Ductility of Seismic Resistant Steel Structures

Ductility of Seismic Resistant Steel Structures

Victor Gioncu and Federico M. Mazzolani

CRC Press is an imprint of the
Taylor & Francis Group, an informa business
A SPON PRESS BOOK

CRC Press
Taylor & Francis Group
6000 Broken Sound Parkway NW, Suite 300
Boca Raton, FL 33487-2742

First issued in paperback 2019

ISBN-13: 978-0-419-22550-8 (hbk)
ISBN-13: 978-0-367-86531-3 (pbk)

British Library Cataloguing in Publication Data
A catalogue record for this book is available from the British Library

Library of Congress Cataloging in Publication Data
A catalog record has been requested

Visit the Taylor & Francis Web site at
http://www.taylorandfrancis.com

and the CRC Press Web site at
http://www.crcpress.com

Ductility of Seismic Resistant Steel Structures

Victor Gioncu and Federico M. Mazzolani

CRC Press is an imprint of the
Taylor & Francis Group, an informa business
A SPON PRESS BOOK

CRC Press
Taylor & Francis Group
6000 Broken Sound Parkway NW, Suite 300
Boca Raton, FL 33487-2742

First issued in paperback 2019

ISBN-13: 978-0-419-22550-8 (hbk)
ISBN-13: 978-0-367-86531-3 (pbk)

British Library Cataloguing in Publication Data
A catalogue record for this book is available from the British Library

Library of Congress Cataloging in Publication Data
A catalog record has been requested

Visit the Taylor & Francis Web site at
http://www.taylorandfrancis.com

and the CRC Press Web site at
http://www.crcpress.com

To
Angela and Silvana

Contents

THE OAK AND THE REEDS

A violent storm uprooted an oak that grew on the bank of a river. The oak drifted across the stream and lodged among some reeds. Amazed that the reeds were still standing, the oak could not help asking them how they had escaped the fury of a storm that had torn him up by the roots. We bent our heads to the blast and it passed over us. You stood stiff and stubborn till you could stand no longer, they said. Moral: Sometimes it is better to bend with forces that are too strong to oppose.

Aesop's Fables

Preface

Earthquakes occur throughout the world. The earth is like a living body in constant motion. Every day small ground movements are registered in some parts of the world, every week a moderate earthquake is reported from some places. At least one significant earthquake causing damage and injury occurs every month, while every year two or three strong earthquakes fill the mass media with dramatic accounts of human and economic losses. Statistically, we can expect that somewhere each year there will be one earthquake of magnitude 8 or greater, 16 of magnitude 7 or greater, 150 of magnitude 6 or greater, and more than 1000 of magnitude 5. The globalisation of the mass media has not only meant that the whole world is informed of earthquake disasters but is also emotionally involved in efforts to save lives in the ruins of buildings. The problems of people in distant places have become everyones problems, and terms such as fault, epicentre, Richter and Mercalli scales, building structure, formerly used only by specialists, are nowadays part of everyday language.

Nowadays, earthquakes are capable of claiming more lives and doing more damage to the built environment than ever before. There are more people and buildings in earthquake-prone areas, which means more buildings, facilities, roads, bridges, dams, and so on are affected by earthquakes each year. Although seismic design has brought progress to engineering practice, there has been a marked increase in financial losses, because rapid and often uncontrolled urbanization and economic development in seismic areas has outpaced the gains from improvement in constructional methods. In any case, there are many old buildings that were erected before or during the early development of seismic design. Many of these were poorly built and are liable to collapse in moderate earthquakes, not just strong ones, killing more people than more recent buildings.

The collapse of both new and old buildings causes large loss of life. The real tragedy is that these human losses are often due not to the earthquakes themselves but to the failure of the construction of the builders. Builders thus become the makes of tools for killing people. Nowadays, for each life lost there is some person or organization that may be held liable for prosecution.

A building is the product of the activity of an interdisciplinary team. The architect is liable if the recently developed seismo-resistant architectural philosophy is ignored. The structural engineer, whether from ignorance or superficiality or weakness in resisting pressure from an architect more interested in the beauty of a building than its safety, may design or approve an unsafe structure. The constructional engineer, due to insufficient control of manufacture or use of poor structural materials, may erect a building that will fall victim to moderate or strong earthquakes. The owner, by poor monitoring and failure to maintain a building properly, can reduce its structural resistance. And last but not least, state or city authorities may be liable because they were slow to incorporate in building codes knowledge recently gained through theoretical and experimental research or through the examination of building behaviour in recent strong earthquakes. Further, lack of official interest in planning the strengthening of old buildings mark them as sure victims of the next earthquake. However, it should also be said that emphasising the deficiencies in the building process in this way holds out the hope that more effective organization of that complex process could substantially reduce the human and financial losses currently being incurred.

Another way of reducing the risk of such losses is improving earthquake-resistant design. In recent times, the defeatist attitude that an earthquake is a fatal force that it is not possible to resist is being transformed by tremendous progress. In the last thirty years, understanding of the nature of earthquakes, the effect of the site soil on the characteristic seismic properties, and of structural response of building subject to seismic waves, has made real advances. The installation of comprehensive instrumentation in high earthquake risk areas has provided a great amount of information about the main characteristics of ground motions. Examination of these records led to the recognition of the differences between near-field and far-field earthquakes, one of the most important recent contributions to advances in seismic design.

The great and costly damage caused by recent strong earthquakes has shown the need to develop a design methodology based on multi-level performance. Different aspects must be considered when a structure is designed for serviceability, susceptibility to damage or ultimate limit states. In this methodological framework, special attention must be paid to ductility as a key factor in resisting partial or global structural failure during very strong earthquakes. Here, ductility is understood as the ability of a structure to sustain large deformations in the plastic range without significant loss of resistance. As in Aesops fable, it is better for a structure to yield under large seismic forces that are too strong to be resisted. Unfortunately, in present design practice, such performance can be attained only through general construction rules. However, real progress in this area is offered by the possibility of ascertaining structural ductility at the same level as for displacement and strength: available ductility, determined at the level of a structural member, must be greater than required ductility, imposed at the level of the structure by an actual near-field or far-field earthquake.

In modern design practice is it generally accepted that steel is an excellent material for seismic-resistant structures because of its performance in terms of strength and ductility: it is capable of withstanding substantial inelastic deformation. In general this is true: the percentage of failure of steel structures has always been very small compared to other constructional materials. But in the last few decades, specialists have recognized that the so-called good ductility of steel structures under particular conditions may be an uncritical dogma that is denied by reality. In the decade 1985 1995, strong earthquakes in Mexico City (1985), Loma Prieta (1989), Northridge (1994), and Kobe (1995) have seriously compromised this ideal image of steel as the perfect material for seismic areas. The performance of steel structures in some cases was very bad, and the same type of damage was caused by different seismic events, clearly showing that there were significant shortcomings in current practice. So now seems the right moment for a critical analysis of progress recently made in conception, design and construction of buildings in seismic areas, to consider the lessons to be learned from these recent dramatic events. Of these lessons, improving the ductility of structures under unfavourable conditions takes a leading place.

The best way to look into the future is to understand the past. However, the international scientific community is also aware of the urgent need to investigate new topics and consequently to improve the current range of provisions for seismic design. The whole framework of modern seismic codes needs a complete review in order to determine and revise the design rules that failed in these recent earthquakes. The challenge for the immediate future is to transfer research achievements into practice, to bridge the gap that has opened between accumulated knowledge and design codes.

Accordingly, this book provides a state-of-the art review of the most advanced issues in the analysis of seismic-resistant steel structures, with the accent on the assessment of structure ductility as the most efficient method of preparing the structure to resist unexpected strong seismic events. At the same time it presents the most recent research results obtained by the authors, which in the near future can be used to improve existing building codes.

In organizing the book, the main idea has been to present the simplest possible formulations, even though these may be no more than approximations of just one phenomenon only, rather than try to elaborate exact specifications. A high degree of exactness is not possible in seismic design because of uncertainties in input data on earthquake characteristics. The most refined and accurate method is useless if the values used in the specifications are not correctly determined.

Chapter 1 begins with a definition of ductility and a description of its place in structural design. Progress in design methods and challenges to building codes since the last strong earthquakes are then presented. Chapter 2 deals with the main lessons to learn from the Michoacan, Northridge and Kobe earthquakes, in which steel structures showed much unexpected damage. Chapter 3 discusses basic elements of design philosophy, such as multi-level design criteria, the modelling of ground motion, and structure conformation and design. Chapter 4 analyses ductility problems at the level of elements and materials, while Chapter 5 deals with the ductility of sections and stubs. Member ductility in a structure is the concern of Chapter 6, which also considers the effect of joints on structural behaviour. Chapter 7 reports the results of state-of-the-art theoretical and experimental research on the main section types used in structures. Finally, Chapter 8 sets out a comprehensive methodology for ductility design, and compares the required ductility for moment-resisting frames with the available ductility determined as the local level, both ductilities being determined using the plastic mechanism method. The Appendix presents the DUCTROT M computer program used to evaluate the rotation capacity of members working in a structure. A CD-ROM containing this program is attached to the back cover of this book.

The authors hope this book will serve as a guide for structural designers seeking to design more economical but safer steel structures, and to open new doors to future developments in the seismic design for all people interested in research and codification. However, they are aware that because the book is the first attempt to analyse the ductility of steel structures from a single, coherent point of view, it is likely to have many shortcomings and make assertions that are open to dispute. They refer readers who encounter these to the words of the wise Chinese author, who said This book would never have appeared if perfection had been awaited.

The present book concentrates on local ductility at the level of members. The authors intend to take their approach further by extending the analysis of ductility to the level of structure, in a book provisionally entitled Global Analysis of Seismic Resistant Steel Structures. They would be grateful for any comments and suggestions about the content of this first book that might help in preparing the second.

The authors are grateful to all colleagues who contributed to the research reported in this book. They would also like to thank those who helped prepare the book for publication, in particular to Emil Danetiu for the illustrations, and Dr Dana Petcu for elaborating the DUCTROT M computer program and the computerized setting of the text.

Victor Gioncu
Federico M. Mazzolani

Notation

Latin Small Letters

a - plate length
a - weld thickness
a - geometrical dimension of buckled shape
a - distance of vertical web stiffner from column face
a - acceleration
a_0 - acceleration amplitude
a_g - ground motion acceleration
a_s - acceleration corresponding to SLS
a_d - acceleration corresponding to DLS
a_u - acceleration corresponding to ULS
b - plate width
b - flange width
b_e - total width of material removed from the flange
b_{eff} - effective width of concrete slab
c - flange semi-width
c_a - coefficient for plastic rotation accumulation
c_r - reduced semi-width of weakened flange
c_T - strain-rate coefficient considering the temperature influence
c_w - strain-rate coefficient considering the welding influence
d - epicentral distance
d - web depth
d_c - effective web depth in compression
d_{fc} - distance from plastic neutral axis to compression flange
d_s - distance from compression flange to web horizontal stiffener
e_0 - nondimensional eccentricity
f_y - nominal yield stress
f_{yc} - corner yield stress
f_{ya} - actual yield stress
f_{yr} - random yield stress
f_{yu} - upper yield stress
f_{yl} - lower yield stress
f_{yf} - flange yield stress
f_{yw} - web yield stress
f_{ysr} - increased yield stress due strain-rate effect
f_{yt} - through-thickness yield stress
f_{ymax} - maximum random value of yield stress
f_{ymin} - minimum random value of yield stress
f_u - ultimate stress
f_{uf} - flange ultimate stress

f_{uw} - web ultimate stress
f_{usr} - increased ultimate stress due strain-rate effect
f_{ut} - through-thickness ultimate stress
f_{umax} - maximum random value of ultimate stress
f_{umin} - minimum random value of ultimate stress
g - gravity acceleration
h - focal depth
h - section depth
h_c - total depth of composite section
h_s - slab thickness of composite section
i - index
j - index
k - index
k - plate buckling coefficient
k - elastic rigidity
l - index
m - mass
m_b - multiplier for plastic buckling moment
m_h - multiplier for strain-hardening moment
m_p - actual moment level related to full plastic moment
m_y - multiplier considering the actual yield stress
m_M - multiplier considering the maximum moment
n - ductility criterium for stubs
n - number of pulses until section fracture initiation
n_b - number of pulses until flange buckling
n_r - number of pulses after buckling until fracture initiation
n_p - actual axial force level related to full plastic axial force
$n(> M)$ - number of events in one year having magnitude greater than magnitude M
p_r - return period
q - distributed beam load
q - behaviour factor, reduction factor
q_μ - strength reduction factor
q_s - overstrength factor
r - radius of flange-web junction
s - multiplier for maximum moment
s - coefficient for strain-hardening effect
t - time
t - thickness
t_f - flange thickness
t_w - web thickness
u - horizontal displacement
v - vertical displacement
v - velocity
v_g - ground motion velocity
v_0 - velocity at the outset of plastic deformations
v_p - velocity of P wave

v_s - velocity of S wave
w - transverse plate displacement
w_a - accumulate transverse plate displacement
w_i - initial plate displacement
w_k - weight of storey k
x - axis
x - distance from the neutral axis to the top of a composite section
x_i - distance of beam inflection point from left end
x_m - distance of beam maximum moment from the left end
z - vertical displacement
z_p - plastic vertical displacement

Latin Capital Letters

A - total section area
A - numerical coefficient for rigid-plastic mechanism behaviour
A_a - reinforcement area for composite section
A_c - corner area
A_f - flange area
A_s - steel profile area for composite section
A_w - web area
B - normalized width-thickness ratio
B - numerical coefficient for rigid-plastic mechanism behaviour
C - numerical coefficient for rigid-plastic mechanism behaviour
C_i - initial cost of building
C_d - damage cost
C_t - total cost
D - damage index under Miner's assumption
D - ground motion duration
E - elastic modulus
E_h - strain-hardening modulus
E_r - reduced elastic modulus
E_s - secant modulus
E_y - energy corresponding to yield strain
E_u - energy corresponding to ultimate strain
F - plate axial loading
F_b - base shear force
F_k - horizontal force acting at storey k
F_{bd} - base shear force for DLS
F_{bs} - base shear force for SLS
F_{bu} - base shear force for ULS
F_d - design plate strength
F_e - elastic plate strength
F_p - plastic plate strength
F_u - ultimate plate strength
G - shear elastic modulus
G_p - shear plastic modulus

H - storey height
H - horizontal component of seismic load
H_k - height of k-storey from base
H_m - height of mass center
H_p - length of column plastic zone
H_s - structure height
I - earthquake intensity
I - moment of inertia
I_d - damage index
I_ω - warping section constant
I_{dg} - global damage index
I_{dm} - member damage index
I_{ds} - storey damage index
K - non-dimensional coefficient of connection stiffness
K - torsional of web stiffness
L - stub length
L - beam span
L_l - left span for standard beam
L_r - right span for standard beam
L_b - length of buckled shape
L_p - length of plastic zone
L_p - loading potential
L_w - length of web buckling shape
M - magnitude
M - bending moment
M_b - bending moment for flange buckling
M_p - full plastic moment
M_l - left end moment of beam
M_r - right end moment of beam
M_u - ultimate moment
M_w - bending moment due to vertical loads
M_y - bending moment for the first yielding
M_{cu} - upper column moment
M_{cl} - lower column moment
M_{pb} - plastic moment of beam
M_{pc} - plastic moment of column
M_{pj} - plastic moment of joint
M_{ph} - moment in strain-hardening range
M_{pf} - moment corresponding to flange plasticization
M_{pl} - left end plastic moment
M_{psr} - right end plastic moment
M_{pN} - reduced plastic moment due to interaction with axial force
M_{pred} - reduced plastic moment by weakening of flanges
M_{max} - maximum moment
M_{uN} - reduced ultimate moment due to interaction with axial force
N - axial force
N_b - buckling axial force

N_p - full plastic axial force
N_f - face axial component
N_c - corner axial component
N_w - axial force from structure weight
N_{ct} - corner torsional axial component
N_{cr} - critical axial load
P - beam concentrate transverse load
S - soil parameter
S_a - acceleration spectral value
S_v - velocity spectral value
S_d - spectral value for DLS
S_s - spectral value for SLS
S_u - spectral value for ULS
S_{el} - elastic spectral value
T - period of vibration
T^0 - temperature (Celsius grade)
T_c - corner period
T_d - period of damaged structure
T_g - natural period of ground motion
T_{cd} - corner period of DLS spectrum
$T(> M)$ - recurrence interval of an earthquake greater than magnitude M
T_{cs} - corner period of SLS spectrum
U - strain energy
U_z - strain energy of plastic zone
U_l - strain energy of yield line
V - total potential energy
V - vertical component of seismic load
V - shear force
V_p - plastic shear force
W - total structure weight
Z - elastic section modulus
Z_p - plastic section modulus
Z_{pr} - reduced plastic section modulus

Greek Letters

α - angle
α - normalized slenderness
α - multiplier of horizontal forces
α - coefficient of pulse asymmetry
α_c - collapse multiplier
α_d - multiplier corresponding to DLS
α_f - fracture rotation of buckled flange
α_y - multiplier for first yield
α_N - numerical coefficient for rigid-plastic stub behaviour
α_M - numerical coefficient for rigid-plastic beam behaviour

β - angle of inclined yield line
β - parameter of buckled shape length
$\beta(T)$ - spectral amplification factor
β_d - spectral amplification factor for DLS
β_s - spectral amplification factor for SLS
γ - parameter of plastic zone length
γ_s - partial safety factor for seismic action
δ - axial shortening of plate and stub
δ - top sway displacement of structure
δ - parameter of web buckled shape
δ_i - initial axial shortening of stub
δ_p - plastic axial shortening of stub
δ_u - ultimate displacement
δ_y - first yield displacement
ϵ - normal strain
$\dot{\epsilon}$ - strain-rate
ϵ_h - strain at the outset of strain-hardening
ϵ_t - total strain
ϵ_u - uniform strain
ϵ_y - yield strain
ϵ_{sh} - strain in hardening range
ϵ_{cu} - ultimate strain of concrete
ζ - numerical coefficient for plate fracture
η - parameter for asymmetry of buckled web shape
η - damper correction factor with reference value 1.0 for 5% viscous damping
θ - rotation
θ_a - accumulated rotation
θ_f - fracture rotation
θ_i - initial rotation
θ_m - rotation corresponding to maximum moment
θ_p - rotation corresponding to formation of plastic hinge
θ_r - ultimate plastic rotation
θ_u - ultimate rotation
θ_y - yield rotation
θ_{rh} - hysteretic plastic rotation
θ_{rk} - kinematic plastic rotation
θ_{rr} - required plastic rotation
θ_{rs} - reduced rotation due to strain-rate effect
θ_{uc} - ultimate rotation for cyclic action
λ_y - lateral beam slenderness
$\bar{\lambda}$ - column normalized slenderness ratio
$\bar{\lambda}_f$ - flange normalized slenderness ratio
$\bar{\lambda}_w$ - web normalized slenderness ratio
μ - ductility
μ_ϵ - material ductility
μ_χ - curvature ductility

μ_θ - rotation ductility
μ_θ - rotation capacity
μ_δ - displacement ductility
μ_E - energy ductility
μ_d - global ductility
μ_r - local ductility
$\mu_{\theta c}$ - rotation capacity for cyclic action
$\mu_{\theta 0.9}$ - rotation capacity for $0.9M_p$ ductility criterium
ν - reduction factor for SLS
ν - Poisson's ratio
ν - damping coefficient
ν_m - left to right plastic moments ratio
ξ - viscous damping ratio expressed in percent
ρ_{ysr} - yield ratio
σ - normal stress
σ_{ysr} - yield ratio for strain rate
σ - standard deviation
σ_b - buckling stress
σ_{cr} - critical stress
τ - shear stress
τ_y - shear yield stress
τ_0 - rupture duration
χ - curvature
χ_b - buckling curvature
χ_h - curvature at the outset of strain-hardening
χ_u - ultimate curvature
χ_y - yielding curvature
χ_{max} - curvature at the maximum moment
ϕ - circular frequency of ground motion
ω - natural circular frequency
Δ - interstorey drift
Δ - plate displacement
Δ_i - plate initial displacement
Δ_c - displacement of compression flange
Δ_t - displacement of tension flange

Abbreviations

al - aluminum
hs - high strength steel
ms - mild steel
cd - constant displacement
id - increasing displacement
rd - random displacement
pd - pulse displacement
ip - inflection point
ph - plastic hinge

FO - full operational
O - operational
LS - life save
LB - local buckling
LT - lateral-torsional buckling
NC - near collapse
SLS - serviceability limit state
DLS - damageability limit state
ULS - ultimate limit state
MRF - moment resisting frame
S-MRF - special moment resisting frame
S-MRFd - special moment resisting frame with limited interstorey drift
G-MRF - global moment resisting frame
CBF - concentrically braced frame
EBF - eccentrically braced frame
DF - dual frame
SSL - space structure layout
PSL - perimetral structure layout
SDOF - single-degree of freedom system
MDOF - multi-degree of freedom system
NDT - nil-ductility temperature
FTP - fracture transition plastic temperature
HD - high ductility class
MD - medium ductility class
LD - low ductility class
SB 1 - standard beam type 1 for moment gradient
SB 2 - standard beam type 2 for quasi-constant moment
GSB - generalized standard beam, including the joint effect
SSB - strengthened standard beam
WSB - weakened standard beam
RS - rolled section
CRS - cold rolled section
CFS - cold formed section
WS - welded section
CoFS - concrete fillet section
MG - moment gradient
CM - constant moment
MT - monotonic test
CT - cyclic test
EC - experimental calibration
PS - parametrical study
BM - beam member
BCM - beam-column member
CP - control parameter
COV - coefficient of variation
PEEQ - effective plastic strain index

1

Why Ductility Control?

1.1 Nature of the Problem

1.1.1 Main Purpose of Seismic Design

From a structural viewpoint, the design term refers to a synthesis of various disciplines of construction science, aiming to create a building that is of such a size and design that the demands for functionality, aesthetics, and resistance are satisfied at the same time in the same measure.

Structural engineering is the science and art of designing and making, with elegance and economy, buildings, bridges, frameworks, and other similar structures so that they can safely resist the forces to which they may be subjected. (Petroski 1985).

The main purpose of structural design is to produce a suitable structure, in which we must consider not only the initial cost, but also the cost of maintenance, damage and failure, together with the benefits derived from the structure function. Thus, the optimum design of a structure requires a clear understanding of the role of each of above aspects and, therefore, requires a general view on the total process (Bertero 1996).

This objective can be achieved by the engineer, involved in designing a specific building, without great difficulty for conventional actions such as dead, live, wind, and snow loads, but with difficulty for exceptional loads produced by natural disasters such as hurricanes, tornadoes, floods, earthquakes, etc. Among these natural disasters, earthquakes are responsible for almost 60 per cent of deaths (Fig. 1.1, IAEE, 1992). In contrast to other natural disasters, which occur above the surface, earthquakes are the consequence of the release, in a very short period of time, of massive energy stored in the interior of the Earth, and, therefore, are very difficult to predict and model for analysis.

Powerful earthquakes are responsible for large losses of life and property. In the past the number of deaths were always very high, and continue to be so up to the present day. From analysis of data regarding losses from previous earthquakes, it is clear that the majority of lives lost was as a consequence of relatively few powerful earthquakes. In fact, during the

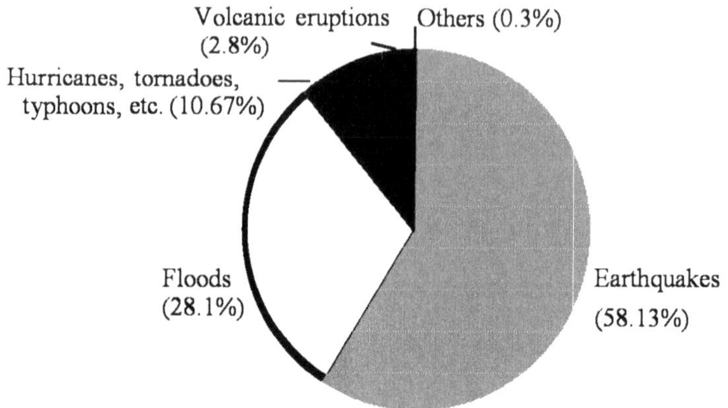

Figure 1.1: Percentage of deaths due to natural disasters

twentieth century, despite 100 events throughout the world of magnitude 7, only 20 earthquakes caused losses of life greater than 10,000 and only 2 greater than 200,000 (Dolce et al. 1995). In 1976, one of the most disastrous earthquakes claimed 250,000 lives in Tangshan, China.

Analysis of these events shows that it is not just ground motion severity that is responsible for this loss of life. Indeed, the main feature of earthquakes is that most human and economic losses are due to failure of human constructions – buildings, bridges, transport systems, dams, etc. – designed and built for ease of travel and comfort of humans. Major losses of life are always concentrated in poor regions with old buildings or with very poorly built constructions. In contrast, major economic losses are localized in rich regions with modern buildings, where, despite few building collapses, the resulting damage is significant and the cost of repairs very high.

We can see that buildings which are poorly designed and constructed suffer much more damage in moderate earthquakes than well-designed and constructed buildings in strong earthquakes. Moreover, these latter buildings should easily be able to withstand severe earthquakes with no loss of life and without severe damage. So, in many cases of building collapse, it is the builders themselves who are responsible for buildings that can kill. This is depressing, but at the same time encouraging, because it seems that earthquake problems are solvable (Bertero 1992), at least theoretically. For a well-designed and erected building, the risk of collapse during a strong earthquake is substantially reduced. It is the duty of building professionals to detect errors which contribute to building collapse and to improve the conceptual design and quality of buildings.

Generally, the engineering approach to design is quantitative and the structural members must be sized to have resistance greater than the actions caused by these events. But the design for the largest credibly imagined loads, resulting from the strongest expected earthquake on the structure site, is unreasonable and economically unacceptable. The design requirements are set at a given level smaller than that associated with the largest possible loads. So, the structures occasionally fail to perform their intended function under these requirements which exceed the design values and, consequently, they may suffer local damage, by the loss of resistance in a single member or in a small portion of the structure. But a properly designed structure must preserve the general integrity, which is the quality of being able to sustain local damage, the structure as a whole remaining stable. This purpose can be achieved by an arrangement of structural elements that gives stability of the entire structural system. The local damaged portions must be able to dissipate a great part of seismic energy, being able to support important deformation in plastic range; the other parts remain in elastic range. The ability of a structure to undergo plastic deformations without any significant reduction of strength, represents the structural ductility, being a measure of the suitable structure behaviour during a severe earthquake. It is easy to understand that, in function of earthquake severity, there are different levels of ductility demands, and the ability to design a good structure is to supply it with sufficient available ductility.

But these excursions in plastic range cause damage, which must be repaired after the events and, in this perspective, the design process against earthquakes becomes a balance between the initial investment and the repair costs after the earthquake. This very difficult design philosophy was the subject of many research works (Waszawski et al, 1996) with very disputable results. So, in spite of the great efforts paid in recent years in solving satisfactorily this problem, recent earthquakes are capable of doing more damage today than ever before. Instead of observing a reduction of damage produced by earthquakes, a marked increase of financial losses results (Fig. 1.2). The years 94 (Northridge earthquake) and 95 (Kobe earthquake) reach the maximum of losses. The reason for this remarkable increase of losses in recent years is due to the fact that many buildings were constructed when Earthquake Engineering had not started, or was in its early stage. But, at the same time, many new buildings were damaged due to the concentration of population and industrialization in high seismic risk regions, the high vulnerability of modern technologies, and even because seismic codes are not infallible.

The last big events of Northridge and Kobe produced enormous damage, but they can be considered minor in comparison with potential losses in big cities like Mexico City, Los Angeles, San Francisco, Tokyo, etc. A recent calculation model predicted losses of US$100-150 billion for "The Big One" earthquake in California and over US$1000 billion in the Tokyo earthquake, if the 1923 earthquake happened today with the same magnitude.

So, the main purpose of modern seismic design is to reduce economic losses, and at the same time to save human life. The control of structure ductility is the key to solve this problem.

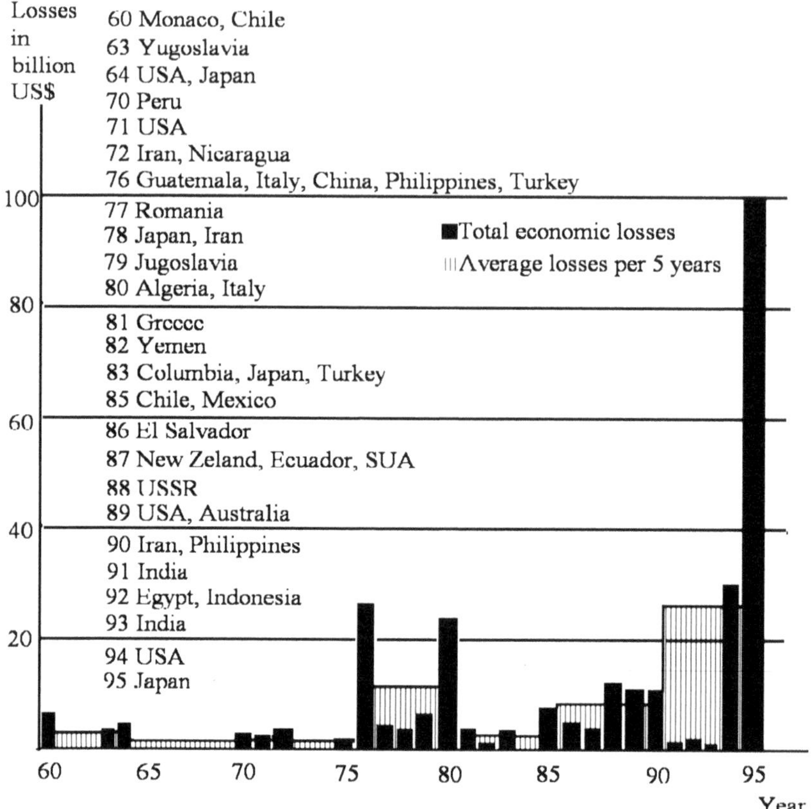

Figure 1.2: Cost of loses due to earthquakes from 1960 to 1995

1.1.2. Main Lessons after the Last Strong Earthquakes

It is very important that some general features of the earthquakes that occurred recently should be seriously analysed:

(i) It is well known that all the attention is concentrated on the regions were earthquakes occurred in the past. At the same time, it can be observed that the major devastating earthquakes occurred in areas where no previous events have been recorded and where the current knowledge would suggest the existence of a quiescent area. The Tangshan-China and Killary-India are examples of such areas. So, there are few, if any, areas of world which are immune to earthquake effects and to which minimum protection can be considered in design. But in any case, all structures must be provided with some ductility properties, as a precaution against an unexpected earthquake.

(ii) Each event is unique, offering new surprises in the vulnerability of buildings affected by earthquakes and showing the great complexity of the phenomenon. Referring only to steel structures, as no serious damage during some major earthquakes was recorded for a long time, so the persuasion that steel structures are a very safe solution for seismic areas have been consolidated among the structural designers. But the 1985 Michoacan earthquake in Mexico City produced the first collapse of a high-rise steel building, due to the difference between required and available ductilities. The 1994 Northridge and 1995 Kobe earthquakes produced many failures in steel moment-resisting frames, which were not expected by the engineering community. These failures were mainly due to the near source position of the structures and also due to very high velocities recorded near the epicenters. These failures dealt a great blow to the use of steel structures in seismic areas, but at the same time there was the start of large activity in research works, in which the control of ductility for severe conditions plays a leading role.

(iii) For a long time the main purpose of seismic design was the protection of the public from loss of life or serious injuries and the prevention of building collapse under the maximum intensity earthquake. The second goal was the reduction of property damage. After each earthquake, attention is concerned only on the performance of structure and, therefore, very little is known about the vulnerability of non-structural elements, secondary structures, contents, installations, equipments, etc., which can produce more economic losses than the structural damage itself, even for a moderate earthquake. Only recently specialists have been informed that to only fulfil the condition of live protection is economically unacceptable. They realized that it is necessary to pay more attention to the reduction of damage of all building elements for all ranges of earthquake intensities. Consequently, a multi-level design approach has been developed and different levels of ductility demand are now used in design process, as a function of the considered seismic intensity.

1.1.3. Required Steps for Control of Ductility Demand

The control of ductility demand requires attention to be given to some important aspects (Fig. 1.3):

(i) *Seismic macro-zonation*, which is an official zoning map to Country scale, based on a hazard analysis elaborated by geologists and seismologists. This map divides the national territory into different categories and provides each area with values of earthquake intensities, on the basis of design spectra. At the same time, this macro-zonation must characterize the possible ground motion types, as a surface or a deep source, an interplate or intraplate fault, etc. The ductility demands are very different for each ground motion type.

(ii) *Seismic micro-zonation*, which considers the possible earthquake sources at the level of region or town, on the basis of common local investigation of seismologists and geologists. The result of these studies is a local map, indicating the positions and the characteristics of the sources, general informations about the soil conditions and design spectra. It is very useful to accompany the time-history accelerograms with very precise

ACTIVITY	SCHEME	AUTHORS	INFORMATION
MACRO-ZONATION	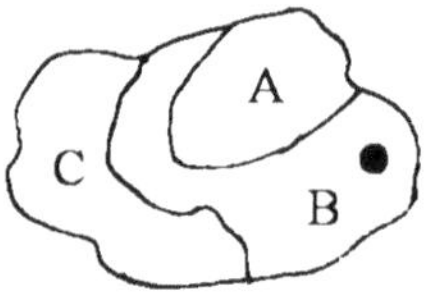 A B C	• geologists • seismologists	• type of possible earthquakes • magnitude • intensities
MICRO-ZONATION		• geologists • seismologists	• source position • magnitude • intensities • attenuation • duration
SITE CONDITION	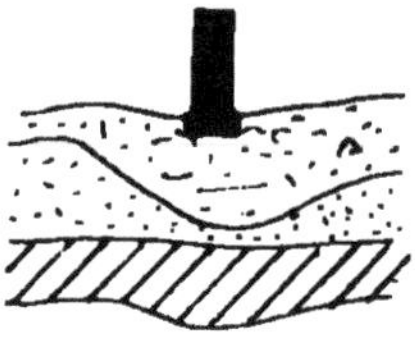	• geologists • geotechnical engineers	• soil stratification • framing in soil type • duration • time-history records • design spectrum
STRUCTURE CHARACTE-RISTICS	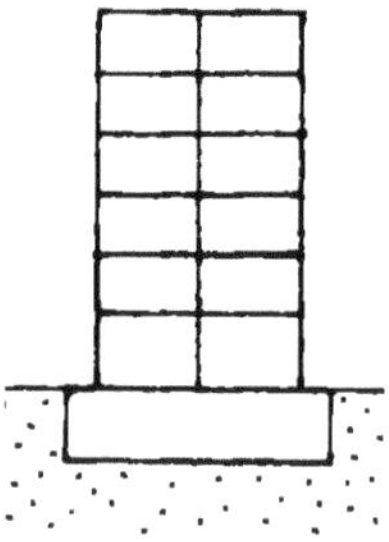	• geotechnical engineers • structural engineers • mechanical engineers • architects • builders • owner	• level of protection • general configuration • materials • foundation type • structure type

Figure 1.3: Steps in control of ductility demand

indications about the place where they have been recorded (directions, distance from epicenters, soil condition, etc.). Recordings such as magnitude, distance from source, attenuation, duration, etc., are directly involved in ductility demand.

(iii) *Site conditions*, established by geologists and geotechnical engineers, from the examination of the stratification under the proposed structure site. This is a very important step, because dramatic changes of earthquake characteristics within a few hundred of meters distance is not an unusual observation during an earthquake. These differences are mainly caused by the different soil conditions. The geotechnical engineers must provide exact information about the framing of local conditions in the soil type, in function of code demands. For soft soil, the ductility demand is more important than for rigid soil. The geotechnical engineers must propose the most suitable design spectrum or time-history records, taking into account of soil conditions.

(iv) *Structure characteristics*, result from the collaboration of geotechnical, structural and mechanical engineers, architects, builders and owners. At this step, the levels of protection are established and the ductility demand is fixed as a function of these levels. This is the main step in the design process; the good or bad behaviour of the structure during a strong earthquake depends on the decisions taken during these discussions. General configuration, structural material, foundation and elevation structural types, technology of erection, etc., result from this activity.

1.2 Evolution Process of Ductility Concepts

In comparison with the other branches of structural engineering, the design of seismic resistant structures is a relatively new branch, with first attempts being developed only at the beginning of this century, and the most important concepts being achieved during the last 40 years. The evolution of design concepts is characterized by a continuous flow of information between the architects, structural engineers, geotechnists and seismologists.

The history of development of ductility concepts is divided into three principal periods (Fig. 1.4).

1.2.1. Early Development

The preliminary design concepts commenced after the severe earthquakes at the beginning of the 20th century. The great builder Gustave Eiffel had the intuition to model the earthquake forces by means of an equivalent wind load. The San Francisco city was rebuilt after the 1906 great earthquake using a 1.4 kPa equivalent wind load. It was not until after the Santa Barbara earthquake in 1925 and the 1933 Long Beach earthquake, that the concept of lateral forces proportional to mass was introduced into practice. The buildings have been designed to withstand lateral forces of about 7.5 percent for rigid soil and 10 percent for soft soil of their dead load. This rule constituted due to the observation that the great majority of well designed and constructed buildings survived strong ground motions, even if they were designed only for a fraction of the forces that would develop if the structure behaved entirely linearly elastic (Fajfar, 1995). In 1943, the Los Angeles

	DESIGN CONCEPT	STRUCTURE MODEL	HORIZONTAL FORCE
DESIGN CONCEPT	Horizontal load as wind	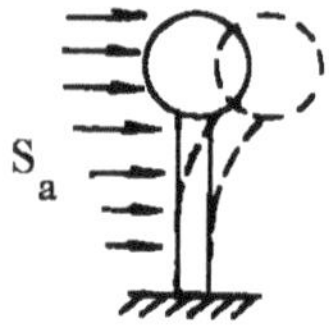 S_a	$S_a = c$
	Influence of mass	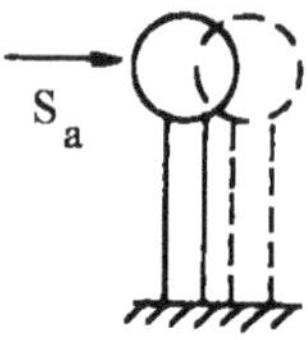 S_a	$S_a = cW$
MODERN DESIGN CONCEPT	Influence of flexibility		$S_a = \frac{cW}{T^{\alpha}}$
	Influence of plastic deformation		$S_a = \frac{S_{el}}{q}$
CONTROL DESIGN	Passive control		$S_a = \gamma_p \frac{S_{el}}{q}$
	Active control	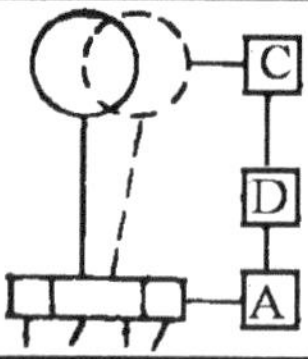 C D A	$S_a = \gamma_a \frac{S_{el}}{q}$

Figure 1.4: Evolution of design concepts

city code recognized the influence of flexibility of structures, and considered the number of structure levels in the design forces. The first provisions where the influence of the fundamental period of structure were introduced, were the San Francisco recommendations, introducing a relation stating that seismic forces are inversely proportional to this period (Bertero, 1992, Popov, 1994).

These preliminary concepts are based on grossly simplified physical models, engineering judgment and a number of empirical coefficients. Influenced by the conventional design concepts, the earthquake actions are considered as static loads and the structures as elastic systems. This simple concept has been the standard design methodology for several decades. There are good reasons for the success of this design approach. This methodology has been well understood by structural engineers because it is relatively easy to be implemented. In most cases this approach helps professional activity, but in some cases, it may lead to inadequate protection (Krawinkler, 1995). Because of these limits new concepts have been developed.

1.2.2. Modern Design Concepts

The beginning of the modern design concepts may be fixed in the 1930s, when the concepts of response spectrum and plastic deformation were introduced to earthquake engineering. The first concept considering the elastic response spectrum was used by Benioff in 1934 and Biot in 1941 (Miranda, 1993). Linear elastic response spectra provide a reliable tool to estimate the level of forces and deformations developed in structures. In 1935 Tanabashi proposed an advanced theory, which suggested that the earthquake resistance capacity of a structure should be measured by the amount of energy that the structure can absorb before collapse. In terms used nowadays, this energy can be interpreted as the dissipated energy through the ductility of structure (Takanashi and Nakashima, 1994).

The first attempts to combine these two aspects, the response spectrum and the dissipation of seismic energy through plastic deformations, was made by Housner (1956, 1959), who made a quantitative evaluation of the total amount of energy input that contributes to the building response, using the velocity response spectra in the elastic system, and, assuming that the energy input, responsible for the damage in the elastic-plastic system, is identical to that in the elastic system (Akiyama, 1985). Housner verified his hypothesis by examining several examples of damage. So, his method proposed a limit design type analysis to ensure that there is sufficient energy-absorbing capacity to give an adequate factor of safety against collapse in the event of extremely strong ground motion. The first study on the inelastic spectrum was conducted in 1960 by Velestos and Newmark. They obtained the maximum response deformation for the elastic-perfectly plastic structure. Since its first application in seismic design, the response spectrum has become a standard measure of the demand of ground motion. Although it is based on a simple single-degree-of-freedom linear system, the concept of the response spectrum has been extended to multi-degree-of-freedom systems, nonlinear elastic systems and inelastic hysteretic systems. The utility of the response spectrum lies on the fact that it gives a simple and direct indication of the overall displacement and acceleration demands of earthquake ground motion, for structures having different period and damping characteristics, without needing to perform detailed numerical analysis.

A new concept was proposed in 1969 by Newmark and Hall, by constructing spectra based on accelerations, velocities, and displacements, in short, medium and long period ranges, respectively. This concept remained a proposal until after the Northridge and Kobe earthquakes, when the importance of velocity and displacement spectra was recognized.

More recently, for structures situated in near-field region of an earthquake, another methodology has been elaborated (Iwan, 1997), based on the drift spectrum of a continuous medium, in opposition with the concept of discrete medium. This concept is based on the observation that the ground motions in near-field regions are qualitatively different from that of the commonly used far-field earthquake ground motions. For near-field earthquakes, the use of the equivalence of multi-degree-of-freedom systems with only one degree-of-freedom gives inaccurate results, because the importance of the superior vibration modes is ignored. So, a new direction of research works for ductility of structures in near-field regions began to be explored.

Since the early 70s a crucial change in seismic design concept has taken place, thanks to the availability of personal computers and the implementation of a great number of programs for structural engineering, which very easly perform static and dynamic analyses in elastic and elasto-plastic ranges. This technological advance allow us to obtain more refined results, and gives to the researchers the perspective to improve the methodology of using the design spectra in current practice, with a more correct calibration of design values. At the same time for important structures, a time-history methodology, using real recorded accelerograms, can be applied and the behaviour of structures under seismic actions can be evaluated in a more precise way, according to the spectrum methodology.

But this concept has been criticized in recent years due to the fact that large deformations, such as those necessary for the building components to provide the required ductility, are associated for strong earthquakes with local buckling, cracking and other damage in structural and non-structural elements, with a very high cost of repairing after each event. In order to minimize this damage, a new approach in seismic design has been developed, mainly based on the idea of controlling the response of the structure, by reducing the dynamic interaction between the ground motion and the structure itself.

1.2.3. Response Control Concept

A significant progress has been recently made in the development and application of innovative systems for seismic protection. The aim of these systems is the modification of the dynamic interaction between structure and earthquake ground motion, in order to minimize the structure damage and to control the structure response. So, this concept is very different from the conventional one, according to which the structure is unable to behave successfully when subjected to load conditions different from the ones it has been designed.

The control of the structural response produced by earthquakes can be done by various means, such as modifying rigidities, masses, damping and providing passive or counteractive forces (Housner et al, 1997). This control is based on two different approaches, either the modification of the dynamic characteristics or the modification of the energy absorption capacity of the structure. In the first case, the structural period is shifted away from the predominant periods of the seismic input, thus avoiding the risk of resonance occurrence. In the second case, the capacity of the structure to absorb energy is enhanced through appropriate devices which reduce damage to the structure (Mele and De Luca, 1995). Both these approaches can be implemented in passive, active or hybrid systems.

(i) *Passive systems* are systems which do not require an external power source. The properties of the structure (period and/or damping capacity) do not vary depending on the seismic ground motion. The base isolation or damping devices serve as the first line of defence against seismic forces, leaving the structure itself and its inelastic reserve of strength as a second defence line. So the structures receive only a part of seismic forces, the rest being dissipated by the behaviour of the devices (Romero, 1995).

(ii) *Active systems* are systems in which an external source of power controls the actuators. Thus, in these systems, the structure's characteristics are modified just as a function of the seismic input. The modifications are obtained by integrating within the structural system a control system consisting in three main components: sensors, interpretation and decision systems/actuators. Thus, the structures are able to determine the present state, to decide in a rational manner on a set of actions which would change its state to a more desirable state and to carry out these actions in a controlled manner and in a short period of time. The goals of active systems are to keep forces, displacements and accelerations of structure below specific bounds, in order to reduce the damage in case of strong earthquakes.

(iii) *Hybrid systems* are systems implying the combined use of passive and active control systems. For example, a base isolated structure is equiped with actuators which actively control the enhancement of its performance.

Recently response control systems, including seismic isolation, have been gradually applied to various structures, e.g. buildings, highway bridges and power plants. The response control systems are utilized not only for the new structures but also for existing structures to retrofit them (Mazzolani et al, 1994a,b).

The response control systems are classified as shown in Fig. 1.5 (ISO 3010, 1998) and illustred in Fig. 1.6. All systems, except active and hybrid control systems, can be classified into passive control systems. The seismic isolation is to reduce the response of the structure by the isolators which are usually installed between the foundation and the structure. In the case of suspended structures, the isolators can be located on the top of the building (Mazzolani, 1986, Mazzolani and Serino, 1997a). Since the isolators elongate the natural period of the structure and dampers increase damping, the acceleration response is reduced, as shown in Fig. 1.7, but the large relative displacement occurs at the isolator installed story. Energy ab-

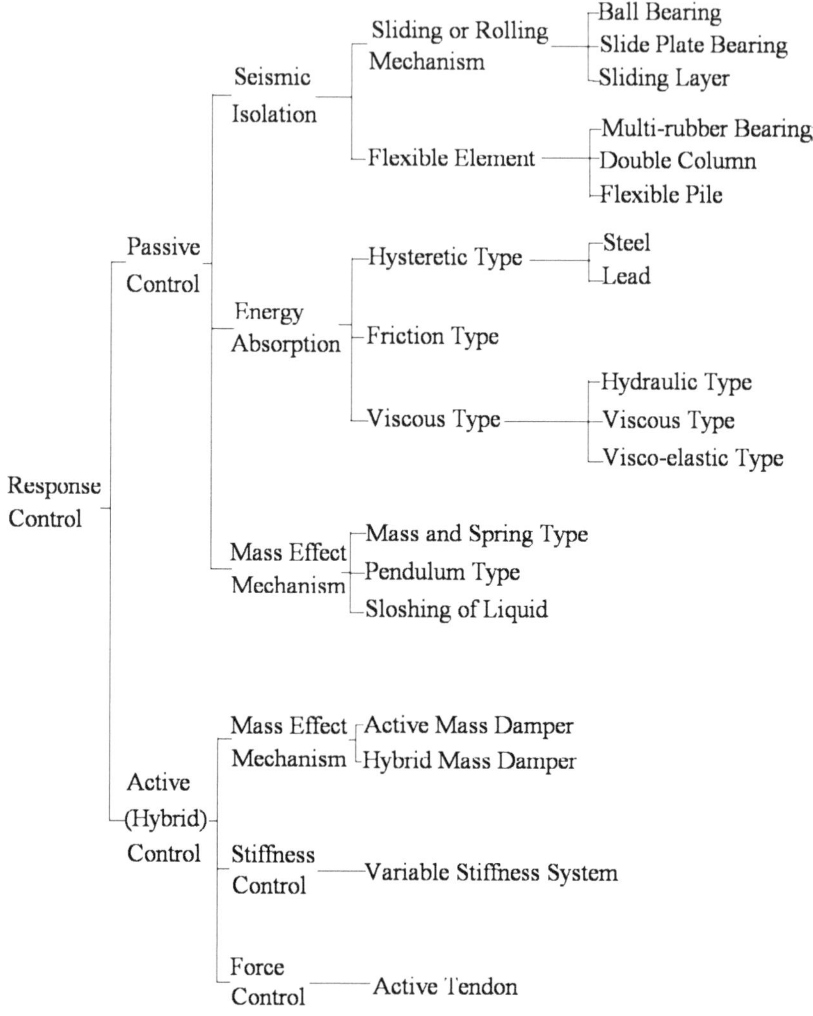

Figure 1.5: Classification of response control systems

sorption devices and additional masses to structure are also used to control the response. The energy absorption devices increase the damping within the structure by plastic deformation or viscous resistance of the devices. In some cases the use of oleodynamic devices can protect the structure from the formation of plastic hinges (Mazzolani and Serino, 1997b). The response of a structure is also reduced by vibration of additional masses and liquid materials. The active response control systems reduce the response of a structure caused by earthquakes and winds using computer controlled additional masses or tendons.

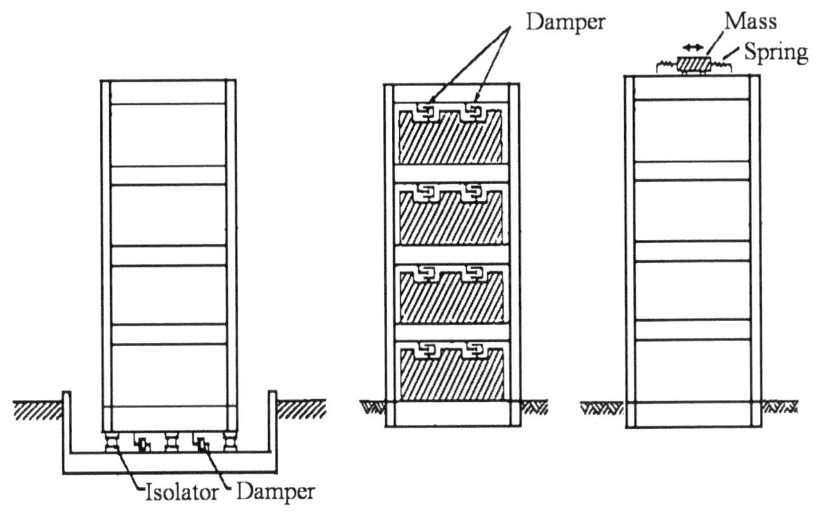

Figure 1.6: Examples of schemes of response control systems

The response control systems are used to reduce floor response and inter-story drift. The reduction of floor response may ensure seismic safety, improve habitability, ease mental anxiety, protect furniture from overturning, etc. The reduction of inter-story drift may decrease the amount of construction materials, reduce damage to non-structural elements and increase design freedom.

During the Northridge and Kobe earthquakes, while many conventional structures suffered excessive damage and even partial or total collapses, some base isolated buildings located in zones close to the epicenter experienced successfully the first severe field tests (De Luca and Mele, 1997).

The concept of response control is a very promising strategy, but there are some limitations in using this system:

-there are situations where more than one source are depicted in the same region and, generally, these sources have different characteristics. It is very difficult to design a control system which has a variable response in function of ground motion type;

-it is not technically possible to design a control system which assures that the structure remains elastic during a strong earthquake. An open question is the behaviour of a structure when it falls in the inelastic range. The development of plastic hinges could in fact reduce the difference in period between fixed base and isolated schemes, reducing the effectiveness of isolators and leading to fast deterioration of dynamic response. In some cases, a sudden increase of damage is observed at some level of acceleration (Ghersi, 1994, Mazzolani and Serino, 1993);

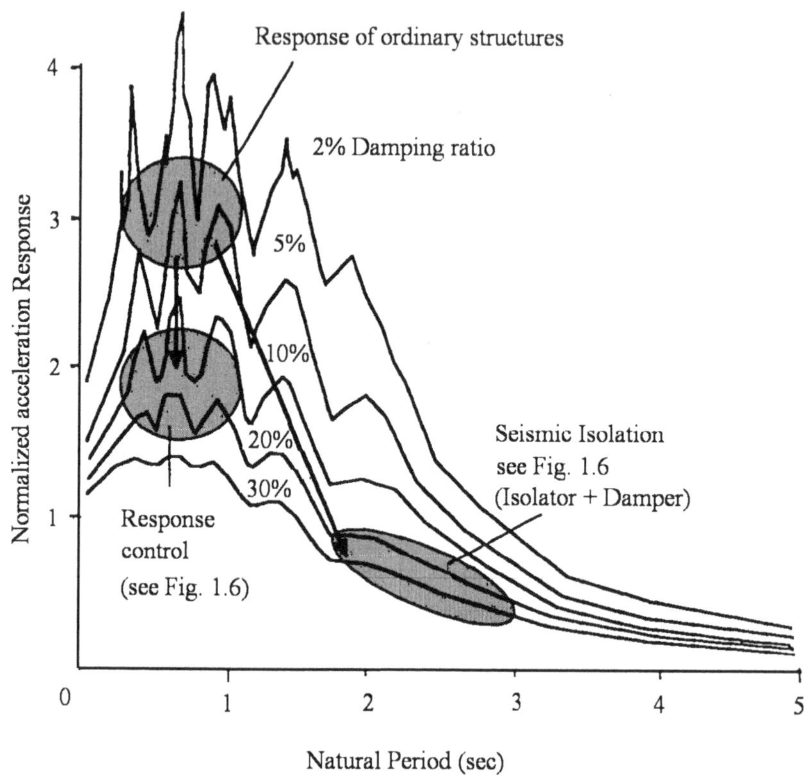

Figure 1.7: Examples of schemes of response control systems

-in case of near-field ground motion, as the energy content and velocity are very high, the required isolator displacements are very large and very often exceed the available displacements of the used isolators. In these cases, a high impact load to the isolated portion of building results (Iwan, 1995);

-when the vertical displacements are very high (as for near-field zones), the efficiency of devices for response control is disputable.

Thus, even in cases of response control, the ductility control remains a very important method of preventing any unexpected behaviour of a structure during severe earthquakes.

1.3 Leading Role of Ductility in Structural Design

1.3.1. Ductility Definition

Before the 1960s the ductility notion was used only for characterizing the material behaviour. After the Housner's studies of earthquake problems and Baker's research works on plastic design, this concept has been extended to a structural level.

In the common practice of earthquake resistant design, the term *ductility* is used for evaluating the performance of structures, by indicating the quantity of seismic energy which may be dissipated through plastic deformations. The use of the ductility concept gives the possibility to reduce the seismic design forces and allows to produce some controlled damage in the structure also in case of strong earthquakes.

In the practice of plastic design of structures, ductility defines the ability of a structure to undergo deformations after its initial yield, without any significant reduction in ultimate strength. The ductility of a structure allows us to predict the ultimate capacity of a structure, which is the most important criteria for designing structures under conventional loads.

The following ductility types are widely used in literature (Fig. 1.8) (Gioncu, 1999):

-material ductility, or deformation ductility, which characterizes the material plastic deformations for different loading types;

-cross-section ductility, or curvature ductility, which refers to the plastic deformations of the cross-section, considering the interaction between the parts composing the cross-section itself;

-member ductility, or rotation curvature, when the properties of members are considered;

-structure ductility, or displacement ductility, which considers the overall behaviour of the structure;

-energy ductility, when the ductility is considered at the level of dissipated seismic energy.

A correlation among these types of ductility exists. The energy ductility is the cumulation of structure and member ductilities; the member ductility depends on cross-section and material ductilities. There are many disputable problems in the above definitions, due to the fact that they have a precise definition and quantitative meaning only for the idealized case of monotonic linear elasto-perfectly plastic behaviour. Their use leads to much ambiguity and confusion in actual cases, where the structural behaviour significantly differs from the idealized one (Bertero, 1988).

A very important value in seismic design is the ductility limit. This limit is not necessarily the largest possible energy dissipation, but a significant changing of structural behaviour must be expected at ductilities larger than these limit ductilities. Two ductility limit types can be defined (Gioncu, 1997, 1998):

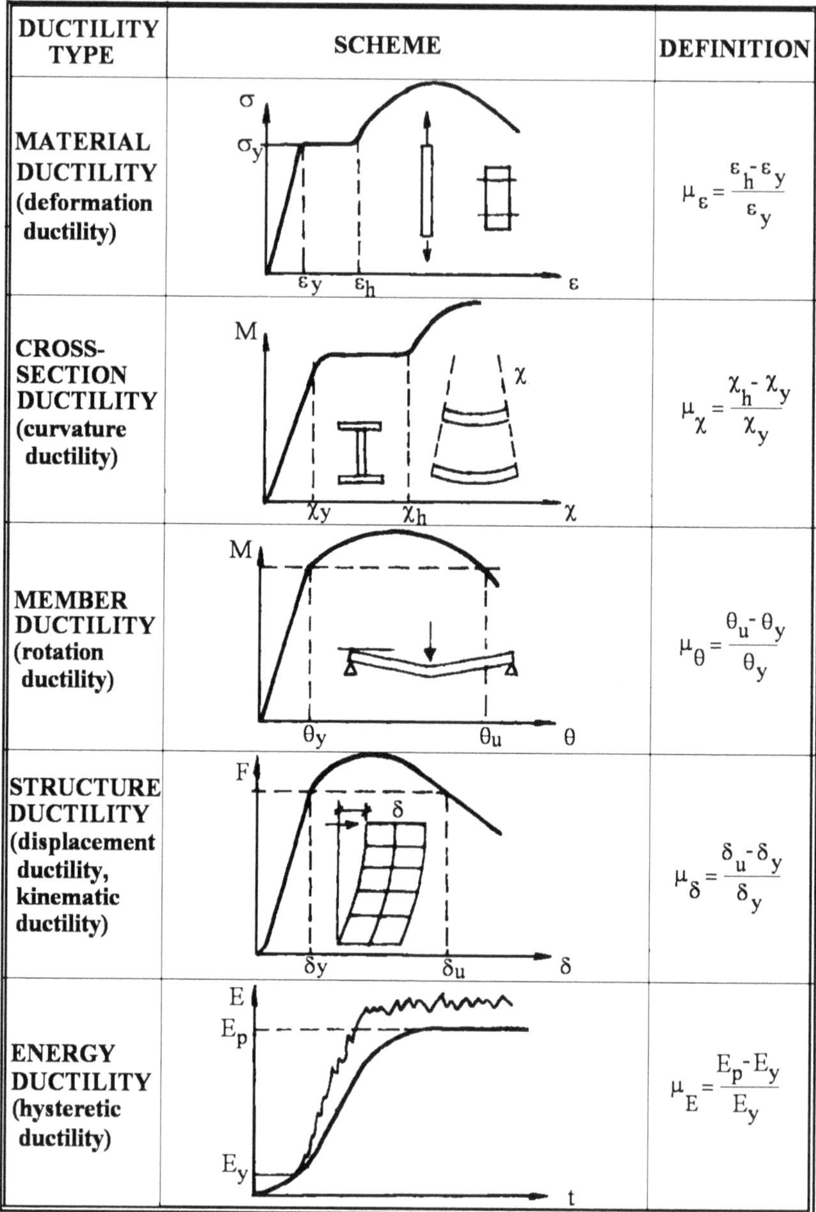

DUCTILITY TYPE	SCHEME	DEFINITION
MATERIAL DUCTILITY (deformation ductility)		$\mu_\varepsilon = \frac{\varepsilon_h - \varepsilon_y}{\varepsilon_y}$
CROSS-SECTION DUCTILITY (curvature ductility)		$\mu_\chi = \frac{\chi_h - \chi_y}{\chi_y}$
MEMBER DUCTILITY (rotation ductility)		$\mu_\theta = \frac{\theta_u - \theta_y}{\theta_y}$
STRUCTURE DUCTILITY (displacement ductility, kinematic ductility)		$\mu_\delta = \frac{\delta_u - \delta_y}{\delta_y}$
ENERGY DUCTILITY (hysteretic ductility)		$\mu_E = \frac{E_p - E_y}{E_y}$

Figure 1.8: Ductility types

-available ductility, resulting from the behaviour of structures and taking into account its conformation, material properties, cross-section type, gravitational loads, degradation in stiffness and strength due to plastic excursions, etc.;

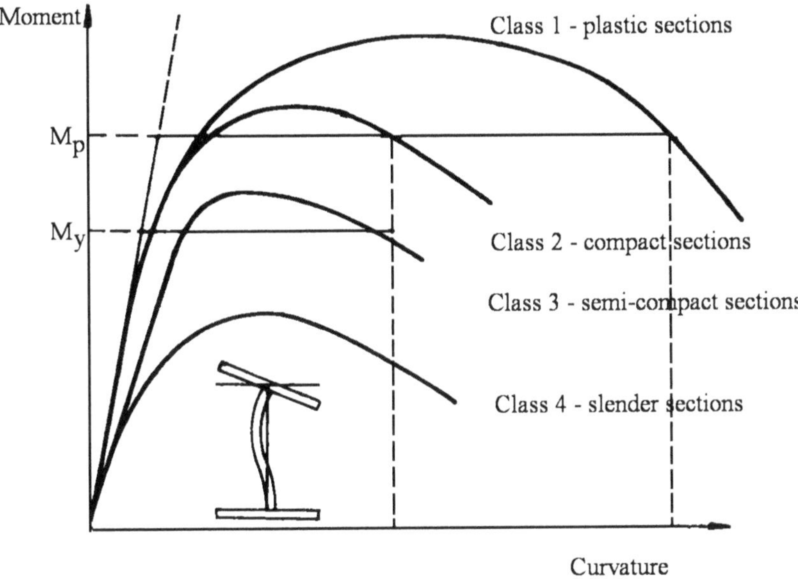

Figure 1.9: Cross-sectional behaviour classes

-required ductility, resulting from the earthquake actions, in which all factors influencing these actions are considered: magnitude, ground motion type, soil influence, natural period of structure versus ground motion period, number of important cycles, etc.

1.3.2. Ductility for Plastic Design

The plastic behaviour of a structure depends upon the amount of moment redistribution. The attainment of the predicted collapse load is related to the position of plastic hinges, where sections reach the full plastic moment, and to the plastic rotation which other hinges can develop elsewhere. Hence, a good behaviour of a plastic hinge requires a certain amount of ductility, in addition to its strength requirement. The plastic rotation capacity is the more rational measure of this ductility.

The basic requirement for plastic analysis of statically undetermined structures is that large rotations (theoretically infinite) are possible without significant changes in the resistant moment. But these theoretical large plastic rotations may not be achieved because some secondary effects occur. The limitation to plastic rotation is usually given by flexural-torsional instability, local buckling or brittle fracture of members. Due to this reduction in plastic rotation, the cross-section behavioural classes are used in design practice (Fig. 1.9):

-class 1 (plastic sections); sections belonging to the first class are characterized by the capability to develop a plastic hinge with high rotation capacity;

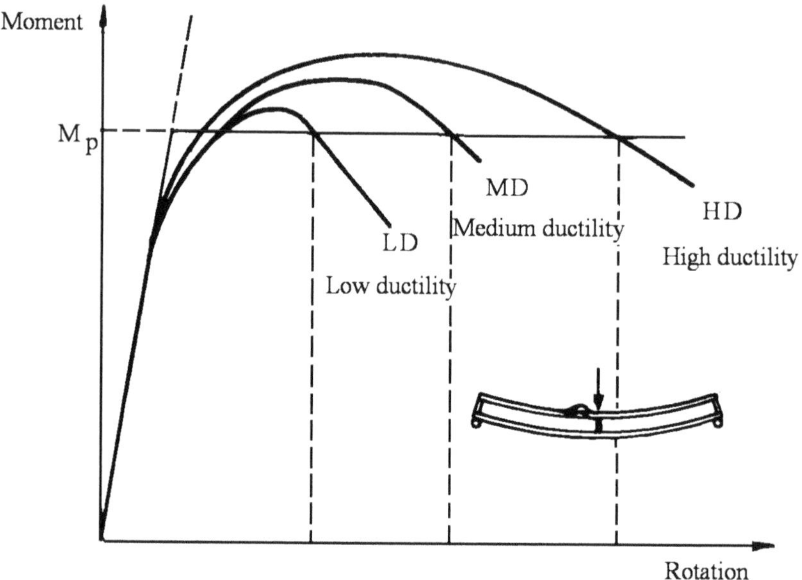

Figure 1.10: Member behavioural classes

-class 2 (compact sections); second class sections are able to provide their maximum plastic flexural strength, but they have a limited rotation capacity, due to some local effects;

-class 3 (semi-compact sections); sections fall in the third class when the bending moment capacity for the first yielding can be attained, without reaching the plastic moment;

-class 4 (slender sections); sections belonging to this class are not able to develop their total flexural resistance due to the premature occurrence of local buckling in their compression parts.

Evidently, only the first two classes have sufficient ductility to assure the plastic redistribution of moments.

This classification is limited at the cross-section level only, so it has many deficiencies. Another more effective classification at the level of a member has been proposed by Mazzolani and Piluso (1993) (Fig. 1.10):

-ductility class HD (high ductility) corresponds to a member for which the design, dimensioning and detailing provisions are such that they ensure the development of large plastic rotations;

-ductility class MD (medium ductility) corresponds to a member designed, dimensioned and detailed to assure moderate plastic rotations;

-ductility class LD (low ductility) corresponds to a member designed and dimensioned according to general code rules which assures low plastic rotations only.

These classifications used for the plastic design are also very useful for earthquake design. But some corrections must be introduced, due to the

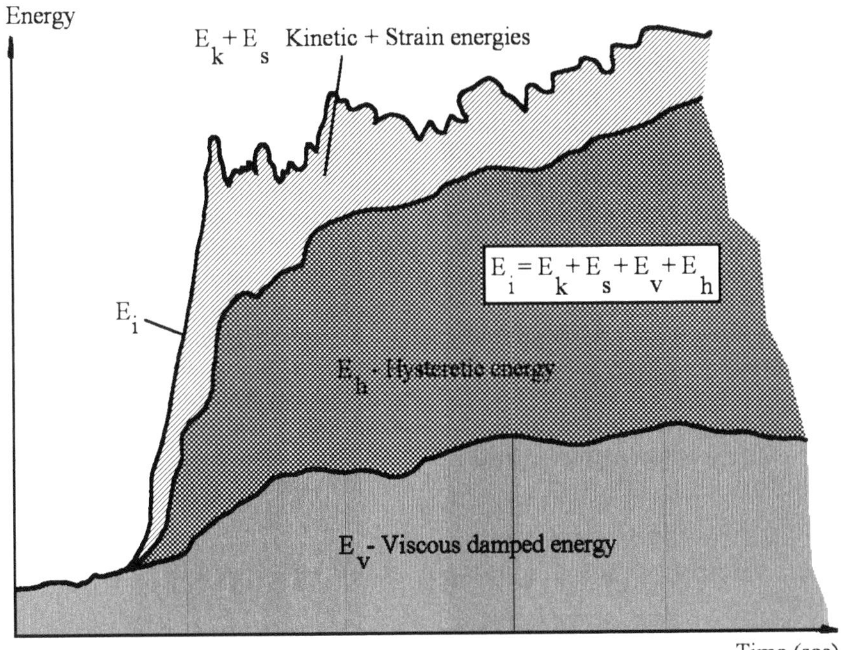

Figure 1.11: Seismic input and energy balance

fact that in the plastic design the loading system is monotonic, while in the case of earthquakes the variation in time is cyclic and an accumulation of plastic deformation occurs.

1.3.3. Ductility for Earthquake Design

The analysis of dynamic responses of structures to a seismic input is based on the application of energy concepts through the use of an energy balance among kinetic energy, recoverable elastic strain energy, viscous damped energy, and irrecoverable hysteretic energy. From Fig. 1.11 it is very clear that at the beginning of an earthquake or for a moderate earthquake, all input energy is balanced by damping. For severe earthquakes, when the input energy is greater than damping, the difference is balanced by hysteretic energy, which implies the ductility of the structure.

Analysing whether it is technically and economically possible to balance the seismic input, the designer may decide to adopt one of the following alternative approaches to protect the structure against severe earthquakes:

-to rely on the elastic behaviour of the structure only;

-to consider the viscous damping given by the non-structural elements and to increase the plastic hysteretic energy, namely to increase the structure ductility, by using appropriate constructional details, but accepting some damage during severe earthquakes;

-to increase the viscous damping, by using some damper devices, which decrease the hysteretic energy, protecting the structure against damage;

-to increase both viscous damping and hysteretic energies;

-to decrease the input energy using base isolation techniques and to balance the remaining energy through elastic vibrations only.

Generally, the absorption of input energy by elastic behaviour only is restricted for those facilities, whose failure may lead to other disasters, affecting man and/or the environment, e.g. nuclear power plants, dams, petrochemical facilities, etc. The common practice is to increase the hysteretic energy as much as possible through inelastic behaviour, using the ductile properties of the structure. Only recently has it been recognized that it is possible to increase the dissipated energy through dissipation devices. But the efficiency of these devices is very doubtful in many situations, so the dissipation of input energy through plastic deformation remains the most realistic measure of protection. So, even in using new control systems, it is absolutely necessary to provide the structure with some given level of ductility.

1.4 Progress in Design Methodology

1.4.1. International Activity

Today we have probably reached the stage when the actual structural performance during strong ground motions can be satisfactorily explained. Even now, the most important effects on the inelastic structural behaviour can be quantified (Fajfar, 1995). This significant progress which has been recently achieved in the earthquake design methodology is due to following factors:

-a great amount of information concerning the features of earthquakes has collected and important databases are operative. For instance, the database for the European area and Middle East of the Imperial College of Science and Technology of London (Ambraseys and Bommer, 1990) contains almost 1000 records for earthquakes of all magnitude and depths. Similar databases exist in Italy, Greece, USA and Japan;

-important activity in macro and micro-zonation has been recently carried out all over the world to identify and characterize all potential sources of ground motions. Important national and international conferences on zonation have been held recently;

-a wide activity in research works concerning the behaviour of structures in seismic areas has been made contemporary. This activity is materialized by a sequence of World Conferences (WCEE), European Conferences (ECEE), and National Conferences, the proceedings of each scientific event contains hundreds of very important papers;

-for steel structures, a sequence of STESSA Conferences on the Behaviour of Steel Structures in Seismic Areas initiated in 1994 in Timisoara (Mazzolani and Gioncu, 1995), followed by the 1997 Kyoto Conference (Mazzolani and Akiyama, 1997) and the 2000 Montreal Conference (Mazzolani and Tremblay, 2000). The proceedings of these Conferences present the state-of-the-art research works for these structures and underline the lessons to be learned from the last great earthquakes;

-important international associations such as the International Association for Earthquake Engineering (IAEE) and the European Association for Earthquake Engineering (EAEE), are involved in promoting the development of research activity. The result was the important initiative to establish the International Decade for Natural Disaster Reduction (IDNDR), with the aim to limit the destruction produced by natural phenomena, among them the earthquakes playing a leading role;

-the interest in the problem of earthquake engineering is manifested also by the European Convention for Constructional Steelwork (ECCS), with the preparation of the first proposal of codification "European Recommendations for Steel Structures in Seismic Zones" in 1988. A manual for using these recommendations for practical purposes has been elaborated by Mazzolani and Piluso, (1993b). Today, this text is incorporated in the Eurocode 8, for the steel buildings Section, with just some editorial changes;

-after each great event, extensive international activity is performed to characterize and understand what happened to buildings. So, after the Northridge earthquake, an SAC Joint Venture research program was elaborated upon by Structural Engineering Association of California (SEAOC), Applied Technology Council (ATC) and California Universities for Research in Earthquake Engineering (CUREE) (Ross, 1995, Krawinkler and Gupta, 1997). After the Kobe earthquake, a JSSC Special Task Committee was organized to analyse the impact of this earthquake on steel building frames (JSSC Technical Report, 1997). The European research project dealing with the "Reability of Moment Resistant Connections of Steel Building Frames in Seismic Areas (RECOS)" has been recently sponsored by the European Community within the INCO-Copernicus Joint Research Project (Mazzolani, 1999). The aim of this project is to examine the influence of joints on the seismic behaviour of steel frames, bringing together knowledge and experience of different specialists from several Countries (Italy, Romania, Greece, Portugal, France, Belgium, Slovenia and Bulgaria). The results of these research works were published by Mazzolani (2000) as editor.

-as a consequence of the significant economic losses that resulted during the last decade, a pressing need has been identified to develop a set of new approaches for improving the civil engineering facilities during severe earthquakes. So, the Structural Engineering Association of California (SEAOC) established the "Vision 2000 Committee" with the aim to develop a conceptual comprehensive framework for seismic design (Bertero et al, 1996, Bertero, 1997). This framework, which is called "performance-based seismic engineering" involves the conception, design, construction and maintenance activities. Consequently, the performance-based seismic engineering is defined as "... consisting of the selection of design criteria, appropriate structural systems, layout, proportioning, and detailing for a structure and its non-structural components and contents, and the assurance and control of construction quality and long-time maintenance, such that at specific levels of ground motion and with defined levels of reliability, the structure

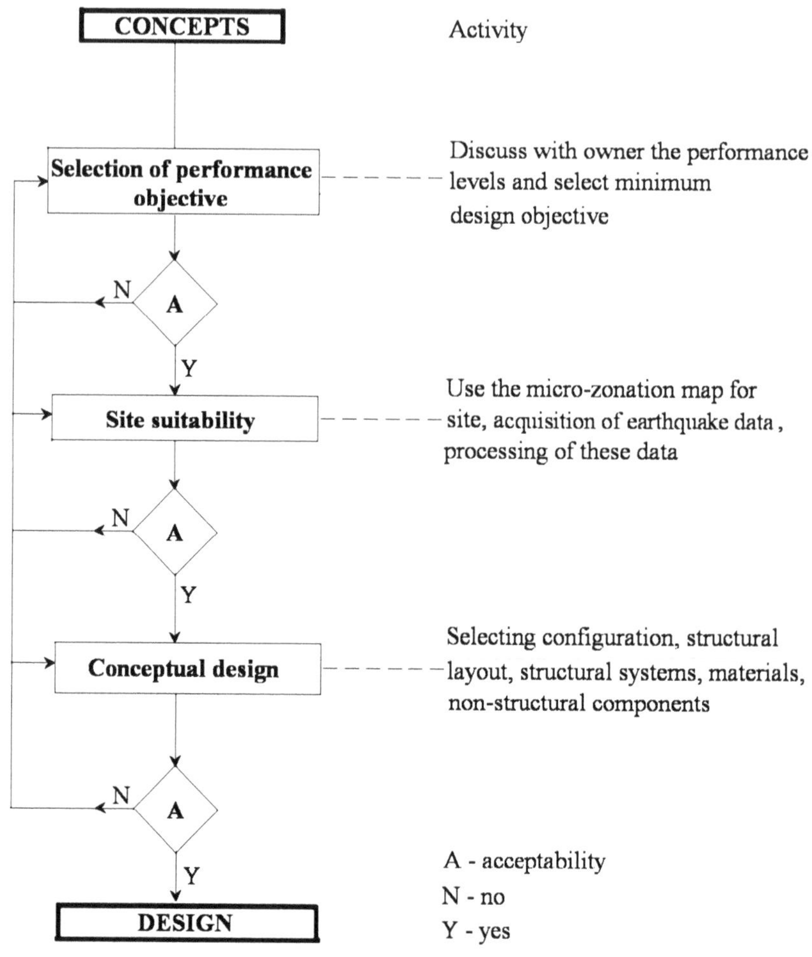

Figure 1.12: Flow chart for the concept development

will not be damaged beyond certain limiting states or other useful limits" (Bertero, 1997).

Thanks to this intense international activity in earthquake engineering, a marked progress in conception, design and construction has been observed recently.

1.4.2. Progress in conceptions

The flow charts for the concept development and its progress are presented in Figs. 1.12 and 1.13 and contain the following steps:

(i) *Selection of performance objectives.* For a long time the main objective of seismic design has been to protect the life losses and consequently to

ACTIVITY	CONVENTIONALCONCEPT	PROGRESS
SELECTED PERFORMANCE	One level performance: - ultimate limit state	Multi-level performance: - fully operational; - operational; - life safety; - near-collapse.
SITE SUITABILITY	Macrozonation: - determination of seismic action based on a single spectrum type	Microzonation information: - magnitude; - return period; - distance from source; - site soil stratification; - attenuation low; - duration.
CONCEPTUAL DESIGN	Traditional concepts based on strength demands	Using the new concepts considering the "rigidity-strength -ductility" triade

Figure 1.13: Progress in conception

satisfy only strength requirements by preventing structural collapse. Thus, the engineers have great difficulties in explaining to the owners what they are buying only according to the minimum code requirements and that in case of strong earthquake the code designed structure could suffer important damage, which must be repaired by using supplementary funds. After the last earthquakes, the level of economic losses has been socially and economically unacceptable, thus the owners have begun to understand that they must accept supplementary cost for additional protection. So, an important conceptual progression is achieved by introducing in design activity the possibility of protecting the structure at different levels of seismic action, in the frame of the so-called multi-level performance concept. Four levels of protection are, therefore, defined: fully operational (serviceable), operational (functional), life safety (damageable), near collapse (preventing collapse). These performance levels are associated to specific probabilities of occurrence : frequent, occasional, rare and very rare. The structural engineer, together with the owner, can establish whether the performance objectives remain at the code level, accepting damage occurrence in case of severe earthquakes or requesting additional protection and accepting paying supplementary costs.

(ii) *Site suitability.* The code provisions give a macro-zonation at the level of Country, which is insufficient for a proper design. The code analysis method is based on a single design spectrum, with some corrections considering the site soil type. This conception is proved to be unsatisfactory in many cases, because it does not catch the actual site feature. So, progress in this subject can be achieved by underlining the great importance of site conditions: actual magnitude, return period for each level of performance, distance from potential source, attenuation low, site soil stratification, direction, duration, etc. If the site conditions are bad from a seismic point of view, the designer can suggest to the owner to change the location or to accept supplementary cost for improving the soil conditions by using specific constructional methodologies.

(iii) *Conceptual design.* The conceptual design consists in the establishment of the general configuration of building (form, regularity, masses and stiffness distribution, gaps, etc.), foundation types (shallow or deep), structural materials (steel, composite, r.c., etc.), structural systems (moment resisting frames, braced frames, dual frames, etc.), joint types (rigid or semi-rigid), non-structural elements (type of interior and exterior walls), etc. Progress in this aspect is due to the large amount of new information obtained after the investigations of the behaviour of structures during the last severe earthquakes, and, from the impressive results of theoretical and experimental research works carried out all over the world. The designers must be conscious of the fact that a good concept design, rather than complex numerical analysis, has permitted many buildings to survive severe earthquake ground motions.

1.4.3. Progress in Design

The flow charts for design process and its progress recorded in the last time are presented in Figs. 1.14 and 1.15, and contain the following steps:

(i) *Preliminary design.* The conventional methodology of preliminary design is mainly based to satisfy the strength performance. But in many cases, the stiffness or ductility demands are more important than strength and the preliminary design must be improved through a series of analyses. Thus, the importance of a proper preliminary design should be overemphasized, because, if the design procedure has a poor preliminary design, the number of iterations for improving it is very high. By introducing the multi-level performance demands, it is necessary to establish, from the beginning, the most drastic requirement and an attempt of optimization is required, in order to smooth these demands. This step requires the evaluation of structure periods, design forces for different performance levels, critical load combinations, torsional effects, inter-storey drifts, required ductilities, expected overstrength, foundation behaviour and, finally, the beam and column preliminary sizing.

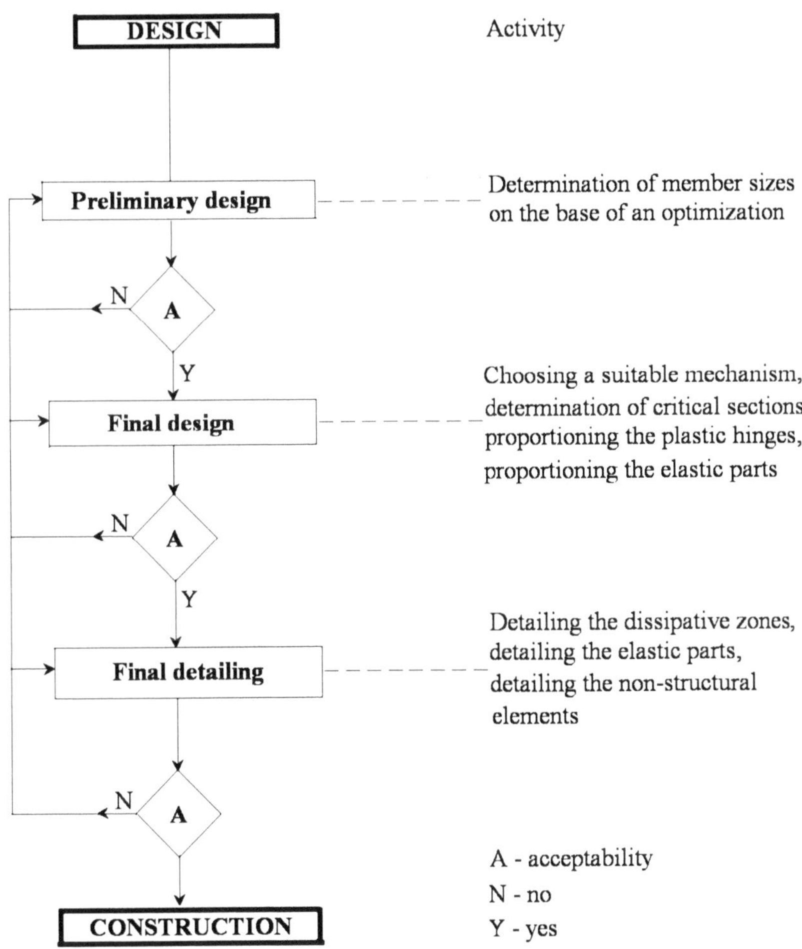

Figure 1.14: Flow charts for the design process

(ii) *Final design.* The conventional methodology considers only the classic verification of internal forces derived from the analysis, but we have to consider that plasticizations are possible everywhere, the pattern of plastic hinges is random, the overall seismic behaviour is difficult to predict and that the local ductility demand cannot be estimated. Due to these shortcomings of the conventional methodology, a new capacity design method is proposed (Pauley and Priestly, 1992, Bachmann et al, 1995). This methodology has been developed for reinforced concrete structures, but its principles can also be very useful for steel structures. Firstly, a complete and admissible plastic mechanism is chosen by locating the potential plastic hinge positions, and, an inelastic redistribution of design actions is carried out. Subsequently, adequate member dimensions are derived, by considering the

ACTIVITY	CONVENTIONAL CONCEPT	PROGRESS
PRELIMINARY DESIGN	On the basis of strength performance	Determination of member sizes on the basis of all level performance and an attempt of an optimization. Calculus of: - structural periods; - design forces for each performance level; - loads combination; - torsional effects; - ductility ratio; - expected overstrength; - foundation behaviour.
FINAL DESIGN	• Dimensions of structural elements • Verification of inter-storey drifts	Design of structural components based on: - choosing a suitable plastic mechanism; - determination of the critical sections; - verification of plastic hinges; - determination of overstrength of plastic hinges; - proportioning of elastic structural parts.
FINAL DETAILING	Few specific details	Improving of details after the last earthquakes for: - assuming a good local ductility; - improving the welded connections.

Figure 1.15: Progress in design

critical sections in the members selected for the eventual development of plastic hinges. The consideration of the overstrength of these critical sections, where plastic hinges occurs, is an important feature of the capacity design method. Finally, the other members or parts of them are designed to resist in elastic range, considering the overstrength of adjacent potential plastic hinges. This procedure ensures that the system may dissipate seismic energy with some local damage, but without global collapse and should allow a very useful design method to obtain the compatibility between ductility demands and available ductilities.

(iii) *Final detailing.* The suitable achievement of design concepts mainly depends on the simplicity of detailing of members, connections and supports. We can say that a good design is the only one which can be constructed (Bertero, 1997). The respect of detailing requirement assures a good behaviour of structure during severe earthquakes, according to the design concept. The failures produced during the ground motions indicate more deficiencies in structure detailing than in structure analysis. However, the code provisions contain only a few details directly involved in the protection against local damage of members and connections. After the last strong earthquakes, a great amount of information has been obtained and real progress in improving the detail conception is now possible.

1.4.4. Progress in Construction

Design and construction are two intimately related phases of the birth of a building. A good design conception is effective only if the building erection is qualitatively good. After each earthquake, field inspection has revealed that a large percentage of damage and failure has been due to poor quality control of structural materials and/or to poor workmanship. The flow charts for constructional aspects and their progress in recent years are represented in Figs. 1.16 and 1.17 and they contain the following steps:

(i) *Quality assurance during construction*, referring to the rules for verification of material qualities and proper workmanship. For instance, the analysis of structural steel specimen tests shows a considerable variation in material characteristics. In view of this variability, many present seismic code provisions specify only the minimum value for yield stress, which can lead to an unsafe design because a random overstrength distribution can modify the global ductility. An upper bound for yield stress must be given in the code and a more severe control of the variation factor is required.

(ii) *Monitoring and maintenance.* In many cases the damage or failure of buildings may be attributed to a lack in monitoring and improper maintenance. Due to this fact, the deterioration of mechanical properties of material and elements undermines the seismic response of the structure. The progress in this field would consist of the elaboration of severe rules for monitoring and maintenance of buildings during their life.

(iii) *Refurbishment, repair and strengthening.* The building may require some functional changing, which claims proper modification of the structural system. If this change is performed without suitable seismic rules, the modified structure could be victim of future earthquakes. On the other hand, there are a lot of buildings which were built many years ago, before the introduction of seismic design. Today, the building industry is looking with particular interest at the restoration, repairing and consolidation of

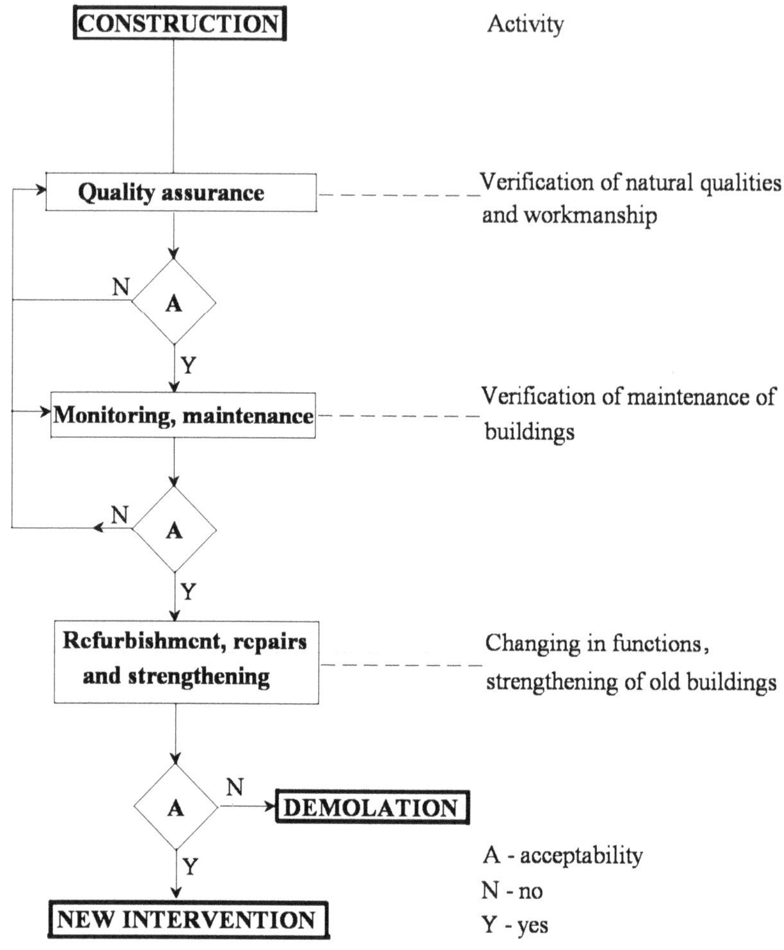

Figure 1.16: Flow charts for constructional aspects

old buildings. In future this industry will be extensively concerned with the refurbishment and strengthening of existing buildings, many of which are very important from an historical and architectural point of view. In all cases steel is an ideal material for refurbishment and important progress is marked in recent years in using some specific technologies (Mazzolani, 1990, 1996, Ivanyi, 1997).

1.5 Progress in Codification

1.5.1. The long way from theory to practice

The general methodology presented in the previous Section is in agreement with the new design philosophy. However, current code design methodology

ACTIVITY	CONVENTIONAL CONCEPT	PROGRESS
QUALITY ASSURANCE	Poor provisions on verification of material quality and workmanship qualification	Very careful control of - material qualities - workmanship qualities
MONITORING AND MAINTENANCE	Without rules	Rules for maintenance and monitoring
REFURBISHMENT, REPAIRS AND STRENGTHENING	Poor rules	Rules for refurbishment repairs and strengthening

Figure 1.17: Progress in construction

fails in realizing goals and objectives of this philosophy (Bertero, 1997). The main reason for this can be explained as follows:

(i) Structural engineers are professionals and not researchers; their activities are driven by the need to deliver the design in a timely and cost effective manner. They may also resist new concepts, unless these concepts are put into the context of their present mode of operation (Krawinkler, 1995). On the other hand, in many cases the research works are performed by professors and researchers who are more interested in publishing their results for their colleagues than in the transmission of new knowledge to those who will apply it (De Bueno, 1996).

(ii) The loading condition on a structure during a major earthquake is very difficult to model. The definition of a design earthquake load inevitably calls for a series of engineering judgments of seismology, safety policy as well as structural engineering matters (Studer and Koller, 1995). For this reason design earthquakes should be elaborated upon in close collaboration between seismologists and engineers. Unfortunately, there are some difficulties in communication between these two professions. Engineers, in designing structures, must rely on proven principles, the engineering approaches being quantitative. In contrast, seismologists, when investigating geologic processes, have the privilege of proposing, testing and discarding erroneous hypotheses (Seeber and Armbuster, 1989). This difference in approach may be in part responsible for difficulties in collaboration.

(iii) The design philosophy for earthquake loads is totally different from the design methodology for the other loading conditions. The admittance of plastic deformations during severe earthquakes implicitly anticipates the occurrence of structure damage, which is not so easily accepted by the building owner. So the most relevant performance criterion for a building structure that has survived an earthquake is the total cost of damage. In this perspective, clear attention to damage control for structural and non-structural elements should be a central concept. This damage control is very difficult to be quantified in a simple manner to be introduced as provision in a design code.

(iv) The implementation of new concepts in codes is constrained by the need to keep the design process simple and verifiable. Today, the progress of computer software has made it possible to predict the actual behaviour of structures subjected to seismic loads. But now the availability of powerful computational tools at relatively low cost does not imply that the most complicated models must always be used. During the elaboration of codes it must be kept in mind that the engineering community tends to be conservative. So the code provisions must always be a compromise between new and old knowledge and procedures, otherwise the new methodology will be rejected by the designers.

(v) Recognizing the need for code development based on a transparent methodology, we must also recognize that it is necessary to underline some dangers of this operation: over-simplification, over-generalization and immediate application in practice of the latest research results, without an adequate period of time during which these results can be verified.

1.5.2. Required Steps in Code Elaboration

Considering the above observations, it is possible to establish the following steps in the elaboration of a performance code (Bertero, 1997):

-establishing a conceptual methodology, which must contain the problems of practical design and construction;

-elaboration of the first draft, in which the methodology is developed in detail in such a way as to be directly applied in practice;

-designing of some buildings by using different regular and irregular configurations, structural layouts and structural systems, which preferably have been designed and constructed according to current seismic codes, and whose response to earthquakes have been recorded or predicted;

-selection of conclusions after using the first draft and establishment of some simplifications of code provisions requested by the designers;

-elaboration of the final draft, in which all the conclusions coming from the previous steps are introduced.

The importance and advantage of developing of code elaboration is due to the fact that it is based on a transparent methodology which considers and checks the selected or desired performance objectives, in a very clear manner for the designer. This methodology assures that the improved provisions can be easily introduced in the code, when new or more reliable data become available, because it is not necessary to change the philosophy or the format of code. Another important advantage is the clear establishment of a program for focused research on the need to improve the code.

In a good codification for seismic design the following major objectives should be considered:

-codification for macro and micro-zonation;

-codification for design of structures, including the foundation design;

-codification for design of non-structural elements.

1.6 Challenges in Design Methodologies

1.6.1. After the Last Severe Earthquakes

For a long time it was generally accepted in design practice that steel is an excellent material for seismic resistant structures, thanks to its performances in strength and ductility at a material level. But very serious alarm signals about this optimistic view arise after the last severe earthquakes of Michoacan (1985), Northridge (1994) and Kobe (1995). Beside a lot of steel constructions which have shown a good performance, at the same time a lot of others exhibited very bad behaviour. For these, the actual behaviour of joints, members and structures has been very different from the design expectation, so the traditionally good performance of steel structures under severe earthquakes has been recognized as a dogmatic principle not always respected in the reality. In many cases the damage occurred when both design and detailing have been performed in perfect accordance with the code provisions, it means that something new happened, which was not foreseen in the design practice (Mazzolani, 1995). The engineers and scientists want to know exactly the reasons of this poor behaviour:

-inaccuracy of ground motion modelling?

-modification of material qualities during severe earthquakes?

-shortcoming in design concept, especially concerning the use of the simplified design spectrum method?

-insufficient code provisions concerning ductility demand?

-shortcoming in accuracy in constructional details?

Today, concerning the measures which are necessary to eliminate the possibility of similar damage occurrance during the future strong earthquakes, the world of specialists is divided. Some of them consider that the actual design philosophy is proper and only some improvement of constructional details, especially concerning welded joints should be enough. But at the same time, there are many other specialists who consider that the abovementioned questions are real problems in the design and some pressing modifications in concept are required.

In the frame of this debate, the challenges in design methodology are presented.

1.6.2. Challenge in Concept

The study of the structural response during an earthquake constitutes an important step for improving the methodologies of analysis and design of structures. In the past, due to the reduced number of records during the severe earthquakes (the famous El Centro record obtained in 1940 was for long time the main information about the time-history of an earthquake), the developed design methods are mainly based on simple hypothesis, with little possibility of verifying their accuracy. Today, due to a large network of instrumentation all over the world, several measurements of ground motions for different distances from the sources and on different site conditions are available. This situation gives the possibility to underline a new very important aspect which was previously neglected in the current concept: the difference in ground motion for near-field and far-field earthquakes. The near-field region of an earthquake is the area which extends for several kilometers from the projection on the ground surface of the fault rupture zone. Because in the past the majority of ground motions were recorded in the far-field region, the current concept refers to this earthquake type only. The great amount of damage during the Northridge and Kobe earthquakes are due to the fact that these towns are situated in a near-field region. So, the ground model, adopted in current design methodology on the basis of ground motions recorded in far-field regions cannot be used to describe in proper manner the earthquake action in near-field regions. The differences are presented in Fig. 1.18 (Gioncu et al, 2000, Mazzolani and Gioncu, 2000):

-the direction of propagation of the fault rupture has the main influence for near-field regions, the local site stratification having a minor consequence. Contrary to this, for far-field regions, soil stratification for travelling waves and site conditions are of first importance;

-in near-field regions, the ground motion has a distinct low-frequency pulse in acceleration time history and a pronounced coherent pulse in velocity and displacement time histories. The duration of ground motion is very short. For far-field regions, the records in acceleration, velocity and displacement have the characteristic of a cyclic movement, with a long duration;

-the velocities in near-field regions are very high. During Northridge and Kobe earthquakes, velocities with values of 150 – 200 cm/sec were recorded at the soil level, while for far-field regions these velocities did not exceed 30 – 40 cm/sec. So, in case of near-field regions the velocity is the most important parameter in design concept, replacing the accelerations which are a dominant parameter for far-field regions;

-the vertical components in near-field regions may be greater than the horizontal components, due to the direct propagation of P waves (see Chapter 3), which reach the structure without important modifications due to soil conditions, their frequencies being far from the soil frequencies.

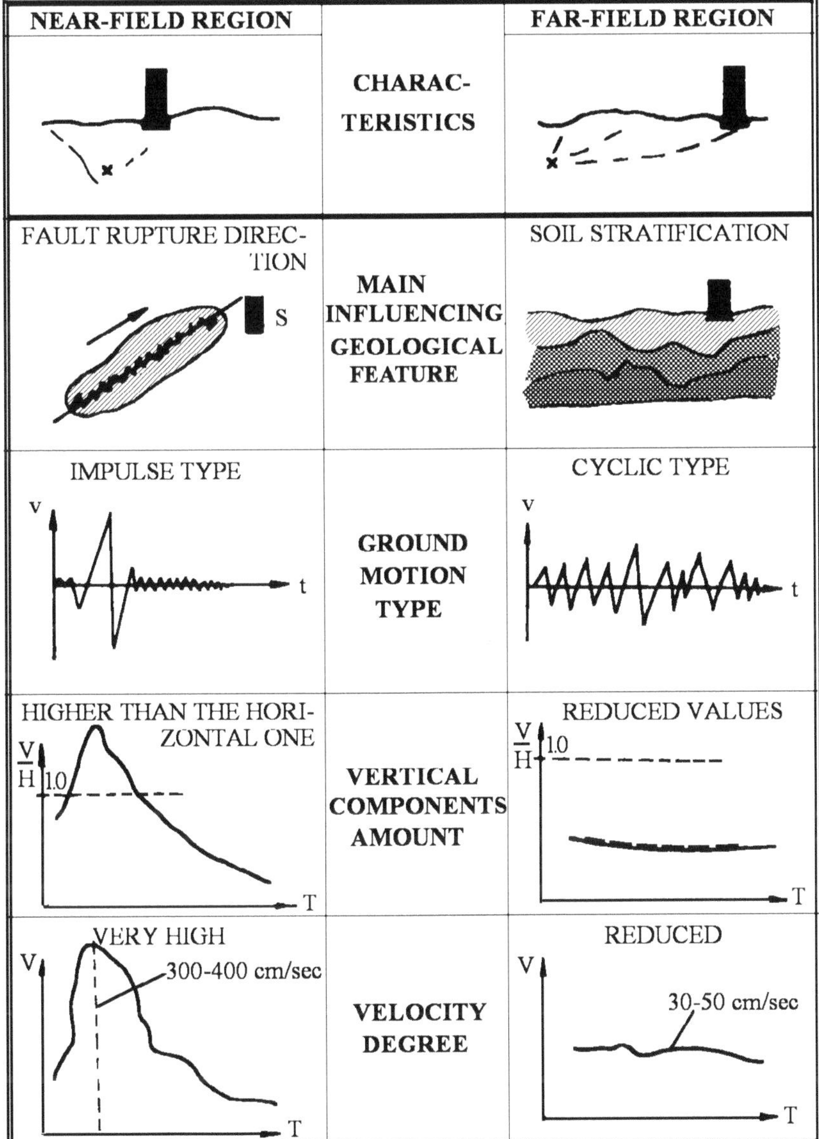

Figure 1.18: Near-field vs. far-field ground motion features

1.6.3. Challenge in Design

As a consequence of the above mentioned differences in ground motions, there are some very important modifications in design concept (Fig. 1.19) (Gioncu et al, 2000):

(i) In near-field regions, due to very short periods of ground motions and due to pulse characteristic of loads, the importance of higher vibration

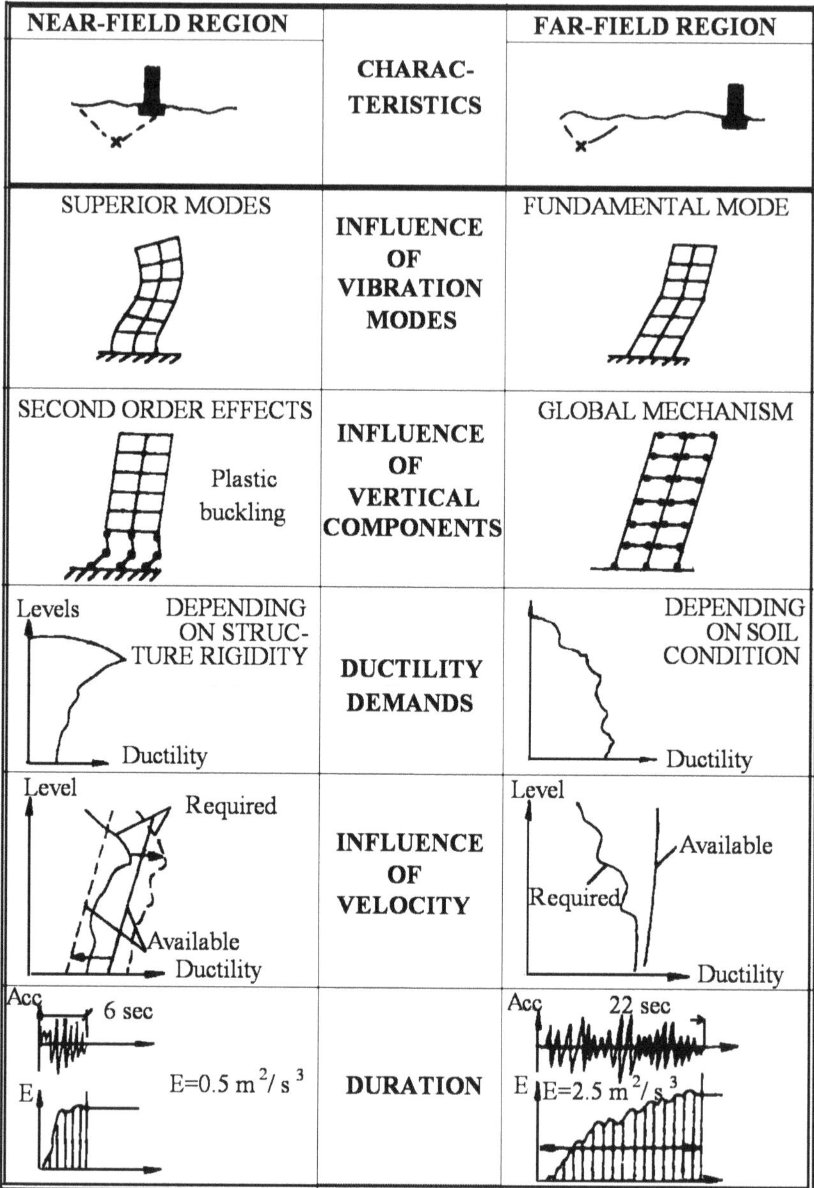

Figure 1.19: Near-field vs. far-field structure behaviours

modes increases, in comparison with the case of far-field regions, where the first fundamental mode is dominant. For structures subjected to pulse actions, the impact propagates through the structure as a wave, causing large localized deformations and/or important inter-storey drifts. In this situation the classic design methodology based on the response of a single-degree-of freedom system characterized by the design spectrum is not sufficient to describe the actual behaviour of structures. Aiming to solve this problem, a continuous shear-beam model is proposed and the design spectrum is the result of the shear strain, produced by the inter-storey drift (Iwan, 1997).

(ii) Due to the concordance of frequencies of vertical ground motions with the vertical frequencies of structures, an important amplification of vertical effects may occur. At the same time, taking into account the reduced possibilities of plastic deformation and damping under vertical displacements, the vertical behaviour can be of first importance for structures in near-field regions. The combination of vertical and horizontal components produces an increase in axial forces in columns and, as a consequence, increases in second order effects.

(iii) Due to the pulse characteristic of actions, developed with great velocity due to the lack of important restoring forces, the ductility demand may be very high. So, the use of the inelastic properties of structures for seismic energy dissipation must be very carefully examined. At the same time, the short duration of ground motions in near-field regions is a favourable factor. A balance between the severity of ductility demand, due to pulse action, and the effect of short duration must be analyzed.

(iv) Due to the great velocity of seismic actions, an increase in yield strength occurs, which means a significant decreasing of available local ductility. Due to this increasing demand, as an effect of the impulse characteristic of loads, together with the decreasing of response, due to the effect of high velocity, the demand-response balance can be broken. The need to determine the ductilities as a function of the velocity of actions is a pressing challenge for research works.

(v) If it is not possible to take advantage of the plastic behaviour of structures, due to this high velocity, it is necessary to consider the variation of energy dissipation through ductile fracture. The fact that many steel structures were damaged by fracture of connections during the Northridge and Kobe earthquakes without global collapse, gives rise to the idea that the local fracture of these connections can transform the original rigid structure into a structure with semi-rigid joints. The positive result of this weakening is the reduction of seismic actions at a level which can be supported by the damaged structure, taking into account that the duration of an earthquake is very short. This is not the case of far-field earthquakes, for which the effect of long duration can induce the collapse of the structure.

1.6.4. Challenge in Construction

Pictures showing failures of structural members, in which details were grossly wrong, are very frequent in the post-earthquake reports. Very often, these mistakes were due to the fact that the detailed analytical calculations were not accompanied by a consistent set of structural details. This reality claims that special parts of codes must be elaborated in order to give constructional requirements for details (Corsanego, 1995). After the joint damage produced during the last earthquakes, when welded connections behaved very badly, the challenge in construction is to establish some provisions to improve the joint behaviour and to eliminate any source of brittle fracture. A lot of important experimental tests (JSSC, 1997) and a very careful examination of the connection behaviour (Miller, 1995) have been performed and the output of these investigations can be used directly in building erection.

1.7 Conclusions

After the careful examination of the state-of-the-art conception, design and construction, by analysing the most important challenge for future developments, it is possible to underline the following conclusions:

-seismic design is one of more intricate activities of the structural engineer due to the impossibility to accurately predict the characteristics of future ground motions that may occur at a given site and the difficulty in evaluating the complete behaviour of a structure when subjected to very large seismic actions. The best way to solve these unknowns is to give to the structure the necessary ductility;

-without any doubt the design concept based on the ductility properties of structures remains for the moment the main method to assure a suitable behaviour against strong earthquakes. The capacity to predict ductility demand and available ductility under seismic loads is a key-point in seismic design;

-even in the case of new concepts of response control, the structures must be provided with some given ductility level to prevent the situation in which the control measures do not react to the seismic input as it has been foreseen in design;

-the methodology for checking the structure ductility must be carried out with the same degree of importance as given for fulfilling the other criteria, as strength, stability and deformation checks. It means that, instead or in addition to the general rules, especially for construction details, some direct analytical formulations must be introduced in practice for ductility verifications.

1.8 References

Akiyama, H. (1985): Earthquake-Resistant Limit-State Design for Buildings. University of Tokyo Press

Ambraseys, N., Bommer, J.J. (1990): Uniform magnitude re-evaluation for the strong-motion database of Europe and adjacent areas. European Earthquake Engineering, No. 2, 3-16

Bachmann, H., Linde, P., Werik, Th. (1995): Capacity design and nonlinear dynamic analysis of earthquake -resistant structures. In 10 th European Conference on Earthquake Engineering, (ed G.Duma), Vienna, 28 August-2 September 1994, Balkema, Rotterdam ,Vol. 1, 11-20

Bertero, V.V. (1988): Ductility structural design. State-of-the art report. In 9th World Conference on Earthquake Engineering, Tokyo - Kyoto, 2-9 August 1988, Vol. VIII, 673-686.

Bertero, V.V. (1992): Major issues and future directions in earthquake -resistant design. In 10th World Conference on Earthquake Engineering, Madrid, 19-24 July 1992, Balkema , Rotterdam, 6407-6444

Bertero, V.V (1996): State-of-the-art report on design criteria. In 11th World Conference on Earthquake Engineering, Acapulco, 23-28 June 1996, CD-ROM Paper 2005

Bertero, V.V (1997): Codification, design, and application. General report. In Behaviour of Steel Structures in Seismic Areas, STEESA 97, (eds. F.M. Mazzolani, H. Akiyama), Kyoto, 3-8 August 1997, 10/17 Salerno, 189-206

Bertero, R.D., Bertero, V.V.,Teran-Gilmore, A. (1996): Performance- based earthquake-resistant design based on comprehensive design philosophy and energy concepts. In 11th World Conference on Earthquake Engineering, Acapulco, 23-28 June 1996, CD-ROM Paper 611

Berz, G., Smolka, A. (1995): Urban earthquake potential: Economic and insurance aspects. In 10th European Conference on Earthquake Engineering, (ed G. Duma) , Vienna, 28 August- 2 September 1994, Balkema, Rotterdam, Vol. 2, 1127-1134

Cosenza, A. (1995): Recent trends in the field of earthquake damage interpretation. In 10th European Conference on Earthquake Engineering, (ed G. Duma) , Vienna, 28 August- 2 September 1994, Balkema, Rotterdam, Vol. 1, 763-771

De Buen, O. (1996): Earthquake resistant design: A view from the practice. In 11th World Conference on Earthquake Engineering, Acapulco, 23-28 June 1996, CD-ROM Paper 2002

De Luca, A., Mele, E. (1997): Base isolation and energy dissipation. General report. In Behaviour of Steel Structures in Seismic Areas, STESSA 97 , (eds F.M.Mazzolani and H.Akiyama),Kyoto, 3-8 August 1997, 10/17 Salerno, 683-699

Dolce, M., Kappos, A., Zuccaro, G., Coburn, A.W. (1995): Report of the EAEE Working Group 3: Vulnerability on risk analysis. In 10th European Conference on Earthquake Engineering, (ed. G.Duma), Vienna, 28 August- 2 September 1994, Balkema, Rotterdam, Vol. 4, 3049-3077

Elnashai, A. (1994): Seismic performance of steel structures: Editorial. Journal of Constructional Steel Research. Special issue, Vol. 29, No. 1-3 , 1-4

Fajfar, P. (1995): Design spectra for new generation codes: Eurocode 8 achieves the half-way mark. In 10th European Conference on Earthquake Engineering (ed. G.Duma), Vienna, 28 August- 2 September 1994, Balkema, Rotterdam, Vol. 4, 2969-2974

Gioncu, V. (1997): Ductility demands. General report. In Behaviour of Steel Structures in seismic Areas, STESSA 97, (eds. F.M.Mazzolani and H. Akiyama) Kyoto, 3-8 August 1997, 10/17 Salerno, 279-302

Gioncu, V. (1998): Ductility criteria for steel structures. In 2nd World Conference on Steel in Construction, San Sebastian, 11-13 May 1998, CD-ROM Paper 220

Gioncu, V. (1999): Framed structures: Ductility and seismic response. General report. In 6th International Conference on Stability and Ductility of Steel Structures, SDSS 99, Timisoara, 9-11 September 1999, Journal of Constructional Steel Research, Vol. 55, No. 1-3, 125-154

Gioncu, V., Mateescu, G., Tirca, L., Anastasiadis, A. (2000): Influence of the type of seismic ground motions. In Moment Resisting Connections of Steel Building Frames in Seismic Areas (ed. F.M.Mazzolani), E& FN Spon, London, 57-92

Ghersi, A. (1994): Non conventional systems for seismic protection. Seminar on Behaviour of Steel Structures in Seismic Areas, Timisoara, 2 July 1994

Housner, G.M. (1956): Limit design of structures to resist earthquakes. The First World Conference on Earthquake Engineering, Berkley, California, 5.1-5.11

Housner, G.M. (1959): Behaviour of structures during earthquakes. Journal of Engineering Mechanical Division, Vol. 85, No. 4, 109-129

Housner, G.M. et al (1997): Structural control: Past, present, and future. Journal of Engineering Mechanics, Vol. 123, No. 9, 897-971

IAEE, (1992): A time for action. World seismic safety initiative. 1 July 1992

Ivanyi, M. (1997): Strengthening and repairing. Seismological considerations. General report. In Behaviour of Steel Structures of Seismic Areas, STESSA 97 , (eds. F.M.Mazzolani and H.Akiyama) Kyoto, 3-8 August 1997, 10/17 Salerno, 907-916

Iwan, W.D. (1997): Drift spectrum: Measure of demand for earthquake ground motions. Journal of Structural Engineering, Vol. 123, No. 4, 397-404

Iwan, W.D. (1995): Near-field considerations in specification of seismic design motions for structures. In 10th World Conference on Earthquake Engineering (ed. G.Duma), Vienna, 28 August- 2 September, Balkema, Rotterdam, Vol. 1, 257-267

Jakob, K.H., Turkstra, C.J.(eds) (1989): Earthquake Hazards and the Design of Constructed Facilities in the Eastern United States, New York, 24- 26 February 1988, Annuals of the New York Academy of Science, Vol. 558, XIII-XIV

Japanese Society of Steel Constructions, JSSC, (1997): Kobe earthquake damage to steel moment connections and suggested improvement. Technical Report, No. 39

Krawinkler, H. (1995): New trends in seismic design methodology. In 10th European Conference on Earthquake Engineering, (ed.G.Duma), Vienna, 28 August- 2 September 1994, Balkema, Rotterdam, 821-830

Krawinkler, H., Gupta, A. (1997): Deformation and ductility demands in steel moment framed structures. In Stability and Ductility of Steel Structures, SDSS 97, (ed. T.Usami), Nagoya, 29- 31 July 1997, Vol. 1, 57-68

Mazzolani, F.M. (1986): The seismic resistant structures of the new Fire Station of Naples, Costruzioni Metalliche, No. 6

Mazzolani, F.M. (1990): Refurbishment and extensions: the case of steel. International Symposium ICSC, Luxemburg

Mazzolani, F.M. (1995): Some simple considerations arising from Japanese presentation on the damages caused by the Hanhsin earthquake. In Stability of Steel Structures, (ed. M.Ivanyi), Budapest, 21- 23 September 1995, Akademiai Kiado, Budapest, vol 2, 1007-1010

Mazzolani, F.M. (1996): Strengthening options in rehabilitation by means of steelwork. Proc. of 5th Int. Colloquium on Structural Stability, SSRC Brazilian Session, Rio de Janeiro, 5- 7 August 1996, 275-287

Mazzolani, F.M. (1999): Reability of moment resistant connections of steel building frames in seismic areas: The first year of the RECOS project. In 2nd European Conference on Steel Structures, (eds. J. Studnicka, F. Wald, J. Machacek), Prague, 26- 29 May 1999, CD-ROM

Mazzolani, F.M. (2000): Moment Resistant Connections of Steel Building Frames in Seismic Areas. Design and Reliability. E&FN Spon, London

Mazzolani, F.M., Akiyama, H. (eds.) (1997): Behaviour of steel structures in Seismic Areas, STESSA 97, Kyoto Conference, 3-8 August 1997 , 10/17, Salerno

Mazzolani, F.M., Gioncu, V. (eds) (1995): Behaviour of Steel Structures in Seismic Areas, STESSA 94, Timisoara Conference, 26 June - 1 July 1994 E&FN Spon, Londra

Mazzolani, F.M., Tremblay R. (eds) (2000): Behaviour of Steel Structures in Seismic Areas, STESSA 2000, Montreal Conference, 21- 24 August 2000, Balkema, Rotterdam

Mazzolani, F.M., Piluso, V. (1993a): Member behaviour classes of steel beams and beam-columns. In Giornate Italiane delle Costruzione in Acciaio, Viareggio, 24-27 October 1993, 405-416

Mazzolani,F.M., Piluso, V. (1993b): Manual on Design of Steel Structures in Seismic Zones. Manual, ECCS Document

Mazzolani, F.M., Piluso, V. (1996): Theory and Design of Seismic Resistant Steel Frames. E&FN Spon, Londron

Mazzolani, F.M., Serino, G. (1993): Most recent developments and applications of seismic isolation of civil buildings in Italy. Int. Conf. on Isolation, Energy Dissipation and Control of Vibrations of Structures, Capri, Italy

Mazzolani, F.M., Serino, G. (1994a): Innovative techniques for seismic retrofit: design methodologies and recent applications. AFPS-ANIDIS Symposium, Nice

Mazzolani, F.M., Zampino, G., Serino, G. (1994b): Seismic protection of Italian monumental buildings with innovative techniques. Int. Workshop IWADBI, Shanton, China

Mazzolani, F.M., Serino, G. (1997a): Top isolation of suspended steel structures: modelling, analysis and application. In Behaviour of Steel Structures in Seismic Areas, STESSA 97, (eds. F.M. Mazzolani and H. Akiyama), Kyoto, 3- 8 August 1997, IO/I7, Salerno, 734-743

Mazzolani, F.M., Serino, G. (1997b): Viscous energy dissipation devices for steel structures: modelling, analysis and application. In Behaviour of Steel Structures in Seismic Areas, STESSA 97, (eds. F.M. Mazzolani and H. Akiyama), Kyoto, 3- 8 August 1997, IO/I7, Salerno, 724-733

Mazzolani, F.M., Gioncu, V. (2000): Seismic Resistant Steel Structures. CISM course, Udine, 18-22 October 1999, Springer, Wien

Mele, E., De Luca, A. (1995): State of the art report on base isolation and energy dissipation. In Behaviour of Steel Structures in Seismic Areas, STESSA 94, (eds. F.M. Mazzolani and V. Gioncu), Timisoara, 26 June- 1 July 1994, E&FN Spon, Londra, 631-658

Miller, D.K. (1995): Northridge: The role of welding clarified. In Seventh Canadian Conference on Earthquake Engineering, Montreal, 573-580

Miranda, E. (1993): Evaluation of site - dependent inelastic seismic design spectra. Journal of Structural Engineering, Vol. 119, No. 5, 1319-1338

Newmark, N.M., Hall, W.J. (1969): Seismic design criteria for nuclear reactor facilities. Proc. of 4th World Conference on Earthquake Engineering, Santiago, Chile , 2 (B-4), 37-50

Pauley, T., Pristley, M.J.N. (1992): Seismic Design of Reinforced Concrete and Masonry Structures, John Wiley and Sons, London

Petroski, H. (1985): To Engineer is Human. The Role of Failure in Successful Design. Mac Millan, Londra

Popov, E.P. (1994): Development of U.S Codes. Journal of Constructional Steel Research, Vol. 29, 191-207

Roesset, J.M. (1992): Modelling problems in earthquake resistant design : Uncertainties and needs. In 10th World Conference on Earthquake Engineering, Madrid, 19- 24 July 1992, Balkema, Rottedam, 6445-6483

Romero, E.M. (1995): Supplementary energy dissipations for maximum earthquake protection of tall buildings structures. The Structural Design of Tall Buildings, Vol. 4, No. 1, 91-101

Ross, A.E. (1995): The Northridge earthquake moment frame failures: Results of phase one of the SAC joint venture research program. The 7th Canadian Conference on Earthquake Engineering, Montreal, 581-608

Seeber, L., Armburster, J.G. (1989): Displacement seismogenic faults and nonstationary seismicity in the Eastern U.S. In Earthquake Hazards and Design of Constructed Facilities in the Eastern U.S., (eds. K.H.Jacobs and C.L Turkstra), New York, 24- 26 February 1988, Annals of New York Academy of Science, Vol. 558, 21-39

Studer, J., Koller, M.G. (1995): Design earthquake: The importance of engineering judgments. In 10th European Conference on Earthquake Engineering , (ed. G.Duma), Vienna, 28 August- 2 September 1994, Balkema, Rotterdam, 167-176

Takanashi, K., Nakashima, M. (1994): Stability considerations on seismic performance of steel structures. Proc. of the SSRC conference, Link between Research and Practice, Bethlehem, 21-22 June 1994, Lehigh University, 119-132

Veletsos, A., Newmark, N.M (1960): Effect of inelastic behaviour on response of simple system to earthquake motions. Proc. 2th World Conference on Earthquake Engineering, Tokyo, 855-912

Waszawski, A., Ghick, J., Segal, D. (1996): Economic evaluation of design codes. Case of seismic design. Journal of Structural Engineering, Vol. 122, No. 12, 1400-1408

2

Learning from Earthquakes Seismic Decade 1985–1995

2.1 Damage of Steel Structures during the Recent Earthquakes

The occurrence of earthquakes, their consequent impact on people and on the facilities they live and work in, the evaluation and interpretation of damage caused by severe ground motions, are the principal items for structural engineers designing buildings in seismic areas. The attempts to find the answer to the question: "why does damage occur, after a wide amount of research work?", is an ethical duty of the specialists. The paradox of structural engineering is that while engineers can learn from the structural mistakes of what not to do, they do not necessarily learn from successes how to do (Petroski, 1985). The failure of a structure contributes more to the evolution of design concepts than structures standing without accidents, on the condition that the engineers have the capability to understand what happened. So, the damage of a structure during an earthquake represents a challenge for structural engineers to improve the design methods.

This aspect can be very well illustrated with the example of steel structures. For a long time it was accepted in design practice that steel is an excellent material for seismic-resistant structures, thanks to its performance in terms of material strength and ductility. To exemplify this good behaviour, in many papers the excellent performance of the Torre Latino-Americana building in Mexico City was mentioned, without showing that the fundamental period of this building is much larger than the predominant period of the Michoachan ground motion and the structure has piles supported by rock. Therefore, the seismic demand was relatively low (Osteraas and Krawinkler, 1989, De Buen, 1996).This is an example of the danger of generalization, mentioned in the previous Chapter. Contrary to this, the last severe earthquakes of Michoacan (1985), Northridge (1994) and Kobe (1995) have seriously compromised this image of steel as the most suitable material for seismic resistant structures. It was a providential sign that in the same place, Mexico City, where the case of Torre Latino-Americana building was assumed as example of good performance of steel structures, the first overall collapse of a steel structure occurred: the Pino Suarez building. The bad performance of joints in steel structures, both in Northridge and Kobe earthquakes, having the same characteristics of damage, shows that there are some general mistakes in design concept. And the fact that in both cases damage arises also when the design and detailing were performed in perfect accordance with the design philosophy and code provisions, amplifies the challenge addressed to structural engineers.

Generally, when such failures occur, the observed damage depends on following factors (Corsanego, 1995):

- general characteristics of the earthquake;
- local soil behaviour;
- seismic vulnerability of buildings;
- incomplete knowledge in seismic behaviour of structures;
- inadequacy of code provisions;
- wrong design, in opposition with code provisions;
- bad construction;
- lack of maintenance.

From the structural engineers' point of view the most important aspects of post-earthquake analysis are the lessons to be learned after each event, having the conviction that the best teacher is the full-scale laboratory of nature. No theory or mathematical model can be accepted, unless they correctly explain what happens in nature (McClure, 1989).

In the following Sections, the best known failures of steel structures during the last earthquakes are presented and a careful examination of the above mentioned factors is carried out. Almost exhaustive references are presented in order to have the possibility of finding supplementary information. Anticipating the results of this analysis, it is necessary to underline that the main conclusions are the existence of some differences between the ductility concept incorporated in codes and the actual ductility demand. So, there are some situations when the code ductility concept does not work (Eisenberg, 1995) and the task of improving these provisions is a crucial problem for structural engineers.

2.2 Michoacan Earthquake

2.2.1. Earthquake Characteristics

On 19 September 1985, a major earthquake of magnitude 8.1 occurred, with an epicenter located in Zacatula City, about 350 km from Mexico City, in the South of Michoacan State, due to the subduction activity of the Mexican Pacific coast. This earthquake was the most severe among a succession of ground motions with magnitudes between 5.8 and 8.1, which were produced in the South-West part of Mexico. Fig. 2.1a shows some of these earthquakes and the area of rupture of 1985 event (Reinoso et al, 1992, Iglesias and Gomez-Bernal, 1992).

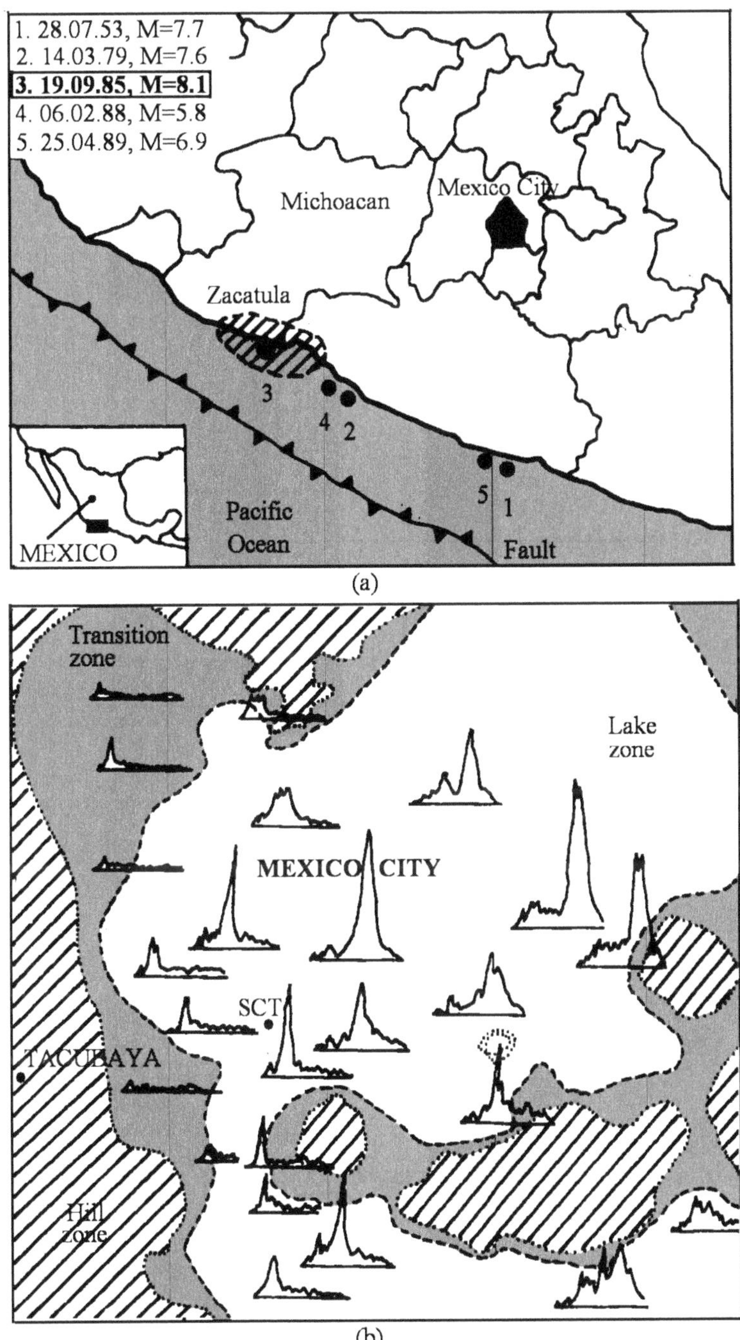

Figure 2.1: Michoacan earthquake: (a) Magnitude and epicenter of some Mexican earthquakes; (b) Response spectra for Mexico City Valley (after Reinoso et al, 1992)

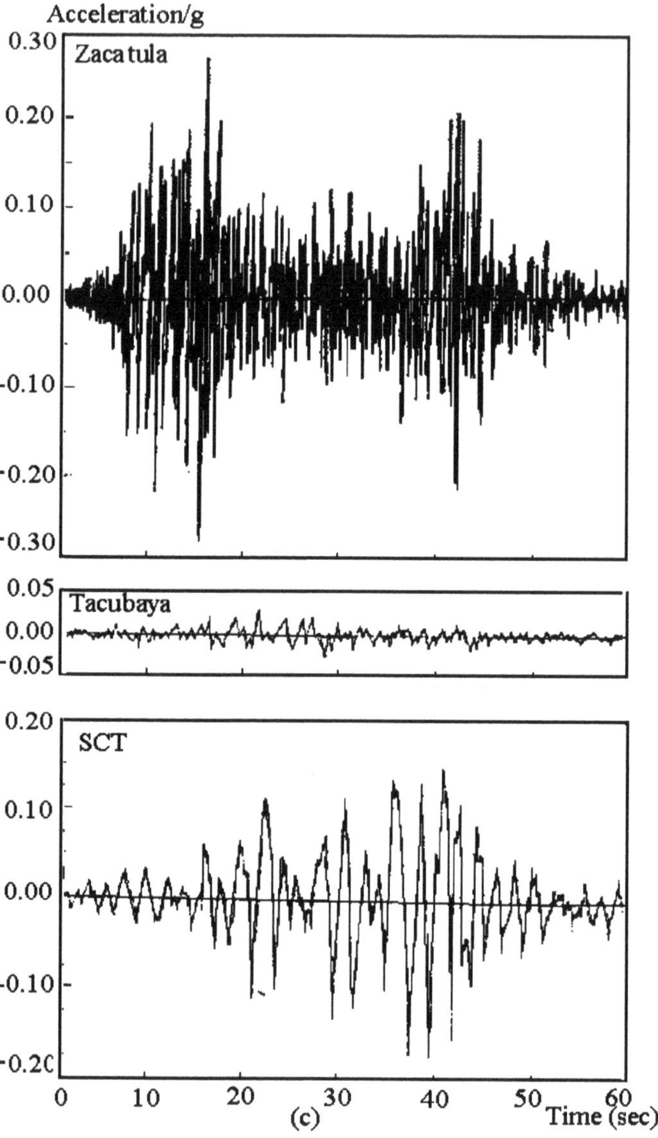

Figure 2.1: (continued) (c) Accelerogram at Zacatula (near epicenter), Tacubaya (hill) and SCT (lake) (after Fischinger, 1997)

Fig. 2.1c presents the accelerations recorded at Zacatula City (near to epicenter) and Mexico City (Tacubaya and SCT Stations). One can see the great differences on peak ground accelerations and natural periods of these three records. The great question of this earthquake is to explain why there were so many important differences in recorded ground motions and why so much damage was produced at such large distance from the epicenter, when normal attenuation laws would suggest that much lower levels of acceleration would be expected at such distance. The first answer results from the fact that the earthquake is of interplate type, with a deep source location, so the area of influence was very large. The characteristics of the soil in Mexico City also played a key role in the disastrous event. The valley of Mexico City is a closed basin which was filled by water and wind-laid transported materials during the ancient periods. Due to the disintegration of rocks, the surrounding hills were gradually eroded and the finest elements were transported by water into the basin (Diaz-Rodriguez, 1995). The subsoil of Mexico City has been divided into three zones from the point of view of foundation engineering: hill zone, transition zone and lake zone (Fig. 2.1b). The lake area consists of a layer of clays of 20 to 30 m deep. The site effects are characterized by large amplification at the resonant frequency of the clay layer. This amplification and the dominant period are also presented in Fig. 2.1b (Reinoso et al, 1992, Chavez-Garcia, 1995, Fischinger, 1997). These periods vary between 0.5 sec for hill zone, up to 5.2 sec for lake area, explaining the amplification of 12.7 times in the some zones (Abbiss, 1989) and the great energy spectra (Fajfar, 1995). The duration was also very different: for the lake area, about 140 sec. and for hill zones, about 30 sec. So, this earthquake was one of most devastating event for structures with long fundamental periods, such as the multi-storey steel moment resisting frames.

2.2.2. General Information about Damage of Steel Structures

Buildings in Mexico City are shaken on average once every two years by an earthquake with magnitude 7 or larger. So it is expected that Mexico City structures be badly affected by degradation not only during one earthquake but of accumulated damage during several earthquakes (Reinoso et al, 2000). More than 100 steel structures were subjected to the 1985 severe test. Among them, 59 buildings were built after 1957, having from 7 to 22 stories (Osteraas and Krawinkler, 1989, Teran-Guilmore and Bertero, 1992). This was the first very important in-site verification of the behaviour of steel structures during a strong earthquake, showing generally a very bad performance. The main cause of this unexpected behaviour was the double resonance phenomena, seismic wave-soil and soil-structure, which gave rise to a required ductility exceeding the normal demand. The influence of higher modes, which were more active than the first one, caused damage on the upper stories and also collisions between adjacent buildings.

In Mexico City the most frequently used steel structures were the moment-resistant frames (MRFs). Typically, this system consisted of box columns (2 channels and cover plates, or four welded plates), H-section columns and beams (either hot rolled or welded) or truss girders built up with angle sections. The MR frames behaved generally well. Of the 41 buildings of this type, one underwent severe structural damage requiring

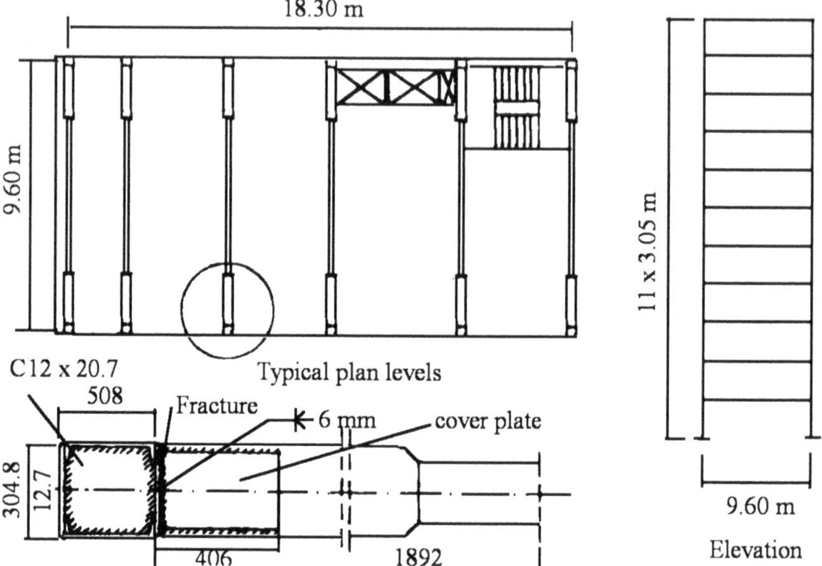

Figure 2.2: Damage of Amsterdam Street building (after Osteraas and Krawinkler, 1989)

partial demolition, one was affected by repairable damages and three sustained minor structural damages. All these damaged buildings were 10 or more stories high, having a long fundamental period. The damage was concentrated at welded beam-column connections or in truss girders, by buckling of compression diagonals.

The second type used in Mexico City area was the steel dual system, some bays of MR frames being braced. Of the 25 buildings of this type, two collapsed totally, one partially, four sustained various degrees of damage, the rest were undamaged. The collapse of the Pino Suarez building is the most famous case.

The third type was the mixed dual systems, consisting of steel frames and concrete shear walls. Of the 6 surveyed buildings, one sustained serious structural damage and one suffered minor structural damage, concentrated primarily in the truss girders.

The 77 Amsterdam Street Building (11-storey building) was built around 1970, and it is a one bay MR frame (Osteraas and Krawinkler, 1989), (Fig. 2.2). The columns consist of two channels and two welded plates and the beams are welded I-section. The damage reported in this building was severe cracking of masonry infill walls in the two longitudinal walls and connection failure in the first four stories of the transverse frames. The connection type constitutes a very weak link, because the only reliable force transfer from beam to column appears to take place from the beam flange splice plate through the full penetration weld to the column cover plate. The filled welds between continuity plate and cover plate were fractured, due to moments generated in the connections by the earthquake and the force transfer from beam to column shifted to vertical welds where cracks immediately ocurred. In these conditions, the connections worked as semi-rigid joints. Despite the large number of inelastic reversals experienced during this earthquake, the deterioration was not sufficient to cause the total collapse of structure due to second-order effects. This is due to the redundancy properties of the structure, which permitted a redistribution of moments. These connection damages were the first alarm signal to structural engineers, confirmed by the next Northridge and Kobe earthquakes, about the wrong behaviour of poor conceived joints and the need to provide connections with a sufficient ductility, assuring a second line of force transfer.

2.2.3. Pino Suarez Building

The Pino Suarez complex shown in Fig.2.3a, comprised five high-rise steel buildings: two identical 15-storey structures(A and E buildings) and three identical 22-storey structures (B, C, and D buildings) (Osteraas and Krawinkler, 1989). The complex is standing on a two level reinforced concrete subway station, which acts as a rigid foundation common for all five buildings. The two first stories are also common.

During the earthquake, the building D collapsed on the building E and buildings B and C were very seriously damaged. Because the building C was close to collapse, it represented a very rare occasion giving the possibility to study a building just before failure, at its ultimate state level. The layout and typical details of buildings C and D are shown in Fig.2.3b. The structural system consists of moment-resisting frames and a bracing system around the service core, consisting in two X braced bays in the transverse direction and one bay in the exterior longitudinal frame. The beams are truss girders, built up with angle sections and plate elements. The trusses are double in the longitudinal direction and single in traverse direction. All columns are welded box sections, built-up with four plates of equal thickness. The braces consist of double T cross-sections, built-up with three plates welded together.

The most evident localized failure observed in building C was the severe local buckling in the fourth storey box columns (Fig. 2.4a). The four plates of the cross-section were separated due to failure of welding, thereby causing significant reduction of column stiffness. The shortening of these columns of about 25 cm was responsible of large deflection of the girder supporting the V-braces (Fig. 2.4b). Buckling was observed also in the X-bracing system. Local failure was present in almost all the truss girders of longitudinal and transverse directions; many of the lacing members buckled (Fig. 2.4c) (Fischinger, 1997).

A very well conducted analysis of this failure has been performed by Cheng et al (1992), Ger and Chang (1992), and Ger et al (1993). Experimental tests were carried out for girders and columns. For transverse truss girders, a ductility factor of 2.3 was obtained, the failure being caused by

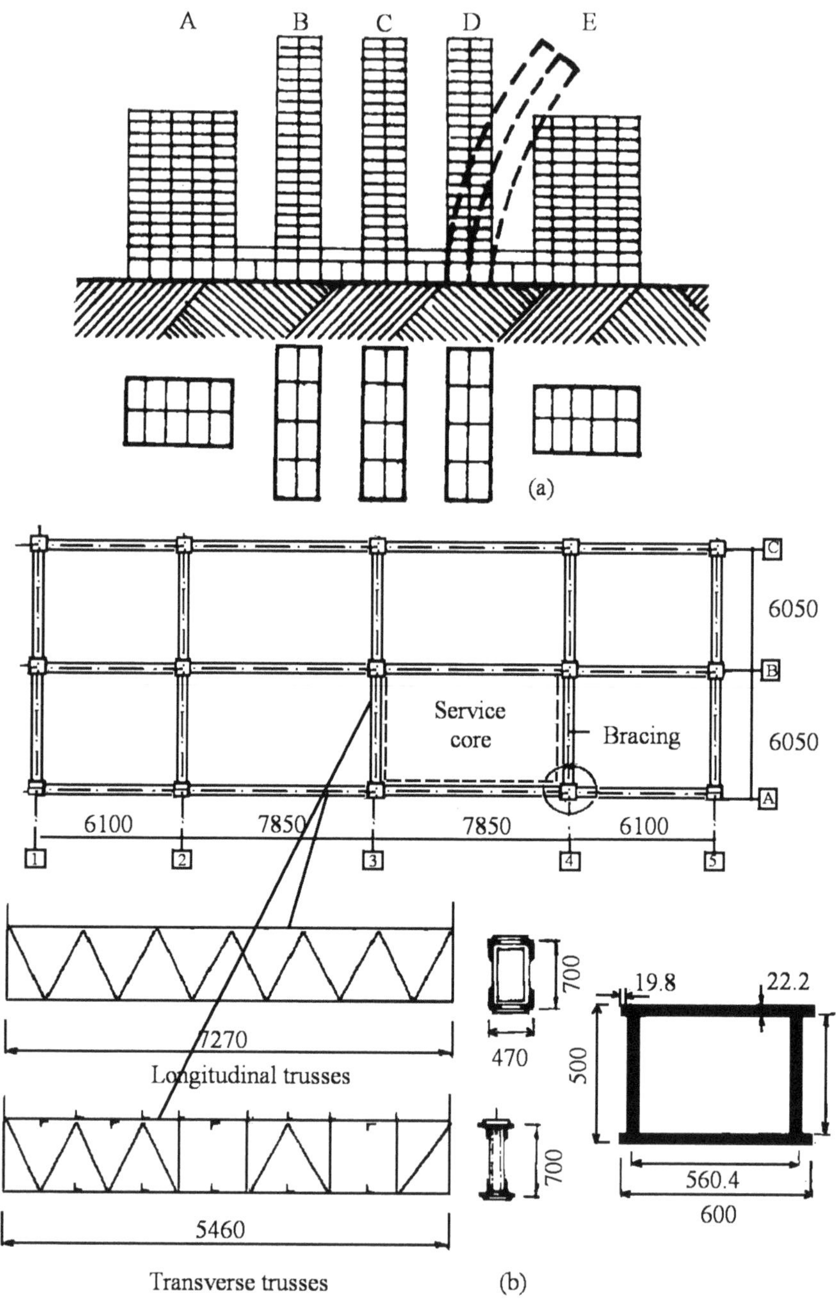

Figure 2.3: Pino Suarez complex (a) Elevation, plan and collapse of building *D*; (b) Plan view and typical framing details (after Osteraas and Krawinkler, 1989, Ger et al, 1993)

Figure 2.4: Collapse of Pino Suarez buildings: (a) Column buckling; (b) Collapse of bracing system; (c) Buckling of truss girder members

buckling of web members. For longitudinal truss girders, only 1.72 and 1.71 ductility factors were obtained, the failure being produced by local buckling and cracks of top chords. For columns, after the local buckling, a very unstable behaviour was observed, with a very bad ductility factor. The column failure was due to high axial force and low moment combination, due to the presence of bracing systems.

According to the code requirements, which do not consider the specific situation of soil in Mexico City and the actual ductility of structural mem-

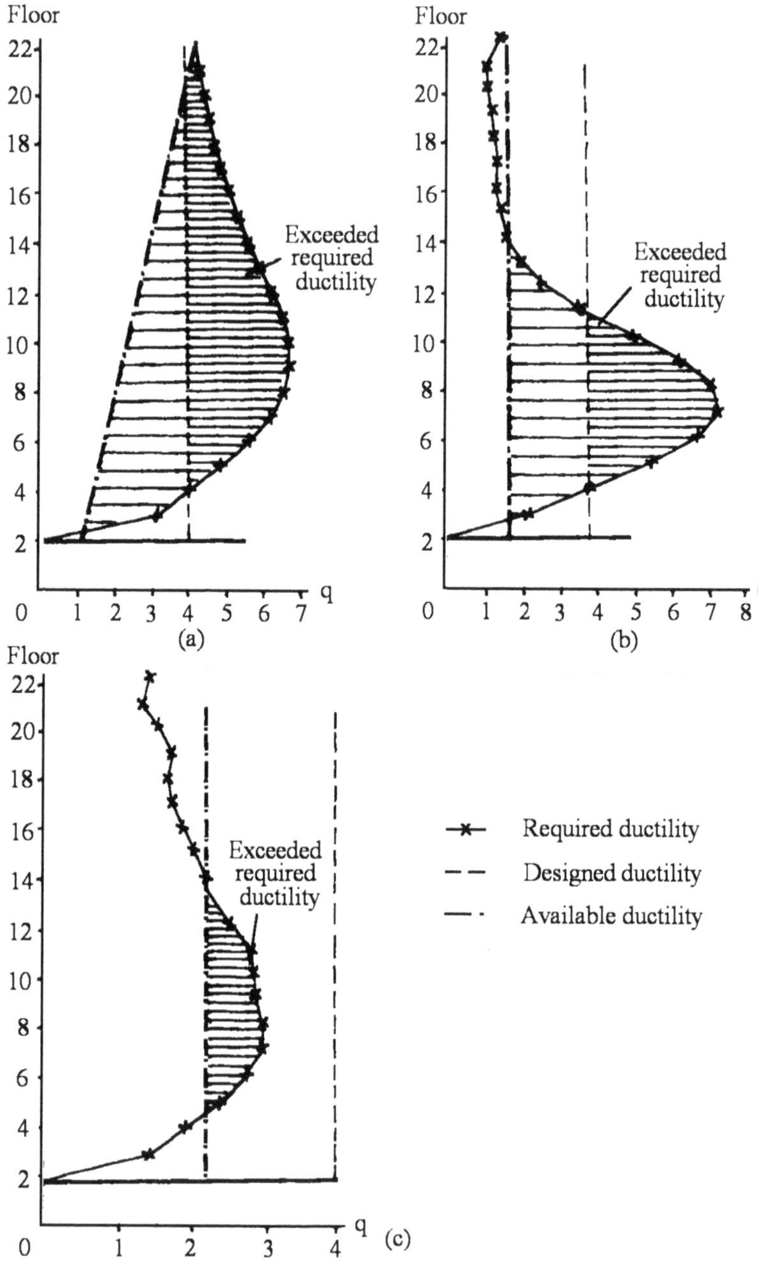

Figure 2.5: Designed, required and available ductilities: (a) For columns; (b) For long direction girders; (c) For short direction girders (after Ger et al, 1993)

bers, the structure was designed for a ductility factor equal to 4. The overall structure analysis, incorporating the specific behaviour of members and the peculiarities of the Mexico City earthquake, allows to evaluate the actual behaviour of the structure and to determine the required ductility for the structural members. Due to soil conditions, the required ductilities were greater than 7 for columns, 7.5 for long direction girders and 3 for short direction girders, the maximum of these ductility demands being obtained at the 9th storey (Fig. 2.5). Comparing these required ductility factors with the experimental values, very large differences may be noted. Thus, it is very clear that the collapse of the structure occurred due to insufficient ductility of columns and girders.

The Pino Suarez building collapse provided an excellent opportunity to underline what may occur if no concordance between required and available ductilities exists.

2.3 Northridge Earthquake

2.3.1. General Description of Californian Earthquakes

The largest earthquakes ever to hit the USA States were centered in the New Madrid seismic zone near Memphis City, Tennessee, in 1811-1812, with the magnitude of 8.6 (Basham, 1989). But the most active seismic regions are along the western shore of the Country, where the Pacific and the North American tectonic plates meet and a system of fault lines has developed (Popov, 1994). California is a part of the circum-Pacific seismic belt, which is responsible for about 80 percent of the world's earthquakes. The West Coast of USA is hit by thousands of shocks every year and earthquakes of destructive magnitude have occurred once a year in the past 50 years. California's crusted surface is crossed by many great fractures or faults, forming lines of weakness in the masses of rock. Some faults are known to be active, while others are presumed to be inactive, but they can give unexpected surprises. The most famous is the San Andreas fault in Southern California, along which the most frequent and dangerous earthquakes occur. The type of seism is an interplate motion, produced at shallow depth, which means that the main characteristic is given by the near-field ground motion type. The most important earthquakes produced in recent years along the San Andreas fault are presented in Fig. 2.6 (Grecu and Moldovan, 1994). Several major urban areas are located alongside major active faults and so could be subjected to near source ground motions from large earthquakes. The San Andreas fault runs 10 km West of downtown San Francisco and extends south to Los Angeles metropolitan region. Among these earthquakes, Loma Prieta and Northridge are the most interesting from the structural engineering point of view.

Due to the fact that a very dense network of instrumentations is now available along the San Andreas fault, a lot of very important information was obtained. About 37,000 well-recorded earthquakes detected on the southern California seismic network between 1981 and 1994 provided data

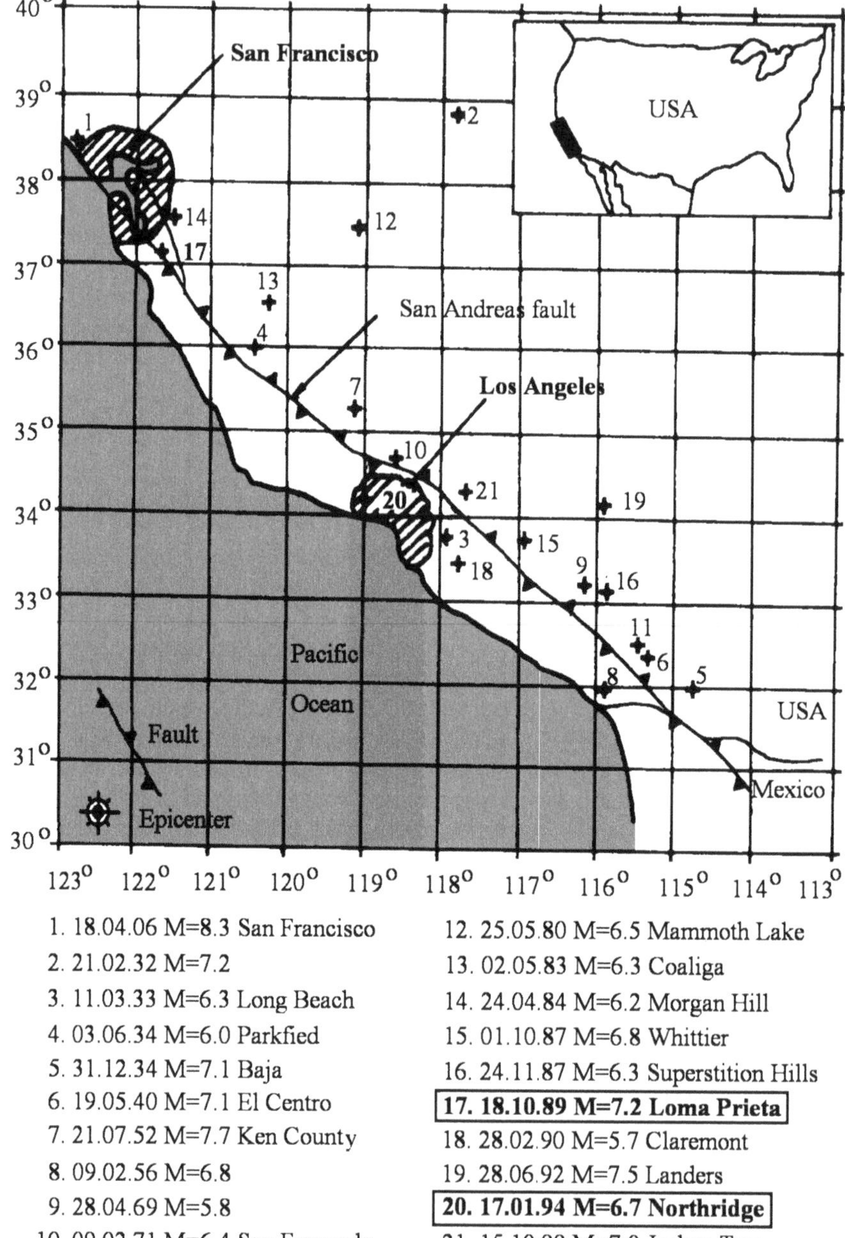

1. 18.04.06 M=8.3 San Francisco
2. 21.02.32 M=7.2
3. 11.03.33 M=6.3 Long Beach
4. 03.06.34 M=6.0 Parkfied
5. 31.12.34 M=7.1 Baja
6. 19.05.40 M=7.1 El Centro
7. 21.07.52 M=7.7 Ken County
8. 09.02.56 M=6.8
9. 28.04.69 M=5.8
10. 09.02.71 M=6.4 San Fernando
11. 15.10.79 M=6.6 Imperial Valley
12. 25.05.80 M=6.5 Mammoth Lake
13. 02.05.83 M=6.3 Coaliga
14. 24.04.84 M=6.2 Morgan Hill
15. 01.10.87 M=6.8 Whittier
16. 24.11.87 M=6.3 Superstition Hills
17. **18.10.89 M=7.2 Loma Prieta**
18. 28.02.90 M=5.7 Claremont
19. 28.06.92 M=7.5 Landers
20. **17.01.94 M=6.7 Northridge**
21. 15.10.99 M=7.0 Joshua Tree

Figure 2.6: Californian earthquakes (after Grecu and Moldovan, 1994)

for calculating the features of these ground motions. During the Northridge event more than 200 strong-motion accelerograms were recorded in the metropolitan area (Magistrale and Zhou, 1996). The main characteristics of these earthquakes were as follows:

(i) *Pulse characteristic.* The analysis of records reveals that the aspect of the time-history variation of ground motions is qualitatively different from the other well known records, for instance the famous El Centro records. Fig. 2.7 shows the 1979 Imperial Valley records, which are typical for a near-field earthquake record. The feature of these records is the low frequency pulses in the acceleration time-history, which translate into the pronounced coherent pulses in velocity and displacement histories.

(ii) *Vertical components of ground motions.* For a long time, the study of earthquake ground motions has been limited to the examination of the horizontal components. The vertical ground motions have been largely ignored, because until last time the recorded earthquakes were far from the source. But during the recent recorded earthquakes near the source, it has been observed that the vertical ground motions are sometimes greater in amplitude than the horizontal components. This remark was very evident in the strong ground motions which were recorded during the 1979 Imperial Valley earthquake (Lew, 1992, Chouw, 2000). From Fig.2.8 it can be observed that the largest maximum vertical accelerations generally occur close to the fault rupture zone. In addition, the vertical movements are associated with frequencies higher than the horizontal ones, showing that different design spectra must be considered. So, for the Californian earthquakes, the vertical actions of the ground motions cannot be ignored.

(iii) *Combination of horizontal and vertical components.* It is generally accepted that the first waves which arrive to the structure are the vertical ones, as shown in the Fig. 2.9a for the Imperial Valley earthquake. But in other cases, as Morgan Hill earthquake, the vertical and horizontal motions are almost exactly coincident in time (Fig. 2.9b). Both cases must be considered in structure analysis, because it is not sure which situation will arise (Elnashai and Papazoglu, 1997).

(iv) *Velocity.* An increase of velocity near-field is marked. Velocities often exceed 150 to 200 cm/sec in areas surrounding the source (Trifunac and Todorovska, 1998). The velocity histories of several Californian records are shown in Fig. 2.10. The ground motion has in the near source field a pronounced coherent pulse in velocity and displacement. The ground motion could be composed by only one pulse (Supersition Hills and Lucerne Valley) or more adjacent pulses (Tabas, El Centro and Loma Prieta). Fig. 2.11 shown a histogram of S-wave velocities of the sites where the seismic surveys were performed (Niwa et al, 1996). One can see that the main velocity is about 200m/sec, values for which the influence of asynchronism in horizontal and vertical ground motions may be important.

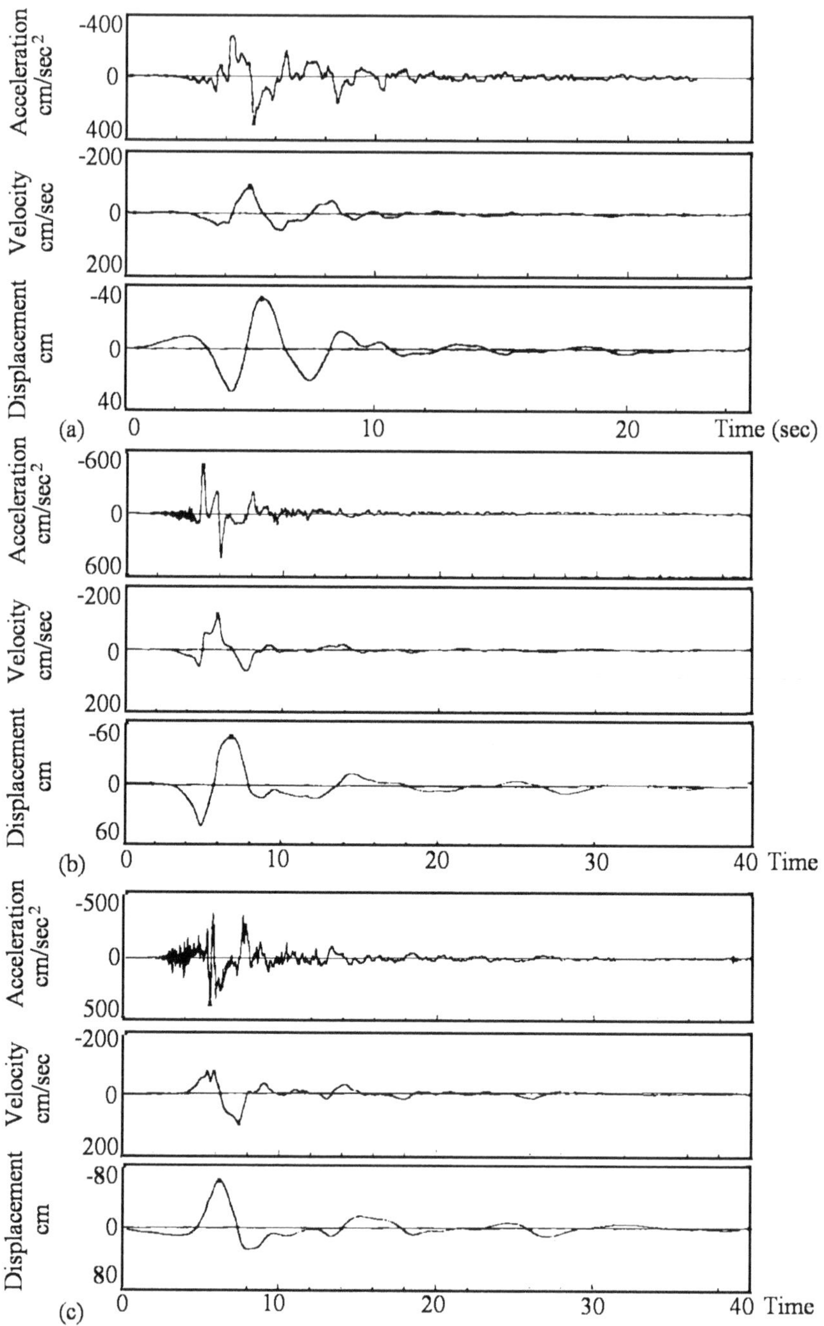

Figure 2.7: Pulse characteristic - Imperial Valley, 1979: (a) Meloland Overpass; (b) Array No. 7; (c) Array No. 5

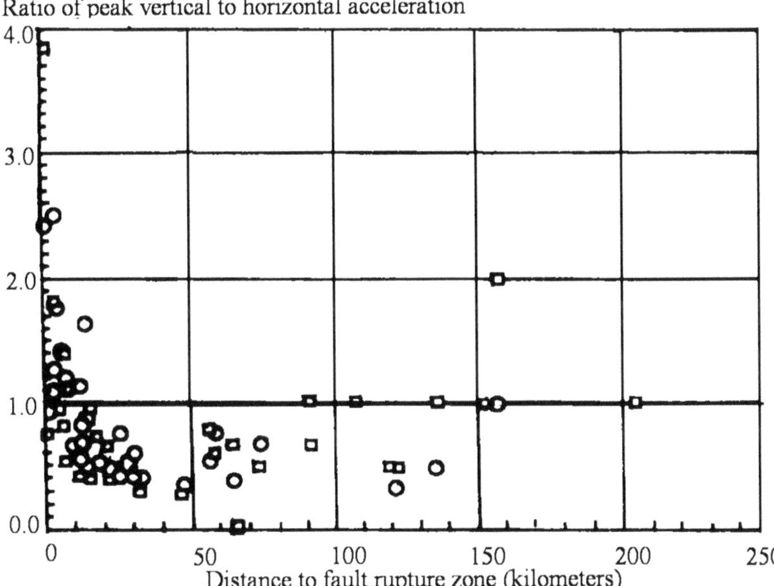

Figure 2.8: Ratio of vertical to horizontal peak ground accelerations for Imperial Valley earthquake, 1979 (after Lew, 1992)

2.3.2. Loma Prieta Earthquake

On 17 October 1989, a major earthquake of 7.1 magnitude occurred at 100km South of San Francisco Bay (Fig. 2.12), at Loma Prieta, situated on the San Andreas fault and in the close vicinity of Hayward fault. If this ground motion is considered belonging to the San Francisco area is, it is an example of intermediate-field type. Examining regional differences in peak of horizontal accelerations, it is very easy to observe that the site is divided in two areas: zone A with an inner angle of 60 degrees and a central direction coincident with the fault line and another area, called zone B. The attenuation of acceleration peaks is less rapid in zone A than in zone B, the direction of propagation being there in the direction of fault (Ejiri et al, 1992). This aspect of directionality of the input motion is not considered in design practice, even when the dominant direction may be significant and should be considered in certain cases.

In spite of the fact that the number of injuries and lives lost was surprisingly low (3,000 injuries and 65 deaths), from the seismological and engineering aspect, the Loma Prieta earthquake was one of largest disasters in the USA history. The economic losses due to damage reached about US$8 billion, due to the high concentration of technology industries and the increased urbanization, which meant that not only buildings but highways, roads, and bridges were also damaged (Bertero, 1994).

The Loma Prieta earthquake has again demonstrated the great influence of local site effects on surface ground motions and the presence of damage

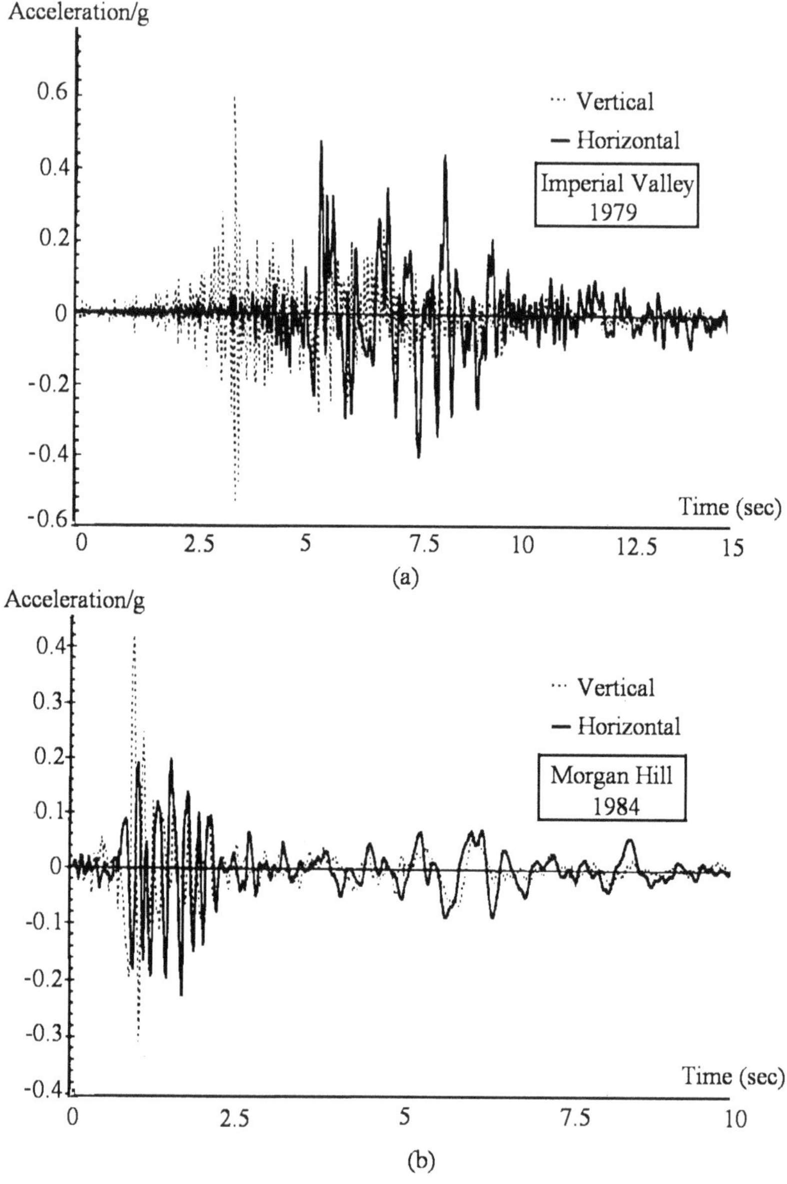

Figure 2.9: Combination of horizontal and vertical components: (a) Imperial Valley, 1979; (b) Morgan Hill, 1984 (after Elnashai and Papazoglu, 1997)

resulting from ground shaking (Seed et al, 1990, Krawinkler and Rahnama, 1992).

The overall steel building performance was good with little to moderate structural damage in buildings. Only some braced frames presented buckled

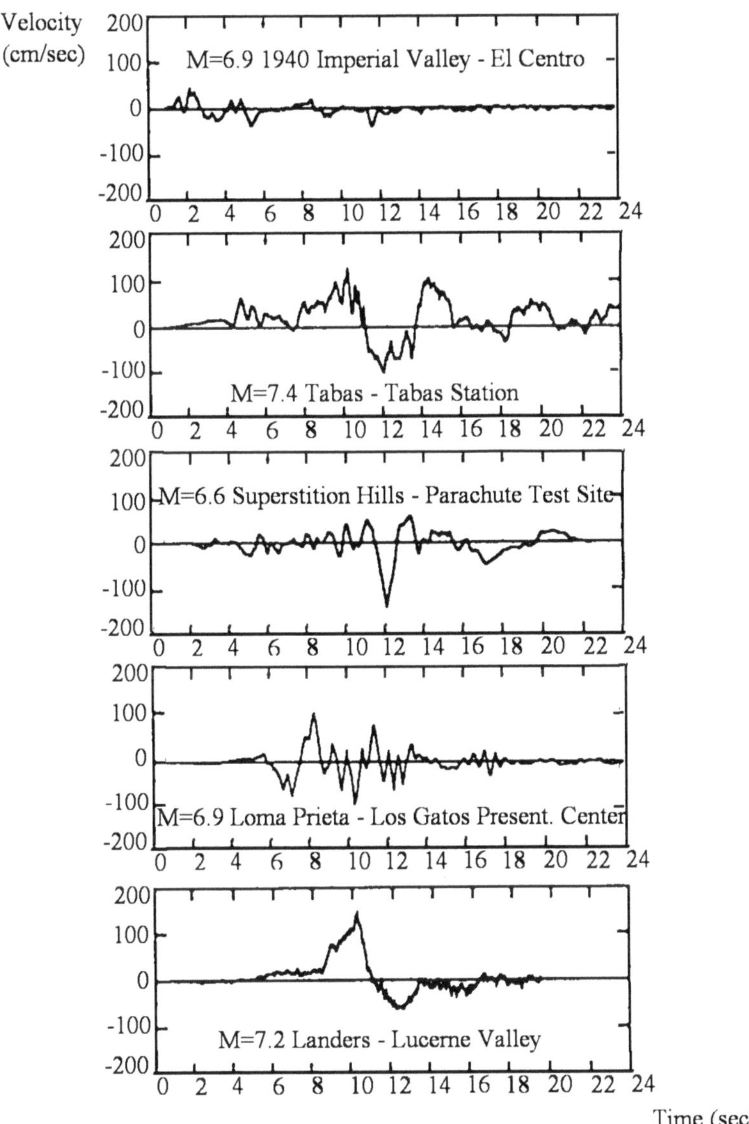

Figure 2.10: Ground velocity of Californian earthquakes (after Hall et al, 1995)

bracings and failed welded connections (Phipps, 1992, Gunturi and Shah, 1992). Contrary to this, the unique and important aspect of Loma Prieta earthquake was the widespread non-structural component damage: partitions, suspended ceilings, curtain walls, facades and cladding, contents such as computer equipment, office furniture, piping, etc. After the earthquake, a great number of buildings survived with little structural damage and no

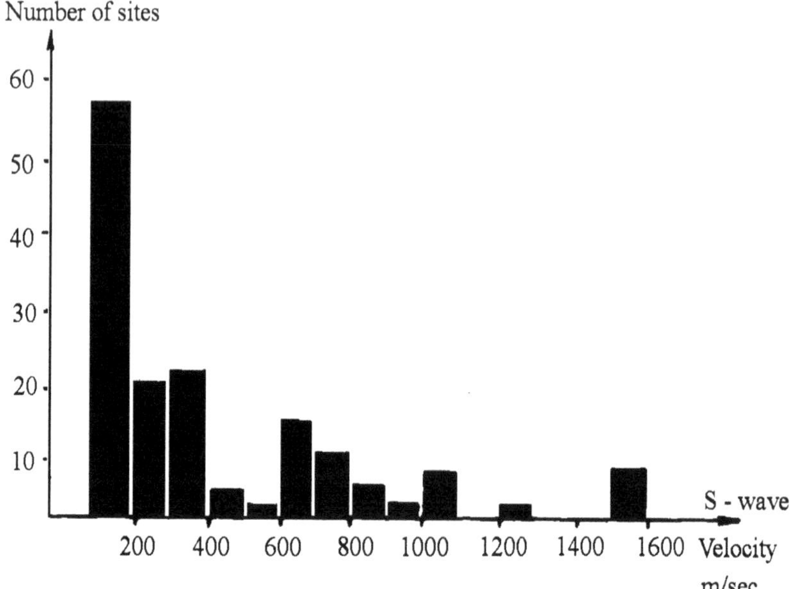

Figure 2.11: Distribution of S-wave velocities for Californian earthquakes (after Niwa et al, 1996)

loss of life, but they were evacuated because of the loss of function due to extensive non-structural damage (Rihal, 1992, Sharpe, 1992).

This situation shows very clearly that:

-today the codes make it possible to design buildings for life safety, but not for limiting monetary damage;

-the public and the owner need to be informed about what can happen to buildings during a severe earthquake, and that the codes are not infallible;

-the seismic codes need to be modified in order to improve the performance of non-structural elements, to reduce the economical losses and to assure the continuation of functionality of contents during the emergency.

Because a lot of buildings in the San Francisco Bay area were instrumented, a large number of records are today available, showing the actual response of buildings during a severe earthquake (Li and Mau, 1997, Celebi, 1992, Chajes and al, 1992). The main conclusions of these analyses are:

-for high-rise buildings, after 20 sec of base excitation, the motion sensors at top of building recorded 120 sec of dynamic response;

-during the first 25 sec the building response exhibited high mode participation and only at the end the response was it dominated by the fundamental mode;

-there are differences between the dynamic characteristics identified from records and the computed ones, due to interaction of soil-structure which was not accounted for in the analysis;

-the free-field motions are influenced by the presence of tall buildings;

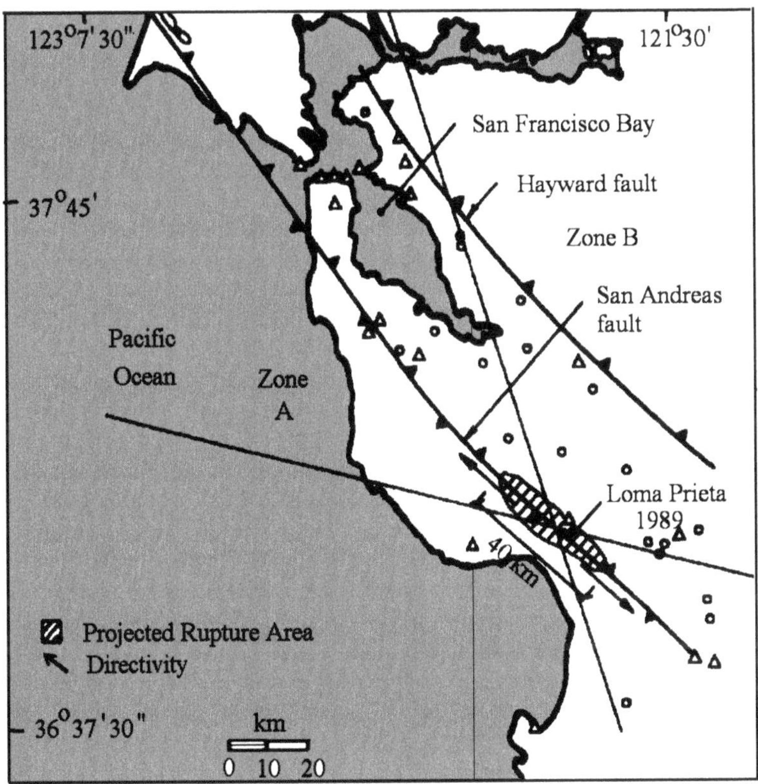

Figure 2.12: San Francisco City and Loma Prieta earthquake (after Ejiri et al, 1992)

-the directionality of earthquake can affect the torsion response of non-symmetric buildings and buildings with wings;

-the damping ratio for steel frame buildings has a range for translation modes from 1 to 6 percent, with the majority falling between 2 and 5 percent;

-the fundamental frequency may vary during an earthquake due to subsequent damage and consequent changes in effective stiffness of both structural and non-structural elements, together with softening in the soil at large strain. This variation of frequency is in strong correlation with drift variation;

-the orthogonal behaviour of structure is significant;

-the non-structural damage is due to the very important site - amplification.

2.3.3. Northridge Earthquake Characteristics

On 17 January 1994, a fairly moderate earthquake, with about 6.7 magnitude, struck a North-western suburb of Los Angeles (Fig. 2.13). Suddenly, the city of Northridge became very well known to structural engineers around the world. This earthquake is the latest of a significant series of seismic events which occurred in the Los Angeles region: San Fernando, 1971, $M = 6.4$; Whittier Narrows, 1987, $M = 5.9$; Sierra Madre, 1991, $M = 5.8$, and belongs to the San Andreas fault deformation.

The Northridge earthquake occurred beneath a heavily urbanized area and caused 57 confirmed fatalities and about 9000 injuries. A preliminary damage estimate is of 30 billion US$ and it is possible to consider this event as the most costly natural disaster in the USA history.

A lot of information about the nature of ground motions was published in journals and in the Proceedings of the European and World Conferences on Earthquake Engineering: 11th WCEE 1996, Acapulco, Citipitioglu and Celebi, Hudson et al, Bozorgnia et al, Mohammadioun and Mohammadioun, Durkin, Iwan, Somerville et al, Saikia, Faccioli, Pinto, Celebi, Doroudian; 11th ECEE 1998, Paris, Elnashai et al, King and Rojahn, Pomonis and Williams; 12th WCEE 2000, Auckland, Chouw; SAC Report 95-03.

The ground motions during the Northridge earthquake were more strongly directed along a North-South axis (Faccioli, 1996, Paret and Attalla, 2000), and it was complicated by the possibility to have at least two shocks, separated by several seconds, so the duration of shaking increased from 6 sec to 25-30 sec (Bonacina et al, 1994). The directivity of rupture has played a great role in the ground motions. Rupture began at the depth of 19km below the surface and propagated up dip toward the North on a plane dipping at about 42 degree from the horizontal (Kataoka and Ohmachi, 1996, Borcherdt, 1996) (Fig. 2.14a). The rupture plan has a length along strike about 18km and up dip width of about 21km; its surface projection is shown in Fig. 2.13a,b. The depths of the bottom (South-West) and top (North-East) edges of the rupture are 20km and 6km respectively, with most of the rupture confined to depths of 12km or more, being categorized as a crustal earthquake. The closed distance of the fault plane to the densely urbanized southern San Fernando Valley is about 17km (Sommerville et al, 1995). Selected accelerations and velocity time histories recorded during the earthquake are shown in Fig. 2.13. The recorded peak velocities are very large near the upper edge of the rupture along the North edge of the San Fernando Valley; the values are much larger than in the epicentral region in the central part of the San Fernando Valley. This is due to the effect of rupture directivity, caused by propagation of the rupture up dip toward the northern edge of the San Fernando Valley (Fig. 2.13a). Thus the densely urbanized southern margin did not experience a rupture directivity effect and most of modern steel structures were not fully tested by the Northridge earthquake. The distinctive characteristics of this directivity were the occurence of large velocity pulses in horizontal direction perpendicular to the strike of the fault. The fault-parallel component was relatively small (Fig. 2.14b).

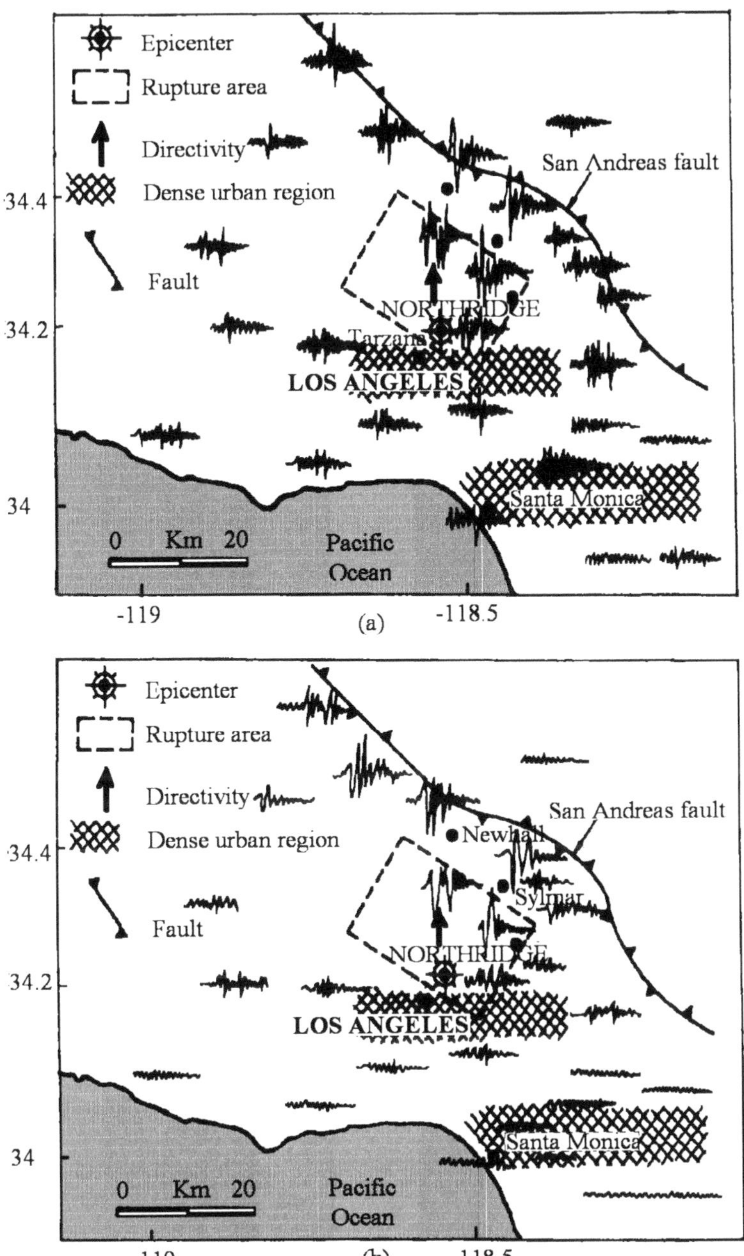

Figure 2.13: Los Angeles City and Northridge earthquake: (a) Ground accelerations - North components; (b) Ground velocities - North components (after SAC, 1995b)

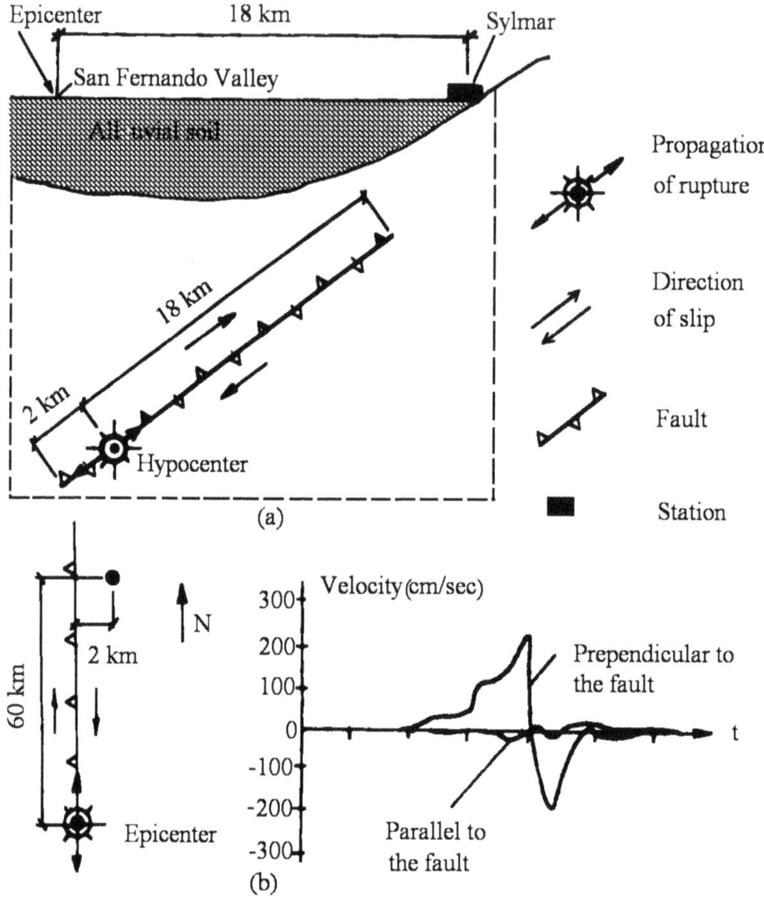

Figure 2.14: Directivity of Northridge earthquake: (a) Vertical directivity; (b) Horizontal directivity (after Hall et al, 1995)

A very dense network of seismic stations, including recording instruments of different soil types, allowed a great number of recorded ground motions. There were also many instrumented buildings, so approximately 120 sets of records now exist, concerning the behaviour of structures during this event (Celebi, 1997).

The acceleration, velocity and displacement time histories for some near field stations are reproduced in Fig. 2.15 (Iwan, 1995). One can see that the ground motions are characterized by velocity and displacement pulses, defined as a ground displacement which is attained rapidly, with a peak velocity of 100cm/sec or greater. This pulse characteristic is associated directly with the fault rupture process (Hall et al, 1995, Hall, 1995). The ground motions recorded at Newhall and Sylmar Stations situated North from the epicenter are higher than the ones recorded at Santa Monica Sta-

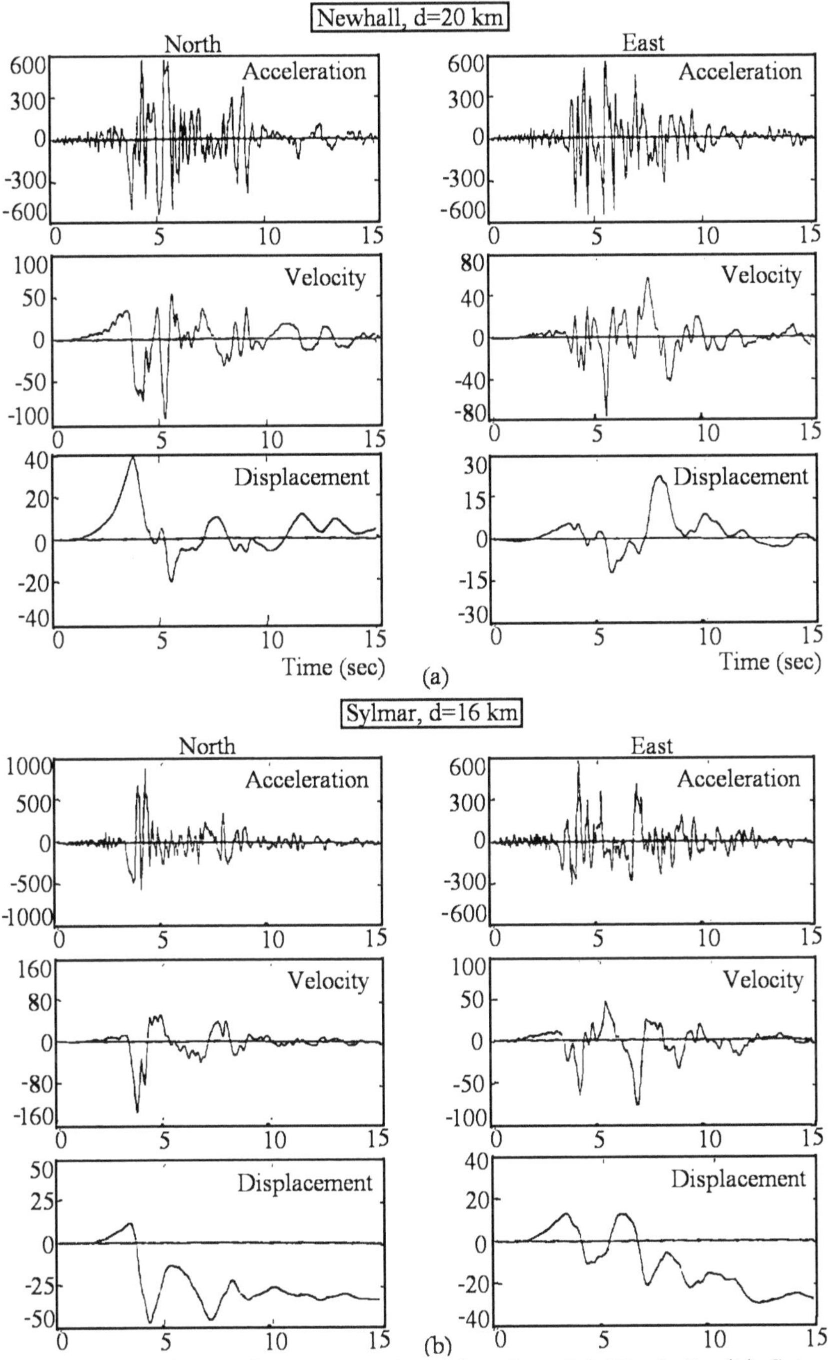

Figure 2.15: Northridge motion time-histories: (a) Newhall; (b) Sylmar (after Iwan, 1995)

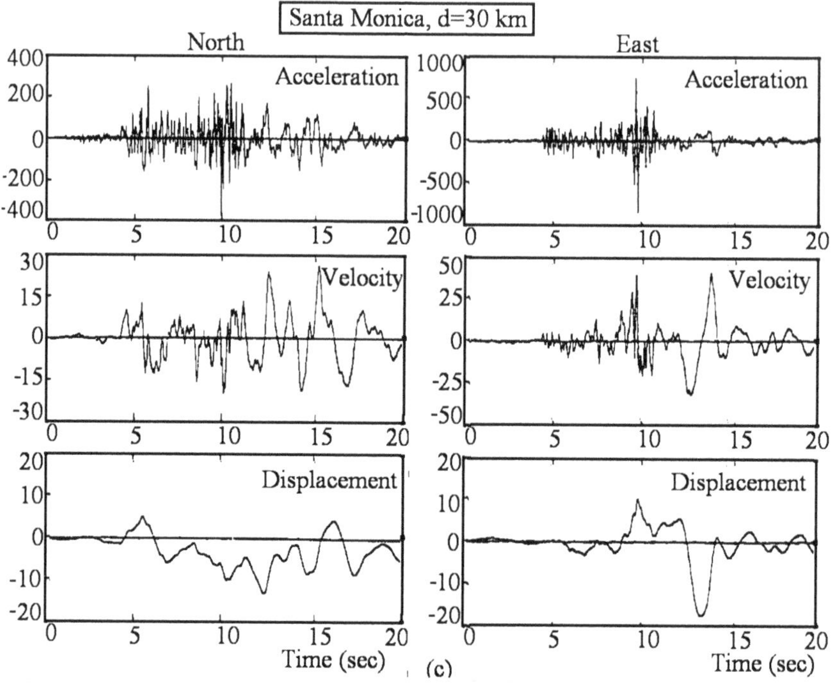

Figure 2.15: (continued) (c) Santa Monica

tion situated in the South part. The ground motions are characterized by few strong pulses (one or two pulses) with long duration (between 1-4sec) (Seco and Pinto, 1996). The largest horizontal peak acceleration of 1.82g was recorded at Tarzana, 5km from the epicenter (Citipitioglu and Celebi, 1996). The peak ground velocities were in the range from 20cm/sec to 150cm/sec, but the areas with very large velocities were small (Trifunac and Todorovska, 1998). The velocity spectrum for Tarzana Station shows great amplifications untill 350-400cm/sec in the field of low vibration periods (Fig. 2.16).

Concerning the vertical ground motion components, the Northridge earthquake provided the most comprehensive set of data ever recorded for a single event. The ratio of the vertical to horizontal spectra are shown in Fig. 2.17, as a function of distance and period. The increase in ratio at small distances from the fault is dramatic (Hudson et al, 1996, Bozorgnia et al, 1996, Mohammadioun and Mohammadioun, 1996).

2.3.4. Damage of Buildings

Because the Northridge earthquake struck a densely populated area of Los Angeles, the damage on existing buildings was very widespread. While Loma Prieta earthquake had the epicenter at about 100 km from San Francisco, the Northridge earthquake shaked Los Angeles directly in a populated

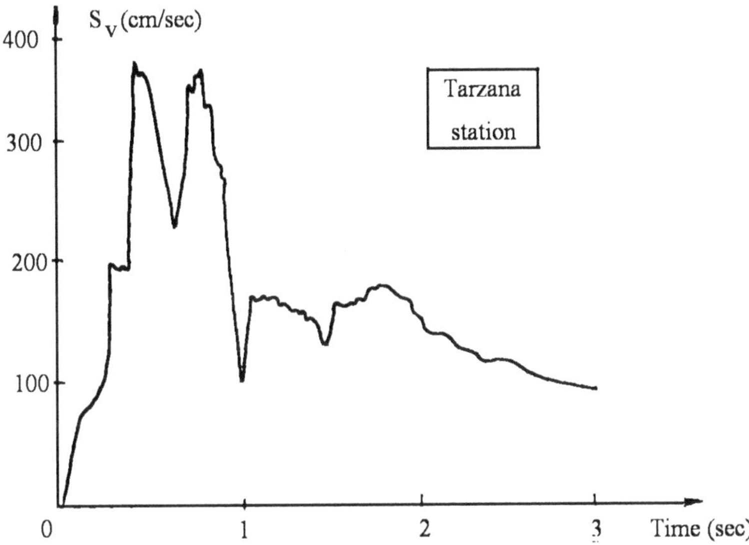

Figure 2.16: Velocity spectrum - Tarzana Station

area. So, the number of collapsed buildings were about 200 units and damaged buildings were about 5600 units. As the development of San Fernando Valley region greatly increased after 1960, most of the buildings were of recent construction and the new buildings were designed according to aseismic codes. Most of the damaged buildings were built before the introduction the UB Code in 1971.

The biggest issue emerging from the Northridge earthquake is the surprisingly poor performance of many steel buildings. The first reports declared that steel structures were unscathed. Titles of papers like "Northridge earthquake confirms the steel superiority" (Vannacci, 1994), were very optimistic. No one death and a single structural collapse, compared to the ones occurred in reinforced concrete structures, contributed to confirm this first impression. Unfortunately, more refined inspections after the earthquake have shown relevant damages in beam-to-column connections in up to 140 steel frame buildings (Miller, 1995). These damages to steel structures were apparently unprecedented. However, this is the only one among the various major earthquakes which occurred in an area containing so large number of modern steel structures. Similar damage may have also occurred in the past, but, due to the difficulties in detecting them, this damage probably remained undetected (Ghosh, 1995).

The main structure types in Los Angeles area are:

-special moment resisting frames, used in the majority of steel buildings, with the conception of strong column-weak beams. Usually columns have a steel grade higher than beams. One can observe that this system was very effective in order to preserve human life and to impede collapse of structure, which are the highest objectives of a proper design. But regarding to the

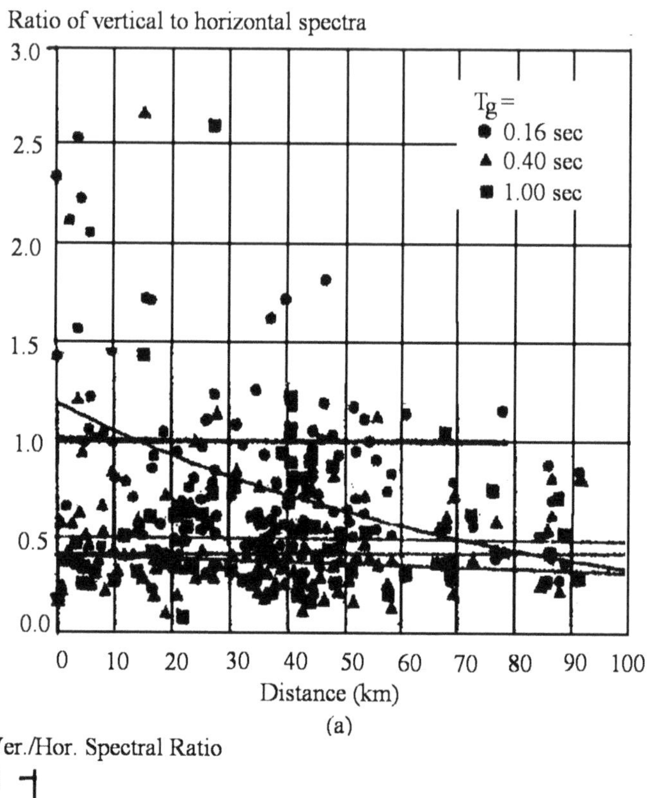

(a)

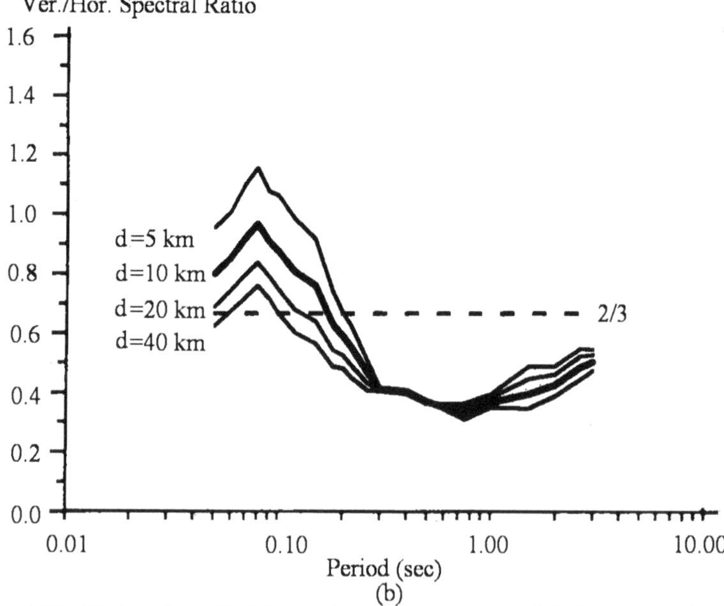

(b)

Figure 2.17: Ratio of vertical to horizontal spectra: (a) Influence of distance (after Hudson et al, 1996); (b) Influence of period (after Bozorgnia et al, 1996)

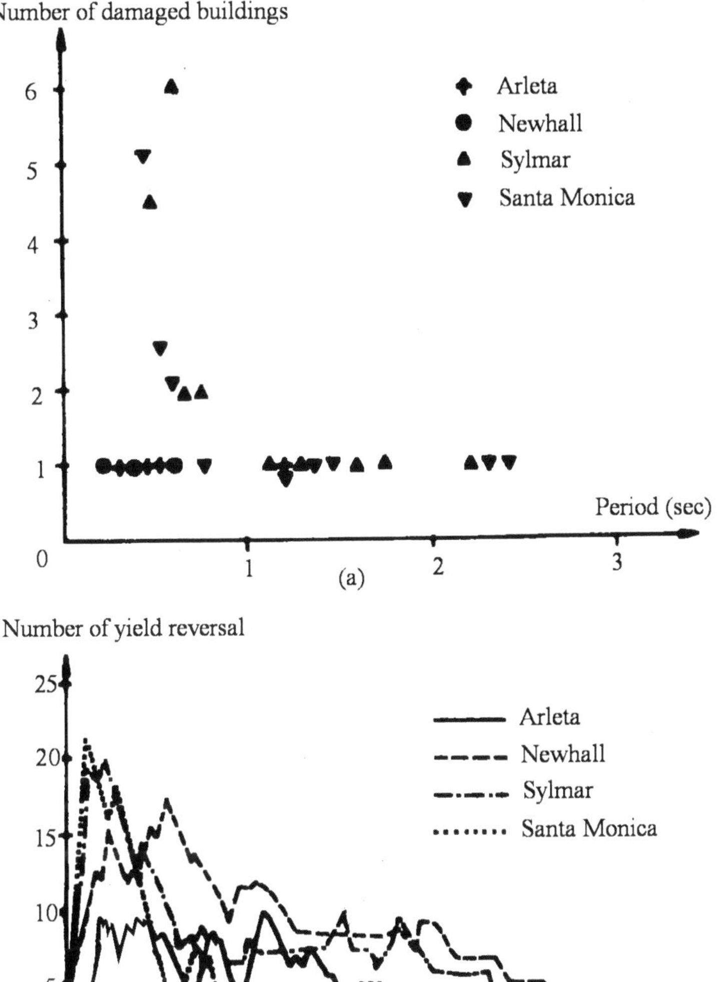

Figure 2.18: Influence of vibration period on damage: (a) Damaged buildings number; (b) Yield reversal number (after Yang and Popov, 1997)

limitation of damage, this system did not behave as expected, because the cost of joint repairs was very high;

-concentrically braced frames, used in some structures; generally these systems performed well.

The types of damage observed in steel structures were mainly failure of anchor bolts, overall buckling of lateral bracing members leading to

Table 2.1: Recorded accelerations and velocities (after Naeim, 1998a)

	Storey member	Acceleration/g		Roof velocity
		base	roof	$cm \cdot s^{-1}$
1	6	0.36	0.47	48
2	19	0.32	0.65	65
3	12	0.22	0.77	–
4	3	0.33	0.97	57
5	52	0.15	0.41	40
6	54	0.14	0.19	34
7	6	0.24	0.48	70
8	7	0.29	0.76	73
9	9	0.18	0.34	45
10	6	0.80	1.71	140

local buckling, in some cases fracture of their ends, and last but not least, cracks in the welded beam-to-column connections of special moment resisting frames. This last case is undoubtedly the most widespread type of failure to occur in steel structures, during the Northridge earthquake (Mazzolani and Piluso, 1997).

A very comprehensive study about the damage on steel structures during this event is due to Yang and Popov (1997). Fig. 2.18 shows the relation between the number of damaged structures in four districts of Los Angeles, (Arleta, Newhall, Sylmar, and Santa Monica) as a function of the structural period. It is very clear that the number of damaged structures having short periods is higher than the one with long periods (33 buildings versus 11). At the same time, the number of yield reversals, defined as a positive yield followed by a negative yield, or vice versa, are reduced for long structural periods. For long periods 3 to 6 yield reversals are produced, while for short periods this number is 10 to 20. The most reduced number of yield reversals for short periods was in the Arleta district, where the amplitude of accelerations was less than in the other districts. The observed trend shows an excellent correlation between the damaged buildings and the number of yield reversals. As a general remark, the displacement ductility factor alone cannot be used to define the damage factor. In fact, the same displacement ductility with larger number of yield reversals must be considered more dangerous than the case with reduced yield reversals.

Due to many instrumented buildings, approximately 120 sets of records now exist concerning the behaviour of structures during this event (Celebi, 1997). This allows the analysis of these buildings by comparing the theoretical and recorded results (De La Llera and Chopra, 1996, Naeim, 1998a,b, 2000, Hall, 1995, Hall et al, 1995, Bertero et al, 1994, Bonacina et al, 1994, King and Rojahn, 2000, Pomonis and Williams, 2000, SAC Reports 95-04, 95-05, 95-06, 95-07).

The main results of the analysed performances are (Naeim, 1998a):

-important amplification of accelerations and velocities are recorded (Table 2.1). The most important amplifications were recorded for building No 10 situated in Sylmar district, the most hitched area of Los Angeles;

-characteristics observed in several structures responses corresponded to the near source pulse action;

-peak vertical ground accelerations were amplified in the structure by a factor ranging from 1.1 to 6.4, due to the fact that vertical structure periods ranged between 0.075 to 0.26 sec correspond to the range of high vertical spectral amplification;

-second and third vibration modes played a more important contribution to the overall response than the fundamental mode;

-torsion of structures significantly contributed to the seismic response.

2.3.5. Moment Connection Fractures

A typical beam-to-column moment connection extensively used for moment resisting frames is presented in Fig. 2.19. Depending on beam and column dimensions, column web stiffeners (continuity plates) may be present or not. The beam web is field bolted to a single shear tab, which is shop welded to the column. The beam flanges are field welded to the column, using complete penetration welds. Web copes are required to accommodate the backup plate at the top flange and to permit making the bevel weld at the bottom flange.

Approximately 140 buildings with welded moment resisting frames were shown to suffer unexpected fractures in or near beam-to-column welds. A typical distribution of fractures in a perimetral frame is presented in Fig. 2.20 (Astaneh-Asl et al, 1995). The fracture distribution on the frame height, especially in the middle zone, has shown the importance of the second and third vibration modes. Typical damage consisted of cracks developed in flange welds, beam flanges, column flanges, column webs and shear plate connection of the beam web. Fig. 2.21 shows typical cracks within and around the full penetration weld connections (Astaneh-Asl et al, 1995, Miller, 1995, Tide, 1995, Engelhardt and Sobol, 1995). The damage was typically confined to the lower flange of beam near connection, the top flange remaining intact. In some cases, in presence of damage in the bottom flange, some damage in the shear tabs occured. Figs. 2.21 a,b illustrate fractures in weld metal, typically occurring near the column flange. The second type of damage is illustrated in Fig. 2.21 c,d, when a fracture initiating at or near the groove weld runs up through the column flange. In some cases the fracture stops within the flange; in other cases, the fracture emerges from the column flange at a distance of several centimeters above the top of the groove weld. Figs. 2.21 e,f illustrate fractures that run across the column flange and in some cases continue into column web. There are some cases when the fracture runs across the full width of column passing through both flanges and the web (Engelhardt and Sobol, 1995).

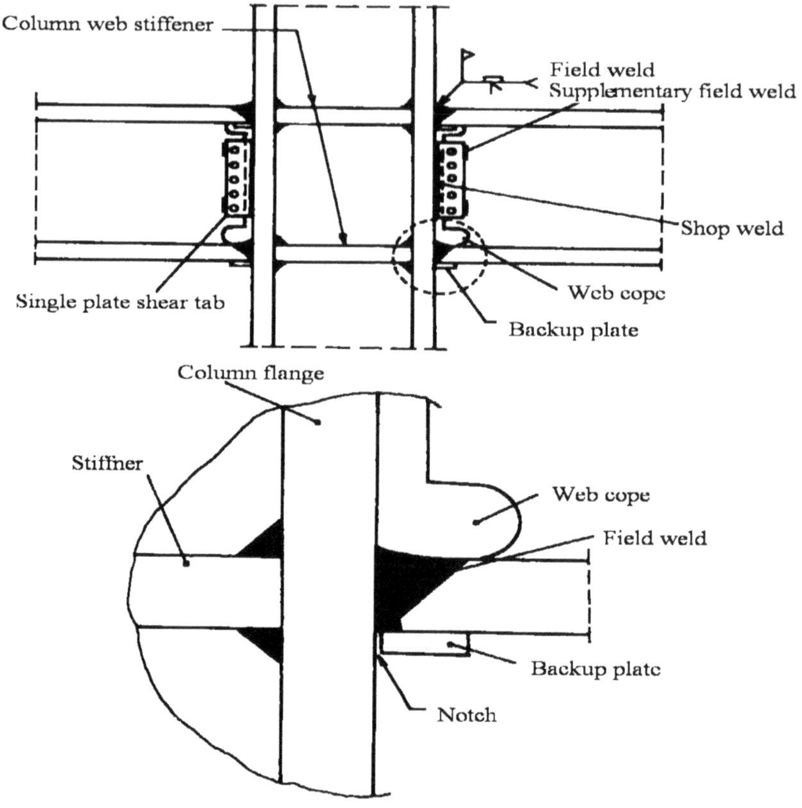

Figure 2.19: Typical beam-to-column connection

This unsatisfactory performance raises some very relevant questions (Daali, 1995, De Buen, 1996):

-are the characteristics of earthquakes relevant as included in the codes?

-do the code specifications correspond to reality?

-is the welding technology guilty of failure?

-are the connection details accurate?

-is the workmanship incorrect, especially in the field of penetration groove welds?

-how the pre-existent cracks in the weld metal or in the adjacent base metal are guilty?

-how the three orthogonal directions residual stresses in the joints, generated during the construction of the structure, including shop and field welding, are guilty?

-are the high velocities producing fast strain rates responsible for the brittle fracture?

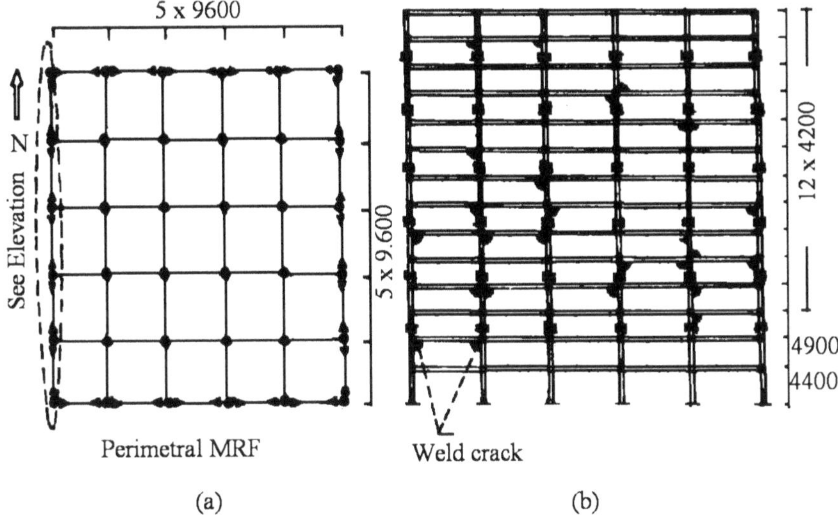

Figure 2.20: Typical location of fractures on the frame height: (a) Floor plan with perimetral MRFs (b) Frame elevation and damage location (after Astaneh-Asl et al, 1995)

Attempts to have responses to these questions were carried out by a lot of research works: Miller (1995), Engelhardt and Sobol (1995, 1997), Tide (1995), Hajjar and Leon (1996), Yang and Popov (1996), Sabol et al (1996), Astaneh-Asl et al (1996), Nahim et al (1996), Housner and Masri (1996), Engelhardt (1996), SAC Reports 95-02, 95-03, Wittaker et al (1998), Popov et al (1998); Special issue of Journal of Structural Engineering (2000): Kunnath and Malley, Peret, Dexter and Melandrez, Liu and Astaneh-Asl, Stojadinovici et al, El-Tawil et al, Chi et al, Gupta and Krawinkler, Maison et al, MacRae and Matheis, Luco and Cornell, Nakashima et al.

Upon examining the damage, it must be mentioned that there was only rare evidence that the plastic hinges actually were formed in beam where expected. Rather, the seismic energy passed directly to the connections, overloading them and causing fractures. First of all, this fact is due to the fact that the Northridge earthquake applied a dynamic load at a very high rate of speed. The yield strength used in design is measured for slowly loaded tensile specimens and the effect of strain rate is to increase this yield strength, in some cases to double it. In these conditions the joints are undersized with respect to the beam section which contrarily is oversized and, therefore, the seismic action cannot produce yielding, being directly transferred to the connection. This can be a rational answer to the first question, with reference to the influence of earthquake characteristics.

Another problem is related to the difference between the minimum and maximum specified yield strength. Generally, in order to obtain a weak-beam strong-column frame, a higher steel grade is used for columns than for beams. It is well known that the difference between the minimum and

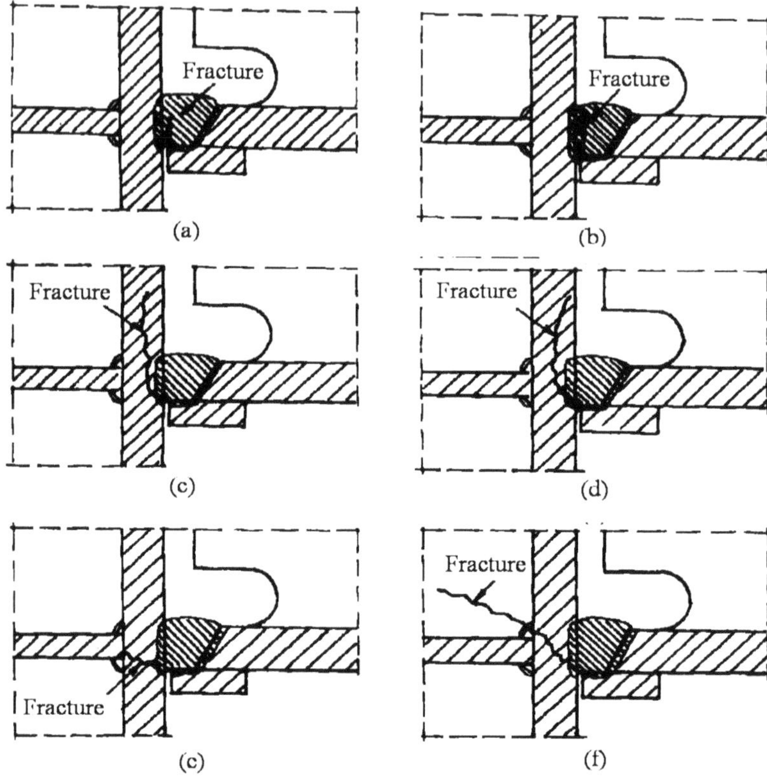

Figure 2.21: Typical connection fracture: (a) and (b) Fractures in weld metal; (c) and (d) Fracture in column flange; (e) and (f) Fracture in column flange and web (after Engelhardt and Sobol, 1995)

maximum values specified for yield strength is much higher as the steel grade is reduced. In many cases the yield strength of the beam material can be very near to the ultimate tensile strength of the column material, and this is an explanation for many fractures which were found in column flanges (Miller, 1995). So, a criticism of the code provisions is concerning the ratio between minimum and maximum yield strength and the possibility of combination of two steel qualities in a structure.

Because the damage was concentrated in the vicinity of the bottom welds, it is reasonable to believe that the technology of welding is guilty for this fact. From the technological point of view, one can observe that it is possible to weld across the full width of the top flange without interruption and that the weld which connects the beam bottom flange to column is more difficult to make, because the beam web prohibits the deposition of a continuous weld along the flange width. The investigations have shown in many cases inadequate fusion, generally concentrated in the portion of the weld directly under the beam web. So, incorrect welding technology can be the reason for the connection damage (Miller, 1995). The fact that all the fractures occur in the bottom flange welds is due to the effect caused by composite floor slabs (Hajjar and Leon, 1996), to the presence of a backing bar which produces an artificial crack and initiates the rupture (Popov et al, 1998), or due to the combination of vertical and horizontal ground motions (Hajjar and Leon, 1996). In order to improve the welding conditions, some additional studies concerning the connection details are necessary. It is imperative that the designer provides access holes with sufficient dimensions to have a good visibility for welding and to facilitate the deposition of weld metal in the respect of the quality requirements. These types of detail are missing in the code specifications.

The investigations concerning the causes of damage in steel buildings during the Northridge earthquake have given rise to a wide discussion within the international scientific community. On one hand, it can be ascribed to the use of field welding so that the poor workmanship is solely to blame, and, therefore, it is necessary to increase the on-site supervision and to improve welding details and procedures. On the other hand, damage causes can be attributed to an excess of seismic loading and to defective design guidance, leading to a rotation ductility supply lower than the earthquake imposed demand. This latter point of view seems nowadays the most supported by the specialists (Elnashai, 1994, Mazzolani and Piluso, 1997, Gioncu, 1999a,b, Gioncu et al, 2000).

2.4 Kobe Earthquake

2.4.1. Earthquake Characteristics

Japan is located geographically where the Eastern edge of Eurasian plate meets the North American plate, Pacific plate and Philippine plate (Fig. 2.22a). The movements of these boundaries are characterized by the Philippine and North-American plates thrusting under Eurasian plate with annual slip rates ranging from 5 to 10cm. This type of plate convergence induces two different types of seismic activity known as interplate and intraplate seismicities. The main seismic Japanese type are the intraplate ground motions, produced by energy release of fault planes. More then 1500 active faults are reported in and around Japan (Yamazaki et al, 1995).

The Kobe event is known also as Hyogoken-Nanbu earthquake, Great Hanshin earthquake or Hanshin-Awaji earthquake. The epicenter was about 200km north of the the major plate boundary between the Philippine plate and Eurasian Plate and about 40km from the Median Tectonic Line, which is a large strike-slip fault zone in South-Western Japan. Numerous active faults exist in that region. In the past, four major intraplate earthquakes occurred in central Japan: Nobi (1891), M = 8, Tango (1927), M = 7.3, Tottori (1943), M = 7.2, Fukui (1948), M = 7.1, earthquakes. Two historically important earthquakes occurred in the vicinity of Kobe City, in 868 and in 1916 (Zhao et al, 1996).

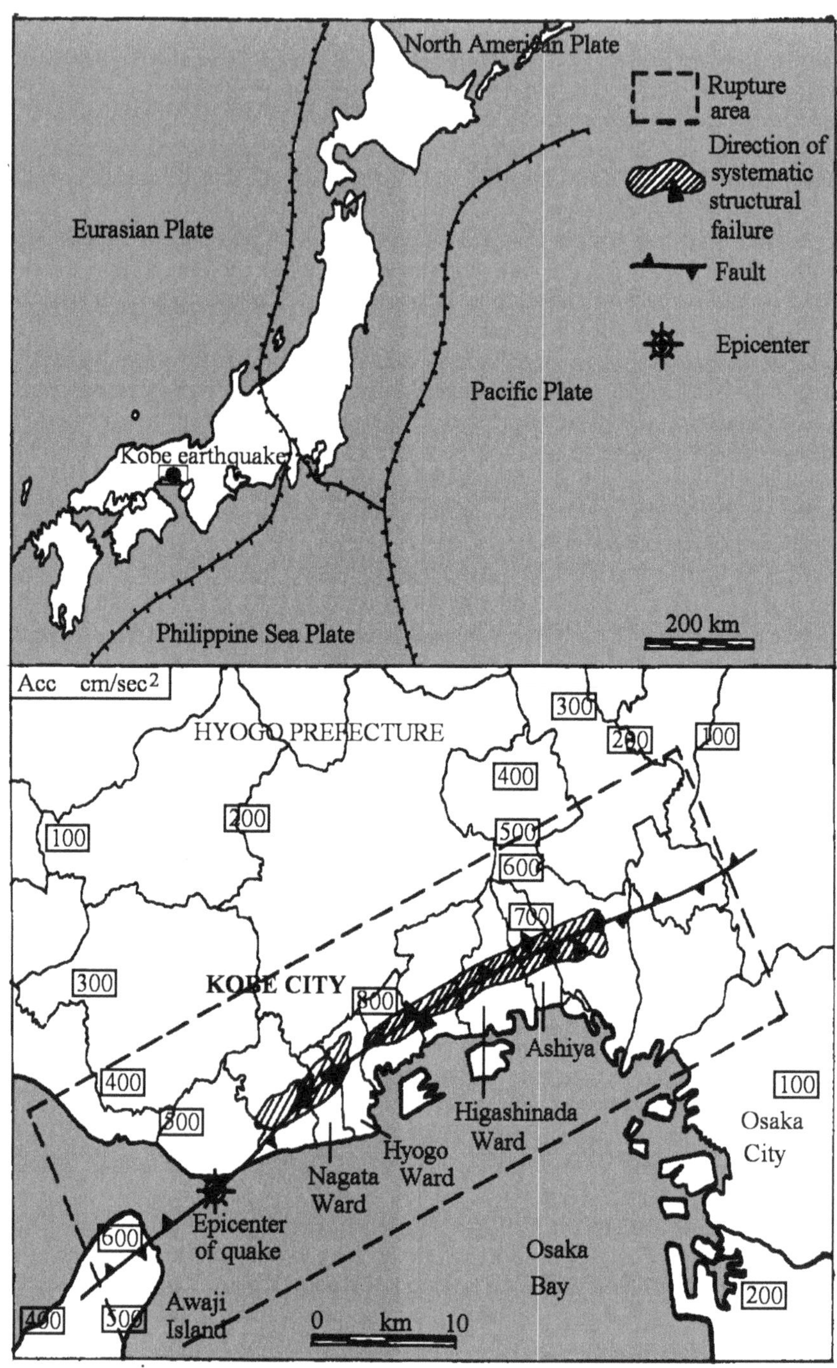

Figure 2.22: Kobe earthquake: (a) Tectonic environment of Japan: (b) Epicenter location and fault rupture of Kobe earthquake

The earthquake, that shook the Southern Hyugo Prefecture on 17 January 1995 measuring 7.2 magnitude, was the most devastating earthquake which struck Japan since the Kanto (Tokyo) event in 1923. The most damage was inflicted to the town of Kobe and the neighbouring towns of Ashiya and Nishinomiya. Kobe, with 1.4 million inhabitants, is the capital of Hyogo Prefecture, and the second largest port in Japan. The most damaged area was along a narrow band straddling the trace of fault rupture, extending from South-West towards North-East (Fig. 2.22b). This strong shaken area stood on the alluvial deposits between the Rokko mountains at North and the coast line at South. Is was believed that the earthquake consisted of three subevents, with the main shock having the above mentioned 7.2 magnitude. The epicenter of the main shock was located below the Awaji Island, 20 km South-West from Kobe downtown, with a focal depth of approximately 14.3 km. The type of fault rupture was strike-slip, similar to the fault ruptures of California. The rupture of the faults is reported to have lasted approximately 11 sec and fault area was about 40 x 10 km. The occurrence of three main shocks within 11 sec each other (generating fault and two co-movement faults, Yao et al, 1996) could have produced a complicated ground motion. The ground motions were amplified by the presence of a thick layer of alluvial soil, which caused an extensive liquefaction over the area of harbour district of Kobe. Most of the buildings and transportation structures damaged during the earthquake were left leaning to the North, indicating that the largest forces were generated perpendicular to the direction where the rupture faults were largest (Azizinamini and Ghosh, 1997, Faccioli, 1996, Pinto, 1996).

The characteristics of source and ground motions of the Kobe earthquake were investigated in many papers: Pinto (1996), Bertero et al (1995), Indirili (1996), Tomazevici and Fischinger (1995), Otani (1997). Many papers were published in the last important Conferences: 11th WCEE (1996), Acapulco, Mochizuchi et al, Borcherdt, Iwan, Inoue et al, Matsushima et al, Midorikawa et al, Niwa et al, Dohi et al, Adam and Takemiya, Faccioli, Pinto, Iwasaki; 11th ECEE (1998), Paris, Elnashai et al, Pomonis and Williams, Kamiyama and Matsukawa, Chouw; 12th WCEE (2000), Auckland, Paret and Attalla, Hori et al, Yamamoto, Wang and Nishimura, Motoki and Seo, Adam et al, Ejiri et al, Fujimoto and Midorikawa, Nakao et al, Takemiya et al.

Similar to the case of the Northridge earthquake, a very dense network of stations with very sophisticated instrumentation was present in the area of Kobe. Thus, a great number of records was obtained from different sites, but unfortunately it seems that no strong motion has been recorded in the most severely shaked areas (Nakashima et al, 1997). For the most strongly shaked district, the peak ground acceleration was equal to 0.835g (Fig. 2.23). From this figure it is evident that even through the duration was not very long, the signal cannot be considered a single impulse type, since several peaks are reaching the maximum accelerations and velocities (De Luca and Mele, 1996).

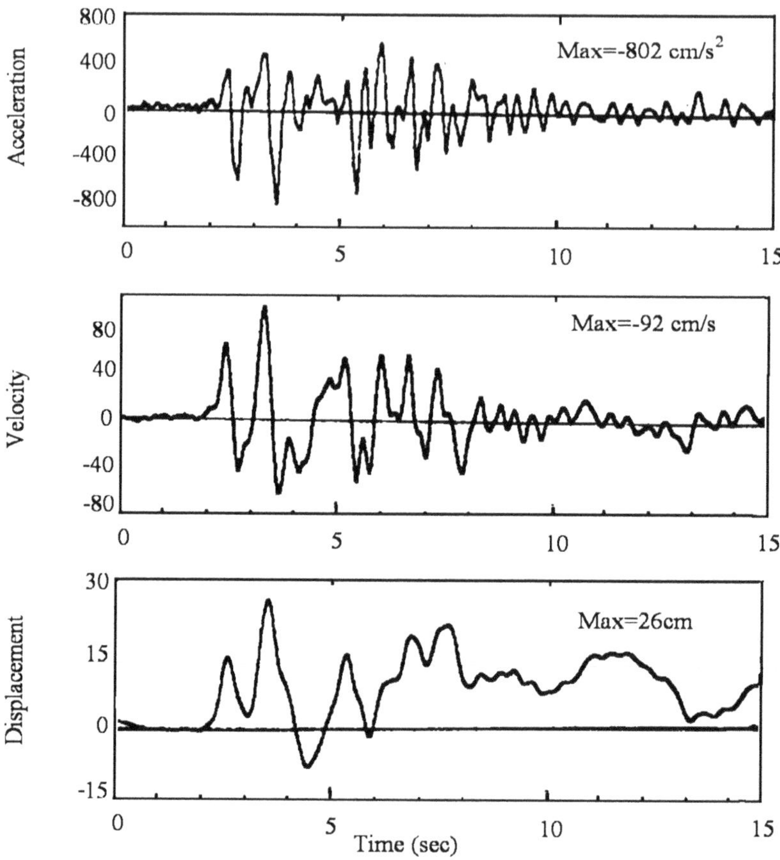

Figure 2.23: Ground motions of Kobe earthquake - JMA Station (after Iwan, 1996)

The acceleration and velocity elastic spectra, computed at a value of damping equal to 2 percent, are presented in Fig. 2.24. A synthetized motion for the different areas, predicted as the maximum possible action, is also depicted (Nakashima et al, 1997). These values are compared with the requirements of the Japanese code. The comparison confirms that the potential of damage of this earthquake was quite large and the code was not targeted toward such a signal. The velocity was very high, being for a damping of 2 percent around 350 cm/sec, and over 500 cm/sec for a damping of 1 percent.

The peak vertical ground accelerations at a distance about 100km to the fault plane were still as large as the horizontal ones, due to poorly compacted soil and the presence of water which was able to transmit compression waves. So the vertical soil motions were not highly damped (Adam et al, 2000).

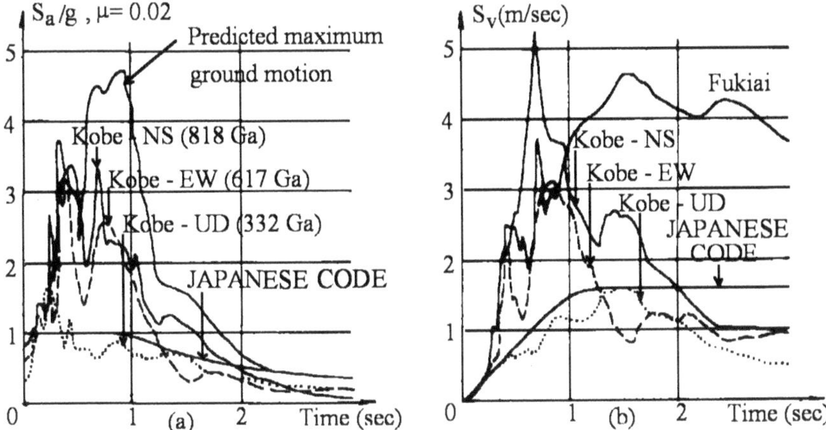

Figure 2.24: Kobe spectra: (a) Acceleration; (b) Velocity (after Kurobane et al, 1996)

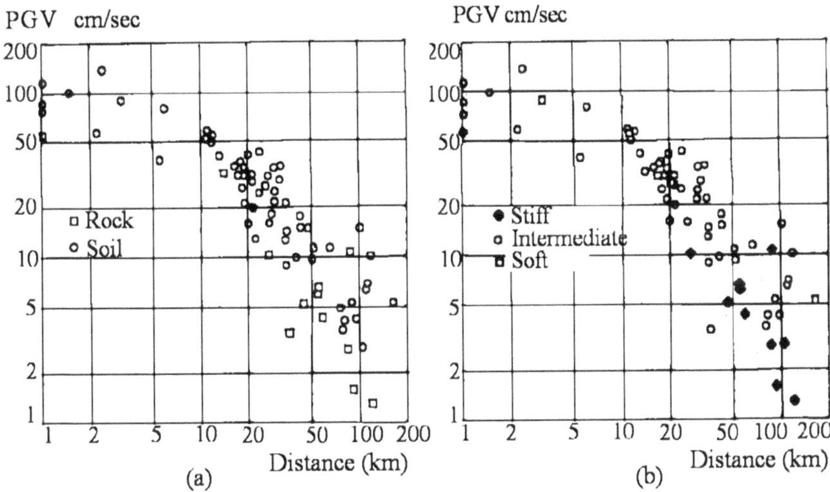

Figure 2.25: Attenuation of peak horizontal velocity: (a) Rock and hard soil sites; (b) Stiff intermediate and soft soil sites (after Midorikawa et al, 1996)

The attenuation of peak horizontal ground velocity is shown in Fig. 2.25. At very close distances, the peak velocities range from 60 to 140cm/sec, indicating that the peak horizontal velocity can be over 200 to 400cm/sec in the epicentral area (Mirodikawa and Matsuoka, 1996). The peak velocities attenuate with the distance. It is very interesting to note that the maximum velocity is attained at 1km from the epicenter (Faccioli, 1996). Due to these high velocities the damage to steel structures were produced for the

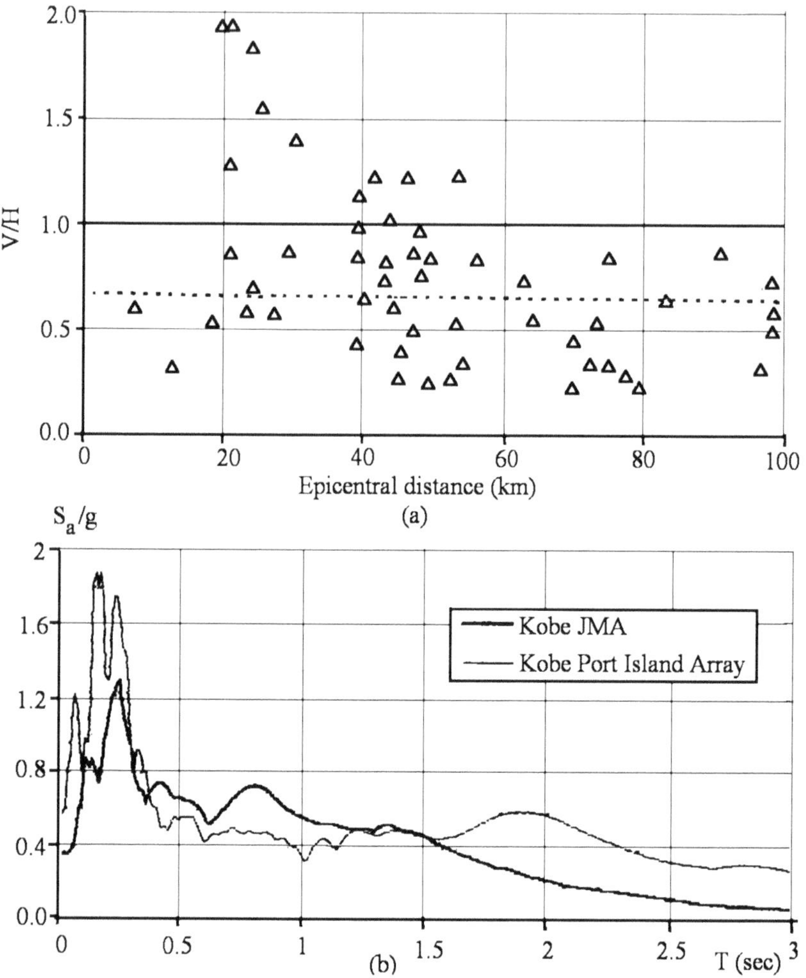

Figure 2.26: Vertical components: (a) Vertical-to-horizontal components ratios; (b) Vertical spectra (after Elnashai et al, 1998)

first or second excursions in plastic range (Kohzu and Suita, 1996), the resulting high strain rate produced an increase of yield stress and a decrease of ductility.

The vertical components were very important in the epicentral area. Many strong motion recordings have exhibited ratios of vertical to horizontal acceleration that frequently exceed unity (Fig. 2.26a) (Elnashai et al, 1998). The spectrum for accelerogram of vertical components is presented in Fig. 2.26b. An important amplification occurs for short periods, while for periods greater than 0.3 sec the amplification is very low.

Because both vertical and horizontal ground accelerations have a large amplitude, the coincidence of their appearance may have significant influence

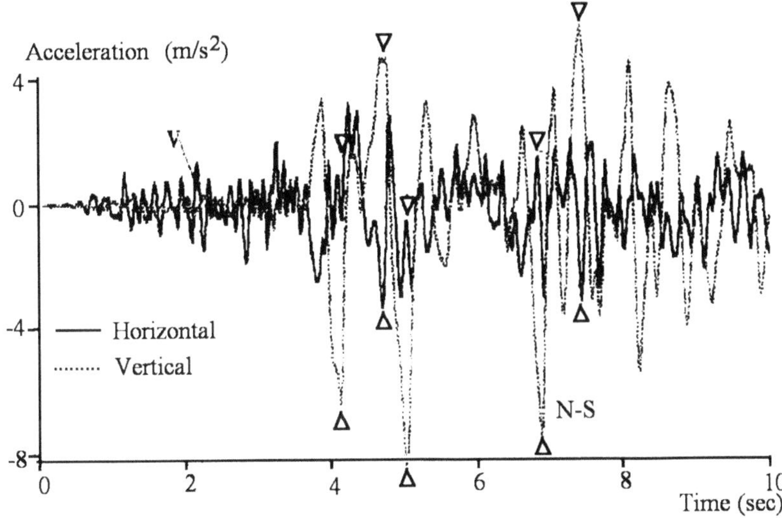

Figure 2.27: Combination of horizontal and vertical component - JMA Station (after Chouw, 1998)

on the behaviour of buildings. In Fig. 2.27 the time coincidence of peak accelerations in the vertical and horizontal direction was indicated by a triangular symbol (Chouw, 1998).

2.4.2. Damage of Steel Structures

Over 100,000 buildings and houses collapsed, over 90,000 buildings were heavily damaged and almost 150,000 buildings were lightly damaged. Among the collapsed or heavily damaged buildings, there were many steel structures. The great impact of this earthquake in the world of structural engineers was enormous, because Japanese engineering was always a reference point in seismic engineering. The delusion was very huge; nobody before could imagine this kind of damage.

The analysis of the damage of buildings was the subject of many papers and reports: Kurobare et al, 1997, Kawaguchi and Hangai (1995), Nakashima et al (1997) Wada et al (1997), Pinto (1996), Tomazevic and Fischinger (1995), Akiyama (1995, 1996), Bertero et al (1995). The damaged buildings were analysed in the papers presented at some Conferences: 11th WCEE (1996), Acapulco, Iwai et al, Inoue et al, Yao et al, Housner and Masri, Nakashima, Akiyama; SSS (1995), Budapest, Igarashi et al, Mazzolani, STESSA (1997), Kyoto, Akiyama and Yamada, De Luca and Mele, Kato et al, Kurobane et al, Tanaka et al, Tanaka and Tabuchi, Yang and Popov; 11th ECEE (1998), Paris, Pomonis and Williams; 12th WCEE (2000), Auckland, Fujimoto and Midorikawa, Usami et al.

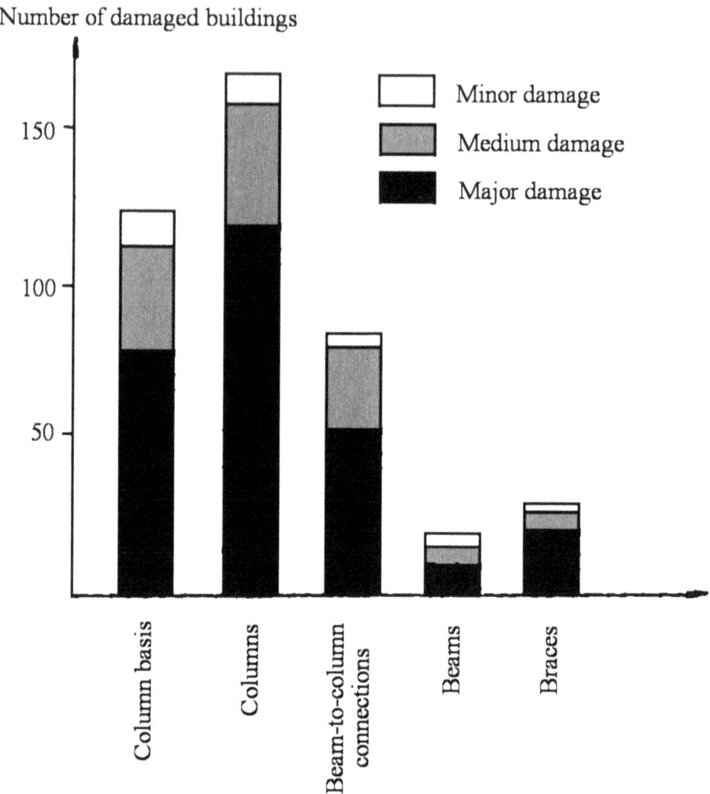

Figure 2.28: Damage of structural elements (after Igarashi et al, 1995)

From the structural point of view, the main characteristics of this earthquake were (Mazzolani, 1995):

-the intensity was much higher than it was considered in the Japanese code;

-the epicenter was very superficial, below an urban habitat and produced very important unexpected vertical quakes, which are not usually considered in the modern seismic codes and very often ignored for normal buildings;

- the strain rate was so high that it is impossible to reproduce in laboratory tests and, therefore, a material like steel, which is considered as a very ductile material, behaved as a brittle one under this very unusual loading condition.

The number of damaged buildings are given in Fig. 2.28, classified according to the structural element where the damage was observed. The definition of "major damage" corresponds to buildings in danger of collapse and "medium damage" to buildings with irreparable damage, but far from collapse. The "minor damage" covers small reparable damage. From the surveyed buildings, 63 percent were found to fall under the category of major damage, 27 percent under medium damage and only 10 percent under minor damage (Igarashi et al, 1995). It can be seen that the cases of damage to columns was most numerous, followed in order by column bases, beam-to-column connections, braces and beams. This astonishing classification shows a reverse situation with respect to the requirements of the European conception, which considers that the damage must arise in beams rather than in columns, the latter remaining unaffected during a severe earthquake. Contrary to this, the Japanese conception allows the occurrence of storey mechanisms, in which damage of columns is involved.

The reasons for this damage may be due to defective design and execution and to the fact that many buildings were designed before the introduction of the new seismic code.

The survey of damage after the Kobe earthquake shows that many steel buildings behaved very well, so no general conclusion about the poor behaviour of steel structures during this event can be drawn. At the same time, there were some cases in which very important damage occurred and the following particular comments result (Igarashi, 1995):

-storey collapse and residual interstorey drifts occured in buildings with little lateral stiffness;

-defective works were very frequently observed for column bases and the failure of these connections led to a major aggravation of structural damage;

-defective works were also observed on connection welding. It was noticed in some cases that filled welds had been implemented instead of butt welds, consequently having an insufficient penetration;

-lowering in structural strength due to corrosion was observed, especially on frames constructed with light-weight cold-formed steel sections;

-elements with thick walls collapsed from dramatic brittle fracture;

-due to the large values of vertical components, an increase of second order effects was produced.

Figs. 2.29 and 2.30 show some failure of structures and structural members during the Kobe earthquake (Fischinger, 1997).

2.4.3. Behaviour of Connections

More than 90 percent of multistorey building frames in Japan use box-section columns, due to their excellent cross-sectional properties to resist biaxial bending. Cold-formed sections with thick walls are the cheapest solution and, therefore, very widely used. The most typical details of beam-to-column connections are shown in Fig. 2.31 (JSSC, 1997). There are basically two types: the first is shop welded and the second site welded. The joints are designed to fulfill the requirements for fully restrained moment connections, according to the Japanese building code. This connection is made through the continuity plate, also called through diaphragms, at the position of beam flanges. They are welded to the diaphragms using single bevel complete penetration groove welds with backup bars. The position of backup bars for shop or site welding are different, because in the first case the welding can be performed in turned position.

Figure 2.29: Failure of some structures in Kobe (after Fischinger, 1997)

From the comparison of fractures or cracks for shop or site welding types, it is clear that the influence of welding conditions plays a very important role, the number of damaged connections of site welding type being 2.6 times greater than the shop welding type.

The fracture modes of beam-to-column connections are studied by Kurobane et al (1997a,b), Kato et al (1997), Tanaka et al (1997), JSSC (1997), Tanaka and Tabuchi (1997). These fracture modes are shown in Fig. 2.32, after JSSC (1997):

- brittle fracture of the base metal of the lower beam flange, initiated at the toe of the weld access hole (Fig. 2.32a);

-brittle fracture of the base metal of the through diaphragm, initiated at the edges of the butt welded joints (Fig. 2.32b);

Figure 2.30: Failure of some structural members in Kobe (after Fischinger, 1997)

-brittle fracture of the butt welded joints initiated at the tack welds of the run-off tab (Fig. 2.32c);

-brittle fracture of the weld metal/heat affected zone of the lower beam flange-to-through diaphragm welded joint (Fig. 2.32d);

-cracks in the weld zone between the column skin plate and through diaphragm (Fig. 2.32e);

-brittle cracks in the column skin plate propagated from the edges of butt welded joints between the lower beam flange and the through diaphragm (Fig. 2.32f).

-cracks in H-beam web initiated at the toe of the weld access hole (Fig. 2.32g);

-cracks in column plate initiated at the site edges of the beam flange (Fig. 2.32h).

For shop welded connections, the most frequent fracture is the one presented in Fig. 2.32a (49.7 percent) and for site welded is the one shown in Fig. 2.32c (42.3 percent).

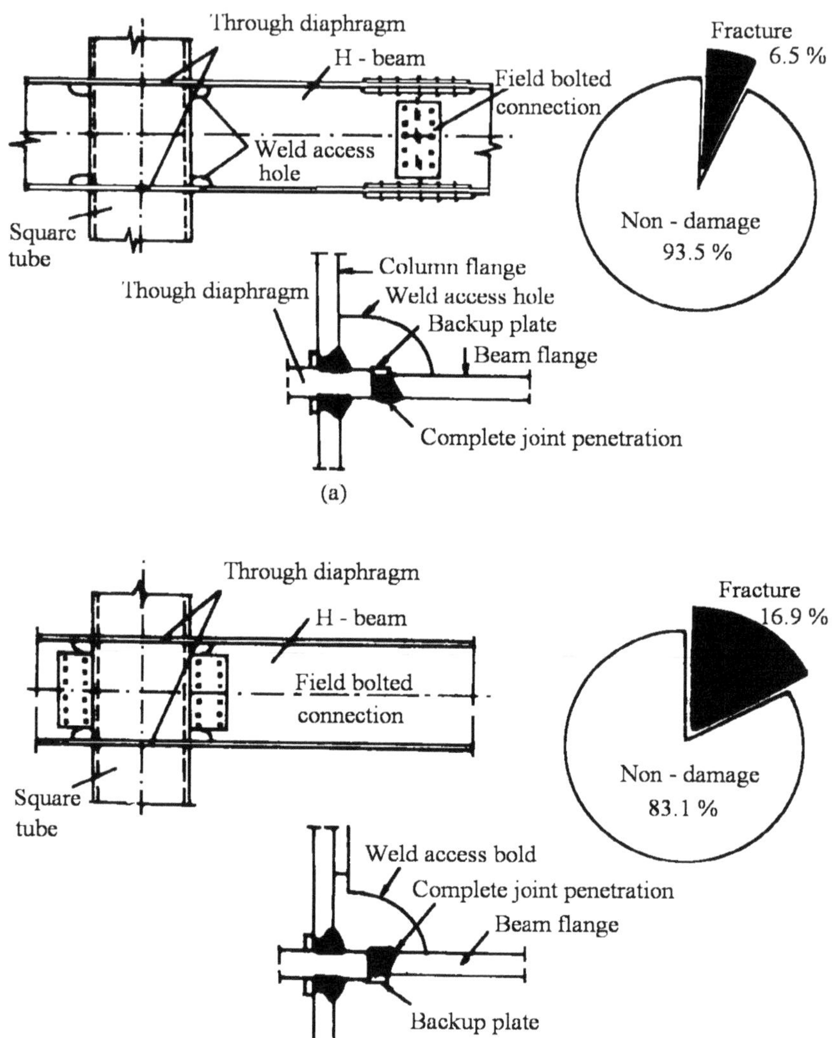

Figure 2.31: Typical beam-to-column connections: (a) Shop welding type; (b) Site welding type (after JSSC, 1997)

Some views of fractured connections are presented in Fig. 2.33 (Fischinger, 1997).

Experimental research works for the structural typologies used in Japan have been done by Kuwamura and Akiyama (1994), showing that the cold press-bracked square-tube members with relatively small wall slenderness have a brittle fracture collapse, due to decrease in ductile elongation capabil-

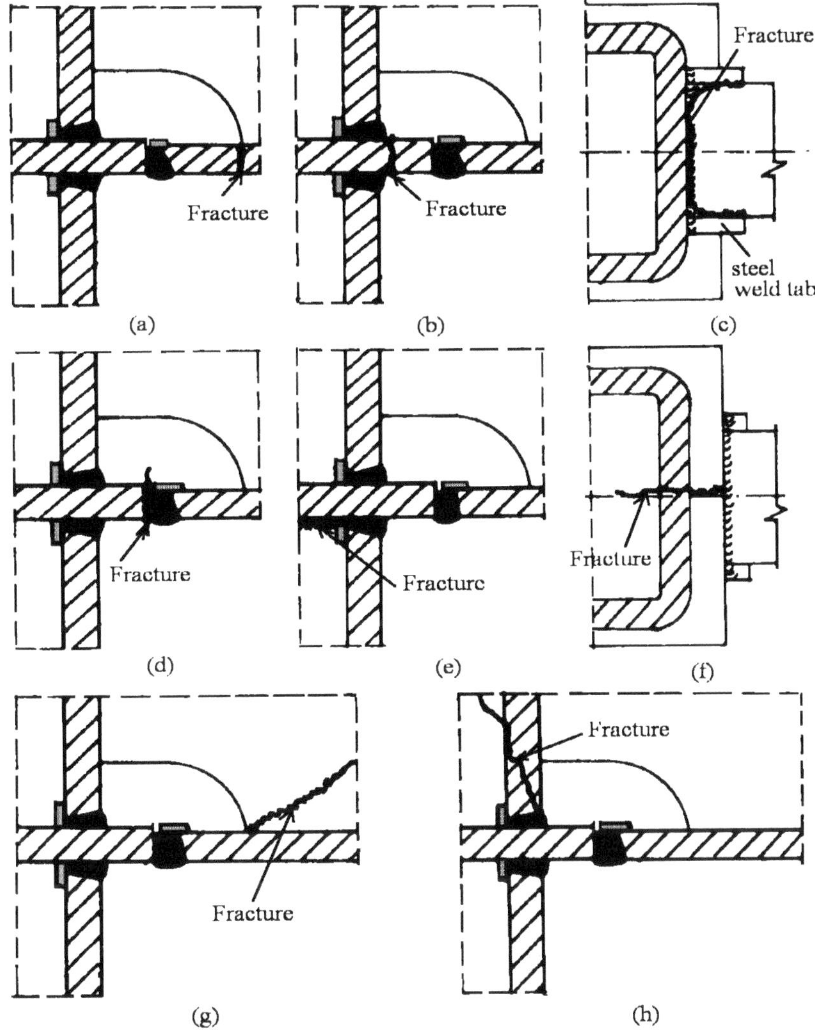

Figure 2.32: Typical fractures: (a), (b), (c) Brittle fracture of lower beam flange; (d), (e) Brittle crackes in welded zones; (f), (g), (h) Brittle cracks in diaphragm, beam or column webs (after JSSC, 1997)

ity and to the considerable reduction in notch toughness in the press-bracked corners. These results, published one year before the Kobe earthquake, were confirmed during the in-site verification of the great event.

After the earthquake a very extensive experimental program has been performed in Japan in order to clarify the reasons for damage during the Kobe earthquake (JSSC, 1997, Terada et al, 1997, Kurobane et al, 1997a,b, Tanaka et al, 1997, Kuwamura, 1997, Nakagomi et al, 1997, Tanaka and Tabuchi, 1997, Usami et al, 1997).

Figure 2.33: Failure of some beam-to-column connections (after Fischinger, 1997)

2.4.4. Ashiyahama Apartment Buildings

The most surprising damage during the Kobe earthquake was the brittle fracture of box-section columns of the Ashiyahama apartment buildings (Kurobane et al, 1996). Ashiyahama is a modern residential town consisting of 51 apartment buildings, located along the seaside of Ashiya and completed at the end of 70s. The number of stories of each building varies from 14 to 29. Among these, 21 buildings with storeys of 19 and 24 suffered tensile fractures, with the exception of few chords in the 14 storey buildings.

The building views, typical plan and structural system are presented in Fig. 2.34. The transverse structure is composed of concentric braced frames, while the longitudinal structure is a mega-structure, composed of trussed girders and columns. Each house unit is made of precast concrete panels, which are mounted in plane and surrounded by steel members, forming a group of 4 storey units (Tomazevic and Fischinger, 1995). The trussed columns have welded box-section chords, which are made of two parallel channels longitudinally welded. The width of column varies from 482 to 544 mm and the thickness from 16 to 47 mm. Only the lower portions of the first storey columns are made from 4 plates welded together. The trussed girders and the bracings are made of wide-flange sections. The steel grade is SM 490.

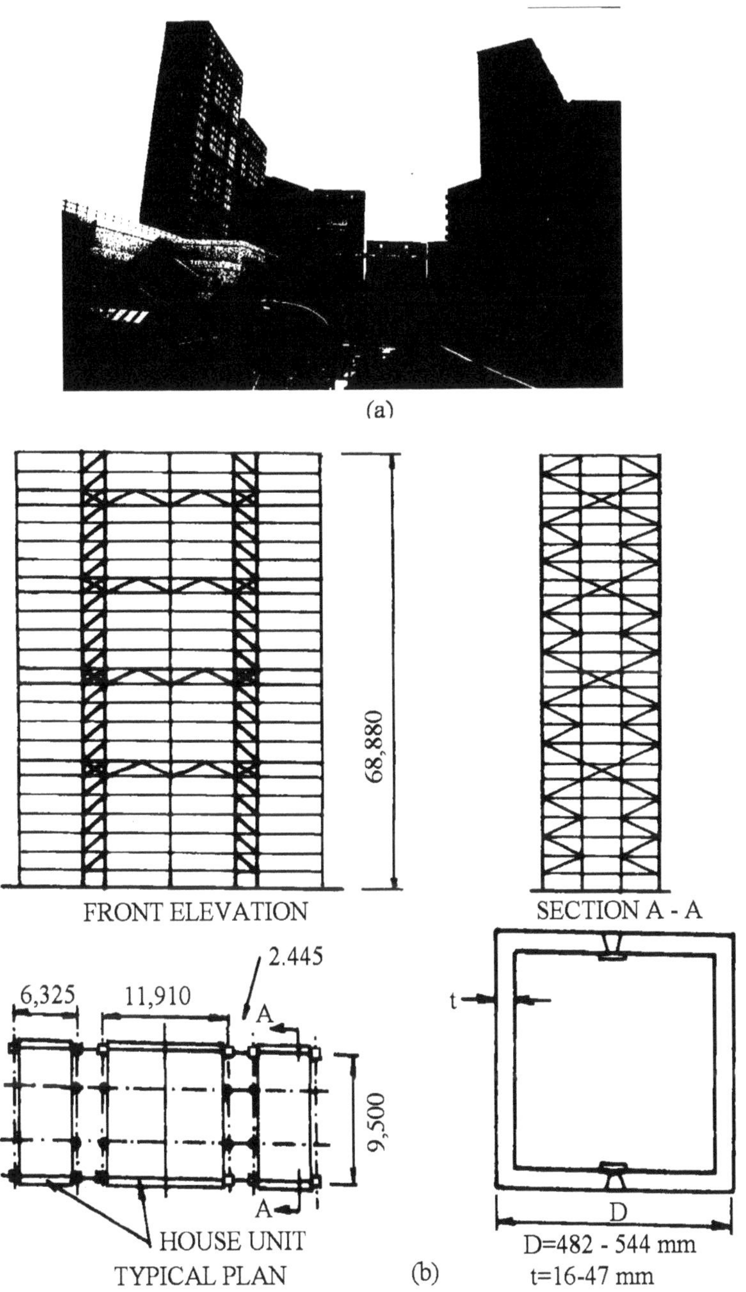

Figure 2.34: Ashiyahama Apartment buildings: (a) Building view (after Fischinger, 1997); (b) Building structure (after Kurobane et al, 1996, 1997a)

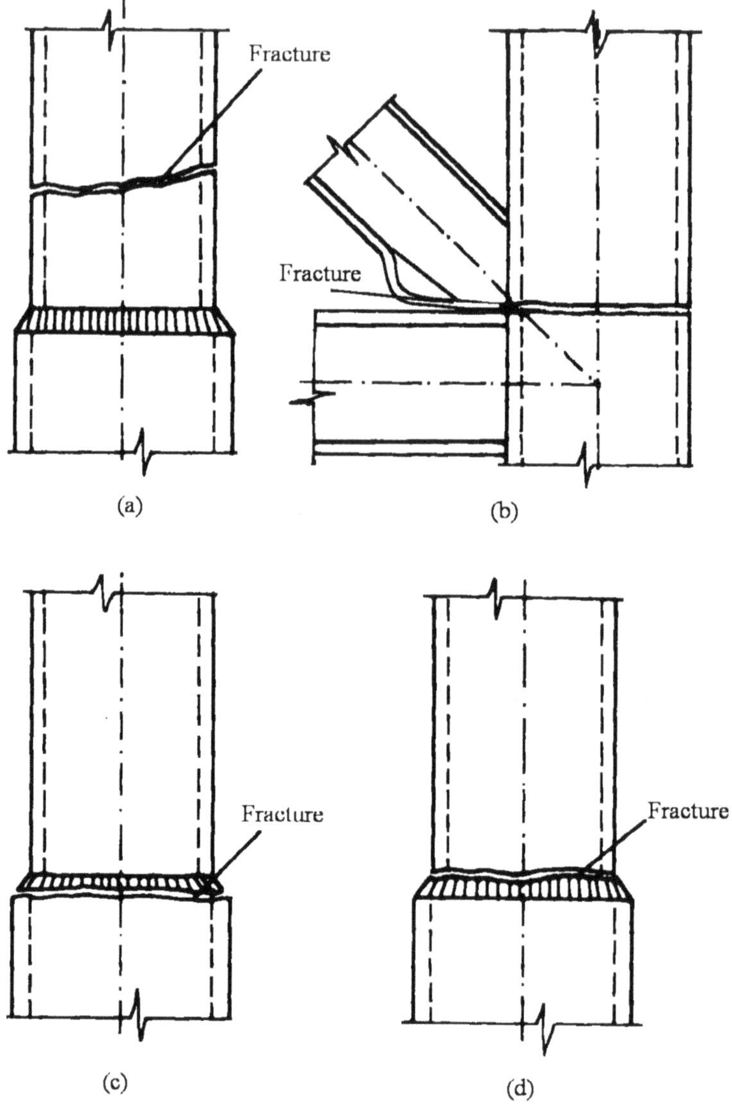

Figure 2.35: Brittle fracture types of trussed columns: (a) Base metal; (b) Chord to brace connection; (c) Groove weld in chord splice; (d) Weld toes of chord splice (after Kurobane et al, 1996)

The brittle fracture occurred in the chord of trussed columns, but never in the girders. Fracture patterns are classified into four main types (Fig. 2.35) (Kurobane et al, 1996):

-mode A, fracture across the full section of the chord at about 300 or 600 mm above the chord splice;

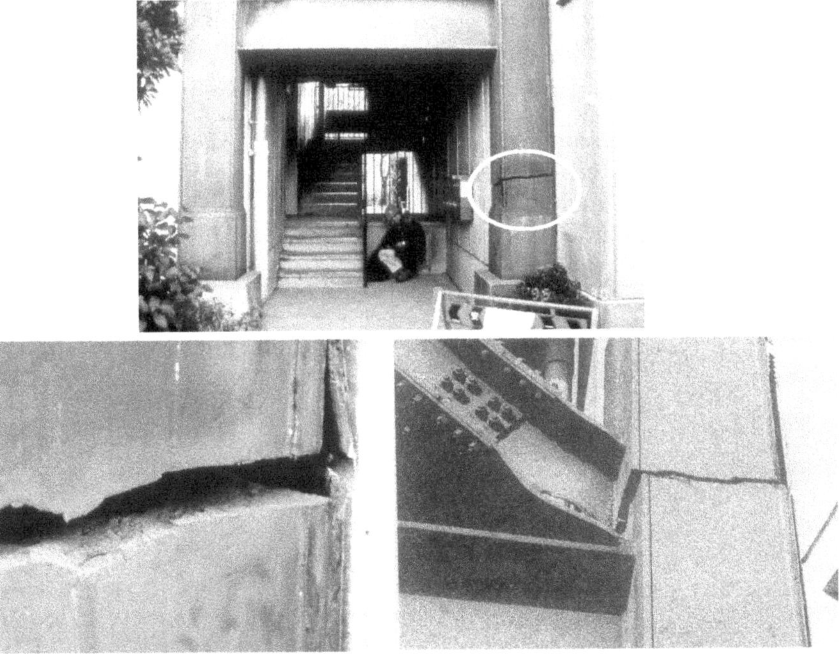

Figure 2.36: Fracture of columns

-mode B, fracture starting from the chord to the brace connection and running through the chord and brace;

-mode C, fracture in the groove welds at the chord splice;

-mode D, fractures initiated at the weld toes of the chord splice, running across the chord full section.

Fig. 2.36 shows some views of fractures of Ashiyahama Apartment Buildings.

All the fractured surfaces appeared rough and from this appearance they can be considered to have occurred just after yielding, without being strained up to the strain-hardening range. Plasticization was witnessed on some of the chord surfaces either in the immediate vicinity of cracks or in regions near the welded connections. Plasticization and local buckling were observed in a few braces. The location of the four types of fracture are summarized in Table 2.2. One can see that the most frequent mode of failure is the C one, but the mode A is the typical failure type at the ground floor. The concentration of some failure at 4, 8 and 11 stories is due, probably, to the influence of high modes of vibrations.

In order to understand what happened, a dynamic analysis of the 24 storey building was done for three ground motions, recorded at Fukiai station (Kurobane et al, 1997a,b). By analyzing the possible causes of brittle fracture, the following factors are considered:

Table 2.2: Locations of fracture occurrences (after Kurobane et al, 1996)

Locations	**A**	**B**	**C**	**D**	
Story	**Base Metal above Chord Splice**	**Chord to Brace Connec- tion**	**Groove Weld in Chord Splice**	**Weld Toes of Chord Splice**	**Total**
24	0	0	0	0	0
23	1	0	0	0	1
22	0	0	0	0	0
21	1	0	0	0	1
13-20	0	0	0	0	0
12	0	1	0	0	1
11	0	0	7	0	7
10	0	0	0	0	0
9	0	0	0	0	0
8	0	0	12	0	12
7	0	0	0	0	0
6	0	2	4	0	6
5	0	2	0	0	2
4	0	0	12	1	13
3	0	2	0	0	2
2	4	2	1	3	10
1	11	0	2	1	14
Total	17	9	38	5	69

-influence of low temperature, which during the earthquake has been estimated to be about 0^0 C in the chords;

-deterioration of material ductility, due to welding and high residual stresses;

-influence of the vertical components;

-influence of high strain rate, due to the impulsive loading, both in horizontal and vertical directions;

-influence of the highest vibration modes.

From these factors, the influence of the strain rate is found to be the most significant influencing factor. But, for the top level of the structure, it is possible that the exceeding of tension axial forces should be another important factor.

Akiyama (1995, 1996) considers also that an important factor of these brittle fractures can be the heavy section of box-columns, with low slenderness of walls. For the above reasons, the applicability of these section types for tall structures must be thoroughly and urgently reviewed.

2.5 Conclusions

A detailed analysis of the effects on steel structures of the last great earthquakes have shown that, generally, the steel structures performed very well, so the reputation of steel as a good material for seismic-resistant structures remains substantially unaffected. But this good performance can be undermined if some behavioural aspects, related to the earthquake loading conditions, are neglected.

The last earthquakes have shaken densely built urban areas, with high economic development, where very different structure types have been affected. The resulting damage has been very varied and a lot of new information concerning the behaviour of steel structures during strong ground motions has been obtained.

The presence of a dense network of stations and numerous buildings with high performance instrumentations has delivered a great amount of records, involving all the factors influencing the phenomena.

On this basis, the last important earthquakes have shown some limitations in present concepts:

-Michoacan earthquake belongs to the far-field ground motion type, where the local soil conditions play the most important effect. The Pino Suarez building collapse has underlined the importance of the proper evaluation of available and required ductility, for which all the influencing factors must be considered in detailed manner;

-Loma Prieta earthquake, due to the very high economic losses caused by the failure of non-structural elements, has shown that a correct design conception, in addition of saving the structure from collapse during severe ground motions, must consider the reduction of losses caused by the damage of non-structural elements;

-Northridge earthquake, which belongs to the near-field ground motion type, have shown that the concept of over-strengthening the connections must be re-examined, by considering all factors influencing the ductility of plastic hinges, like impulse type of loading, large strain-rate and vertical components;

-Kobe earthquake was of the same type as the Northridge earthquake. The similarities between the damage of steel structures for the two events show that there are serious mistakes in present basic concepts, which must be urgently ruled out.

Due to the superior performance associated to steel structures in past earthquakes, much of the available research works has been concentrated to reinforced concrete structures. After the last earthquakes, it is clear that also steel structures justify the increased research investments, faced to identify and, hopefully, to eliminate the possible situations in which steel structures behave deficiently.

2.6 References

Abbiss, C.P. (1989): Seismic amplification in Mexico City. Earthquake Engineering and Structural Dynamics, Vol. 18, 79-88

Adam, M.A., Schmid, G. Chouw, N. (2000): Investigation of ground motions and structural responses in near field due to incident waves. In 12th World Conference on Earthquake Engineering, Auckland, 30 January- 4 February 2000, CD-ROM 1313

Adam, M.A., Takemiya, H. (1996): Seismic wave amplification in Kobe during Hyogo-ken Nanbu earthquake. In 11th World Conference Earthquake Engineering, Acapulco, 23-28 June 1996, CD-ROM 1895

Akiyama, H., Yamada, S. (1995): Damage of Structures in the Hyogoken-Nanbu earthquake. In EASEC 95, Gold Cost Australia, 1-12

Akiyama, H. (1996a): Damage of structures in the Hyogokeen - Nanbu earthquake. In La Cita Sicurra: Terremoti, Eruzioni e Protezione Civile, Napoli, Messina, 10-13 Febbraio 1996, 1-13

Akiyama, H. (1996b): Damage of steel buildings observed in the 1995 Hyogoken Nanbu earthquake. In 11th World Conference on Earthquake Engineering, Acapulco, 23-28 June 1996, CD-ROM 2144

Akiyama, H., Yamada, S. (1997): Seismic input and damage of steel moment frames. In Behaviour of Steel Structures in Seismic Areas, STESSA 97, (eds. F.M.Mazzolani, H. Akiyama) , Kyoto, 2-8 August 1997,10/17 Salerno, 789-800

Astaneh-Asl, A., Shen, J.H, D'Amore, E. (1995): Seismic vulnerability of welded WMRF damages during 1994 Northridge earthquake. Giornale Italiane dela Costruzione in Acciaio, Riva del Garda, 15-18 Octobre 1995, Vol. 1, Ricerco Tecnico e Sperimentale, 69-79

Astaneh-Asl, A., Modjtahedi, D., McMullin, K.M., D'Amore, E. (1996): Seismic safely of damaged steel moment frames. In 11th World Conference on Earthquake Engineering, Acapulco, 23-28 June 1996, CD-ROM 1014

Azizinami, A., Ghosh, S.K. (1997): Steel reinforced concrete structures in 1995 Hyogoken-Nanbu earthquake. Journal of Structural Engineering, Vol. 123, No. 8, 986-992

Basham P.W. (1989): Earth-science issue of seismic hazards assessment in Eastern North America. In Earthquake Hazards and the Design of Construction Facilities in the Eastern United States, (eds. K.H. Jacob and C.J.Turstra), New York, 24-26 February 1988, Annals of the New York Academy of Science, Vol. 558, 1-10

Bertero, V.V. (1994): Major issues and future directions in earthquake - resistant design. In 10th World Conference on Earthquake Engineering, Madrid, 19-24 July 1992, Balkema, Rotterdam, 6407-6444

Bertero, V.V. (1995): Seismological and engineering aspects of the January 17, 1995 Hyokoken-Nanbu (Kobe) earthquake. Report UCB/EERC-95/10, University of California, Berkeley, November 1995

Bertero, V.V., Anderson J.C., Krawinkler, H. (1994): Performance of steel building structures during the Northridge earthquake. Report UBC/EERC-94/09, University of California, Berkeley

Bonacina, G., Indirili, M., Negro, P. (1994): The January 17, 1994 Northridge earthquake. Special Publication No. I.94

Borcherdt, R.D. (1996): Strong ground motions generated by the Northridge and Hanshin- Awayi earthquake of January 17, 1994 and 1995: Implication for site-specific design factors. In 11th World Conference on Earthquake Engineering, Acapulco, 23-28 June 1996, CD-ROM 1246

Bozorgnia, Y., Niazi, M. (1993): Distance scaling of vertical and horizontal response spectrum of the Loma Prieta earthquake. Earthquake Engineering and Structural Dynamics, Vol. 22, 695-707

Bozorgnia, Y., Niazi, M. (1995): Characteristics of the free field vertical ground motions during the Northridge earthquake. Earthquake Spectra, Vol. 11, No. 4, 515-525

Bozorgnia, Y., Niazi, M., Campbell, K.W. (1996): Relation between vertical and horizontal response spectra for Northridge earthquake. In 11th World Conference on Earthquake Engineering, Acapulco, 23-28 June 1996, CD-ROM 893

Celebi, M.V. (1992): Highlights of Loma Prieta responses of four tall buildings. In 10th World Conference on Earthquake Engineering, Madrid, 19-24 July 1992, Balkema, Rotterdam, 4039-4044

Celebi, M.K. (1996): Unique ground motions recorded during the Northridge (California) earthquake of January 17, 1994 and implications. In 11th World Conference on Earthquake Engineering, Acapulco, 23-28 June 1996, CD-ROM 2148

Celebi, M. (1997): Response of Olive Hospital to Northridge and Whittier Earthquakes. Journal of Structural Engineering, Vol. 123, No. 4, 389-396

Cajes, M.J., Yang, C.Y., Zhang, L. (1992): Stability of 47-story office building with active controls. In 1992 Annual Technical Session, Earthquake Stability Problems in Eastern North America, Pittsburg, 6-7 April 1992, 237-247

Charez-Garcia, F.J. (1995): A Mexican experience in seismic microzonation. In 10th European Conference on Earthquake Engineering, (eds G.Duma), Vienna, 28 August- 2 September 1994, Balkema, Rotterdam, 2553-2558

Cheng, F.Y., Lu, L.W., Ger, J.F. (1992): Observations on behavior of a tall steel building under earthquake excitations. SSRC 1992 Annual Technical Session, Earthquake Stability Problems in Eastern North America, Pittsburg, 6-7 April 1992, 15-26

Chi, W.M., Deierlein, G.C., Ingraffea, A. (2000): Fracture toughness demands in welded beam-column moment connections. Journal of Structural Engineering, Vol. 126, No. 1, 88-97

Chouw, N. (1998): Effect of strong vertical ground motions on structural responses. In 11th European Conference on Earthquake Engineering, Paris, 6-11 September 1998, CD-ROM 608

Chouw, N. (2000): Performance of structures during near-source earthquakes. In 12th World Conference on Earthquake Engineering, Auckland, 30 January-4 February 2000, CD-ROM 0368

Citipitioglu E., Celebi, M. (1996): Should the design response spectra be revised as a result of Northridge earthquake motions? In 11th World Conference on Earthquake Engineering, Acapulco, 23-28 June 1996, CD-ROM 44

Corsanegro, A. (1995): Recent trends of earthquake damage interpretation. In 10th European Conference on Earthquake Engineering, (ed. G.Duma), Vienna, 28 August- 2 September 1994, Balkema, Rotterdam, Vol. 1, 763-771

Daali, M.L. (1995): Damage assessment in steel structures. In 7th Canadian Conference on Earthquake Engineering, Montreal, 517-524

De Buen, O. (1996): Earthquake resistant design: A view from the practice. In 11th World Conference on Earthquake Engineering, Acapulco, 23-28 June 1996, CD-ROM 2002

De la Llera J.C., Chopra A.K. (1996): Analysis of the behaviour of buildings during the 1994 Northridge earthquake. In 11th World Conference on Earthquake Engineering, Acapulco, 23-28 June 1996, CD-ROM 220

De Luca, A., Mele, E. (1996): The recent Kobe earthquake: General data and lessons learned on steel structures. Report

De Luca, A., Mele, E. (1997a): The lessons learned from the Northridge and Hyogoken-Nanbu earthquakes: Inelastic requirements of perimetral frames. In Behavior of Steel Structures on Seismic Areas, STESSA 97, (eds. F.M. Mazzolani, H.Akiyama), Kyoto, 2-8 August 1997, 10/17 Salerno, 801-810

De Luca, A., Mele, E. (1997b): Seismic behaviour of steel frame buildings. Comparison of perimeter and spatial frame structures. In Giornate Italiane della Costruzione in Acciaio, Ancona, 2-5 October 1997, Stato della Ricerca, 257-268

Dexter, R.J., Melendrez, M.I. (2000): Through-thickness properties of column flanges in welded moment connections. Journal of Structural Engineering, Vol.126, No. 1, 24-31

Diaz-Rodriguez, J.A. (1995): Dynamic properties and mineralogy of soft clay. In 10th European Conference on Earthquake Engineering, (ed.G.Duma), Vienna, 28 August- 2 September 1994, Balkema, Rotterdam, Vol. 1, 447-452

Dohi, H., Kawano, M., Matsuda, S. (1996): Response spectra of ground motion above earthquake fault. In 11th World Conference on Earthquake Engineering, Acapulco, 23-28 June 1996, CD-ROM 1772

Doroudian M., Vucetic, M., Martin G.R. (1996): Development of 3-dimensional geotechnical data for Los Angeles seismic microzonation. In 11th World Conference on Earthquake Engineering, Acapulco, 23-28 June 1996, CD-ROM 2148

Durkin, M.E. (1996): Casualty patterns in the 1994 Northridge, California earthquake. In 11th World Conference on Earthquake Engineering, Acapulco, 23-28 June 1996, CD-ROM 979

Eisenberg, J.M. (1995): Recent strong earthquakes evidences against ductility concepts. In 10th European Conference on Earthquake Engineering, (ed. G.Duma), Vienna, 28 August- 2 September 1994, Balkema, Rotterdam, Vol. 2, 1339-1346

Ejiri, J., Fujimori, T., Nakayama, T., Goto, Y., Yasui, Y. (1992): Bedrock motion characteristics during Loma Prieta earthquake in the nord-west area from the epicenter. In 10th World Conference on Earthquake Engineering, Madrid, 19-24 July 1992, Balkema, Rotterdam, 1003- 1008

Eriji, J., Goto, Y., Toki, K. (2000): Peak ground motion characteristics of 1995 Kobe earthquake and an extracted simple evaluation method. In 12th World Conference on Earthquake Engineering, Auckland, 30 January- 4 February 2000, CD-ROM 1659

Elnashai, A. (1994): Comments on the performance of steel structures in the

Northridge (Southern California) earthquake of January 1994, New Steel Construction, Vol. 2, No. 5

Elnashai, A.S., Papazoglou, A.S. (1997): Procedure and spectra for analysis of RC structures subjected to strong vertical earthquake loads. Journal of Earthquake Engineering, Vol 1, No 1, 121-155

Elnashai, A.S., Bommer, J.J., Martinez-Pereira A. (1998): Engineering implications of strong motion records from recent earthquakes. In 11th European Conference on Earthquake Engineering, Paris, 6-11 September 1998, CD-ROM 59

El-Tawil, S, Mikesell, T., Kunnath, S.K. (2000): Effect of local details and yield ratio behaviour of FR steel connections. Journal of Structural Engineering, Vol. 126, No. 1, 79-87

Engelhardt, M.D. (1996): Damage to steel moment frames observed in the 1994 Northridge earthquake. In 11th World Conference on Earthquake Engineering, Acapulco, 23-28 June 1996, CD-ROM 2139

Engelhardt, M.D., Sabol, T.A. (1995): Lessons learned from the Northridge earthquake: Steel moment frame performance. Symposium on A New Direction in Seismic Design, Tokyo, 9-10 October 1995, 1-14

Engelhardt, M.D., Sabol, T.A. (1997): Seismic-resistant steel moment connections: Development since the 1994 Northridge earthquake. Progress in Structural Engineering and Materials. Construction Research Communication Limited, 68-77

Faccioli, E. (1996): On the use of engineering seismology tools in ground shaking scenarios. In 11th World Conference on Earthquake Engineering, Acapulco, 23-28 June 1996, CD-ROM 2007

Faccioli, E., Tolis, S.V., Borzi, B., Elnashai, A.S., Bommer, J.J. (1998): Recent developments in the definition of the design action in Europe. In 11th European Conference on Earthquake, Paris, 6-11 September 1998, CD-ROM 14

Fajfar, P. (1995): Elastic and inelastic design spectra. In 10th European Conference on Earthquake Engineering, (ed. G.Duma), Vienna, 28 August- 2 September 1994, Balkema, Rotterdam, Vol. 2, 1169-1178

Fischinger, M. (1997): EASY, Earthquake Engineering Slide Information System. FGG-IKPIP Institute Slovenia

Fujimoto, K., Midorikawa, S. (2000): Isoseismal map of the 1995 Hyogoken Nanbu earthquake. In 12th World Conference on Earthquake Engineering, Auckland, 30 January- 4 February 2000, CD-ROM 1670

Ger, J.F., Cheng, F.Y. (1992): Collapse assessment of a tall building damaged by 1985 Mexico earthquake. In 10th World Conference on Earthquake engineering, (ed. G.Duma), Vienna, 28 August- 2 September 1994, Balkema, Rotterdam, 51-59

Ger, J.F., Cheng, F.Y., Lu, L.W. (1993): Collapse behavior of Pino Suarez building during 1985 Mexico City earthquake. Journal of Structural Engineering, Vol 119, No 3, 852-870

Ghosh, S.K. (1995): Recent and impending major changes in US Seismic Codes. In 7th Canadian Conference on Earthquake Engineering, Montreal, 983-991

Grecu, V., Moldovan, A. (1994): Geostatistical methods for seismic hazard estimation and earthquake prediction. In XXIV General Assembly of European

Seismological Commission, Athens, Greece, University of Athens, Vol. 1, 345-355

Gunturi, S.K., Shah, H.C. (1992): Building specific damage estimation. In 10th Conference on Earthquake Engineering, Madrid, 19-24 July 1992, Balkema, Rotterdam, 6001-6006

Gupta, A., Krawinkler, H. (2000): Behaviour of ductile SMRFs at various seismic hazard levels. Journal of Structural Engineering, Vol. 126, No. 1, 98-107

Hajjar, J.F.,Leon, R.T. (1996): Effect of floor slabs on the performance of SMR connections. In 11th World Conference on Earthquake Engineering, Acapulco, 23-28 June 1996, CD-ROM 656

Hall, J.F. (1995): Parameter study of the response of moment-resisting steel frame buildings to near-surface ground motions. SAC Report 95-05

Hall, J.F., Heaton, T.H., Halling, M.W., Wald, D.J. (1995): Near-source ground motion and its effects on flexible buildings. Earthquake Spectra, Vol. 11, No. 4, 569-605

Hori, N., Yamamoto, S., Yamada, M. (2000): Source analysis of near-field earthquake records observed in rock sites. In 12th World Conference on Earthquake Engineering, Auckland, 30 January- 4 February 2000, CD-ROM 0698

Housner, G.W., Masri, S.F. (1996): Structural control researh issues arising out of the Northridge and Kobe earthquakes. In 11th World Conference on Earthquake Engineering, Acapulco, 23-28 June 1996, CD-ROM 2009

Hudson, R. L., Skyers, B.D., Lew, M. (1996): Vertical strong motion characteristics of the Northridge earthquake. In 11th World Conference on Earthquake Engineering, Acapulco, 23-28 June 1996, CD-ROM 728

Igarashi, S., Nakashima, S., Kadoy, A. (1995): Damage to Steel Structures due to the Great Hanshin Earthquake. In Stability of Steel Constructions, (ed. M.Ivanyi), Akademiai Kiado, Budapest, 21-23 September 1995, Vol. 2, 985-992

Iglesias, V., Gomez-Bernal, A. (1992): Seismic zonation of Mexico City. In 10th World Conference on Earthquake Engineering, Madrid, 19-24 July 1992, Balkema, Rotterdam, 6215-6220

INCEDE ERS, Kobe net (1999): Joint raport on 1995 Kobe earthquake. INCEDE Report 1999-03

Indirli, M. (1996): Il terremoto di Great Hanshin-Awaji, Giappone, 17 Gennaio, 1995. ANIIS-GLIS Rapporto 04/96, ENEA, Bologna

Inoue, R., Susuki, S., Kudo, K., Takahanshi, M., Sakaue, M.(1996): A detailed survey on the damage of buildings during the Hyogoken Nanbu (Kobe) earthquake with references to aftershock data and geology. In 11th World Conference on Earthquake Engineering, Acapulco, 23-28 June 1996, CD-ROM 1253

Ishihara, K., Yoshida, K., Kato, M. (1997): Characteristics of lateral spreading in liquefied deposits during the 1995 Hanshin-Awaji earthquake. Journal of Earthquake Engineering, Vol. 1, No. 1, 23-55

Iwan, W.D. (1995): Drift demand spectra for selected Northridge sites. SAC Report 95-05

Iwan, W.D., Chen, X. (1995): Important near-field ground motion data from the Landers earthquake. In 10th European Conference on Earthquake Engineering, (ed. G.Duma), Vienna, 28 August- 2 September 1994, Balkema,

Rotterdam, 229-234

Iwai, S., Susuki, Y., Kakumoto S. (1996): GIS application of damage data management on buildings and urban facilities on the January 17 1995 Hyogoken-Nanbu earthquake. In 11th World Conference on Earthquake Engineering, Acapulco, 23-28 June 1996, CD-ROM 851

Iwan, W.D. (1996): The drift demand spectrum and its application to structural design and analysis. In 11th World Conference on Earthquake Engineering, Acapulco, 23-28 June 1996, CD-ROM 1116

Iwan, W.D. (1997): Drift spectrum: Measure of demand for earthquake ground motion. Journal of Structural Engineering, Vol. 123, No. 4, 397-404

Iwasaki, Y. (1996): Strong ground motion during the Kobe earthquake of January 17, 1995. In 11th World Conference on Earthquake Engineering, Acapulco, 23-28 June 1996, CD-ROM 2149

Japanise Society of Steel Structures (1997): Kobe earthquake damage to steel moment connections and suggested improvement. JSSC Technical Report, No 39

Kamiyama, M., Matsukama, T. (1998): Non-linear response inthe downhole strong motion records at Port Island during the 1995 Kobe earthquake, Japan. In 11th European Conference on Earthquake Engineering, Paris, 6-11 September 1998, CD-ROM 93

Kaneto, T., Mikani, T., Hayashikawa, T., Matsui, Y. (1992): Directional behaviour of strong motion ground motions during the Loma Prieta earthquake. In 10th World Conference on Earthquake Engineering, Madrid, 19-24 July 1992, Balkema, Rotterdam, 605-610

Kataoka, S., Ohmachi, T. (1996): Synthetic earthquake motion of 2-D irregular ground in near field. In 11th World Conference on Earthquake Engineering, Acapulco, 23-28 June 1996, CD-ROM 1248

Kato, B., Morita, K., Maruoka, Y., Sugimoto, H., Teraoka, M. (1997): Seismic damage of steel beam-to-column rigid connections in the 1995 Hyogoken-Nanbu earthquake. Fabrication. In Behaviour of Steel Structures in Seismic Areas, STESSA 97, (eds. F.M.Mazzolani, H.Akiyama), Kyoto, 2-8 August 1997, 10/17 Salerno, 811-820

Kawaguchi, K., Hangai, Y. (1995): Report on Spatial Structures Damaged by the 1995 Great Hansin Earthquake. Bulletin of Earthquake Resistant Structure Research Center, No 28, Sept., 69-76

King, S.A., Rojahn, C. (1998): Performance of buildings near-strong motion recording sites in the 1994 Northridge earthquake. In 11th European Coference on Earthquake Engineering, Paris, 6-11 September 1998, CD-ROM 66

Kohzu I., Suita, K. (1996): Single or few excursions failure of steel structural joints due to impulsive shocks in the 1995 Hyogoken Nanbu earthquake. In 11th World Conference on Earthquake Engineering, Acapulco, 23-28 June 1996, CD-ROM 412

Krawinkler, H., Osteraas, J. (1990): Comportamento delle costruzioni in acciaio durante il terremoto del Messico del 1985. Costruzioni Metalliche, No. 2, 97-107

Krawinkler, H., Rahnama, M. (1992): Effect of soft soil on design spectra. In 10th World Conference on earthquake Engineering, Madrid, 19-24 July 1992, Balkema, Rotterdam, 5841-5846

Krawinkler, H., Gupta, A. (1997): Deformation and ductility demands in steel moment frame structures. In Stability and Ductility of Steel Structures, SDSS 97, (ed. T.Usami), Nagoya, 29-31 July 1997, Vol. 1, 57-68

Kunnath, S.K., Malley, J.O. (2000): Structural forum. Seismic behaviour and design of steel moment frames: Aftermath of the 1994 Northridge earthquake. Journal of Structural Engineering, Vol. 126, No 1, 5-9

Kurobane, Y., Ogawa, K., Veda, C. (1996): Kobe earthquake damage to high-rise Ashiyahama apartment buildings: Brittle tensile fracture of box section columns. Tubular Structures VII, (eds. I.Farkas and K.Jarmai), Miskolc, 28-30 August 1996, Balkema, Rotterdam, 277-284

Kurobane, Y., Azuma, K., Ogawa, K. (1997a): Brittle fracture in steel building frames. Comparative study of Northridge and Kobe earthquake damage. In International Institute of Welding. Annual Assembly, San Francisco, 1-30

Kurobane, Y., Wang, B., Azuma, K., Ogawa, K. (1997b): Brittle fracture in steel frames. In Behaviour of Steel Structures in Seismic Areas, (eds. F.M. Mazzolani and H.Akiyama), Kyoto, 2-8 August 1997, 10/17 Salerno, 833-844

Kuwamura, H., Akiyama, H. (1994): Brittle fracture under repeated high stresses. Journal of Constructional Steel Research, Vol. 29, 5-19

Lew, M. (1992): Characteristics of vertical ground motions recorded during recent California earthquakes. In 10th European Conference on Earthquake Engineering, Madrid, 19-24 July 1992, Balkema, Rotterdam, 573-576

Li, Y., Mau, S.T. (1997): Learning from recorded earthquake motion of buildings. Journal of Structural Engineering, Vol. 123, No. 1, 62-69

Liu, J., Astaneh-Asl, A. (2000): Cyclic testing of simple connections. Journal of Structural Engineering, Vol. 126, No. 1, 32-39

Luco, N., Cornell, A.C. (2000): Effects of connection fractures on SMRF seismic drift demands. Journal of Structural Engineering, Vol. 126, No. 1, 127-136

MacRae, G.A., Mattheis, J. (2000): Three-dimensional steel building response to near-fault motions. Journal of Structural Engineering, Vol. 126, No. 1, 127-136

Magistrale, H., Zhou, H.W. (1996): Lithologie control of the depth of earthquakes in Southern California. Science, Vol. 273, 639-642

Maison, B.F., Rex, C.O., Lindsey, S.D., Kasai, K. (2000): Performance of PR moment frame buildings in UBC seismic zones 3 and 4. Journal of Structural Engineering, Vol. 126, No 1, 108-116

Mahin, S.A., Hamburger, R.O., Malley, J.O. (1996): An integrated program to improve the performance of welded steel frame buildings. In 11th World Conference on Earthquake Engineering, Acapulco, 23-28 June 1996, CD-ROM 1114

Matsushima, Y., Myslimaj, B.(1996): A comparison between nonlinear responses of structures subjected to 1995 Hyogoken Nanbu earthquake and 1993 Kushiro-oki earthquake. In 11th World Conference on Earthquake Engineering, Acapulco, 23-28 June 1996, CD-ROM 1428

Mazzolani, F.M. (1995): Some simple considerations arising from Japanise presentation on the damages caused by the Hanshin earthquake. In Stability of Steel Structures, (ed. M.Ivavyi), Budapest, 21-23 September 1995, Akademiai Kiado, Budapest, Vol. 2, 1007-1010

Mazzolani, F.M., Piluso, V. (1997): The influence of the design configuration on the seismic response of moment - resisting frames. In Behaviour of Steel Structures in Seismic Areas, STESSA 97, (eds. F.M.Mazzolani and H.Akiyama), Kyoto, 2-8 August 1997, 10/17, Salerno, 444-453

Mc Clure, F.E. (1989): Lessons learned from recent moderate earthquake. In Earthquake Hazard and the Design of Constructed Facilities in the Eastern United States, (eds. K.H. Jacob, C.S. Turkstra), New York, 24-26 February 1988, Annuals of New York Academy of Science, Vol. 558, 251-258

Midorikawa, S., Si, H., Matsuoka, M. (1996): Empirical analysis of peak horizontal velocity for Hyogo-Ken Nanbu Japan earthquake of January 17, 1995. In 11th World Conference on Earthquake Engineering, Acapulco, 23-28 June 1996, CD-ROM 1564

Miller, D.K. (1995): Northridge: The Role of Welding Clarified. In 7th Canadian Conference on Earthquake Engineering, Montreal, 573-580

Mochizuchi T., Amakuni, K, Takao, M. (1996): What is the 1995 great Hanshin-Awayi earthquake disaster? In 11th World Conference on Earthquake Engineering, Acapulco, 23-28 June 1996, CD-ROM 877

Mohammadioun, B. (1997): Nonlinear response of soils to horizontal and vertical bedrock earthquake motion. Journal of Earthquake Engineering, Vol 1, No 1, 93-119

Mohammadioun, G., Mohammadioun, B.(1996): Vertical / horizontal ratio for strong ground motion in the near field and soil non-linearity. In 11th World Conference on Earthquake Engineering, Acapulco, 23-28 June 1996, CD-ROM 899

Motoki, K., Seo, K. (2000): Strong motion characteristics near the source region of the Hyogoken-Nanbu earthquake from analysis of the directions of structural failures. In 12th World Conference on Earthquake Engineering, Auckland, 30 January- 4 February 2000, CD-ROM 0959

Naeim, F. (1998a): Performance of 20 extensively-instrumented buildings during the 1994 Northridge earthquake. The Structural Design of Tall Buildings, Vol. 7, 179-194

Naeim, F. (1998b): Research overview: Seismic response of structures. The Structural Design of Tall Buildings, Vol. 7, 195-215

Naeim, F. (2000): Learning from structural and nonstructural seismic performance of 20 extesively instrumented buildings. In 12th World Conference on Earthquake Engineering, Auckland, 30 January- 4 February 2000, CD-ROM 0217

Nakagima, T., Fujita, T., Minami, K. (1997): Welding connections on slop assembling recommendations for beam - end details in steel structures. In Behaviour of Steel Structures in Seismic Areas, STESSA97, (eds. F.M. Mazzolani and H.Akiyama), Kyoto, 2-8 August 1997, 10/17, Salerno, 632-639

Nakao, Y., Tamura, K., Kataoka, S. (2000): Effects of earthquake source parameters on estimated ground motion. In 12th World Conference on Earthquake Engineering, Auckland, 30 January- 4 February 2000, CD-ROM 1895

Nakashima M. (1996): Damage to steel buildings observed in the 1995 Hyogoken Nanbu earthquake. In 11th World Conference on Earthquake Engineering, Acapulco, 23-28 June 1996, CD-ROM 2143

Nakashima, M., Yamao, K., Minami, T. (1997): Post-earthquake analysis of steel buildings damaged during the 1995 Hyogoken-Nanbu earthquake. In Stability and Ductility of Steel Structures, SDSS 97, Nagoya, 29-31 July 1997 723-730

Nakashima, M., Minami, T., Mitani, I. (2000): Moment redistribution caused by beam fracture in steel moment frames. Journal of Structural Engineering, Vol. 126, No. 1, 137-144

Negro, P., Bonacina, G., Indirli, M. (1994): Il terremoto di Northridge (Los Angeles) del 17 Gennaio 1994. CTIE Atti del 10 Congresso CTE sulle Nuova Tecnologia, Edilizia per L' Europa, Milano, 3-5 Nov. , Vol. 1, 237-245

Niazi, M. Bozorgnia, Y. (1992): Behaviour of near source vertical and horizontal response spectra at SMART-1 array, Twain. Earthquake Engineering and Structural Dynamics, Vol. 21, 37-50

Nigdor, R.L., Madura R. M. (1996): Collection and achiving of code accelerograph data from the Northridge earthquake. In 11th World Conference on Earthquake Engineering, Acapulco, 23-28 June 1996, CD-ROM 738

Niwa, M., Ohno, S., Takanashi, K., Takemura, M. (1996): Estimation of peak accelerations and response spectra on rock for the 1995 Hyogoken Nanbu earthquake, Japan. In 11th World Conference on Earthquake Engineering, Acapulco, 23-28 June 1996, CD-ROM 1716

Ohno, S., Konno, T., Abe, K., Masao, T. (1996): Method of evaluationg horizontal and vertical earthquake ground motion for aseismic design. In 11th World Conference on Earthquake Engineering, Acapulco, 23-28 June 1996, CD-ROM 1791

Osteraas, J., Krawinkler, H. (1989): The Mexico earthquake of September 19, 1985; Behaviour of steel structures. Earthquake Spectra, Vol. 5, No. 1, 51-88

Otani S. (1997): Development of performance-based design methodology in Japan. In Seismic Design Methodologies for the Next Generation of Codes, (eds. P. Fajfar and H. Krawinkler), Bled, 24-27 June 1997, Balkema, Rotterdam, 59-67

Papaleontiou, C., Roesset, J.M. (1993): Effect of vertical acceleration on the seismic response of frames. In Structural Dynamics, EURODYN 93, (eds. Moan et al), Balkema, Rotterdam, 19-26

Paret, T.P., Attalla, M.R. (2000): Changing perception of the extent of damage to welded steel moment frames in the Northridge earthquake. In 12th World Conference on Earthquake Engineering, Auckland, 30 January- 4 February 2000, CD-ROM 0054

Peret, F.T. (2000): The W1 Issue. I: Extent of weld fracturing during Northridge earthquake. II: UT reability for inspection of T-joints with backing. Journal of Structural Engineering, Vol. 126, No 1, 10-18, 19-23

Phipps, M.T., Jirsa, J.O., Picado, M., Karp, R. (1992): Performance of high technology industries in the Loma Prieta earthquake. In 10th World Conference on Earthquake Engineering, Madrid, 19-24 July 1992, Balkema, Rotterdam, 85-89

Pinto P.S.S. (1996): Considerations on the geotechnical behaviour of structures during earthquakes. In 11th World Conference on Earthquake Engineering, Acapulco, 23-28 June 1996, CD-ROM 2145

Petroski, H. (1989): To Engineering is Human. The Role of Failure in Successful

Design. MacMillan, London Limited

Pinto, A.V. (1996): The Kobe earthquake (January 17th 1995) damage to R/C structures. Performance of Civil Engineering Structures, COST 1, 21-44

Pomonis, A., Williams, M.S. (1998): Lessons learned from the EEFIT missions to the Northridge and Kobe earthquakes. In 11th European Conference on Earthquake Engineering, Paris, 6-11 September 1998, CD-ROM 69

Popov, E.E. (1994): Development of U.S seismic code. Journal of Constructional Steel Research, Vol. 29, 191-207

Popov, E.E., Yang, T.S., Chang, S.P. (1998): Design of steel MRF connections before and after 1994 Northridge earthquake. Engineering Structures, Vol. 20, No. 12, 1030-1038

Reinero, E. Perez-Rocha, L.E., Arciniego, A., Ordaz, M. (1992): Prediction of response spectra at any site in Mexico City. In 10th World Conference on Earthquake Engineering, Madrid, 19-24 July 1992, Balkema, Rotterdam, 767-77

Reinoso, E., Ordaz, M., Guerrero, R. (2000): Influence of strong ground motion duration in seismic design of structures. In 12th World Conference on Earthquake Engineering, Auckland, 30 January- 4 February 2000, CD-ROM 1151

Rihal, S.S. (1992): Correlation between recorded buildings data and non- structural damage during the Loma Prieta earthquake of October 17, 1989, selected studies. In 10th World Conference on Earthquake Engineering, Madrid, 19-24 July 1992, Balkema, Rotterdam, 73-78

Roeder, C.W. (1997): Column cracking in steel moment frames. In Stability and Ductility of Steel Structures, SDSS 97, Nagoya, 28-30 August 1997, Vol. 2, 793-800

Roesset J.M. (1994): Modelling problems in earthquake resistant design: Uncertainties and needs. In 10th World Conference on Earthquake Engineering, Madrid, 19-24 August 1992, Balkema, Rotterdam, 6445-6483

SAC (1995a): Evaluation, repair, modification and design of steel moment frames. Report 95-02

SAC (1995b): Characterization of ground motions during the Northridge earthquake of January 17, 1994. Report 95-03

SAC (1995c): Analytical and field investigations of buildings affected by Northridge earthquake of January 17, 1994. Report 95-04

SAC (1995d): Parametric analytical investigation of ground motion and structural response, earthquake of January 17, 1994. Report 95-05

SAC (1995e): Surveys and assessment of damage to buildings affected by the Northridge earthquake of January 17, 1994. Report 95-06

SAC (1995f): Case studies of steel moment frame building performance in the Northridge earthquake of January 17, 1994. Report 95-07

SAC (1996): Interim guide lines. Advisory No 1: Evaluation, repair, modification and design of welded steel moment frame structures. Report 96-03

Saikia, C.K. (1996): Prediction of design ground motions for large earthquakes. In 11th Conference on Earthquake Engineering, Acapulco, 23-28 June 1996, CD-ROM 1931

Seed R.B. et al (1990): Preliminary report on the principal geotechnical aspects of the October 17, 1979, Loma Prieta earthquake. Report UBC/EERC-90/05,

University of California, Berkeley, April 1990

Sharpe, R.L. (1992): Acceptable earthquake damage on design performance. In 10th World Conference on Earthquake Engineering, Madrid, 19-24 July 1992, Balkema, Rotterdam, 5891-5894

Sobol T.A., Engelhardt M.D., Aboutaha, R.S., Frank, K.H. (1996): Overview of the AISC Northridge moment connection test program. In 11th World Conference on Earthquake Engineering, Acapulco, 23-28 June, 1996, CD-ROM 857

Somerville, P., Graves, R., Saikia, C. (1995): Characterization of ground motions of the Northridge earthquake of January 17, 1994. SAC Report 95-03

Somerville, P., Graves, R., Saikia, C. (1996): Estimation of strong motion time histories experienced by steel buildings during the 1994 Northridge earthquake. In 11th World Conference on Earthquake Engineering, Acapulco, 23-28 June 1996, CD-ROM 1178

Stojadinovic, B., Goel, S.C., Lee, K.H., Margarian, A.G., Choi, J.H. (2000): Parametric tests on unreinforced steel moment connections. Journal of Structural Engineering, Vol. 126, No. 1, 40-49

Tanaka, T., Tabuchi, M. (1997): Fracture on SHS column to H-beam connections by Hyogoken - Nanbu earthquake. In Behaviour of Steel Structures in Seismic Areas, STESSA 97, (eds. F.M.Mazzolani and H.Akiyama), Kyoto, 2-8 August 1997, 10/17, Salerno, 866-873

Takemiya, H., Miyagawa, G., Kagawa, Y. (2000): Synthesis of seismic motions in view of source to site: The Kobe case. In 12th World Conference on Earthquake Engineering, Auckland, 30 January- 4 February 2000, CD-ROM 2018

Tanaka, A., Kato, B., Kanetko, H., Sakamoto, S., Takahashi, Y., Teraoka, M. (1997): Seismic damage of steel beam-to-column rigid connections. Evaluation from statical aspect. In Behaviour of Steel Structures in Seismic Areas, STESSA 97, (eds. F.M.Mazzolani and H.Akiyama), Kyoto, 2-8 August 1997, 10/17 Salerno, 856-865

Terada, T., Yabe Y., Mase, S., Sakamoto, S., Toshio, U. (1997): Structural behaviour of steel beam-to-column connections subjected to dynamic loads. In Behaviour of Steel Structures in Seismic Areas, STESSA 97, (eds. F.M. Mazzolani and H.Akiyama), Kyoto, 2-8 August 1997, 10/17, Salerno, 656-663

Teran-Gilmore A., Bertero, V.V. (1992): Performance of tall buildings during the 1985 Mexico City earthquakes. Report UBC/EERC 92/17, University of California, Berkeley

Tide, R.H.R. (1995): Fracture of beam-to-column connections under seismic load. Northridge, California Earthquake January 17, 1994. In Habitat and High-Rise. Tradition and Innovation, (eds. L.S. Beedle and D. Rice) Amsterdam, 14-19 May 1995, 1273-1286

Tomazevic, M., Fischinger, M. (1995): Observazioni sul territorio di Hyogoken - Nanbu del 17 Gennaio 1995. Ingegnirie sismica, anno XII, No. 2, 3-19

Trifunac, M.D., Todorovska, M.I. (1998): Nonlinear soil response as a natural passive isolation mechanism. The 1994 Northridge, California, earthquake. Soil Dynamic and Earthquake Engineering, Vol. 17, 41-51

Usami, T., Kaneko, H., Kimura, M., Une, H., Kushibe A. (1997): Real scale model tests on flange fracture behaviour of beam adjacent to beam to column joint and the seismic resistance after repairing and strengthening. In Behaviour of Steel Structures in Seismic Areas, STESSA 97, (eds. F.M.Mazzolani and H.Akiyama), Kyoto, 2-8 August 1997, 10/17, Salerno, 955-962

Usami, T., Teshigawara, M., Kitagawa, Y., Kawase, H. (2000): Evaluation of strong ground motion and building damage in the 1995 Hyogo-Ken Nanbu earthquake. In 12th World Conference on Earthquake Engineering, Auckland, 30 January- 4 February 2000, CD-ROM 2038

Vannacci, G. (1994): 17 Gennaio '94 : Il terremoto di Northridge conferma le superiorita dell' acciaio. Costruzioni Metalliche, No. 5, 9-12

Wada, A., Iwata, M., Huang, Y. H. (1997): Seismic design trend of tall steel buildings after the Kobe earthquake. Post SMiRT Conference on Seismic Isolation, Passive Energy, Dissipation and Control of Vibrations of Structures. Taormina, 25-27 August 1997, 251-269

Wang, H., Nishimura, A. (2000): On seismic motion near active faults based on seismic records. In 12th World Conference on Earthquake Engineering, Auckland, 30 January- 4 February 2000, CD-ROM 0853

Watanabe, M. (1999): Kobe earthquake. In Earthquake Resistant Engineering Structures (eds. G. Oliveto and C.A. Brebbia), Catania, June 1999, WIT Press, Southampton, 778-783

Whittaker, A., Gilani, A., Bertero, V.V.: Evaluation of pre-Northridge steel moment-resisting frame joints. The Structural Design of Tall Buildings, Vol. 7, 263-283

Yamamoto, S. (2000): Spatial distribution of strong ground motion considering asperity and directivity of fault. In 12th World Conference on Earthquake Engineering, Auckland, 30 January- 4 February 2000, CD-ROM 0700

Yamazaki, F., Meguro, K., Tong, H. (1995): General review of recent five damaging earthquakes in Japan. Bulletin of Earthquake Resistant Structure Research Center, No. 28, September 1995, 7-23

Yang, F.Y. (1993): Collapse behaviour of Pino Suarez Building during 1985 Mexico City Earthquake, Journal of Structural Engineering, Vol. 119, ST 2, 852-870

Yang, T.S., Popov, E.E. (1996): Analytical studies of pre-Northridge steel moment-resisting connections. In 11th World Conference on Earthquake Engineering, Acapulco, 23-28 June 1996, CD-ROM 661

Yang, T.S., Popov, E.E. (1997): Comparison of MRF connection damage due to Northridge quake with theory. In Behaviour of Steel Structures in Seismic Areas, STESSA 97, (eds. F.M.Mazzolani and H.Akiyama), Kyoto, 2-8 August 1997, 10/17, Salerno, 897-904

Yao, S., Murayama, K., Suga, M., Nishida, K.,Kusumi, H. (1996): The structural damage near active fault in Hyongoken Nanbu earthquake. In 11th World Conference on Earthquake Engineering, Acapulco, 23-28 June 1996, CD-ROM 1258

Zhao, D., Kanamori, H., Negishi, H., Wiens, D. (1996): Tomography of the source area of the 1995 Kobe earthquake: Evidence for fluids at the hypocenter?. Science, Vol. 274, 1891-1894

3

Basic Design Philosophy

3.1 Performance Based Seismic Design

3.1.1. Multi-Level Seismic Design Criteria

The structure design procedure on the basis of multi-level criteria is not a new concept. Under gravity, live, snow, wind loads, the limit state design considers the service and ultimate levels. In the case of seismic loading, the declared intent of building codes is to produce buildings capable of achieving the following performance objectives (Fajfar, 1998):

-to resist minor earthquakes without significant damage;

-to resist moderate earthquakes with repairable damage;

-to resist major earthquakes without collapse.

However, as a rule, the majority of codes consider explicitly only one performance objective, defined as protection, in cases of rare major earthquakes, of occupants against injury or loss of life. Criteria for structure checking to minor or moderate earthquakes that may occur relatively frequently in the life of the building are not specified explicitly. A review of 41 codes elaborated all over the world shows that 38 are based on just one level, the principal design being concentrated on strength requirements (Bertero, 1997). During recent earthquakes, including those of California and Japan, structures in conformance with the modern seismic codes performed as expected: the loss of lives was minimal. However, the economic loss due to sustained damage was substantial. Earthquakes in urban areas have demonstrated that the economic impact of physical damage, loss of function and business interruption was huge and the damage control must become a more explicit design consideration. So, in the recent years, a new philosophy for building design has been discussed within the engineering community, adopting a performance based seismic design philosophy. In the United States, the Vision 2000 Document provides the foundation concepts for this approach (Bertero, 1996a,b). The goal of this design philosophy is to produce structures that have predictable seismic performances under multiple levels of earthquake intensity (Leelataviwat et al, 1999).

Performance based seismic design is defined as "consisting of selection of design criteria and structural systems such that at the specified levels of ground motion and with defined levels of reliability, the structure will not be damaged beyond certain limiting states or other useful limits" (Bertero and Bertero, 2000). So a comprehensive performance based design involves several steps:

-selection of the performance objectives;

-definition of multi-level design criteria;

-specification of ground motion levels, corresponding to the different design criteria;

-consideration of a conceptual overall seismic design;

-options for a suitable structural analysis method;

-carrying out comprehensive numerical checking.

The subject of performance based seismic design was in the interest of many reports and papers published in journals: Teran-Gilmore (1998), Leelataviwat et al (1999), Kennedy and Medhekar (1999). They were the subject of many Conference papers: 11th WCEE, Acapulco (1996), Bertero, Bertero et al, Krawinkler, Holmes, Hamburger, Otani; Next Generation of Codes, Bled (1997), Bertero, Hamburger, Jirsa, Krawinkler, Otani, Poland and Hom, Ghobarah et al; 11th ECEE, Paris (1998), Calvi, Fajfar, Tassios; 12th WCEE, Auckland (2000), Bertero and Bertero, Yamawaki et al, Fujitani et al, Priestley.

3.1.2. Performance Levels

The need to erect buildings in seismic areas has led to the development of a particular design philosophy. The basic principle of this philosophy consists of considering that it is not economically justified that, in a seismic active area, all structures should be designed to survive the strongest possible ground motion, without any damage. It is more reasonable to take the point of view that the structures must exceed a moderate earthquake without damage, but, in a rare event of very strong ground motion, damage would be tolerated as long as structural collapse can be prevented. The following cases are exceptions from this concept (Gurpinar, 1997):

-constructions whose failure may lead to other disasters affecting man or environment, e.g. nuclear power plants, dams, petrochemical facilities;

-buildings which are required in emergency plans, e.g. hospitals, communication centers;

-life-line facilities, e.g. communication lines, pipelines, bridges;

-housing containing irreplaceable articles, e.g. museums.

For these buildings, specific seismic design criteria are developed, which do not allow damage under common actions, but they are out of the subject of this book.

In the case of usual buildings in seismic areas the code provisions for earthquake design cannot guarantee a safe structure or no damage during a very strong earthquake. Each building has some weakness not covered by the design and, when a strong earthquake occurs, this weak point results in serious damage.

So, the first step in the performance basic design philosophy is to define an acceptable damage level due to an earthquake, and this is the purpose of the design code. There is no general agreement on this damage level, but there are some general accepted criteria for determining these performances (Krawinkler, 1995):

(i) *Life safety*, is the primary requirement. The loss of life and the injuries in a building due to an earthquake are usually caused by the collapse of the building components. The evaluation of the number of deaths and injuries as an economic damage in an optimization process, as it is suggested in some studies (Warszawski et al, 1996), poses very difficult ethical problems.

(ii) *Collapse prevention*, is directly related to the prevention of loss of lives, injuries and damage of the contents of buildings. The structure can undergo serious damage during the major earthquakes, but it must stand after the ground motion.

(iii) *Reparable damage*. A distinction is made between structural damage which cannot be repaired and damage which can be repaired. Irreparable damage is a specific subject for individual engineering judgement of experts. The damage refers both to structure and to non-structural elements.

(iv) *Acceptable business interruption*, can be appreciated by the owner. In some cases the value of the business is more important than the value of the buildings themselves and the interruption of this activity is intolerable. If the owner of a building wishes to avoid the cost of interruptions, it is necessary to do more than the minimum requirement of design codes. By using stronger and stiffer designs, it is possible to reduce, or even eliminate, the building function interruption after a strong earthquake, but this means that a more expensive structure results.

(v) *Acceptable loss of architectural and historical heritage*, in which artistic valuables and masterpieces are destroyed at the same time with the building collapse.

The main goal of seismic design and requirement is to protect life and structure collapse. However, more recent earthquakes with extensive non--structural element collapses, interruption of functionality for many buildings, evacuation of people, losses in workplaces for varying periods of time, monetary losses, have shown that the above mentioned goal is not sufficient for a proper design methodology. So, the concept of more than one level of earthquake intensity must be adopted as a basic design philosophy.

3.1.3. Economic Evaluation of Performance Levels

There is a minimum level of protection demanded by society and introduced in codes in order to safeguard buildings against collapse which endangers human life. But some owners want options for providing live safety protection for the inhabitants of their facilities beyond the minimum required by society (Krawinkler, 1997). With the development of the multi-level performance based seismic design, property owners now have choices, that were never available before, to conform the performance of their building to their needs. The buildings where the value of business is more important than the value of the buildings themselves (hi-tech manufactures, movie studios, banks, etc) have been able to design and upgrade their facilities to maintain continued operation, even after a major earthquake (Poland and Hom, 1997). For these owners an economic evaluation of performance levels is required, because they desire sound judgement on costs and benefits of earthquake protection.

After the negotiations between designers and owners for establishing the performance objectives, the calibration procedure could be based on the minimisation of the objective function (Warszawski et al, 1996, Pinto, 1998), which is the cost function C_t, given by:

$$C_t = C_i + C_d \tag{3.1}$$

where C_i is the initial cost and C_d is the damage cost. The initial cost may depend on the site conditions C_{si}, structural elements C_{st}, architectural finishings C_{fi}, equipments C_{eq}:

$$C_i = C_{si} + C_{st} + C_{fi} + C_{eq} \tag{3.2a}$$

The damage costs may be separated into five components, namely repair/-replacement C_{re}, loss of contents C_{co}, economic loss due to business interruption C_{bu}, cost of injuries C_{in}, cost of life loss C_{li}:

$$C_d = C_{re} + C_{co} + C_{bu} + C_{in} + C_{li} \tag{3.2b}$$

A higher level of performance corresponds to higher initial cost, but also to higher life-cycle benefits. A higher load-carrying capacity of structural members can be obtained by using larger dimensions, stronger materials or more advanced construction technologies (Warszawski et al, 1996). All these measures usually entail higher cost, but higher structural performance will result in greater reliability and lower expected damage costs for the same seismic load. The search for an optimal solution must therefore determine the initial and damage costs associated with various performance levels, and select the one with the optimum cost (Fig. 3.1). The minimum of total cost function corresponds to a target for design.

3.1.4. Multi-Level Design Approaches

A building can be subjected to low, moderate, or severe earthquakes. It may cross these events undamaged, can undergo slight, moderate or heavy damage, may be partially destroyed or can collapse. These levels of damage depend on the earthquake intensities. The low intensity earthquakes occur frequently, the moderate earthquakes more rarely, while the strong earthquakes may occur once or maximum two times during the structure life. It is also possible that a devastating earthquake will not affect the structure during its life.

In these conditions, the checks, required to guarantee a good behaviour of a structure during a seismic attack, must be examined in the light of

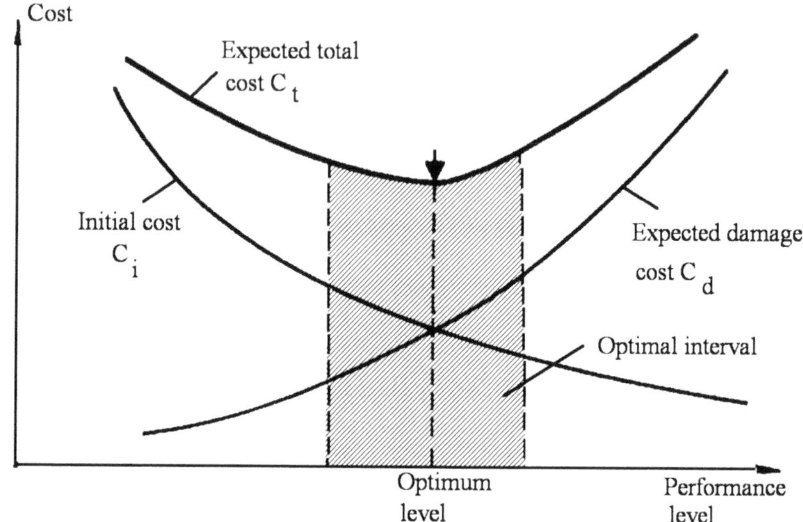

Figure 3.1: Optimal performance level

a multi-level design approach. Thus, it is very rational to establish some limit states, as a function on the probability of damage occurrence, both for structure and non-structural elements (Bertero, 1996a,b, 1997, Bertero et al, 1996, Mazzolani and Piluso, 1997). These limit states are presented in Fig. 3.2, in function of both structure and non-structural elements. Three categories of damage may occur during the structure behaviour (Tassios, 1998):

-damage of contents, which depends on the accelerations of each storey and on the integrity of the non-structural and structural components of the building;

-damage of non-structural components as a result of earthquake induced deformations, the interstorey drifts being the most important characteristic for this damage type;

-damage of structural elements produced by local and/or overall buckling of members, brittle fracture of members or joints. The damage index is proposed to be a criterium for the damage level.

In the seismic load–top–sway displacement curve, there are three very important points: limit of elastic behaviour without any damage, limit of damage with major damage and limit of collapse, for which the structure is at the threshold of breakdown. In function of taking different limit states for structure and non-structural elements, some multi-level approaches are possible:

(i) *Four levels design approach.* The performance-based seismic engineering, elaborated by the Vision 2000 Committee of SEAOC (1995) and ATC (1995), consists of a selection of appropriate systems, layout and detailing of a structure and non-structural components and contents, so that at specified levels of ground motion and defined levels of reliability, the structure

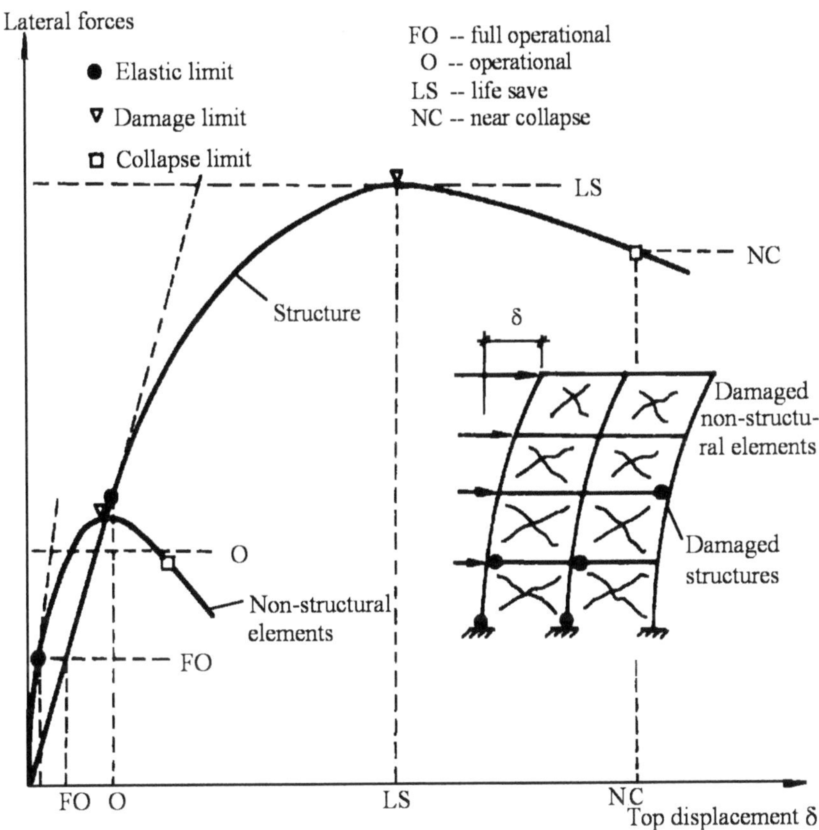

Figure 3.2: Performance levels

will not be damaged, beyond certain limit states. The performance levels have been defined for four levels as a combination of damage of structure and non-structural elements, building facilities and required repairs (Table 3.1).

A classification of performances in five levels is proposed by Yamawaki (2000).

The relation between the four limit states and the four probabilities of occurrence are presented in Fig. 3.3, as a performance objective matrix. The minimum objectives of an earthquake design are illustrated in the figure as the diagonal line. The unacceptable performance are values under these minimum objectives. In the same way, there are enhanced objectives, if the owner consents to a supplementary payment for providing better performance, or lower risk than the one corresponding to the minimum objectives (Bertero, 1997).

(ii) *Three level design approach.* Verification at the three levels has been proposed by Bertero and Bertero (1992), Mazzolani and Piluso (1993, 1996), Mazzolani et al (1995), Sharpe (1992):

Table 3.1: Seismic performance levels according to 1. ATC; 2. SEAOC; 3. Yamawaki

	Designation	Overall	Structure	Repairs
1	1.Operational 2.Fully operational 3.Keep function	• No damage to almost all functions • Facilities continue in operation	• No damage in structure • No visible residual deformations	Without repairs
2	1.Immediate occupancy 2.Functional 3.Keep major functions	• Light damage to non-structural components • Main facilities continue in operation, non-essential facilities are interrupted, but can resume immediately	• Very minor structural damage • No residual deformation • Structure retains all its original stiffness and strength	• Slight repairs for non-structural components • Without repairs for structure
3	1.Life safety 2.Life safe 3.Life safety	• Significant non-structural component damage • Activity is interrupted • Building remains accessible to emergency activity	• Significant structural damage • Structure loses its original stiffness and strength, but retains some lateral strength against collapse	• Repairs of non-structural components • Immediate repairs for structural elements
4	1.Collapse prevention 2.Near collapse 3.No guarantee for life safety	• Non-structural components are completely damaged and present a falling hazard • No entry into building is permitted	• Serious damage in structural members • Substantial loss of structural strength • Structure supports only the gravity loads • Partial collapse is probable, but not overall collapse	• Experts decide if the building should be demolished or can be repaired • Repair is probably not practical

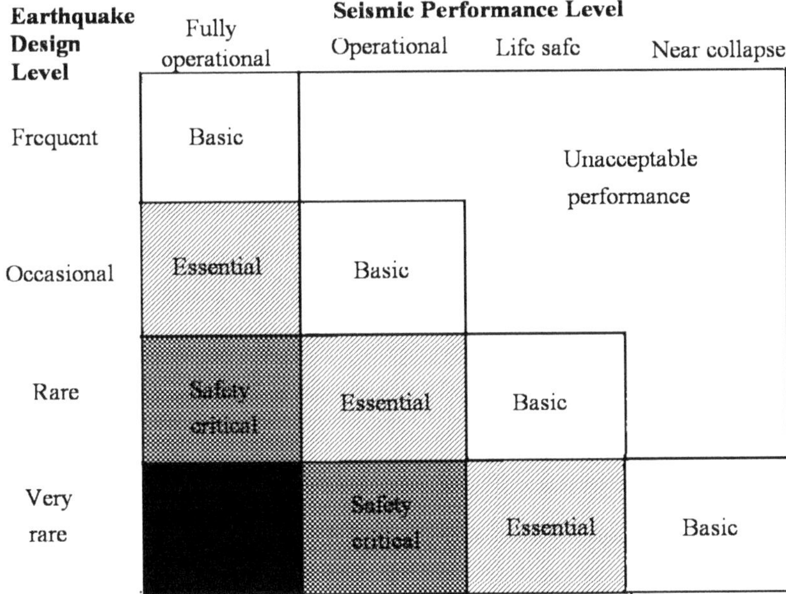

Figure 3.3: Earthquake occurrence versus seismic performance levels (after Bertero, 1996a,1997a)

-*serviceability limit state* (SLS), for frequent earthquakes; the corresponding design earthquake is called a service earthquake. This limit state imposes that the structure, together with non-structural elements, should suffer minimum damage and the discomfort for inhabitants should be reduced to the minimum. So, for this level, the structure must remain within the elastic range or it may suffer unimportant plastic deformations;

-*damageability limit state* (DLS), for occasional earthquakes; this limit state considers an earthquake intensity which produces damage in non-structural elements and moderate damage in the structure, which can be repaired without great technical difficulties;

-*survivability-ultimate limit state* (ULS), for earthquakes which may rarely occur, represents the strongest possible ground shaking. For these earthquakes, both structural and non-structural damage are expected, but the safety of inhabitants has to be guaranteed. In many cases damage is so substantial that structures are not repaired and demolition is the recommended solution.

(iii) *Two levels of design approach.* Although it is recognized that the ideal methodology would use four or three levels for design, today current code methodologies and seismic design philosophy may be based on just two levels (Eurocode 8):

-*serviceability limit state*, for which structures are designed to remain elastic, or with minor plastic deformations and the non-structural elements remain undamaged or have minor damage;

-*ultimate limit state*, for which structures exploit their capability to deform beyond the elastic range, the non-structural elements being partially or totally damaged.

In the Eurocode 8, the accelerations corresponding to the serviceability limit state are given as a fraction of the corresponding ones for the ultimate limit state. Generally, this methodology cannot assure a controlled damage, also because the determination of this relationship is not clearly assigned in code.

3.1.5. Coherent Strategy for Seismic Design

Within the context of performance-based design a structure is designed so that, under a specified level of ground motion, the performance of the structure is within prescribed bounds. To achieve these levels of verification, the seismic design problem is laid out through required-available formulation:

$$\text{Required capacity} \leq \text{Available capacity} \tag{3.3}$$

Currently the required-available pairs of three mechanical characteristics are considered in seismic design: rigidity, strength and ductility:

$$\text{Required rigidity} \leq \text{Available rigidity} \tag{3.4a}$$

$$\text{Required strength} \leq \text{Available strength} \tag{3.4b}$$

$$\text{Required ductility} \leq \text{Available ductility} \tag{3.4c}$$

These verifications must be incorporated in the performance based seismic design.

When the number of performance levels is discussed, one must recognize that the use of four levels is the most rational proposal and that two levels represent the minimum acceptable option. Because it is questionable to ask design engineers to perform many verifications, it seems more rational to introduce no more than three levels of verifications: serviceability, damageability and ultimate (survivability) levels.

Considering that the rigidity verification is a problem of the serviceability level, the strength verification is related to the damageability level and ductility verification is a problem of survivability, the relation between verification types and performance levels are illustrated in Table 3.2. There are basic verifications and optional ones.

In a coherent strategy for seismic design, the structure must be verified for rigidity at serviceability level, for strength at damageability level and for ductility at survivability level. So, the basic verifications of rigidity, strength and ductility triade must be performed at different levels of seismic loads. Other verifications at each level, as presented in Table 3.2, are only optional or not necessary, in the function of building importance or earthquake intensities.

The relationship between performance levels and the other parameters of structure-seismic design are shown in Table 3.3 and Fig. 3.4. They show that:

Table 3.2: Verification vs performance levels

Verification / Performance	Rigidity	Strength	Ductility
Serviceability	Basic	Optional*	–
Damageability	Optional*	Basic	Optional*
Survivability	–	Optional*	Basic

* This verification is not necessary for current buildings and for low earthquakes

Table 3.3: Approaches of structure seismic design

Perfor-mance	Earth-quake intensity	Avoided failure	Verifi-cation	Analysis method	Analysis object
Servi-ceabi-lity	Low	Non-structural element	Rigidity	Elastic	Storey drift
Dama-geabi-lity	Moderate	Local collapse	Strength	Elasto-plastic	Cross-section capacity
Survi-vability	Severe	Global collapse	Ductility	Kinematic	Rotation capacity

-for *serviceability level*, under frequent low earthquakes, the strategy calls for the elastic response of a structure. The lateral deformations are limited for the interstorey drift limits, given for the non-structural elements. Due to their integrity, interaction structure-non-structural elements must be considered. The basic verification refers to the structure rigidity and the strength verification (condition to elastic behaviour) is only optional;

-for *damageability level*, under occasional moderate earthquakes, an elasto-plastic analysis must be performed. The basic verification refers to the structure member strengths, verifications for rigidity or ductility being optional. The non-structural elements are partially damaged, so the analysis must consider only the structure behaviour, without any interaction with the non-structural elements;

-for *ultimate (survivability) level*, under rare severe earthquakes, a kinematic analysis, which considers the behaviour of possible formed plastic mechanisms, must be performed. The basic verification refers to the ductility, the strength verification being only optional. The design strategy refers to the control of the formation of a preselected plastic mechanism and the rotation capacities of plastic hinges. The non-structural elements are completely damaged.

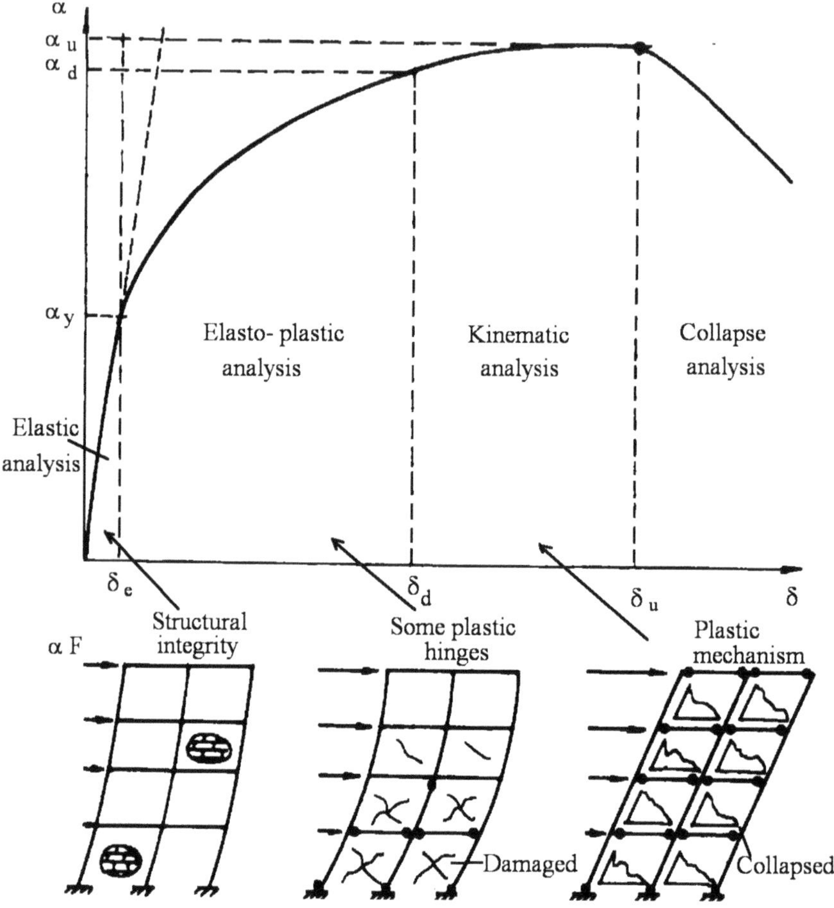

Figure 3.4: Structure response for different load levels

One of the most important problems of this multi-level design approach is the optimization of solutions. If the design is performed without any conception, it is possible that one of the limit states will dominate the sizing. A structural solution able to satisfy two or more requirements simultaneously represents a case of optimum design.

3.1.6. Definition of Earthquake Design Levels

To be effective for design, the performance levels must be translated into seismic action values in term of design magnitude and accelerations. There are two different cases:

(i) For *given structures* the accelerations corresponding to different performance levels can be determined from the analysis of their behaviours. Such evaluation has been performed by Mazzolani and Piluso (1997) for the frames of Fig. 3.5a and for different sizing criteria:

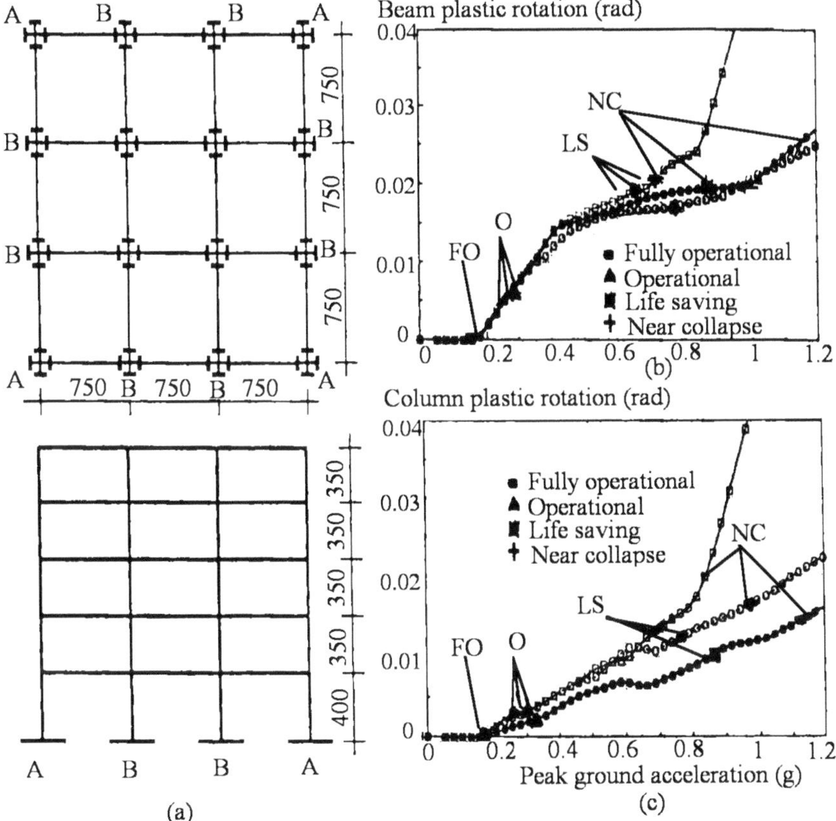

Figure 3.5: Peak ground accelerations corresponding to the different performance levels: (a) Analysed structure scheme; (b) Maximum beam plastic rotation; (c) Maximum column plastic rotation (after Mazzolani and Piluso, 1997a)

S-MRF: special moment resisting frame sized by means of the code hierarchy criterium;

S-MRFd: special moment resisting frame with limited interstorey drift;

G-MRF: global moment resisting frame which guarantees the formation of a global mechanism.

The determined accelerations for the four performance levels are given in Fig. 3.5b and Table 3.4.

The values of peak accelerations are strongly influenced by the structure typology. For S-MRF one can see that the values of peak accelerations may be classified in two groups, the first for fully operational and operational levels and the second for life saving and near collapse levels, observation which can justify the use of two performance levels only, proposed by the codes. Contrary to this, for the other structure types, the differences in peak accelerations for the different performance levels are very important. But

Table 3.4: Predicted values of the peak ground acceleration for different performance levels (after Mazzolani and Piluso, 1997a)

Performance	S-MRF	S-MRFd	G-MRF
Fully operational	0.15 g	0.16 g	0.16 g
Operational	0.22 g	0.27 g	0.31 g
Life saving	0.66 g	0.78 g	0.86 g
Near collapse	0.67 g	0.98 g	1.15 g

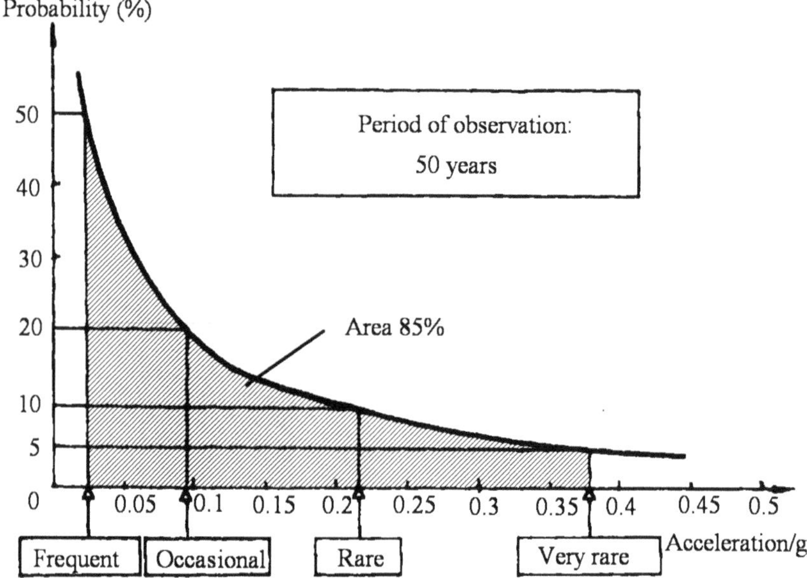

Figure 3.6: Probability acceleration function

for fully operational and operational performance levels there are not very great differences, showing that these two limits can be unified for practical design in the above proposed three performance levels.

(ii) For *design of new structures* the situation is very different, because the structure designers need values for ground motion characteristics. There are two proposals to determine these characteristics on the basis of:

-*Probability acceleration (or magnitude) function* (Fig. 3.6), determined for a prescribed probability. Design earthquakes are determined in function of probability of excedence in 50 years (Ghobarah et al, 1997, Hamburger, 1996), which is considered the period of building maintenance without any structural intervention:

-frequent, for a probability of 50 per cent;

-occasional, for a probability of 20 per cent;

Table 3.5: Recurrence periods (years)

Performance levels	Author	Frequent	Occasional	Rare	Very rare
Four levels	SEAOC Vision 2000 (1995)	43	72	475	970
	Bertero and Bertero (1996)	10	30	450	900
	Bertero and Bertero (2000)	30	75	475	970
Three levels	Pauley et al (1990)	10-50	50-200	150-1000	
Two levels	Kenedy and Medhekan (1999)	–	50	475	
	Wen (1996)	10		475	
	Pauley and Priestly		50		

-rare, for a probability of 10 per cent;

-very rare, for a probability of 5 per cent.

-*Recurrence period.* The level of acceleration or magnitude is determined in function of recurrence periods. Some proposals to determine these periods in function of number of performance levels and earthquake frequency are presented in Table 3.5.

If for rare and very rare earthquakes there are no contradictions between recurrence periods (970 years being considered as the strongest possible earthquake, but this is only a statistic value because there are no determined values for so long period), for frequent and occasional earthquakes there are very diverging proposals, from 10 to 200 years. This fact is due to the difficulties in decision to choose of a rational criterium for non-damage limit states and to a very arbitrary definition of damage level. For very active sources, as the ones from California, the choice of a long recurrence period (50–70 years) seems to be a rational proposal. But for some with rare manifestations it seems that a short recurrence period (10–30 years) is an economic proposal.

In function of established recurrence periods, the magnitude of earthquakes corresponding to each performance level may be determined from the recurrence–magnitude relationships (Lungu et al, 1997):

-for low magnitude values, the Gutenberg–Richter may be used:

$$\log n\ (\geq M) = a - bM \tag{3.5a}$$

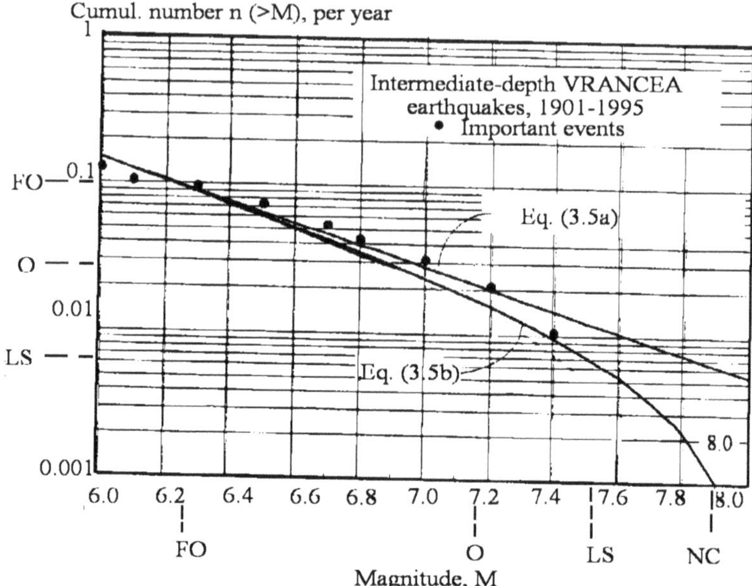

Figure 3.7: Magnitude recurrence relationship – Vrancea source (after Lungu et al, 1997)

-for high magnitude values, the Hwang–Huo modified Gutenberg–Richter relationship is:

$$n\ (\geq M)\ =\ e^{\alpha-\beta M}\frac{1-e^{-\beta(M_{max}-M)}}{1-e^{-\beta(M_{max}-M_0)}} \tag{3.5b}$$

where $n(\geq M)$ represents the mean number of events in one year having a magnitude equal to or greater than magnitude M, a, b, α, β coefficients which have to be fitted to the recorded data, M_{max}, M_0, the maximum credible magnitude of source and the threshold magnitude, respectively.

The mean recurrence interval (in years) of an earthquake of magnitude greater or equal to is the reverse of the number $n(\geq M)$:

$$T\ (\geq M)\ =\ \frac{1}{n\ (\geq M)} \tag{3.6}$$

The relationships (3.5) are presented in Fig. 3.7 for the Vrancea earthquake (for Bucharest). So, the following magnitudes as a function of performance levels result:

-fully operational for frequent earthquakes (10 years), $M = 6.24$;

-operational, for occasional earthquakes (50 years), $M = 7.2$;

-live safety, for rare earthquakes (475 years), $M = 7.9$;

-near collapse, for very rare earthquakes (970 years), $M = 8.0$.

For design practice, these magnitudes must be translated in accelerations, using the attenuation relationship presented in the next section.

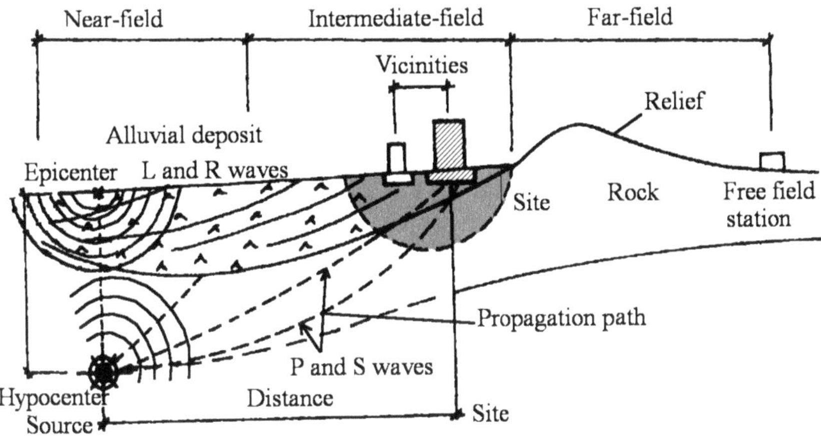

Figure 3.8: Earthquake parameters

-*Approximate evaluation.* If no one of above methods is available, due to lack of adequate recorded data, there is a proposal of Mazzolani et al (1995) for establishing the accelerations corresponding to performance levels:

-serviceability, $0.4A$;

-damageability, A;

-survivability, $1.6A$;

where A is the acceleration determined for a recurrence period of 50 years.

3.2 Site Ground Motions

3.2.1. Earthquake Parameters

The tectonic earthquakes result from motion between a number of large plates comprising the crust of earth or lithosphere. The plate motions are very slow, may be ranging in strain rate of 10^{-13} to 10^{-16}/sec (Men et al, 1998). Relative plate motion at the fault interface is constrained by friction and/or asperities. However, the strain energy accumulates in the plates sometimes overcomes any resistance and causes the slip between the two sides of the fault, producing the earthquake (Scawthorn, 1997). The location of initial dynamic rupture is termed as *hypocenter*, while the projection on the surface of the earth directly above the hypocenter is termed as *epicenter*. Other terminology includes *near-field* (near-source area), which is the area around the epicenter, and *far-field*, which is the area beyond near-field. Energy is radied through the earth by body waves (P and S waves) and surface waves (L and R waves) (Fig. 3.8).

One can define three categories of sources (Yao et al, 1996):

-*seismic generating sources*, are the first to produce the fault rupture and make the first seismic waves;

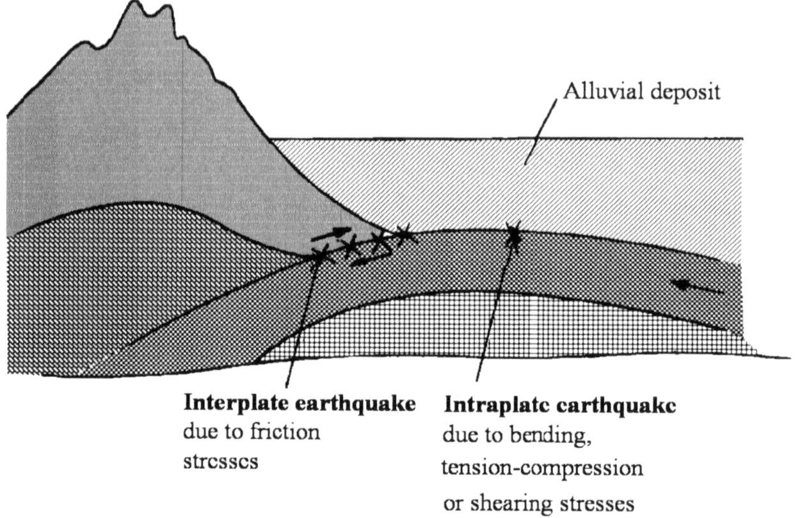

Figure 3.9: Earthquake types

-*co-movement sources*, move by their own tectonic energy triggered by the seismic wave of the generating source;

-*pseudo seismic sources*, are formed not by their own energy, but by large seismic waves produced by other sources.

Data about ground motions during an earthquake which the structures are exposed to, are fundamental for the seismic behaviour evaluation. Prediction of these ground motions of site soil subjected to an earthquake is a very difficult problem, due to the complexity of geological and geotechnical conditions. Without proper data, all analysis performed by structural engineers would be based only on assumptions. It is very well recognized that the irregularities in occurrence of earthquakes makes difficult to have exact values for seismic actions, as the structural engineers like to use. Some general proper information can be obtained, by examining the factors influencing ground motions. The site ground motions are influenced by the following factors (Fig. 3.8):

-source characteristics;

-propagation path effects;

-site soil effects;

-building effects.

3.2.2. Source Characteristics

The main characteristics of source are:

(i) *Earthquake type*, which can be classified in function of the movement of geological faults (Chandler et al, 1992, Jankulovski et al, 1996) (Fig. 3.9):

-*interplate earthquakes*, produced by sudden relative movement of two

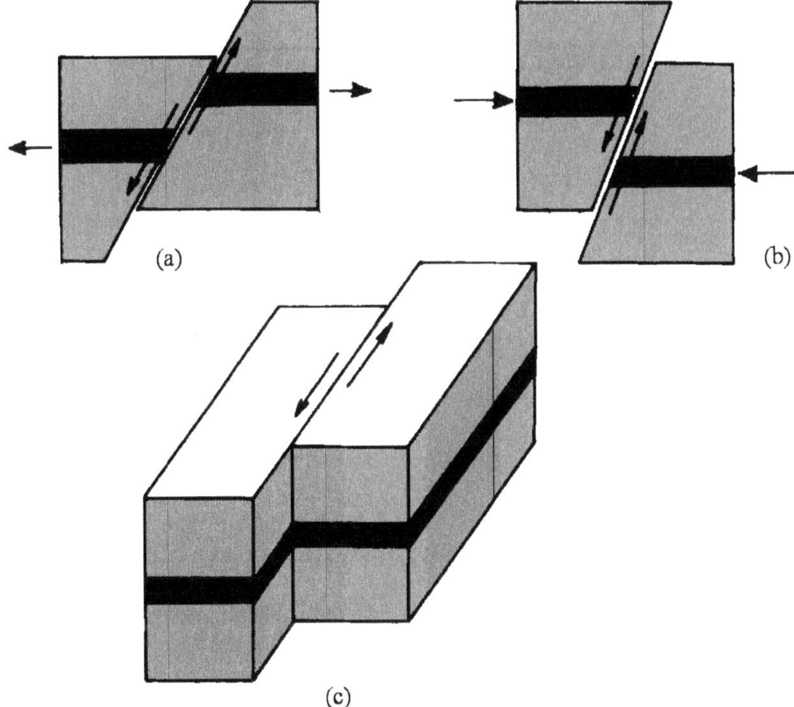

Figure 3.10: Fault types: (a) Normal fault (tension); (b) Reversal fault (compression); (c) Strike–slip fault (shear)

adjacent tectonic plates, at their boundaries. Earthquakes originating on plate boundaries are often strike–slip events (Fig. 3.10), they may produce rupture on a very large area and last for a considerable time. So, these earthquakes are characterized by large magnitudes, long durations, influences on a wide area;

-*intraplate earthquakes*, associated with the relative slip across geological faults within a tectonic plate, produced by bending and shearing. The maximum credible earthquakes which are produced by these faults are, generally, smaller in magnitude, frequency of occurrence, peak-ground acceleration and area of influence in comparison with the interplate earthquakes. At the same time, these earthquakes are characterized by low period and pulse action type. The intraplate earthquakes occur in areas that are tectonically relatively stable and geologically more uniform than in case of interplate earthquakes.

Generally, the positions of interplate faults are known, due to their manifestations during the historical time and to the fact that the boundaries of tectonic plates are very well defined. Contrary to this, the positions of in-

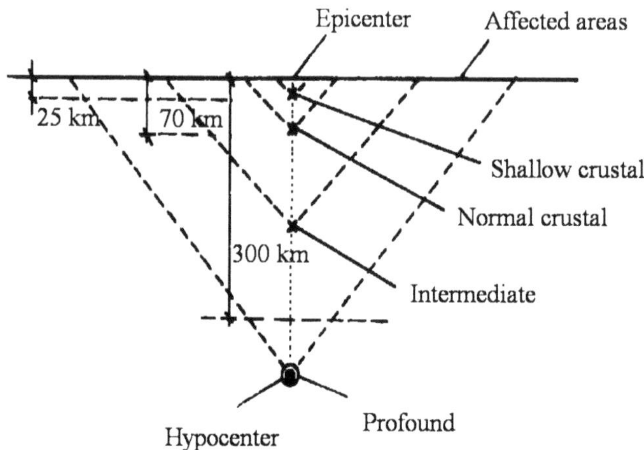

Figure 3.11: Focal depths and earthquake types

traplate faults are more difficult to be detected, due to the fact that there are some which were inactive for many years (hidden faults), but potentially they are active.

The Californian, Japanese and New Zealand earthquakes are interplate ground motions, while the European (with the exception of some Romanian, Greek and Italian events), Canadian, and Australian earthquakes are intraplate ground motions.

Another classification refers to the direction of stresses, which can be normal, reversal or strike–slip (Scawthorn, 1997),(Fig. 3.10).

(ii) *Spatial description* refers to the focal depth, rupture area, and directionality:

-*Focal depth* has a considerable influence on the earthquake behaviour. In function of this depth there are the following types (Fig. 3.11):

-shallow crustal earthquakes, having the hypocenter situated at the depths from 0 to 25 km;

-normal crustal earthquakes, with depths from 25 to 70 km;

-intermediate earthquakes, with depths from 70 to 300 km;

-profound earthquakes, for depths over 300 km.

The area affected by the earthquake is directly influenced by the focal depth. As the depth is higher, the area is larger.

More than 700 events, collected from Europe and adjacent areas, have been classified into the range of 10 km (Fig. 3.12). The majority of the records are from 0 to 25 km, 60 per cent are in the range from 4 to 14 km (Ambraseys, 1995a, 1997, Ambraseys and Bommer, 1991). In Europe, the crustal events are the most frequent and this finding is very important, due to the particularities of this kind of earthquake. Intermediate earthquakes are characteristic for some Romanian, Greek and Italian events only.

-*Rupture area.* The fault geometry and size of the rupture area determines the distribution of the origins of possible ruptures. In reality, the

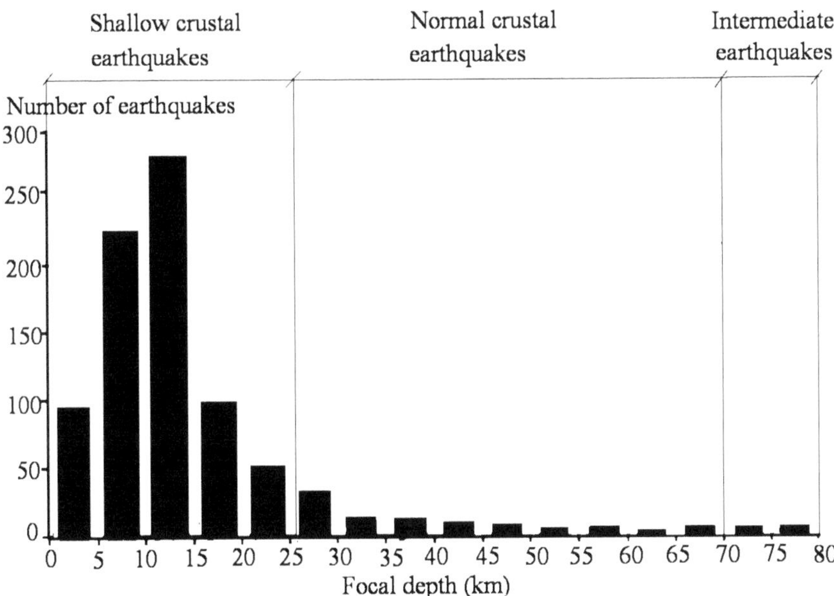

Figure 3.12: Distribution of European earthquake in terms of focal depth (after Ambraseys, 1995a)

earthquake sources are irregular surfaces or volumes. They can be modelled by plains, lines or points. Generally, it is accepted that for the analysis it is more useful to work with a rupture area, characterized by fault length and width (Fig. 3.13). In the rupture area a dislocation occurs, which is a motion of one side relative to the other side. The range of velocity of this dislocation is between 2–3 km/sec (Trifunac, 1992).

-*Directivity.* The examination of wave propagation shows that the main ground motion of the earthquake is dominated by the direction of the rupture fault (Fig. 3.13). The variation of acceleration peaks on two orthogonal directions can vary by a factor of 10, depending on the distance from the epicenter and the velocity peaks and by a factor of about 5, in function of the fault rupture process (Bertero, 1994).

(iii) *Magnitude/intensity.* Earthquakes are very complex phenomena and the design process requires to quantify the damage power of the possible events. Prior to the invention of modern scientific instruments, earthquakes were qualitatively measured by their intensity, as the effects on the buildings and human behaviour. With the development of seismometers the measure of earthquake power became possible using the notion of magnitude. While an earthquake has many intensity values in different surface points, it has only one magnitude value. The intensity–magnitude conversion is given by a relationship:

$$M = aI + b \tag{3.7}$$

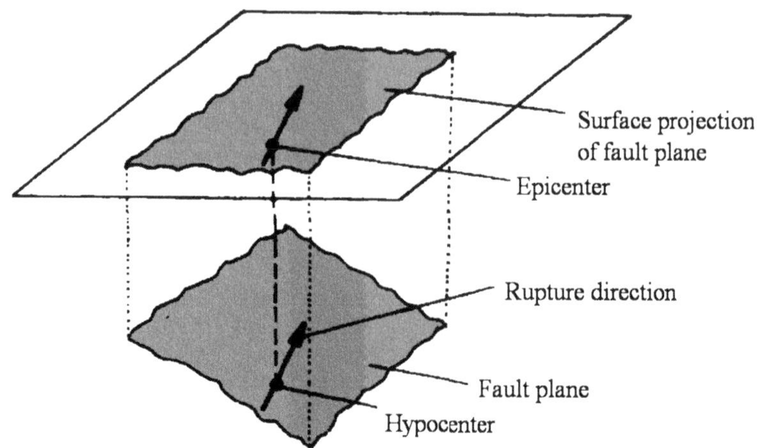

Figure 3.13: Rupture area and directivity

where a and b are numerical coefficients, with different values for each earthquake.

The magnitude/intensity is usually assessed in two ways (Studer and Koller, 1995):

-estimating of maximum dimensions of the future rupture, by empirically relating these dimensions to the magnitude. This approach is commonly applied in areas where surface faulting is observed. It is important to include the earthquake potential for hidden faults;

-considering the size of historical and recorded earthquakes associated to the source and with tectonically analogous sources. Historical periods of observation can be divided into instrumental and pre-instrumental periods. Of course, the first case is more exact, but the pre-instrumental observations cover a much longer period, giving useful information on the occurrence of large events.

Fig. 3.14 shows the results of the analysis performed by Ambraseys and the Bommer (1991a) for European ground motion records (842 for horizontal and 721 for vertical accelerations), concerning the distribution of surface magnitudes. The more frequent horizontal ground motions have magnitudes from 4 to 5, and for the vertical ones from 3.5 to 4.5.

The following framing is generally accepted for relating seismicity to magnitude:

-very low seismicity: $M < 3.0$;

-low seismicity: $M = 3.0 - 4.5$;

-moderate seismicity: $M = 4.5 - 6.5$;

-high seismicity: $M > 6.5$.

(iv) *Temporal characteristics*, which are described by:

-*Duration of rupture*, which depends on the amount of released energy, defined by magnitude, on the geometry of rupture area and on the speed of the rupture process. This duration can be approximated by an exponential

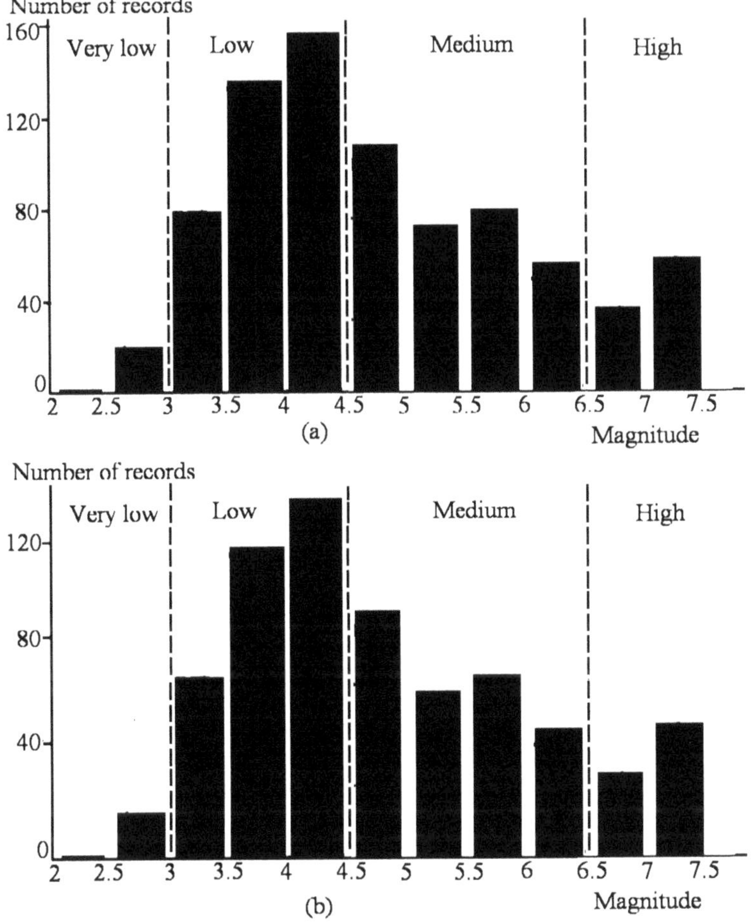

Figure 3.14: Distribution of European earthquakes in terms of magnitude: (a) Horizontal magnitude; (b) Vertical magnitude (after Ambraseys and Bommer, 1991a)

function (Trifunac and Novikova, 1992):

$$\tau_0 = \alpha e^{\gamma M} \tag{3.8}$$

where α and γ are numerical coefficients.The duration of rupture varies from 0.2 to 30 sec. The duration of rupture depends on the number of events involved in the earthquake. It is necessary to consider the possibility of multi-events, i.e two or more separate ruptures, not necessarily in the same fault, leading to a significant increase in the duration of strong motions at the epicenter area, in comparison with the duration of a single event (Bertero, 1994).

-*Recurrence period*, represents the time between two events with the same magnitude, i.e. the rate of occurrence or return period of a given

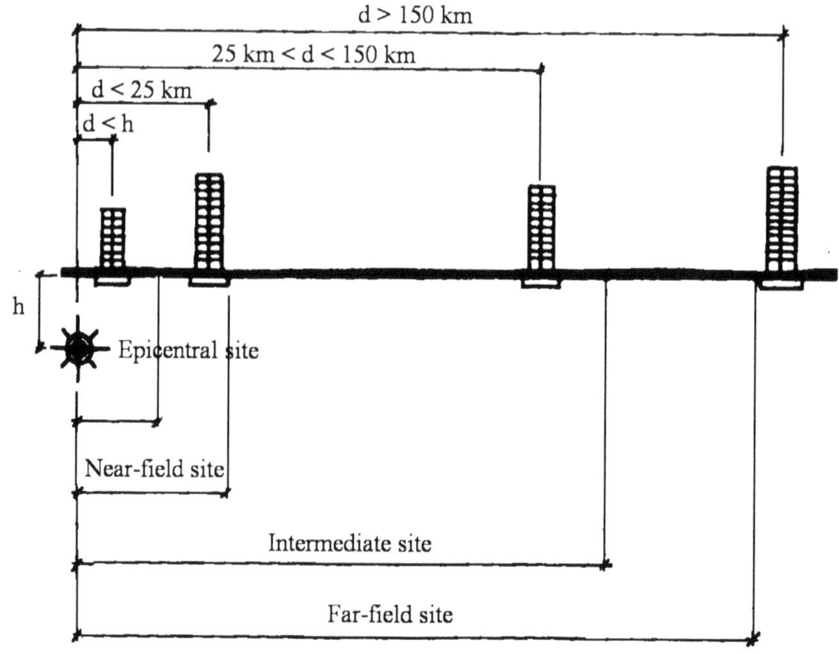

Figure 3.15: Earthquake types

motion. Fig. 3.7 shows the number of events in one year having a magnitude equal to or greater than a given magnitude. For low and moderate earthquakes, the Gutemberg–Richter relation is available, while for strong earthquakes, the Hwang–Huo relation is proposed (Lungu et al,1997).

3.2.3. Propagation Path Effects

The main factors related to the influence of propagation path effects are:

(i) *Site-source distance.* If the site is far from the source, the distance from site to the epicenter is an adequate measure. But for very close distance, the definition is more complex, due to the length and depth of rupture and requires a generous dose of expert judgment (Mohammadioun, 1995). Various definitions have been used, as epicentral or hypocentral distances. The most used is the site distance to the closed point on the surface projection of the rupture.

The site-source distance plays a leading role in the design of structures and a classification is absolutely necessary. In function of this distance, the following classification may be considered (Fig. 3.15):

-*epicentral site*, including the area around the epicenter, generally with a radius equal to the source depth;

-*near-field site*, including area within a distance of 25–30 km from the epicenter;

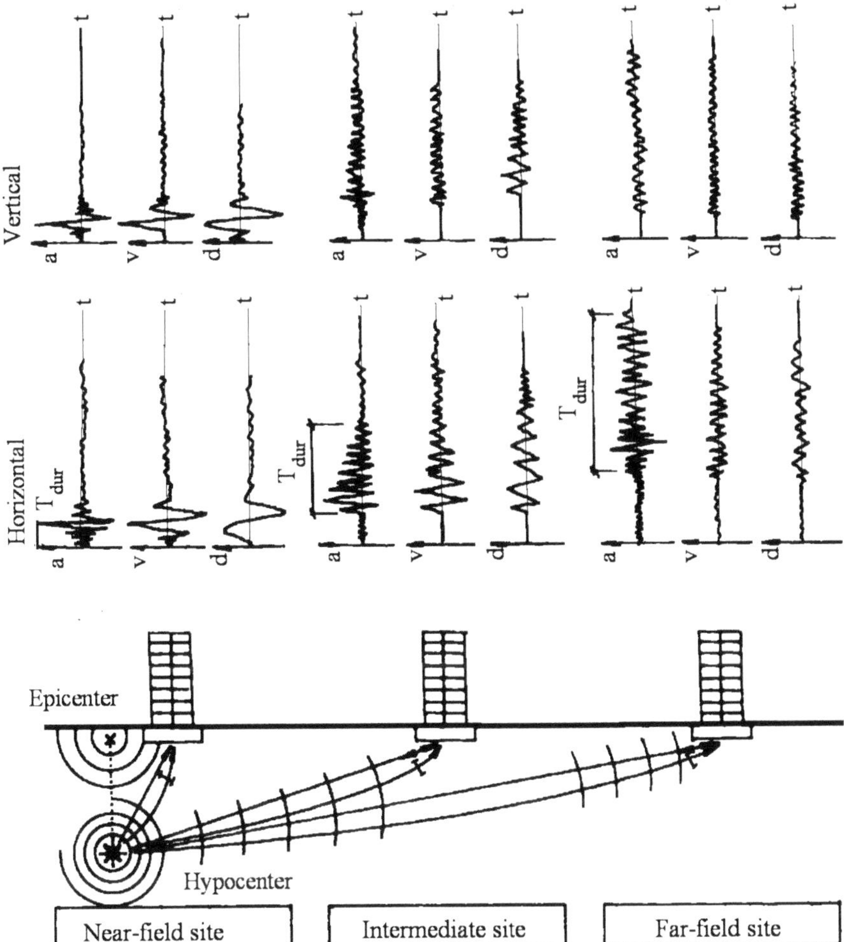

Figure 3.16: Ground motions for near, intermediate and far-field sites

-*intermediate-field site*, including area within a distance of about 150 km from the epicenter;

-*far-field site*, including area with distances more than 150 km.

This classification is available only for shallow sources, because for deep sources which affect very large areas, the classification is more intricate.

The above classification is crucial in defining the characteristics of the earthquake motions. Fig. 3.16 shows an idealized situation for a uniform soil between site and source. In this case, the conclusions are the following:

- there are important differences between the accelerations and velocities recorded in the near-field, intermediate-field, and far-field sites. The records in near-fields are characterized by low-frequency pulses in the acceleration time history, which coherently produces pulses in the velocity and displace-

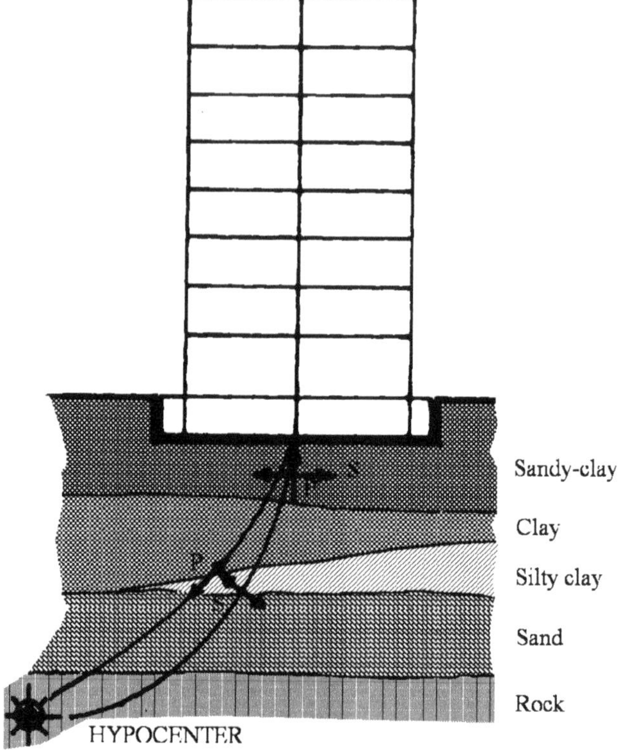

Figure 3.17: P and S waves action

ment time histories. These seismic loads are known also as fling actions. The records in far-field sites have a cyclic characteristic with many peaks in accelerogram records;

-the vertical peaks for accelerations and velocities may be greater than the horizontal ones, in near-field sites, while for far-field sites the horizontal ground motions are always more important;

-the duration of the earthquakes in near-field sites is shorter than the duration in far-field sites. These differences may be attributed to the ray paths of direct waves, which have the tendency to assume a vertical direction due to the reduction of wave velocities, as far as the surface layers are concerned (Fig. 3.17). Then, P-wave motions are thought to be dominant on the vertical direction, while the S-wave motions are thought to be dominant on the horizontal direction (Akao et al, 1992). Always P-waves are characterized by higher frequencies compared with S-waves. It is very well known that higher frequencies attenuate more rapidly with the distance than lower frequencies. Consequently, the far-field site motions are characterized by important horizontal components and low frequency energy, while the near-field motions are characterized by important vertical components and high frequency energy (Chandler, 1992).

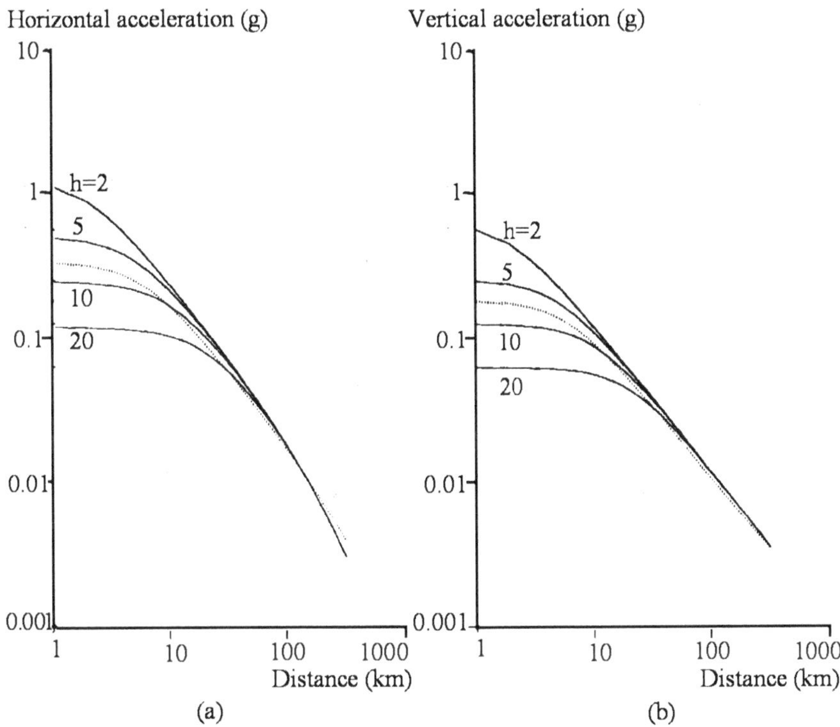

Figure 3.18: Attenuation of peak ground acceleration as a function of epicentral distance: (a) Horizontal attenuation; (b) Vertical attenuation (after Ambraseys and Bommer, 1991)

(ii) *Attenuation of ground motions.* A considerable number of attenuation laws, predicting the ground motions in terms of magnitude and distance and in some cases other additional factors, have been derived for different parts of the world. It was found that the attenuation of peaks of horizontal and vertical accelerations, in g, is given by:

$$\log PGA = c_1 + c_2 M + c_3 M + c_3 d + c_4 h + \sigma \tag{3.9}$$

where PGA is the peak ground acceleration, M, magnitude, d, epicentral distance, h, focal depth, σ, standard deviation and $c_1 \ldots c_4$, numerical coefficients determined from the recorded data.

In Fig. 3.18, the attenuation laws for horizontal and vertical ground motions are presented after Ambraseys (1995a,b), Ambraseys and Bommer (1991) for European strong motion data. The information is related to records obtained from 865 earthquakes, and the attenuation laws refer to horizontal and vertical accelerations, for different magnitudes and focal depths. One can see that for the near-field sites the focal depth has a very important effect, while for intermediate and far-field sites small differences are observed. Reduced attenuations are observed for near-field sites, and

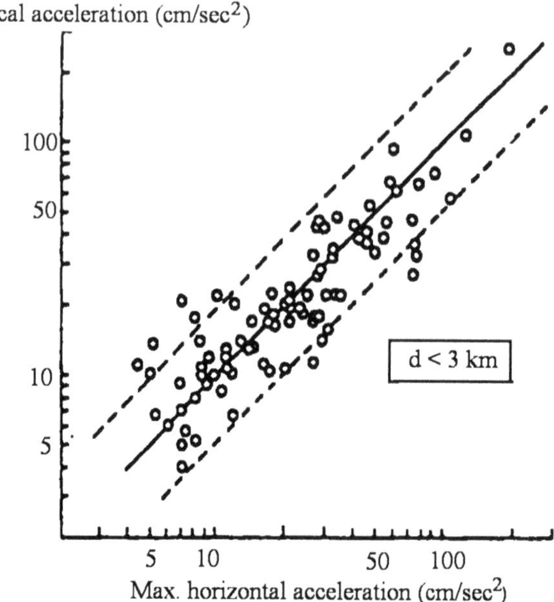

Figure 3.19: Maximum horizontal and vertical accelerations (after Kudo et al, 1992)

important reductions in acceleration are produced for intermediate and far-field sites. In near-field sites the vertical components can be greater than the horizontal ones. So, for epicentral distances within 3 km, the number of vertical components greater than the horizontal ones is almost equal to the opposite situation (Kudo et al, 1992), (Fig. 3.19). The ratio between vertical and horizontal components as a function of the source distance is given in Fig. 3.20 for different values of magnitude (Elnashai and Papazoglou, 1997).

(iii) *Travelled soil conditions.* The propagation-path effect depends on the percentage of the path travel through rock or through soft sediments (Fig. 3.21). Deviations from an uniform horizontally layered crust model occur along the path of the waves propagating from the fault to the site. These deviations are produced due to the topography of basement rock, the path being a collection of sedimentary basins with alluviums, separated by irregular basement rock, forming mountains and geological and topographical irregularities. Using a map showing these distributions of rock, soft rock and alluvium, it is possible to characterize the transmission path for the fault site (Trifunac and Novikova, 1995). The travelled soil can be classified in three classes (Erdik, 1995): hard rock (metamorphic and crystalline rock, older sedimentary deposits), soft rock (sedimentary deposits of tertiary age), and alluvium. Fig. 3.22 indicates the amplification of peak-ground acceleration on alluvium and soft rock travelled soil with respect to hard rock soil. Results indicate that an attenuation occurs for alluvium soil

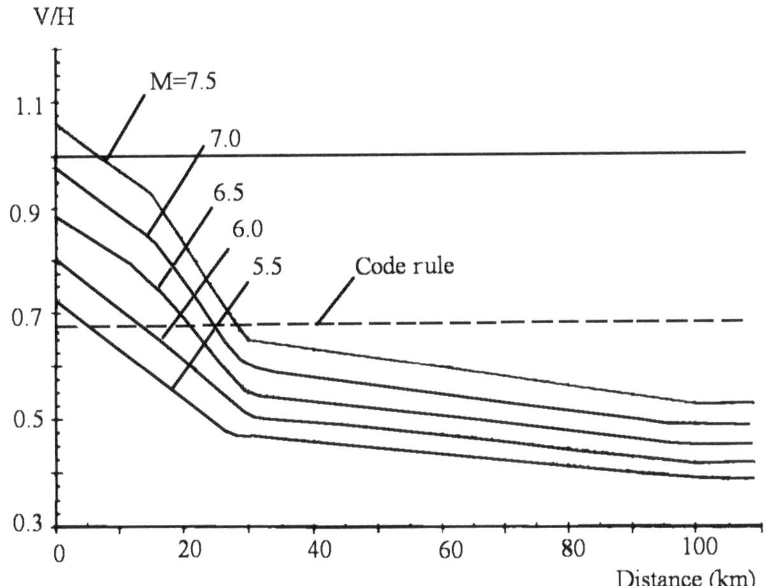

Figure 3.20: Vertical to horizontal ratio attenuation (after Elnashai and Papazoglou, 1997)

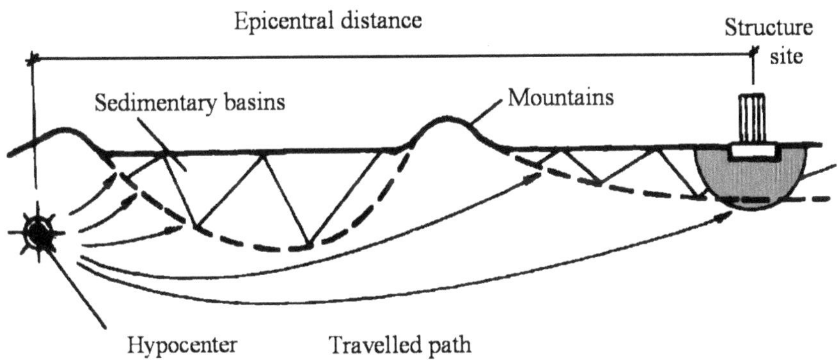

Figure 3.21: Travelled soil conditions

at short distances, while the reverse is true for long distance. Amplification is observed on all distances in the rock.

(iv) *Increasing of duration.* The initial rupture duration increases due to propagation-path effects for an uniform horizontal layer crust, in function of soil conditions. The maximum increase is a period of 0.2 secs and can be about 0.2 sec/km for alluvium and 0.08 sec/km for stiff soil (Fig. 3.23) (Trifunac and Novikova, 1995). Deviations from an uniform horizontal layered crust model depend on the percentage of the path travelled through rocks or soft sediments. The increase in duration may be determined using this percentage. But the actual value of duration is more complex, because

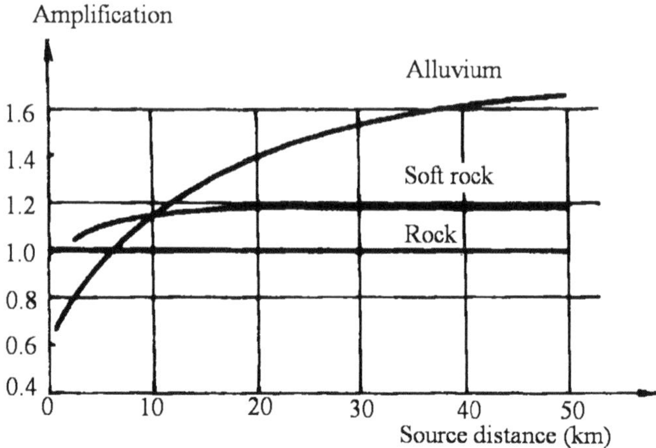

Figure 3.22: Amplification due to travelled soil conditions (after Erdik,1995)

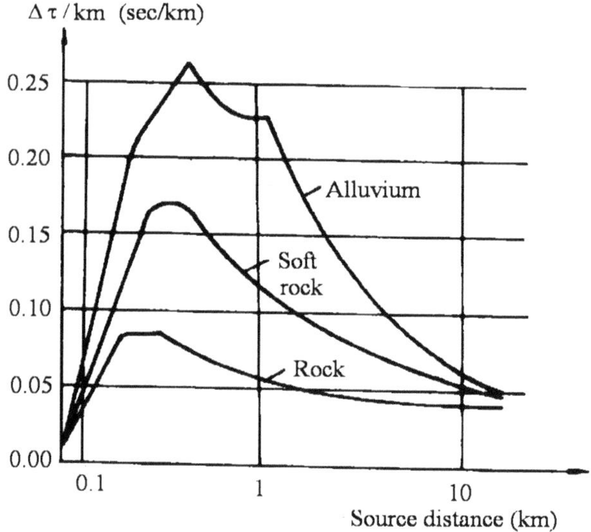

Figure 3.23: Duration increasing due to travelled soil conditions (after Trifunac and Novikova, 1995)

it depends also on the reflections of the waves rut surface and boundaries of sedimentary basin, which produce a prolongation of duration, in function of the depth and lengths of sediment zones. Generally, one can observe that the increase in duration is proportional to the epicentral distance. The durations are short near the earthquake source and long far from the source. A simple relation is proposed by Trifunac and Brady (1975) and presented in Fig. 3.24:

$$D = 10^{M/2-2} + d\left(\frac{1}{v_s} - \frac{1}{v_p}\right) \tag{3.10}$$

where v_s and v_p are velocities (km/sec) of S and P waves, respectively.

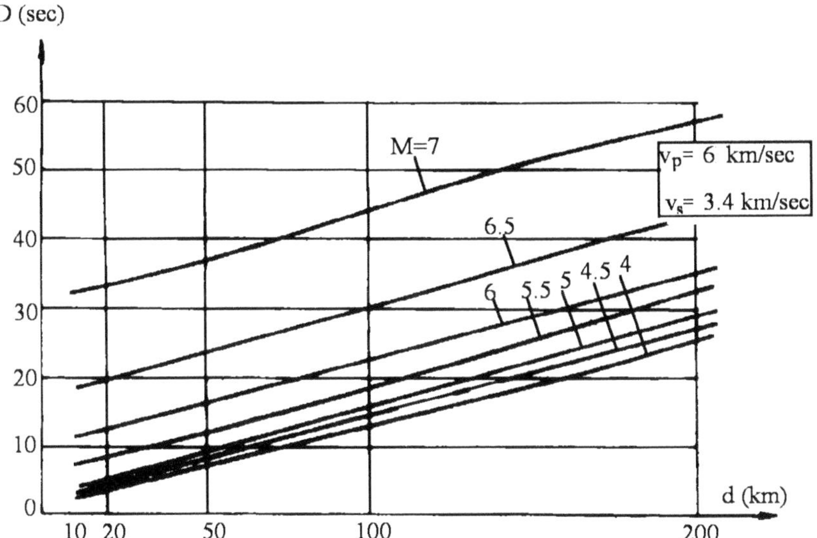

Figure 3.24: Duration increase in function on magnitude (after Trifunac and Brady, 1975)

3.2.4. Site Soil Effects

Site soil effect can be responsible for microseismic variation, which can be more important than the epicentral distance influence. The ignorance or inappropriate consideration of site effects, which can still be encountered in practice, is, therefore, a professional negligence (Studer and Koller, 1995). The influence of site can be expressed by means of the following aspects:

(i) *Local soil profile.* Each soil media has variable geologic and physico-mechanic characteristic, what marks its complexity. The main characteristic is the period of vibration of soil media, which can influence the spectra of earthquakes. Generally, the soil profile is composed by multi-layers (Fig.3.17), with different mechanical properties and thickness. The alternance of layers is another very important factor. Due this complexity, the effect of local soil conditions is a problem not yet solved, although valuable results have been obtained until now.

The classification of soil conditions contain two, three and more soil types. A general classification refers to rock, alluvium and soft soil sites (Miranda, 1993). The most usual and accurate classification considers the following categories:

- Class A: rock or stiff deposits of sand, gravel or over-consolidated clay, up to several tens of meters thick, characterized by shear wave velocity of 400–800 m/sec at a depth of 10 m.

- Class B: alluvium composed by deep deposits of medium dense sand, gravel or medium stiff clays with thickness from several tens to many hundreds of meters, characterized by shear wave velocity of 200–400 m/sec at a depth of 10 m.

- Class C: soft soil with loose cohesionless soil deposits or deposits with predominant soft-to-medium stiff cohesive soils, characterized by shear velocity below 200 m/sec.

More details of this classification may be necessary to conform better with special soil conditions (Erdik, 1995). If site specific data are available, it is possible to refine this classification and to consider more site variables.

(ii) *Peak-ground acceleration.* Seismic waves may be amplified or attenuated due to the following factors (Mahammadioun, 1997):

-contrasting impedance between the bedrock and superficial layers, due to the conservation of energy;

-resonance phenomena if the dominant period of the incident wave is close to the fundamental period of the superficial layers:

-topographical effects.

As an orientation, the fundamental periods for some soil types are:

-rock < 0.3 sec;

-thin over-consolidated alluvium 0.3–0.5 sec;

-thick consolidated alluvium 0.3–0.7 sec;

-thin unconsolidated alluvium 0.5–1.0 sec;

-thick unconsolidated alluvium 0.8–1.6 sec;

-thick soft alluvium 1.5–3.5 sec.

A seismic wave which arrives at the rock base site with some fundamental period can be amplified or deamplified, as a function of the site stratification and the frequency contents (Fig. 3.25). Thus, it is interesting to know the records at the basic rock, at different soil layers, and at the free surface. As it can be observed in Fig. 3.25, the shape of recorded accelerations has great variations, according to the position in vertical and horizontal directions. A clear amplification results in vertical direction, due to the effect of stratification, the accelerations at the free surface are far greater than the depth recorded ones. On the other hand, great differences result on recorded accelerations on the free surface at short distances. These differences are the result of variations in the stratification and the thickness of layers.

The effect of site soil conditions is presented in Fig. 3.25. It is clear that for moderate earthquakes, the soft soil increases the fundamental period only. Contrary to this, in the case of strong earthquakes, a very important amplification occurs, with an important increase in fundamental period. Fig. 3.25 clearly shows the great difficulties to define a prototype of acceleration for a given area.

(iii) *Ratio a/v.* A common approach in current design practice characterizes the earthquake effect by using the acceleration peaks only. However, due to a great amount of records recently available, it becomes apparent that the use of a single parameter, the acceleration peak, is inadequate. Two earthquakes, having the same acceleration peak, can give dramatically different responses, because the peak-ground velocities are neglected. So, a very important factor, which characterizes the ground motion, is the peak

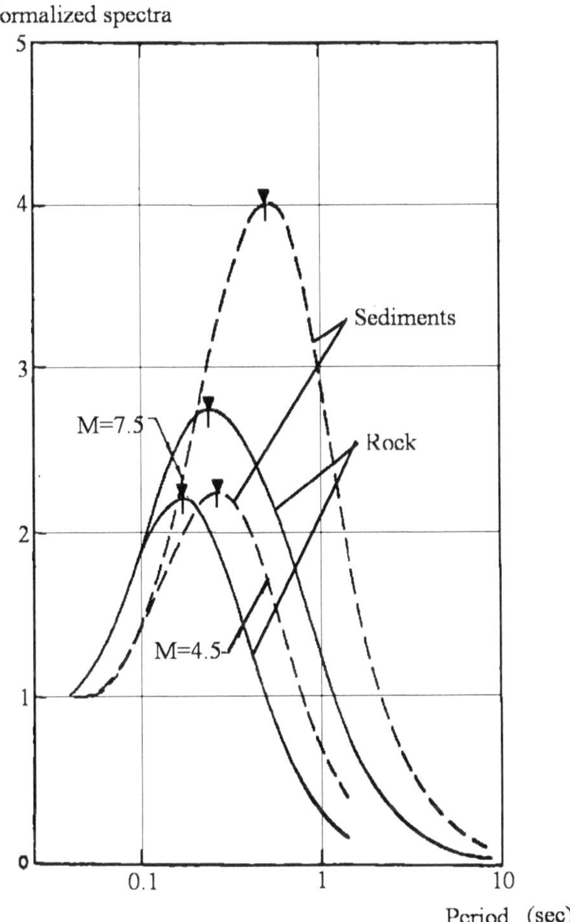

Figure 3.25: Influence of site soil in function of magnitude (after Trifunac, 1990)

ground acceleration-to-velocity ratio a/v. This ratio can be framed into three groups (Zhu et al, 1988):

-high values > 1.2 g/(m/sec);

-intermediate values 0.8–1.2 g/(m/sec);

-low values < 0.8 g/(m/sec).

The highest a/v ratios occur in near-to-source sites and the lowest in far-to-source sites. The amplifications for different values of a/v ratios are presented in Fig. 3.26 (Tso et al, 1993). In the short period range, the high a/v records have larger values when compared to the intermediate and low values. For periods longer than 0.6–1.0 sec the situation is changed.

(iv) *Dominant period*, is a very important parameter to classify ground motions. Generally, near the source, the records show very short periods, while far from the source, the records are characterized by long periods.

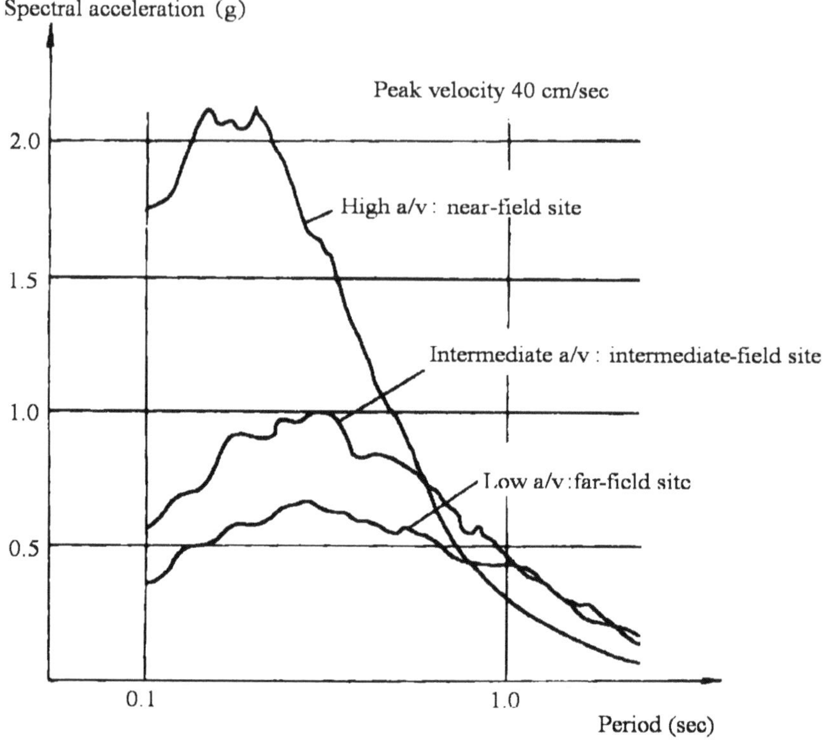

Figure 3.26: Influence of acceleration–velocity ratio (after Tso et al, 1993)

These differences are caused by the difference between the fundamental periods of P and S waves. The P waves, which dominate the near-source fields, are characterized by short periods, while the S waves, are very important for far-source fields, have long periods. The dominant period can be determined on the site, using the soil frequencies in some places (Fig. 3.27). It it very clear that, at relatively short distances, there is a great variation of frequencies, depending on the soft sediment thickness.

The dominant periods can be classified in four groups (Yayong and Minxian, 1990):

-very short period < 0.1 sec;

-short period 0.1–0.4 sec;

-medium period 0.4–0.8 sec;

-long period > 0.8 sec.

(v) *Duration.* Due to the site conditions, the duration of strong ground motion can be prolonged. For a site located on a deep sedimentary layer with frequency of about 0.5 Hz the duration is increased of 3–4 sec, while for soft soil with frequency of about 1 Hz, the increasing is of 5–6 sec. Generally, the rock soil has no effect in the increasing of earthquake duration. The ground motion duration is related also to the a/v ratio. For high a/v values, the duration is shorter compared with low values.

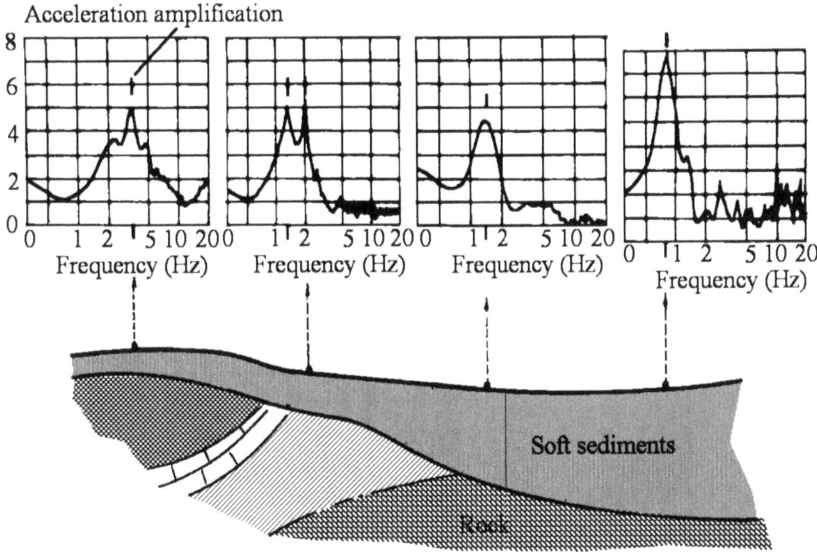

Figure 3.27: Dominant frequencies in function of soft sediments

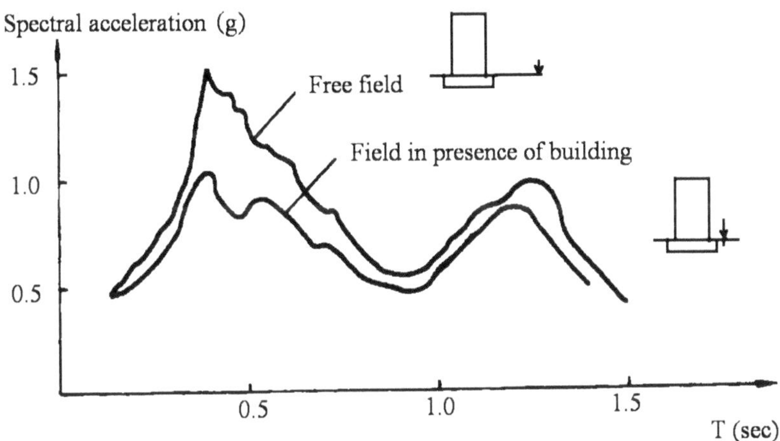

Figure 3.28: Influence of building weight

3.2.5. Building Effects

The free-field motions are usually considered as the source of the structure motions. However, there is the evidence that the actual ground motions which shock the structure depend on the building presence and on the nearby constructions.

(i) *Building weight.* It is very important to have records in free-field conditions, measured at a convenient distance away from the building and at

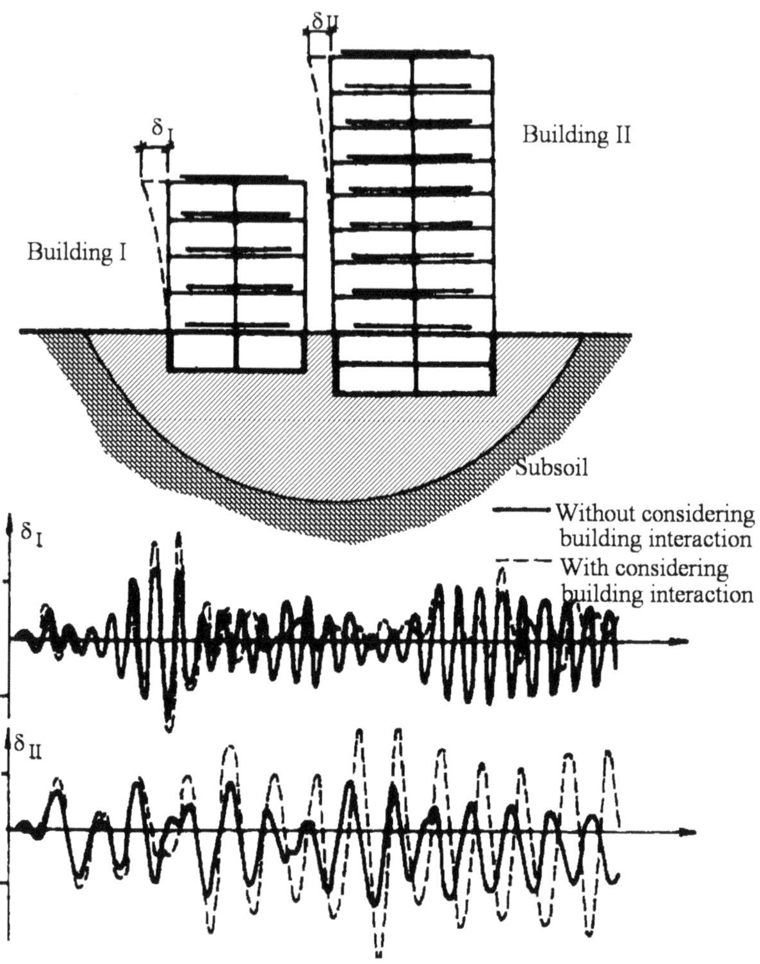

Figure 3.29: Interaction of two buildings (after Chouw and Schmid, 1995)

its base (Fig. 3.28). Due to the actual building weight, the soil is overloaded, becoming stiffer and this produces a change in the seismic behaviour of soil. The effect is to reduce the acceleration peak (Pitilakis, 1995). But for very soft soils it is possible to observe a relative increase in acceleration peaks at high frequencies, due to the building weight.

(ii) *Nearby buildings.* The response of two nearby buildings are plotted in Fig. 3.29. The response of the first building is influenced by the presence of the second building, due to the fact that both have an influence on soil ground motions (Celebi, 1995). For analysing this situation, it is necessary to consider the whole system of the two buildings, foundations and the

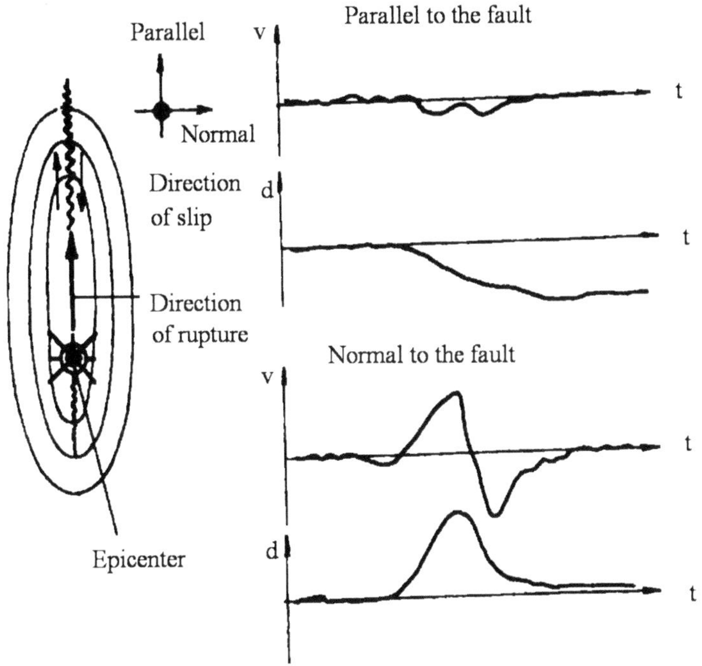

Figure 3.30: Normal and parallel to the fault components

common subsoil. The case was studied by Chouw and Schmid (1995), with and without considering the interaction between the two nearby buildings. One can see that the influence is important, the amplitudes of motions are increased due to the presence of the nearby buildings. So, in order to obtain a realistic response, the presence of other nearby buildings must be considered.

3.2.6. Assessing Near-Field Earthquakes

During the last 30 years an ever increasing database of recorded earthquakes has indicated that dynamic characteristics of the ground motions can vary significantly among different recording stations that are located in the same area. This is particularly true for stations located near the epicentral region. Some recent recorded earthquakes have shown that near the source the ground motions can be very strong, stronger than the seismic input used in present codes. This is due to the fact that the main dynamic characteristics are very different from that of recorded ones far from the source. Near-fault ground motions, which have caused the dramatic damage in the Northridge and Kobe eartquakes, are characterized by a short duration impulsive motion, clear directivity (the fault normal components are more severe than fault parallel components, Fig. 3.30), important vertical components and high velocity of both horizontal and vertical components. The ground motions are strongly influenced by the source characteristics. So, in some cases, the motions contain high-peak accelerations and short-pulse periods (0.2–1 sec), which are known as acceleration spikes. In other cases, there are lower peak accelerations with longer pulse periods (1–3 sec).

Recently there has been increased interest in the assessment of design-based earthquake ground motions for near-field earthquakes: Anderson and Bertero (1987), Hall et al (1995), Naeim (1998), Midorikawa (1995). Many papers were published during international conferences: 10th WCEE (1992), Madrid, Akao et al, Kudo et al, Gariel; 10th ECEE (1995), Vienna, Iwan and Chen, Iwan; 11th WCEE (1996), Sommervill et al, Jankulowski et al, Kataoka and Ohmachi, Yao et al; STESSA 97 (1997), Mylonakis and Reihorn, Nakashima et al; 12th WCEE (2000) Auckland: Chouw, Yamamoto, Aagaard et al, Hori et al, Wang and Nishimura, Motoki and Seo, Adam et al, Yuan et al, Umeruma et al, Nakao et al, Erdik and Durukal, Alavi and Krawinkler.

3.3 Ground Motions Modelling

3.3.1. Basic representations

The ground motion is very difficult to predict in terms of its time of occurrence, intensity and duration. The definition of the spatial characteristics, like acceleration, velocity and displacement in three directions, together with the temporal ones, like period and duration, is also considerably difficult. With respect to earthquake engineering, there are three ways in which ground motions can be modeled:

-time history representation;

-response spectra representation;

-spectral density function representation.

3.3.2. Recorded Time-History Representation

The main problems of using the time-history representation are:

(i) *Realization of network instruments.* The earthquake engineering always involves numerous problems which cannot be solved without the existence of a network of instruments for recording the ground motion characteristics. For each station it is necessary to know the characteristics of understating soil, the characteristics of the soil traveling path and the distance of the source-station. Data on the ground motion during an earthquake, where structures can be exposed are fundamental for the seismic hazard evaluation and definition of design parameters. The selection of station locations must make it possible to obtain records on most characteristics in seismic areas. The recording instruments must be placed in free fields, at the structure bases and at some stories of buildings.

(ii) *Recording.* The instrumentation must be able to record accelerations, velocities and displacements in 3D. These must be protected against seismic effects in the same way as the black boxes for aircrafts.

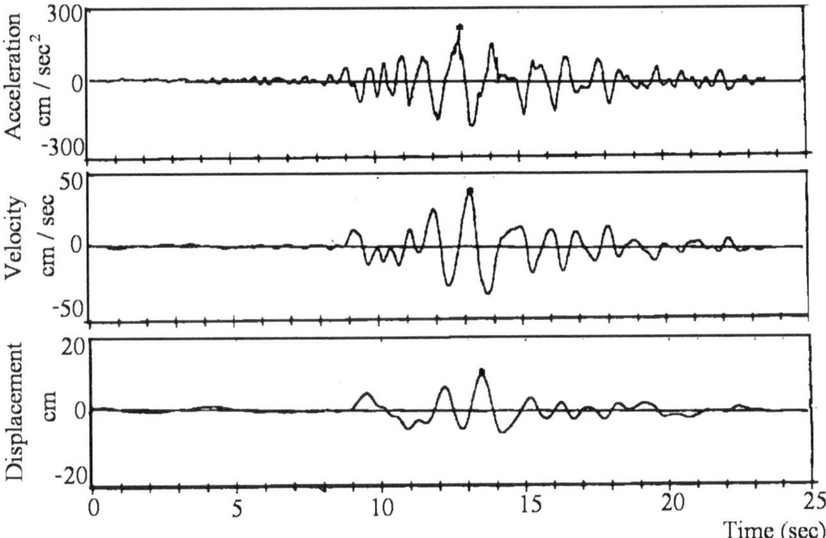

Figure 3.31: Digitized ground motions – Loma Prieta, Hyatt Regency Lobby, West component (after Lee, 1995)

(iii) *Data processing.* The digitilization of recorded ground motions must convert the curve coordinates into numbers. The computer program is used to read and plot the data in order to allow the starting of a structural analysis program. An example is the digitilization of the Loma Prieta (Hyatt Regency Lobby) earthquake (Lee, 1995), (Fig. 3.31). With the rapid development of personal computers and the availability of scanners, the automatic digitilization of accelerograms is easy and efficient.

(iv) *Data bank.* Each earthquake prone country must assess a data bank which represents the medium for transferring the recorded and digitized accelerograms for storage, to be at disposal for researchers and designers. The recorded ground motions must be classified in function of source type, the distance of the source station, station soil site characteristics, in order to be useful to similar places. So, the designer can select the most representative time-history records, corresponding to the involved site. There are some very well organized data banks. In Italy, Enea–Enel data bank collected more than one thousand records, for earthquakes which have occurred in Italy during the last twenty years (Calderoni et al, 1995). A similar data bank exists in Greece–Thessaloniki, organized by ITSAK. In England, the ICSTM (Imperial College of Science, Technology and Medicine) organized a data bank for Europe and surrounding areas (Ambraseys and Bommer, 1991b). Very important data banks exist also in USA (Abrahamson and Litehiser, 1989) and Japan (Fukushima and Tanaka, 1990), which have been enriched after the last great events with a lot of new records.

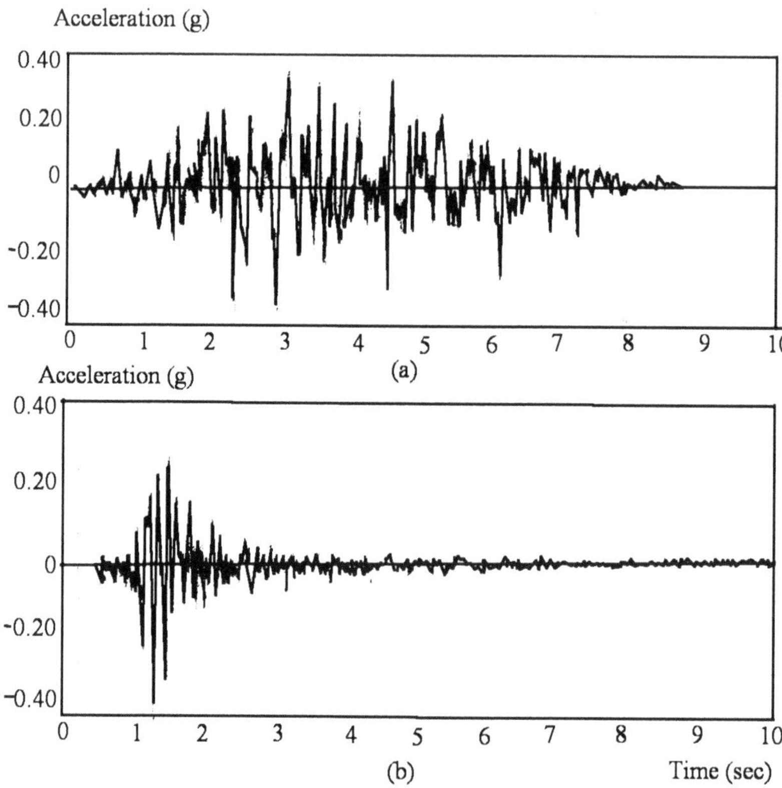

Figure 3.32: Recorded accelerogram types: (a) Far-field record; (b) Near-field record (after Combescure et al, 1998)

There are two types of recorded accelerograms (Combescure et al, 1998) (Fig. 3.32):

-for far-field records, the duration is long and the number of cycles is high;

-for near-field records, the accelerograms have higher peaks, but the period and duration are shorter.

A special case of recorded accelerations are the pulse-ground motion type (Sasani and Bertero, 2000), which are very destructive, in spite of their small magnitude. Generally, this type is a characteristic for near-field records (Loma Prieta and Kobe, Fig. 3.33a,b), but they were recorded at far-field sites due to the soil conditions (Vrancea–Bucharest, Fig. 3.33c).

3.3.3. Artificially Generated Time-History Representation

The time-history representation gives the possibility of a very good description of the ground motion during a particular earthquake, but it is question-

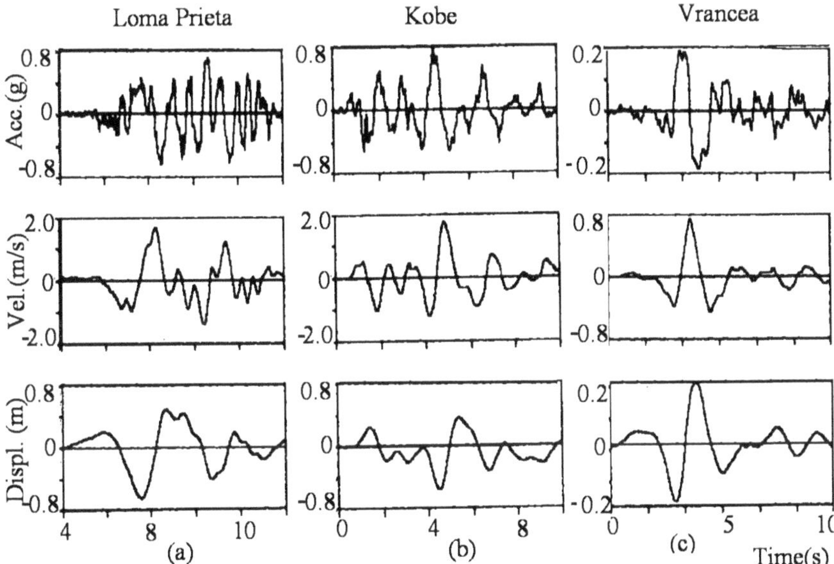

Figure 3.33: Pulse type ground motions: (a) Loma Prieta–Los Gatos Station; (b) Kobe–Takatori Station; (c) Vrancea–INCERC Bucharest Station (after Sasani and Bertero, 2000)

able whether it should be considered as a prediction for the next earthquake on the same site. This remark is supported by the observation that considerable differences have been obtained among ground motions recorded at the same place during different earthquakes, originating more or less from the same source zone (Sandi, 1995). Taking into account of this aspect, generally the codes ask the use of minimum of 5 records for the structure analysis. In spite of the large number of available records, it is very difficult to have a sufficient number of accelerograms with characteristics similar to those expected at the site of the structure. On the other hand, the large variability in the characteristics of ground motions, like peak acceleration, period, duration, etc, which are able to influence in very large measure the structure response, produces an important scattering, without the certainty that these responses represent the actual ones.

So, as an alternative to the recorded accelerograms, the use of artificially generated accelerograms may be a very convenient way (Mezzi et al, 1991). The time-history representation can be created by the combination of actual recording portions, to better preserve the relevant contents of the actual accelerograms (Krawinkler, 1995, Studer and Koller, 1995). The used time-history records are not necessarily the representation of a physically realistic earthquake. One must mention that for all loading types for buildings (live load, wind, snow, etc), the codes allow the use of a conventional equivalent loading system, which can be the envelope of the actual one. So, the artificial time-history representation should be considered in the same way as the envelope of all possible seismic loadings, corresponding to the duration,

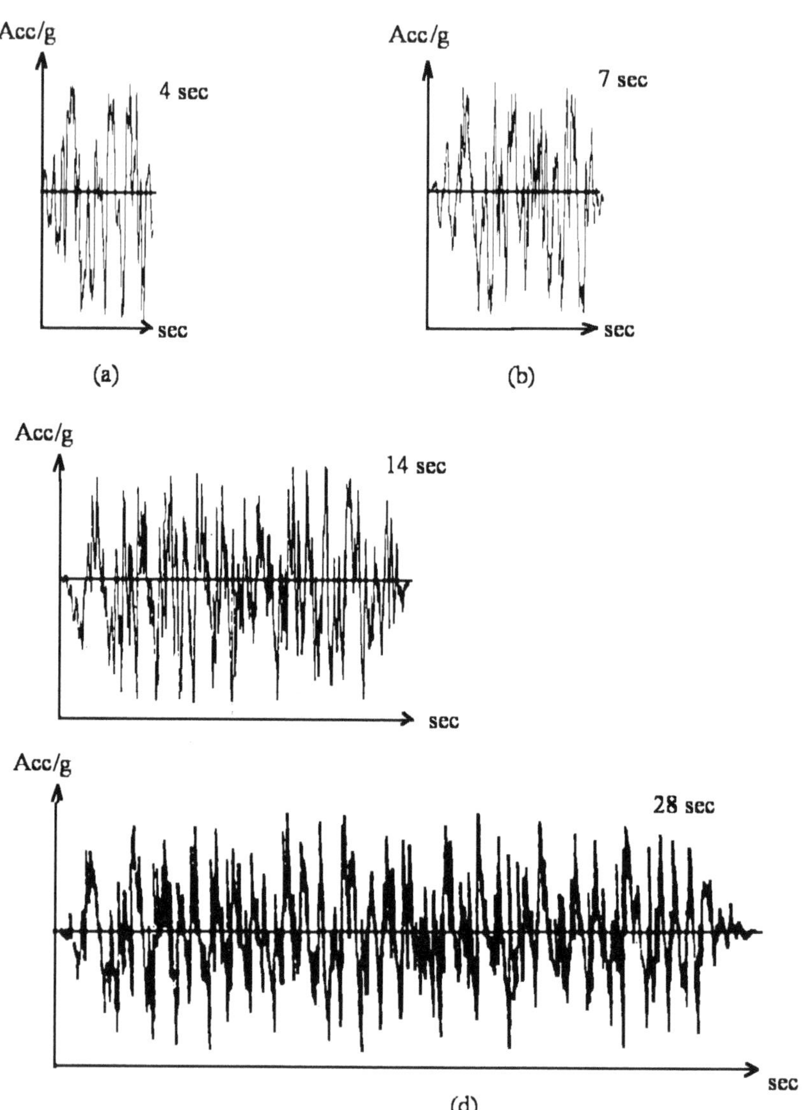

Figure 3.34: Artificially digitized accelerograms (after Background Document for EC8, 1988)

maximum acceleration and dominant period. There are two methods to build an artificial time-history representation:

-*digitalization method*, in which some artificial ground motions are numerically simulated, with a time step of 0.01 sec. Fig. 3.34 presents the artificial accelerograms proposed in the Background Document for EC8 (1988), for different durations: 4 sec, 7 sec, 14 sec, and 28 sec;

-*analytical method*, based on the Fourier series functions and an envelope exponential function, determined in relation with the ground motion characteristics. For different earthquake types, it is possible to use different durations, number of cycles, periods, etc. The acceleration may be defined by the following expression (Lam et al, 1996):

$$a_g(t) = I(t) \sum_{n=1}^{N/2} A_n \sin\left(\frac{2\pi n t}{T} + \varphi_n\right) \qquad (3.11a)$$

where $I(t)$ is the normalized intensity envelope given by:

$$I_t = I_0\left(e^{-\alpha t} - e^{-\beta t}\right) \qquad (3.11b)$$

In (3.11a) T is the length of the record and N the number acceleration time steps with regular intervals given by vibration period T_g (Fig. 3.35):

$$N = \frac{T}{T_g} \qquad (3.11c)$$

An artificial generated accelerogram is proposed by Reinoso et al (2000) for pulse-type ground motions (Fig. 3.36):

$$a_g(t) = a_{max} \sin\frac{2\pi t}{T_1} \sin\frac{2\pi t}{T_2} \qquad (3.12)$$

where a_{max} is the peak acceleration, T_1, the dominant soil period and T_2, the duration of strong ground motion.

Masri and Safford (1980) have proposed a pulse technique to stimulate dynamic actions, in which the continuous variation of actions due to ground motions is simulated by pulse forces (Fig. 3.37). This methodology gives the idea to use artificially generated accelerograms containing only the main pulse part of the records. In order to control the possibility to use this methodology, the Vrancea–Bucharest accelerogram (1977) is used, by dividing the 17 sec main duration in four time intervals, denoted from a to d (Fig. 3.38a): three of 5 sec and one of 2 sec (Ifrim et al, 1986, Gioncu, 2000). The a interval corresponds to the beginning, b represents the major ground motion characterized by a pulse acceleration, the intervals c are the consequences of the main shock. Fig. 3.38b shows the importance of each accelerogram fragment, considering each interval as an independent earthquake. It is clear that the response is dominated by the interval b, which corresponds to the pulse acceleration. The superposing of the response for the different intervals is presented in Fig. 3.38c. Curve A corresponds to interval a, curve B to $a + b$, curve C to $a+b+c$ and curve D the sum of all intervals. One can see that for periods ranged between 0.2 and 1.6 sec, the curves B, C and D practically coincide. So the use of the main pulse can only describe the behaviour of structure behaviour. This example shows that for pulse-type earthquakes, the accelerogram recordings may be simulated using an idealized ground motion composed by a simple pulse. Sasani

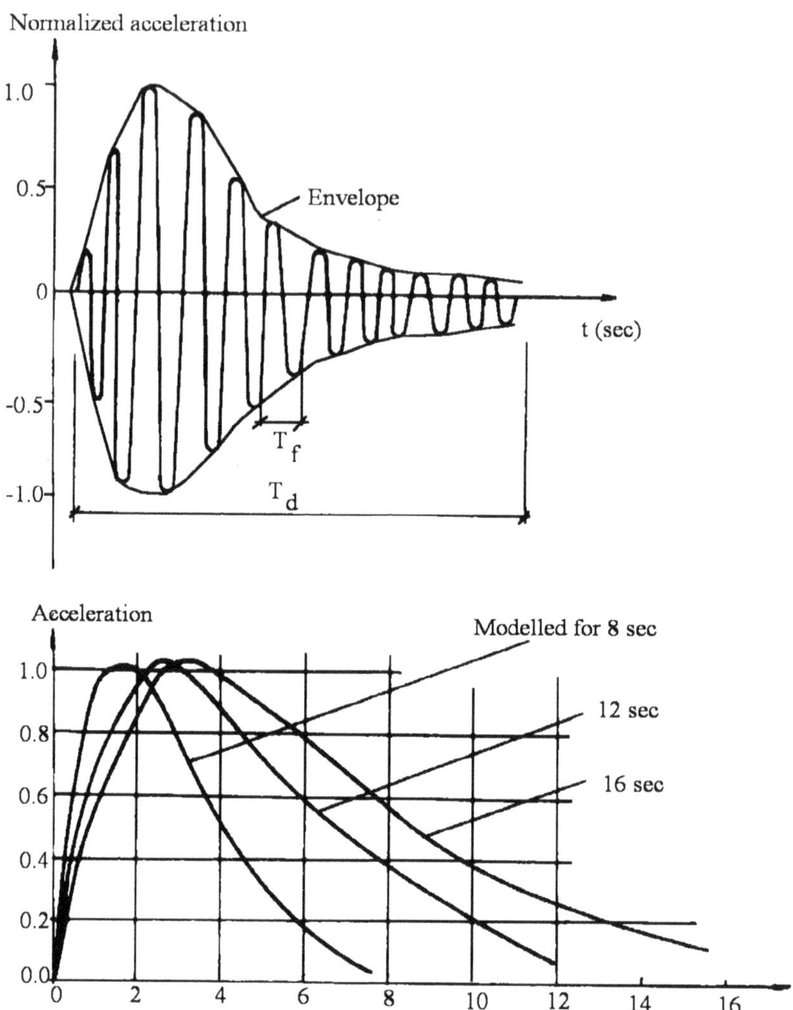

Figure 3.35: Analytic response: (a) Accelerogram; (b) Envelopes for different record lengths (after Lam et al, 1996)

and Bertero (2000) suggest the shapes of these pulses for normal fault and parallel fault ground motions (Fig. 3.39). As it can be seen, for normal fault ground motions the displacement at the end of motion are set equal to zero, while for parallel fault ground motion, there are permanent displacements caused by relative displacement of ground at two sides of strike–slip faults.

The simple pulse-type action is used for the study of near-field earthquakes by Hall et al (1995), Mylonakis and Reinhorn (1997), Nakashima et al (1997), Alavi and Krawinkler (2000), Gioncu (2000), Gioncu et al

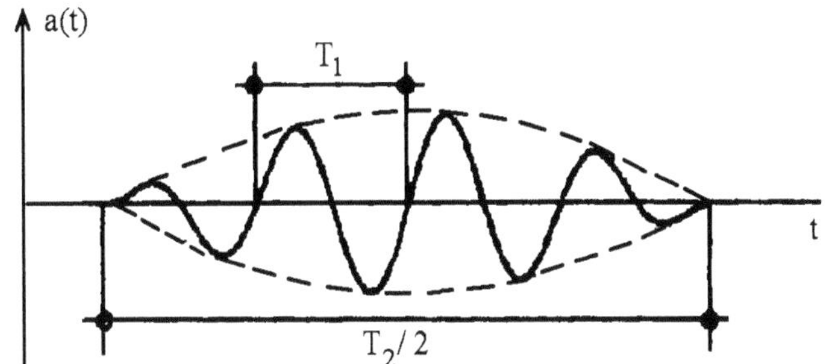

Figure 3.36: Synthetic accelerogram (after Reinoso et al, 2000)

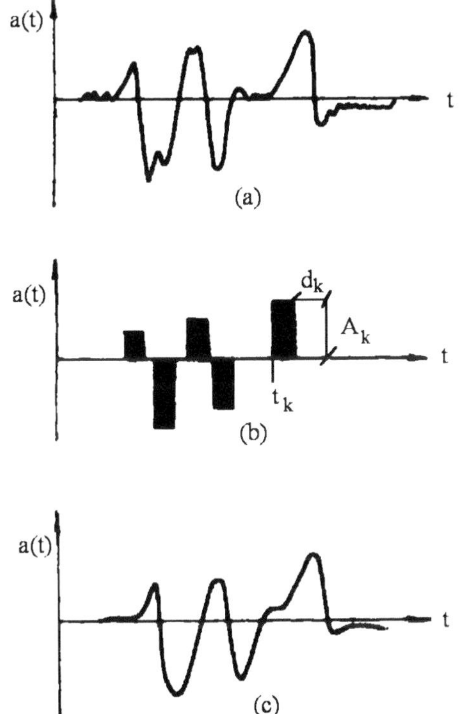

Figure 3.37: Pulse technique for simulating artificial accelerograms: (a) Recorded accelerogram; (b) Input pulse train; (c) Digitized accelerogram (after Masri and Safford, 1980)

(2000), Mateescu and Gioncu (2000), Tirca (2000). Fig. 3.40a presents the acceleration, velocity and displacement time history for one pulse, which is fully defined by two parameters, the pulse period T_p and the maximum ground acceleration a_{max}. The cases of asymmetric pulses, two or more ad-

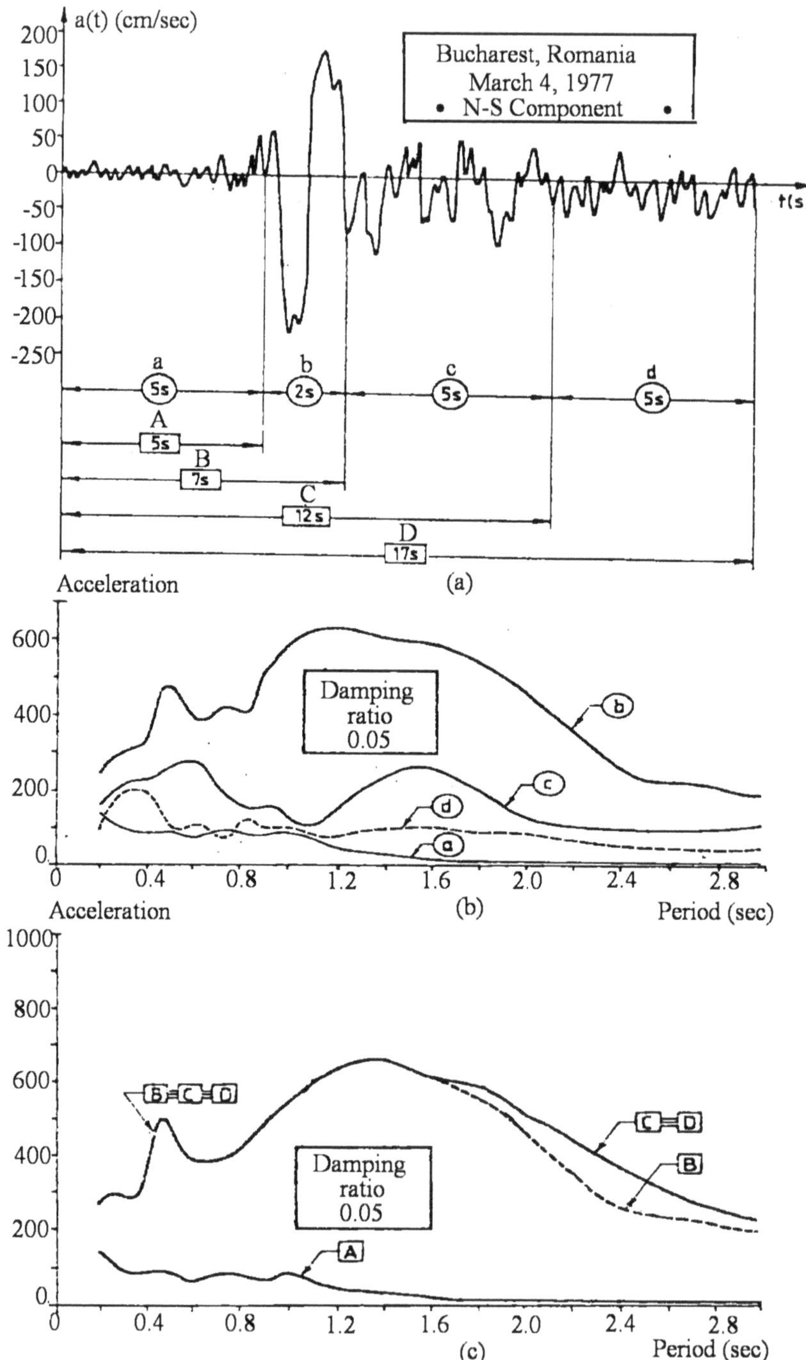

Figure 3.38: Vrancea–Bucharest accelerogram, 1977:(a) The dividing of the accelerogram in fragment of time; (b) Contribution of each fragment; (c) Sequential seismic response spectra (after Ifrim et al, 1986)

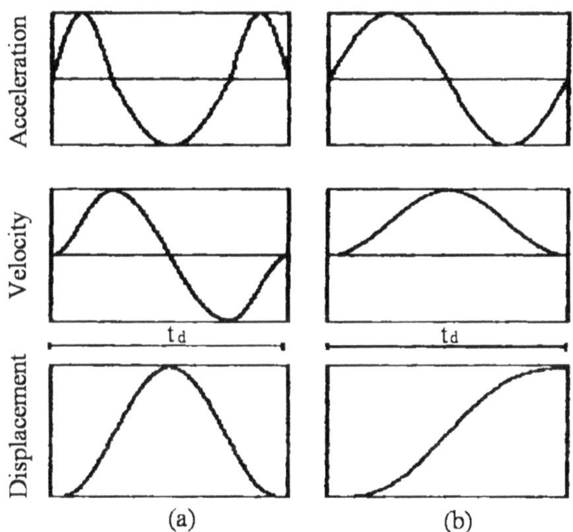

Figure 3.39: Pulse ground motions: (a) Normal to fault; (b) Parallel to fault (after Sasani and Bertero, 2000)

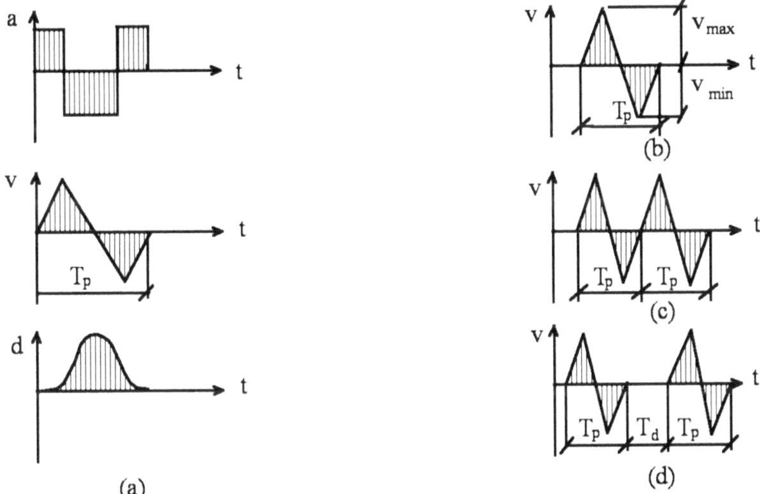

Figure 3.40: Simple pulse ground motions: (a) Symmetric pulse; (b) Asymmetric pulse; (c) Two adjacent pulses; (d) Two distinguished distant pulses (Gioncu et al, 2000)

jacent pulses and two distinguished distant pulses (Fig. 3.40b) are studied by Gioncu et al (2000).

Thus, the use of the artificial time-history representation seems to be a promising way for the structure design.

3.3.4. Response Spectra for SDOF

A common approach in current seismic design practice is to characterize ground motion and earthquake effects by the response spectra. Today, response spectra form the basis for the evaluation of the seismic design forces in the majority of seismic codes. Natural or artificial accelerograms, velocities and displacements, can be used for constructing response spectra, for a given earthquake. There are some steps in this construction:

(i) *Elastic response spectra.* After digitizing an accelerogram of a particular earthquake (Fig. 3.41a) and assuming a numerical value for period and damping, the response of a single-degree-of-freedom SDOF, elastic system can be calculated (Fig. 3.41b). The dynamic motion is applied at the base of a cantilever, which models the case of a structure restrained at the ground. The complete history of response for this elastic system can be computed. The maximum values for accelerations, velocities and displacements are determined. By repeating the above process for a great number of SDOF systems, for a given value of damping, a plot of the response spectrum can be obtained (Fig. 3.41c). Due to the resonance effect, the spectra have the tendency to amplify the ground motions in some range of period. The aim of these spectra is to emphasize this amplification.

There are two possibilities to represent a spectrum:

-tripartite plots which allow for the presentation of all response parameters together (Fig. 3.42a);

-separate plots for accelerations, velocities and displacements (Fig. 3.42b), in which for a given structure period, the corresponding maximum values of the involved characteristics are obtained.

(ii) *Probabilistic approach.* Strong ground motions are time process affected by a high degree of uncertainty, deriving from many sources. Therefore, the modelling of a seismic excitation turns out to be a random process. So, from the processing of the recorded data, the average value and the standard deviation can be determined. The proposed values are the average plus one standard deviation (Fig. 3.43). But for normal structures, it is generally considered adequately conservative to use mean values. The mean plus one standard deviation value can be used only for special structures (Fajfar, 1995a). The parameters having a specified probability of non-excedance at site during structure lifetime are (Lungu et al, 1997):

-peak-ground acceleration, velocity or displacement (PGA, PGV, PGD), obtained from ground-motion records;

-effective peak acceleration or velocity (EPA, EPV), which attempts to compensate for the inadequacy of the actual single peak to describe the potential of the ground motion. EPA is the average of the maximum ordinates of elastic-acceleration-response spectra in the period range from 0.1 to 0.5 sec, divided by a mean value of 2.5 (for 5 per cent of damping). EPV is the same average of the maximum ordinates of elastic velocity response spectra in the period range of 0.8 to 1.2 sec, divided by the same mean value of 2.5.

(iii) *Smoothed spectra.* Because the actual elastic spectra present many cusps which can not be used for design practice, the constructed elastic spectra must be smoothed (Fig. 3.43). So, the actual variation is substituted

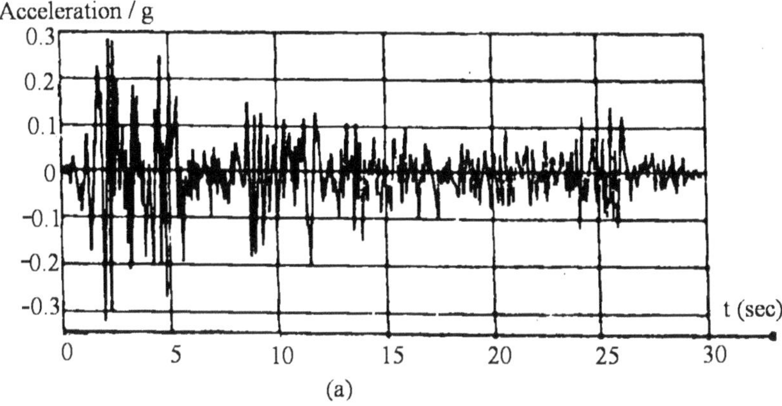

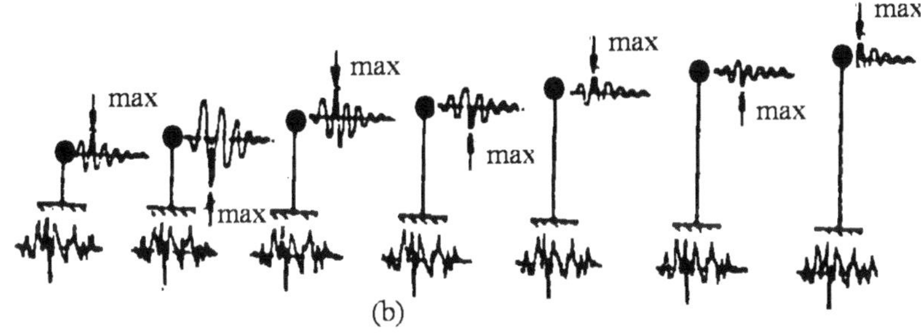

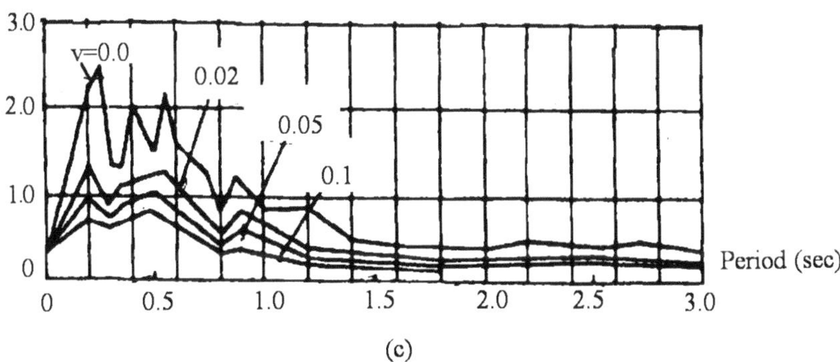

Figure 3.41: Elastic response spectra: (a) Acceleration; (b) SDOF system; (c) Resulted spectra

by a combination of linear and hyperbolic variations. A very important characteristic of the smoothed spectra is the *corner period*, which is equivalent to the predominant period of ground motion (Lam et al, 1998). It can be determined also as the period corresponding to peak acceleration associated with the maximum relative velocity spectrum (Miranda, 1993).

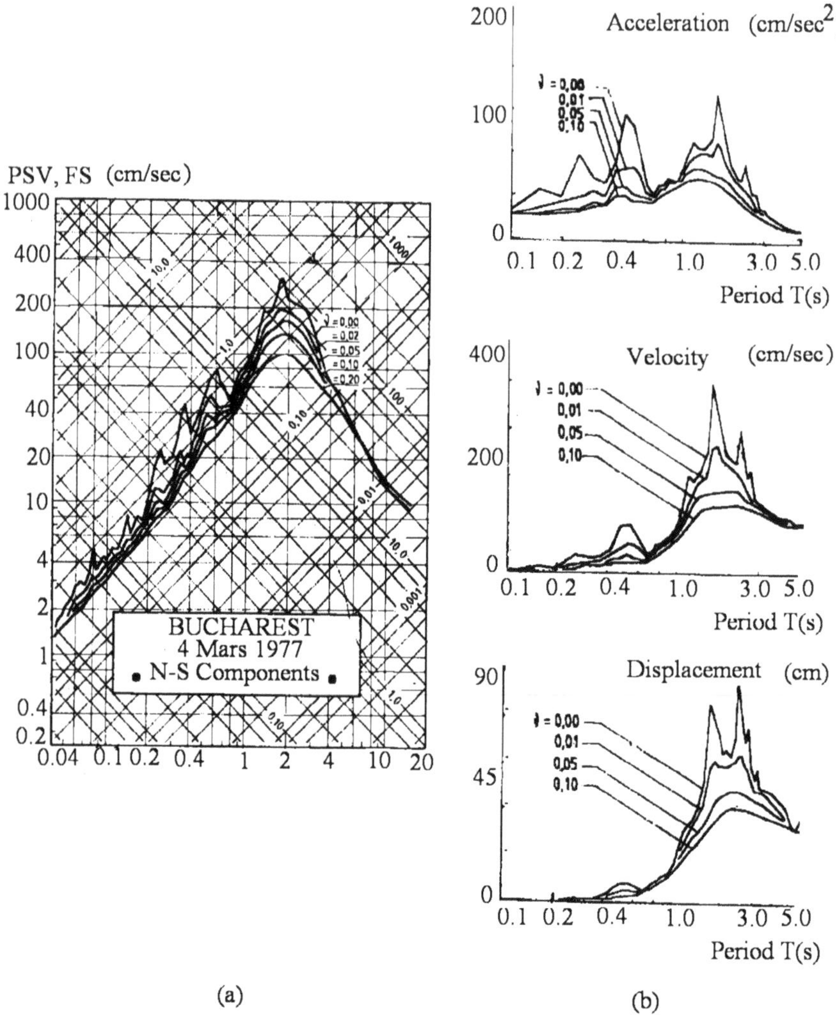

Figure 3.42: Spectrum types: (a) Tripartite plot; (b) Separate plots (after Ifrim, 1984)

The database containing 81 strong accelerogram records obtained from both interplate and intraplate regions around the world and measured for rock, stiff and soft soils, within 60km from epicenter and for accelerations 0.1 to 0.6g, are used for determining the dominant corner periods (Fig. 3.43b). One can see that the majority of earthquakes have corner periods in the interval 0.1 to 0.5 sec. The cases of Mexico City and Vrancea–Bucharest long corner periods (1.5–3.0 sec) are some exceptions due to very bad soil conditions.

(iv) *Normalized spectra.* It is customary to employ a design spectrum with the general form shown in Fig. 3.44, in which the spectrum is normalized to the value of peak-ground acceleration. It is very important to no-

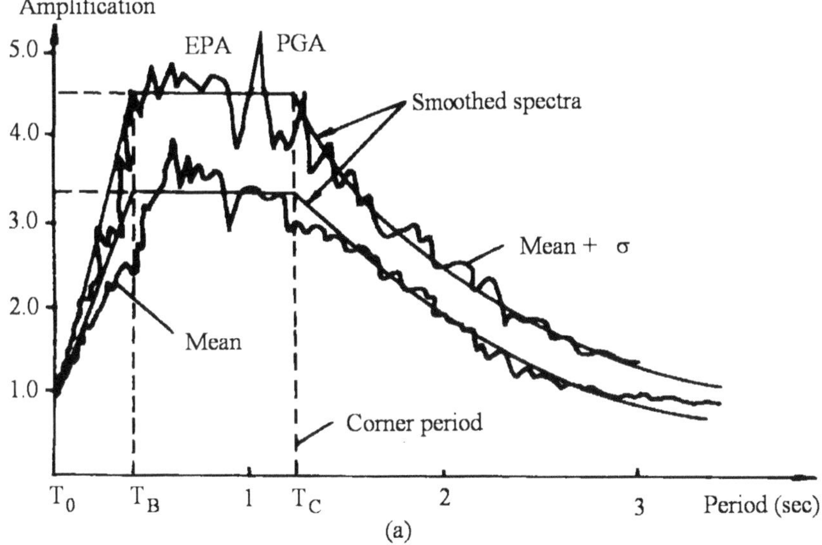

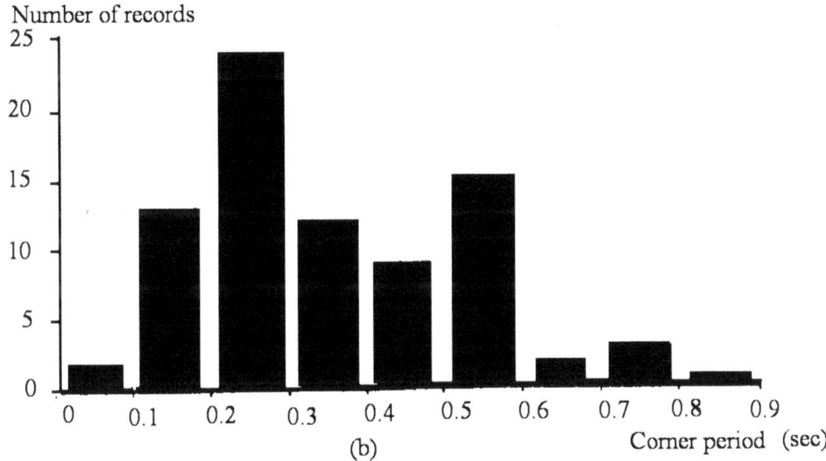

Figure 3.43: Smoothed spectra: (a) PGA and EPA values; (b) Corner period distribution (after Lam et al, 1998)

tice the great difference between the interplate and intraplate earthquakes, which must be reflected into the elastic spectra (Lam et al, 1996).

(v) *Inelastic-response spectra.* Elastic-response spectra are derived by amplifying peak-ground motions for acceleration, velocity and displacement, due to the resonance effect. The inelastic response spectra are obtained by reducing the elastic design spectra, taking into account the structure capacity to dissipate seismic energy. The inelastic spectra can be obtained in the following ways (Mahin and Bertero, 1981, Lai and Biggs, 1980):

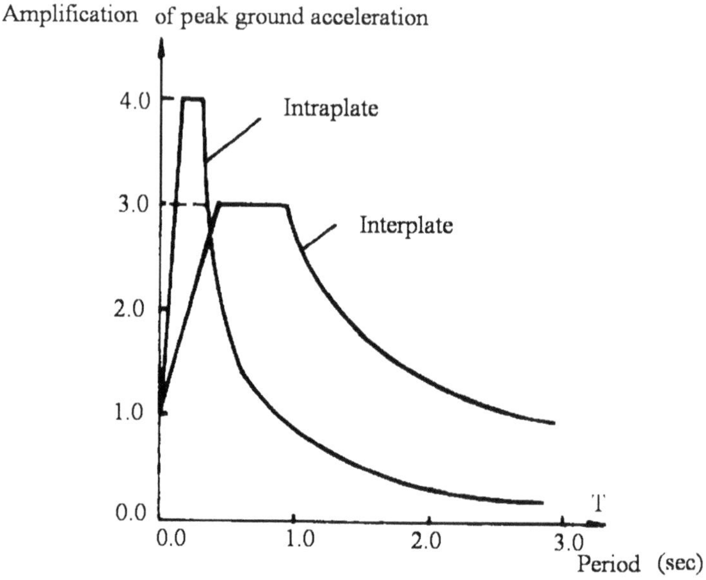

Figure 3.44: Normalized spectra for interplate and intraplate earthquakes (after Lam et al, 1998)

-by reducing the elastic-response spectra with an empirical and period-independent coefficient, known as the reduction factor q, intended to consider the inelastic deformation of the structure (Fig. 3.45a);

-by reducing the elastic-response spectra with a variable coefficient in function of the structure period (Fig. 3.45b). There are many proposals for this coefficient (see Section 3.7);

-by using an elasto-plastic SDOF system subjected to a given ground motion, the inelastic-response spectra can be determined directly (Fig. 3.45c). In this way, the inelastic spectrum considers the local site conditions, period of vibrations, etc. Of course, the last method is the most proper, but for practice design purposes it is too expensive. So, a combination of the last two methods is required, the direct method being used for the calibration of the reduction factor q in function of the structure period.

(vi) *Data bank for spectra.* In the same way as for the time-history representation, it is necessary to store all the spectra in a data bank, in order to keep them at disposal for designers. Several acceleration response spectra for different strong earthquakes are presented in Fig. 3.46 (Popov, 1994). One can see that the very well known and often used El Centro event is modest in comparison with the most recent earthquakes.

Another alternative to present the design spectrum is the ***capacity spectrum*** (Krawinkler, 1995, Reinhorn, 1997). Instead to use the acceleration–period curve, a seismic–force–capacity curve (acceleration versus displacement) is plotted (Fig. 3.47). The structure period T is represented by radial lines. The advantage of this representation is that both strength and dis-

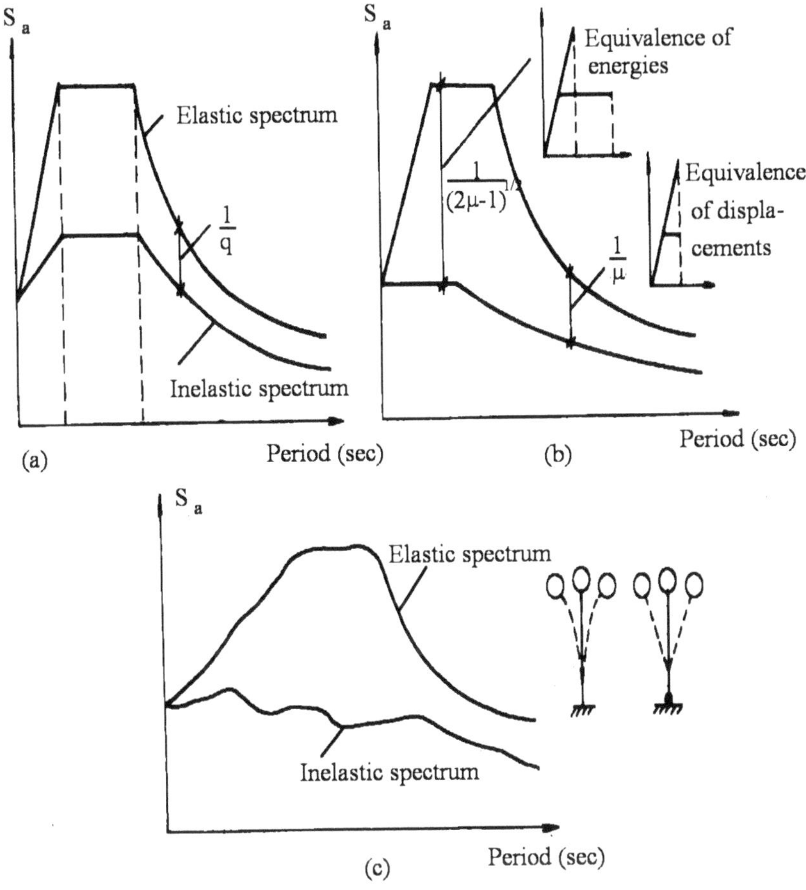

Figure 3.45: Inelastic spectra: (a) Using a constant q factor; (b) Using a variable q factor; (c) Direct determination using SDOF elsto-plastic system

placement demands are evident from a single graph. For elastic forces, the elastic displacement demand and the structure period may be determined. For inelastic behaviour, at the reduced force, the inelastic displacement is obtained at the horizontal branch of the capacity curve with inelastic reduced spectra. At the same time, the reduced period due to inelastic deformation may be determined.

The actual design philosophy uses the spectral force (acceleration) based design method, which is a force-based approach. In the last time the limitations of this method are widely discussed in literature, being proposed a new methodology based on the ***displacement spectra*** (Medhebar and Kennedy, 2000, Tolis and Faccioli, 1999, Fajfar, 1995, 1996, Bertero and Bertero, 2000, Priestley, 1997, 1998, 2000, Calvi, 1998, Tassios, 1998, Faccioli et al, 1998,

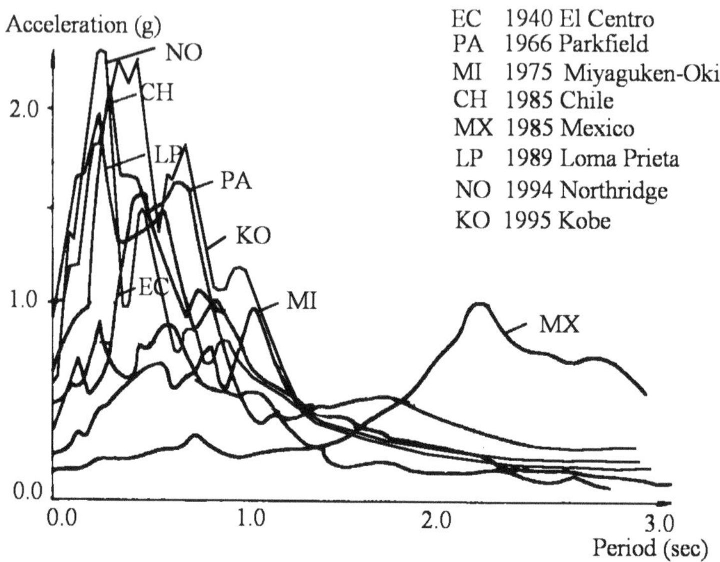

Figure 3.46: Response spectra for different strong earthquakes

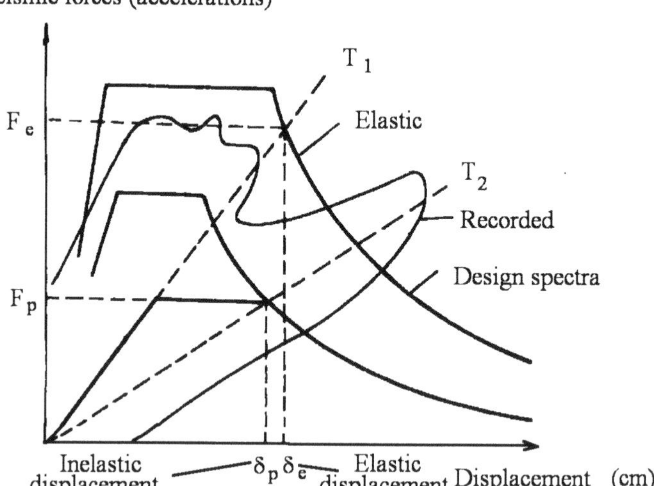

Figure 3.47: Capacity spectrum

Moehle, 1996, Bommer and Elnashai, 1999). Fig. 3.48a shows the main displacement spectra determined from records, the main+ σ band and the design displacement spectra. One can see that for the actual periods the variation-displacement period is linear (Fig. 3.48b).

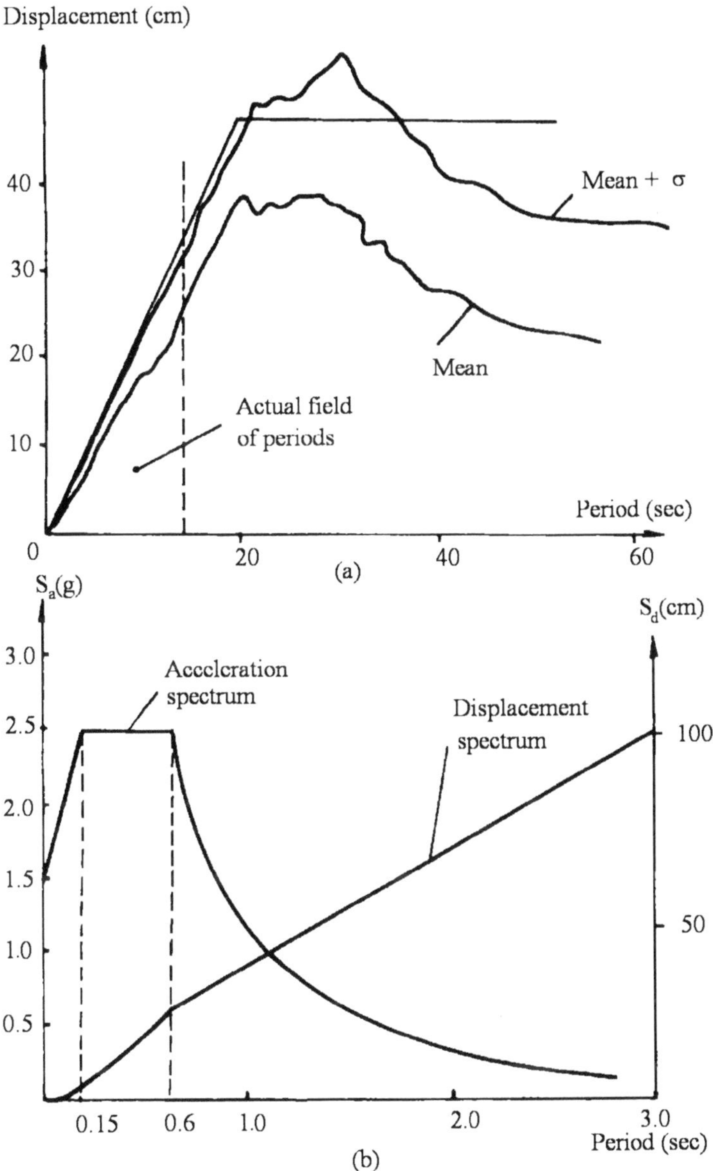

Figure 3.48: Displacement-based spectra: (a) Elastic design spectra; (b) Elastic acceleration and displacement spectra

Due to its simplicity, the response spectra representation provides a very reliable tool to estimate the ground-motion effects on the ordinary structures. The criticism on the use of this representation in the present codes is based on the following points:

-the assumption that only one design spectrum is sufficient to describe the earthquakes to be used for seismic design purposes. As several earthquake records are available, it is clear that the use of a single design spectrum is inadequate, because some of earthquake ground motions can have the shape of response spectra dramatically different from the standard design spectrum (Zhu et al, 1988);

-on the same site, the spectra for distinct ground motion levels, required for a multi-level design approach, are very different. The use of the same spectra for serviceability and ultimate states is a shortcoming in the design methodology;

-the spectrum for near to source is very different from the one recorded far from epicenter, being influenced by pulse type of ground motion;

-near to the source, the spectra for vertical ground motions are very different from the ones for horizontal ground motions and the use of just a fraction of the last spectrum is an inadequate methodology;

-the damping effects are different in function of the levels of the considered earthquake. For serviceability limit state, a maximum value of 2 per cent must be considered, while for the ultimate limit state, an average value of 5 per cent can be used;

-the duration of earthquake, which is a very important factor, is completely ignored;

-the influence of a/v ratio is not considered;

-the impossibility to consider in proper ways the effect of the superior modes of vibrations, which is a very important aspect for multi-storey buildings located in near-to-source site.

Despite these criticisms, the use of response spectra remains the most practical method for the usual design.

3.3.5. Response Spectra for MDOF

The response spectrum method is based on the behaviour of a single-degree-of-freedom, SDOF, elastic linear system. Due to its simplicity, the method has been extended to multi-degree-of-freedom, MDOF, systems, nonlinear elastic and inelastic hysteretic systems, using some equivalence coefficients (Fig. 3.49a). For many earthquake types, the response spectrum is an adequate measure of demand. This is the case when the demand of ground motion can be specified in term of the dominant response, which is the first mode of vibration. These ground motion types are characteristic of far-field sites. The SDOF system can model very well the displacements of the MDOF structure (Iwan, 1995, 1997).

However, there are some cases where the SDOF systems are inadequate when describing the structure displacements and this is the case of near-field sites for an earthquake. The earthquake action has a characteristic impulse, and the displacements propagate through the structure as waves, causing large localized interstorey drifts (Fig. 3.49b). The superior vibration modes have an increasing effect, which can not be detected using the response spectrum for SDOF systems.

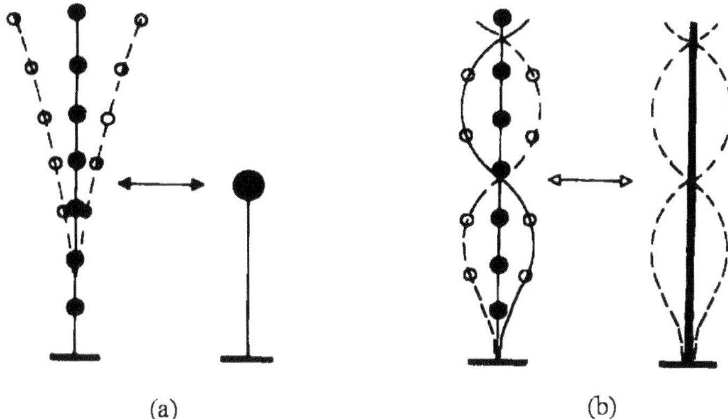

Figure 3.49: Modelling of multi-degree of freedom systems: (a) Cantilever model; (b) Shear-beam model

So, a different approach must be used in this situation and Iwan (1994, 1997) has proposed a continuous shear–beam model (Fig. 3.49b) and, as a design criterion, the interstorey drift ratio (interstorey displacement-to-storey height). A comparison among drift demands for the Northridge, Kobe and El Centro events are presented in Fig. 3.50. The drift spectra have features which are similar to those of response spectra. They tend to reduce for both short and long periods. Typically there are one or more distinct period ranges for which the values of spectra are greater than the ones for other periods. As far as the damping ratio increases, the spectrum values decrease and the shape becomes smooth. There is demand for the Northridge earthquake to have two peaks in the period interval from 0.5 to 2.0 sec, ranging from 3.5 to 5.3 per cent. For the Kobe earthquake, the peak occurs for periods ranging from 0.7 to 1.5 sec, with values from 4.5 to 8 per cent. The El Centro earthquake, frequently used as a standard earthquake type, has a reduced drift demand.

Iwan (1995, 1997) compared the drift results from response spectra and the drift spectra. For short periods the differences are small, but for longer structural periods, the results diverge significantly, the response spectra underestimating the drift demand.

3.3.6. Power Spectral Density Function

The response spectrum approach is the most widely used in design practice, because of its simplicity. But some difficulties in using this method arise from the problems related to the combination with other directions of shaking and to the application to multiple degrees of freedom systems. Another problem is related to the fact that the usual earthquake intensity is measured by means of the acceleration peak, but in many cases this peak does

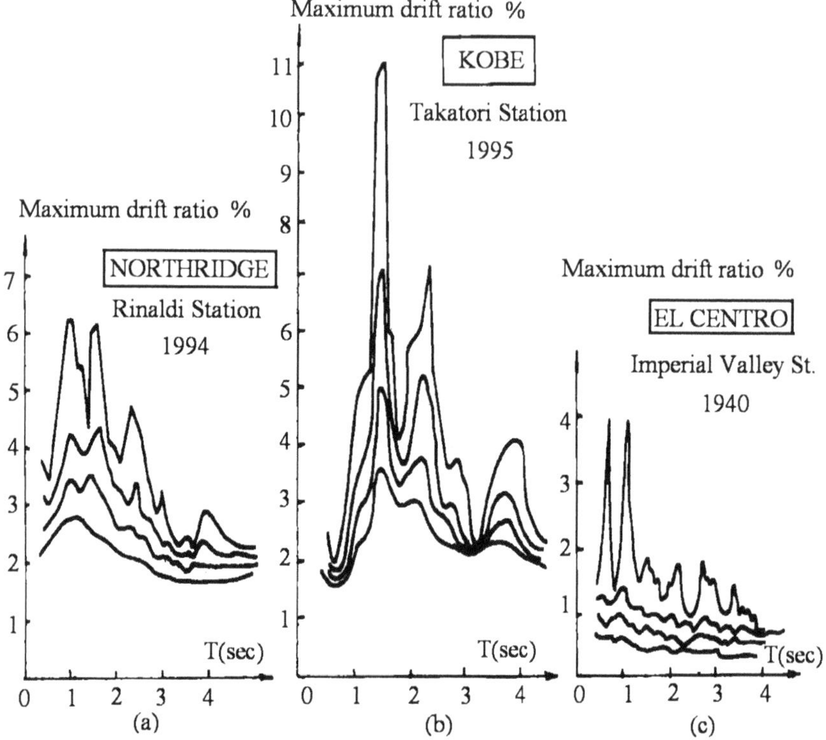

Figure 3.50: Interstorey drift spectra (after Iwan, 1997)

not represent the energy of the input motion and so the response spectrum is not a direct description of the motion itself.

In the last few years there has been an increasing interest in utilizing the spectral density function as a monitor of the energy at the fundamental period of interest in the ground motion time history.

The power-spectral-density function describes the energy content of the signal, as a function of frequency and thus it is a representation of the distribution of power over the frequencies that make up a given motion (Christian, 1989). The ground motion is represented as a random process (Fig.3.51).

Unfortunately, this very interesting approach is widely discussed in literature, but somewhat less used in practice.

3.4 Conceptual Design

3.4.1. Seismo-Resistant Architecture

From the analyses of damage caused by earthquakes it is possible to conclude that in general buildings having a proper configuration survive with minor

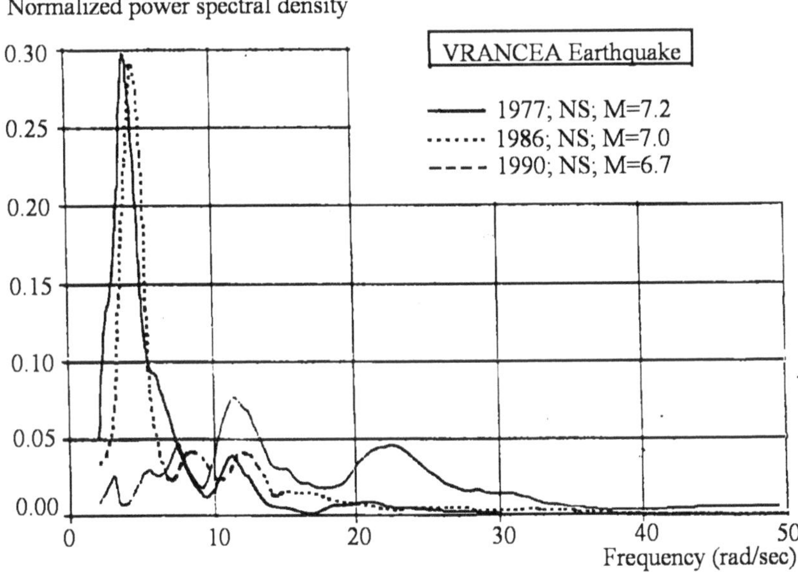

Figure 3.51: Power spectral density function (after Lungu et al, 1997)

damage, whereas buildings not properly conformed and/or poorly erected may collapse. Also the use of sophisticate computer programs, as recently developed for the analysis of complex structures, gives the same answer: badly conceived structures in the preliminary design can never survive. So, the concern of a good general configuration plays a crucial role in seismic design.

Because these general configuration problems belong to the field of architecture, a *Seismo-Resistant Architecture Philosophy* has developed over the last few years (Arnold, 1996, De Buen, 1996, Giuliani et al, 1996, Giuliani, 2000). This philosophy refers to an "architecture engaged in the necessity of optimizing the design and materialisation processes of human sites located in highly risky seismic zones, based on the compatibility of interrelationships among its components or interacting subsystems, during the seismic actions" (Giuliani et al, 1996). For constructions, "the seismo-resistant architecture deals with the interaction of each subsystem of the building during seismic shaking in order that the architectural project does not originate structural wrong adjustments which would decrease the seismo-resistant capacity of the building" (Giuliani, 2000).

In order to obtain a good solution for a seismo-resistant structure the architect–engineer relationships must be very good. The collaboration refers to the general configuration of building, consisting in:

(i) *Regularity*. Regular structures may be defined as those having a nearly uniform distribution throughout their height of storey strength, stiffness, weight and dimensions, and being approximately symmetrical for

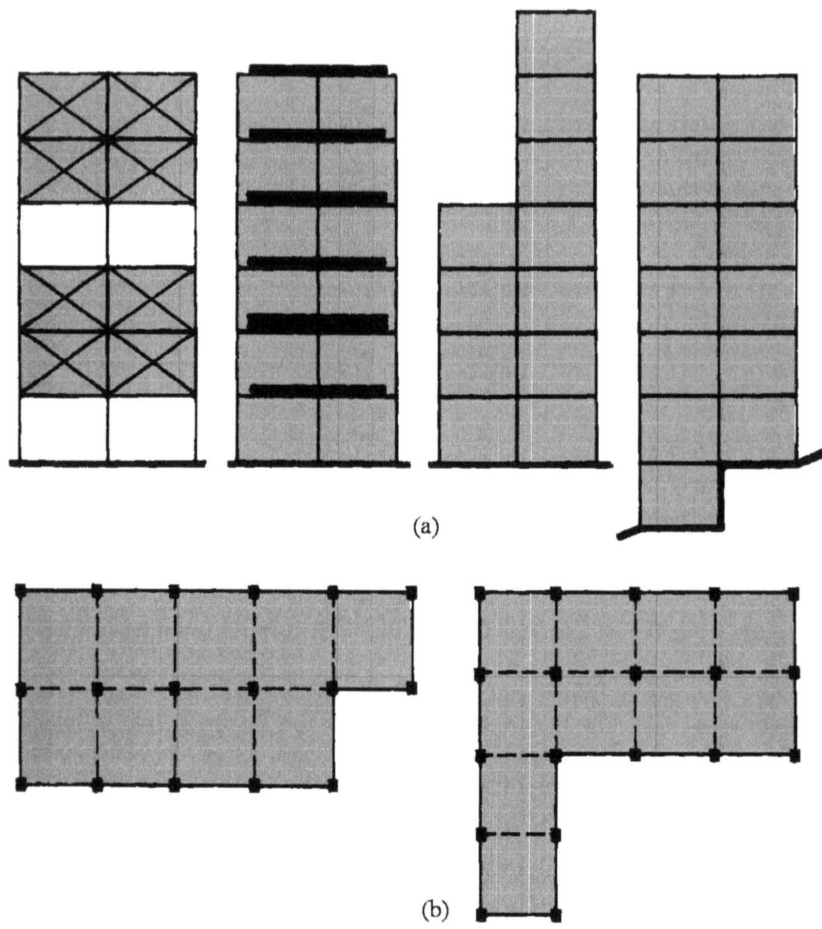

Figure 3.52: Irregular structures: (a) Vertical irregularities; (b) Horizontal irregularities

weight and stiffness in their plane. Modern codes consider the regularity as a pre-requisite for a good seismic behaviour (Rutenberg et al, 1996). Vertical irregularities may be introduced by stiffness and strength non-uniform distribution, due to existing soft and weak storeys, to abrupt changes in the masses of adjacent storeys, to setback and offset of the structure, or to irregularities in ground topography (Fig. 3.52a). Horizontal irregularities are produced by re-entrant corners, presence of different structure types, and floor plan shape (Fig. 3.52b). The problem of the influence of set back on the seismic behaviour of steel structures has been deeply analysed (Guerra et al, 1990, Mazzolani and Piluso, 1996a,b, 1997b), leading to the conclusion that geometrical irregularity does not always mean structural irregularity and the adopted design criteria play the most important role.

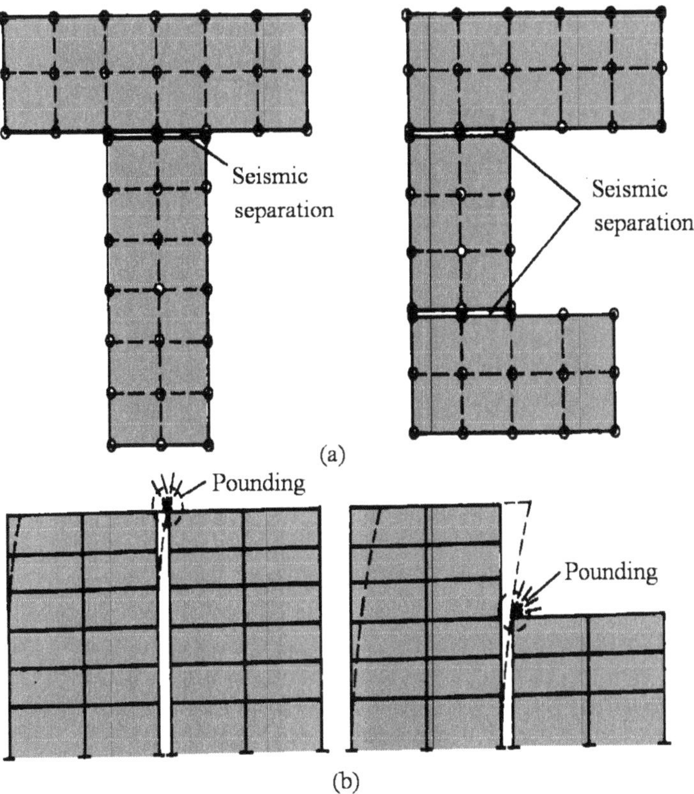

Figure 3.53: Seismic separation: (a) Horizontal separation; (b) Vertical separation – structure pounding

(ii) *Seismic separation.* The concept of plan regularity is related to the ability a structure has to vibrate without torsion. The ideal shape of this demand is the square or at least rectangular one, where the centroid and rigidity centers coincide. If the architectural conditions impede a regular solution, it is desirable to use separation gaps, which divide the building into independent parts, having a regular seismic behaviour (Fig. 3.53a). In this case, the seismic separation gap must be a sufficient size to avoid the earthquake induced pounding of the two adjacent buildings (Fig.3.53b). During the recent important earthquakes, the insufficient separation sizes produced major damages.

(iii) *Staircase emplacement.* The position of the staircase in a building plays a very important role, because it represents an irregularity both in horizontal and vertical planes (Fig. 3.54). Always the staircase must be protected by rigid anti-fire walls, so a rigid core is introduced in the structure. The different rigidity in comparison with the rigidity of floor structure and the presence of the intermediate landing are the factors which produce a source of asymmetry in the building, generally causing torsions.

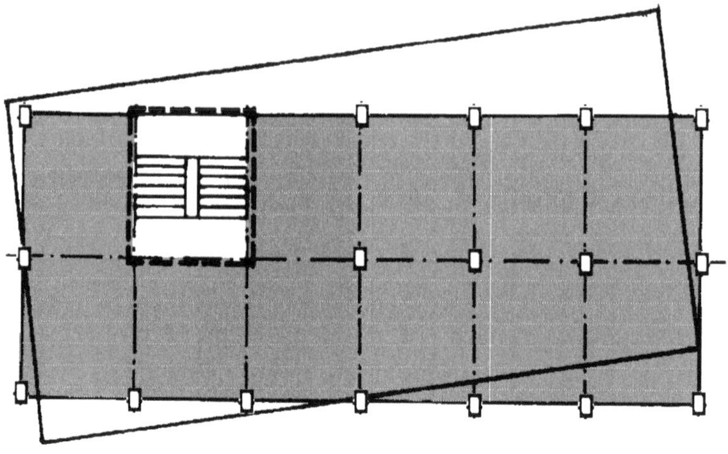

Figure 3.54: Staircase emplacement

(iv) *Non-structural elements.* These elements are attached to floors, roofs and frames, without being part of the main structural system; they react when subjected to large seismic forces, and try to resist them, according to their own structural characteristics (Villaverde, 1997). These non-structural elements may be classified in architectural components (partitions, parapets, cladding systems, suspended ceilings, etc.), mechanical and electrical equipments (storage tanks, pressure vessels, piping systems, computer and data-acquisition systems, fire protection systems, boilers, etc.) and building contents (bookshelves, storage racks, furniture, etc.). From the structural view point, the partition walls play the most important role. Masonry infilled panels or light-gauge steel panels can be frequently found as interior and exterior partitions. Since they are normally considered as architectural elements, their presence is often ignored by engineers. However, even they are considered non-structural elements, they tend to interact with the bounding frames, when the structure is subjected to strong earthquake loads. This interaction may or not be beneficial to the structure performance. If their position is regular, the infilled panels may be used as a means to strengthen the existing frames and to improve the structure performance (Mehrabi et al, 1996, Mazzolani and Piluso, 1990, 1996a). Otherwise, an irregular and uncontrolled distribution of panels may contribute to unreliable behaviour, by introducing an abrupt change in stiffness and strength, leading to torsional effects (Fig. 3.55). A recent European project sponsored by ECSC has been particularly devoted to the analysis of the influence of cladding panels on the seismic behaviour of steel frames (De Matteis et al, 1995, 1998, Mazzolani et al, 1996a,b,c, 1997a,b), leading to very detailed conclusions related to the type of sandwich panel and the connection systems.

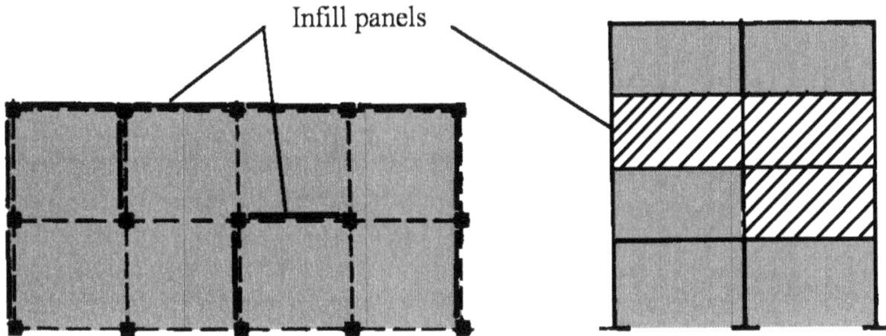

Figure 3.55: Irregular non-structural element distribution

3.4.2. Structural Configuration

For structural configuration two basic aspects must be ensured:

-interaction of all structural components in taking over the seismic actions;

-correspondence between actual structure behaviour and that expected in design analysis regarding rigidity, strength and ductility.

Central for the successful design of a structure in a seismic area is an adequate solution for the three components: foundations, floors, and structure system.

(i) *Foundations.* The design of foundations is a very difficult problem, due to the variability of site and soil conditions. In seismic behaviour, foundations have a dual role in terms of loading transmission: to capture the ground motions, to transfer them to the structure, to receive the inertial forces from the structure and to return them to the soil.

Foundations may be classified into two main categories:

-shallow foundations, as isolated, combined, strip footings and mats;

-deep foundations, as piles and drilled piers or caissons.

From the seismic view point, there are two main ways which can improve their structural behaviour:

-inter-connecting all the footing elements into a single mass in order to restrain any relative movements of these elements. The main assumption in design philosophy considers that all points at the building base move in phase. But in some cases, especially near the source and on soft soil, the asynchronical movement of foundations is possible (Trifunac, 1997). The connecting beams may reduce or eliminate these differential movements (Fig. 3.56):

-over-strengthening the foundations in comparison with the super-structure elements. The main problem for foundation design is whether they should be conceived as ductile or not. The code provisions require foundations to be designed only for gravity and for reduced seismic loads. Although, in reality, larger forces will develop in the structure during a major earthquake, so foundations must be ductile in code conception also. Since the foundations have to carry the entire structure and taking into account

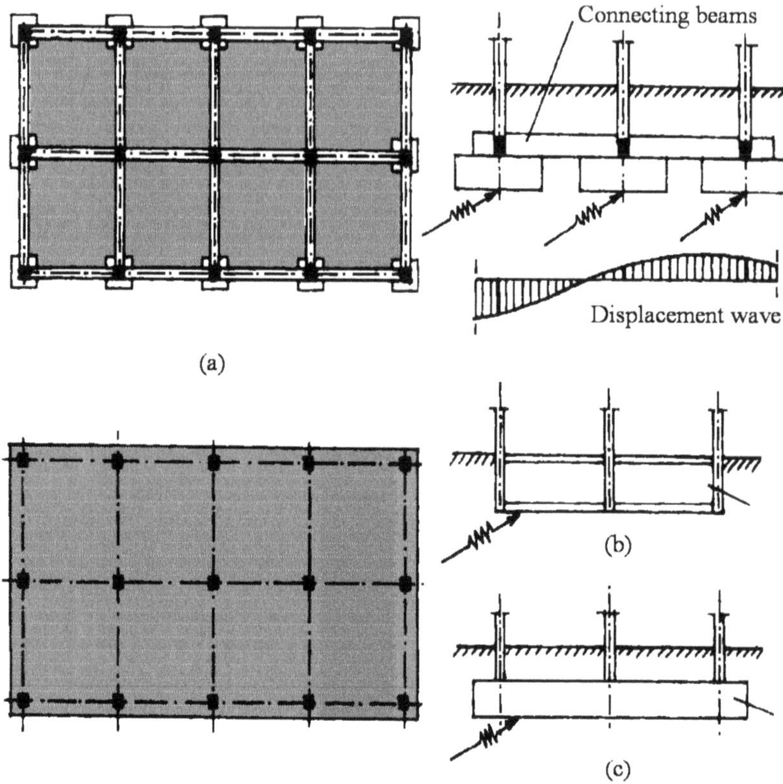

Figure 3.56: Foundation types: (a) Isolated foundations; (b) Box foundations; (c) Slab foundations

that the eventual repairs of damage are very expensive, it is clear that they should be designed conservatively, in order to allow them to undergo inelastic deformations, for avoiding the foundation collapse before the one of the super-structure (Ali, 1995). However, some detailing rules to assure a minimum level of foundation ductility are required.

(ii) *Floors.* Virtually, quite all the modern multi-storey steel framed buildings use trapezoidal sheetings supported by the floor structure. Most of them are designed as composite slabs, the steel decks being utilized as tensile reinforcement for the slab (Fig. 3.57). The structural designer selects the depth and thickness of the steel deck and the depth of concrete slab, by considering the vertical loads only. But from the earthquake view point, there are some other aspects to be considered:

-floor structures must work in horizontal plane as a diaphragm with sufficient rigidity to distribute the seismic forces among all vertical elements and to assure a rigid motion of the floors in horizontal direction. This aspect is crucial in case the structural system is composed by different structural types or in case of asymmetric structures and asynchronical seismic loads

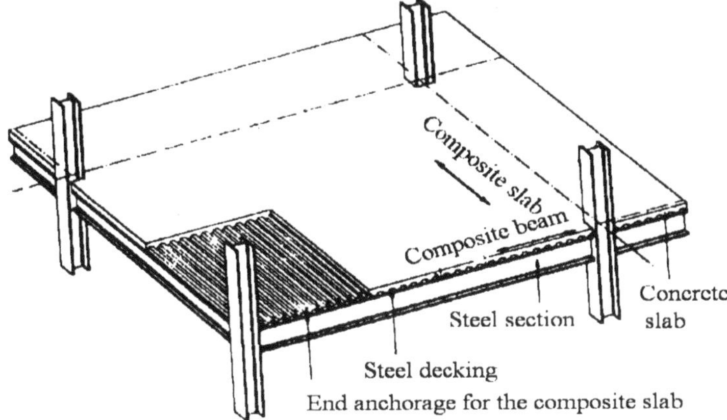

Figure 3.57: Floor system

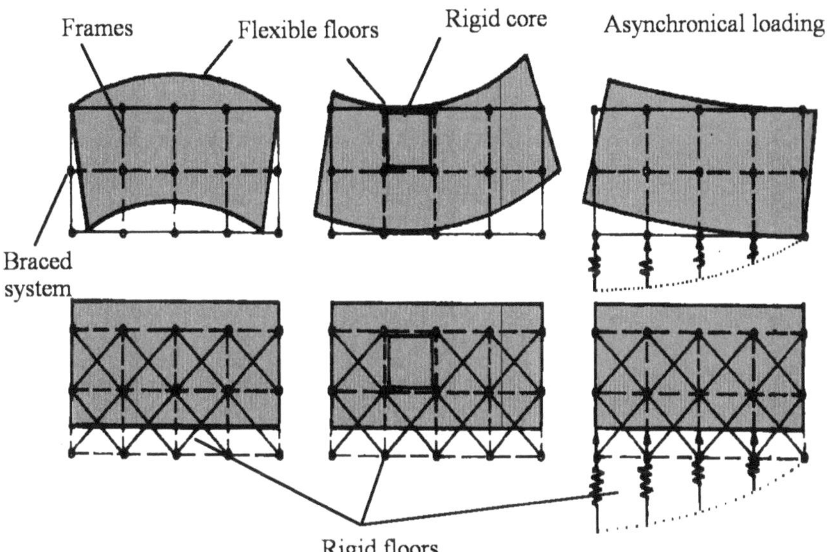

Figure 3.58: Floor horizontal diaphragm

(Fig. 3.58). The flexibility of floors has a significant effect on the distribution of seismic forces among all vertical elements. The increasing of diaphragm stiffness of floors leads to an important reduction of force concentrations in vertical elements and therefore to a significant improvement of seismic behaviour (Lopez et al, 1995);

-openings in floors, unavoidable due to some functional demands, must be limited and controlled in order to avoid the reduction of diaphragm stiffness and the break of the structure along weaker structure lines (Fig. 3.59);

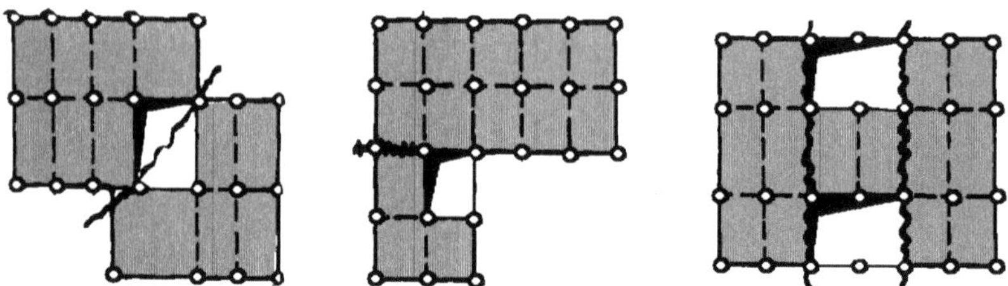

Figure 3.59: Avoided openings in floors

-floors must remain elastic during the strong earthquakes, to assure a correct force distribution among all the structure system. So, the floors must be over-strengthened in comparison with the vertical elements.

(iii) *Structure system.* The choice of the structure system plays the leading role in the structural design. From the seismic point of view, the structural typologies of seismic-resistant steel structures can be classified in two ways:

-according to the type and number of dissipate zones they are: moment resisting frames (MRF), concentrically braced frames (CBF), eccentrically braced frames (EBF) and dual frames (DF) (Mazzolani and Piluso, 1994) (Fig. 3.60a);

-according to the in plane configuration of the structure they can have: a space structure layout (SSL) or a perimetral structure layout (PSL), (Mazzolani and Piluso, 1997c),(Fig. 3.60b).

3.5 Structural Analysis Methods

3.5.1. Structure Modelling

A correct design of steel structures supposes a proper modelling of the behaviour of materials, elements, connections and structure. It is important to develop models capable of simulating the hysteretic behaviour during an earthquake. These models refer to the following aspects:

(i) *Materials.* For the uniaxial constitutive law (Fig. 3.61a), the most important aspect is related to the behaviour in the strain hardening range, because the stress state belongs to this field in the case of strong earthquakes. For the cyclic constitutive law it is very important to model the hysteretic phenomenon.

(ii) *Members.* According to the concept that the structure must dissipate the seismic energy input, some critical zones of members should be selected for the eventual development of plastic hinges (Fig. 3.61b). These zones must be appropriately detailed enabling the development of a sufficient rotation capacity, which is evaluated using the moment–rotation law for monotonic or cyclic loads;

(iii) *Joints.* The modelling of joint behaviour is performed through a moment–rotation curve. Fig. 3.61c shows some different behaviours, belonging to the range of quasi-perfectly rigid, semi-rigid or flexible (pinned).

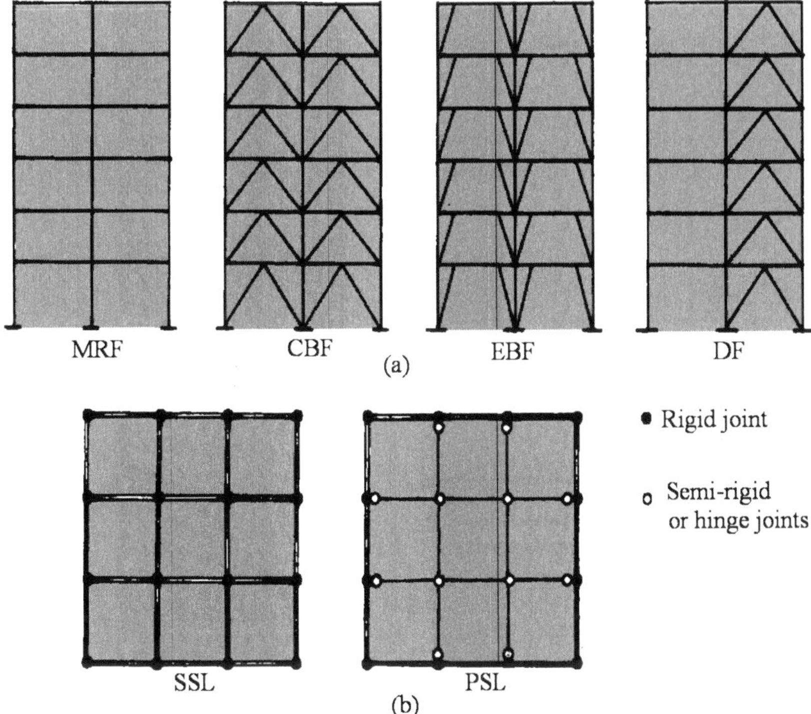

Figure 3.60: Structural systems: (a) Vertical configurations; (b) Plane configurations

(iv) *Structure.* There are three possibilities to model the structure behaviour (Fig. 3.62) by using: a system with N-degree-of-freedom (MDOF system), an equivalent system with a single-degree-of-freedom (SDOF system) or an uniform shear-beam system.

(v) *Soil-structure interaction.* The soil-structure interaction must take into account the possibility of having different horizontal and vertical motions of supports (Fig. 3.63a), modification of the natural period of structure due to interaction with the soil (Fig. 3.63b), changing of the base motions in comparison with the motions in free field (Fig. 3.63c), increasing of effective damping due to the difference between the tendency of regular structure motions and the chaotic motions of soil.

(vi) *Spatial behaviour.* The determination of internal forces in different components of the structure is generally performed using plane frames. Due to the torsion effects, caused by the static eccentricity (distance from the mass center to the center of rigidity) and the accidental eccentricity (caused by asynchronical ground motions) (Fig. 3.64), the values of forces in some parts of the structure increases with respect to the value resulting from plane modelling. Due to the inelastic behaviour, the rigidity center position can change and the effect of torsion significantly increases. So, spatial behaviour modelling must be introduced in design practice.

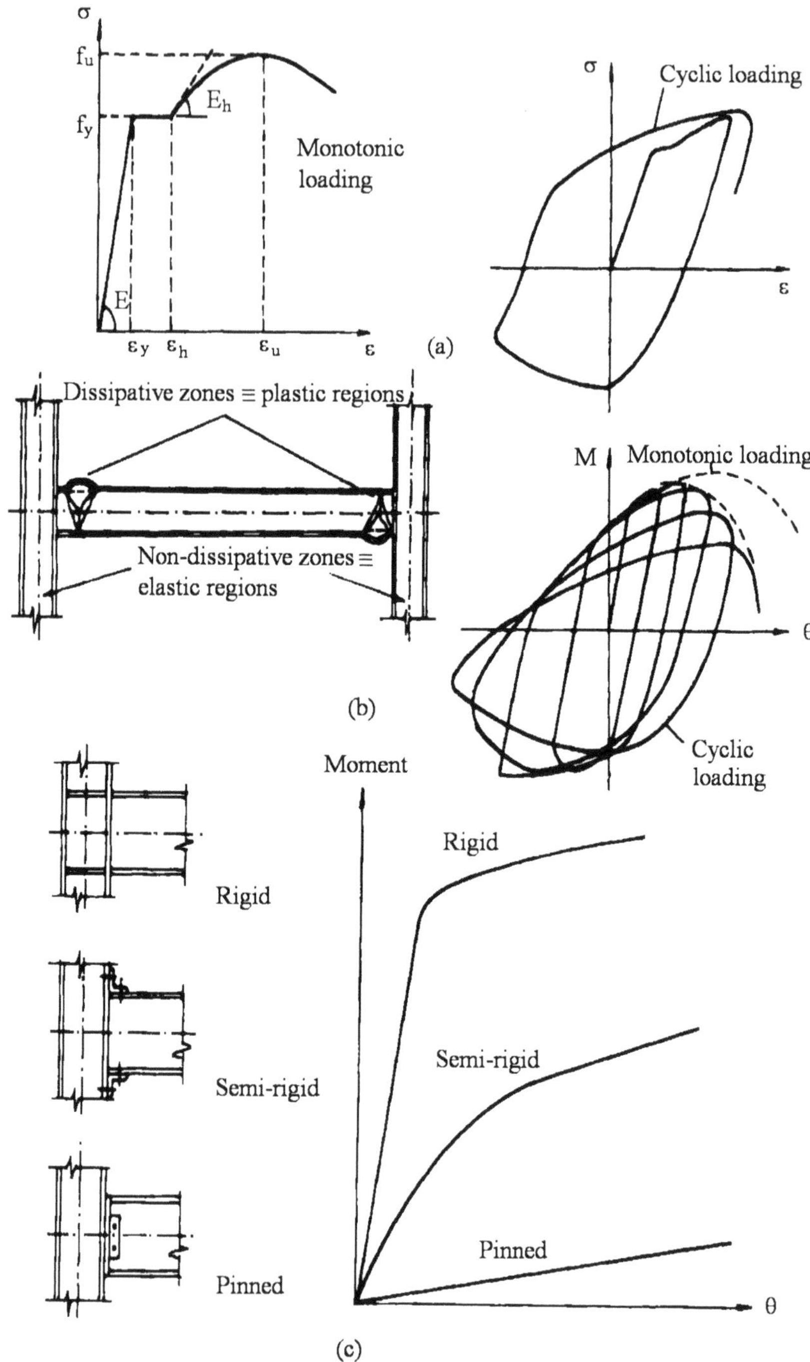

Figure 3.61: Structure modelling: (a) Material modelling; (b) Member modelling; (c) Joint modelling

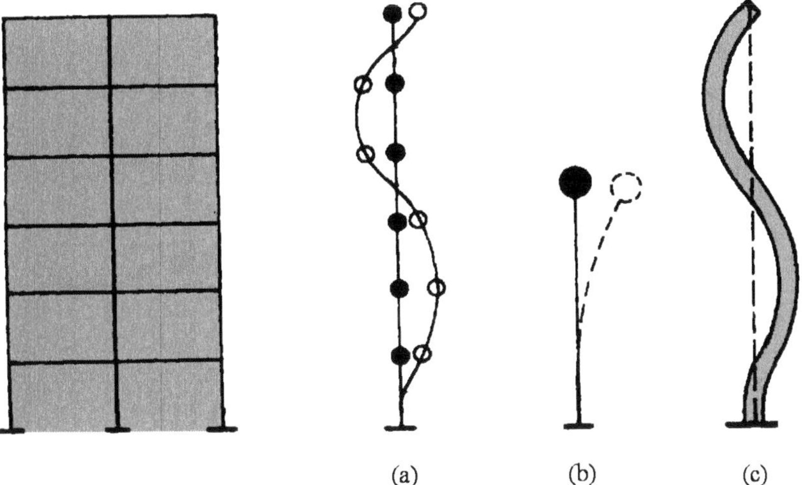

Figure 3.62: Modelling types: (a) MDOF system; (b) SDOF system; (c) Uniform shear-beam system

(vii) *Non-structural elements.* Infilled walls, as cladding panels or masonry infilled panels, which increase significantly the lateral stiffness of the building, shall be taken into account. The most usual modelling is the proposal to replace the infilled panels by an equivalent pin-jointed X bracing (Fig. 3.65) (Madan et al, 1997, Mazzolani and Piluso, 1990, 1996a, Mazzolani et al, 1996a,b,c).

3.5.2. Methods of Analysis

The stress and strain due to the seismic forces can be evaluated by means of different methods which provide different degrees of accuracy. Depending on the nature of the considered variables, the methods of analysis can be classified as deterministic or probabilistic. In deterministic analyses, specific values of the structure properties are used. In the probabilistic analyses, probability distributions of these properties are used, permitting to quantify the realibility of the results (Riddell and De la Llera, 1996). Based on the level of structure response, linear elastic or linear or nonlinear inelastic models are used (Fig. 3.66). The model may consider three components of ground motion and their spatial variations, or, as it is very common in most analyses in practice, just a single component.

(i) *Deterministic linear elastic methods* which consider the following procedures (Riddell and De la Llera, 1996):

-*Equivalent lateral force analysis* is the oldest and simplest analysis procedure and used mostly for current building design. The method is based on the assumption that the structural behaviour is governed by the fundamental period of vibration and the corresponding modal shape (Fig. 3.67a). The stress analysis is usually conservative for low to medium high buildings with a regular conformation. The distribution of statically applied hori-

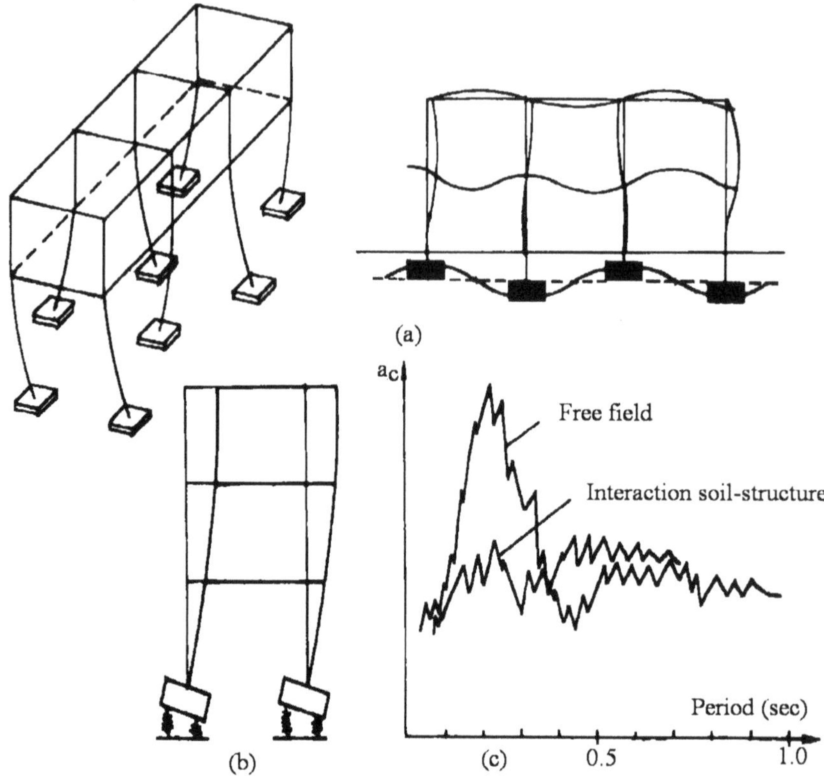

Figure 3.63: Soil-structure interaction: (a) Horizontal and vertical interaction; (b) Modelling of soil-structure interaction; (c) Influence of soil-structure interaction

zontal forces is close to the first vibration mode only, what represents a great simplification. The characteristics of ground motions are described by means of a linear elastic response spectrum. It defines the accelerations which have to be sustained by structure when it is designed in order to remain in elastic range under the severe expected earthquake. But the structures are usually designed by assuming that earthquake input energy during severe ground motions is dissipated through inelastic deformations. As a consequence, design spectra are obtained by scaling the linear design response spectra by means of a reduction factor, namely q-factor, which takes into account the energy dissipation capacity of the structure. The method has many shortcomings, so its use is limited only for simple structures as low and medium rise buildings with regular heightwise and planwise distribution of mass and rigidity and in areas with reduced seismicity.

-*Mode superposition method.* This method is widely recognized as a powerful method for calculating the dynamic linear response of elastic damped structural systems. The method is attractive, because the response of multi-degree-of-freedom systems (MDOF) is expressed as the superposition of modal response, each modal response being determined from the spectral

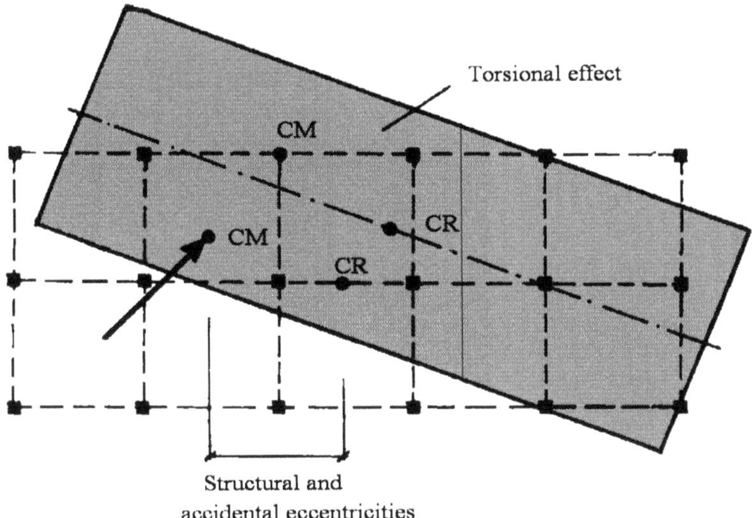

Figure 3.64: Spatial behaviour

analysis of single-degree-of-freedom systems (Chopra, 1996) (Fig. 3.67b). The advantage of this method consists on the possibility to implement only the modes which show a significant contribution to the response (Lopez and Cruz, 1996). The main problem of this method is the calculation of the frequencies for each mode, which varies during the earthquake due to the change of building stiffness (formation of plastic hinges, damage of structure and non-structural elements) and soil stiffness (softening of soil at large strain, soil-structure interaction). Modal analysis leads to the response history of the structure to a specified ground motion: however, the method is usually in conjuction with a response spectrum. In such cases, the maximum modes are computed and combined by available modal superposition rules to obtain an estimation of the expected mean-maximum response of the structure. If an elastic spectrum is used, the method is exact; however, the method is often used together with a design spectrum whose ordinates have been reduced to account for inelastic behaviour. The use of a constant q behaviour factor should be regarded as an imperfect way of representing the ratio of elastic to inelastic strength for a given ductility. Due to these aspects this method has a limit of availibility.

-*Elastic time-history analysis.* It is a linear elastic dynamic response analysis to a base excitation performed by direct integration in time of the equations of motion (Fig. 3.67c). One interesting advantage of such a procedure is that the relative signs of response quantities are preserved in the response histories, opposite to the mode superposition method in which the signs are lost. This is particularly important when interaction effects are considered in designs among stress resultants. The principal disadvantage of this elastic procedure is that it gives limited insight into the inelastic response of structure under severe earthquake loading.

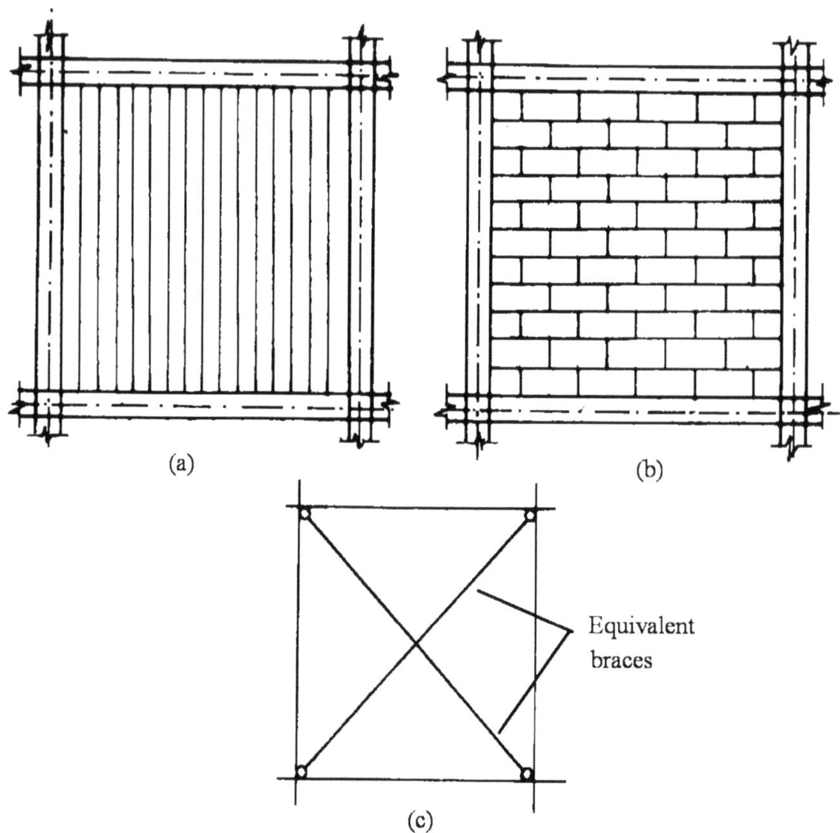

Figure 3.65: Infilled wall types: (a) Cladding panel; (b) Masonry panel; (c) Equivalent braces

(ii) *Deterministic linear or nonlinear inelastic methods* which consider the following procedures (Riddell and De la Llera, 1996):

-*Plastic analysis.* This procedure was initially developed for the analysis of steel-frame structures to take advantage of the redistribution of stresses which take place in ductile redundant structures for loads beyond the elastic limit. Plastic analysis is used well to establish a collapse mechanism in a structure subjected to a loading pattern of seismic actions (Fig. 3.68a). So, this procedure provides a framework to understand the ultimate behaviour of a structure.

-*Push-over analysis.* Following this method the structure is subjected to static incremental lateral loads, which are distributed along the height of the structure (Fig. 3.68b). A nonlinear inelastic static analysis is performed, and an inelastic load–deflection response curve is obtained by controlling the displacement at the top of the structure. The method is relatively simple to implement, and provides information on the strength, deformation and ductility of the structure and the distribution of demands. This permits to identifying the critical members likely to reach limit states during the

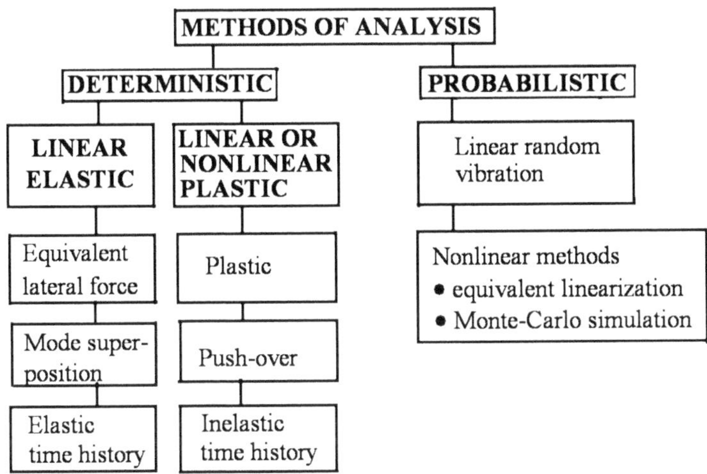

Figure 3.66: Classification of analysis methods

earthquake, for which attention should be given during the design and detailing process. But this method contains many limiting assumptions which neglect the variation of loading patterns, the influence of higher modes, the effect of resonance. In spite of these deficiencies, the push-over method can provide a reasonable estimation of the global deformation capacity, especially for structures which primarily respond according to the first mode (Krawinkler, 1995).

-Inelastic time-history analysis. This method is based on the direct numerical integration of the motion differential equations. In this aim, different algorithms can be adopted, in which the elasto-plastic deformation of the structure elements must be considered. The analysis may be performed by using real or simulated ground motions. The variation of displacements at diverse levels of a frame are presented in Fig. 3.68c. One can see that an amplification due to resonance effect, an increasing of motion duration and a tendency of regularization of movements result as far as the level increases from the bottom to the top. This method is the only able to describe the actual behaviour of a structure during an earthquake. But the great problem of this method is the choice of a proper accelerogram record, knowing the great variability in function of soil and the source distance. Therefore, it is essential to use more record types or to use an artificial accelerogram, in which all the principal characteristics are contained. A complete three-dimensional analysis is possible today, although important processing time and data storage are required. Since these hardware limitations reduce every day, this procedure, which now is mostly for academic use, should soon become customary in engineering practice. Also, the great difficulty for nonlinear analysis is the use of mathematical models for beam, beam-column, joint and brace behaviour and the modelling of the hysteretic load-deformation behaviour of these elements. Limit or collapse states need

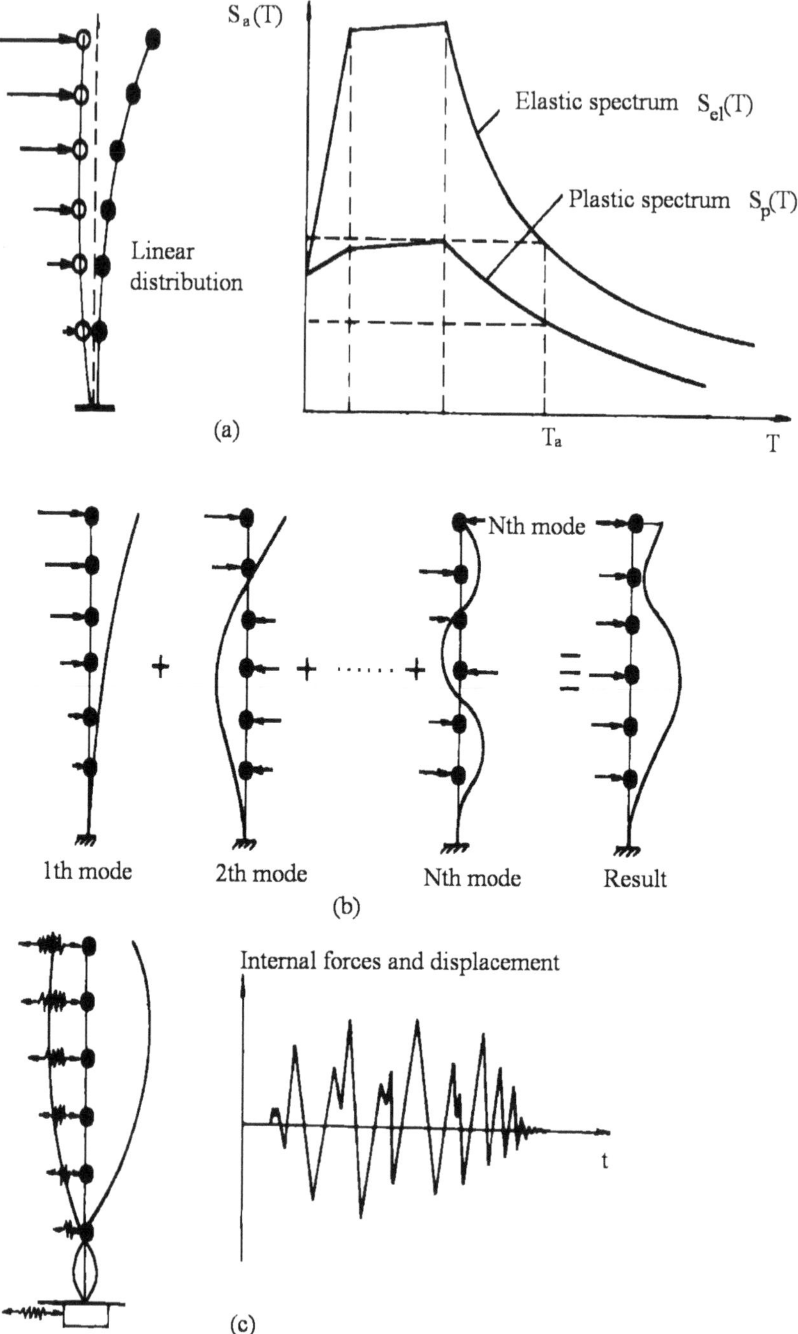

Figure 3.67: Linear elastic analysis methods: (a) Equivalent lateral forces; (b) Mode superposition; (c) Time history

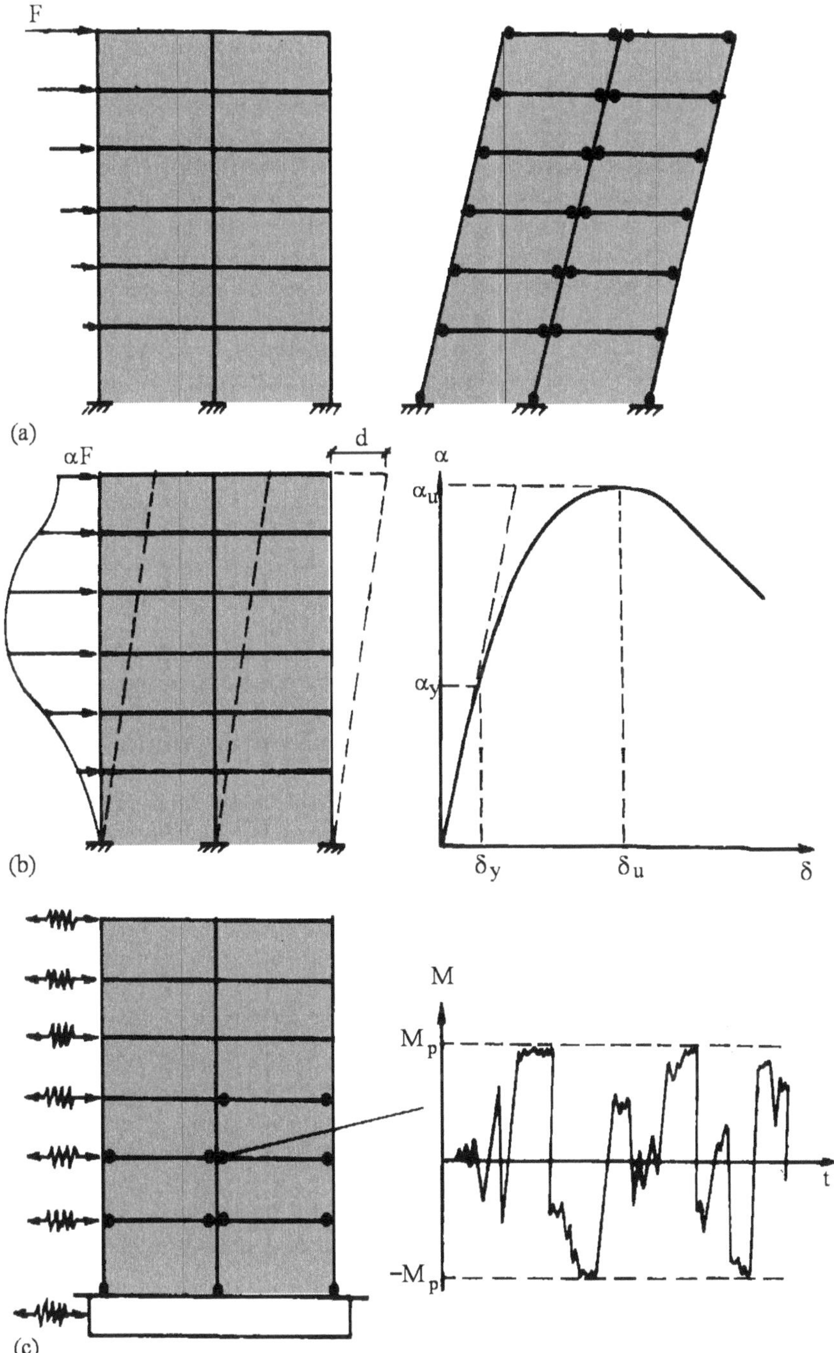

Figure 3.68: Inelastic analysis methods: (a) Plastic; (b) Pushover; (c) Time history

to be identified and defined including fracture of welds and local plastic buckling. Furthermore, a realistic nonlinear structural model should include the foundations and surrounding soil.

(iii) *Probabilistic methods.* Probabilistic methods in earthquake analysis and design of structures provide an extra dimension in the interpretation and solution of the problem (Riddell and De la Llera, 1996). For computed internal forces, stresses, deformations or any other response characteristics a probability is associated to computed values. This becomes essential in earthquake problems where important uncertainties exist in the input of ground motion and properties of the soil, materials and structures. The used methods are: linear random vibration analysis, and nonlinear methods (Riddell and De la Llera, 1996). Among these methods, a very powerful procedure is the Monte-Carlo simulation. The main advantage of this method is its similarity to the procedures used in deterministic analyses. Indeed, Monte-Carlo simulation is just a repetition or collection of deterministic analyses, each of them for a realization of the random input process. Its main drawback is the higher computational cost since many analyses have to be repeated in order to obtain a good statistic representation.

However, in spite of the great theoretical and practical value of these methods, their use today is limited in practice. Several reasons might explain this situation (Riddell and De la Llera, 1996):

-probabilistic procedures are more complex than the deterministic ones;

-existing codes do not establish a definite probability of exceedance of engineer designing;

-the computational tools for stochastic analysis in practice are more scarce than those for deterministic analysis.

Adding to these reasons, the fact that there are also many unclear solutions even for deterministic problems, one can consider that the probabilistic methods are a very promising tool for the future design methodologies.

3.6 Design Methods

3.6.1. Conceptual Methodology for Design

A coherent conceptual methodology was elaborated by Bertero (1994), which takes into account the simultaneous demands for rigidity, strength and ductility from the early beginning of the design procedure. The proposed methodology leads to a rational and transparent design procedure, which is divided into two main phases (Fig. 3.69). The first phase is the preliminary design in which the main characteristics of a structure are established. The second phase is devoted to the final design of a structure, in which the design delivers details for the construction phase.

3.6.2. Preliminary Design

The first phase covers the acquisition and processing of data needed for design (Fig. 3.70): soil profile and topography of site, information about

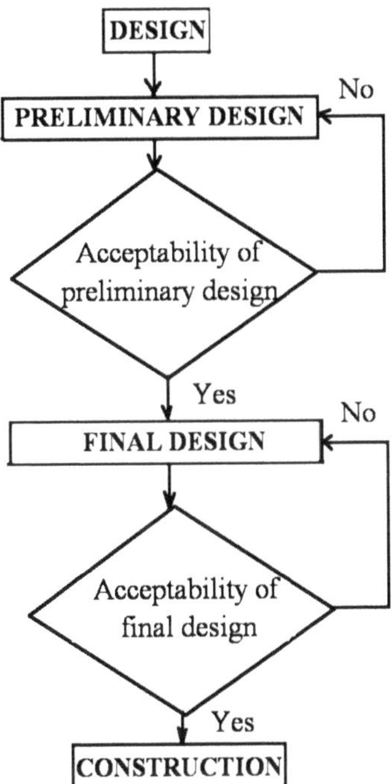

Figure 3.69: Flow chart for structural design

possible sources, return periods for different levels of ground motion, characteristic periods, durations, etc. In the second phase of preliminary design the data is translated into design values as accelerograms, spectra, etc. In the third phase the general configuration of building, structural system, structural material, non-structural elements and contents are established. The fourth phase consists in the preliminary sizing of structure members considering all performance demands. The preliminary design is an iterative process, additional modifications being necessary from the beginning of design in order to arrive at a solution that is as close as possible to the desired final design.

3.6.3. Final Design

Building analysis usually reveals that preliminary design is satisfactory and only minor changes in the member sizes are necessary in the final design. For this design phase, the most frequently used procedures are summarized next (Riddell and De la Llera, 1996):

-*Lateral strength design.* This is the most common seismic design approach used today. It is based on providing the structure with the minimum lateral strength to resist seismic loads, assuming that the structure will behave adequately in the nonlinear range. For this reason only some simple

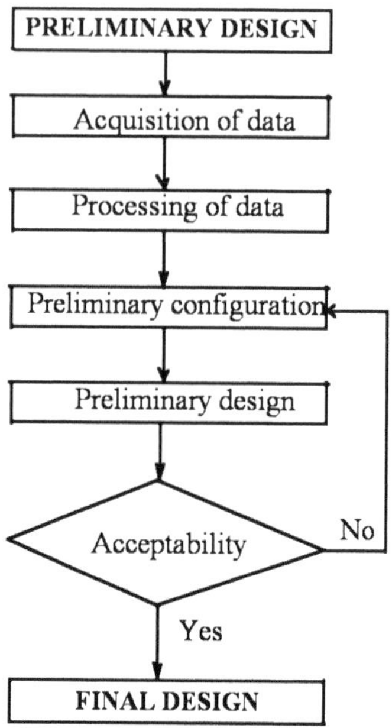

Figure 3.70: Flow chart for preliminary design

constructional detail rules are provided to be satisfied, as material ductility, member slenderness, cross-sectional classes, etc.

-*Displacement based design.* Recognizing that damage in structures subjected to earthquakes is the result of excessive deformations, this method operates directly with deformation quantities and therefore gives better insight on expected performance of structures, rather than simply providing strength in the above lateral strength design approach.

-*Capacity design.* Based on the current seismic code principles that accept severe excursions into inelastic range, the method seeks protection of resisting structures against collapse by providing adequate ductility capacity to critical plastic hinge locations. By this approach a collapse mechanism, for a given code type lateral force distribution, is chosen in order to ensure that plastic deformations will occur only in the selected zones. Such zones are then designed and detailed so that ductility demands are met, while all other structural parts are provided with sufficient overstrength as to remain elastic; thus, plastic deformations are confined to the predetermined locations.

-*Energy based design.* Energy concepts in seismic design have been recently emphasized as a mean toward better appraisal of earthquake demands on structures (Akiyama, 2000). The basic equation of interest is the energy balance between total input energy and the energies dissipated by viscous damping and inelastic deformations.

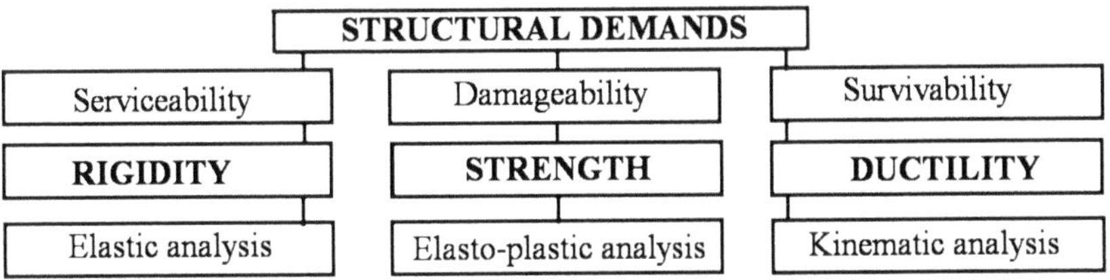

Figure 3.71: Design criteria for three performance levels

3.7 Design Criteria for Three Performance Levels

3.7.1. Rigidity, Strength and Ductility Triade

In the conventional practice for non-seismic loads, structures are designed only for two demands, strength and rigidity, which correspond to a good structural performance. The strength checking, including stability, is related to the ultimate limit state, assuring that the force level developed in structures remain in the elastic field, or some limited plastic deformations can occur in agreement with the design assumption. The rigidity checking is, generally, related to the serviceability limit state, for which the structure displacements must remain in some limits which assure that no damage occurs in non-structural elements. Although it is recognized that damage is also due to deformations, strength checking plays the leading role in designs for conventional loads.

Contrary to this, in case of earthquake design, a new demand must be added to the two above mentioned ones, that is the ductility demand. The survivability of a structure under strong seismic actions relies on the capacity to deform beyond the elastic range, and to dissipate seismic energy through plastic deformations. So, the ductility check is related to the control of whether the structure is able to dissipate the given quantity of seismic energy considered in structural analysis or not.

The new approach in the structural design under seismic loads, known as *capacity design conception*, requires the verifications of three demands: rigidity, strength and ductility (Fig. 3.71). The rigidity is checked for serviceability limit state, strength for the damageability limit state and ductility for the survivability limit state. It is very important to notice that for each limit state different accelerograms or spectra must be used due to the fact that the ground motion characteristics change by increasing the magnitude or intensity. This is due to the differences in the seismic wave site soil interaction. The main feature of these ground motions are (Fig. 3.72):

-for serviceability limit states, the low earthquake magnitude has short periods and duration, but the amplification for short periods is different compared with strong earthquakes;

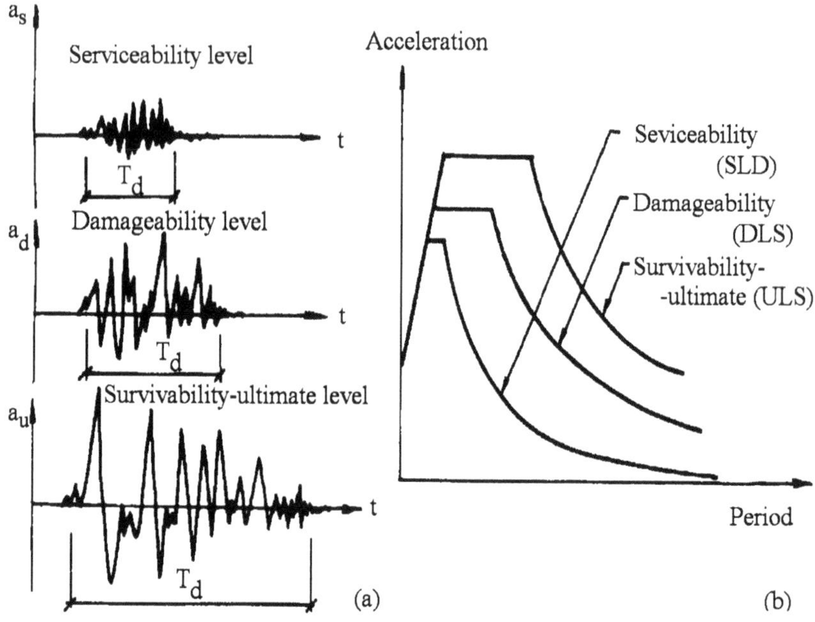

Figure 3.72: Ground motion characteristics for the three performance levels: (a) Accelerograms; (b) Elastic spectra

-for damageability limit states, the earthquakes have a severe magnitude and the periods and durations are longer compared with the serviceability limit state;

-for survivability limit states under very severe earthquakes (the larger one expected on the site), the characteristic periods (corner periods) and duration are long due to the interaction with the site soil.

So the ratio between the magnitudes corresponding to the different performance levels depend on the structure's natural periods. To use the same pattern of spectra for all these limit states, changing the peak accelerations only, as it has been suggested in the code provisions, is a mistake in the design methodology.

As the feature of these demands is rather different from the conventional checking for non-seismic loads, the main problems of these verifications are briefly presented in the following.

3.7.2. Rigidity Design Criterion

Traditionally, strength checking is considered by designer as the primary goal of design process. In the last time, due to social and economic impact of the loss of functionality in buildings, there is a particular focus on the control of damage through the rigidity checking. This operation must be done for the performance levels, mentioned in Section 3.1.

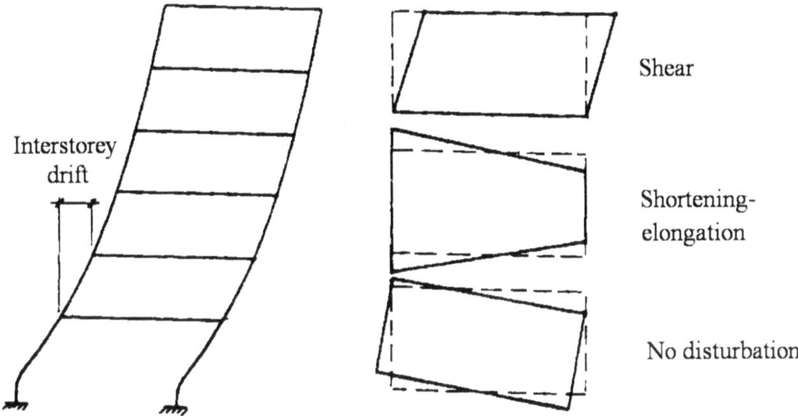

Figure 3.73: Interstorey drift types

The rigidity control of a structural system is important for the following reasons (Bertero, 1992):

-to maintain architectural integrity for moderate earthquakes and to avoid dangerous damage of non-structural elements during severe earthquakes;

-to limit structural damage and the increased instability phenomena, due to second order effects;

-to avoid human discomfort under minor or even occasional moderate earthquakes.

For the structural damage, the displacement of the top of a structure can provide a good indication, but it cannot adequately reflect the damage of non-structural elements, which are more dependent on the relative displacement between two storeys, the so-called interstorey drift (Fig. 3.73). This is the result of a shear racking or shortening-elongation deformation, but there are some cases for which the drift occurs with no distorsion, for significant foundation rocking (see Pisa Tower). Because only shear deformations produce important damage in non-structural elements (Fig. 3.74), these must be separated from the deformations due to shortening-elongation. The tangential storey drift index is (Mayes, 1995):

$$R_T = \frac{1}{H}(u_3 - u_1) + \frac{1}{L}(v_3 + v_4 - v_1 - v_2) \tag{3.13}$$

in which L is the bay width and H is the storey height. The first term on the right-hand side of above equation is the conventional storey drift ratio and the second is the correction applied to each bay accounting for the slope of floors above and below the storey.

In order to protect the non-structural elements, the rigidity design criterion is laid out through a required-available formulation:

$$\text{Required rigidity} \leq \text{Available rigidity} \tag{3.14a}$$

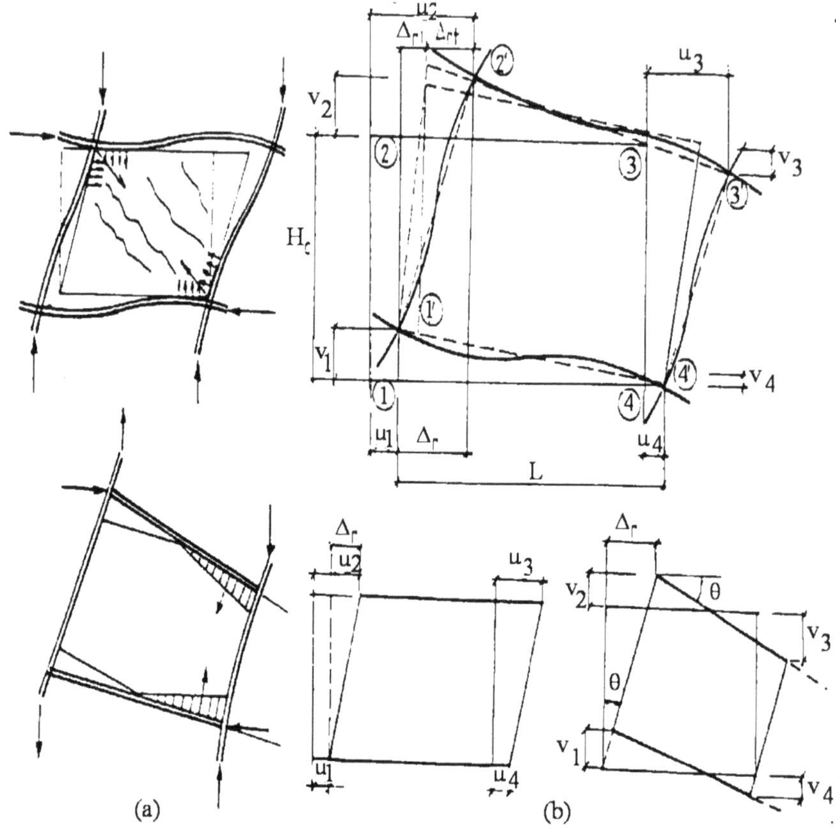

Figure 3.74: Tangential interstorey drift: (a) Damage of non-structural elements; (b) Geometrical definition

The relationship may be used also in the form of:

$$\text{Determined interstorey drift} \leq \text{Interstorey drift limit} \qquad (3.14b)$$

The required rigidity may be determined in one of two ways:

-direct determination of interstorey drift from an elastic structure analysis using a time-history record or an elastic spectrum for the earthquake corresponding to serviceability limit states;

-indirect determination of interstorey drift from the spectra corresponding to ultimate state, using a reducing factor. This methodology, currently used in codes, is very simple, but has the shortcomings mentioned in Section 3.7.1.

Due to the fact that in the serviceability limit state the steel frames work with the undamaged infilled walls, the modelling of the frame must consider an equivalent diagonal strut (Saneinejad and Hobbs, 1995, Mazzolani et al, 1996a,b,c, 1997a,b) (Fig. 3.75) with the characteristics determined in function on geometrical and mechanical properties of infilled walls.

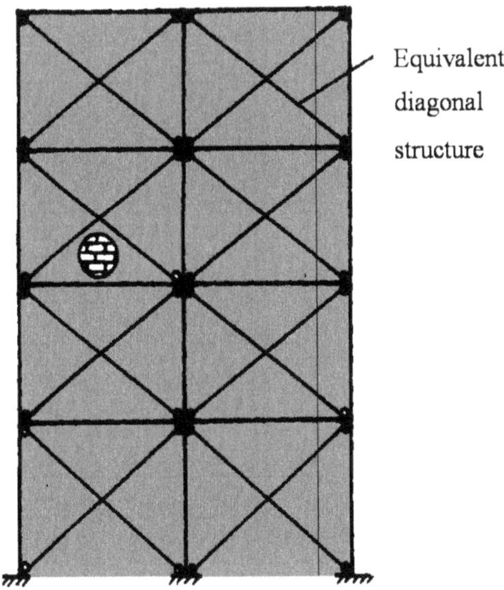

Figure 3.75: Structure modelling for infilled walls

Table 3.6: Rigidity design criterion

Rigidity	
Required	**Available**
• Earthquake type • Structure type • Elastic sprectrum	• Deformation limits of non-structural elements • Isolation of non-structural elements

Due to the reduced level of seismic loads and the protection demands for structure and non-structural elements, the analysis of structure must be completed in the linear elastic range.

The factors influencing the rigidity design criterion are presented in Table 3.6.

(i) *Required rigidity* depends on:

-*Earthquake type*. The interstorey drifts depend on the ground motion type. The far-field regions are characterized by structure deformations for which the first vibration mode is dominant, so the drifts have the maximum values in the lower part of the structure(Fig. 3.76a). The structure deformations for ground motions in the near-field regions are qualitatively quite different. In this case, the second and third modes are dominant, so the maximum of interstorey drifts occur at the top of structures (Fig. 3.76b). At the same time and consequently, in the first case the plastic hinges take

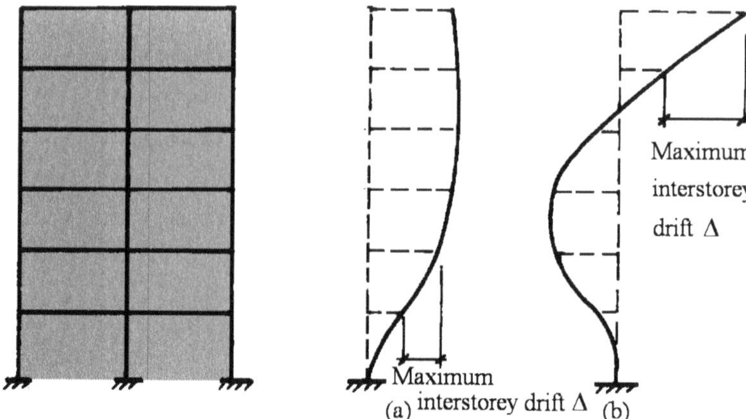

Figure 3.76: Influence of vibration modes on the interstorey drift: (a) First vibration mode; (b) Superior vibration modes

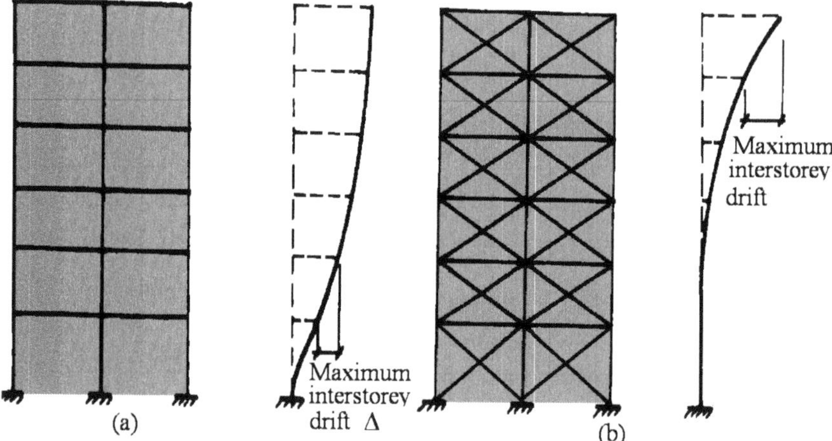

Figure 3.77: Dominant interstorey drifts: (a) MR frames; (b) CB frames

place in lower parts, while in the second case they are concentrated in the top of the structure. The presence of these plastic hinges produces an additional increasing of interstorey drifts in these zones.

-*Structure type.* The lateral displacements of multi-storey buildings due to lateral loading can be regarded as being caused by two deformation types: flexural and shear-racking deformations. Generally, both displacement types occur in structures, but the shear-racking deformations tend to be predominant in moment resisting frames, while the flexural deformations are usually dominant for braced frames (Fig. 3.77). The maximum and the position of interstorey drift depend on the displacement type. The shortening-elongation deformations can cause significant shear racking in

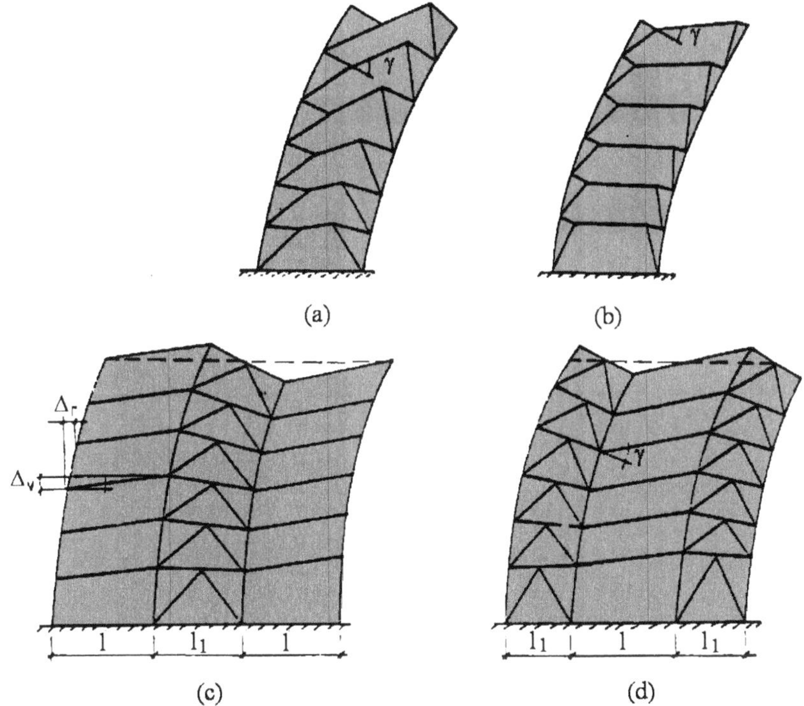

Figure 3.78: Displacement in braced frames: (a,b) Eccentric braced frames; (c,d) Dual frames

the case of eccentric braced frames or dual frames (Fig.3.78), where the damage of non-structural elements is produced by vertical displacements, resulted from deformations of bracing systems.

(ii) *Available rigidity* depends on:

-*Deformation limit of non-structural elements.* The damage of non-structural elements, when traditional materials are used, may be defined according to the following categories (Fig. 3.79):

-minor damages with cracks of 0.2–1 mm that do not significantly affect the element serviceability, and the repair is very easy to be performed;

-important damages with cracks of 1–2 mm, but they do not affect the safety of the elements and the cost of repair is important but supportable;

-very important damages with cracks larger than 2 mm which produces partial or total collapse of elements, so they must be replaced with new ones.

Fig. 3.79 provides some guidelines in unserviceability which occur at different interstorey drift levels (Galambos and Ellingwood, 1986). Precise serviceability limits depend on many factors which are very difficult to define in the design practice. Information concerning the values of available limits for different non-structural elements are given by Mayes (1995) and Nair (1995). Unfortunately, the stage of knowledge on the deformation limits

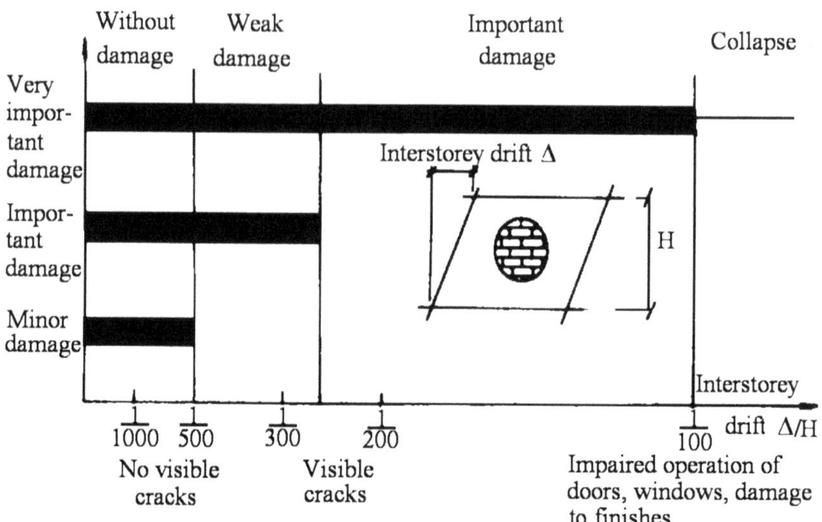

Figure 3.79: Interstorey drift limits for different limit states

of non-structural elements to racking distorsion does not appear to have advanced in the recent years. One of the most important criteria which, in many cases, can determine the structure sizing, is based on the obsolete information.

-*Isolation of non-structural elements from building structures.* In the most modern buildings nowadays an attempt is made to isolate the exterior cladding from the structure due to seismic loading. The interior walls are detailed in such a way as to isolate them from structure deformations. If this is done, deformation limits of the non-structural elements would not be mandatory in establishing the lateral rigidity criterion for structure and only the building contents protection remains a criterion for interstorey drift limitation. In some special cases, the cladding connectors are designed to dissipate energy when the structure is subjected to a strong earthquake (Pinelli et al, 1992).

3.7.3. Strength Design Criterion

As a result of the currently used seismic design philosophy for buildings that accept structural damages in the event of severe earthquake ground motions, structures are likely to experience significant inelastic excursions. Thus, design lateral forces are much lower than those required to maintain structures in the elastic range. The strength criterion is traditionally emphasized as a primary goal of seismic design. So, the structure strength must satisfy the equation:

$$\text{Required strength} \leq \text{Available strength} \tag{3.15}$$

Table 3.7: Strength design criterion

Strength	
Required	**Available**
• Earthquake type • Structure type • Design spectrum	• Randomness of yield stress • Increase of cross-section dimensions • Hardering effects • Moment redistribution • Effect of strain rate • Cumulative plastic deformation

The two terms of this equation are determined from the seismic action and structure configuration, respectively. The factors influencing the strength design criterion are presented in Table 3.7.

(i) The *required strength* may be determined using one of the analyses methods presented in Section 3.5.2 depending on:

-*Earthquake types*, refer to the intensity, periods, durations, etc. The differences between near-field and far-field earthquake types must be considered. So the influence of short duration pulse-ground motions produced with great velocity versus long duration with important number of cyclical plastic excursions must be analysed in function of the structure position related to expected source position;

-*Structure types*, consider the differences in behaviour of moment resisting frames, concentrically and eccentrically braced frames, dual frames. These differences consist in predominance of bending moment or axial forces on the structure behaviour in the plastic range.

-*Design spectrum*, is derived from the elastic acceleration spectrum, using the reduction factor q (called also behaviour factor) determined in function of structure type, overstrength and natural period of structure.

The elastic spectrum is established for a single-degree-of-freedom system (SDOF), corresponding to the elastic behaviour of this model (Fig. 3.80a). From this elastic spectrum results the elastic strength F_e (Fig. 3.80b). A structure which relies on energy dissipation through inelastic deformations should have an *actual strength* F_p:

$$F_p = \frac{F_e}{q_\mu} \tag{3.16}$$

where q_μ is the strength reduction factor corresponding to the structure ductility. So an inelastic spectrum can be obtained by reducing the elastic one with the reduction factor q_μ. The design strength F_d is, as a rule, lower than the actual strength, mainly due to the overstrength, which is an inherent property of a well designed, detailed and constructed structure.

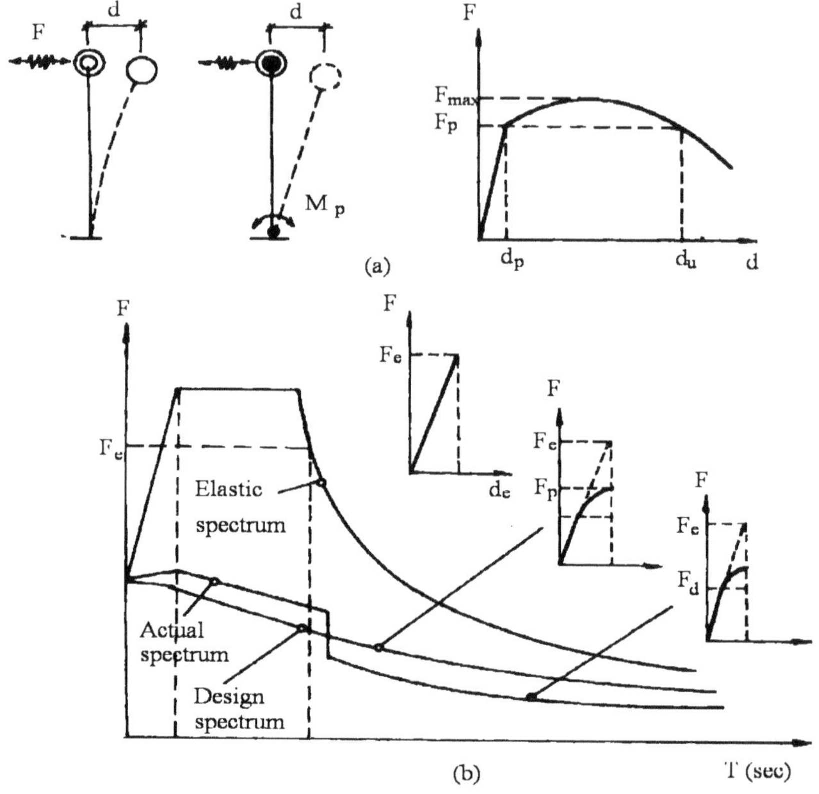

Figure 3.80: Inelastic spectrum: (a) SDOF elasto-plastic system; (b) Elastic, actual and design spectra

Results of the *design strength*:

$$F_d = \frac{F_e}{q} \tag{3.17}$$

where q is the design strength reduction factor, which considers also the overstrength. Introducing the overstrength factor q_s:

$$q_s = \frac{F_p}{F_d} \tag{3.18}$$

it results:

$$q = q_\mu \cdot q_s \tag{3.19}$$

Thus the total strength reduction factor q

$$q = \frac{F_e}{F_d} \tag{3.20}$$

is the product of the ductility factor q_μ and the overstrength factor q_s.

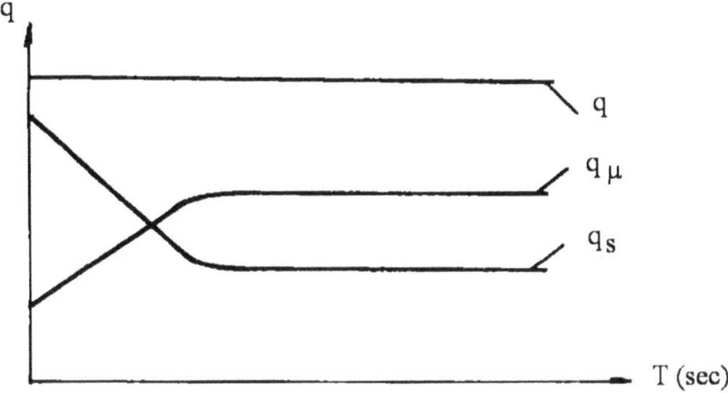

Figure 3.81: Reduction factors q, q_{μ} and q_s

The *ductility factor* q_{μ} depends on the structure type and the local ductility of the cross-sections and members, level and type of loadings, etc. The *overstrength factor* may come from the internal force redistribution, higher material strength than those specified in the design, strain hardening, member oversize, strain rate, multiple loading combinations, effect of non-structural elements, etc. Having in mind that the ductility factor q_{μ} decreases in the short period region, while the overstrength factor increases in this region, from the point of view of the structural design practice, it appears to be a reasonable approximation to use a constant period independent behaviour factor q (Fajfar, 1995) (Fig. 3.81).

(ii) The *available strength* depends on the characteristics of selected dissipative zones which must develope plastic hinges. If the strength of these zones is adequately provided, brittle fractures or other undesirable failure modes are prevented. The strength of these dissipative zones depends on the following factors:

-*randomness of the yield stress.* In the resistance to problems, the lower bound values of yield stress are usually considered. In reality the actual material strength is greater that the one specified in the codes. This difference can produce an increase of section strength;

-*increase of cross-section dimensions.* This factor refers to the use of standard shapes which only in very rare cases correspond to the design required section. Generally, the designer uses a section greater than the one resulted from the analysis. As a consequence, an overstrength is obtained;

-*hardening effect.* Due to the gradient bending moments, the plastic hinges work in the hardening range, producing an increasing of plastic moment;

-*moment redistribution.* Due to the success of plastic hinges occurring in statically indeterminate structures, the design moment can reduce or increase;

-*cumulative plastic deformations.* The structure strength depends on the number of inelastic cycles, sequences and relative amplitudes.

It is clear that a combination of maximum values of these effects is not realistic, but a possible increase in strength of 50–150 per cent can be expected. So, the verification of the strength of a plastic hinge must consider the existence of an important overstrength.

The steps in the determining the available strength are as follows:

-*establishing the suitable plastic mechanism.* The concept of a desirable member hierarchy in the dissipating mechanism is universally recognized as a procedure to assure that plastic hinges develop in beams rather than in columns (Fig. 3.82a). A new design method has been proposed by Mazzolani and Piluso (1996, 1997), by controlling the failure mechanism types (Fig. 3.82b). The most suitable plastic mechanism is the beam–hinge model and obtaining this mechanism type must be the main goal of the strength design;

-*identifying the critical sections* where the plastic deformations have the task to dissipate the seismic energy input. These zones may be the ends of members or the joints (Fig.3.82c);

-*sizing and detailing the plastic regions* in order to provide the critical sections to have the capacity to dissipate the seismic energy;

-*sizing and detailing the elastic regions* in order to assure that no accidental plastic deformations can occur in some regions which are not designed for this possibility (Fig. 3.82d).

3.7.4. Ductility Design Criterion

The ductility is the capacity of structure to undergo high plastic deformations in some predefined locations, ensuring a protection against structure collapse during severe earthquakes. Recent development of advanced design and concepts, as the ones included in the capacity design method (Bachmann et al, 1995), is based on the objective to provide a structure with sufficient ductility, in the same way as for rigidity and strength. For this, a comprehensive and transparent methodology for direct ductility control must be developed. This purpose may be achieved if it has satisfied the ductility criterion (Bertero, 1988, Mazzolani and Piluso, 1996a, Gioncu, 1997, 1998):

$$\text{Required ductility} \leq \text{Available ductility} \tag{3.21}$$

where the required ductility is determined from the global behaviour of structure and the available ductility from the local elastic deformations. So, the ductility design criterion may be transformed in the required–available relation:

$$\text{Required local ductility (resulted from global analysis)} \leq$$
$$\leq \text{Available local ductility} \tag{3.22}$$

These two ductilities are defined as:

-*global ductility*, determined at the level of the whole structure (Fig. 3.83):

$$\mu_d = \frac{\delta_u}{\delta_y} \tag{3.23}$$

where δ_u and δ_y are the elasto-plastic and yielding top sway displacements, respectively;

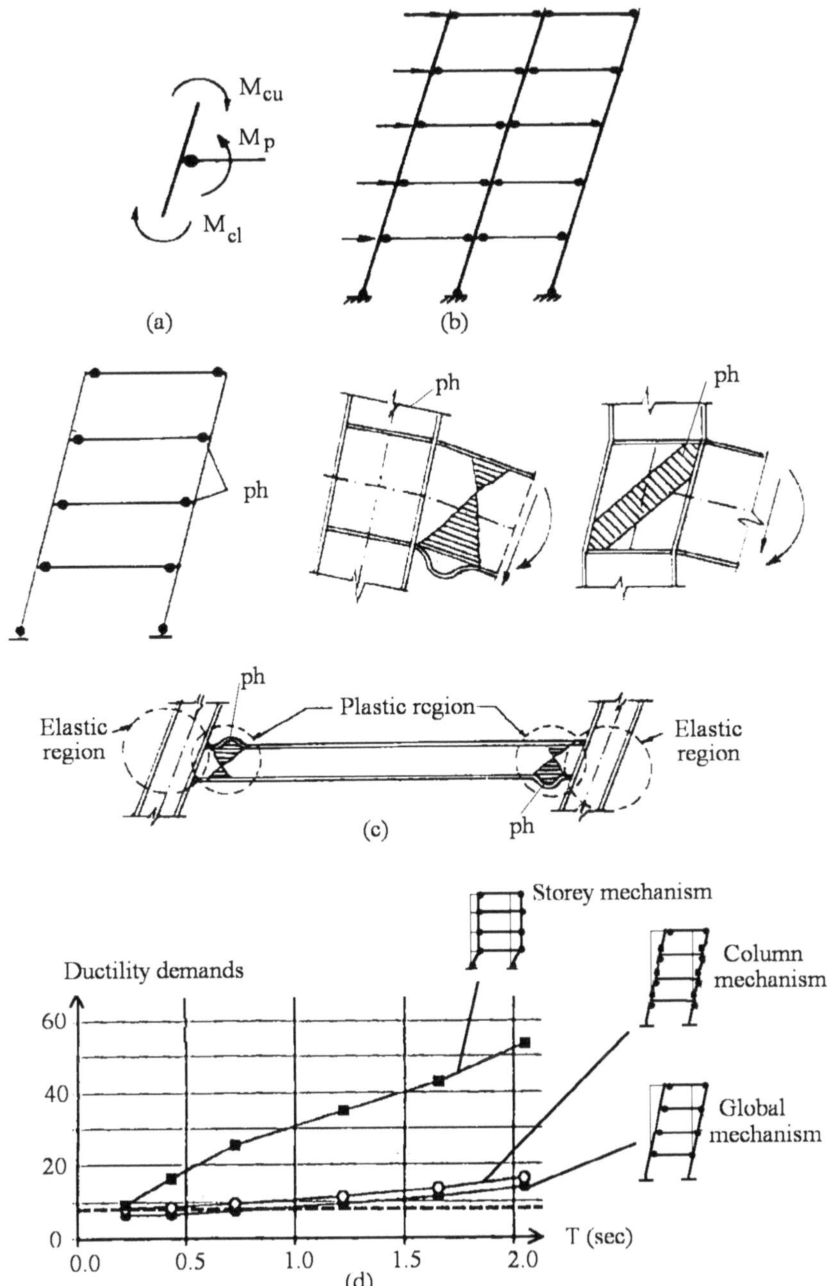

Figure 3.82: Steps in determining the available strength: (a) Member hierarchy; (b) Plastic mechanism; (c) Identifying the critical sections; (d) Influence of plastic mechanism type

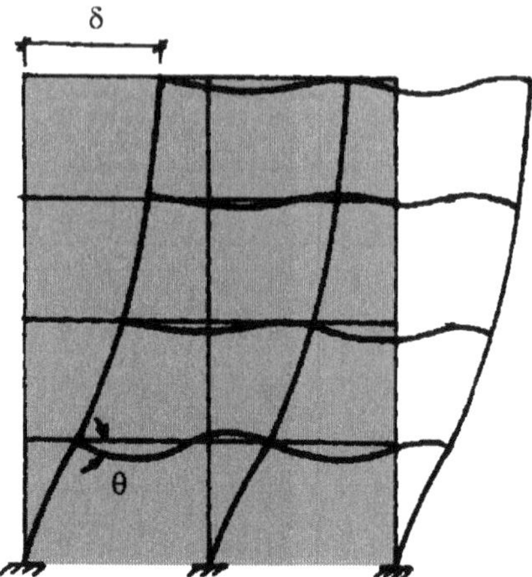

Figure 3.83: Ductility characteristics

-*local ductility* determined at the level of structure member:

$$\mu_r = \frac{\theta_p}{\theta_y} \tag{3.24}$$

where θ_p, and θ_y are the plastic and yielding rotation of a plastic hinge, respectively.

The relation between these two ductilities is (Cosenza, 1987):
-for multi-storey buildings:

$$\mu_d = 1 + \frac{2}{3}\mu_r - 2\left(\frac{\alpha_u}{\alpha_y} - 1\right) \tag{3.25}$$

-for single-storey buildings:

$$\mu_d = 1 + \mu_r \tag{3.26}$$

In relation (3.25) α_u and α_y are the multiplier of horizontal forces for ultimate and first yielding limit states, respectively.

The main factors influencing the ductility criterion are presented in Table 3.8.

(i) *Required ductility.* Factors influencing the required ductility are presented in detail in the Table 3.9 (Gioncu, 1997, 1998).

For global ductility the hierarchy is at the level of source, site, foundation and structure. For assessment of global ductility it is necessary to gather information on characteristics of possible earthquakes. With this purpose we must underline the importance of engineering judgement from a practical point of view and the collaboration of seismologists, geotechnists and structural engineers for establishing the most characteristic design earthquakes for the given site.

Table 3.8: Ductility design criterion

Ductility	
Required	**Available**
• Maximum possible earthquake • Foundation • Structure type • Mechanism type	• Material • Cross-section • Member • Joint • Connections

Table 3.9: Factors influencing global ductility

PREDICTED GLOBAL DUCTILITY	
MAXIMUM GROUND MOTIONS	**STRUCTURE RESPONSE**
Source • Earthquake type • Focal depth	Foundations • Foundation type • Base isolation
Distance from source • Near and far-field • Attenuation	Structure system • Structure type • Collapse mechanism
Site • Soil profile • Amplification • Duration	Non-structural elements • Interaction • Damage limits • Collapse limits
REQUIRED DUCTILITY	

The estimation of required global ductility is based on the ultimate top sway displacement of structure. Unfortunately, there are no standard definitions, accepted by all the specialists, with respect to required ductility. An approximate estimation is given by the relation between the global ductility μ_d and the ductility factor q_μ. Several proposals for the determination of ductility factor q_μ have been made. The first belongs to Newmark and Hall (1982), being related to the period T_0, which is the characteristic period of the elastic spectrum, representing the limit period between short periods with a constant acceleration and medium periods with a constant velocity (Fig. 3.84). A period $T_0 = 0.5$ sec was proposed by Newmark and Hall. The relations between ductility factor and global ductility are:

-for $T > T_0$, also named the displacement amplification region, for which the elastic and inelastic peak displacements are equal, q_μ factor only slightly depends on the period T:

$$q_\mu = \mu_d \qquad (3.27a)$$

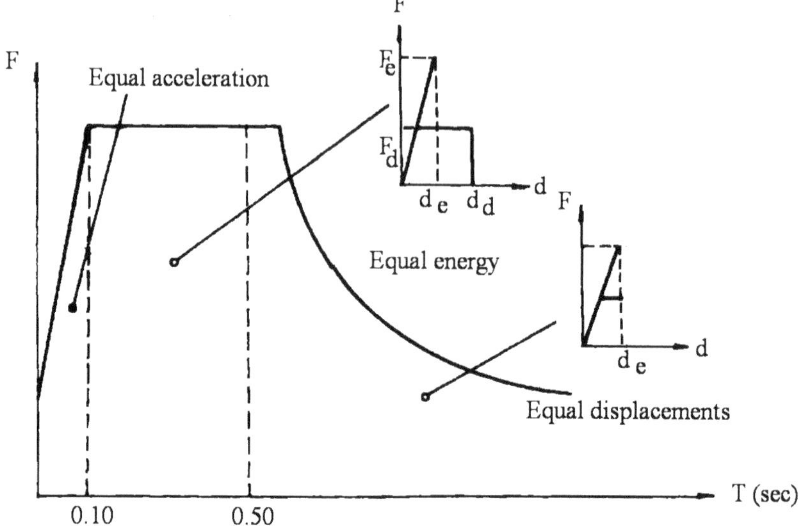

Figure 3.84: Influence of period on the determination of q_μ-factor

-for $0.1 < T < T_0$, also named the acceleration amplification region in which the maximum elastic and elasto-plastic energies are equal, it results that ductility factor strongly depends on global ductility:

$$q_\mu = (2\mu_d - 1)^{1/2} \tag{3.27b}$$

-for $T < 0.1$ it results that accelerations are equal:

$$q_\mu = 1 \tag{3.27c}$$

Because equation (3.27a) always gives larger values than (3.27b) and q_μ resulting from this equation is greater than 1, in the $q_\mu - T$ relationship there are some discontinuities (Fig. 3.85).

Another proposal for determining q_μ in function of T is the one of Vidic and Fajfar (1995):

-for $T > T_0$

$$q_\mu = \mu_d \tag{3.28a}$$

-for $T < T_0$

$$q_\mu = (\mu_d - 1)\frac{T}{T_0} + 1 \tag{3.28b}$$

where

$$T_0 = 0.65\mu^{0.3}T_c \tag{3.29a}$$

$$T_c = 2\pi\frac{c_v v_g}{c_a a_g} \tag{3.29b}$$

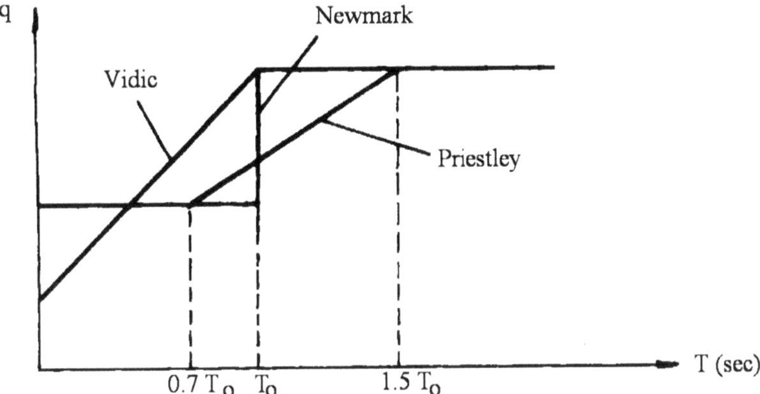

Figure 3.85: Various proposals for $q_\mu - T$ relationships

In these equations c_v and c_a are amplification factors for velocity and acceleration, respectively, obtained from statistical studies (c_v = 1.44-2.60, c_a = 1.98-2.70). The ratio v_g/a_g is a very important characteristic of ground motion. There are the following values (Fajfar, 1995):

$$\frac{v_g}{a_g} = \begin{array}{l} 0.025 \text{ sec} - \text{very stiff soil} \\ (0.05\text{-}0.07) \text{ sec} - \text{rock} \\ (0.12 - 0.13) \text{ sec} - \text{alluvium} \\ 0.3 \text{ sec} - \text{soft soil} \end{array} \tag{3.30}$$

One can see that a simple bilinear variation for reduction factor is an improvement value in comparison with the Newmark proposal.

Another proposal for relating the ductility factor and global ductility is the one of Priestley and Calvi (1991) and Priestley (1997):

-for $T \geq 1.5T_0$:

$$q_\mu = \mu_d \tag{3.31a}$$

-for $0.7T_0 < T < 1.5T_0$:

$$q_\mu = 1 + (\mu_d - 1)\frac{T}{1.5T_0} \tag{3.31b}$$

-for $T < 0.7T_0$

$$q_\mu = (2\mu_d - 1)^{1/2} \tag{3.31c}$$

For this proposal, the variation of the ductility factor is three linear (Fig. 3.85). An analysis of the possibility to use these relationships for practice was performed by Shinouza and Moriyama (1989), using a large number of numerical tests for structures with 1, 3, 5 and 10 storeys (Fig. 3.86) and different earthquake types. The mean, mean $\pm\sigma$ and mean $\pm 2\sigma$ values are considered. Relations (3.27) are also ploted, showing that they tend

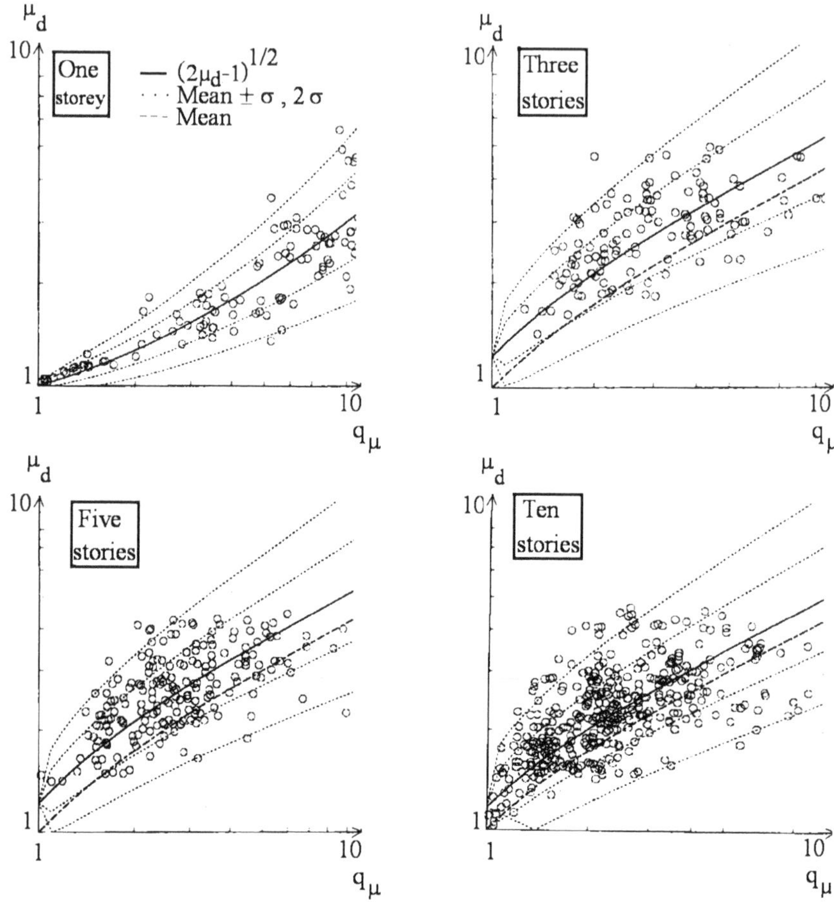

Figure 3.86: Numerical tests for $q_\mu - \mu_d$ relationships (after Shinouza and Moriyama, 1989)

to underestimate the maximum nonlinear required ductility. The results suggest that a modified relationship (3.27b) may be used:

$$q_\mu = \varepsilon(2\mu_d - 1)^{1/2} \tag{3.32}$$

where ϵ must be determined in function of global and damping ratios.

(ii) *Available ductility.* Generally, it is accepted in design that the ductility of a structure is measured by the rotation capacity of members. This evaluation is very difficult due to the great number of factors (Table 3.10),(Gioncu, 1997, 1998), influencing the actual behaviour of members which belong to structures with a complex behaviour themselves.

Thus, it is very important to simplify the analysis, by substituting the actual member with a simple member behaving in a similar way. This member is the *standard beam*, so named because both theoretical and experimental tests presented in the technical literature are performed on this

Table 3.10: Factors influencing the available ductility

LOCAL DUCTILITY	
ELEMENT	**JOINTS**
Material • steel grade • yield ratio • randomness • strain-rate	Joint panel • joint panel type • shear mechanism • crushing mechanism
Cross-section • section type • wall slenderness • wall interactions	Column flanges • column type • plastic mechanism
Members • strain-hardening • buckling • axial forces • cyclic loads	Connections • connection type • plastic local mechanism • cycling loading • strain-rate
AVAILABLE DUCTILITY	

beam (Gioncu and Petcu, 1997). The span of the standard beam is determined by the inflection points, and the actual structure can be replaced by a combination of different standard beams. The available ductility of the structure can be determined on the basis of the ductility of the standard beam (see Chapter 6):

-if the rotation capacity is limited by the effects of local plastic buckling, the formula to calculate the *available rotation capacity* is given by (Fig. 3.87a):

$$\mu_r = \frac{\theta_r}{\theta_p} = \frac{\theta_u}{\theta_p} - 1 \qquad (3.33a)$$

where θ_r is the ultimate plastic rotation, θ_p, the rotation corresponding to the first plastic hinge and θ_u, the total ultimate rotation.

-if the collapse is produced by a local failure, due to brittle fractures, (Fig. 3.87b), the available rotation capacity is given by:

$$\mu_f = \frac{\theta_f}{\theta_p} - 1 \qquad (3.33b)$$

where θ_f is the rotation corresponding to the local failure by fracture.

The relations (3.25) and (3.27) were used by Mazzolani and Piluso (1993b) in order to establish an approximate relation between q_μ factor and local ductility:

$$q_\mu = \frac{2}{3}\mu_r + 1 \qquad (3.34)$$

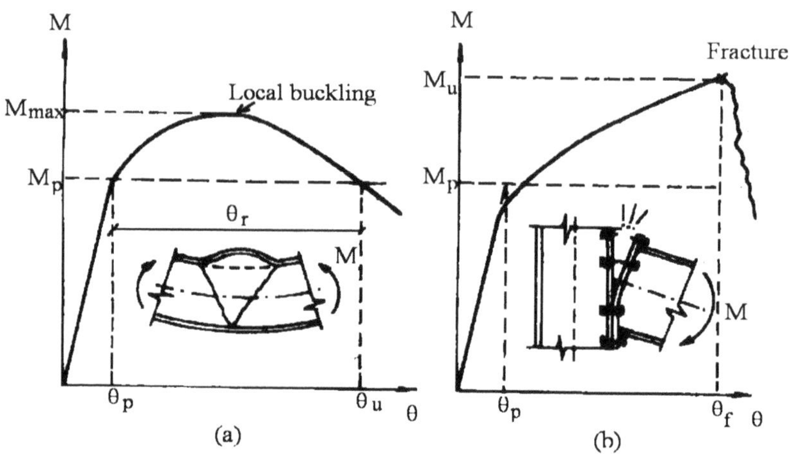

Figure 3.87: Available ductility: (a) Rotation limited by local plastic buckling; (b) Rotation limited by fracture

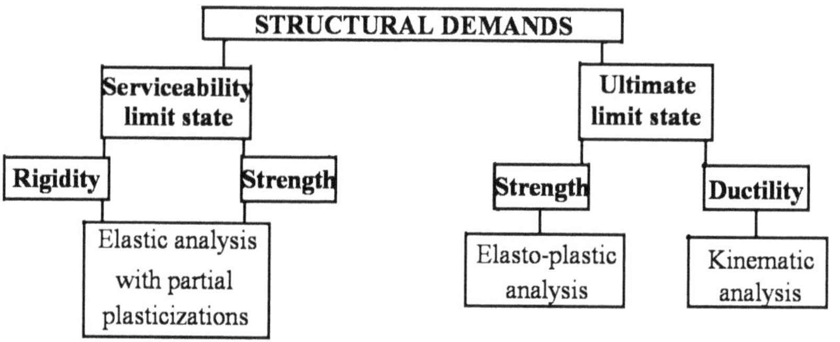

Figure 3.88: Design criteria for two performance levels

3.8 Design Criteria for Two Performance Levels

3.8.1. Structural Demands for Two Levels

Although the most rational design process is to consider three performance levels, modern codes today consider only two levels: serviceability and ultimate limit states. This seismic design philosophy is connected to the traditional design methodology for other loading types, for which two levels are also considered. But for seismic loadings there are some features in design criteria which must be underlined. The structural demands, the limit states, the connection of these two limit states with the rigidity, strength and ductility triade and the analysis methods are presented in Fig. 3.88. In this methodology, checking the two demands must be performed for each limit state.

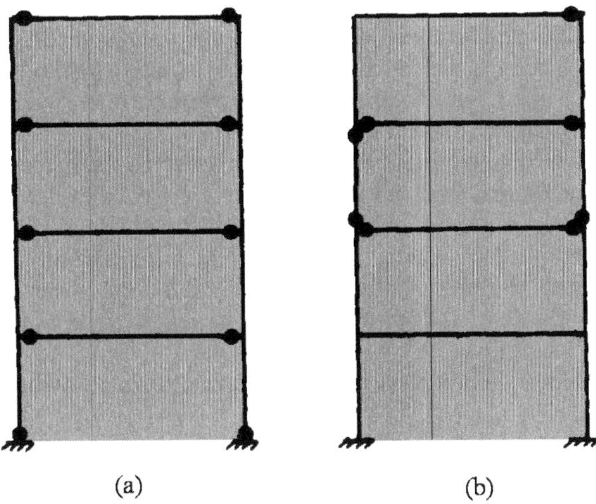

(a) (b)

Figure 3.89: Plastic mechanism: (a) Designed; (b) Unexpected

3.8.2. Structure Design

At each level there are some important design strategies:

(i) *Serviceability limit state.* This limit must be analysed through the following steps:

-defining the characteristics of seismic loads corresponding to the serviceability limit state according to the philosophy of the two levels of performance:

-checking if the structure works in the elastic range for these seismic loads. Only some number of zones may undergo a reduced plastic deformation;

-evaluating the performance of non-structural elements by considering the attachment configuration, dynamic interaction between structure and non-structural elements and the non-classical damping effects of these non-structural elements (Soong, 1995);

-comparing the structure interstorey drift obtained by an elastic analysis with the available deformity of non-structural systems.

So, to satisfy these demands, a pair of design criteria must be checked:

$$\text{Determined interstorey drift} \leq \text{Limit interstorey drift} \qquad (3.35a)$$

$$\text{Determined strength} \leq \text{Elastic limit strength} \qquad (3.35b)$$

(ii) *Ultimate limit state.* In this limit state the steps are as follows:

-defining the characteristics of seismic loads corresponding to ultimate limit states, according to the philosophy of the two performance levels;

-choosing the structure configuration where the beam hinge plastic mechanism (global mechanism) is obtained; the member and joint dimensions must assure that no unexpected plastic mechanism occurs (Fig. 3.89);

-structure analysis in elasto-plastic range, using one of the methods presented in Section 3.5; the analysis refers to strength and ductility too;

-identifying the critical sections, named as dissipative zones, where the plastic hinges have the task to dissipate the seismic energy input, as a consequence of the chosen plastic mechanism;

-sizing and detailing the plastic regions in order to enable the critical sections to have strength and rotation capacity to dissipate the seismic energy;

-sizing and detailing of elastic regions which must exhibit a nominal strength higher than the one in plastic regions. The aim of this step is to assure that no accidental plastic hinges occur in some regions which are not designed for this possibility.

So, to satisfy these demands, a pair of design criteria must be checked:

$$\text{Required strength} \leq \text{Available strength} \qquad (3.36a)$$

$$\text{Required ductility} \leq \text{Available ductility} \qquad (3.36b)$$

3.8.3. Correlation Among Design Criteria

Structural steel moment-resisting frames, MRFs, submitted to lateral forces must be proportional to satisfy both serviceability and ultimate limit states. For this purpose the above design criteria must be checked. The correlation between these criteria are examined for 3, 6 and 9 storey structures (Tirca et al, 1997a,b).

(i) *Selected frames and parameters.* Some typical structures were selected for numerical tests (Fig. 3.90). For the 6 and 9 storey buildings, eccentically braced frames, EBFs, with long link beams are also examined. For beams IPE profiles and for columns HEB profiles are used. The design spectra for ultimate and seviceability limit states are presented in Fig. 3.91. The following parameter values are used:

-earthquake intensity coefficient α, depending on site zone: low seismicity, 0.12, 0.14; medium seismicity, 0.20, 0.24; high seismicity, 0.28, 0.32;

-behaviour factor q considering the frame ductility: low ductility, 4; medium ductility, 6; high ductility, 8;

-plastic mechanism types: local plastic mechanism (hierarchy criterium), global plastic mechanism;

-interstorey drift criterion for the reduction factor ν (Fig. 3.91b): limited reduction, 2; moderate reduction, 3; important reduction 4.

(ii) *Numerical test results.* The results for the 3, 6, 9 storey MRFs are presented in Figs. 3.92–3.94 in function of earthquake intensity α and of the depth of HEB column profiles, for four different design criteria: S-strength, M-mechanism, D-ductility and I-interstorey drift. For 6 and 9 storey frames, the interstorey drift criterion for high seismicity, led to profiles exceeding the practical dimensions. Therefore, the solution of using eccentrically braced frames has been studied. Fig. 3.95 shows the interstorey drift criterion for EBFs in comparison with the MRFs, in function of steel weight for the

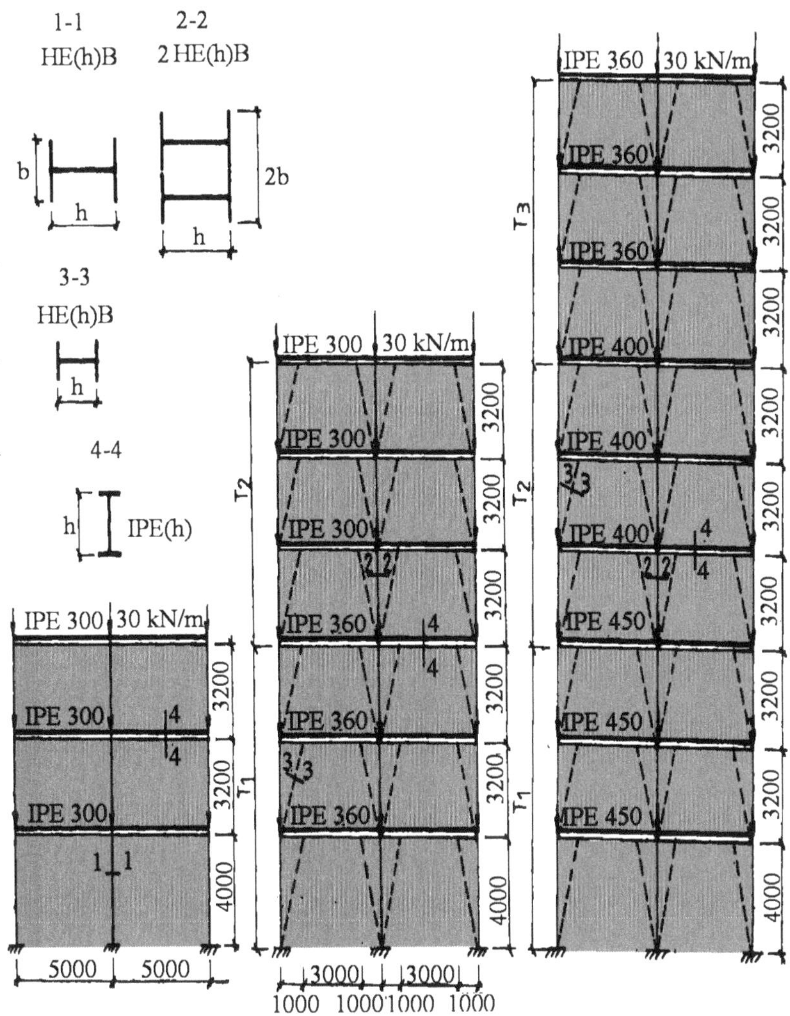

Figure 3.90: Examined frames

columns of first storey. One can see that EBFs with long link beams represents a very good solution to improve the lateral displacements of MRFs in high seismicity areas (Tirca and Gioncu, 1999).

(iii) *Discussion of results.* Examining the numerical results, some very important conclusions regarding the design criteria correlation can be pointed out:

-the strength criterion is not an important requirement for frames with moderate height, but it becomes determinant for tall frames;

-the mechanism criterion depends only on storey numbers and the beam capacities, and generally it is not too severe. For low seismicity areas, a

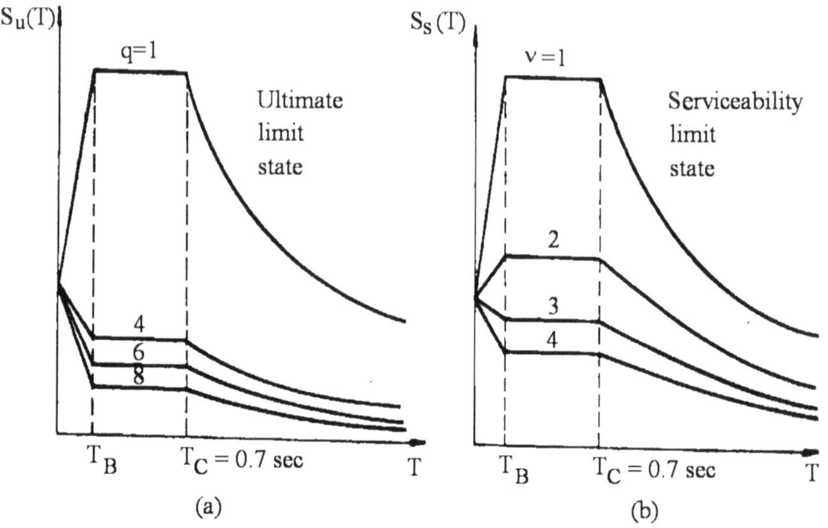

Figure 3.91: Design spectra: (a) Strength criterium; (b) Interstorey drift criterium

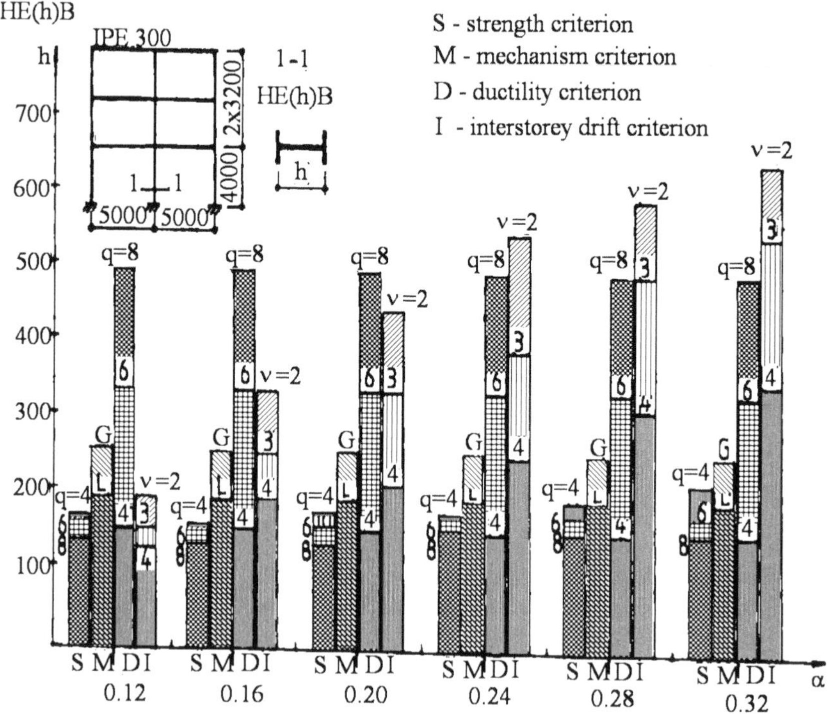

Figure 3.92: Influence of different design criteria for 3 storey MRF

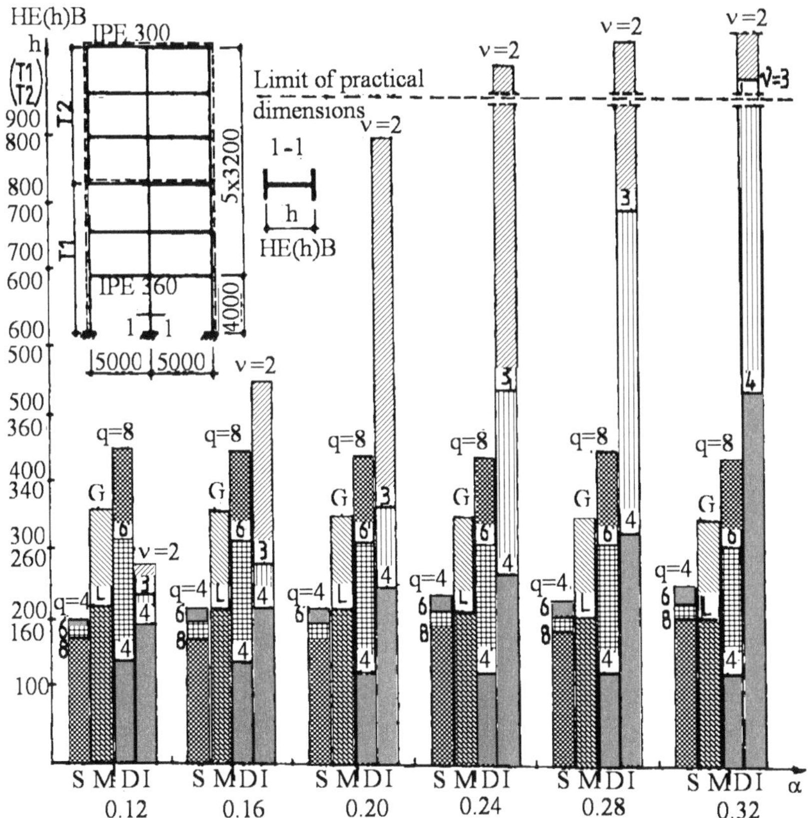

Figure 3.93: Influence of different design criteria for 6 storey MRF

local mechanism, associated with a reduced value of behaviour factor q can be accepted;

-the ductility criterion is essentially influenced by the value of the q factor. For a high values of q, the ductility demand is also high. So, a rational balance between these two factors for ductility and action reduction must be considered. For frames with a reduced number of storeys, a reduced value $q = 4$ is recommended. Only for tall structures it is suitable to consider the high values $q = 6$ or 8;

-the interstorey drift criterion essentially depends on the value of reduction coefficient ν, which defines the serviceability limit state. Values $\nu = 3$ or 4 are more reasonable that the value $\nu = 2 - 2.5$ proposed by EC8.

(iv) *Suggestions for design practice.* The analysis of the correlation among the design criteria has evidence that the principal factors influence the MRF sizing. From the results of this analysis the following suggestions can be derived:

-for structures in low seismicity zones, the interstorey drift criterion is not a determinant and the choice of the ν factor is not very important. The

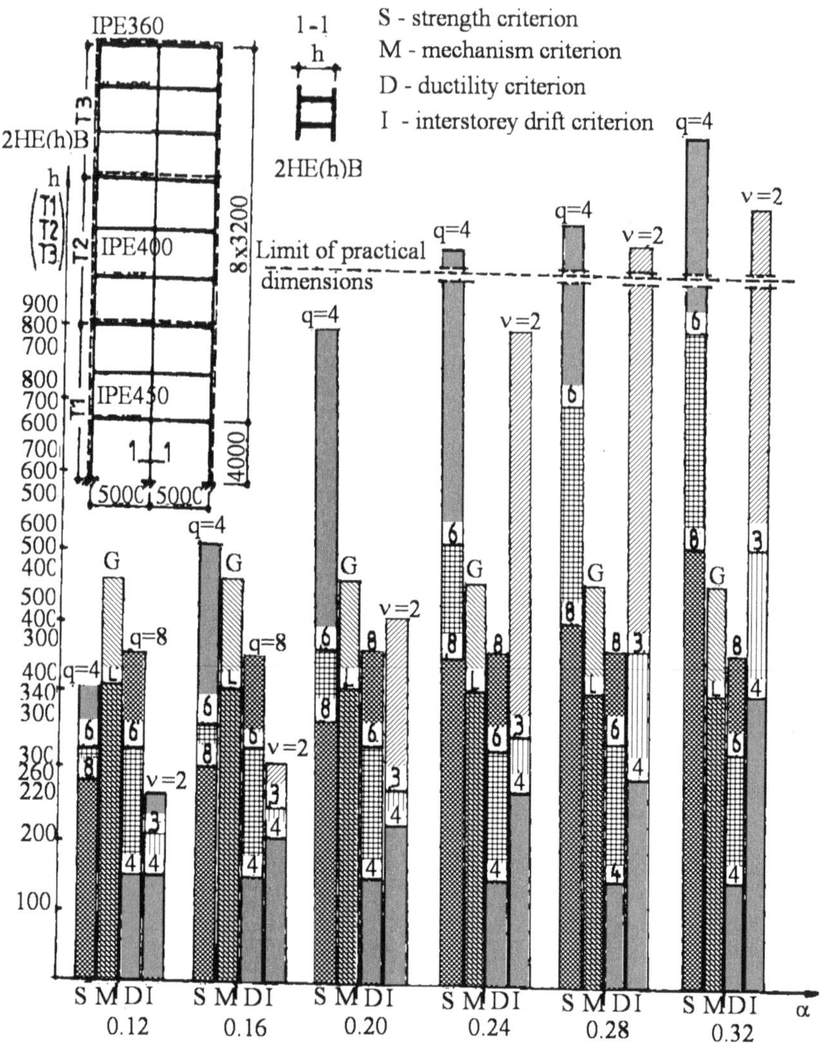

Figure 3.94: Influence of different design criteria for a 9 storey MRF

use of a reduced value for the behaviour factor $q = 4$, associated with a local plastic mechanism, can lead to a suitable solution;

-for structures in medium seismicity zones, one can use values $q = 6$ to 8, for which the global mechanism must be imposed. For the interstorey drift criterion it is recommended to use $\nu = 3$ to 4;

-for structures in high seismicity zones, both strength and intersorey drift criteria become very important. It is recommendable to use a high value $q = 8$ associated with the global mechanism and values $\nu = 3$ to 4. A very good solution to limit the lateral displacements of MRFs is the use of eccentrycally braced frames with long link beams.

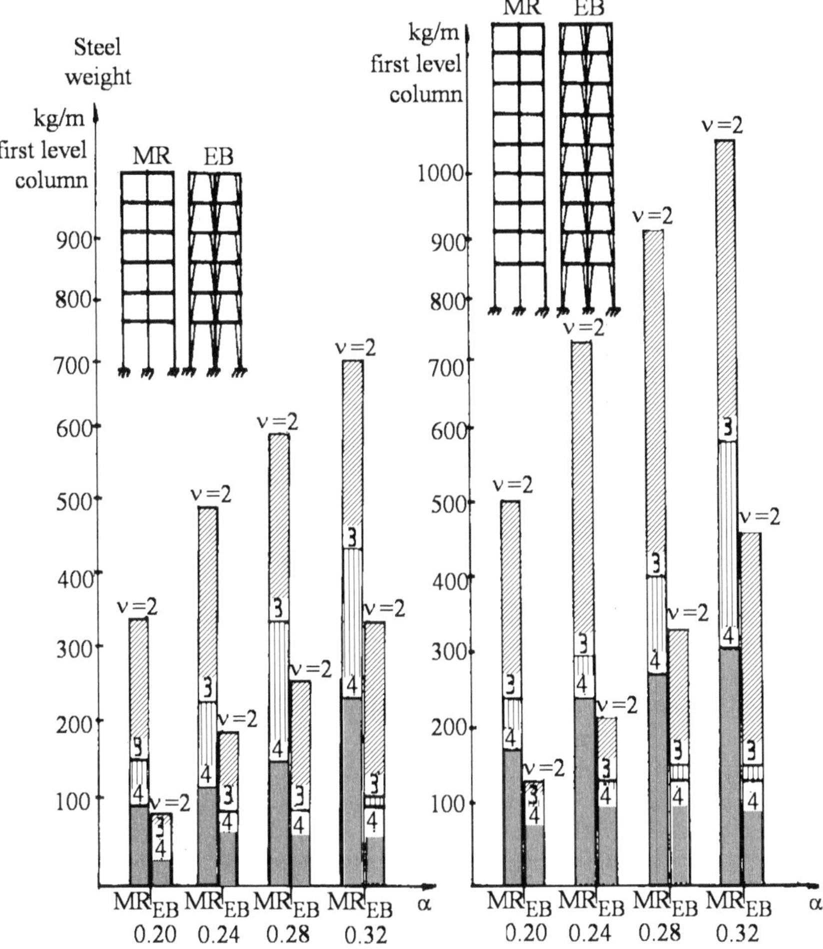

Figure 3.95: Comparison of interstorey drifts for MRF and EBF

3.9 Conclusions

The basic design philosophy presented in this chapter underlines some new important aspects derived from the experience of the last strong earthquakes, which must be introduced in the design practice:

-in order to reduce economic losses, it is advisable to utilise the multi-level design approach. The performance-based seismic design is a very important concept, but its implementation in practice has a long way to go. The more rational approach is to use four performance levels, as in the Vision 2000 proposal. But the most practical design methodology is to limit these levels to a maximum of three levels: serviceability, damageability and ultimate (survivability);

-today, when the practice of seismic design is based on a single level, one can admit that the first step is to arrive to a rational procedure using the minimum of two performance levels, serviceability and ultimate limit states;

-a coherent strategy for seismic design must consider the verification of structure rigidity for serviceability levels, the strength for damageability levels and the ductility for ultimate (survivability) levels. So, the verification of structure ductility is incorporated in the design process at the same level as the usual verification of rigidity and strength;

-the characteristics of earthquakes, like type, source, traveling path, site stratification, building effect, are factors which cannot be neglected in a proper design process. There are important differences between the ground motions in the near-field and far-field regions, which have been neglected until now in design practice. The effects of pulse characteristics on near-field earthquakes (high velocity, reduced number of cycles, short duration, etc) must be introduced in the design practice;

-the artificial time-history representation of ground motions seems to be a promising way for a proper structure design. Instead of a random very complicated representation, the artificial earthquake description allows consideration of only the main characteristics of ground motions and structure response;

-the seismo-resistant architecture philosophy developed recently considers the main general configuration which assures a good seismic response of structures;

-the different methods of structure analysis must be used in function of structure and earthquake importance;

- the method of design plays a crucial role in the seismic design. The new methodologies, of the capacity design, assures a good seismic behaviour during strong earthquakes;

-a proper checking of design criteria must consider all factors influencing the structure behaviour. The verification of the required–available relations is a way to a proper structure design;

-the application of the design criteria for MRFs has shown that there are general rules, and they must be used case by case as a function of earthquake intensity and structure configuration.

3.10 References

Aagaard, B.T., Hall, J.F., Heaton, T.H. (2000): Sensitivity study of near-source ground motion. In 12th World Conference on Earthquake Engineering, Auckland, 30 January- 4 February 2000, CD-ROM 0722

Abrahamson, N.A., Litehiser, J.J. (1989): Attenuation of vertical peak acceleration. Bulletin of the Seismological Society of America, Vol. 80, No. 4, 549-580

Adam, M., Schmid, G., Chouw, N. (2000): Investigation of ground motions and structural responses in near-field due to incident waves. In 12th World

Conference on Earthquake Engineering, Auckland, 30 January- 4 February 2000, CD-ROM 1313

Ali, M.M. (1995): Earthquake-resistant design of foundations. In 7th Canadian Conference on Earthquake Engineering, Montreal, 179-186

Akao, Y., Fukushima, S., Mizutani, M. (1992): Probing characteristics of vertical strong motions. In 10th World Conference of Earthquake Engineering, Madrid, 19-24 July 1992, Balkema, Rotterdam, 583-588

Akiyama, H. (2000): Method based on energy criteria. In Seismic Resistant Steel Structures: Progress and Challenge, (eds. F.M. Mazzolani, V. Gioncu), CISM course, Udine, 18-22 October 1999, Springer, Wien, 101-158

Alavi, B., Krawinkler, H. (2000): Consideration of near-fault ground motion effects in seismic design. In 12th World Conference on Earthquake Engineering, Auckland, 30 January- 4 February 2000, CD-ROM 2665

Ambraseys, N.N. (1995a): The prediction of earthquake peak ground acceleration in Europe. Earthquake Engineering and Structural Dynamics, Vol. 24, 467-490

Ambraseys, N.N. (1995b): Reappraisal of the prediction of ground acceleration in Europe. EAEE Working Group report. In 10th European Conference of Earthquake Engineering, (ed.G.Duma) , Vienna, 28 August- 2 September 1994, Balkema, Rotterdam, Vol. 4, 3041-3048

Ambraseys, N.N. (1997): Surface-wave magnitude calibration for European region earthquakes. Journal of Earthquake Engineering, Vol. 1, No. 1, 1-22

Ambraseys, N.N., Bommer, J.J. (1991a): Database of European strong ground motions records. European Earthquake Engineering, No. 2, 18-37

Ambraseys, N.N., Bommer, J.J. (1991b): The attenuation of ground accelerations in Europe. Earthquake Engineering and Structural Dynamics, Vol. 20, 1179-1202

Anagnostopoulos, S.A. (1995): Earthquake induced pounding: State of the art. In 10th European Conference of Earthquake Engineering, (ed.G.Duma), Vienna, 28 August- 2 September 1994, Balkema, Rotterdam, Vol. 2, 897-905

Anastasiadis, A., Gioncu, V., Mazzolani, F.M. (2000): Toward a consistent methodology for ductility checking. In Behaviour of Steel Structures in Seismic Areas, STESSA 2000, (eds. F.M. Mazzolani, R. Tremblay), Montreal, 21-24 August 2000, Balkema, Rotterdam, 443-453

Anastasiadis, A., Gioncu, V., Mazzolani, F.M. (2000): New trends in the evaluation of available ductility of steel members. In Behaviour of steel Structures in Seismic Areas, STESSA 2000, (eds. F.M. Mazzolani, R. Tremblay), Montreal, 21-24 August 2000, Balkema, Rotterdam, 3-10

Anderson J.C., Bertero, V.V. (1987): Uncertainties in establishing design earthquakes. Journal of Structural Engineering, Vol. 113, No. 8, 1709-1725

Arnold, C. (1996): Architectural aspects of seismic resistant design. In 11th World Conference on Earthquake Engineering, Acapulco, 23-28 June 1996, CD-ROM 2003

ATC -Applied Technology Council (1995): Guidelines and commentary for seismic rehabilitation of buildings. ATC Report 33.03, Redwood City, California

Bachmann, H., Linde, P., Wenk, T. (1995): Capacity design and nonlinear dynamic analysis of earthquake-resistant structures. In 10th European Conference on Earthquake Earthquake (ed. G. Duma), Vienna, 28 August- 2 September 1994, Balkema, Rotterdam, 11-20

Bertero, V.V. (1988): Ductility based structural design. State of the art report. In 9th World Conference on Earthquake Engineering, 2- 9 August 1988, Tokyo- Kyoto, , Vol. VIII, 673-686

Bertero, V.V. (1994): Major issues and future direction in earthquake-resistant design. In 10th World Conference on Earthquake Engineering, Madrid, 19-24 July 1992, Balkema, Rotterdam, 6407-6444

Betrero, V.V. (1996a): State-of-the-art in design criteria. In 11th World Conference on Earthquake Engineering, Acapulco, 23-28 June 1996, CD-ROM 2005

Bertero, V.V. (1996b): The need for multi-level seismic design criteria. In 11th World Conference on Earthquake Engineering, Acapulco, 23-28 June 1996, CD-ROM 2120

Bertero, V.V. (1997a): Codification, design, and application. General report. In Behaviour of Steel Structures in Seismic Areas, STESSA 97, (eds. F.M. Mazzolani, H.Akiyama), Kyoto, 2-8 August 1997, 10/17 Salerno, 189-206

Bertero, V.V. (1997b): Performance-based seismic engineering: A critical review of proposed guidelines. In Seismic Design Methodologies for the Next Generation of Codes, (eds. P. Fajfar, H. Krawinkler), Bled, 24-27 June 1997, Balkema, Rotterdam, 1-31

Bertero, R.D., Bertero, V.V. (1992): Tall reinforced concrete buildings: Conceptual earthquake resistant design methodology. Report No. UCB/EERC-92/16, University of California at Berkley

Bertero, R.D., Bertero, V.V., Teran-Gilmore, A. (1996): Performance based earthquake - resistant design based on comprehensive design philosophy and energy concepts. In 11th World Conference on Earthquake Engineering, Acapulco, 23-28 June 1996, CD-ROM 611

Bertero, R.D., Bertero, V.V.(2000): Application of a comprehensive approach for the performance-based earthquake-resistant design buildings. In 12th World Conference on Earthquake Engineering, Auckland, 30 January- 4 February 2000, CD-ROM 0847

Bommer, J.J., Elnashai, A.S. (1999): Displacement spectra for seismic design. Journal of Earthquake Engineering, Vol. 3, No. 1, 1-32

Bozorgnia, Y., Niazi, M. (1995): Characteristics of free-field vertical ground motion during the Northridge earthquake. Earthquake Spectra, Vol. 11, No. 4, 515-525

Calderoni, B., Rinaldi, Z., Ghersi, A. (1995): Influence of overstrength on the seismic behaviour of steel frames. In Steel Structures -EUROSTEEL 95, (ed. A.Kounadis), Athens, 18- 20 May 1995, Balkema, Rotterdam, 119-126

Calvi, G.M. (1998): Performance-based approaches for seismic assessment of existing structures. In 11th European Conference on Earthquake Engineering, Paris, 6-11 September 1998, Balkema, Rotterdam, Invited Lectures, 3-19

Celabi, M. (1995): Free-field motions near buildings. In 10th European Conference on Earthquake Engineering, Viena, 28 August- 2 September 1994, Balkema, Rotterdam, 215-221

Chandler, A.M., Hutchinson, G.L., Wilson, J.L. (1992): The use of interplate derived spectra in intraplate seismic region. In 10th World Conference on

Earthquake Engineering, Madrid, 19- 24 July 1992, Balkema, Rotterdam, 5823-5827

Chopra, A.K. (1996): Modal analysis of linear dynamic systems: Physical interpretation. Journal of Structural Enginering, Vol. 122, No. 5, 517-527

Chouw, N. (2000): Performance of structures during near-source earthquakes. In 12th Conference on Earthquake Engineering, Auckland, 30 January- 4 February 2000, CD-ROM 0368

Chouw, N., Schmidt, G. (1995): Influence of soil-structure interaction on pounding between buildings during earthquakes. In 10th European Conference on Earthquake Engineering, Vienna, 28 August- 2 September 1994, Balkema, Rotterdam, 553-558

Christian, J.T. (1989): Generating seismic design power spectral density functions. Earthquake Spectra, Vol. 5, No. 2, 351-561

Combescure, D., Queval, J.C., Sollogoub, P., Bonnici, D., Labbe, P. (1998): Effect of near-field earthquake on a R/C bearing wall structure. Experimental and numerical studies. In 11th European Conference on Earthquake Engineering, Paris, 6-11 September 1998, CD-ROM 162

Commission of EC (1988): Background Documents for Eurocode 8, Vol. 2, Design Rules. Report EUR 12266

Cosenza, E. (1987): Duttilita globale delle strutture sismo-resistenti in accaio. Ph D Thesis, University of Napoli

De Buen, O. (1996): Earthquake resistant design: A view from the practice. In 11th World Conference on Earthquake Engineering, Acapulco, 23-28 June 1996, CD-ROM 2002

De Matteis, G., Landolfo, R., Mazzolani, F.M. (1995): On the shear flexibility of corrugated shear panels. Steel Structures, Journal of Singapore Structural Steel Society, 1, 103-111

De Matteis, G., Landolfo, R., Mazzolani, F.M. (1998): Diaphragm effect for industrial steel buildings under earthquake loading. In 2nd World Conference on Constructional Steel Design, San Sebastian, 11-13 May 1998, CD-ROM 401

Elnashai, A.S., Papazoglou, A.J. (1997): Procedure and spectra for analysis of RC structures subjected to strong vertical earthquake loads. Journal of Earthquake Engineering, Vol. 1, No. 1, 121-155

Elwood, K.J., Wen, Y.K. (1996): Evaluation of dual-level design approach for the earthquake resistant design of buildings. In 11th World Conference on Earthquake Engineering, Acapulco, 23-28 June 1996, CD-ROM 609

Erdik, M. (1995): Developments on empirical assessment of the effects of surface geology on strong ground motion. In 10th European Conference on Earthquake Engineering, (ed.G.Duma), 28 August -2 September 1994, Balkema, Rotterdam, Vol. 4, 2593-2598

Erdik, M., Durukal, E. (2000): Assessment of design basis earthquake ground motions for near-fault conditions. In 12th World Conference on Earthquake Engineering, Auckland, 30 January- 4 February 2000, CD-ROM 2361

Faccioli, E., Tolis, S.V., Borzi, B., Elnashai, A.S., Bommer, J.J. (1998): Recent development in the definition of the design seismic action in Europe. In 11th European Conference on Earthquake Engineering, Paris, 6-11 September 1998, CD-ROM 14

Fajfar, P. (1995): Elastic and inelastic design spectra. In 10th European Conference on Earthquake Engineering, (ed.G.Duma), Vienna, 28 August- 2 September 1994, Balkema, Rotterdam, Vol. 2, 1169-1178

Fajfar, P. (1996): Design spectra for new generation of codes. In 11th World Conference on Earthquake Engineering, Acapulco, 23-28 June 1996, CD-ROM 2127

Fajfar, P. (1998): Trends in seismic design and performance evaluation approaches. In 11th European Coference on Earthquake Engineering, Paris, 6-11 September 1998, Balkema, Rotterdam, Invited Lectures, 237-249

Fujitani, H., Tani, A., Aoki, Y., Takanashi, I.(2000): Performance levels of building structures against the earthquake (Concept of performance-based deign standing on questionaires). In 12th World Conference on Eartquake Engineering, Auckland, 30 January- 4 February 2000, CD-ROM 1682

Fukushima, Y., Tanaka, T. (1990): A new attenuation relation for peak horizontal acceleration of strong ground motions in Japan. Bulletin of the Seismological Society of America, Vol. 80, No. 4, 357-783

Galambos, T.V., Ellingwood, B. (1985): Servicebility limit states: Deflection. Journal of Structural Engineering, Vol. 112, No. 1, 67-84

Gariel, J.C. (1992): Near-fault site effects: Some theoretical results. In 10th Conference on Earthquake Engineering, Madrid, 19-24 July 1992, Balkema, Rotterdam, 721-726

Gioncu, V. (1997): Ductility Demands. General Report. In Behaviour of Steel Structures in Seismic Areas , STEESA 97, (eds. F.M. Mazzolani, H. Akiyama), Kyoto, 2- 8 August 1997, 10/17 Salerno, 279-302

Gioncu, V. (1998): Ductility Criteria for Steel Structures. In 2nd World Conference on Steel in Construction, San Sebastian, 11-13 May 1998, CD-ROM 220

Gioncu, V. (1999): Framed structures: Ductility and seismic response. General report.In Stability and Ductility of Steel Structures, SDSS 99, Timisoara, 9-11 Sepember 1999, Journal of Constructional Steel Research , Vol. 55, No. 1-3, 125-154

Gioncu, V. (2000): Design criteria for seismic resistant steel structures. In Seismic Resistant Steel Structures. Progress and Challenge, (eds. F.M. Mazzolani, V. Gioncu), Udine CISM course, 18-22 October 1999, Springer, Wien, 19-99

Gioncu, V., Petcu, D. (1997): Available rotation capacity of wide-flange beams and beams-columns. Part 1 , Theoretical Approaches. Part 2, Experimental and Numerical Tests. Journal of Constructional Steel Research, Vol 43, No 1-3, 161-217, 219-244

Gioncu, V., Mateescu, G., Tirca, L., Anastasiadis, A. (2000): Influence of the type of seismic ground motions. In Moment Resisting Connections for Steel Building Frames in Seismic Areas, (ed. F.M.Mazzolani), E& FN Spon, London, 57-92

Giuliani, H. (2000): Seismic resistant architecture: A theory for the architectural design of buildings in seismic zones. In 12th World Conference on Earthquake Engineering, Auckland, 30 January- 4 February 2000, CD-ROM 2456

Giuliano, H., De Acosta, R., Yacante, M.I., Campora, A.M., Giuliano H.L. (1996): Seismic resisting architecture on building scale: A morphological answer. In 11th World Conference on Earthquake Engineering, Acapulco, 23-28 June

1996, CD-ROM 1067

Ghobarah, A., Aly, N.M., El-Attar, M. (1997): Performance level criteria and evaluation. In Seismic Design Methodoligies for the Next Generation of Codes, (eds. P. Fajfer, H. Krawinkler), Bled, 24-27 June 1997, Balkema, Rotterdam, 207-215

Guerra, C.A., Mazzolani, F.M., Piluso, V. (1990): On the seismic behaviour of irregular steel frames. In 9th European Conference on Earthquake Engineering, Moskow, September, 11-16

Gunturi, S.K., Shah, H.C. (1992): Building specific damage estimation. In 10th World Conference on Earthquake Engineering, Madrid, 19-24 July 1992, Balkema, Rotterdam, 6001-6006

Gurpinar, A. (1997): A review of seismic safety considerations in the life cycle of critical facilities. Journal of Earthquake Engineering, Vol. 1, No. 1, 57-76

Hall, J.F., Heaton, T.H., Halling, M.W., Wald, D.J. (1995): Near-source ground motion and its effects on flexible buildings. Earthquake Spectra, Vol. 11, No. 4, 565-605

Hamburger, R.O. (1996): Implementing performance based seismic design in structural engineering practice. In 11th World Conference on Earthquake Enginnering, Acapulco, 23-28 June 1996, CD-ROM 2124

Hamburger, R,O. (1997): Defining performance objectives. In Seismic Design Methodologies for the Next Generation of Codes, (eds. P. Fajfar, H. Krawinkler), Bled, 24-27 June 1997, Balkema, Rotterdam, 33-42

Holmes, W.T. (1996): Seismic evaluation of existing buildings: State of the practice. In 11th World Conference on Earthquake Engineering, Acapulco, 23-28 June 1996, CD-ROM 2008

Hori, N., Yamamoto, S., Yamada, M. (2000): Source analysis of near-field earthquake record observed in rock sites. In 12th World Conference on Earthquake Engineering, Auckland, 30 January- 4 February 2000, CD-ROM 0698

Ifrim, M. (1984): Dynamics of Structures and Earthquake Engineering (in Romanian), Editura Didactica si Pedagogica, Bucuresti

Ifrim, M., Macavei, F., Demetriu, S., Vlad, I. (1986): Analysis of degradation process in structures during the earthquakes. In 8th European Conference on Earthquake Engineering, Lisbon, 65/8-72/8

Iwan, W.D. (1995): Near-field considerations in specification of seismic design motions for structures. In 10th European Conference on Earthquake Engineering, Vienna, 28 August- 2 September 1994, Balkema, Rotterdam, Vol. 1, 257-267

Iwan, W.D. (1997): Drift spectrum: Measures of demand for earthquake ground motions. Journal of Structural Engineering, Vol. 123, No. 4, 397-404

Iwan, W.D., Chen, V. (1995): Important near-field ground motion data from the Landers earthquake. In 10th European Conference on Earthquake Engineering, Vienna, 28 August- 2 September 1994, Balkema, Rotterdam, 229-234

Jankulovski, E., Sinadinovski, C., McCue, K. (1996): Structural response and design spectra modelling: Results from intraplate earthquakes in Australia. In 11th World Conference on Earthquake Engineering, Acapulco, 23-28 June 1996, CD-ROM 1154

Jirsa, J.O. (1997): Opportunities and challenges. Development of performance-sensitive engineering. In Seismic Design Methodologies for Next Generation of Codes, (eds. P. Fajfar, H. Krawinkler), Bled, 24-27 June 1997, Balkema, Rotterdam, 43-46

Kataoka, S., Ohmachi, T. (1996): Synthetic earthquake motion of 2-D irregular ground in near-field. In 11th World Conference on Earthquake Engineering, Acapulco, 23-28 June 1996, CD-ROM 1248

Kennedy, D.J.L., Medhekar, M.S. (1999): A proposal strategy for seismic design of steel structures. Canadian Journal of Civil Engineering, Vol. 26, 564-571

Krawinkler, H. (1995): New trends in seismic design methodology. In 10th European Conference on Earthquake Engineering, (ed. G.Duma), 28 August- 2 September 1994, Balkema, Rotterdam, 821-830

Krawinkler, H. (1996a): A few basic concepts for performance based seismic design. In 11th World Conference on Earthquake Engineering, Acapulco, 23-28 June 1996, CD-ROM 1133

Krawinkler, H. (1996b): Introduction to special theme session on seismic design criteria. In 11th World Conference on Earthquake Engineering, Acapulco, 23-28 June 1996, CD-ROM 2119

Krawinkler, H. (1997): Researh issues in performance based seismic engineering. In Seismic Design Methodologies for the Next Generation of Codes, (eds. P. Fajfar, H. Krawinkler) Bled, 24-27 June 1997, Balkema, Rotterdam, 47-58

Kudo, K., Sakane, M., Wang, Z. (1992): Some features of near-field strong ground motions in the central Japan. In 10th World Conference on Earthquake Engineering, Madrid, 19-24 July 1994, Balkema, Rotterdam, 589-592

Kuribayashi, E. (1987): Earthquake Engineering for Practicing Engineers. IABSE Periodica 3, August, 41-60

Lai, S.P., Biggs, J.M. (1980): Inelastic response spectra for aseismic building design. Journal of Structural Division, Vol. 106, ST 6, 1295-1310

Lam, N., Wilson, J., Hutchinson, G. (1996): Building ductility demand: Interplate versus intraplate earthquakes. Earthquake Engineering and Structural Dynamics, Vol. 25, 965-985

Lam, N., Wilson, J., Hutchinson, G. (1998): The ductility reduction factor in the seismic design of buildings. Earthquake Engineering and Structural Dynamics, Vol. 27, 749-769

Lee, V.W. (1995): Automatic digitization and data processing of strong - motion accelerograms: State-of-the art review. In 10th European Conference on Earthquake Engineering, (ed. G.Duma), 28 August- 2 September 1994,, Balkema, Rotterdam, Vol. 1, 183-194

Lee, H.S. (1996): Revised rule for concept of strong-column weak-girder design. Journal of Structural Engineering, Vol. 122, No. 4, 359-364

Leelataviwat, S., Goel, S.C., Stojadinovic, B. (1999): Toward performance-based seismic design of structures. Earthquake Spectra, Vol. 15, No. 3, 435-461

Lopez, O.A., Cruz, M. (1996): Number of modes for the seismic design buildings. Earthquake Engineering and Structural Dynamics, Vol. 25, 837-855

Lopez, O.A., Anichiario, W., Genatios, C., Raven, E. (1995): The influence of floor plan shape on the earthquake response of buildings. In 10th European Conference on Earthquake Engineering, (ed. G. Duma), Vienna, 28 August- 2 September 1994, Balkema, Rotterdam, Vol. 2, 935-940

Lungu, D., Cornea, T., Aldea, A., Zaiceanu, A. (1997): Basic representation of seismic action. In Design of Structures in Seismic Areas, Eurocode 8, Worked Examples, (eds. D.Lungu, F.M.Mazzolani, S.Saridis), Tempus Phare Project 01198, Bridgeman, Romania, 9-60

Madan, A., Reinhorn, A.M., Mander, V.B., Valles, R.E. (1997): Modelling of masonry infill panels for structural analysis. Journal of Structural Engineering, Vol. 123, No. 10, 1295-1302

Mahin, S.A., Bertero, V.V. (1981): An evaluation of inelastic seismic design spectra. Journal of Structural Division, Vol. 107, ST 9, 1777-1795

Masri, S.F., Safford, F.B. (1980): Optimization procedure for pulse-simulated response. Journal of Structural Division, Vol. 107, ST 9, 1743-1761

Mateescu G., Gioncu, V. (2000): Member response to strong pulse seismic loading. In Behaviour of Steel Structures in Seismic Areas. STESSA 2000 (eds. F.M. Mazzolani, R. Tremblay), Montreal, 21-24 August 2000, Balkema, Rotterdam

Mazzolani, F.M. (2000): Steel structures in seismic zones. In Seismic Resistant Steel Structures: Progress and Challenge, (eds. F.M. Mazzolani, V. Gioncu) CISM course, Udine, 18-22 October 1999, Springer Verlag, Wien, 1-17

Mazzolani, F.M. (2000): Design of moment resisting frames. In Seismic Resistant Steel Structures: Progress and Challenge, (eds. F.M. Mazzolani, V. Gioncu) CISM course, Udine, 18-22 October 1999, Springer Verlag, Wien

Mazzolani, F.M., Piluso, V. (1990): Skin effect in pin-jointed structures. Ingegneria Sismica, VII, No. 3

Mazzolani, F.M., Piluso, V. (1993a): Design of Steel Structures in Seismic Zones. ECCS Manual. TC13, Seismic Design Report

Mazzolani, F.M., Piluso, V. (1993b): Member behavioural classes of steel beams and beam-columns. In XIV Congreso CTA, Viaregio, 24-27 October 1993, Riccerca Teorica e Sperimentale, 405-416

Mazzolani, F.M., Piluso, V. (1995): Seismic design criteria for moment resisting steel frames. In Steel Structures. Eurosteel 95, (ed. A.N. Kounadis), Athens, 18-20 May 1995, Balkema, Rotterdam, 247-254

Mazzolani, F.M., Piluso, V. (1996a): Theory and Design of Seismic Resistant Steel Frames. E&FN Spon, London

Mazzolani, F.M., Piluso, V. (1996b): Behaviour and design of set-back steel frames. In 1st European Workshop on the Seismic Behaviour of Asymmetric and Set-back Structures, (eds. R.Ramasco, A. Rutenberg), Anacapri, 4-5 October 1996, 279- 297

Mazzolani, F.M., Piluso, V. (1997a): Plastic design of seismic resistant steel frames. Earthquake Engineering and Structural Dynamics, Vol. 26, 167-191

Mazzolani, F.M., Piluso, V. (1997b): A simple approach for evaluating performance levels of moment-resisting steel frames. In Seismic Design Methodologies for the Next Generation of Codes, (eds. P.Fajfar, H. Krawinkler), Bled, 24-27 June 1997, Balkema, Rotterdam, 241-252

Mazzolani, F.M,, Piluso, V. (1997c): The influence of design configuration on the seismic response of moment- resisting frames. In Behaviour of Steel Structures in Seismic Areas, STESSA 97, (eds. F.M. Mazzolani, H. Akiyama), Kyoto, 3-8 August 1997, 10/17 Salerno, 444-453

Mazzolani, F.M., Georgescu, D., Astaneh-Asl, A. (1995): Safety levels in seismic design. In Behaviour of Steel Structures in Seismic Areas, STESSA 94, (eds. F.M.Mazzolani, V. Gioncu), Timisoara, Romania, 495-506

Mazzolani, F.M., De Matteis, G., Landolfo, R. (1996a): The stiffening effect of cladding panels on steel buildings: The ECCS research project in progress. In European Workshop on Thin- Walled Structures (eds. K. Rykaluck, H. Pasternak), Krzyzowa-Kreisau, 25-34

Mazzolani, F.M., Delponte, R., De Matteis, G., Landolfo, R. (1996b): Monotonic and cyclic tests on sandwich panels in shear. In Earthquake Performance of Civil Engineering Structures, (ed. A.V. Pinto). ELSA Joint Research Center, European Commission, Special Publication No. 1.96.56

Mazzolani, F.M., De Matteis, G., Landolfo, R. (1996c): Analytical models for cladding panels under monotonic and cyclic shear loads. In Stability Problems in Designing, Construction and Reabilitation of Metal Structures (eds. R.C. Batista, E.M. Batista, M.S. Pfeil), SSRC/IC Brasil, Rio de Janeiro, 5-7 August 1996, 163-174

Mazzolani, F.M., Piluso, V. (1997a): A simple approach for evaluating performance levels of moment-resisting frames. In Seismic Design Methodologies for the Next Generation of Codes, (eds. P. Fajfar, H. Krawinkler), Bled, 24-27 June 1997, Balkema, Rotterdam, 241-252

Mazzolani, F.M., Piluso, V. (1997b): Review of code provisions for vertical irregularity. In Behaviour of Steel Structures in Seismic Areas, STESSA 97, (eds. F.M. Mazzolani, H. Akiyama), Kyoto, 2-8 August 1997, 10/17 Salerno, 250-257

Mazzolani, F.M., De Matteis, G., Landolfo, R. (1997a): Dynamic behaviour of sandwich diaphragms in simple pin-jointed steel frames. In Behaviour of Steel Structures in Seismic Areas, STESSA 97, (eds. F.M. Mazzolani, H. Akiyama), Kyoto, 3-8 August 1997, 10/17 Salerno, 434-443

Mazzolani, F.M., De Matteis, G., Landolfo, R. (1997b): Hysteretic behaviour of fastening in sheeting. In Behaviour of Steel Structures in Seismic Areas, STESSA 97, (eds. F.M. Mazzolani, H. Akiyama), Kyoto, 3-8 August 1997, 10/17 Salerno, 622-631

Mayes, R.L. (1995): Interstory drift design and damage control issues. The Structural Design of Tall Buildings, Vol. 4, No. 1, 15-25

Medhekar, M.S., Kennedy, D.J.L. (2000): Displacement-based seismic design of buildings. 1 Theory, 2 Applications. Engineering Structures, Vol. 22, 201-209, 210-221

Mehrabi, A.B., Shing, B., Schuller, M.P., Noland, J.L. (1996): Experimental evaluation of masonry-infilled RC frames. Journal of Structural Engineering, Vol. 122, No. 3, 228-237

Men, F.L., Cui, J., Chen, W.H. (1998): Estimation of influence of earthquake mechanism on ground motion and wave propagation. In 11th European Conference on Earthquake Engineering, Paris, 6-11 September 1998, CD-ROM 104

Mezzi, M., Radicchia, R., D'Ambrisi, A. (1991): Use of artificially generated accelerograms in the seismic analysis of non linear structures. In Earthquake Protection of Buildings, Ancona, 6-8 June 1991, 113B-121B

Midorikawa, S. (1995): Ground motion intensity in epicentral area. In New Direction in Seismic Design, Tokyo, 9-10 October 1995, 247-250

Miranda, E. (1992): Evaluation of site-dependent inelastic seismic design spectra.

Journal of Structural Engineering, Vol. 119, No. 5, 1319-1338

Miranda, E. (1993): Site-dependent strength-reduction factors. Journal of Structural Engineering, Vol. 119, No. 12, 3503-3519

Moehle, J.P. (1996): Displacement-based seismic design criteria. In 11th World Conference on Earthquake Engineering, Acapulco, 23-28 June 1996, CD-ROM 2125

Mohammadioun, G. (1995): Calculation of site-adapted reference spectra from statistical analysis of an extensive strong motion data bank. In 10th European Conference on Earthquake Engineering, (ed. G.Duma), Vienna, 28 August- 2 September 1994, Balkema, Rotterdam, Vol. 1, 177-181

Mohammadioun, G. (1997): Nonlinear response of soils to horizontal and vertical bedrock motion. Journal of Earthquake Engineering Vol.1, No 1, 93-119

Motoki, K., Seo, K. (2000): Strong motion characteristics near the source region of the Hyogoken-Nanbu earthquake from analysis of the directions of structural failures. in 12th World Conference on Earthquake Engineering, Auckland, 30 January - 4 February 2000, CD-ROM 0959

Mylonakis, G., Reinhorn, A.M. (1997): Inelastic seismic response of structures near fault: An analytical solution. In Behaviour of Steel Structures in Seismic Areas, STESSA 97, (eds. F.M. Mazzolani, H. Akiyama), Kyoto, 3-8 August 1997, 10/17 Salerno, 82-89

Naeim F. (1998): Researh overview: Seismic response of structures. The Structural Design of Tall Buildings, Vol. 7, 195-215

Nair, R.S. (1995): Stiffness and serviceability issues in tall steel buildings. In Habitat and the Hige-Rise, (eds. L.S.Beedle, D.Rice), Proc. of the Fifth World Congress, Amsterdam, 14-19 May 1995, 1171-1183

Nakao, Y., Tamura, K., Kataoka, S. (2000): Effects of earthquake source parameters on estimated ground motions. In 12th World Conference on Earthquake Engineering, Auckland, 30 January- 4 February 2000, CD-ROM 1895

Nakashima, M., Mitani, T., Tsuji, B.(1997): Control of maximum and cumulative deflections in steel building structures combined with hysteretic dampers. In Behaviour of Steel Structures in Seismic Areas, STESSA 97,(eds. F.M. Mazzolani, H. Akiyama), Kyoto, 3-8 August 1997, 10/17 Salerno, 744-751

Newmark, N.M., Hall, W.V. (1982): Earthquake spectra and design. EERI Berkeley

Otani, S. (1996): Recent developments in seismic design criteria in Japan. In 11th World Conference on Eartquake Engineering, Acapulco, 23-28 June 1996, CD-ROM 2124

Otani, S. (1997): Development of performance-based design methodology in Japan. In Seismic Design Methodologies for the Next Generation of Codes, (eds. P. Fajfar, H. Krawinkler), Bled, 24-27 June 1997, Balkema, Rotterdam, 59-67

Pauley, T., Bachmann, H., Moser, K. (1990): Erdbenbemessung von Stahlbetonhochbauten. Birkhausen Verlag, Basel

Pauley, T. Priestley M. J. N. (1992): Seismic Design of Reinforced Concrete and Masonry Buildings. John Wiley&Sons Inc., New York

Pinelli, J.P., Moor, C., Craig, J.I., Goodno, B.J. (1992): Experimental testing of ductile cladding connections for building facades. The Structural Design of Tall Buildings, Vol. 1, No. 1, 57-72

Pinto, A.V. (1998): Achievents of the COST-C1 seismic group. In Control of

Semi-Rigid Behaviour of Civil Engineering Structural Connections. Liege, 17-19 September 1998, 349-358

Pitilakis, K.D. (1995): Seismic microzonation practice in Greece: A critical review of some important factors. In 10th European Conference on Earthquake Engineering, (ed. G.Duma), Vienna, 28 August- 2 September 1994, Balkema, Rotterdam, Vol. 4, 2537-2545

Poland, C.D., Hom, D.B. (1997): Opportunities and pitfalls of performance based seismic engineering. In Seismic Design Methodologies for Next Generation of Codes, (eds. P. Fajafar, H. Krawinkler), Bled, 24-27 June 1997, Balkema, Rotterdam, 69-78

Popov, E.P. (1994): Development of U.S Seismic Codes. Journal of Constructional Steel Research, Vol. 29, 191-207

Priestley, M.J.N. (1997): Displacement-based seismic assessment of reinforced concrete buildings. Journal of Earthquake Engineering, Vol. 1, No. 1, 157-159

Priestley, M.J.N. (1998): Displacement-based approaches to rational limit states design of new structures. In 11th European Conference on Earthquake Engineering, Paris, 6-11 September 1998, Invited Lectures, 317-335

Priestley, M.J.N. (2000): Performance based sesmic design. In 12th World Conference on Earthquake Engineering, Auckland, 30 January- 4 February 2000, CD-ROM 2831

Priestley, M.J.N., Calvi, G.M. (1991): Towards a capacity-design assessment procedure for reinforced concrete frames. Earthquake Spectra, Vol. 7, 413-437

Reinhorn, A.M. (1997): Inelastic analysis techniques in seismic evaluations. In Seismic Design Methodologies for Next Generation of Codes, (eds. P. Fajfar, H. Krawinkler), Bled, 24-27 June 1997, 277-287

Reinoso, E., Ordaz, M., Guerrero, R. (2000): Influence of strong ground motion duration in seismic design of structures. In 12th World Conference on Earthquake Engineering, Auckland, 30 January- 4 February 2000, CD-ROM 1151

Riddell, R., DeLa Llerra, J.C. (1996): Seismic analysis and design: Current practice and future trends. In 11th World Conference on Earthquake Engineering, Acapulco, 23-28 June 1996, CD-ROM 2010

Roesset, J.M. (1994): Modelling problems in earthquake resistant design: Uncertainties and needs. In 10th World Conference on Earthquake Engineering, Madrid, 19-24 July 1992, Balkema, Rotterdam, 6445-6483

Rutenberg, A., Chandler, A.M., Ramasco, R. (1996): EAEE Task Group TG 8: Behaviour of irregular and complex structures: Agenda. In European Workshop on the Seismic Behaviour of Asymmetric and Setback Structures, (eds. R.Ramasco, A.Rutenberg), Anacapri, Italy , 7-5 October 1996, 3-18

Sandi, H. (1995): Some considerations on the needs of the future development of earthquake resistant design practice and codes. In Behaviour of Steel Structures in Seismic Area, STESSA 94, (eds. F.M. Mazzolani, V.Gioncu), Timisoara, 26June- 1 July 1994, E&FN Spon, London, 507-516

Saneinejad, A., Hobbs, B. (1995): Inelastic design of infilled frames. Journal of Structural Engineering, Vol. 121, No. 4, 634-650

Sasani, M., Bertero, V.V. (2000): Importance of severe pulse-type ground motions in performance-based engineering. Historical and critical review. In 12th

World Conference on Earthquake Engineering, Auckland, 30 January- 4 February 2000, CD-ROM 1302

Scawthorn Ch. (1997): Earthquake Engineering. In Handbook of Structural Engineering, (ed. W.F. Chen), CRC Press, New York, 5.1-5.83

SEAOC- Structural Engineers Association of California, (1995): Vision 2000- A framework for performance based design. Sacramento, California

Shape, R.L. (1992): Acceptable earthquake damage on designed performance. In 10th World Conference on Earthquake Engineering, Madrid, 19-24 July 1992, 5891-5894

Shinouza, M., Moriyama, K. (1989): An assessment of uncertainties in seismic response. In Earthquake Hazards and the Design of Constructed Facilities in the Eastern United States, (eds. K.H.Jacob, C.J. Turkstra), New York, 24-26 February 1988, Annals of the New York Academy of Sciences, Vol. 558, 234-250

Soong, T.T. (1995): Seismic behaviour of nonstructural elements: State-of-the art report. In 10th European Conference on Earthquake Engineering, (ed. G. Duma), Vienna, 28 August- 2 September 1994, Balkema, Rotterdam, Vol. 3, 1599-1606

Studer, J., Koller, M.G. (1995): Design earthquake: The importance of engineering judgments. In 10th European Conference on Earthquake Engineering, (ed. G.Duma), Vienna, 28 August- 2 September 1994, Balkema, Rotterdam, Vol 1, 167-176

Somerville, P.G., Smidth N.F., Abrahamson, N.A. (1996): Accounting for near-fault rupture directivity effects on the development of design ground motions. In 11th World Conference on Earthquake Engineering, Acapulco, 23-28 June 1996, CD-ROM 711

Tassios, T.P. (1998): The seismic design: State of practice. In 11th European Conference on Earthquake Engineering, Paris, 6-11 September 1998, Balkema, Rotterdam, Invited Lectures, 255-267

Teran-Gilmore, A. (1998): A parametric approach to performance-based numerical seismic design. Earthquake Spectra, Vol. 14, No. 3, 501-518

Tirca, L. (2000): Behaviour of MRFs subjected to near-source earthquakes. In Behaviour of Steel Structures in Seismic Areas, STESSA 2000 (eds. F.M. Mazzolani, R. Tremblay), Montreal, 21-24 August 2000, Balkema, Rotterdam, 635-642

Tirca, L., Gioncu, V.(1999): Ductility demands for MRFs and LL-EBFs for different earthquake types. In Stability and Ductility of Steel Structures, SDSS 99, (eds. D. Dubina, M. Ivanyi). Timisoara, 9-11 September 1999, Elsevier, Amsterdam, 429-438

Tirca, L., Gioncu, V., Mazzolani, F.M. (1997a): Influence of design criteria for multistorey steel MR frames. In Behaviour of Steel Structures in Seismic Areas, STESSA 97, (eds. F.M. Mazzolani, H. Akiyama), Kyoto, 3-8 August 1997, 10/17 Salerno, 266-275

Tirca, L., Gioncu, V., Mazzolani, F.M. (1997b): Design criteria demands for earthquake-resistant steel structures. In 8th Conference on Steel Structures (ed. M. Ivan), Timisoara, 25-28 September 1997, Mirton, Timisoara, Vol. 2 558-567

Tolis, S.V., Faccioli, E. (1999): Displacement design spectra. Journal of Earthquake Engineering, Vol. 3, No. 1, 107-125

Trifunac, M.D. (1990): How to model amplification of strong earthquake motions by local soil and geologic site conditions. Earthquake Engineering and Structural Dynamics, Vol. 19, 833-846

Trifunac, M.D. (1992): Earthquake source variables for scaling spectral and temporal characteristics of strong ground motions. In 10th European Conference on Earthquake Engineering, (ed. G.Duma), Vienna, 28 August- 2 September 1994, Balkema, Rotterdam, 2585-2520

Trifunac, M.D. (1997): Relative earthquake motion of building foundations. Journal of Structural Engineering, Vol. 123, No. 4, 414-422

Trifunac, M.D., Brady, A.G. (1975): A study on the duration of strong earthquake ground motion. Bulletin of the Seismological Society of America, Vol. 65, 581-626

Trifunac, M.D., Noricova, E.I. (1995): State of the art review on strong motion duration. In 10th European Conference on Earthquake Engineering, (ed. G.Duma), Vienna, 28 August- 2 September 1994, Balkema, Rotterdam, Vol 1, 131-140

Tso, W.K., Zhu, T.J., Heidebrecht, A.C. (1993): Seismic energy demands on reinforced concrete moment resisting frames. Earthquake Engineering and Structural Dynamics, Vol. 22, 533-545

Umemura, H., Sakai, Y., Minami, T. (2000): Estimation of response of structures under near field ground motions considering inhomogeneous faulting. In 12th Worl Conference on Earthquake Engineering, Auckland, 30 January- 4 February 2000, CD-ROM 1810

Villaverde, R. (1997): Seismic design of secondary structures: State of the art. Journal of Structural Engineering, Vol. 123, No. 8, 1011-1019

Wang, H., Nishimura, A. (2000): On seismic motion near active faults based on seismic records. In 12th World Conference on Earthquake Engineering, Auckland, 30 January- 4 February 2000, CD-ROM 0853

Warsazarski, A., Gluck, J., Segal, D.(1996): Economic evaluation of design codes-case of seismic design. Journal of Structural Engineering, Vol. 122, No. 12, 1400-1408

Yamamoto, S. (2000): Spatial distribution of strong ground motion considering asperity and directivity of fault. In 12th World Conference on Earthquake Engineering, Auckland, 30 January- 4 February 2000, CD-ROM 0700

Yamawaki, K., Kitamura, H., Tsuneki, Y., Mori, N., Fukai, S. (2000): Introduction of performance-based design. In 12th World Conference on Earthquake Engineering, Auckland, 30 January- 4 February 2000, CD-ROM 1511

Yao, S., Murayama, K., Suga, M, Nishida, K., Kusumi, H. (1996): The structural damage near-fault active fault in Hyougoken Nanbu earthquake. In 11th Conference on Earthquake Engineering, Acapulco, 23-28 June 1996, CD-ROM 1258

Yayong, W., Mirixian, C. (1990): Dependence of structural damage in the parameters of earthquake strong motions. European Earthquake Engineering, Anno IV, No. 1, 13-23

Yuan, X., Meng, S., Shi, Z., Sun, R. (2000): A procedure for evaluation of differential settlements of buildings during earthquakes. In 12th World Conference on Earthquake Engineering, Auckland, 30 January- 4 February 2000, CD-

ROM 1465

Zhu, T.J., Tso, W.K., Heidebrecht, A.C. (1988): Effect of peak ground a/v ratio on structural damage. Journal of Structural Engineering, Vol. 114, No. 5, 1019-1037

4

Material and Element Ductilities

4.1 Erosion of Native Properties

In normal conditions the steel, as a structural material, offers high strength, stiffness and ductility, being regarded as an ideal material for structures in seismic areas. These prerequisites for good performance may be defined as native properties.

On the other hand, in non-normal conditions, as produced during an earthquake attack, this potential good performance may be lost, due to the erosion of the native properties. At the material level, the erosion is due to the effect of loading conditions, as high-velocity and cyclic actions. At the element level, this erosion is induced by the random variability of element strength and by the effect of local buckling (Fig. 4.1). The erosion of steel native prerequisites is also caused by other factors and the consequent reduction in ductility will be the subject of the following chapters.

Thus, the native properties of steel cannot be assumed as nominal values to evaluate the response of a structure loaded by seismic action, but their degradation must be taken into account.

4.2 Material Ductility

4.2.1. Selecting Material Promising Good Ductility

The selection of initial material properties plays a leading role in a proper design of a structure in a seismic area. A good selection of material assures that, after the ductility erosion due to different factors, the available ductility remains greater than the required ductility. A bad selection, without considering all the factors producing erosions, may generate a discrepancy between the two ductilities (Fig. 4.1).

This aspect is very real, because the development of steel production and the applications of steel in civil and industrial buildings have the tendency

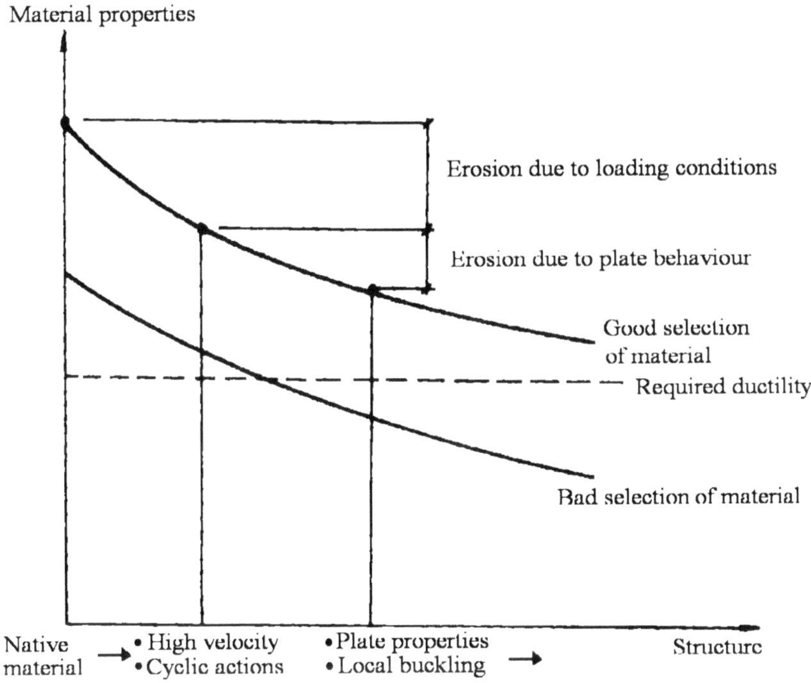

Figure 4.1: Erosion of native properties

to introduce high strength materials (Hemerich and Milcic, 1992, Fukumoto, 1994). Recent advances in important structures require use of thick plates with high strength, high fracture toughness and high corrosion resistance. All these requirements give rise to a decreasing of steel plastic properties and in practice the basic problem in using new kinds of steel is to determine the actual erosion of the material ductility in the behavioural conditions of a structure loaded by an earthquake.

4.2.2. Main Factors Influencing the Steel Properties

The steel properties are subdivided into mechanical and chemical (Kuwamura, 1997a). The former are related to strength and deformation, represented by yield stress, tensile strength, yield ratio, elongation capacity and fracture type. The chemical properties are related to the composition, as content of different elements, in order to obtain a desired alloying effect, as well as the weldability.

(i) *Mechanical properties.* The main factors influencing the material properties, which may give an important contribution to the erosion of the native steel qualities may be summarized as:

-high-velocity loading;
-cyclic loading;
-plastic strain accumulation.

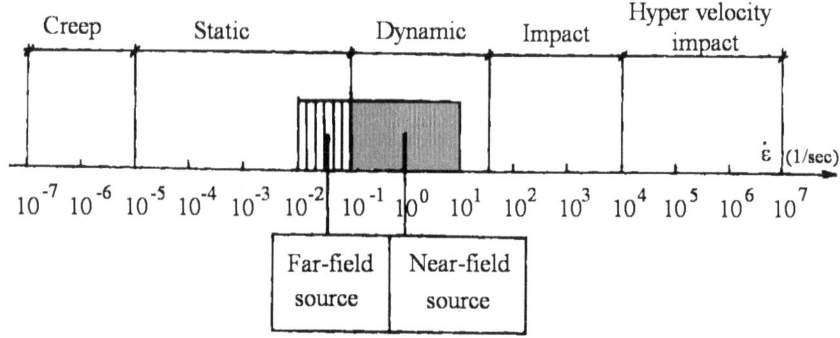

Figure 4.2: Classification of the dynamic effects

Referring to the influence of velocity, which may be expressed by the strain-rate $\dot{\varepsilon}$, Fig. 4.2 shows a classification of the dynamic effects (Kaneko, 1997). For strain rates smaller than $10^{-1}/sec$ the inertial forces can be disregarded, while for greater values, the inertial forces play an important role in the structural behaviour. For strain-rates smaller than $10^{-5}/sec$, a creep rate of steel is studied; in the range within $10^{-5}/sec$ and $10^{-1}/sec$ it is possible to consider that the loading is static; between $10^{-1}/sec$ and $0.5 \times 10^2/sec$ the behaviour is dynamic; the interval of $0.5 \times 10^2/sec$ and $10^4/sec$ is the field of the impact phenomena, while for rates greater than $10^4/sec$, hyper velocity impact occurs. The problem of framing an earthquake strain-rate in this classification is an open one, because the corresponding value depends on many very complex factors, such as ground motion velocity, amplification due to structure behaviour, damper due to non-structural elements, etc. The general accepted range is considered between $10^{-2}/sec$ and $10^{-1}/sec$ (Uang and Bondad, 1996, Suita et al, 1998). But recently, after the new data obtained during the last great events when the characteristics of the impulse loading in the field near to sources are emphasized, very high strain-rate, up to $10^1/sec$ may be recorded during severe earthquakes (Gioncu, 2000).

Referring to the influence of cyclic loading, the first problem is to decide if the seismic action can be studied using cyclic loading. The seismic ground motions have a very chaotic history, depending on many factors, while the cyclic loadings are normally regular, according to the research requirements (Fig. 4.3). Recognizing that some very important differences exist between the two loading types, it must be admitted that today it is not possible, for practical purposes, to use the characteristics of the actual ground motions in design. Due to this fact, it is indispensable to study the effects of seismic loading through the cyclic load approach. The main problem is to select the most appropriate form of the loading cycle which characterizes the actual ground motions in the best way.

During the earthquake the structure is submitted to a limited number of loading cycles. Due to this fact, many research works have classified the failure of members in a structure as due to low-cycle fatigue, considering this phenomenon as an extension of the classical high-cycle fatigue. During

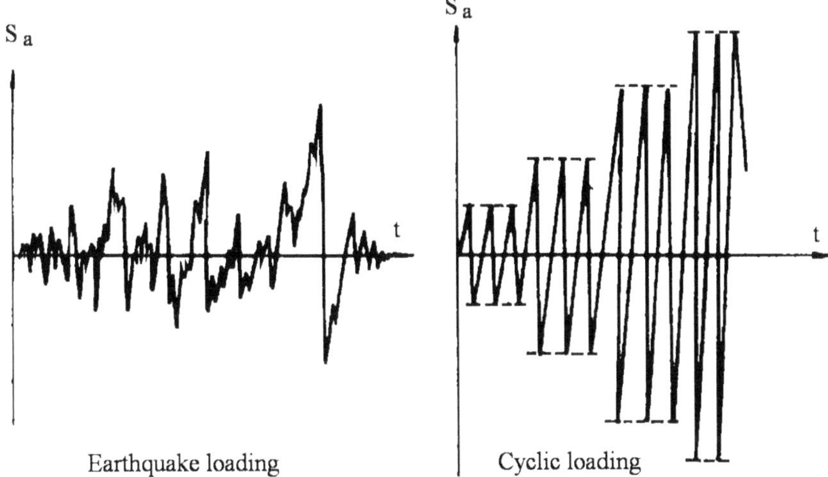

Figure 4.3: Earthquake and cyclic loadings

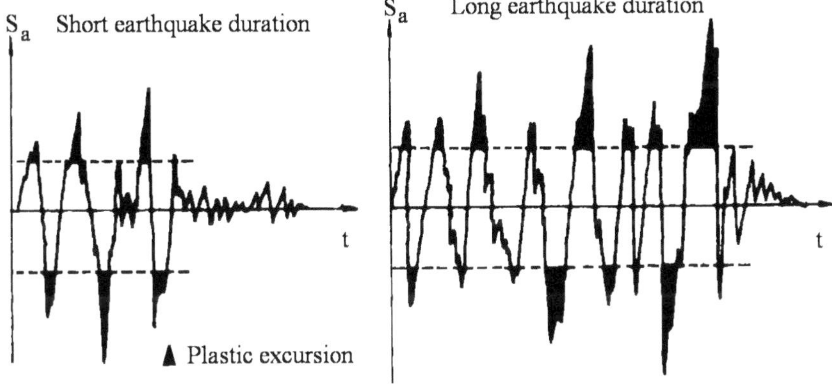

Figure 4.4: Influence of earthquake duration

an earthquake with short duration, less than five cycles of large plasticity occur, while for an earthquake with long duration no more than 20 cycles induce plastic excursion (Fig. 4.4). Therefore, classifying failures that occur under repeated large deformations as belonging to the category of fatigue failure is questionable. The low cycles with high plastic deformations cause an accumulation of these deformations along the yield lines, inducing cracking or rupture in the buckled plates, rather than a reduction of material strength, as in the case of failure under high cycle fatigue.

(ii) *Chemical properties.* Taking into account the content of different elements, the steel used for structural applications may be classified into two fundamental groups (Mazzolani and Piluso, 1995, EN10020, 1988):

-non-alloy steels;

-alloy steels.

Table 4.1: Limit content between non-alloy and alloy steels

Element	Limit content (%)	Element	Limit content (%)	Element	Limit content (%)
Al	0.10	B	0.008	Bi	0.10
Co	0.10	Cr	0.30	Cu	0.40
La	0.05	Mn	1.65	Mo	0.08
Nb	0.06	Ni	0.30	Pb	0.50
Se	0.10	Si	0.50	Te	0.10
Ti	0.05	V	0.10	W	0.10
Zr	0.05	Other(*)	0.05		
(*) Exception is made for C, P, S, N					

Table 4.2: Weldable fine grain steels: limit content between quality and special steels

Element	Limit content (%)	Element	Limit content (%)	Element	Limit content (%)
Cr	0.50	Cu	0.50	La	0.60
Mn	1.80	Mo	0.10	Nb	0.08
Ni	0.50	Ti	0.12	V	0.12
Zr	0.12	Others	see Table 4.1		

Non-alloy steel are those in which the content of elements added in order to obtain a desired effect is less than the one provided in Table 4.1. Alloy steels are the ones in which at least one element has a content equal or greater than the one provided in Table 4.1. Weldable fine grain steels have to be considered as quality alloy steels having a content of alloying elements less than the one provided in Table 4.2. Special alloy steels are characterized by a precise chemical composition, providing particular properties.

4.2.3. Native Steel Performances

The response of material in a structure is a very intricate problem, which must be reduced to simple one, by emphasizing its native performance, useful for the design engineer practice. The uniaxial tension test under monotone loading is conventionally recognized as a unified basis to compare structural materials. But it must be underlined that the result of the axial test never represents the actual behaviour of steel in a structure (Bjorhovde,

1998). The good ductility of steel determined by tension tests does not certainly assure a good ductility at the structural level. This difference in behaviour was for a long time the main cause of great confusion in the appreciation of steel as an ideal material for seismic-resistant structures.

(i)*Main mechanical properties.* The definitions of the most important mechanical properties, which interest structural engineering applications, are the following (Mazzolani and Piluso, 1995):

-*yield point* is the stress corresponding to the plateau in the stress-strain curve; in other words, it is the stress value at which an increase of strain occurs without an increase in stress. In the case of steels that do not exhibit a distinct yielding behaviour, reference is made to the *yield stress* which is defined as the stress providing, after the unloading phase, a residual plastic deformation equal to 0.2 %. *Yield strength* is often used as a generic term to denote either the yield point or the yield stress;

-*ultimate strength* or tensile strength is the maximum stress reached in the tensile test;

-*modulus of elasticity* (Young's modulus) is the stress to strain ratio at levels for which stress is linearly proportional to strain. For all the structural steels the modulus of elasticity can be assumed equal to 205000 N/mm^2;

-*initial strain-hardening modulus* is the slope of the stress-strain curve at the beginning of the strain-hardening range. The initial strain-hardening modulus and the strain at which strain-hardening begins are particularly important for evaluating the deformation capacity of structural steel members;

-*yield strain* is the strain corresponding to the yield strength;

-*uniform strain* is the strain attained at the maximum point of the engineering stress-strain curve, which corresponds to the beginning of necking;

-*total strain* is the maximum strain attained in the stress-strain curve immediately before fracture occurs;

-*Poisson's ratio* is the ratio between the absolute value of transverse strain and axial strain in elastic range; it can be assumed equal to 0.3. Its value increases as far as strain increases and reaches the maximum value 0.5 after the occurrence of complete yielding.

The tension test is performed on specimens with circular or rectangular cross-section, the latter being obtained by cutting some structural elements (Fig. 4.5a).

The stress-strain curve may be determined in two ways: displacement-controlled or load-controlled tensile tests.

(ii)*Displacement-controlled test* in which the displacement is increased at a constant rate until the specimen is fractured. The form of the stress-strain curve for a specimen of typical low-carbon structural steel is shown in Fig. 4.5b. The strain-rate proposed by the RILEM testing method is $0.5 \times 10^{-4}/sec$ until plastic flow occurs, after which a strain-rates of $10^{-4}/sec$ is recommended for the remainder curve (Kato et al, 1990). Due to the low strain rate, this test can be considered as a static one. The general shape of the stress-strain curve is typically represented by four regions corresponding to elastic, plastic, strain-hardening and necking ranges.

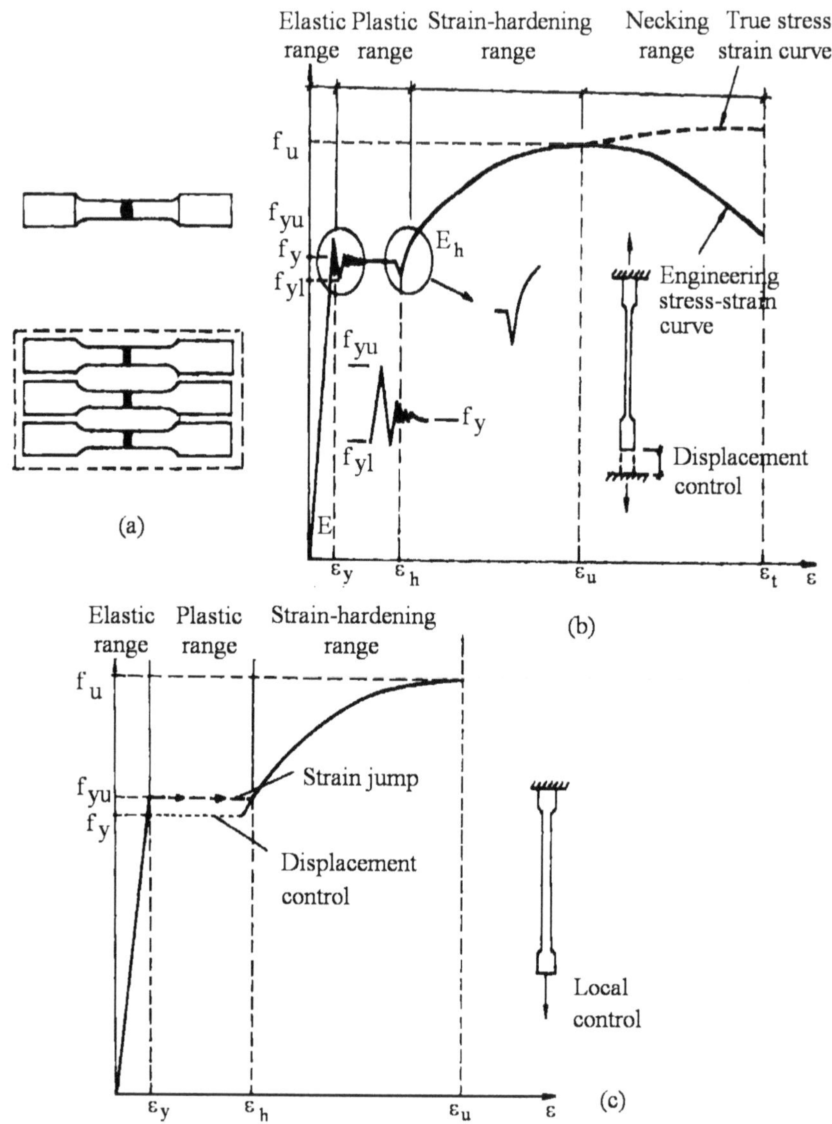

Figure 4.5: Stress-strain curve for structural steel: (a) Specimen types; (b) Displacement-controlled test; (c) Load-controlled test

- The ***linear elastic region*** is defined for $0 \leq \varepsilon < \varepsilon_y$. The tensile stress-strain relationship is given by a straight line $\sigma = E\varepsilon$, where E represents the initial elastic modulus. The yield strain ε_y is given by $\varepsilon_y = f_y/E$.

- The ***yield plateau***, defined for $\varepsilon_y \leq \varepsilon \leq \varepsilon_h$, is assumed to be horizontal, but, as Fig. 4.5b shows, this is not actually the case, because small stress fluctuations arise with upper or lower values of yield stress. The upper

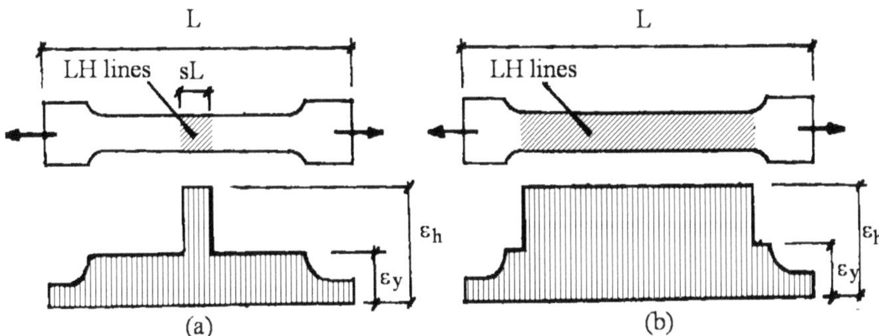

Figure 4.6: Luders-Hartmann lines: (a) Limited plastic zones; (b) Extended plastic zones

yield stress, f_{yu}, is the maximum attaining stress in elastic range, after which an increase of strain occurs, without an increase of stress. The lower yield stress, f_{yl}, is the lowest level of yield stress, immediately following the upper value. Using the dislocation theory to explain the yielding of steel, the upper yield stress mobilizes dislocations and the lower yield stress maintains the movement of the dislocations (Bjorhovde, 1998). The yield stress f_y is the average stress during the actual yielding in the plastic range. It is necessary to underline that the decreasing of stress from upper to lower yield stress is possible only in case of displacement-controlled test.

At yield point, regular line systems appear on the surface of steel specimens, indicating that the plastic strain is concerted in some bands (Fig. 4.6a), their traces on the surface being referred as Luders-Hartmann (LH) lines (Lay, 1965a, Szabo and Ivanyi, 1995). This behaviour is due to the slip of crystalline planes in the direction of the critical shear stress. Generally, the direction of slip plane coincides with the maximum shear stress direction, at 45° inclination from specimen axis. The LH lines occur on the specimen surface only in a limited zone, where the strain traverses all the plastic plateau, reaching the hardening strain ε_h, while in the remained zones, the strain corresponds to yield one. Therefore, the behaviour of specimens is inhomogenous, having discontinuous characteristics. The overall strain is

$$\varepsilon = \frac{\delta L}{L} = (1-s)\varepsilon_y + s\varepsilon_h \tag{4.1}$$

where sL measures the length of slip planes. The increasing of strain in plastic range is due to the formation of new slip planes, until the entire specimen reaches the hardening strain (Fig. 4.6b). So, the yield plateau does not correspond to a continuous deformation, but it is characterised by a succession of deformation jumps from yield strain to hardening strain.

-The *strain-hardening region* is defined for $\varepsilon_h < \varepsilon < \varepsilon_u$. The point at which the yield plateau ends and strain-hardening begins is not clearly identified, because a dip generally occurs at the end of the yield plateau before strain-hardening initiates (Dodd and Restrepo-Posada, 1995). This dip is followed by a steep increase that suddenly changes the slope into relative smooth strain-hardening range, at the beginning of the hardening strain, ε_h. The curve slope in this point is referred to as hardening modulus E_h. The strain-hardening range is extended until the maximum tensile stress, f_u, corresponds to the uniform strain ε_u. At this point the stress-strain curve has a zero slope and the necking range begins.

-The ***post-ultimate stress*** or ***strain-softening region*** is defined for $\varepsilon > \varepsilon_u$. In this region an apparent decrease in the engineering stress-strain curve occurs, due to the specimen necking. In reality, if the actual area is used for determining the actual stress, a continuous increasing of stress is obtained (true stress-true strain curve). The stress corresponding to the specimen fracture is f_t and the strain is ε_t. Due to the fact that this range cannot be used in practice, the ultimate stress f_u and uniform strain ε_u mark the end of the practical useful region of the stress-strain curve.

(iii)*Load-controlled test*, in which the load is increased at a constant rate until the specimen fractures. The corresponding stress-strain curve is presented in Fig. 4.5c. The first elastic range until the upper yield stress is identical as for displacement-controlled test. When the stress reaches the upper yield stress, a decreasing to lower yield stress occurs and a continuous formation of slip bands can be observed in the case of displacement-controlled tests. But in the case of load-controlled tests this decreasing is impossible and thus all the bands slip directly until the hardening range with great velocity. This produces a jump of strain from ε_y till ε_h, local strain variation between these two limits being physically impossible. The test stops at the ultimate stress and uniform strain, because it is not possible to obtain the decreasing of load in the necking range.

(iv)*Behaviour in compression.* Generally, if the effect of buckling is neglected, it is considered that the monotonic stress-strain curve for compression is equal and opposite to the curve in tension. There are some proposals to modify the tension curve, but the differences are not so high. So, the two curves shall be considered equal in the following Sections.

(v)*High strength steel.* The above native properties of steel are presented for mild carbon steel with yield strength in the range of 230-350 N/mm^2. Due to the advances in fabrication technology, the recent trend leads to built structures using higher strength steel, having yield strengths in the range of 400-700 N/mm^2. The effect of the increasing of yield strength on the stress-strain curve is presented in Fig. 4.7. It is clear that the stress-strain behaviour of high strength steels is considerably different from that of low strength steel, exhibiting lower material deformability. For this reason high strength steels must be used with caution for moment-resistant structures in seismic areas.

(vi)*Structural steels.* The principal material properties which determine the structural ductility are: the yield stress, the length of yield plateau, the strain-hardening characteristics and the ratio between yield stress and tensile strength.

The yield stress of structural steel is dependent on the chemical composition of the alloy and the method of manufacturing. These variables are

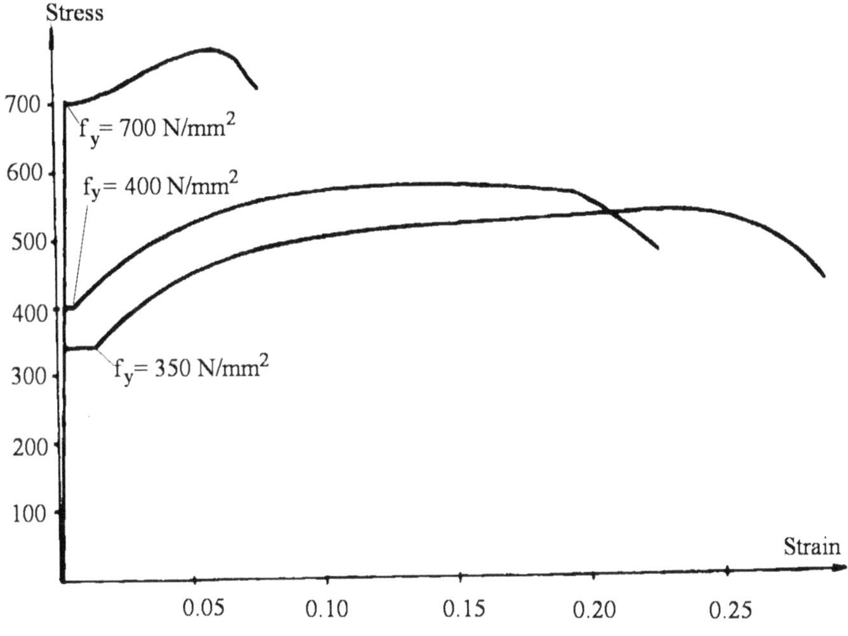

Figure 4.7: High strength steels

reflected in the designated grade of the steel. According to Eurocode 3, Chapter 3 "Materials" and to EN 10025 code, the structural steels usually adopted for civil engineering structures are FeE 235 (Fe 360), FeE 275 (Fe 430), FeE 355 (Fe 510). The nominal values of yield stress and ultimate strength are given in Table 4.3; they are the minimum values guaranteed by the producer. But the actual yield stress and tensile strength values are much higher than those guaranteed and a tendency to increase this difference exists, due to the improving technology of steel production. If this is a very good situation for strength problems, it is not true for seismic applications. This discrepancy and the implications for seismic design will be in detail discussed in the next Sections.

The strain-hardening property is mainly defined by the initial strain-hardening modulus, E_h, for which no standard procedure of measuring exists, so various values are proposed. Generally, its value is evaluated as a fraction of the elastic modulus, say 30 to 50 times smaller. In addition there are two very important parameters: the value of uniform strain, ε_u, which marks the end of hardening range and the total elongation at rupture, ε_t, which limits the necking range.

(vii)*Material ductility*, named also kinematic ductility, can be defined as the ratio between the hardening strain ε_h and the yield strain ε_y:

$$\mu_h = \frac{\varepsilon_h}{\varepsilon_y} = 10 \div 12 \tag{4.2a}$$

Table 4.3: Stress-strain characteristics

Steel	t mm	f_y N/mm^2	f_u N/mm^2	ε_y %	ε_h %	ε_u %	ε_t %	E_h N/mm^2
FeE235	<40	235	360	0.115	1.41	14.0	25.0	5500
(Fe 360)	>40	215	340	0.105				
FeE275	<40	275	430	0.134	1.47	12.0	22.0	4800
(Fe 430)	>40	255	410	0.124				
FeE355	<40	355	510	0.173	1.70	11.0	20.0	4250
(Fe 510)	>40	355	490	0.163				

or as the ratio between the uniform strain, ε_u, and the yield strain ε_y

$$\mu_u = \frac{\varepsilon_u}{\varepsilon_y} = 60 \div 120. \tag{4.2b}$$

One can see that the material ductility is very high, but the ductility of steel in a structure can be eroded at very low values. So, the material ductility can give an unreliable representation of the actual ductility.

(viii) *Yield ratio.* Due to the fact that structures are subjected to internal force gradients along the member axis, a good ductility is provided for low values of the ratio between yield stress and tensile strength:

$$\rho_y = \frac{f_y}{f_u} \tag{4.3}$$

The relation between yield ratio and yield stress is shown in Fig. 4.8. Low yield ratio is considered for $\rho_y < 0.75$ and height yield ratio for $\rho > 0.75$ (Iwatsubo et al, 1997). The yield ratio increases with the increasing of yield stress. As a consequence, ductility is substantially impaired by the elevated yield ratio. A yield ratio which assures a good behaviour is within 0.5–0.7. From Fig. 4.8 one can see that the used steels frame in this condition. The problem of yield ratio is very important for recent steel production, in which a tendency exists to realize high strength steels with tensile strength greater than 600 N/mm^2 and yield ratio greater than 0.9. Therefore, these high strength steels show a very poor structural ductility and this is the reason why they are not recommended to be used for seismic-resistant structures. However, recent developments in steel production obtained high strength steels with yield ratio of about 0.75 (Kuwamura, 1988, Iwatsubo et al, 1997), which can be considered acceptable for structures in seismic areas.

4.2.4. Idealisation of Stress-Strain Curve

For analytical and numerical studies, the actual stress-strain curve is often idealized by means of classical rigid-plastic, elasto-plastic or three-linear

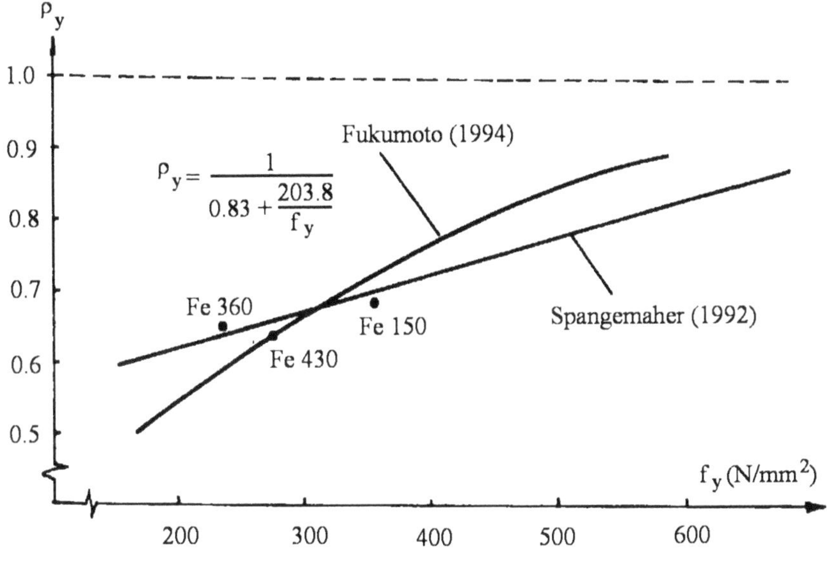

Figure 4.8: Yield ratio

models (Fig. 4.9a). However these models have limited applications in cases of ductility problems where the strain-hardening and fracture ranges play an important role.

For materials with a sharp knee at yielding, as the carbon steel used for ductile structures, the nominal stress-strain relationship can be interpreted in different ways:

(i) *Menegotto-Pinto model*, which is proposed by RILEM Recommendations (Fig. 4.9b), the strain-hardening being described by a nonlinear curve (Rondal, 1998):

$$\sigma = E\varepsilon \qquad 0 < \varepsilon \leq \varepsilon_y \qquad (4.4a)$$

$$\sigma = f_y \qquad 0 < \varepsilon \leq \varepsilon_y \qquad (4.4b)$$

$$\sigma = E\varepsilon / \left[1 + \left(\frac{\varepsilon}{\varepsilon_0}\right)^R\right]^{1/k} \qquad \varepsilon_h < \varepsilon \leq \varepsilon_m \qquad (4.4c)$$

$$\sigma = f_u \qquad \varepsilon_m < \varepsilon < \varepsilon_u \qquad (4.4d)$$

where:

$$\varepsilon_y = \frac{f_{yu}}{E}; \qquad \varepsilon_0 = \frac{\gamma f_u}{E}; \qquad \varepsilon_m = f_u/E\left[1 - \left(\frac{f_u}{E\varepsilon_0}\right)^R\right]^{1/k} \qquad (4.5a-c)$$

R and γ being numerical coefficients determined from experimental data.

(ii) *Model with straight line for hardening range* (Fig. 4.9c) (Boeraeve et al, 1993). The maximum stress is reached for strain value:

$$\varepsilon_{hs} = \varepsilon_h + \frac{f_u}{E_h} \qquad (4.6)$$

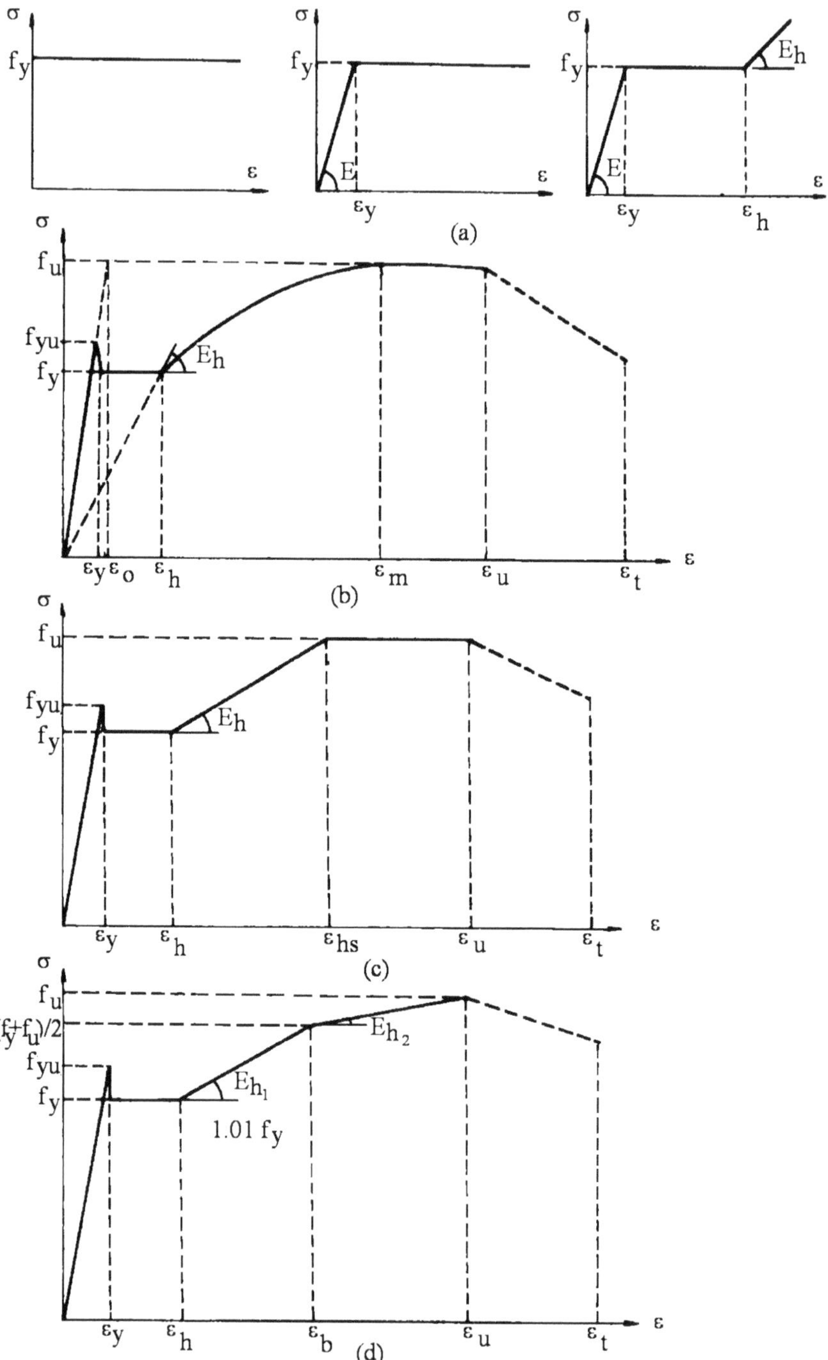

Figure 4.9: Idealisation of stress-strain diagram: (a) Classical models; (b) Menegotto-Pinto model; (c), (d) Multi-linear models

(iii) *Bi-linear model for hardening range* (Fig. 4.9d), the broken point being defined by (Galambos, 1999):

$$\sigma_b = \frac{f_y + f_u}{2}; \qquad \varepsilon_b/\varepsilon_y = 10\ldots40 \qquad (4.7a, b)$$

the lower and upper limits for strains corresponding to high-performance and conventional steels, respectively.

4.2.5. Influence of High-Velocity Loading

The previous monotonic tensile tests correspond to quasi-static loading. Because of the dynamic nature of the structural response to seismic excitations, the accuracy of models derived from these quasi-static tests is uncertain. Thus, it is a crucial problem to clarify how the loading rate affects the material response. The loading-rate effect during an earthquake is now-a-days a topic of many discussions. In some research works this effect is considered negligible (Mahin et al, 1972, Wallace and Krawinkler, 1989), especially for earthquakes which occurred before the Northridge and Kobe events, where moderate velocities were recorded. But after these very important and special earthquakes, where the recorded velocities have been very high, many specialists consider that the loading-rate may be a possible cause of the unexpected bad behaviour of steel structures (Gioncu et al, 2000, Kurobane et al, 1996, Kohzu and Suita, 1996).

There are two types of loading-rate effects, which correspond to strain-rate and stress-rate, respectively. In laboratory tests, the strain-rate is used, while the loading conditions in situ are characterized by stress-rate.

(i)*Strain-rate influence.* The first research work concerning the effect of strain-rate on the behaviour of metals was performed by Morrison (1932), Quinney (1934) and Manjoine (1944). Manjoine's tests were conducted at room temperature for strain-rates from $9.5 \times 10^{-7}/sec$ to $3 \times 10^{2}/sec$, with testing durations between 24 h to a fraction of second (Restrepo-Posada et al, 1994). These results reproduced in Fig. 4.10a indicate a very important increasing of yield stress with an increase of strain-rate, especially for strain-rates greater than $10^{-1}/sec$. The increase of ultimate tensile is moderate, the influence of strain-rate being less important than for yield stress (Fig. 4.10b). Consequently, the yield ratio, defined by equation (4.3), increases as far as the strain-rate increases, with the tendency to reach the value 1 (Fig. 4.10c). So, a reduction of material ductility occurs, especially for strain-rates greater than $10^{-1}/sec$.

More recent results (Wright and Hall, 1964, Rao et al, 1996, Leblois, 1972,Kaneta et al, 1986, Soroushian and Choi, 1987, Fujimoto et al, 1988, Nakagomi and Tsuchihashi, 1988, Kassar et al, 1992, Kassar and Yu, 1992, Obata et al, 1996, Kaneko, 1996, Nakamura et al, 1999) have confirmed the previous results of Manjoine. More detailed research works have shown that the modulus of elasticity is not influenced by the strain-rate variation and the upper yield stress is more strain-rate sensitive than the lower stress

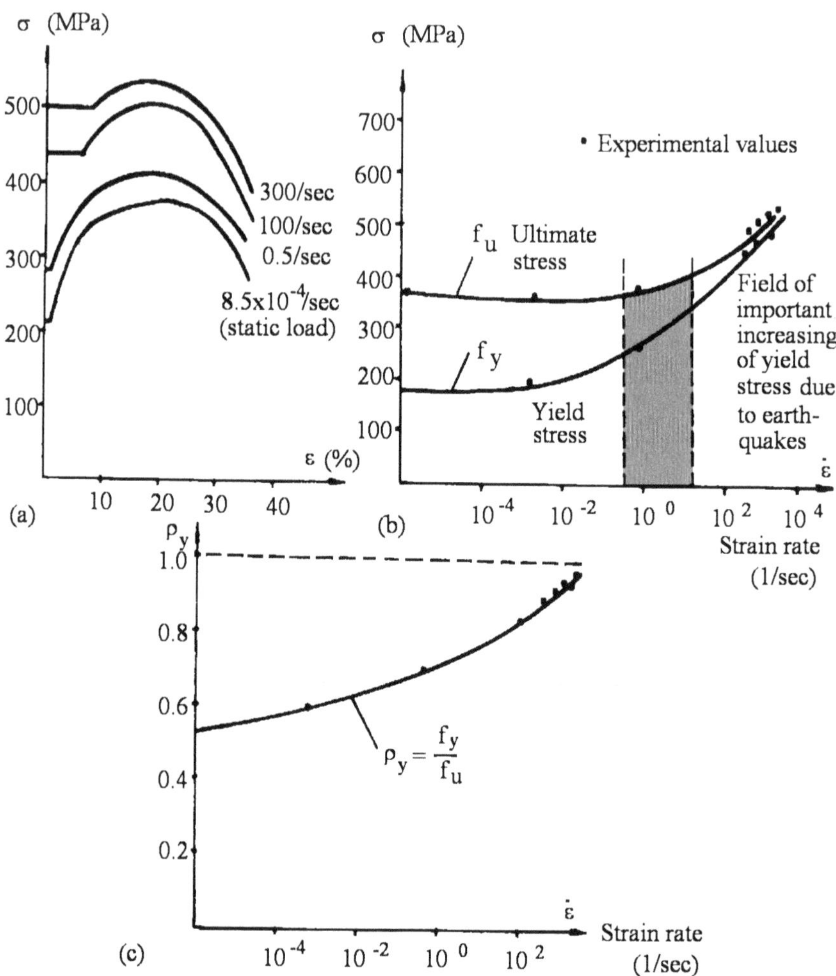

Figure 4.10: Manjoine's strain-rate tests: (a) Stress-strain diagrams; (b) Influence of strain-rate on yield and ultimate stresses; (c) Influence of strain-rate on yield ratio (after Restrepo-Posada et al,1994)

(Fig. 4.11a,b). The upper over lower ratio as a function of strain-rate is presented in Fig. 4.12 (Leblois, 1972). The ultimate tensile strength is also shown in Fig. 4.11c; its increasing has a slower rate when compared with the increase of yield stress. The increasing of strain-rate produces a rapid increase of yield plateau, but contrarily a slow increase of the ultimate strain. No important differences were observed by analyzing the strain-effect on structural steel, reinforcing steel for concrete, wires, cold-formed members and sheeting. Contrary to this, the galvanising treatment amplifies the influence of strain-rate, due to the increasing of mechanical properties (Kaneta et al, 1986).

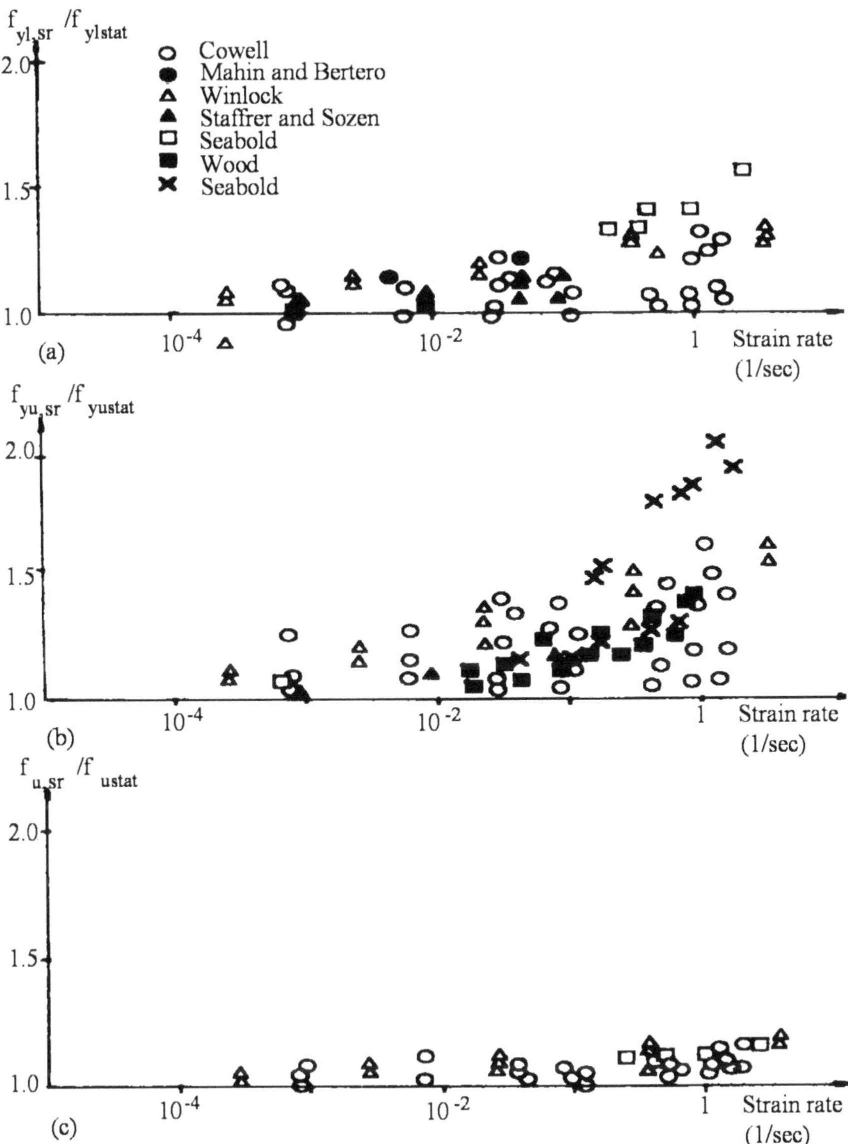

Figure 4.11: Experimental results on strain-rate influence: (a) Lower yield stress; (b) Upper yield stress; (c) Ultimate strength (after Soroushin and Choi, 1987)

(ii) *Stress-rate influence.* The results of Wright and Hall (1964) are presented in Fig. 4.13, where the influence of stress-rate is shown together with the influence of temperature. In this case the most important mechanical property is the upper yield stress. One can see an important increasing in

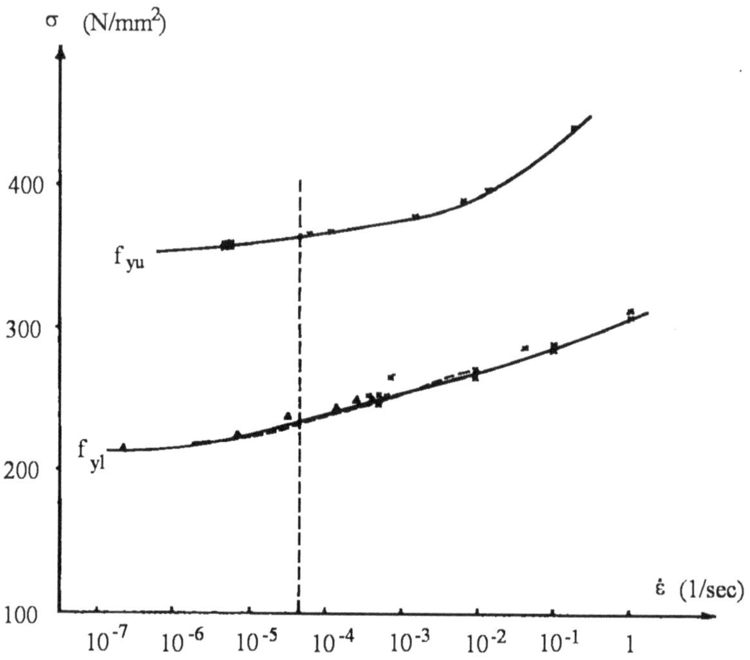

Figure 4.12: Upper and lower yield stress (after Leblois, 1972)

this yield stress with the increasing of stress-rate, in similar way as for the strain-rate.

(iii) *Constitutive proposed laws.* Different constitutive laws modeling the influence of strain-rate $\dot{\varepsilon}$ are proposed:

-Wright and Hall (1964): $10^{-6} < \dot{\varepsilon} < 10^{3}$

$$\frac{f_{ysr}}{f_y} = 1 + 2.77\exp[0.162(\log\dot{\varepsilon} - 3.74)] \tag{4.8a}$$

-Rao et al (1966): $0 < \dot{\varepsilon} < 1.4 \cdot 10^{0}$

$$\frac{f_{ys}}{f_y} = 1 + 0.021(\varepsilon)^{0.26} \qquad : A36 \tag{4.8b}$$

-Soroushian and Choi (1987): $10^{-4} < \dot{\varepsilon} < 10^{1}$

$$\begin{aligned} \frac{f_{ysr}}{f_y} &= 1.46 - 4.51 \times 10^{-7} f_y + (0.0927 - 9.20 \times 10^{-7} f_y) \log\dot{\varepsilon} \\ \frac{f_{usr}}{f_u} &= 1.15 - 7.71 \times 10^{-7} f_y + (0.04969 - 2.44 \times 10^{-7} f_y) \log\dot{\varepsilon} \end{aligned} \tag{4.8c}$$

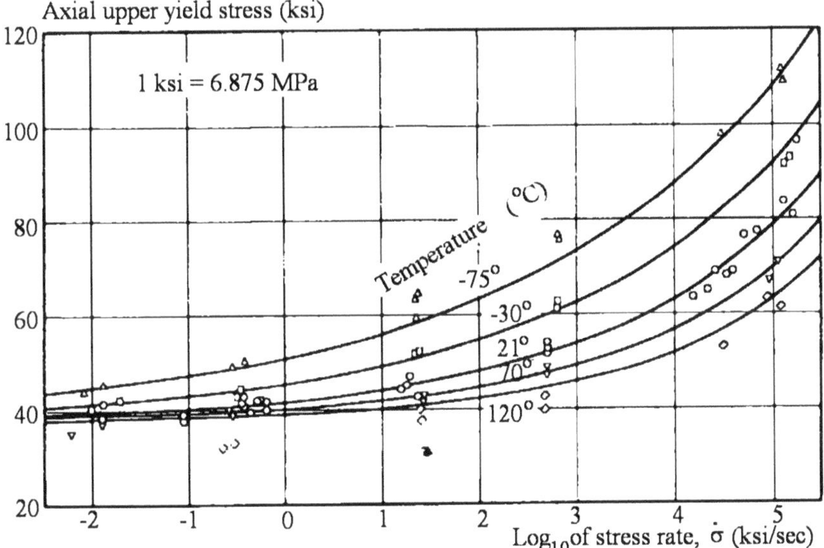

Figure 4.13: Stress-rate influence (after Wright and Hall, 1964)

-Nakagomi and Tsuchihashi (1988): $10^{-4} < \dot{\varepsilon} < 10^{1}$

$$\begin{aligned} \frac{f_{ysr}}{f_y} &= 1.378 + 0.1251 \log \dot{\varepsilon} + 0.010 (\log \dot{\varepsilon})^2, \\ \frac{f_{usr}}{f_u} &= 1.030 + 0.231 \log \dot{\varepsilon} + 0.004 (\log \dot{\varepsilon})^2 \end{aligned} \tag{4.8d}$$

-Wallace and Krawinkler (1989): $10^{-4} < \dot{\varepsilon} < 10^{1}$

$$\frac{f_{ysr}}{f_y} = 0.973 + 0.45 \dot{\varepsilon}^{0.53} \tag{4.8e}$$

-Kassar and Yu (1992): $10^{-4} < \dot{\varepsilon} < 10^{0}$

$$\begin{aligned} \frac{f_{ysr}}{f_y} &= 1.289 + 0.109 \log \dot{\varepsilon} + 0.009 (\log \dot{\varepsilon})^2, \quad (f_y = 320\text{N/mm}^2) \\ \frac{f_{ysr}}{f_y} &= 1.104 + 0.302 \log \dot{\varepsilon} + 0.002 (\log \dot{\varepsilon})^2, \quad (f_y = 495\text{N/mm}^2) \end{aligned} \tag{4.8f}$$

-Kaneko (1996): $10^{-4} < \dot{\varepsilon} < 10^{1}$

$$\begin{aligned} \frac{f_{ysr}}{f_y} &= 1 + \frac{21}{f_y} \log \frac{\dot{\varepsilon}}{\dot{\varepsilon}_0} \qquad \dot{\varepsilon}_0 = 10^{-4}/sec \\ \frac{f_{usr}}{f_u} &= 1 + \frac{7.4}{f_u} \log \frac{\dot{\varepsilon}}{\dot{\varepsilon}_0}, \qquad (\text{N/mm}^2) \end{aligned} \tag{4.8g}$$

In these relations f_{ysr} is the strain-rate yield stress. A comparison among some of these different laws is presented in Fig. 4.14.

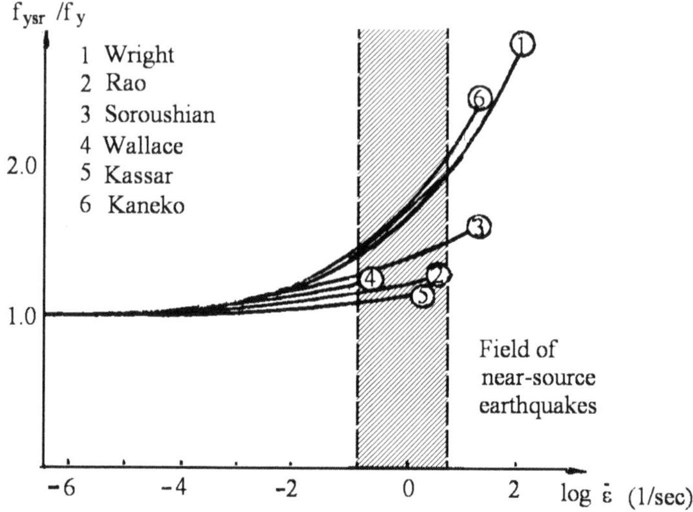

Figure 4.14: Comparison between different strain-rate laws

(iv) *Temperature influence.* The low temperature has a great influence on the upper yield stress, which increases as the temperature decreases (Fig. 4.15). Experimental tests performed with different strain-rates and temperatures (Nakagomi and Tsuchihashi, 1988) have shown that the influence of temperature on the yield stress increases with the increasing of strain-rate, but in the range of +20°C to −20°C the influence is not high. The influence of temperature variation is greatly reduced on the ultimate stress.

A relationship for determining the influence of strain-rate and temperature on yield stress is given by Wright and Hall (1964):

$$\frac{f_{ysr}(T^o)}{f_y} = 1 + 2.77e^{-1.3\left(1+\frac{T^o}{273}\right)(1-0.267\log\dot{\varepsilon})} \tag{4.9}$$

This relation is presented in Fig. 4.16 for temperatures +20°C; 0°C; -20°C. A comparison of upper yield stress values obtained from equation (4.9) and from some experimental results (Nakagomi and Tsuchihashi, 1988) shows a very good correspondence.

In cases where the structure works in conditions where low temperature can occur (structure in open air) the strain-rate influence must be considered by an increasing the yield stresses:

$$f_{ysr}(T^o) = c_T f_{ysr} \tag{4.10}$$

A value of $c_T = 1.1$ is proposed for unprotected buildings for the variation of temperature in the range of +20°C to −20°C. For temperatures out of this range, the values obtained from relation (4.9) may be used.

(v) *Welding influence.* The influence of welding on the material behaviour was the main cause of brittle fracture in connections during the

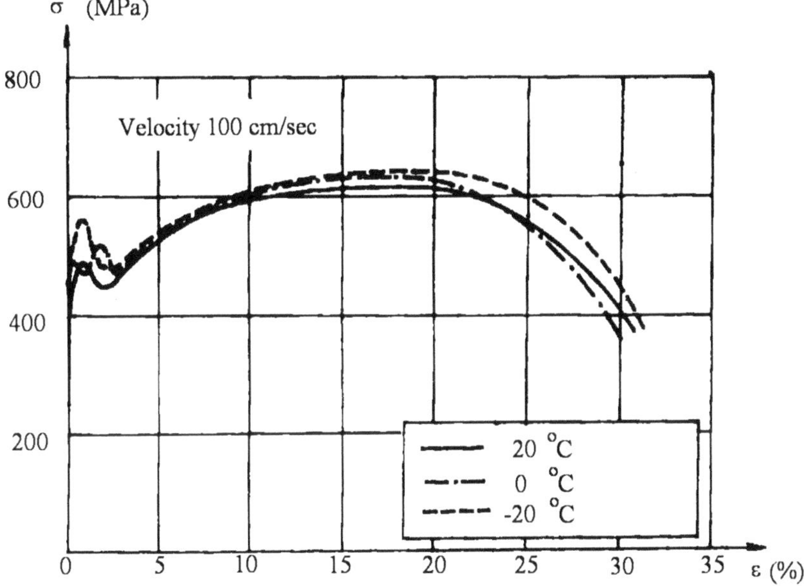

Figure 4.15: Influence of temperature (Nakagomi and Tsuchihashi, 1988)

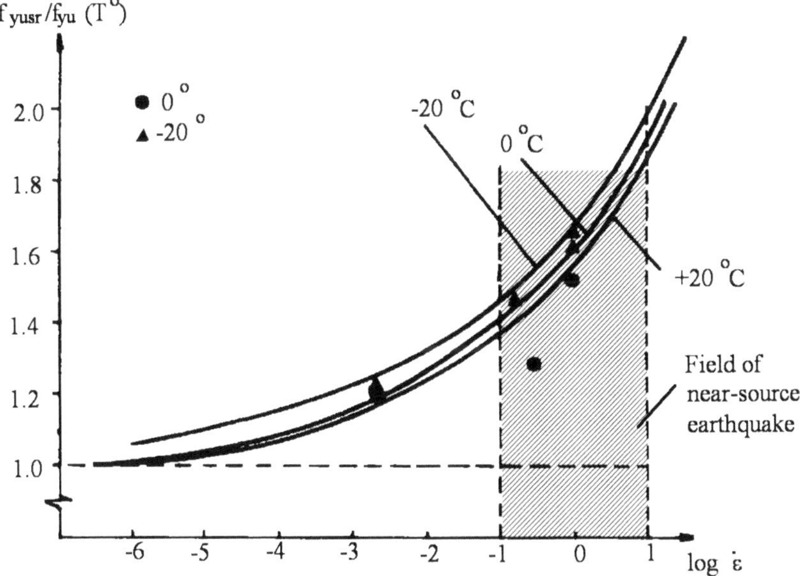

Figure 4.16: Influence of strain-rate and temperature

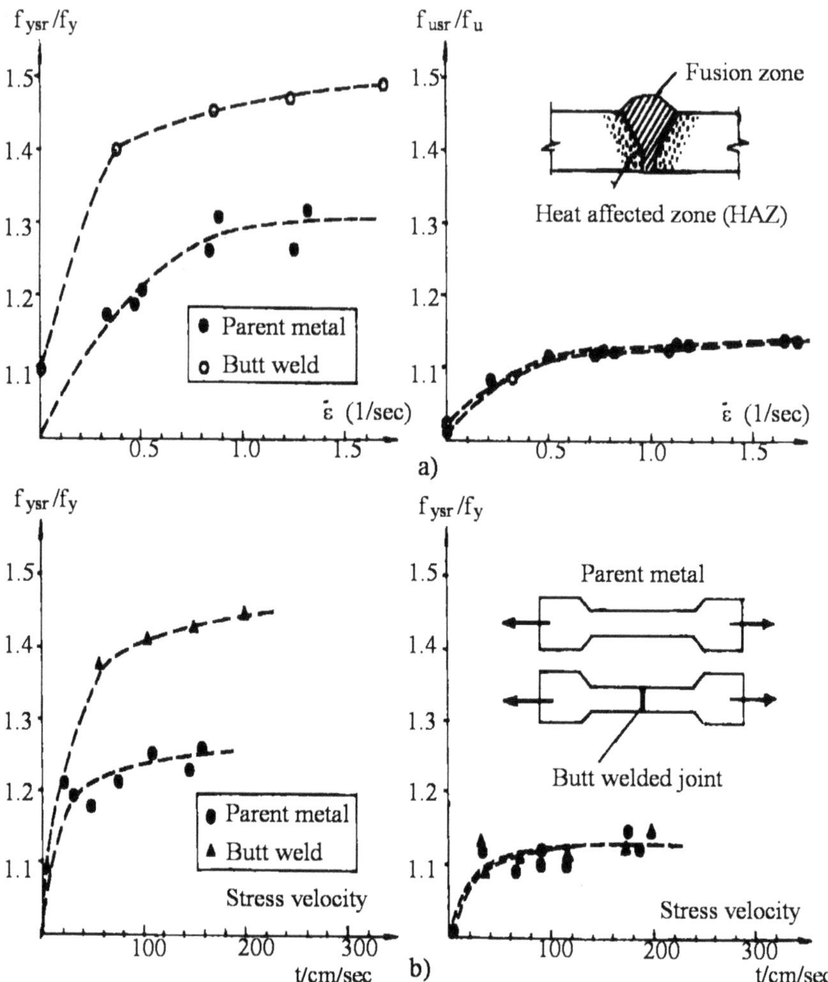

Figure 4.17: Influence of welding: (a) Influence of strain-rate (after Kohzu and Suita, 1996); (b) Influence of stress-rate (after Kaneta et al, 1986)

last important seismic events. As a consequence, many research works have been devoted to this subject (Kaneta et al, 1986, Kohzu and Suita, 1996) by using unwelded and welded specimens. A welded joint basically consists in material, fusion zone and heat affected zone (HAZ). The HAZ plays the most important role on the behaviour of a welded joint due to the crystalline modifications induced by welding.

Under static loading this influence is not particularly significant, but it is very important in the case of loading-rate (Fig. 4.17). A remarkable increasing of yield stress of butt welded joints is observed in comparison with the parent steel. At the same time, the ultimate strength is not sensitive to strain-rate. These observations show that the yield ratio of butt welded joints is greater than the one of parent steel. Similar results have been obtained by Beg et al (2000). So, the welding influences the sensitivity to the loading-rate and the material ductility. Therefore, for welded sections an increase of strain-rate influence must be considered:

$$f_{ysr}(w) = c_w f_{ysr} \tag{4.11}$$

where the correction value c_w is proposed on the basis of experimental data.

(vi)*Influence on yield ratio.* The influence of strain-rate on the yield ratio, using the relationships (4.8), may be determined from one of the following proposals:

-Soroushian and Choi (1987):

$$\frac{\rho_{ysr}}{\rho_y} = c_T c_w \frac{1.46 + 0.0925 \log \dot{\varepsilon}}{1.15 + 0.00496 \log \dot{\varepsilon}} \tag{4.12a}$$

-Nakagoni and Tsuchihashi (1988):

$$\frac{\rho_{ysr}}{\rho_y} = c_T c_w \frac{1.378 + 0.1251 \log \dot{\varepsilon} + 0.010(\log \dot{\varepsilon})^2}{1.030 + 0.231 \log \dot{\varepsilon} + 0.004(\log \dot{\varepsilon})^2} \tag{4.12b}$$

-Kaneko (1996):

$$\frac{\rho_{ysr}}{\rho_y} = c_T c_w \frac{1 + \dfrac{21}{f_y} \log \dfrac{\dot{\varepsilon}}{\dot{\varepsilon}_0}}{1 + \dfrac{7.4}{f_y} \log \dfrac{\dot{\varepsilon}}{\dot{\varepsilon}_0}} \tag{4.12c}$$

the coefficients c_T and c_w being determined from the relationships (4.10) and (4.11), respectively. No very important differences may be noticed among the results obtained using these relationships. The increasing of yield ratio due to high strain-ratio has a very bad influence on the members ductility, significantly reducing the capacity of seismic energy dissipation, especially in the strain-rate range of 10^{-1}/sec to 10^{1}/sec, corresponding to the velocities of near-field earthquakes.

(vii)*Influence of strain-rate during an earthquake.* In conclusion, examining the theoretical and experimental results, one can state that the effect of strain-rate is not very important in case of values of strain-rate less than 10^{-1}/sec, which corresponds to the majority of earthquakes, because the maximum increasing of yield stress is about 20 percent. Contrary to this, in the case of strain-rates greater than 10^{-1}/sec, which occur for earthquakes near the source, the strain-rate effect can be determinant.

4.2.6. Influence of Cyclic Loading

When a specimen, which has been plastically deformed in tension, is unloaded and subsequently reloaded in compression, the stress-strain curve in the new loading phase deviates from the linearity at a stress level far below the one corresponding to the yield point of the virgin material. This phenomenon is known as the Bauschinger effect (Mazzolani, 1974, Dodd and

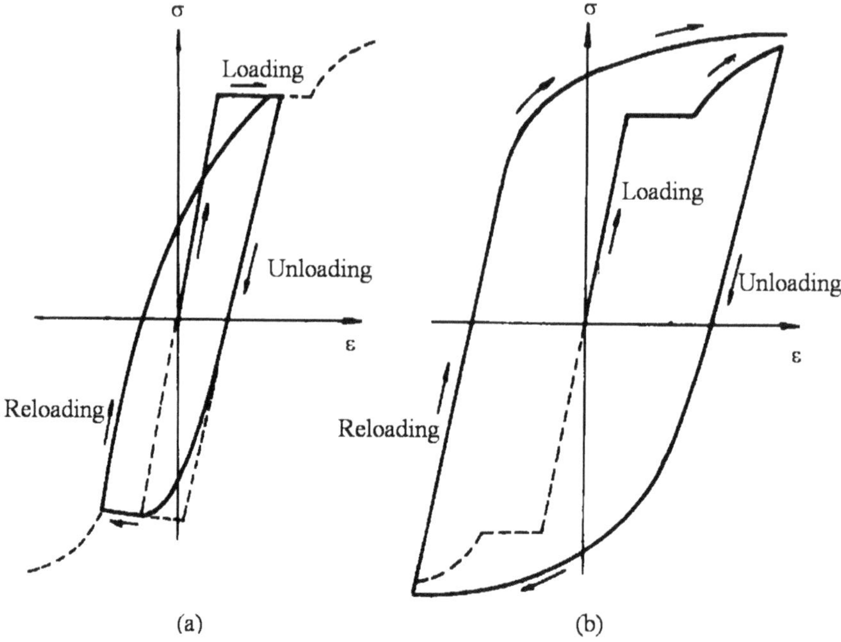

Figure 4.18: Cyclic strain-stress curves: (a) Reversal within yield plateau; (b) Reversal from strain-hardening region

Restropo-Posada, 1995, Mazzolani and Piluso, 1995) and depends on the position of stress reversal. The Bauschinger effect depends on the carbon content, lower carbon steels having a stiffer effect than higher carbon steels.

(i)*Reversal within yield plateau regions.* This is only a theoretical case, because, during an important earthquake, the structure is loaded in strain-hardening region. Fig. 4.18a illustrates the typical stress-strain behaviour of a steel specimen after a load reversal of a point of the yield plateau region. The first part of the unloading branch can be approximated by a straight line parallel to the elastic loading, with the same slope corresponding to the elastic modulus E. In reality, the unloading modulus decreases and depends on plastic strain (Dodd and Restrepo-Posada, 1995), but the differences are not so high ($E_r = (0.85 \div 1.00)E$). In the reversal stress, a stiffness softening gradually occurs due to the earlier yielding. If the specimen is loaded in tension again, then the stress-strain curve will deviate from the linear response far below the virgin yield stress, but it will be able to reach the stress and strain obtained in the first tension loading cycle. The stress-strain curve for the subsequent cycles will follow the same path of the first cycle, provided that the maximum strains are not greater than the ones reached in the first cycle.

(ii)*Reversal from strain-hardening region.* This is the usual case for the behaviour of structures under severe earthquakes. When a reversal starts from the strain-hardening, the shape of the reversal curve is independent

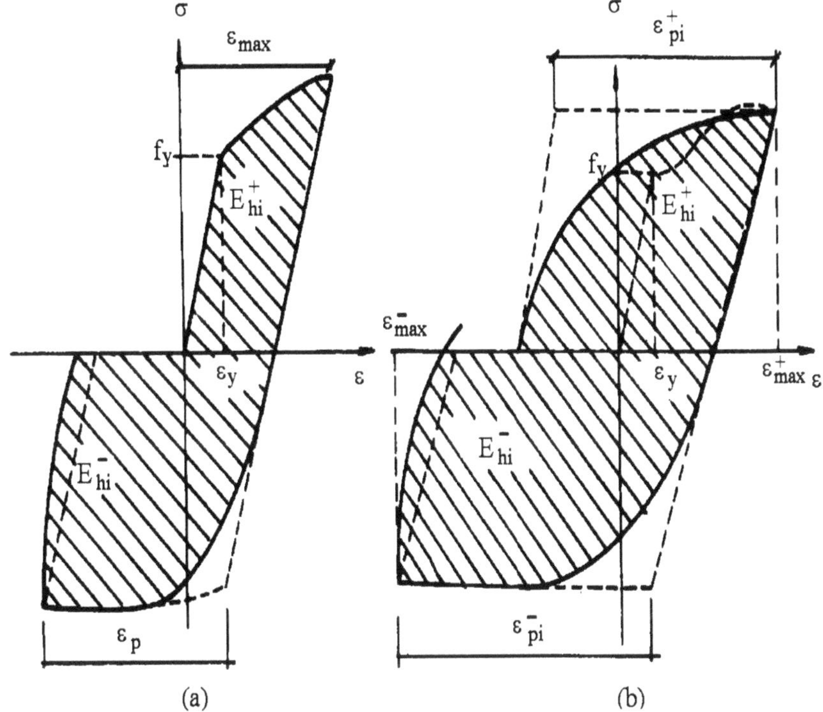

Figure 4.19: Hysteretic energy: (a) First cycle; (b) i cycle

of the virgin stress-strain curve; the yield plateau effectively disappears in both directions (Fig. 4.18b). The reversal curve depends on the maximum tensile load which has been reached.

(iii)*Hysteretic ductilities.* The loading, unloading and reloading curves form a hysteretic loop (Fig. 4.19). The area included within these curves is the dissipated energy. Figure 4.19a shows the first cycle, while the Fig. 4.19b the ith cycle. The kinematic ductility given by the equations (4.2) is appropiate only for one cycle. For deformation histories characterized by many cycles, the use of energy criterion seems more realistic. Two energy criteria have been adopted (Mazzolani and Piluso, 1996):

-hysteretic ductility in one direction:

$$\mu_h = \frac{E_{h,max}}{f_y \varepsilon_y} + 1 \tag{4.13a}$$

where

$$E_{h,max} = \max(E_h^+, E_h^-) \tag{4.13b}$$

-total hysteretic ductility:

$$\mu_h = \frac{E_{h,total}}{f_y \varepsilon_y} + 1 \tag{4.14a}$$

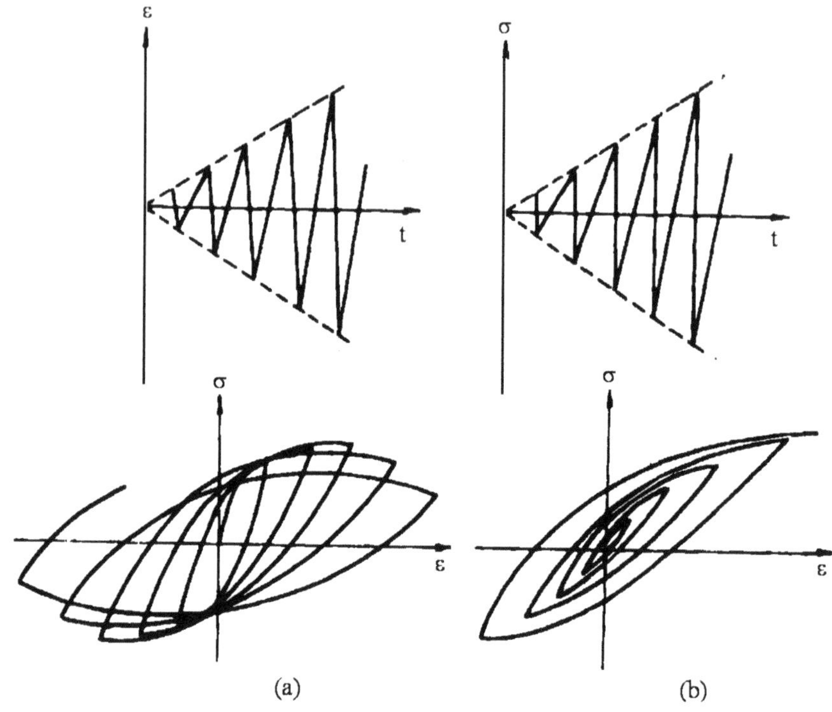

Figure 4.20: Cyclic tests: (a) Strain control; (b) Stress control

where

$$E_{h,total} = E_h^+ + E_h^- \tag{4.14b}$$

(iv) *Control parameters of cyclic test.* As in the case of a monotonic tensile test, the cyclic test can be performed by means of the control of strain or stress. In the first case, under constant periodicity of strain, a continuous degradation of stress occurs for a given level of strain (Fig. 4.20a). In the second case, when the stress is under control, a degradation of strain is produced for a given level of stress (Fig. 4.20b). The control of strain is characteristic for laboratory tests, while the stress control is more realistic for interpreting the behaviour of structures under seismic actions.

(v) *Cyclic loading types.* The cyclic loading types generally used in research works are summarized in Fig. 4.21. The first case is characterized by a fixed imposed value of the control parameter CP (strain or stress) with repetition up to failure (Fig. 4.21a). This cyclic loading type is used for determining the low cycle fatigue effects. The second type is characterized by an increasing of the control parameter after each cycle, or after three cycles, according to ECCS Recommendations (Fig. 4.21b). This loading type is generally recommended for the analysis of structures under ground motions corresponding to earthquakes far from source. The last type is characterized by a continuous decreasing of the control parameter (Fig. 4.21c),

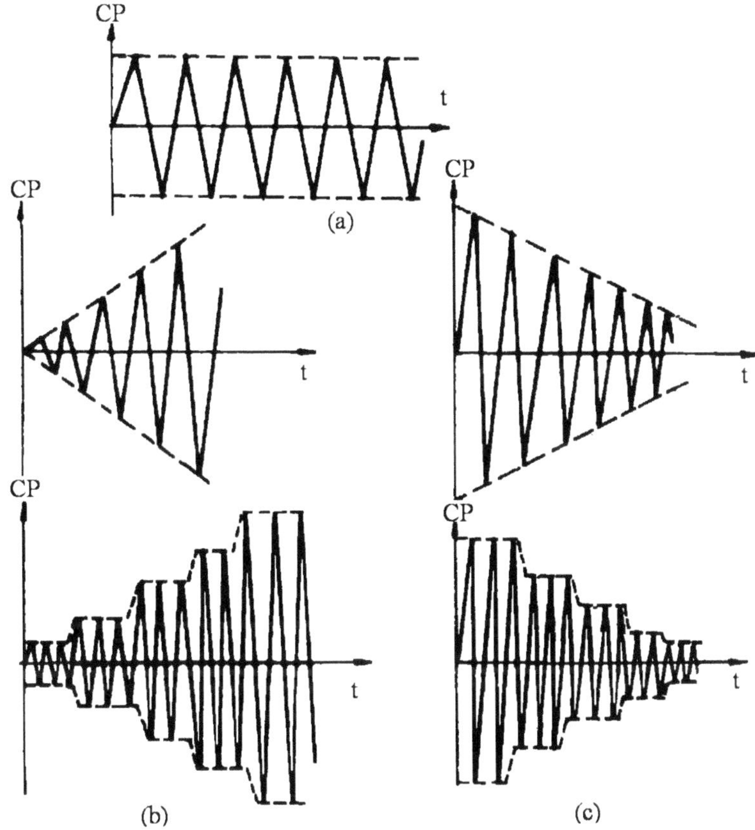

Figure 4.21: Cyclic loading types: (a) Fixed imposed control parameter; (b) Increasing of the control parameters; (c) Decreasing of the control parameters

following the same rule as for the second case. This is the loading history characteristic for the structural behaviour under earthquakes near to source. It is worth reminding ourselves that cyclic loading tests are performed following many other types of laws, very different from the above mentioned ones and their definition is an open problem in the standardization of tests (De Martino and Manfredi, 1994).

(vi)*Skeleton curve.* The concept of skeleton curve has been adopted to compare the ductility and energy dissipation capacity of cyclic loads having different loading histories (Kato et al, 1988, Suita et al, 1996). For this purpose it is very useful to compare the specimen cyclic behaviour with the monotonic one. The skeleton curve is constructed by using the restoring force versus deformation curves which derive from a cyclic loading test (Fig. 4.22). The portion of a force-deformation curve that exceeds the maximum force achieved in previous loading cycles is defined as a new skeleton portion and this portion is added to the skeleton curve already constructed, with

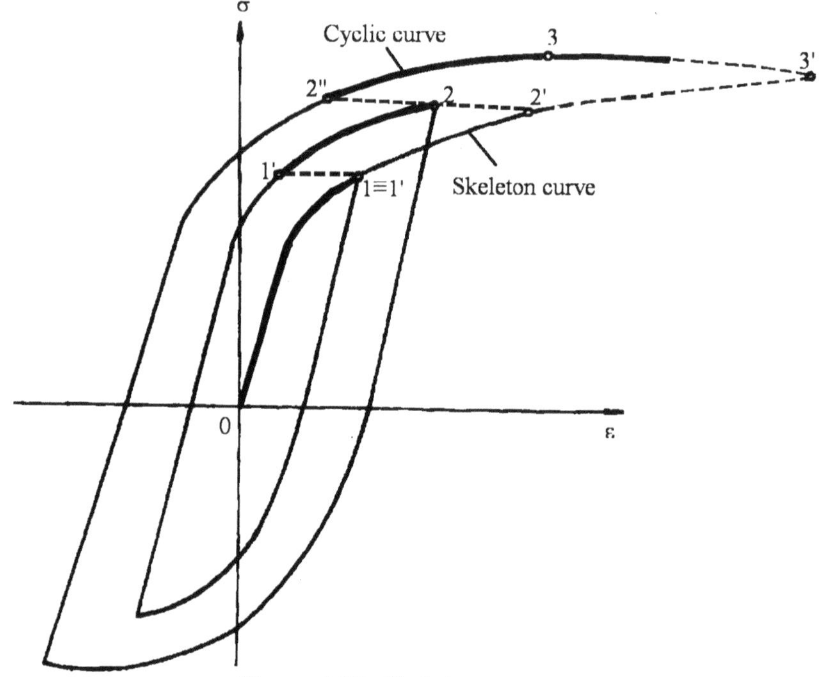

Figure 4.22: Skeleton curve

the horizontal origin shifted to the end of the already built skeleton curve. This treatment is repeated for each cycle and the maximum deformation obtained is defined as the skeleton deformation capacity. This curve may be constructed for both positive and negative loading directions. The skeleton curve shows how the monotonic curve can be used to predict the cyclic behaviour.

(vii)*Influence of cycle velocity.* As for monotonic loading, during the cyclic loads an important increase of yield stresses is observed as a function of strain-rate, having the effect to reduce the material ductility. In case of cyclic loading a favorable effect for ductility behaviour occurs due to the increasing of specimen temperature. Indeed, during the dynamic loading tests a significant temperature rise is observed (Suita et al, 1998) (Fig. 4.23). The temperature continuously rises during loading, gradually decreases during a pause and then rises again during the next loading. It is very clear that this temperature increase depends on the strain-rate and number of cycles. During the experimental test, the maximum temperature reached during each cycle was about 10^oC. Therefore, considering 6–8 the number of important cycles during an earthquake, the temperature of a member can rise by about 60°–80^oC. The increasing of yield stresses for strain-rates of 10^{-2}/sec to 10^1/sec and temperatures increasing from 20^0C to 100^0C is presented in Fig. 4.24, using equation (4.9). One can see that the influence of temperature increasing due to the cyclic action on the reduction of strain-rate effect is not as high as expected.

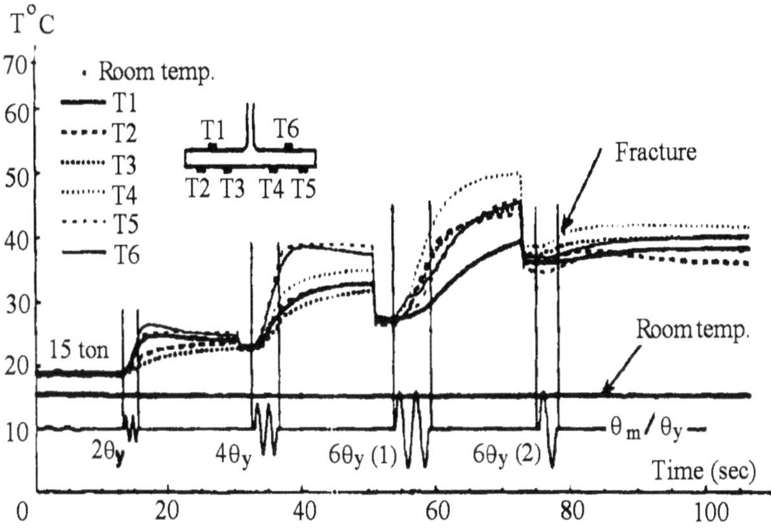

Figure 4.23: Increasing of specimen temperature during the cyclic tests (after Suita et al, 1998)

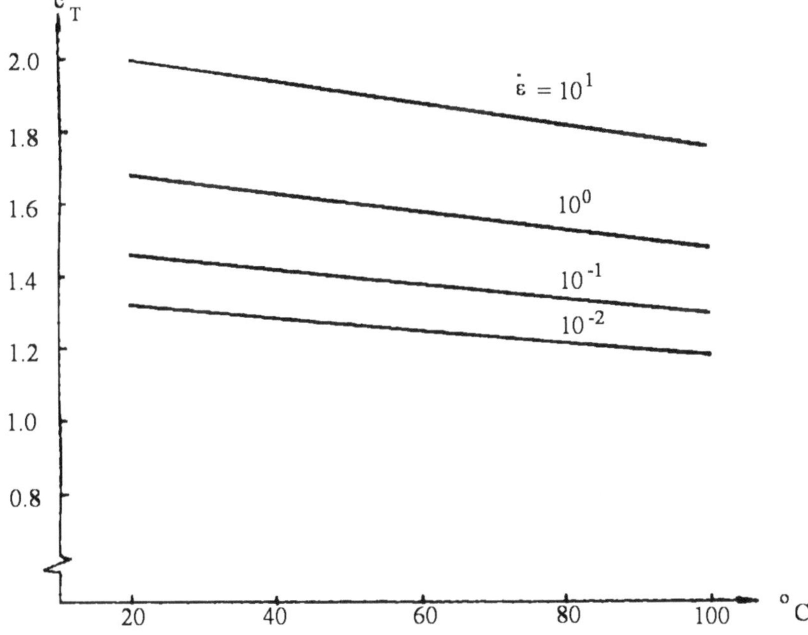

Figure 4.24: Influence of specimen temperature and strain-rate

4.2.7. Ductile, Brittle and Fatigue Fractures

Steel structures subjected to strong earthquakes can collapse either by ductile or brittle fracture. The mode of fracture is governed by stress level, strain-rate, temperature and inherent toughness of constructional steel. In addition, the structures contain geometric discontinuities and fabrication imperfections such as holes, cuts, abrupt change of sections, porosities, slag inclusions, as well as others. These discontinuities and imperfections are regions of stress concentrations and act like notches (Barsom, 1992, Barsom and Rolfe, 1987).

Thus, in addition to the native material properties, the structural engineer must be aware of the effect of material notches, cracks and other imperfections in behaviour of structural components. These difficulties can be solved using the methodologies of *Mechanics of Fracture*, in which the behaviour and strength of solids having crack discontinuities are studied. Unfortunately, the structural engineers do not have sufficient knowledge in this field and, in many cases, they ignore these aspects, considering that the responsibility of fracture belongs to steel manufacturers and fabricators (Kuwamura and Akiyama, 1994). This is the reason why a lot of confusion exists at the level of structural engineering in the field of steel fractures.

From the point of view of material mechanics, *fracture* is produced by the creation of new surface areas within a body, because of irreversible thermodynamic processes (Nica, 1981). It consists of breaking the cohesive bonds of microparticles, hence the accumulation of microcracks. Fracture can be characterized as:

-*ductile*, occurring after extensive plastic deformations with significant energy dissipation, accompanied by slow crack propagation;

-*brittle*, where a very rapid development of cracks occurs with low energy dissipation.

The great problem for a design engineer is to determine the conditions for which a ductile fracture is transformed into a brittle fracture, in order to avoid the last failure type during strong earthquakes. The failure of steel structure due to these ground motions may occur either due to a single plastic excursion, or after more plastic excursions.

(i)*Single plastic excursion failure*, in which the fracture takes place as in the case of monotone test, when the available strength is exceeded by the first plastic deformation. Generally, it is the case of structures subjected to seismic actions which are near to source. In this case the load is of impulse type and the first cycle can produce the structure collapse.

Under a single plastic load, observations on tests of steel members reveal that the fracture is composed of three sequential phases: the first is the initiation of a ductile crack, then the stable growth of the crack and finally the sudden propagation of the crack in a brittle manner (Kuwamura, 1996, 1997a,c, Kuwamura and Yamamoto, 1997). This kind of fracture is released by ductile cracks at strain concentration. A conservative criterion for safety against the brittle fracture can be established by preventing the ductile cracks. The initiation of ductile cracks is governed by three physical factors, that is the average strain, ε, the peak of three-axial stresses, τ_{peak}, and the

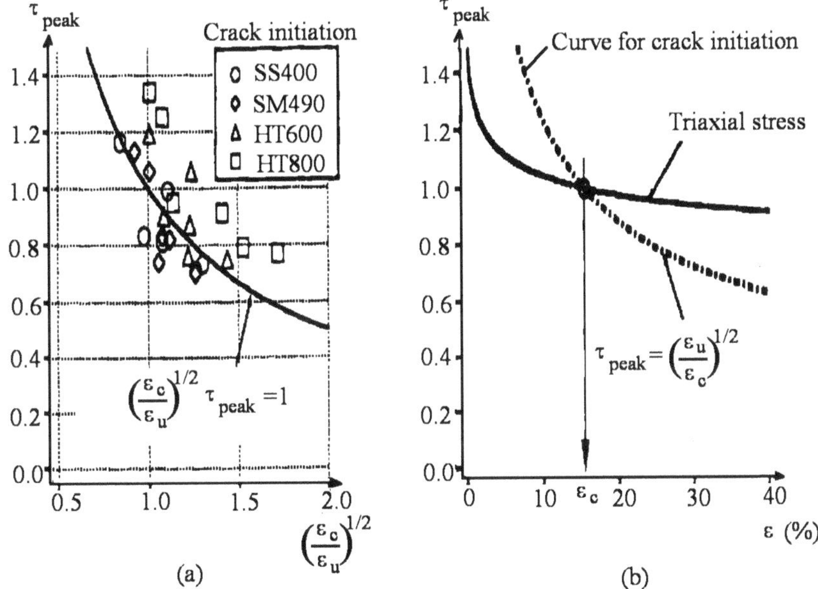

Figure 4.25: Ductile crack initiation stress: (a) Criterion for ductile crack initiation; (b) Scheme to determine crack initiation strain (after Kuwamura, 1996, 1997a)

uniform strain capacity of material, ε_u (Kuwamura, 1997b), in which the strain rate effect must be considered:

$$\varepsilon = \frac{\varepsilon_u}{\tau_{peak}^2} \tag{4.15}$$

This empirical formula was derived from experimental results (Fig. 4.25a). The value of peak, τ_{peak}, may be determined according to the actual detail by means of FEM. The intersection of this curve with the one from equation (4.7) gives the strain at the onset of ductile behaviour (Fig. 4.25b).

Generally, the fracture in a single excursion is a ductile one, but there are some cases in which brittle fracture may occur. One instance is the case of cold formed profiles. It is well known that during the cold press-bracket of a plate the strains exceed the hardening strain and in cases of loading after some time an increasing of yield stress occurs, due to the so call *strain aging effect* (Restrepo-Posada et al, 1994) (Fig. 4.26a). But at the same time that the yield stress increases an accompanying dramatic decreasing of material ductility occurs (Fig. 4.26b). The experimental tests on the cold press-bracket profiles (Kuwamura and Akiyama, 1994) have shown that the strength is noticeably increased at the centre of the corner surface in comparison with those in the flat region. These mechanical changes are associated with fibrous cracks in outer surface. But the elongation capacity is remarkably decreased and consequently an important decreasing of material ductility also occurs (Fig. 4.26b). This effect depends on the thickness

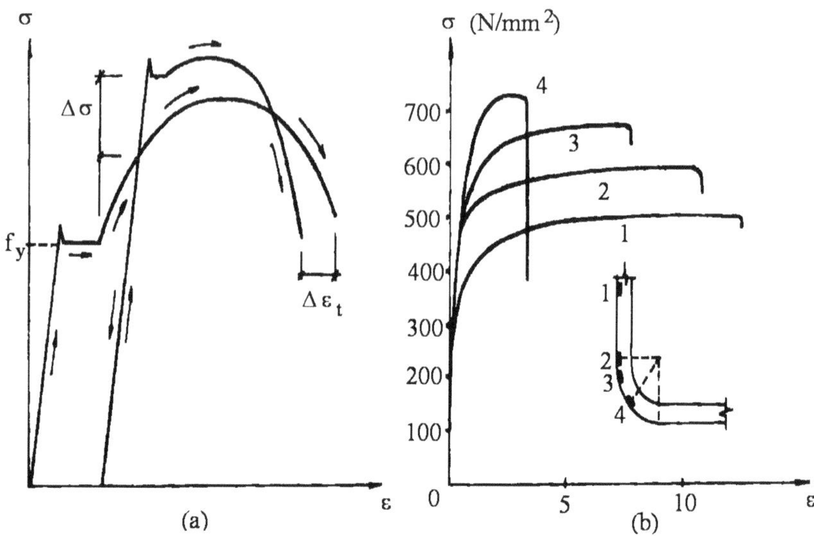

Figure 4.26: Influence of strain aging effect: (a) Increasing of yield stress; (b) Influence of temperature on fracture toughness (after Kuwamura and Akiyama, 1994)

of the plate, with changes more pronounced for thicker plates. Due to the reduction of the elongation capacity in the corner of cold-formed profiles, the ductile fracture may be converted into brittle fracture.

Another factor influencing the fracture type is related to *specimen temperature* and strain-rate of loading. This influence is reflected in the load-displacement curves (Fig. 4.27). For some values of temperature and strain-rate, the ductile behaviour of the specimen is transformed into a brittle one. The study of this phenomenon began due to some brittle failures which happened to the welded steel ships, especially in the Artic Sea. For practical purposes the change in behaviour is studied using the Charpy V-notch specimen subjected to impact loading (Barsom and Rolfe, 1987, Barsom, 1992, Bruneau et al, 1998). The result is the notch toughness and the temperature from which the ductile behaviour is tranformed in a brittle fracture. *Nil-ductility temperature* (NDT) is the highest temperature at which a specimen fails in a purely brittle maner. For rolled H-profiles, Akiyama (1999) has shown that NDT is about -95^oC, while for welded H-sections, NDT is determined to be -65^oC, the differences being given by some defects in welding manufacture.

A very important problem for structural design is to determine the criteria to eliminate brittle fracture. The temperature for which the danger of this brittle fracture is eliminated is the fracture-transition plastic (FTP) temperature. The experimental results (Akiyama, 1999), for strain-rate range from 5×10^{-1}/sec to 9×10^{-1}/sec, have shown that:

$$FTP \approx NDT + (40^o \ldots 60^o)C \tag{4.16}$$

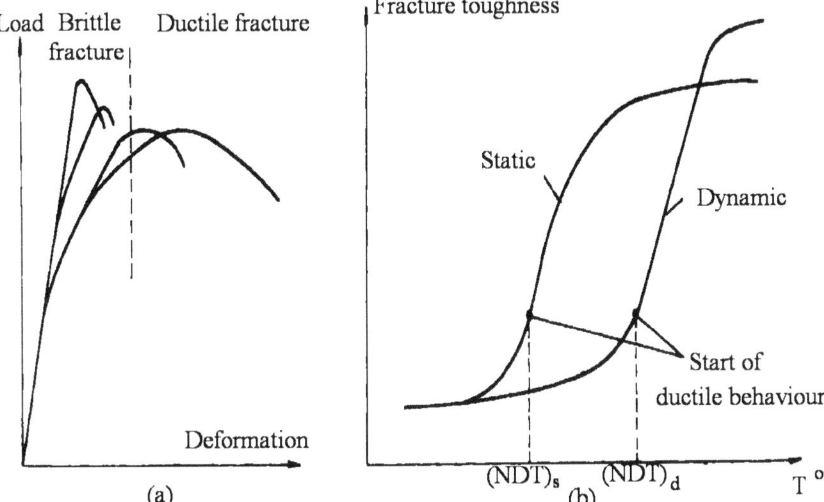

Figure 4.27: Ductile and brittle fracture: (a) Load-displacement curve; (b) Influence of temperature on fracture toughness (after Barsom, 1992)

Under temperatures beyond FTP, the material ductility can be adequately estimated on the basis of tensile strength. One can see that for rolled profiles the relation (4.16) is satisfied by the normal temperature conditions. Contrary to this, for welded sections, the relation (4.16) indicates temperatures of about -20^oC to 0^oC , very close to possible temperatures, when the structure works in open air.

The influence of strain-rate on the FTP temperature is studied by Barsom (1992). The shift temperature from static to strain-rate loading is proposed by:

$$T^o_{shift} = \frac{5}{9}\left[\left(150 - \frac{f_y}{6.895}\right)(\dot{\varepsilon})^{0.17} - 32\right) \qquad T^o \text{ in C, } f_y \text{ in MPa} \quad (4.17)$$

For steel with $f_y = 330$ MPa, for strain rate from 10^{-1} to 10^1/sec, the temperature shift results from 20^oC to 60^oC, in very good agreement with the experimental values determined by Akiyama.

So, for large energy input developed during a very short time and at reduced temperature, brittle fracture of welded structures working in open air can occur. These were the conditions of the Kobe earthquake, hence the brittle fractures observed at Ashiyahama buildings can be understood.

(ii) *More plastic excursion failure*, in which collapse occurs due to cumulative effects of a number of excursions in plastic range. This failure type is characteristic for structures in the situation of intermediate or far field ground motions, where the collapse is dominated by cyclic loading.

This failure type is tributary to the studies on high cycle fatigue, characterized by a large number of cycles with stress range in elastic field. It is very well known that the tensile strength of materials decreases with the

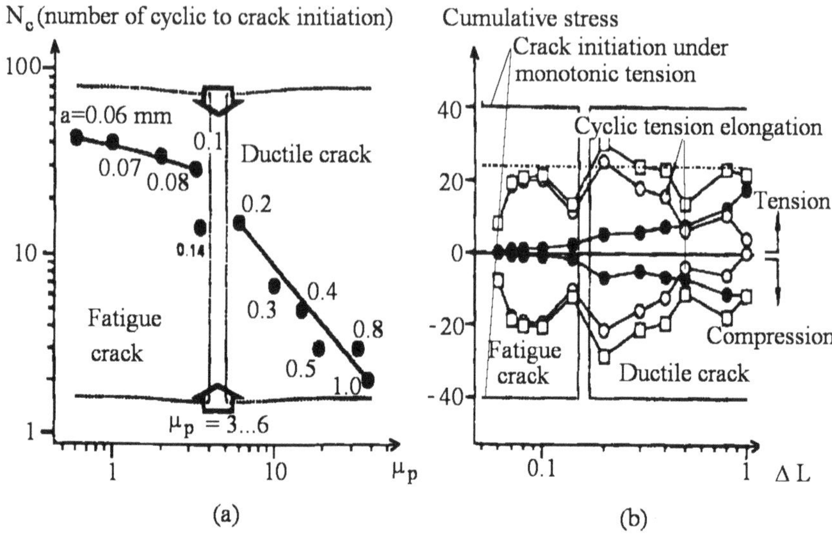

Figure 4.28: Fatigue and ductile cracks: (a) Fatigue to ductile cracks; (b) Cumulative strains preceding crack initiation (after Kuwamura, 1997a)

number of cycles, which is from 10^6 to 10^7. The fatigue effect is the decreasing of strength from 1/3 up to 2/3 of the static fracture strength. A similar reduction is observed for a small number of cycles (up to 10^2), but in the field of large plastic deformations, so the phenomenon is called *low cycle fatigue*. The corresponding reduction in strength in cases of seismic loading is also attributed to this low cycle fatigue by many authors (Cosenza and Manfredi, 1992, Yamada, 1992, Ballio and Castiglioni, 1994a,b, 1995). But the number of plastic cycles in cases of seismic actions is only from 5 to 20. Therefore, Park et al, 1997, propose to call this phenomenon as *very low cycle fatigue*. Due to reduced numbers of plastic cycles, the classification of this failure type, as belonging to the category of fatigue failure, is questionable.

A study on the transition between fatigue and fracture has been performed by Kuwamura (1997a), leading to the conclusion that considering the fracture under a very large plastic strain as fatigue is doubtful, due to the fact that the picture of the fractographs are very different for fatigue and plastic cycles. Fig. 4.28a shows the relation between the number of cycles, N_c, and the plastic ductility, obtained experimentally. One can see that the experimental data may be classified into two groups and fit two different lines. The intersection of the two lines corresponds to the transition point from fatigue to ductile cracks obtained from the fractographs. The transition point lies between 3 and 6 values of ductility, or between 0.6% and 1.2% in plastic strain amplitude. In the area of ductile crack predominance, the crack initiation life is less than 20 cycles, which corresponds to the maximum seismic number of cycles. Thus, the crack induced by an earthquake is not a fatigue crack, but a ductile crack and the fracture in steel under cyclic plastic strains is not due to the fatigue. The main conclusion of Kuwamura's study is that the failure of members during a seismic action is due to the plastic strain accumulation until the ductile crack initiation arises. Fig. 4.28b shows the cumulative strain up to crack initiation, in relation to elongation amplitude. One can see that the cumulative strain capacity under plastic cycles in the ductile crack range seems to be basically equal to the monotonic strain capacity, when the accumulation is due only to tension field. This observation is crucial, because the plastic strain capacity preceding the crack onset under cyclic seismic loading can be predicted from the crack initiation strain obtained under monotonic loading. So, the monotonic tensile test could be used to qualify steel for seismic applications (Kuwamura, 1997a).

Another design problem is to eliminate brittle fracture during cyclic loading. If for monotonic loading the condition (4.16) of FTP is satisfied, no problem occurs for cyclic loading due to important rises of temperature, observed during the experimental tests (Suita et al, 1998, Akiyama, 1999).

4.3 Element Ductility

4.3.1. Main Factors Influencing the Element Ductility

The mechanical properties obtained from static and dynamic tests on the material are generally kept also for the elements. But there are some differences which will be presented in the following Sections.

(i)*Behaviour in tension and compression* (Fig. 4.29). At the level of material there are not important behavioural differences in tension and compression. Contrary for element there are very important differences which influence its ductility, in particular:

-the tension behaviour is dominated by the random variability of mechanical properties;

-the compression behaviour is characterized by the buckling in plastic or strain-hardening ranges.

(ii)*Element type* (Fig. 4.29b). The most used members for steel structures have single or double T, channel, angle and hollow cross-sections. All these sections are composed by two element types:

-supported edge plate;

-one free edge plate.

(iii)*Loading type* (Fig. 4.29c). The element loading is always in its plane and can be:

-uniform axial compression or tension;

-bending;

-bending and axial force.

4.3.2. Through-Thickness Properties of Structural Steels

During the inspection of steel buildings after the Northridge and Kobe earthquakes and experimental tests on beams and connections, many cracks

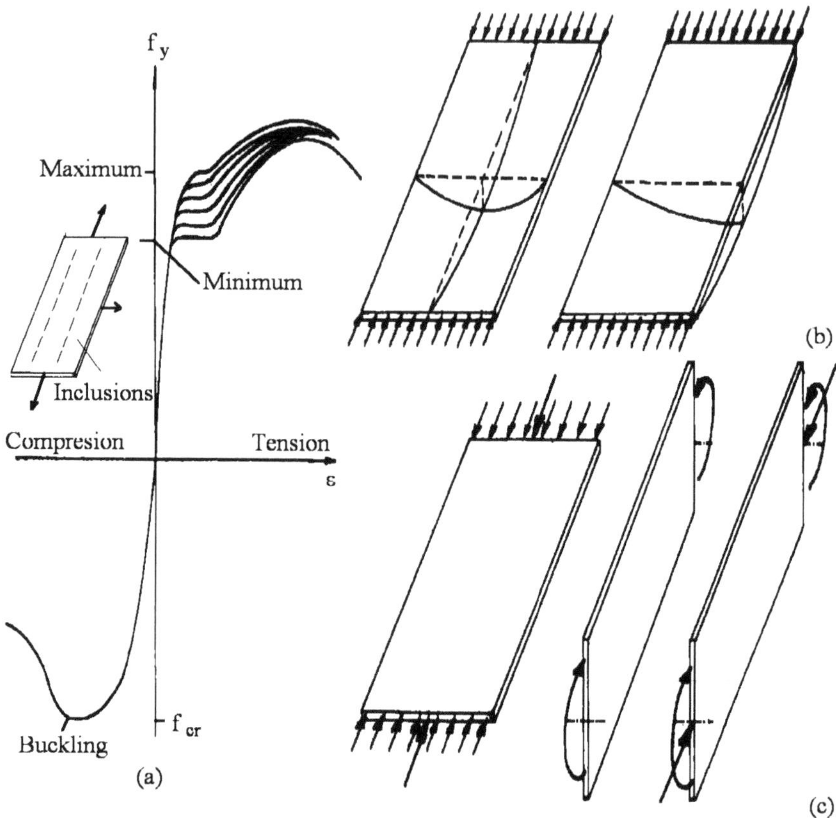

Figure 4.29: Factors influencing the element ductility: (a) Behaviour in tension and compression; (b) Element type; (c) Loading type

propagated along the flanges have been revealed. This is due firstly to the state of the stress and strain, but at the same time can be attributed to the *anisotropy* of steel, especially in through-thickness direction. Barsom and Korvink (1998) have shown that the differences in the longitudinal, transverse and through-thickness properties are caused by the nonhomogeneous distribution of inclusions and voids. Deformable manganese sulfide inclusions are the major contributions to steel anisotropy, because their effect is most severe in the through-thickness direction. The examination of tension tests (Barsom and Korvink, 1998) have shown that the yield stress and tensile strength in the longitudinal and transverse directions are essentially identical. An increase in data scatter is observed when these values are compared with the through-thickness values. A conservative through-thickness for yield stress and tensile strength can be derived from the longitudinal values:

$$f_{yt} = 0.90 f_{yl} \tag{4.18a}$$

$$f_{ut} = 0.80 f_{ul} \tag{4.18b}$$

An important effect of this anisotropic property of steel is the *lamellar tearing*, which is a stepped internal cracking pattern parallel to a plate surface. This lamellar tearing is caused by weld restrain and welding residual tensile stresses and normal stresses occur just below the weld heat affected zone. Three basic factors govern the susceptibility of a plate to lamellar tearing: material properties (steel making without strict control of inclusions and voids), connection geometry (filled weld without residual tensile stress control) and welding conditions (without adequate welding techniques and heat control). Therefore, detailing, welding, fabrication and design should be selected and implemented to minimize stress and strain in the through-thickness direction.

In spite of these observations concerning the through-thickness properties, the review of fractography performed on connection fractures from the 1994 Northridge earthquake indicated that none of these fractures could be attributed to through-thickness properties (Dexter and Melendrez, 2000). After 1970, steel that is susceptible to lamellar tearing has not been produced in the USA, because the producers have controlled the large non-metallic inclusions that caused this fracture. But in many experimental tests the lamellar tearing phenomenon has been the cause of specimen rupture.

4.3.3. Random Variability of Plate Mechanical Properties

The tensile tests on specimens extracted from plate elements may show a very important variability of mechanical properties related to the steel grade and plate thickness. For strength checking, the ultimate limit design is based on the plastic moment, determined on the basis of the lower limit of yield stress. This is the reason why the material regulations only guarantee this lower value, without any specification regarding the maximum value. But in the case of ductility checking, the limit value is safely obtained by using the upper bound of yield stress, leading therefore to a very large variation in the actual ductility in comparison with the design value. The yield stress variation largely depends on chemical compositions and on production conditions and, consequently, the upper value can be controlled in the same way as the lower one.

The random variability of steel mechanical properties of sections and plates as a function of steel grade and thickness has been studied at Naples University (Mazzolani et al, 1990, 1993, Calderoni et al, 1995, Mazzolani and Piluso, 1996, Piluso, 1992). The statistical data are provided by the data collected in recent years during testing and certification activity on steels produced in various Countries of the European Community. Fig. 4.30 shows the random variability with thickness of yield stress, tensile strength and yield ratio for Fe 360(a), for Fe 430(b) and for Fe 510 (c) steels. The yield stress versus ultimate elongation relationships are also shown. One can see that the scatter of all values is very high, especially for yield stress. The influence of thickness on the yield stress is also very important. The yield stress for thinner plates is higher than the one for thicker plates.

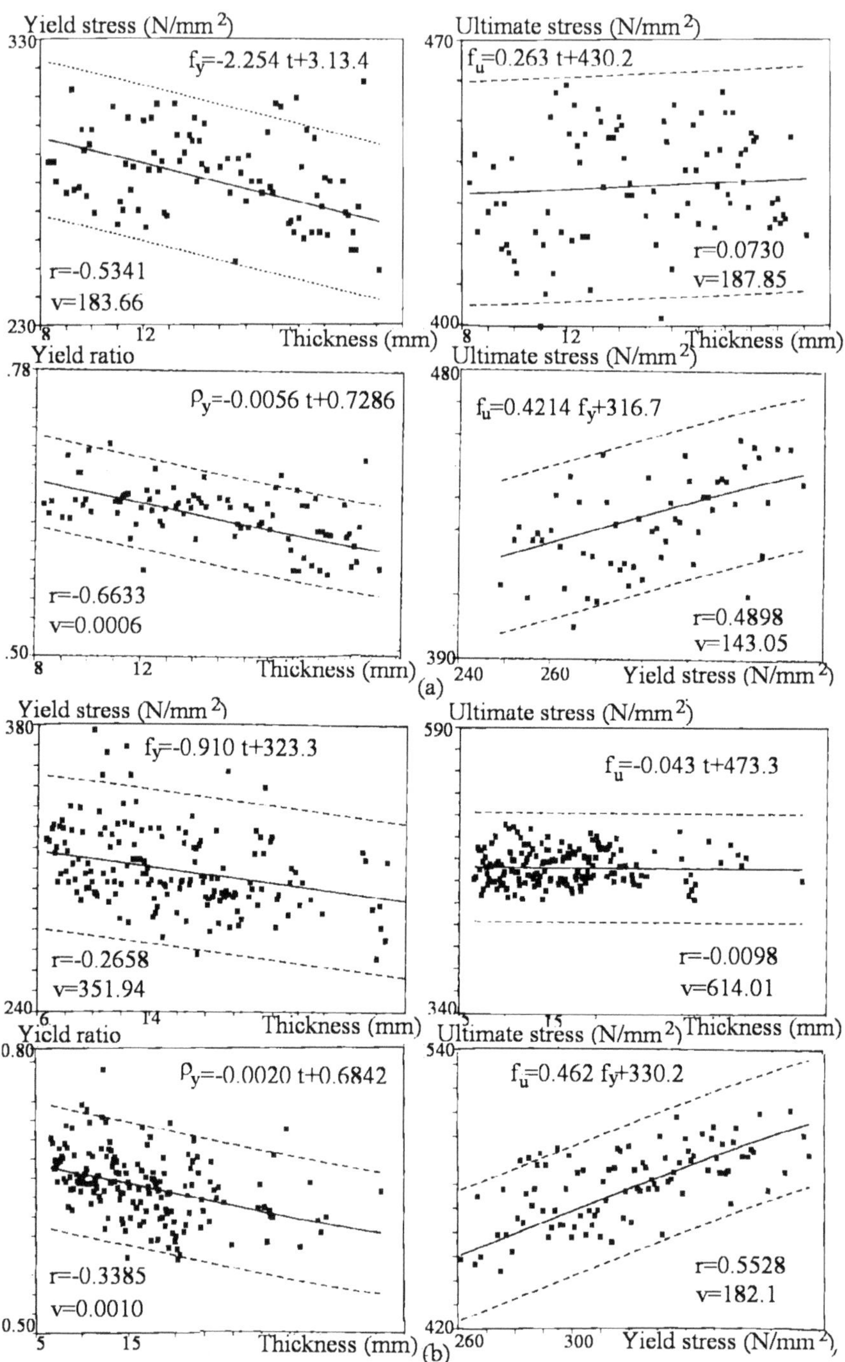

Figure 4.30: Random variability of mechanical properties: (a) Fe 360; (b) Fe 430 (after Mazzolani et al, 1990,1993)

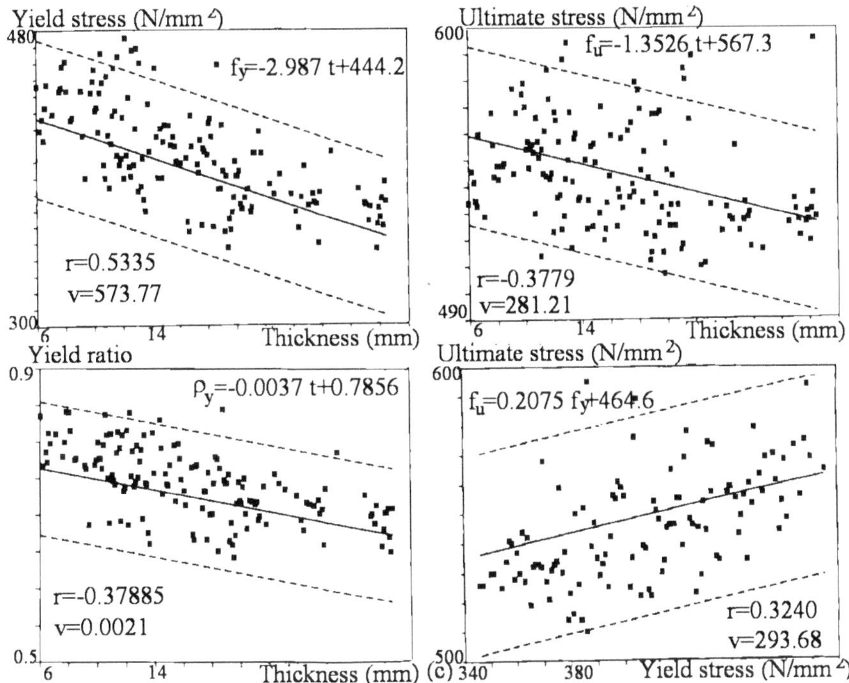

Figure 4.30 (continued): Random variability of mechanical properties: (c) Fe 510

The relationships between mechanical properties are (in N/mm^2):
-Fe 360:

$$\begin{aligned} f_y &= 313 - 2.25t \pm 25 \\ f_u &= 430 + 0.263t \pm 27.5 \\ \rho_y &= 0.729 - 0.0056t \pm 0.05 \end{aligned} \tag{4.19a-c}$$

-Fe 430:

$$\begin{aligned} f_y &= 323 - 0.910t \pm 37.5 \\ f_u &= 473 - 0.043t \pm 50 \\ \rho_y &= 0.684 - 0.002t \pm 0.07 \end{aligned} \tag{4.20a-c}$$

-Fe 510:

$$\begin{aligned} f_y &= 444 - 2.987t \pm 47.5 \\ f_u &= 567 - 1.353t \pm 33 \\ \rho_y &= 0.785 - 0.0037t \pm 0.095 \end{aligned} \tag{4.21a-c}$$

The sign + is for the upper value, while – is for the lower one. One can see, for instance for Fe 360 steel, that for $t = 22$ mm the lower value of yield stress results $f_{yl} = 235$ N/mm^2, which corresponds to the nominal value of yield stress. The upper value is $f_{yu} = 288$ N/mm^2, 22% greater than the

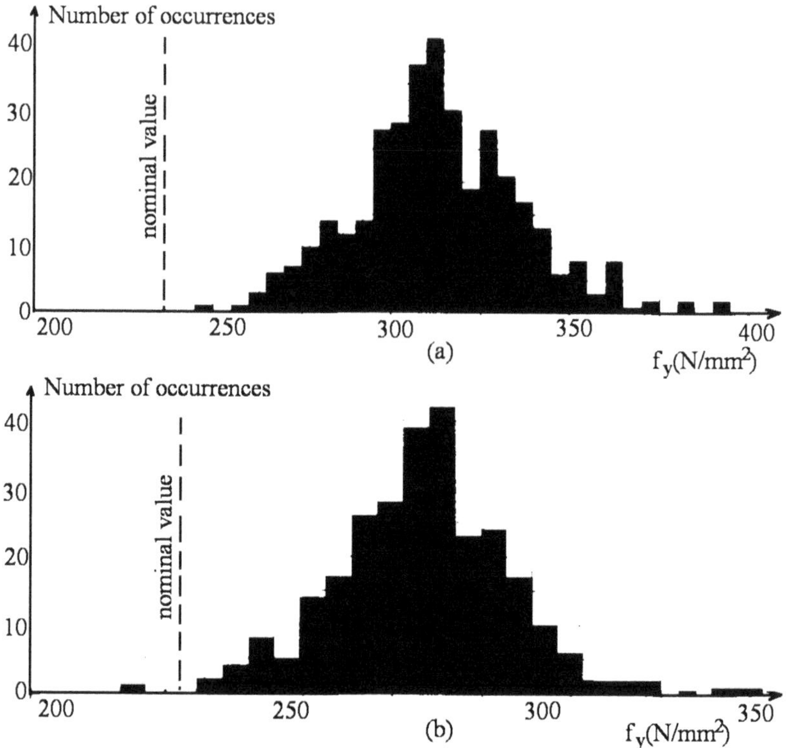

Figure 4.31: Yield stress histograms: (a) $8 < t < 15$ mm; (b) $40 < t < 80$ mm (after Croce et al, 1997a,b)

nominal value. For $t = 8$ mm, the corresponding values are 270 N/mm^2 and 320 N/mm^2 respectively, which is 36% greater than the nominal value. These variations cannot be neglected in design practice.

The analysis of the yield stresses and tensile strengths for three different producers has been performed at Milano University (Agostoni et al, 1993), with the same results as from the above mentioned Naples conclusions.

Similar studies were performed at Pisa University by Croce et al (1997 a, b), by collecting a large set of experimental test results for a total of about 22000 steel specimens from European production. The data have been subdivided in terms of producer (5 different producers), steel grades, type of rolling (hot and cold rolled), type of production (hot rolled profiles, laminated plates and coils), plate thickness (5-80mm). All these factors have great influence on the yield stress, so, it is very difficult to establish some general relationship. But the most important conclusion is that the mean and upper values of yield stress are much higher than the expected values (Fig. 4.31).

The random variability of plate mechanical properties is studied by Fukumoto (1994, 2000) for the Japanese production. For SM 400 C steel

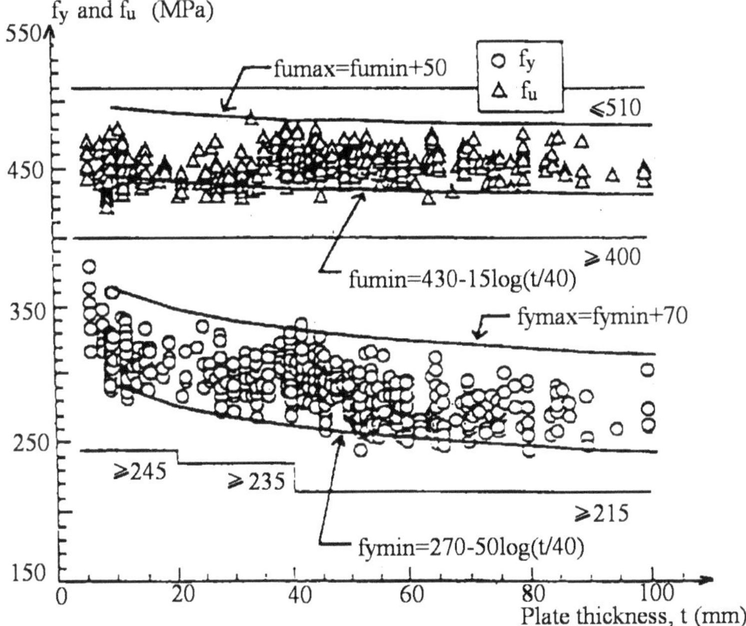

Figure 4.32: Random variability for Japanese steels (after Fukumoto, 2000)

(corresponding to European Fe360), Fig. 4.32 represents the results of experimental values in function of plate thickness. Fitted lower and upper bounds for the actual mechanical properties are (t in mm):

$$f_{y\,min} = 350 - 50\log t \quad \text{(t, mm)}$$

$$f_{y\,max} = f_{y\,min} + 70 \tag{4.22a, b}$$

$$f_{u\,min} = 455 - 15\log t$$

$$f_{u\,max} = f_{u\,min} + 50 \tag{4.23a, b}$$

All these research works have shown that the actual values of yield stress are higher than the nominal values. These findings are in favour of the strength checking, but they are detrimental to the ductility checking, due to the fact that the available ductility decreases with the increasing of yield stress. Table 4.4 summarizes the deterioration of the local ductility factor μ of compressed plates versus actual yield stress f_a to nominal yield stress f_y ratio and slenderness parameter b/t (for nominal yield stress f_y = 235 MPa). A considerable decrease can be seen in ductility factor for compact plates (Fukumoto, 2000).

So, the decrease of design ductility of elements must be prevented by controlling the variability of actual yield stresses and in ductility analysis it is necessary to use an increased nominal value for yield stress.

Table 4.4: Deterioration of ductility factor in plate elements(after Fukumoto, 2000)

b/t f_a/f_y	**20**	**23**	**28**
1.00	10.4	6.0	3.3
1.15	7.6	4.8	3.0
1.45	4.1	3.6	2.6

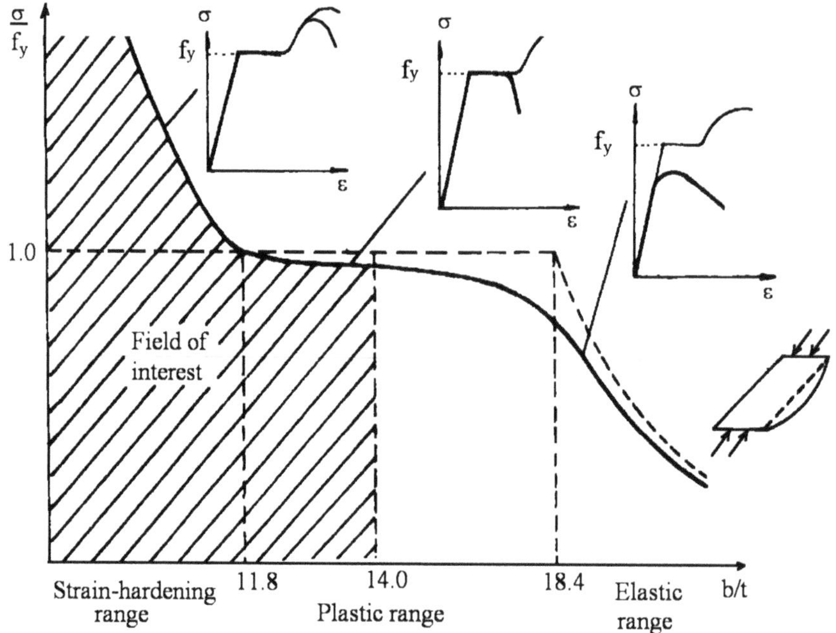

Figure 4.33: Plate buckling

4.3.4. Plastic Buckling of Plates

The problem of buckling of plated structures is very popular and continuous research activity produced a great amount of publications. A complete presentation of this subject is beyond the scope of this book and, therefore, only the particular aspects of the collapse strength directly related to the plate ductility are reviewed here.

(i) *Buckling in strain-hardening range.* The three buckling types of a rectangular plate with one free edge are presented in Fig. 4.33: elastic buckling, plastic buckling in the yielding range and buckling in strain-hardening range. For seismic design purposes the field of interest is the domain of plastic and strain-hardening buckling, when the instability occurs after reaching the yield stress.

The analysis of plastic buckling of plates has a long historical background (Inoue and Kato, 1993). There are three theories used for determining the plastic critical loads.

(ii) *Deformation theory*, in which the elastic buckling relation is used, but the elastic modulus is changed with the secant modulus (McDermott, 1969). This theory deals with metal with continuous stress-strain curves whenever yield flow does not occur. This is not the case of mild steel, for which important discontinuities occur between elastic, plastic and strain-hardening ranges.

(iii) *Incremental theory*, for plates working in strain-hardening range. Haaijer (1957) considers that steel is homogeneous, but with an orthotropic behaviour, because the modulus corresponds to the strain-hardening range in the axial direction, while the modulus is elastic in the transverse direction. The solution of the differential equations, by using an approximate deflection surface for the buckling zone, gives the following relationships:

-plate with one free edge, the opposite one being:

- hinged

$$\sigma_{cr} = \left[\frac{\pi^2 D_x}{k}\left(\frac{b}{l}\right)^2 + G\right]\left(\frac{t}{b}\right)^2; \quad l = L \qquad (4.24a, b)$$

- fixed

$$\sigma_{cr} = \left[0.769(D_x D_y)^{1/2} - 0.020(D_{xy} + D_{yx}) + 1.712G\right];$$

$$\frac{L}{d} = 1.46\left(\frac{D_x}{D_y}\right)^{1/4} \qquad (4.25a, b)$$

-plate supported on two edges:

- hinged

$$\sigma_{cr} = \frac{\pi^2}{12}\left(2(D_x D_y)^{1/2} + D_{xy} + D_{yx} + 4G\right)\left(\frac{t}{b}\right)^2; \quad \frac{L}{b} = \left(\frac{D_x}{D_y}\right)^{1/4} \qquad (4.26a, b)$$

- fixed

$$\sigma_{cr} = \frac{\pi^2}{12}\left[4.554(D_x D_y)^{1/2} + 1.237(D_{xy} + D_{yx}) + 4.943G\right)\left(\frac{t}{b}\right)^2;$$

$$\frac{L}{b} = 0.664\left(\frac{D_x}{D_y}\right)^{1/2} \qquad (4.27a, b)$$

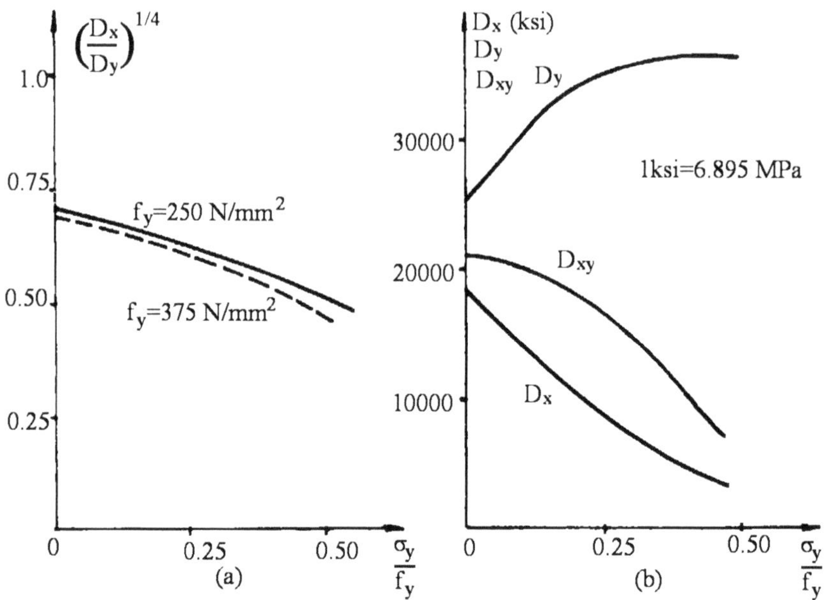

Figure 4.34: Plate rigidity variation: (a) Influence of ratio δ_y/f_y; (b) Variation of D_x, D_y, D_{xy} moduli (after Haaijer, 1957)

where the moduli are:

$$D_x = \frac{E_x}{1-\nu_x\nu_y}, \quad D_y = \frac{E_y}{1-\nu_x\nu_y};$$

$$D_{xy} = \frac{\nu_x E_x}{1-\nu_x\nu_y}, \quad D_{yx} = \frac{\nu_x E_y}{1-\nu_x\nu_y}, \tag{4.28a-d}$$

E_x, E_y being the tangent moduli in x, y directions, respectively, G, the shear modulus, ν_x, ν_y, the Poisson's coefficients in x, y directions. The modulus D_x is reduced in comparison with the modulus D_y due to the influence of strain-hardening. The ratio $(D_x/D_y)^{1/4}$ is presented in Fig. 4.34a for two steel grades in relation with the variation of the ratio σ_y/f_y. Fig. 4.34b shows the variations of the D_x, D_y, D_{xy} moduli. From these figures one can conclude that the influence of the steel grade is small. From experimental tests, it is suggested that the following values of the moduli can be selected:

$D_x = 20800\ N/mm^2$ (3000 ksi)
$D_y = 227600\ N/mm^2$ (32800 ksi)
$D_{xy} = 56200\ N/mm^2$ (8100 ksi)
$G = 16600\ N/mm^2$ (2400 ksi)

As an application, the ϵ_{cr} is plotted in Fig. 4.35 for a plate with a free edge. One can clearly see that for $b/t < 8$ the plastic buckling occurs in the strain-hardening range.

Fig. 4.36 shows the half-wave buckling lengths in function of the restraint coefficient of the edges. For plates with one free edge, this length is within

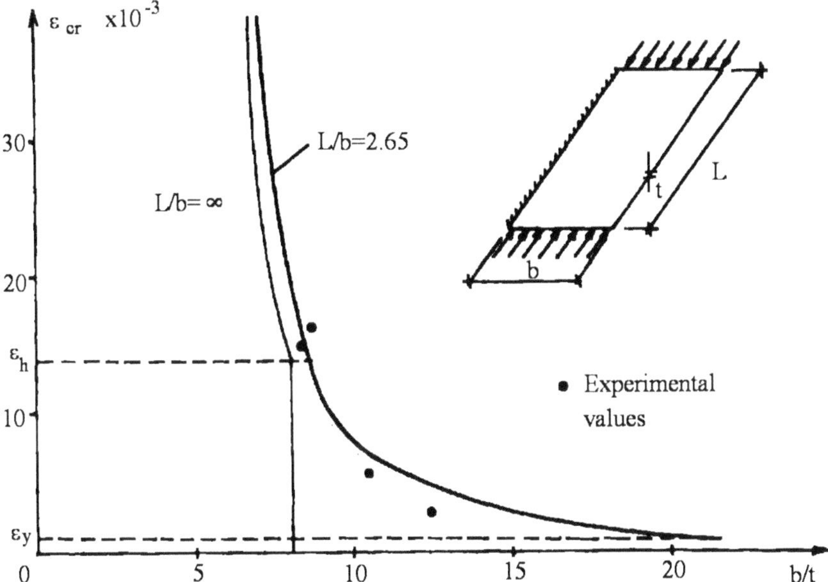

Figure 4.35: Critical strain for a plate with one free edge (after Haaijer, 1957)

(0.8 ÷ 3.0)b, while for plates supported along two edges the range is (0.365-0.55)b.

Similar solutions were established for one free edge plate (Inoue and Kato, 1993) and for a supported edge plate (Inoue, 1994):

-one free edge plate:

$$\sigma_{cr} = \frac{\pi^2}{12}\left[\frac{G_p}{2} + E_h\left(\frac{b}{l}\right)^2\right]\left(\frac{t}{b}\right)^2 \tag{4.29}$$

-supported edge plate:

$$\sigma_{cr} = \frac{\pi^2}{12}\left[2G_p + E_h\left(\frac{b}{l}\right)^2\right]\left(\frac{t}{b}\right)^2 \tag{4.30}$$

where G_p and E_h are the shear and axial modului in the strain-hardening range. The equations (4.29, 4.30) are similar to the ones established by Haaijer. It is very important to observe that the critical load for one free edge plate may be obtained from the supported edge plate by replacing the width b with the corresponding width b/2. This shows that half of the supported edge plate works exactly like the one free edge plate (Fig. 4.37).

From the equations (4.29 and 4.30) one can see that the most important part of the buckling critical load belongs to the term multiplied by G_p. So, the evaluation of a proper value for the shear modulus in plastic and strain-hardening ranges is a crucial problem. This is possible in the following theory of slip planes.

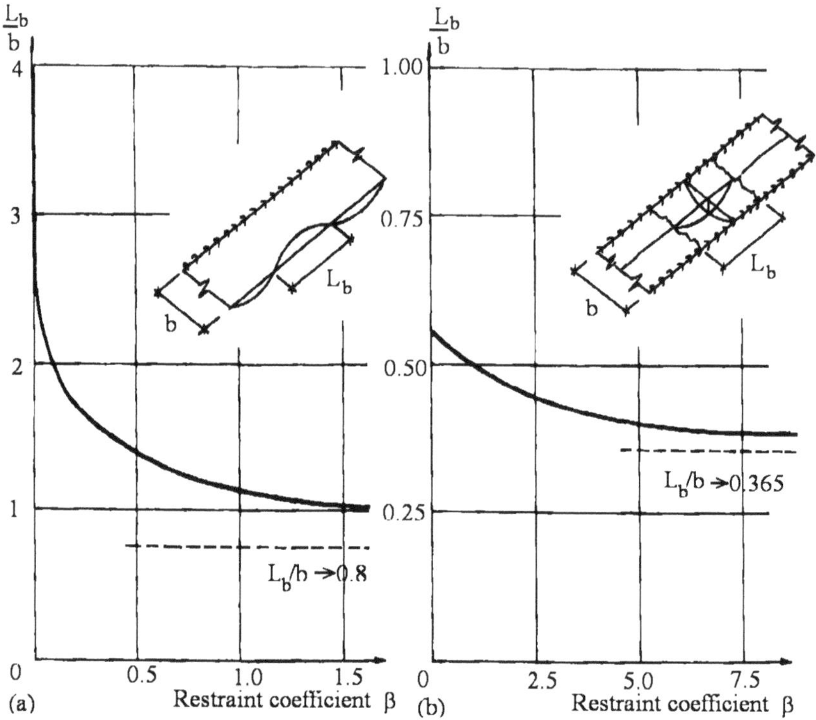

Figure 4.36: Length of buckled shape: (a) One free edge plate; (b) Supported edge plate (after Haaijer, 1957)

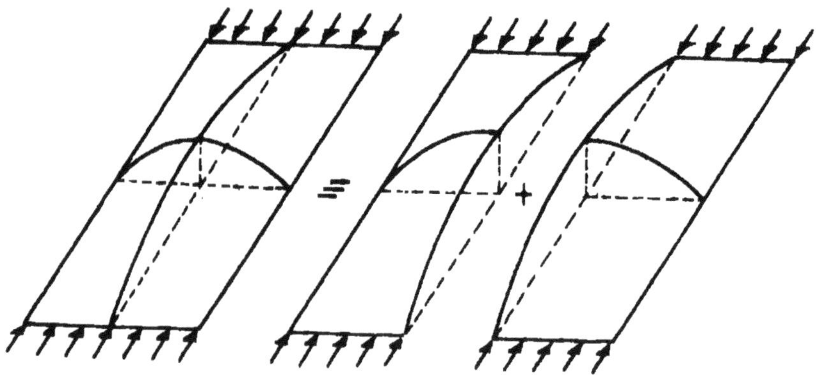

Figure 4.37: One free edge and supported edge plate equivalence

(iv) *Slip plane theory.* Lay (1965) presented a model for shear response of an axially loaded steel member, based on considerations regarding the origin of plastic deformations, namely slips in planes with large shear stresses. Using this theory, an analytical relationship between plastic and elastic moduli is determined:

$$G_p = \frac{2}{1 + \frac{1}{4}\frac{E}{E_h(1+\nu)}}G \tag{4.31}$$

Taking into account the value of E_h given in Table 4.3, results $G_p = (0.20 \div 0.25)G$. This reduction is more drastic than the one proposed by Inoue and Kato (1993): $G_p = 0.5G$. This difference can explain the fact that all experimental values from Inoue and Kato tests are less than the analytical predictions. The Lay's slip plane theory has been extensively used in the inelastic buckling analysis, with good agreement with the experimental results. This is the reason why this theory will be used in this book. However, the Lay's theory has been criticized by Moller et al (1997), because, after these authors, it contains some theoretical faults and the agreement between theory and experimental results can be considered as purely circumstantial.

4.3.5. Behaviour of Plates under Cyclic Loading

Stiffness, strength and ductility of steel plates deteriorate under cycle loading as for the material properties. The behaviour of a rectangular plate in-plane loaded, subjected to a series of cyclic elasto-plastic large deflections was studied by Zhang et al (1997). The plate is hinged at the two opposite edges and has initial deflections

$$w_i = 0.001b \sin\frac{\pi x}{b} \sin\frac{\pi y}{b} \tag{4.32}$$

and a residual stress distribution caused by welding (Fig. 4.38a).

(i) *Cyclic behaviour under constant strain amplitudes.* Fig. 4.38 shows the plate behaviour when the constant strain amplitude is four times the material yield strain. In all cases, the cyclic loading started from the compression side and the compression is considered positive. It is observed that the maximum load capacity in compression side decreases and the residual deflection increases, as far as the number of cycles increases. The reduction of load carrying capacity in compression is due to the sensitivity to the initial deflection, which is the residual deflection at the start of compressive loading, corresponding to the former cycle. One can see that the cyclic loading has a small influence on the tensile capacity of the plate. Due to this erosion of compressive load capacity, the plate ductility decreases continuously during the cyclic loading.

(ii) *Cyclic behaviour under various strain amplitude.* The strain amplitudes are increasing $\epsilon/\epsilon_y = 2, 4, 6, 8$, and for each value a group of three cycles are performed, using the loading history suggested by ECCS (Fig. 4.39). The loading carrying capacity at the compression side decreases as the strain amplitude increases, the reduction being more important than under monotone loading. This is due to the fact that the initial deflection for each cycle increases due to the former cycle being greater than the one

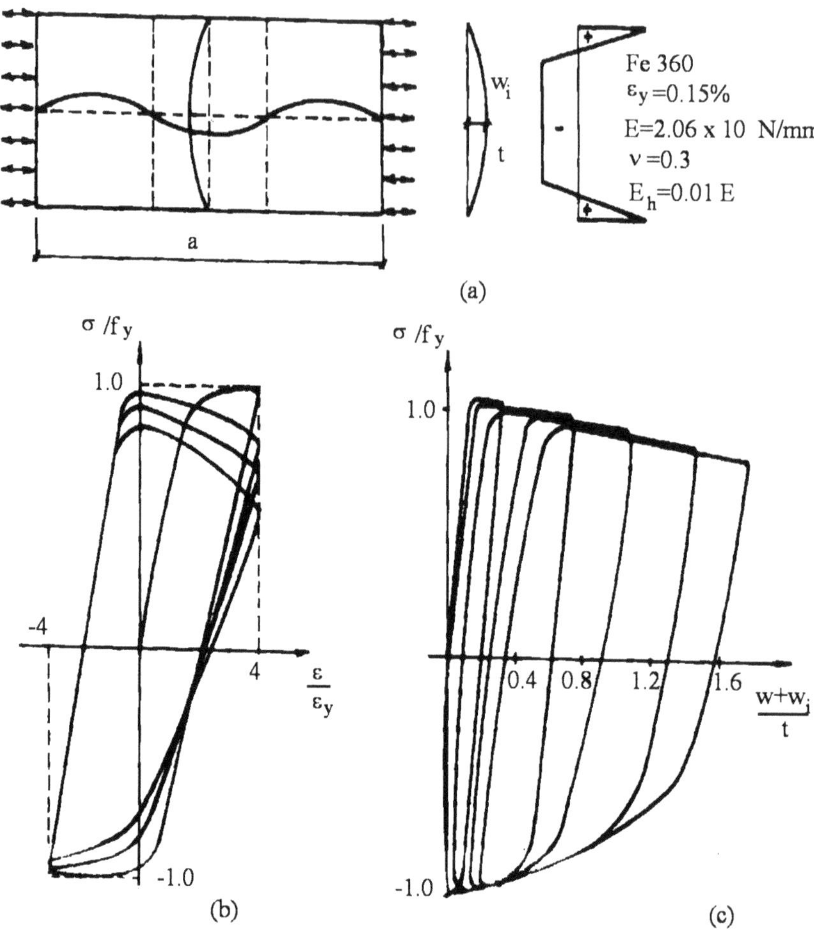

Figure 4.38: Cyclic behaviour under constant strain amplitude: (a) Plate with geometrical imperfections and residual stresses; (b)and (c) Behaviour of plate (after Zhang et al, 1997)

corresponding to monotonic loading. Consequently, the plate ductility may be drastically reduced during a cyclic loading.

(iii) *Accumulation of plastic deflections.* Figs. 4.38b and 4.39a show the non-dimensional stress-strain relationship for cyclic loading; Figs 4.38c and 4.39b show how the deflection modes change as far the number of cycles increases. It is evident that the initial deflection at each new cycle is greater that the one of the later cycle (Fig. 4.40a), due to the residual plastic deflection.

One can see that after the buckling occurrence, the opposite load cannot change the direction of deflection, each cycle producing the increasing of the initial deflection for the next cycle. This fact causes a decreasing of the

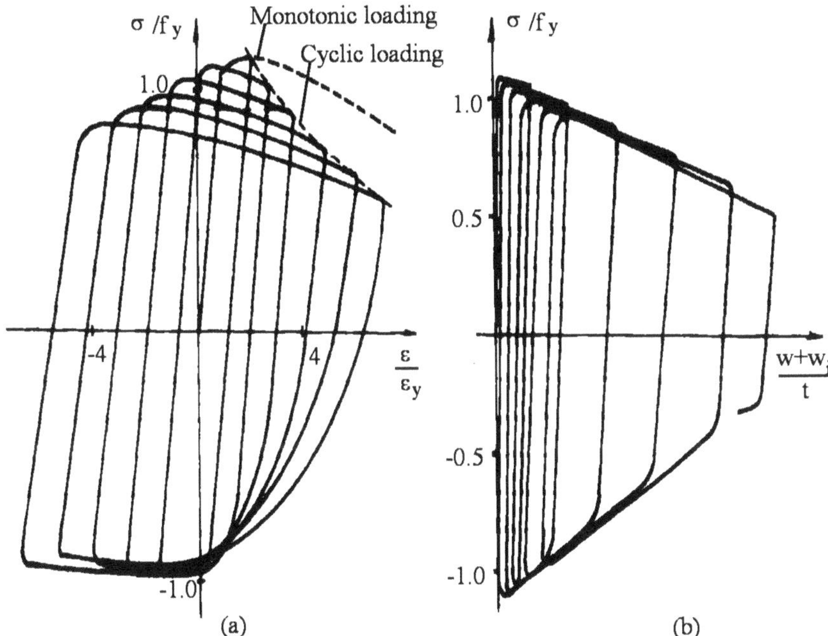

Figure 4.39: Cyclic behaviour under various strain amplitude: (a) Stress-strain relationship; (b) Stress-central deflection relationship (after Zhang et al, 1997)

maximum load capacity (Fig. 4.40b), in the same way as the influence of geometrical imperfections act on the critical load in elastic fields.

Therefore, it is very clear that the effect of cyclic loading with a reduced number of cycles, as in the case of earthquakes, is not a problem of fatigue, being only an effect of plastic deflection accumulation. This is especially undeniable when the structure is situated near to source, where the number of cycles is small, but the amplitudes of ground motions are very high.

(iv) *Asymmetry in cyclic behaviour.* A very important aspect concerning the accumulation of plastic deflections refers to the behaviour of a plate as a part of a cross-section. In the monotonic loading, due to local buckling of the plate, a part of the width becomes ineffective, so the neutral axis is shifted towards the tensioned flange. This produces different deformations at the two opposite plates, the compressive strain being larger than the tensile one (Fig. 4.41a). So, the compressive strain increases with the increase of curvature, while the tensile strain remains constant. So an asymmetry in strains over the section occurs (Vayas, 1997).

In the case of cyclic loading, this asymmetry in deformations is preserved, the compression strains always being much greater than the tension strains (Fig. 4.41b). So, the tension stresses cannot straighten the buckled surface and each cycle starts with an initial deformation produced by the former cycle. This initial deformation increases with each cycle.

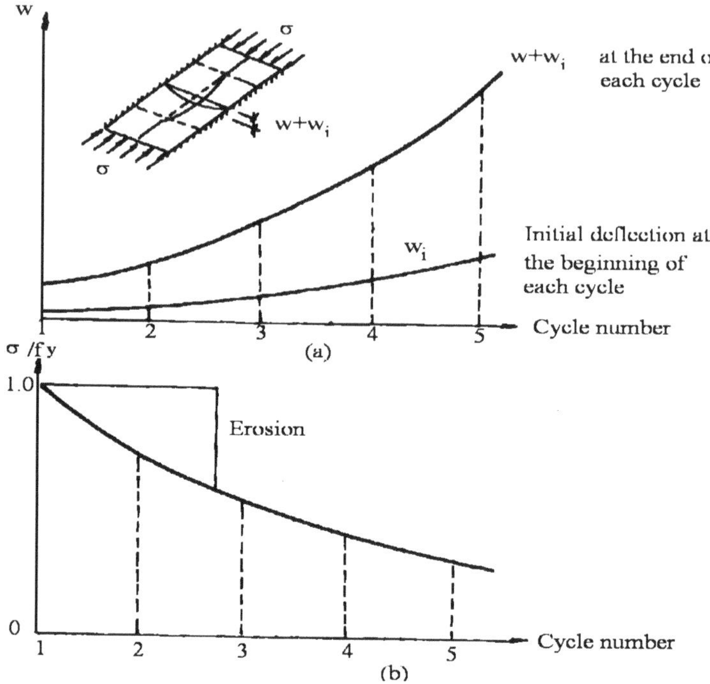

Figure 4.40: Accumulation of plastic deflection: (a) Increasing of central deflection; (b) Erosion of stress

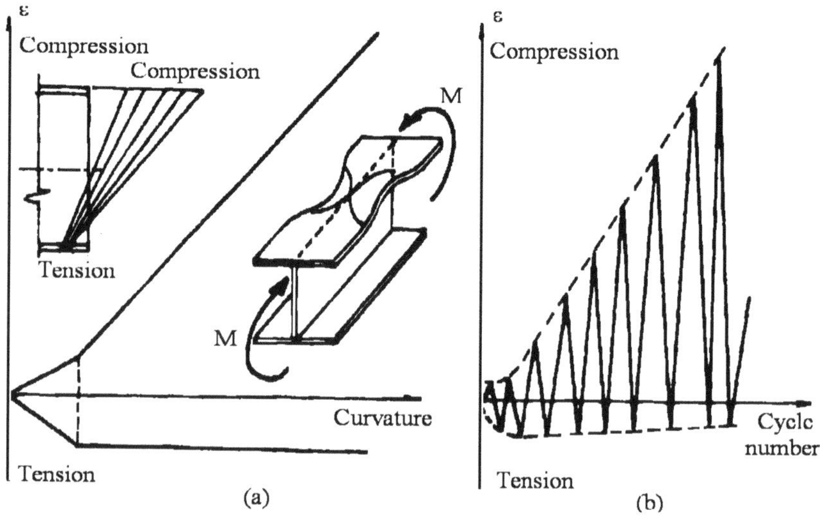

Figure 4.41: Asymmetry in cyclic behaviour: (a) Monotonic loading; (b) Cyclic loading

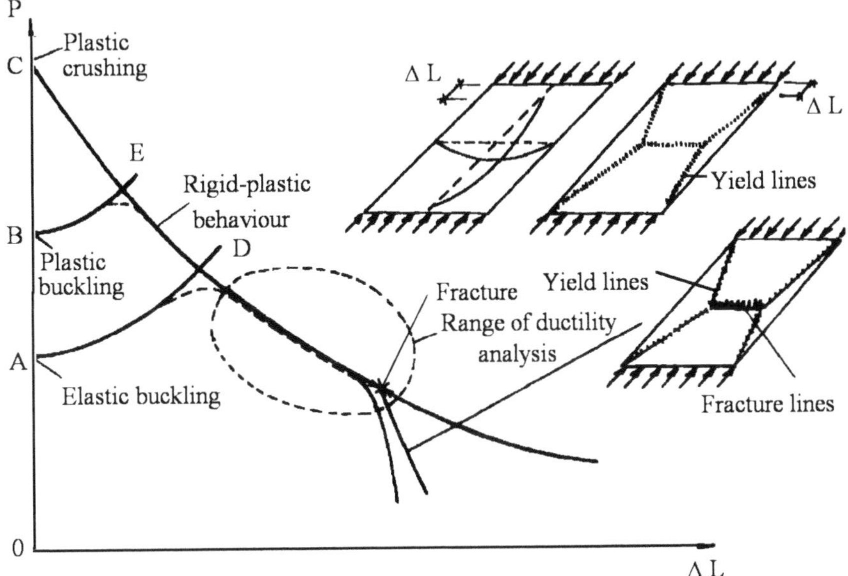

Figure 4.42: Plate behaviour at collapse

4.4 Plastic Collapse Mechanisms

4.4.1. Plate Behaviour at Collapse

The plate ductility must be determined in the range of large plastic deformations. The use of the plastic theory of plates in this field gives rise to very difficult problems. The FEM numerical approach is certainly the best solution, but the computer time and, therefore, the cost are very high, due to the need of a very dense network in the buckled surface. Cost apart, the method of plastic collapse mechanisms seems to be the most suitable for determining the plate ductility. Consequently the analysis of the collapse behaviour of a plate is a step of great importance when considering its ductility.

The beginning of the plate collapse behaviour may be introduced by elastic buckling, by plastic buckling or by crushing (Fig. 4.42), but only for the last case the critical load corresponds to the ultimate load. Both elastic and plastic bucklings have a stable post-critical behaviour, during which the plate may be loaded in the buckled configuration. Contrary to this, the post-crushing behaviour is always unstable. So, the true collapse load is only obtained when a plastic mechanism is formed and an unloading behaviour occurs.

A very practical method is the estimation of ultimate load as the intersection of two load-displacement curves. The first one represents the post-buckling curve, considering also the geometrical imperfections. The second curve characterizes the variation of collapse load with the geometry change in the kinetic plastic mechanism. The intersection gives the points D and E, corresponding to elastic or plastic buckling, respectively. The actual value of the collapse load is always less than these intersection points, due to the presence of geometrical imperfections. This scatter becomes smaller for a thick plate, for which plastic buckling occurs. For determining the ductility in compression, the plate behaviour at collapse can be adequately described by the plastic collapse mechanism.

It is well known that the plastic analysis is based on the fulfillment of the following conditions: yield, equilibrium and mechanism. The satisfying of the first two gives the lower bound value of the collapse load. When both equilibrium and mechanism conditions are satisfied, the upper bound value of the collapse load is obtained (Moy, 1985). So, when the plastic mechanism is used for evaluating the load carrying capacity, this approach leads in general to an overestimation of the load carrying capacity. But in particular the range of interest for the evaluation of plate ductility is far from the maximum load value, where the actual and the plastic mechanism behaviours are very close. In conclusion, no important differences, therefore, exist between the ductility values, determined using either the intricate plasticity theory or the very simple method of plastic collapse mechanism.

During the plastic crushing, some yield lines have more pronounced plastic rotation than the other lines and the bending moment along these lines exceeds the plastic moment values, working in the strain-hardening range. For some mechanism rotations the material strength limit is reached and an important crack occurs, forming a fracture line (Fig. 4.42). The behaviour curve corresponding to this new plastic mechanism type, formed by both yield and fracture lines, intersects the rigid-plastic curve, showing a dramatic decreasing of carrying capacity.

4.4.2. Plastic Collapse Mechanism Types

The plastic mechanism of a plate is mainly composed of a series of rigid parts, separated by yield lines, along which plastic bending occurs. The rigid parts of the plate transmit the internal actions. The yield lines are the deformable parts, where the plastic flow takes place and the external energy is absorbed and dissipated.

The rigid-plastic analysis is used to determine the plastic collapse mechanism which corresponds to the collapse load. To this purpose it is necessary to know some rules for build-up of a proper plastic mechanism. There are the following mechanism types (Murray, 1995, Kotelko, 1996a,b):

(i) *True and quasi-mechanisms.* If the plastic mechanism is composed only by yield lines, the mechanism is called *true* (Fig. 4.43a). If the plastic mechanism involves also plastic zones in which membrane yieldings occur, its name is *quasi-mechanism* (Fig. 4.43b). This differentiation is very important, because the quasi-mechanism involves a larger amount of energy than the true mechanism, this aspect being very important in the ductility calculation.

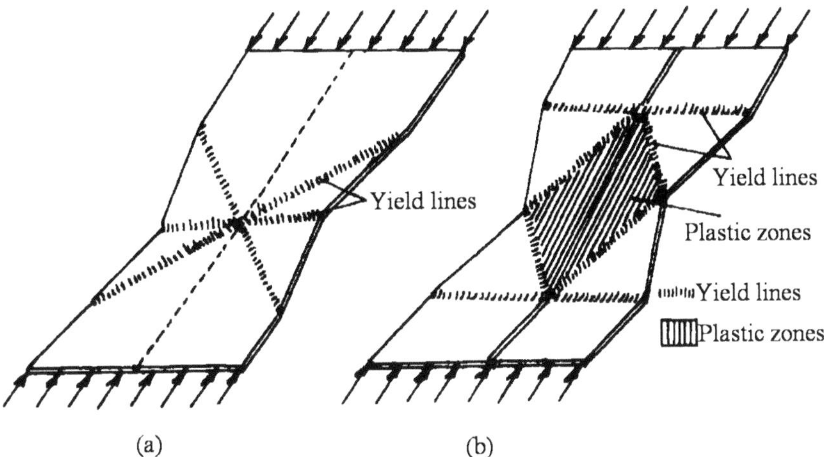

Figure 4.43: Mechanism types: (a) True mechanism; (b) Quasi-mechanisms

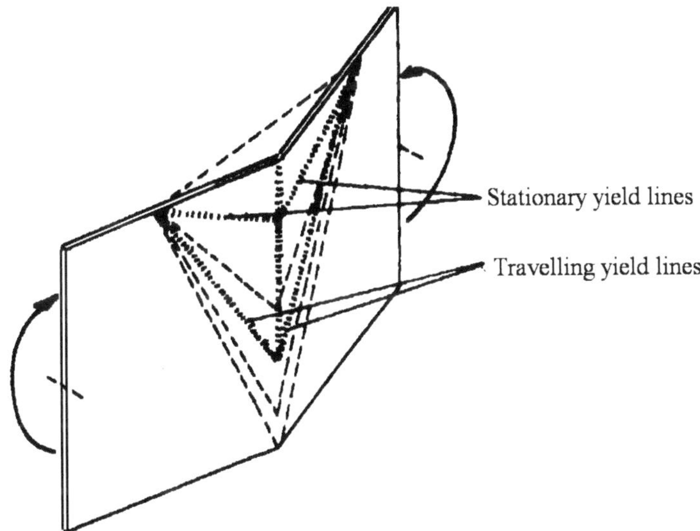

Figure 4.44: Stationary and travelling yield lines

(ii) *Mechanisms with stationary or/and travelling yield lines.* There are yield lines which are determined by the plastic buckling pattern, being influenced by plate configuration, loading system and initial imperfections. During the rotation of plastic mechanisms, these yield lines remain in a fixed position, being the stationary yield lines. Contrary to this, some other yield lines, formed after the occurrence of the buckling shape, change their position, trying to obtain the best mechanism pattern. These are the traveling yield lines (Fig. 4.44).

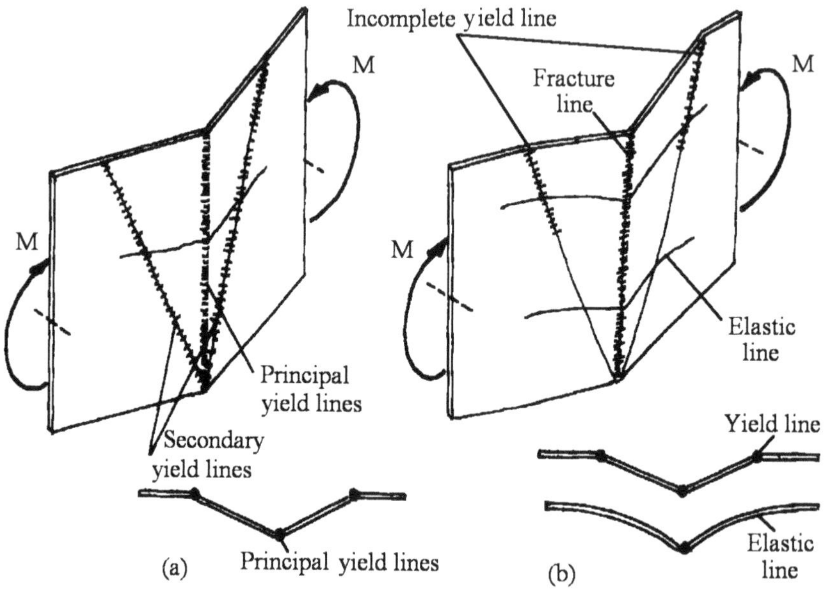

Figure 4.45: Mechanism types: (a) Line types; (b) Incomplete and fracture mechanisms

From the examination of member or connection failure during the last great earthquakes and experimental tests, it was clear that, in addition to the above-mentioned mechanism types, there are also the following plastic mechanisms:

(iii) *Incomplete mechanisms.* In a plastic mechanism there are principal and secondary yield lines (Fig. 4.45a). The principal yield lines must be always present in the plastic mechanism pattern. Contrary to this, the secondary yield lines may be also partially formed, a part of the bended line working in elastic range. From the experimental evidence, the plastic mechanisms of compressed plates are usually complete, while the plastic mechanisms formed in bended plates may be incomplete.

(iv) *Fracture mechanisms.* For large rotations or for steel with large yield ratio, some yield lines turn into the fracture ones and the new formed mechanism is composed by two line types: yield and fracture lines. These mechanisms describe in the best way the ultimate carrying capacity and ductility of the structural members (Fig. 4.45b).

4.4.3. Rigid-Plastic Analysis

The plastic collapse mechanism is composed of rigid parts and local plastic lines or zones. The admissible collapse mechanism can be primarily determined by considering the plate shape, edge conditions and loading systems. The work of a collapsed plastic mechanism implies that the larger amount of energy is absorbed in the small area of plastic lines or zones, so the elastic deformations of the other parts can be neglected. The rigid-plastic analysis is based on the principle of the minimum of the total potential energy. The total potential energy functional V, is defined as:

$$V = U - L_p \tag{4.33}$$

where U is the strain energy (internal potential energy) and L_p is the loading potential (external potential energy). The first variation of energy is equal to the virtual work. The principle of the minimum of the total potential energy states that for equilibrium the first derivative vanishes. Here, only the displacement field is subject to variation. In case of plastic mechanisms described by the displacement d_i, the principle takes the simple form for equilibrium:

$$\frac{\partial V}{\partial d_i} = 0 \tag{4.34}$$

The strain energy is given by the plastic mechanism work:

$$U = \sum U_{ls} + \sum U_{sl} + \sum U_z \tag{4.35}$$

U_{ls}, U_{lt}, U_z being the strain energies corresponding to stationary and travelling yield lines and the plastic zones, respectively. For yield lines only the rotations are involved, while for plastic zones only the axial deformations are considered:

$$U_{ls,t} = \sum_i M_{pi}\theta_i l_i, \quad U_z = \sum_j N_{pj}\varepsilon_j A_j \tag{4.36a, b}$$

where M_{pi} is the plastic moment along the i line, θ_i, the rotation of this line, l_i, the length of i line, N_{pj}, the axial forces of plastic zone j, ϵ_j, the axial deformation of this plastic zone, A_j, the area of plastic zone j.

The loading potential is given by:

$$L_p = \int_\delta F_k d\delta_k + \int_\theta M_l d\theta_l \tag{4.37}$$

F_k and M_l being the external forces and moments acting in the element, δ_k and θ_l, the displacement and rotation under the external forces and moments, respectively.

Taking into account (4.34):

$$F_k = \frac{\partial U}{\partial \delta_k}; \quad M_k = \frac{\partial U}{\partial \theta_l} \tag{4.38a, b}$$

A special treatment must be applied when the plastic mechanism contains travelling lines. Because the strain energy for these lines is a function of a geometrical parameter, χ, the pattern of collapse mechanism must be obtained by minimizing equation (4.35):

$$U_{min} = \min_{\rightarrow\chi} U \tag{4.39}$$

From this condition χ_m resultant the load-deformation relation can be determined for rigid-plastic mechanisms (Fig. 4.46). This curve intersects the one corresponding to primary elastic behaviour.

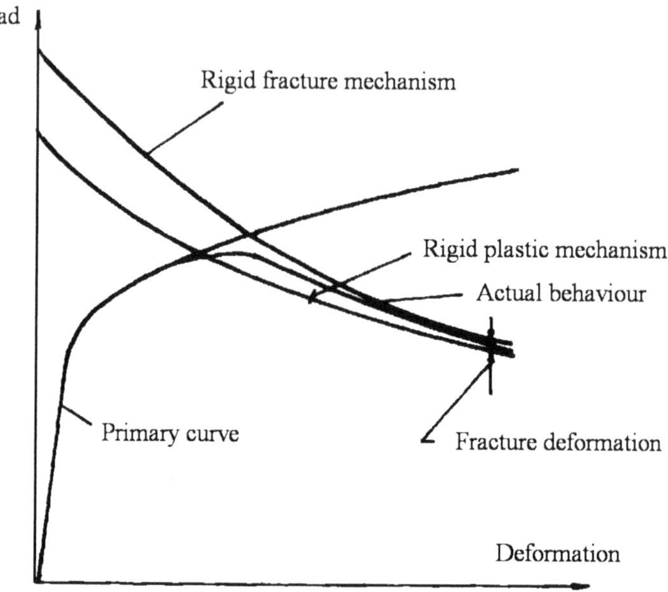

Figure 4.46: Rigid plastic and rigid fracture analysis

4.4.4. Plastic Moment Capacity of the Yield Lines

The evaluation of plastic moment capacity from equation (4.36a) is of substantial importance when the plastic mechanism approach is applied.

(i) *Plastic capacity at yield plateau.* The plate is assumed to be composed of a rigid perfectly plastic material, for which the Tresca yield criterion is adopted. The yielded material, which is responsible for all deformations, is concentrated in regions whose rigid boundaries form a 45° angle with respect to the longitudinal axis (Fig. 4.47a). From this figure it is easy to recognise that

$$\varepsilon_1 = \frac{\theta}{2}; \quad \varepsilon_2 = -\frac{\theta}{2} \tag{4.40a, b}$$

The normal stresses σ_1 and σ_2 are related through the yield condition:

$$\sigma_1 - \sigma_2 = f_y \tag{4.41}$$

The power of dissipation per volume unit is given by:

$$P = \sigma_1\varepsilon_1 + \sigma_2\varepsilon_2 + \sigma_3\varepsilon_3 = (\sigma_1 - \sigma_2)\frac{\theta}{2} = f_y \cdot \frac{\theta}{2} \tag{4.42}$$

By integrating the strain energy over the yielded volume V, the strain energy is given by

$$U = V\int_0^{\varepsilon} f_y d\varepsilon = V f_y \varepsilon = (1 - 2\delta + 2\delta^2)bt^2 f_y \cdot \frac{\theta}{2} \tag{4.43}$$

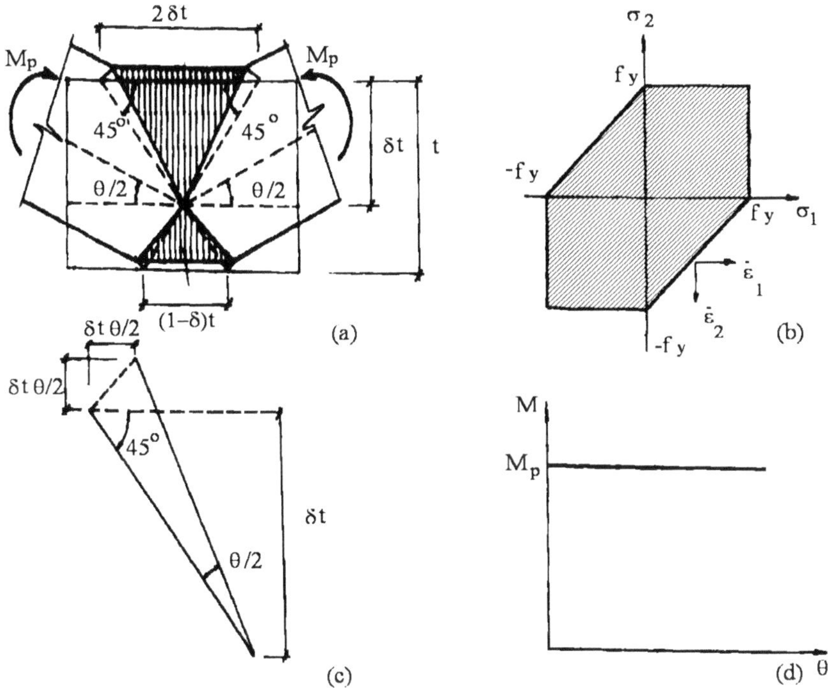

Figure 4.47: Plastic moment capacity of yield lines: (a) Plastic mechanism; (b) Tresca yield criterium; (c) Mechanism rotation; (d) Moment-rotation curve

Using the relation (4.38b), it results:

$$M = \frac{1}{2}(1 - 2\delta + 2\delta^2)bt^2 f_y \tag{4.44}$$

By differentiating with respect to δ in the aim to obtain the minimization of equation (4.44), the value $\delta = 1/2$ is obtained, from which the well known value of full plastic moment results:

$$M_p = \frac{bt^2}{4} f_y \tag{4.45}$$

For axially loaded plates, the full plastic axial capacity or the crushing capacity, is given by:

$$N_p = bt f_y \tag{4.46}$$

When both bending moment and axial force act on a plate, the reduced plastic moment is given by (Moy, 1985):

$$M_{pN} = M_p \left[1 - \left(\frac{N}{N_p}\right)^2\right] \tag{4.47}$$

This equation is presented in Fig. 4.48.

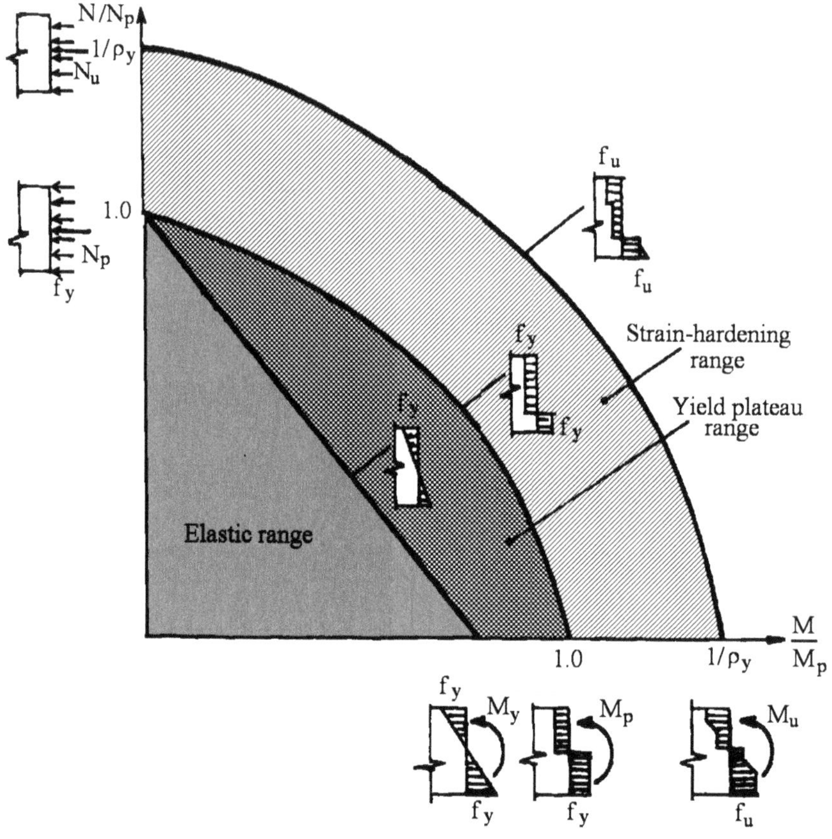

Figure 4.48: Moment-axial force interaction

(ii) *Moment capacity of inclined yield lines.* In many cases the yield lines are inclined with the angle β with respect to the stress direction (Fig. 4.49a). The presence of shear and twist can have an influence on the plastic moment capacity.

The plastic moment capacity of an inclined yield line under axial force is studied by Murray (1973), Zhao and Hancock (1993a,b). In the last papers, some simplified formulae are proposed:

$$M_{p\beta} = \frac{f_y t^2 b}{4}\left[1 - \left(\frac{\sigma}{f_y}\right)^2\right]\frac{f(\beta)}{\cos\beta} \tag{4.48}$$

where $M_{p\beta}$ is the plastic moment capacity of inclined yield line and $f(\beta)$ is a function of the angle β.The following expressions are given (β in radians):

- linear function:

$$f(\beta) = 1.0 - 0.119\beta \text{ where } \beta \in (0, \pi/6) \tag{4.49a}$$

$$f(\beta) = 1.379 - 0.842\beta \text{ where } \beta \in (\pi/6, \pi/2)$$

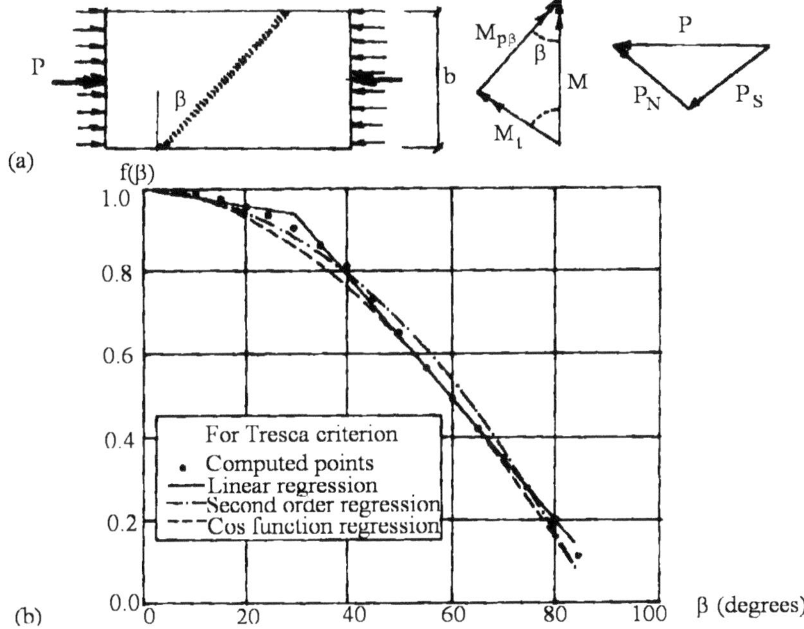

Figure 4.49: Moment capacity of inclined yield lines: (a) Line inclination; (b) Inclination function (after Zhao and Hancock, 1993a,b)

- second order function:

$$f(\beta) = 1.0 - 0.420\beta^2 \tag{4.49b}$$

- cosine function:

$$f(\beta) = \cos\beta \tag{4.49c}$$

Fig. 4.49b shows a comparison between the Tresca criterion and the curves corresponding to the above functions. It is interesting to note that according to equation (4.49c), the plastic moment capacity, given by (4.48) is independent of the angle β. It means that equations (4.47) and (4.48) are coincident, despite the fact that $M_{p\beta}$ (4.48) acts along the inclined yield line, while M_{pN} (4.47) acts across the width of the plate, which coincides with the direction of yield line ($\beta = 0$). Hence, the reduced plastic moment capacity along a yield line under axial force can be expressed by a simple formula which is independent of the inclination angle.

4.5 Fracture Collapse Mechanisms

During both strong earthquakes and experimental tests some cracks may be noticed along the yield lines (Fig. 4.50) due to reaching of ultimate strain. These cracks may occur both in cases of monotonic and cyclic loadings, being caused by the accumulation of plastic rotations.

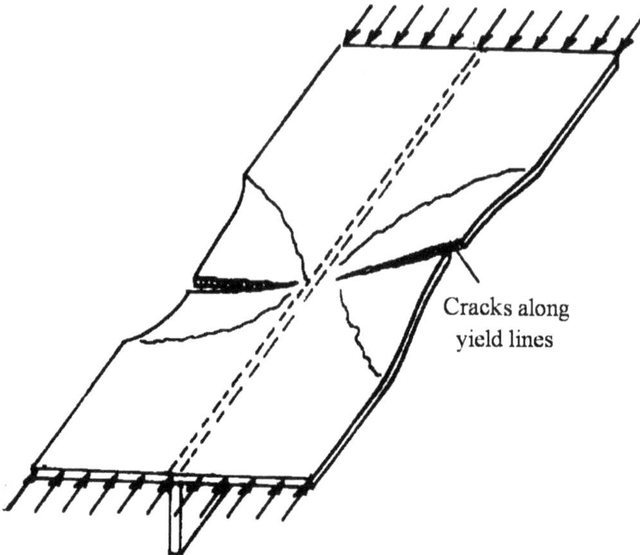

Figure 4.50: Cracking of yield lines

4.5.1. Ultimate Moment Capacity

Taking into account that no stress peak occurs during bending and fractures may occur when the uniform strains are reached at the section extremities. In the same way as for yielding, the ultimate moment is given by:

$$M_u = \int \sigma_y dA = \int_{e+ep+sh+fp} \sigma_y dA \tag{4.50}$$

where the integration over the cross-section is divided in four parts: elastic, plastic (yield plateau), strain-hardening and failure plateau, in accordance with the idealized stress-strain relationship (see Fig. 4.9b). The term corresponding to failure plateau is dominant and, therefore, the others may be neglected. So, a simple relation results:

$$M_u = f_u \frac{bt^2}{4} = \frac{1}{\rho_y} M_p \tag{4.51}$$

where ρ_y is the yield ratio defined by the relation (4.3). The ultimate axial load is given by

$$N_u = f_u bt = \frac{1}{\rho_y} N_p \tag{4.52}$$

If both bending moment and axial force act on a plate, the reduced ultimate moment is given by a similar relation to (4.47):

$$M_{uN} = M_u \left[1 - \left(\frac{N}{N_u}\right)^2\right] = \frac{1}{\rho_y} M_p \left[1 - \rho_y^2 \left(\frac{N}{N_p}\right)\right] \tag{4.53}$$

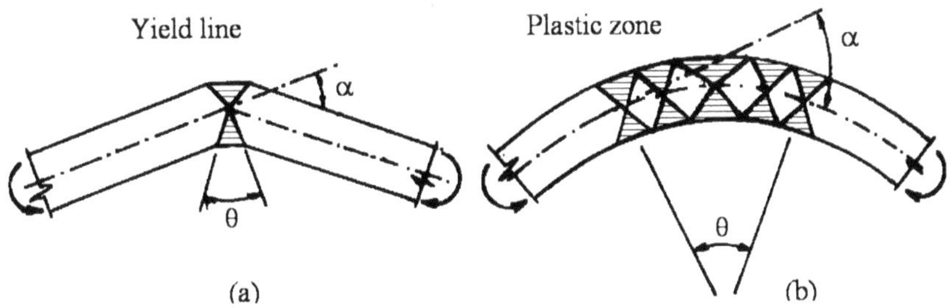

Figure 4.51: Plastic deformation along the yield lines: (a) Concentrated; (b) Distributed

This equation is presented also in Fig. 4.48. One can see that the hardening range is directly dependent on the yield ratio. Due to assumed hypothesis, the collapse load is proportional to the plastic moment.

4.5.2. Rotation of Fracture Lines

The method of plastic collapse mechanisms is based on the approximation that plastic rotation is concentrated into a single section (Fig. 4.51a). The fracture rotation can result from the equation (4.40)

$$\alpha_f = 2\varepsilon_u = 0.22 \div 0.28 \tag{4.54}$$

where ϵ_u is the uniform strain (see Table 4.3). But this condition is too severe, because in reality the rotation is distributed along a plastic zone (Fig. 4.51b) and the fracture rotation is usually larger than the values obtained from relation (4.54). A simplified relation is proposed by Kotelko (1996b):

$$\alpha_f = 2n\varepsilon_u \tag{4.55}$$

where n is an empirical coefficient, determined experimentally or by means of a minimization procedure. The values $n = 3 \div 4$ fit very closely to the experimental results, which means that the plastic zone in reality is composed by three or four yield lines.

Taking into account this aspect, the length of the plastic zone can be determined from Fig. 4.52:

$$L_p = \left(\frac{M_{uN}}{M_{pN}} - 1\right) L = \left(\frac{1}{\rho_y} - 1\right) L \tag{4.56}$$

where equation (4.47) and (4.51) are used. If an idealized stress-strain diagram of Fig. 4.9d type is used, the curvature of the plastic zone is:

$$\chi_u = \frac{2}{\zeta}\frac{\alpha_f}{L_p} = \frac{1}{1/\rho_y - 1} \cdot \frac{\alpha_f}{\zeta L} \tag{4.57}$$

where L is the distance between the two inflection points of the plastic mechanism. It results in the fracture rotation

$$\alpha_f = \zeta \left(\frac{1}{\rho_y} - 1\right) \frac{L}{t} \cdot \varepsilon_u \tag{4.58}$$

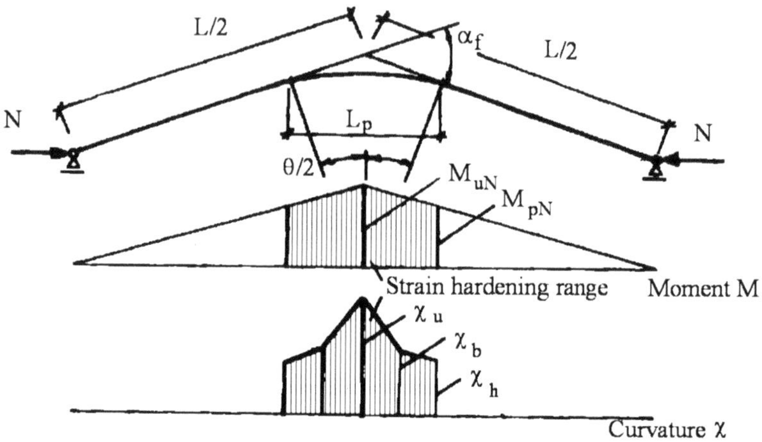

Figure 4.52: Fracture rotation

where

$$\zeta = 1 + 2\frac{\varepsilon_b}{\varepsilon_u} + \frac{\varepsilon_h}{\varepsilon_u} \tag{4.59a}$$

$$\varepsilon_b = \frac{\chi_b t}{2}\ ; \quad \varepsilon_h = \frac{\chi_h t}{2}\ ; \quad \varepsilon_u = \frac{\chi_u t}{2} \tag{4.59b-d}$$

For Fe 360 and $L/t = 6$, eq. (4.59) gives $\alpha_f = 6.38\epsilon_u$, which corresponds very well to the values proposed by Kotelko. One can see that this ultimate rotation depends on yield ratio, length of plastic mechanism and steel quality. Fig. 4.53 shows the influence of European, USA and Japanese steel qualities on the fracture rotations. A very important reduction of fracture rotation may be noticed when the yield stress and yield ratio increases, showing the importance of steel quality in the protection against premature collapse.

4.5.3. Rigid Fracture Analysis

The rigid-plastic analysis presented in Section 4.4.2 considers that all the yield lines work in the plastic range with the plastic moment M_p. But when the load increases until the collapse, the moments along the yield lines have different values, as functions of rotations: some work in plastic range, others in strain-hardening. Contrary to the rigid-plastic analysis, the moment values are different, depending on the rotation amount: ultimate moment for fracture lines, moment in strain-hardening range for some lines and plastic moment for yield lines (Fig. 4.54). The same situation may be related to axial loads. Therefore, the terms of strain energy given by equations (4.36a,b) must be evaluated according to the corresponding moments:

$$U_{ls,t} = m_i \sum_i M_{pi}\theta_i l_i; \qquad U_z = n_j \sum_j N_{pj}\varepsilon_s A_j \tag{4.60}$$

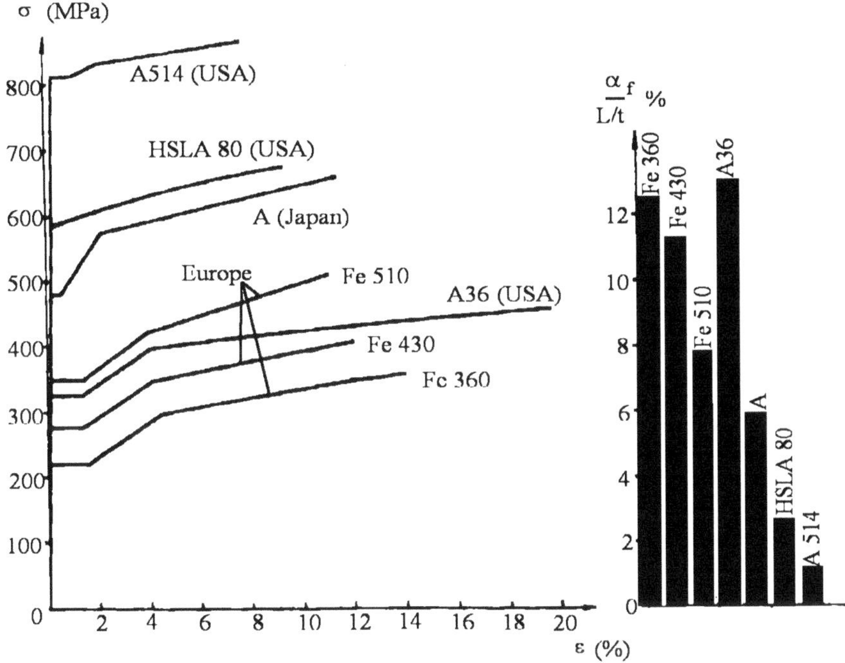

Figure 4.53: Influence of steel quality

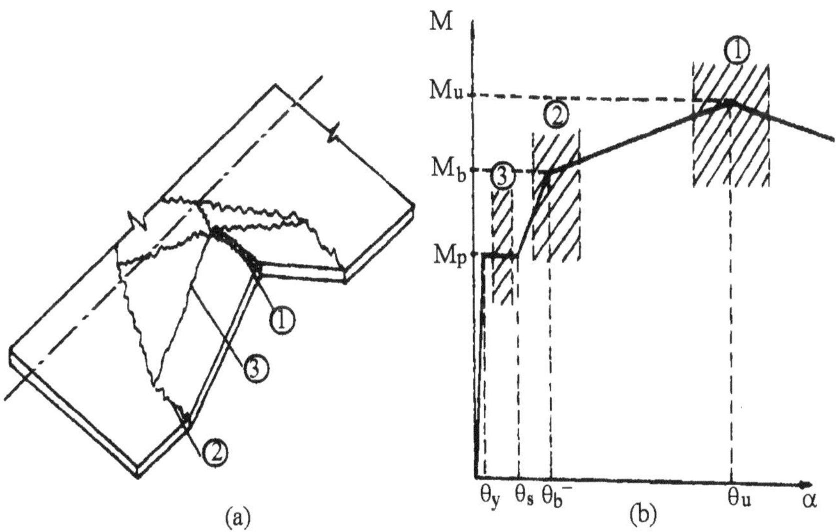

Figure 4.54: Local fracture mechanism: (a) Yield and fracture lines; (b) Moments corresponding different plastic lines

where

$$m_i = \frac{M_i}{M_{pi}}; \qquad n_j = \frac{N_j}{N_{pj}} \tag{4.61a, b}$$

M_i and N_j are the actual values of moment and axial forces along the yield lines i and on the plastic zones j, respectively. For M_i and N_j values one can use the values corresponding to ultimate, strain-hardening or plastic ranges, as function of rotation values α. Using the relations (4.38) and (4.39) a new load-deformation relation results for rigid-fracture analysis. The actual curve is situated between the rigid-fracture and rigid-plastic curves. The ultimate rotation is given by the fracture rotation (4.59).

4.6 Plastic and Fracture Mechanisms of Plates

4.6.1. Plastic Mechanisms for Compression Plates

The behaviour of plastic mechanisms depends on the edge conditions and loading type.

(i) *Free edges* (Fig. 4.55a). The plate has an initial geometrical imperfection which produces the plate initial shortening Δ_{ir}. The shortening is due to the load is Δ_r. The relationship between the two longitudinal (Δ_r and Δ_{ri}) and transversal (Δ and Δ_i) displacements are:

$$\Delta_i = (2a\Delta_{ri})^{1/2}; \quad \Delta + \Delta_i = [2a(\Delta_r + \Delta_{ri}]^{1/2} \tag{4.62a, b}$$

Being:

$$\Delta_{ri} = a\left(1 - \cos\frac{\theta_i}{2}\right) \approx \frac{a}{2}\left(\frac{\theta_i}{2}\right)^2 \tag{4.63}$$

the initial rotation results:

$$\frac{\theta_i}{2} = \left(\frac{2\Delta_{ri}}{a}\right)^{1/2} \tag{4.64}$$

Similarly, the total rotation is given by

$$\frac{\theta_i + \theta}{2} = \left(2\frac{\Delta_r + \Delta_{ri}}{a}\right)^{1/2} \tag{4.65}$$

The strain energy is:

$$U = 2M_{pN}\left(\frac{\theta_i + \theta}{2} - \frac{\theta_i}{2}\right) = 2\sqrt{2}M_{pN}\left[\left(\frac{\Delta_r + \Delta_{ri}}{a}\right)^{1/2} - \left(\frac{\Delta_{ri}}{a}\right)^{1/2}\right] \tag{4.66}$$

From (4.38a) it results:

$$F = \frac{dU}{d\Delta_r} = \sqrt{2}\frac{M_{pN}}{\sqrt{a}}\left[\frac{1}{(\Delta_r + \Delta_{ri})}\right]^{1/2} = 2M_{pN}\frac{1}{\Delta + \Delta_i} \tag{4.67}$$

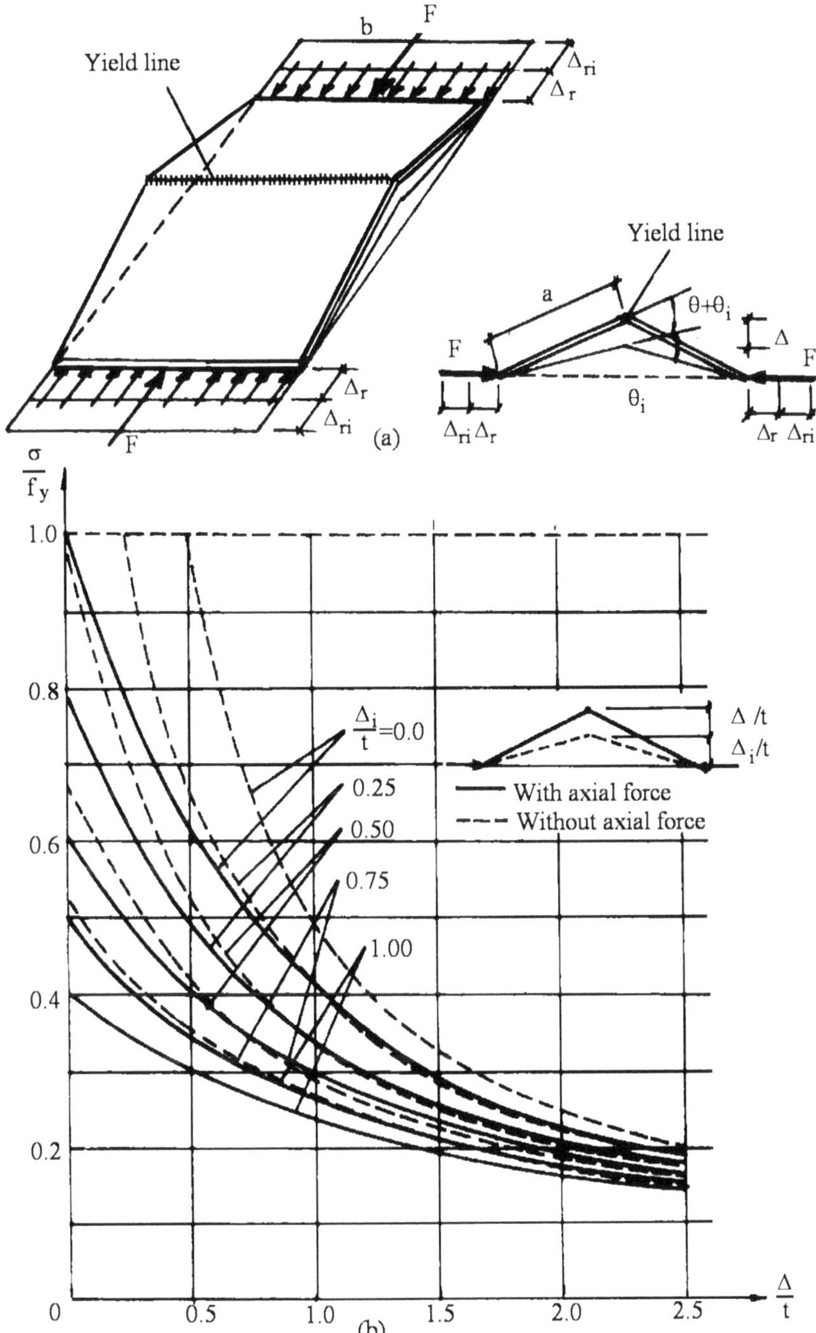

Figure 4.55: Plastic mechanism for plate with free edges: (a) Plate configuration; (b) Stress-strain relationship

For an ideal plate without imperfections, it results:

$$F = 2M_{pN}\frac{1}{\Delta} \tag{4.68}$$

From equations (4.67) and (4.68) it is easily observed that the initial and loading deformations have the same effect on the decreasing of the element capacity, and the sum of these two deformations produces the total degradation of axial force.

From eqs. (4.67) and (4.47) the following relation is obtained:

$$\frac{F}{F_p} = \frac{\sigma}{f_y} = \left[\left(\frac{\Delta + \Delta_i}{t}\right)^2 + 1\right]^{1/2} - \frac{\Delta + \Delta_i}{t} \tag{4.69}$$

If the effect of axial forces is neglected, it results:

$$\frac{F}{F_p} = \frac{\sigma}{f_y} = \frac{1}{2}\frac{t}{\Delta + \Delta_i} \tag{4.70}$$

The equations (4.69) and (4.70) are presented in Fig. 4.55b. The difference between the two curves becomes negligible as far as the $\frac{\Delta}{t}$ ratio increases.

If the fracture behaviour is considered by using equation (4.53) instead of (4.47) the result is:

$$\frac{F}{F_u} = \rho_y\frac{F}{F_p} \tag{4.71}$$

This equation is presented in Fig. 4.56. The fracture displacement results from (4.59)

$$\frac{\Delta_f}{t} = \frac{1}{4}\theta_f\frac{l}{t} = \frac{1}{2}\left(\frac{1}{\rho_y} - 1\right)\left(\frac{l}{t}\right)^2\varepsilon_u \tag{4.72}$$

The values of δ_f/t are marked in Figure for L/t = 6, 8, 10, by assuming ρ_y = 0.653 and ϵ_u = 0.14 which correspond to Fe 360 steel grade.

(ii) *Supported edges.* For supported edge plates, under in-plane compression, four types of local plastic mechanisms are proposed in the literature: pyramidal, wedge, roof and flip-disc shapes.

The *pyramidal shape mechanism* (Fig. 4.57a), is studied by Kragerup (1982), Rondal and Maquoi (1985). The post-plastic behaviour is governed by the equation

$$\frac{F}{F_p} = \frac{\sigma}{f_y} = \frac{1}{2}\left\{1 - \frac{2\Delta + \Delta_i}{t} + \left[1 + 4\left(\frac{\Delta + \Delta_i}{t}\right)^2\right]^{1/2}\right\} \tag{4.73}$$

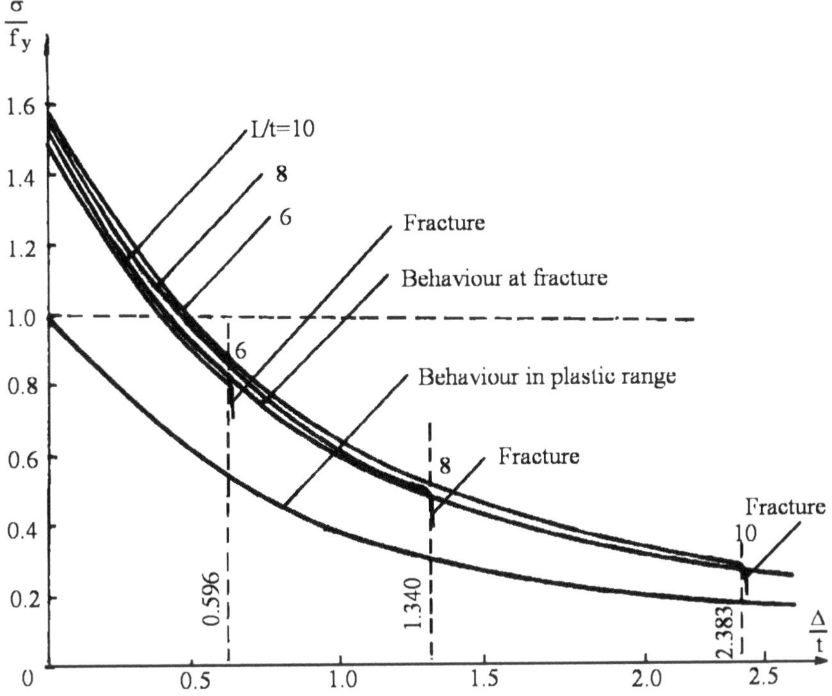

Figure 4.56: Fracture behaviour for plate with free edges

The post-plastic unloading curve resulting from this equation is presented in Fig. 4.57b. One can see that the ultimate load of an uniaxially compressed plate is approximately found as the intersection of two load-displacement curves. The first curve represents the elastic pre-collapse response, while the second one is characteristic of the kinetically admissible plastic mechanism.

The *wedge shape mechanism* (Fig. 4.58a) is studied by Korol and Sherbourne (1972), Sherbourne and Korol (1972). The geometrical parameters are the inclination of diagonal hinges, α, and the length of plastic mechanism, 2d. Considering the diagonal yield lines as travelling ones, the unloading post-crushing curves are obtained for various widths of plate to thickness ratios. As expected, the post-crushing load derivable from the mechanism solution decreases with plastic rotation θ. The load versus transverse displacement may be approximated by the equation:

$$\frac{F}{F_p} = 1.00 - 0.355\frac{\Delta + \Delta_i}{t} + 0.056\left(\frac{\Delta + \Delta_i}{t}\right)^2 - 0.003\left(\frac{\Delta + \Delta_i}{t}\right)^3 \quad (4.74)$$

being almost completely insensitive to plate width b. Equation (4.74) is depicted in Fig. 4.58b.

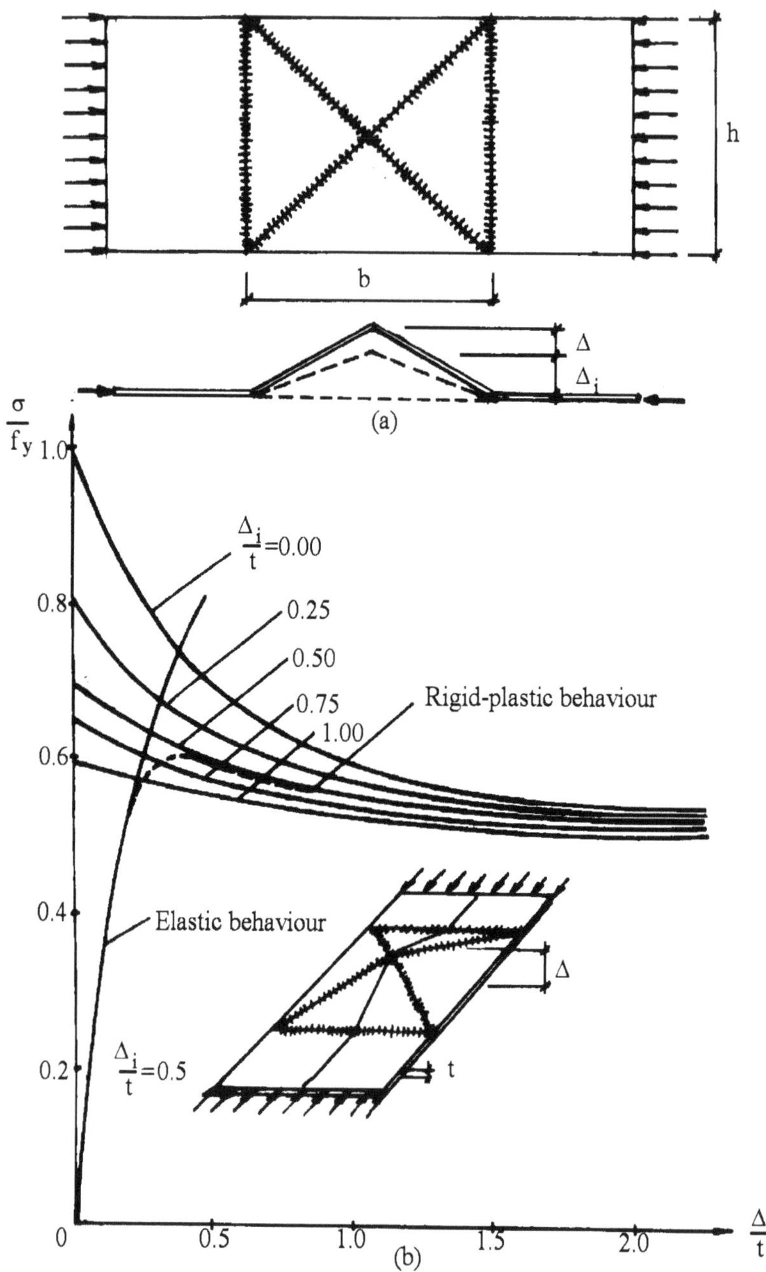

Figure 4.57: Pyramidal shape mechanism for plate with supported edges: (a) Mechanism configuration; (b) Stress-strain relationship

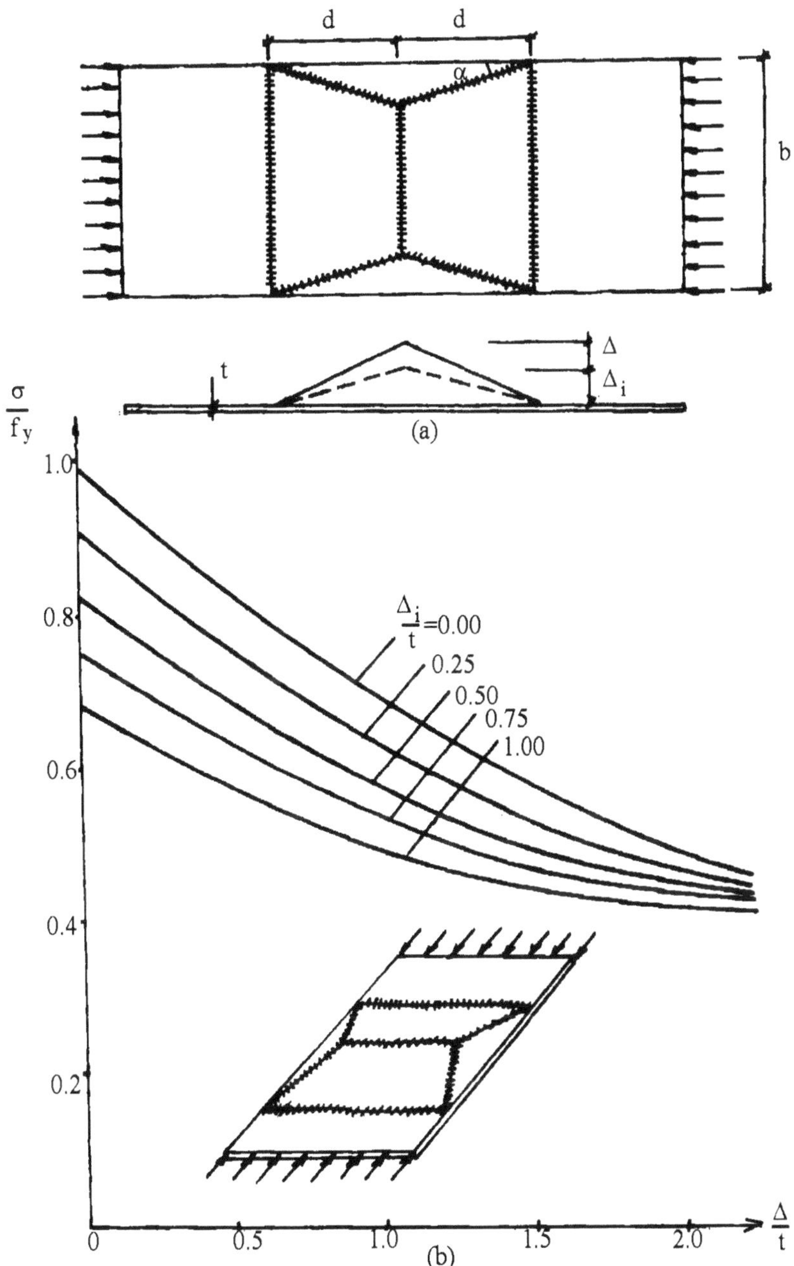

Figure 4.58: Wedge shape mechanism for plate with supported edges: (a) Mechanism configuration; (b) Stress-strain relationship

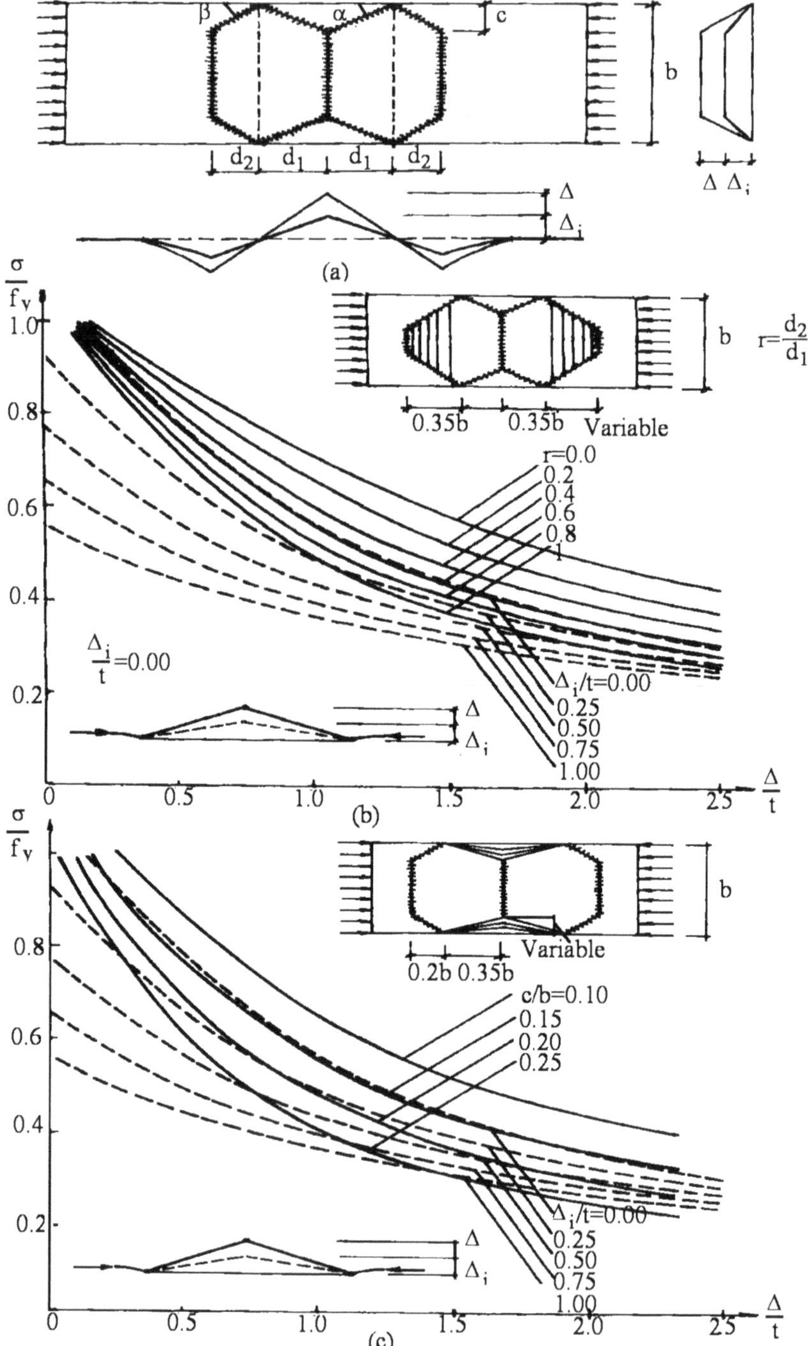

Figure 4.59: Roof mechanism for plates with supported edges: (a) Mechanism configuration; (b), (c) Parametric study and imperfection influence

For *roof mechanism* detailed in Fig. 4.59a, the average axial stress is given by the equation (Mahendran, 1997, Mahendran and Murray, 1991):

$$\frac{\sigma}{f_y} = \left(1 - 2\frac{c}{b}\right)\left\{\left[(1+r^2)\left(\frac{\Delta+\Delta_i}{t}\right)+1\right]^{1/2} - (1+r)\frac{\Delta+\Delta_i}{t}\right\}$$

$$+\frac{c}{b}\left\{\left[\frac{4(1+r)^2}{k^2}\left(\frac{\Delta+\Delta_i}{t}\right)^2+1\right]^{1/2} - \frac{2(1+r)}{k}\frac{\Delta+\Delta_i}{t}+\right.$$

$$+\frac{1}{\dfrac{2(1+r)}{k}}\frac{1}{\dfrac{\Delta+\Delta_i}{t}}\ln\left\{\left[\frac{4(1+r)^2}{k^2}\left(\frac{\Delta+\Delta_i}{t}\right)^2+1\right]^{1/2}+\right.$$

$$\left.\left.+\frac{2(1+r)}{k}\frac{\Delta+\Delta_i}{t}\right\}\right\} \tag{4.75}$$

where:

$$k = \text{cosec}^2\alpha + \text{cosec}^2\beta \tag{4.76a}$$

$$r = \frac{d_2}{d_1} \tag{4.76b}$$

The parameters of the plastic mechanism are α, c and r. The results of a parametric study are presented in Figs. 4.59b,c. One can see that the variation of the parameters does not give a minimum value for stresses, showing that the yield lines are not travelling ones. In this situation only the length and pattern of the buckled surface may give indications about the mechanism dimensions. The influence of geometrical imperfections is also presented in Fig. 4.59b,c.

For *flip-disc mechanism* (Fig. 4.60a), a parabolic shape is considered. The average axial stress has been obtained using the Simpson's rule (Mahendran, 1997):

$$\frac{\sigma}{f_y} = \frac{1}{6}\left[1 - 2\frac{\Delta+\Delta_i}{t} + 2\left(\frac{\Delta+\Delta_i}{t}\right)^2+1\right]^{1/2} - \frac{6\dfrac{\Delta+\Delta_i}{t}}{1+4\dfrac{a^2}{b^2}}+$$

$$+4\left[\left(\frac{3}{2}\frac{\dfrac{\Delta+\Delta_i}{t}}{1+\dfrac{4a^2}{b^2}}\right)^2+1\right]^{1/2} \tag{4.77}$$

The basic geometrical parameter is the d/b ratio. The results of a parametric study and the influence of imperfections are shown in Fig. 4.60b.

A comparison between the four mechanism types are presented in Fig. 4.61 for ideal and actual plates. One can see that the roof and flip-disc mechanisms give the most unfavourable values. Therefore, only these two forms must be considered for practical purposes.

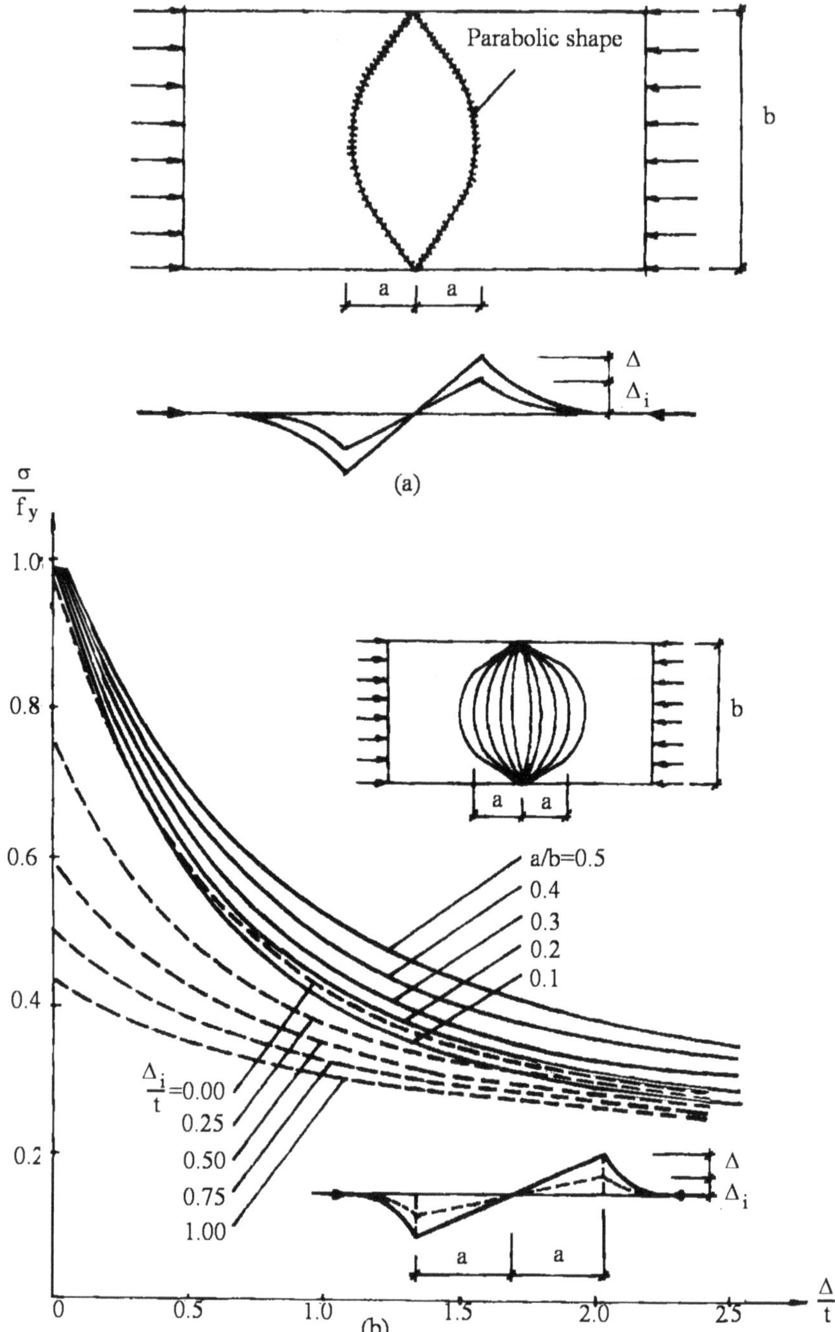

Figure 4.60: Flip-disc mechanism: (a) Mechanism configuration; (b) Parametric study and imperfection influence

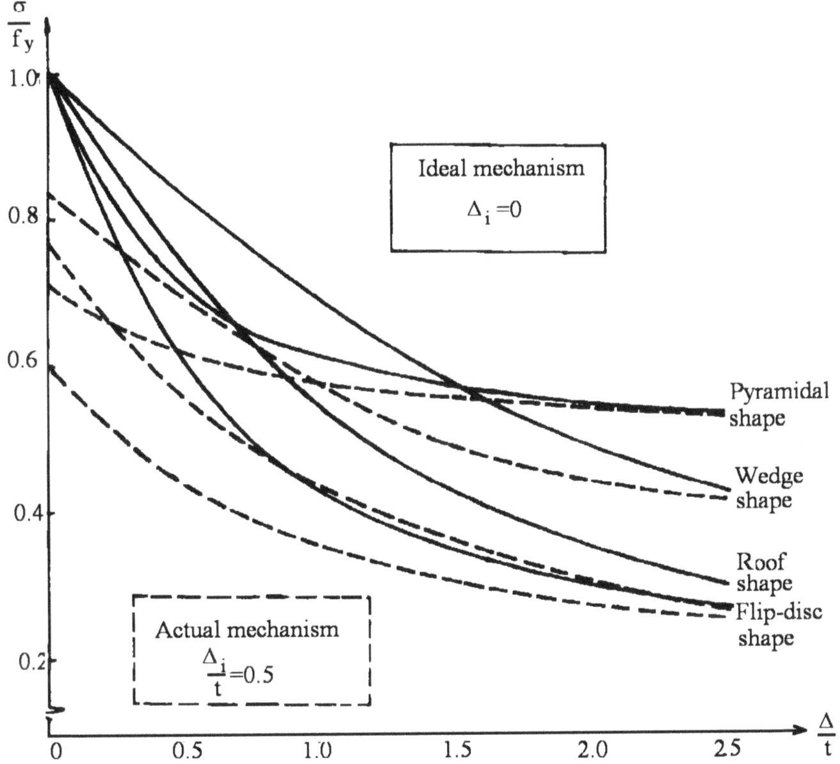

Figure 4.61: Comparison between different mechanism types

The decisive mechanism type results from the comparison of the intersection point of the elastic curve for imperfect plates and the rigid-plastic curve for the mechanism behaviour (Fig. 4.62). So, the ultimate strength of plate depends on the size of imperfections and the mechanism type. Fig. 4.63 indicates, after Mahendran (1997) and Mahendran and Murray (1991), the change of mechanism type when the plate becomes thinner, by increasing the b/t ratio, or when the magnitude of geometrical imperfection Δ_i/t increases. One can see that thicker plates (say b/t ratio of 20–40), which usually do not have higher levels of initial imperfections, would develop the roof mechanism. However, plates with moderate b/t ratios would develop both mechanism types, depending on the magnitude of initial imperfections.

In order to verify these aspects an intensive experimental analysis has been performed (Mahendran, 1997). Approximately 80% of experimental results agree with the analytical prediction, only roof or flip-disc mechanisms occurring. The remaining 20% refer to the cases when flip-disc mechanisms were observed instead of roof mechanisms, due to larger specimen imperfections. So, the general conclusion of this study is that experiments verified the analytical predictions.

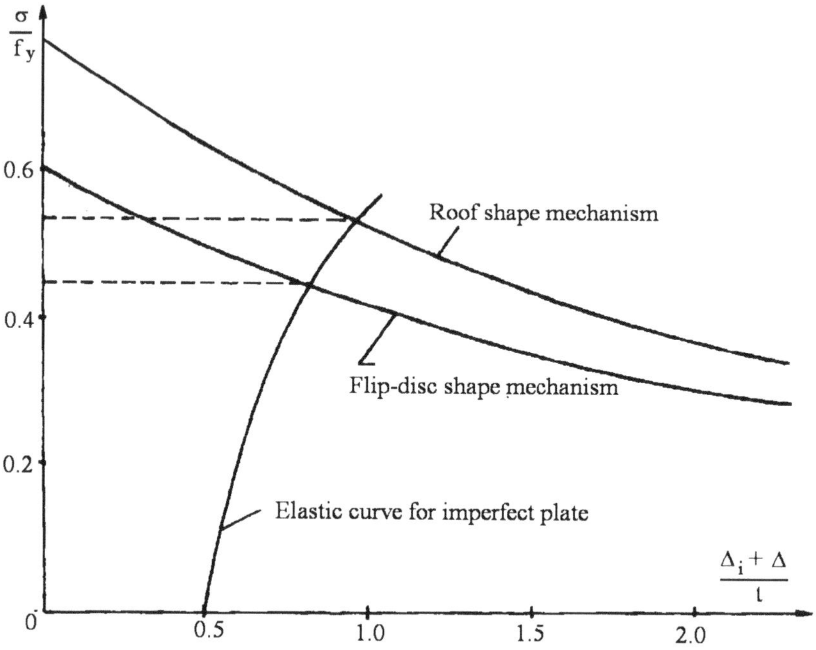

Figure 4.62: Determination of decisive mechanism type

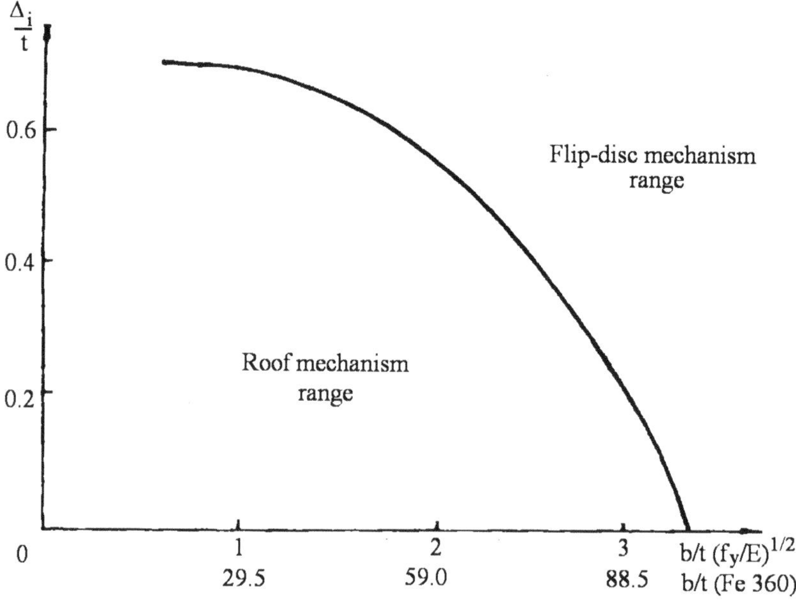

Figure 4.63: Roof and flip-disc ranges

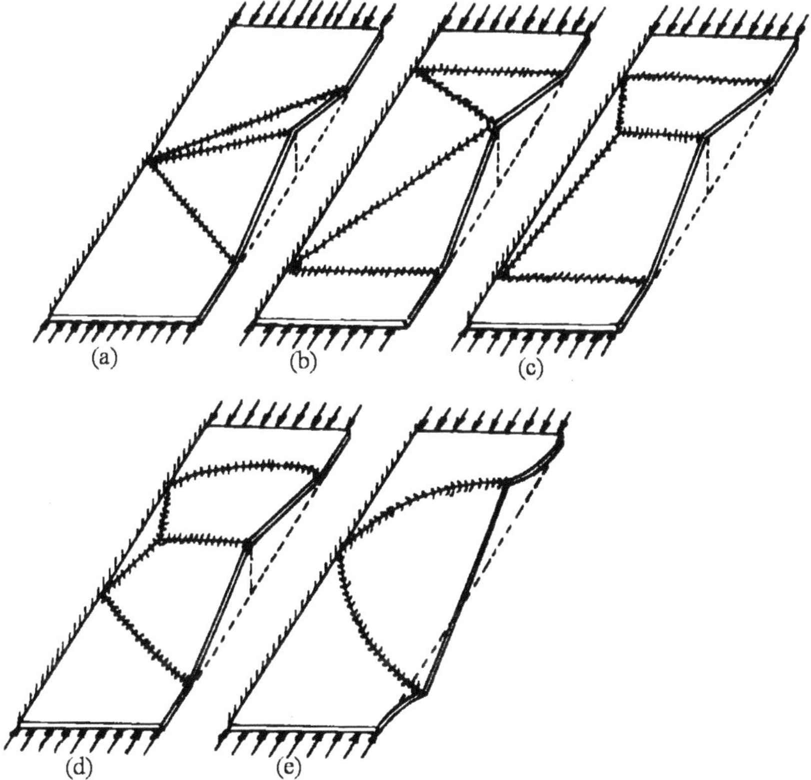

Figure 4.64: Plate with one free edge: Mechanism types

For plates used in seismic-resistant structures, which are generally thick with small geometrical imperfections, it is very important to note that the roof mechanism type seems to be the most common form of collapse.

(iii) *One free edge.* As it is shown in Section 4.3.4, the critical load for supported edge plates is the same as the critical load of one free edge with half width. If the mode of buckling wave for supported edge plates is divided along its centre line in the longitudinal direction, the mode of each plate is equal to that of a plate with one free edge. Conversely, if connecting two modes of such type at their free edges, the obtained mode is the same as that of a plate supported along its four edges (Inoue, 1994). This observation is very important because many results obtained for supported edge plates may be used for the study of one free edge plate behaviour.

There are five types of plastic mechanisms which may be considered (Fig. 4.64): diagonal semi-pyramidal, semi-pyramidal, semi-wedge, semi-roof and semi-flip-disc shapes.

The *diagonal semi-pyramidal shape mechanism* (Fig. 4.65a) is studied by Murray (1986) and Feldmann (1994). The post-crushing curve is given

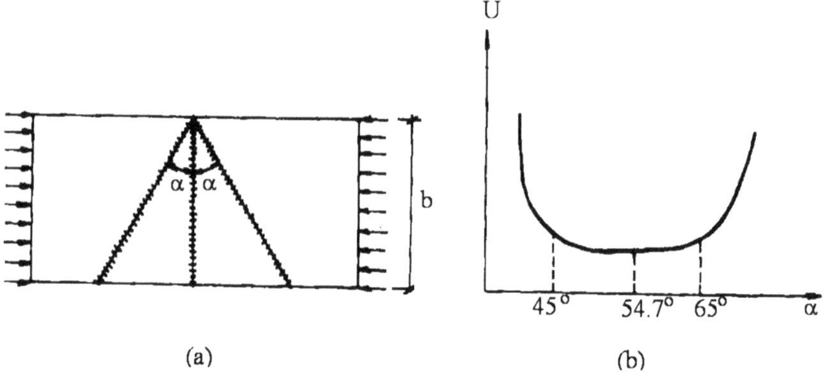

Figure 4.65: Diagonal semi-pyramidal mechanism: (a) Mechanism configuration; (b) Potential function minimizing

by:

$$\frac{\sigma}{f_y} = \frac{1}{2}\left\{ \left[\frac{4}{k^2}\left(\frac{\Delta + \Delta_i}{t}\right)^2 + 1 \right]^{1/2} - \frac{2}{k}\frac{\Delta + \Delta_i}{t} \right.$$
$$\left. + \frac{k}{2}\frac{t}{\Delta + \Delta_i} \ln \left\{ \left[\frac{4}{k^2}\left(\frac{\Delta + \Delta_i}{t}\right) \right]^{1/2} + \frac{2}{k}\frac{\Delta + \Delta_i}{t} \right\} \right\} \tag{4.78}$$

where:

$$k = 1 + \operatorname{cosec}^2 \alpha \tag{4.79}$$

The geometric parameter is the angle α of the yield lines. A minimum potential energy is obtained for $\alpha = 54.7^o$, but the function is practically constant within 45° and 65°, so the value $\alpha = 45^o$ may be used for practical purposes (Fig. 4.65b). A simple equation can be obtained from (4.78):

$$\frac{\sigma}{f_y} = 1 - 0.25\frac{\Delta + \Delta_i}{t} \tag{4.80}$$

For the other four mechanism types the equations (4.73, 4.74, 4.75, 4.77) are available, respectively. A comparison between the five mechanism types is presented in Fig. 4.66. As expected, the conclusions reached for supported plates are also available for one free edge plates, the semi-roof and semi-flip-disc mechanisms being the only ones which must be considered for practical purposes.

4.6.2. Plastic Mechanisms for Bended Plates

(i) *Mechanism types.* Two mechanism types are studied by Park and Lee (1996), being composed by three or four yield lines (Fig. 4.67a). The plastic rotation θ is considered as process parameter and the dimensions of mechanisms as geometrical parameters. The minimum of plastic work gives the following geometrical parameters:

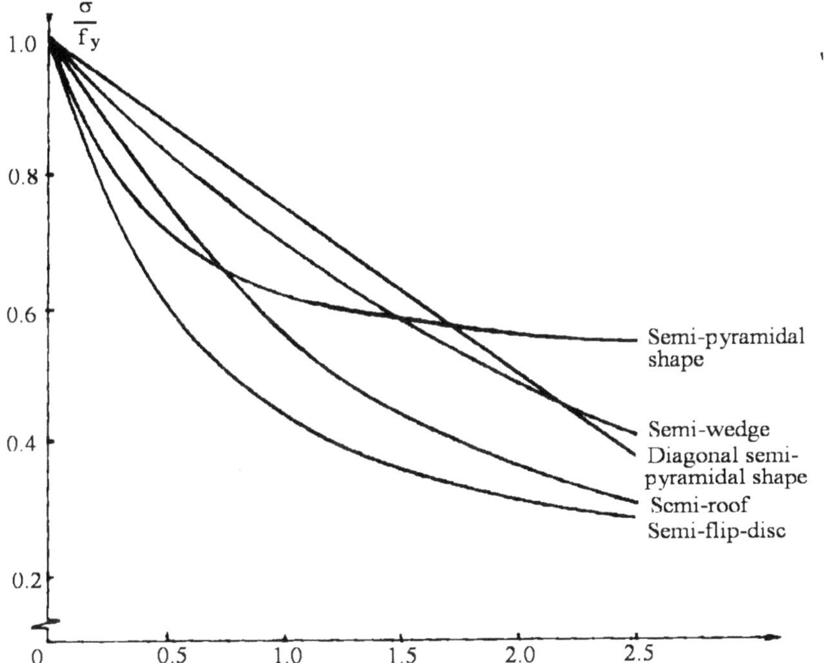

Figure 4.66: Comparison between different mechanism type for plate with one free edge

-mode I: $L_p = 1.64$ h; $\alpha = 39.35^o$
-mode II: $L_{p1} = 1.74$ h; $\alpha_1 = 33.02^o$; $L_{p2} = 0.44$ h $\alpha_2 = 23.75^o$

The moment-rotation curves for the two collapse modes are presented in Fig. 4.67b. One can see that the second mode gives the most unfavourable behaviour, but the differences are insignificant.

The plastic mechanism corresponding to mode I is also studied by Feldmann (1994), for which the post-crushing curve is

$$\frac{M}{M_p} = \frac{1}{6}k^2 \left(\frac{t}{\Delta}\right)^2 \left\{ \left[\frac{4}{k^2} \left(\frac{\Delta + \Delta_i}{t}\right)^2 + 1 \right]^{3/2} - 1 - \frac{8}{k^3} \left(\frac{\Delta}{t}\right)^3 \right\} \quad (4.81)$$

where k is given by (4.79).

The case of plate with geometrical imperfections is presented in Fig. 4.68. From geometrical conditions it results:

$$\Delta_{1i} = h\frac{\theta_i}{2}; \quad \Delta_i = h\frac{\theta}{2} \quad (4.82a-d)$$

$$\Delta_i = \left(2bh\frac{\theta_i}{2}\right)^{1/2}; \quad \Delta = (2bh)^{1/2} \left[\left(\frac{\theta + \theta_i}{2}\right)^{1/2} - \left(\frac{\theta_i}{2}\right)^{1/2} \right]$$

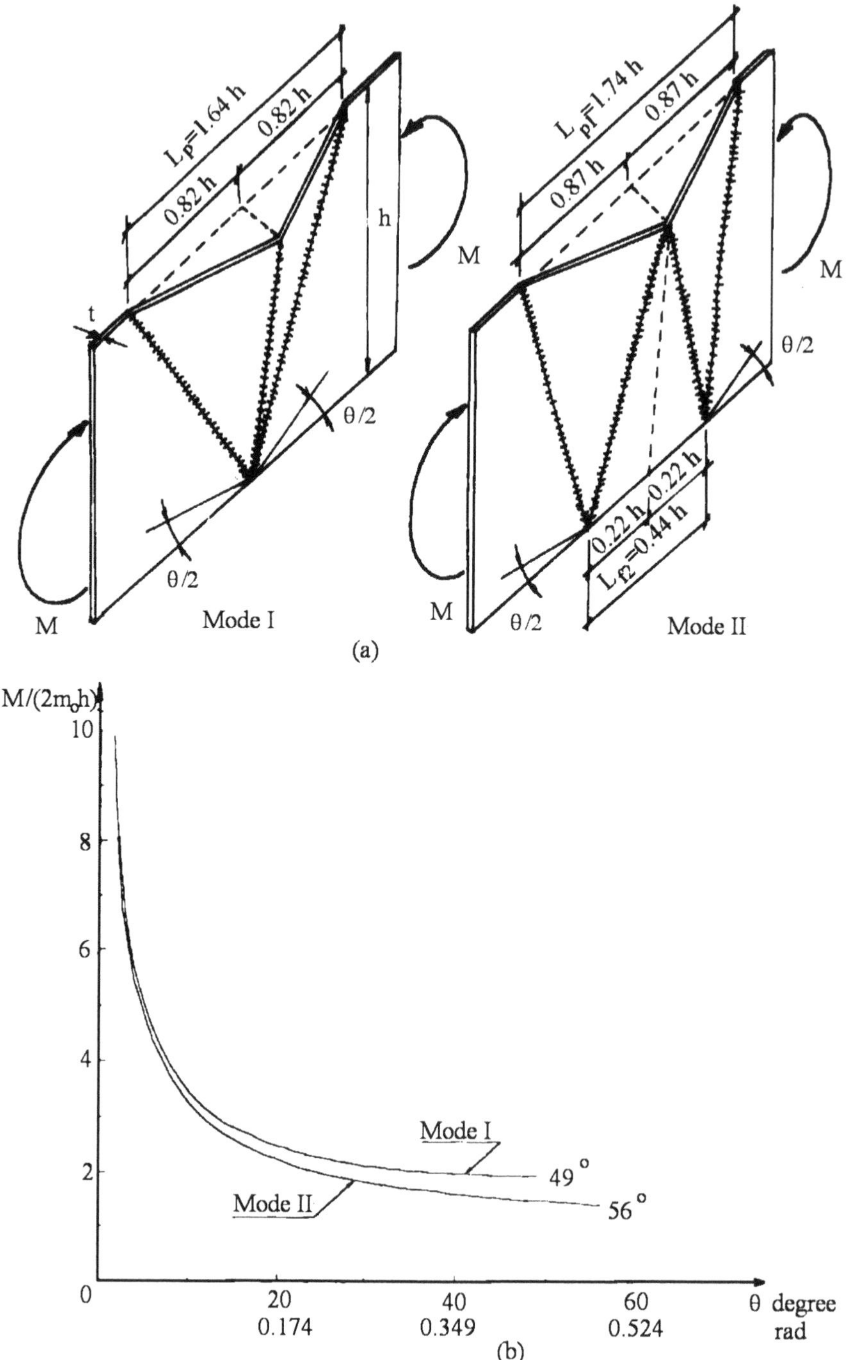

Figure 4.67: Plastic mechanism for bended plates: (a) Mechanism configuration types; (b) Moment-rotation relationship (after Park and Lee, 1996)

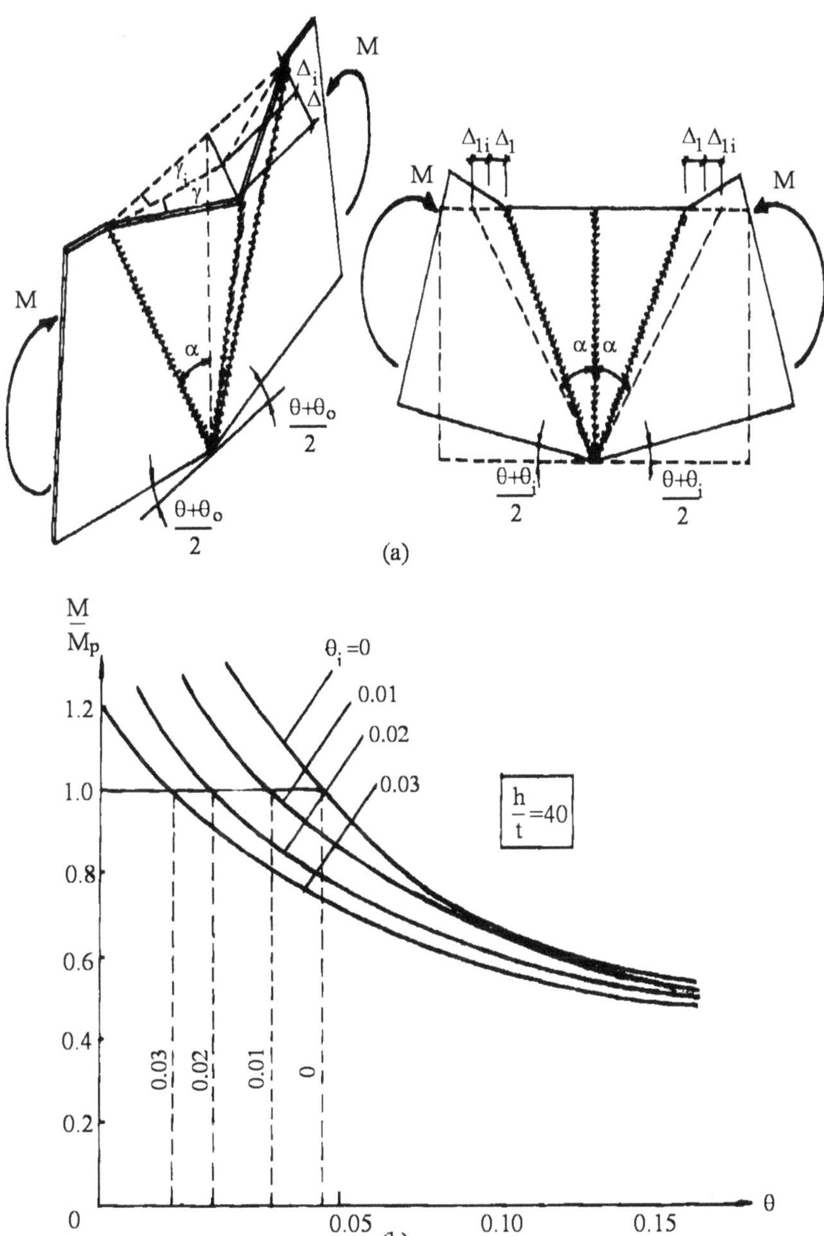

Figure 4.68: Influence of imperfections: (a) Mechanism configuration; (b) Moment-rotation relationship

The strain energy is given by

$$U = 2f_y\frac{t^2}{2}h\left(\frac{2}{\tan\alpha}\right)^{1/2}\left(1+\frac{1}{\cos^2\alpha}\right)\left[\left(\frac{\theta+\theta_i}{2}\right)^{1/2}-\left(\frac{\theta_i}{2}\right)^{1/2}\right] \quad (4.83)$$

which from (4.38b) provides:

$$M = \frac{dU}{d\theta} = 2f_y h\frac{t^2}{2}k(\alpha)\frac{1}{\left(\frac{\theta+\theta_i}{2}\right)^{1/2}} \quad (4.84)$$

and

$$\frac{M}{M_p} = 2\frac{t}{h}k(\alpha)\frac{1}{\left(\frac{\theta+\theta_i}{2}\right)^{1/2}} \quad (4.85)$$

where:

$$k(\alpha) = \left(\frac{2}{\tan\alpha}\right)^{1/2}\left(1+\frac{1}{\cos^2\alpha}\right) \quad (4.86)$$

The minimum of function (4.86) results for $\alpha = 0.6847$ (39.23^o), which corresponds to the value obtained by Park and Lee (1996) for mode I. For this angle, $k = 4.1736$ and:

$$\frac{M}{M_p} = 8.347\frac{t}{h}\frac{1}{(\theta+\theta_i)^{1/2}} \quad (4.87)$$

For $h/t = 40$ and $\theta_i = (0\text{-}3)\text{x}10^{-2}$ the curves (4.87) are presented in Fig. 4.68b. One can see that the presence of geometrical imperfections reduces the rotation capacity, determined at the level M = M_p.

(ii) *Influence of incomplete mechanisms.* For the bended plate, the influence of incomplete mechanism is presented in Fig. 4.69. If the secondary yield lines are incompletely formed, the plastic moment decreases as a linear in function of the ratio between the length of incomplete and complete yield secondary lines (Anastasiadis et al, 2000).

(iii) *Influence of strain-rate.* The influence of strain-rate on a bended plate has been studied for the plastic mechanism presented in Fig. 4.70. The fracture rotation for a simple mechanism is given by relation (4.59). In this case the fracture rotation refers to the angle γ and the length is considered the distance between two inflection points ($L \simeq$0.81h):

$$\gamma_f = 2\left(\frac{1}{\rho_y}-1\right)\cdot\frac{0.81h}{t}\cdot\varepsilon_u \quad (4.88)$$

and the relation between θ_f and γ_f results from:

$$\theta_{fsr} = \frac{\tan\alpha}{4}\gamma_f^2 = 0.531\cdot\left(\frac{1}{\rho_y}-1\right)^2\cdot\left(\frac{h}{t}\right)^2\cdot\varepsilon_u^2 \quad (4.89)$$

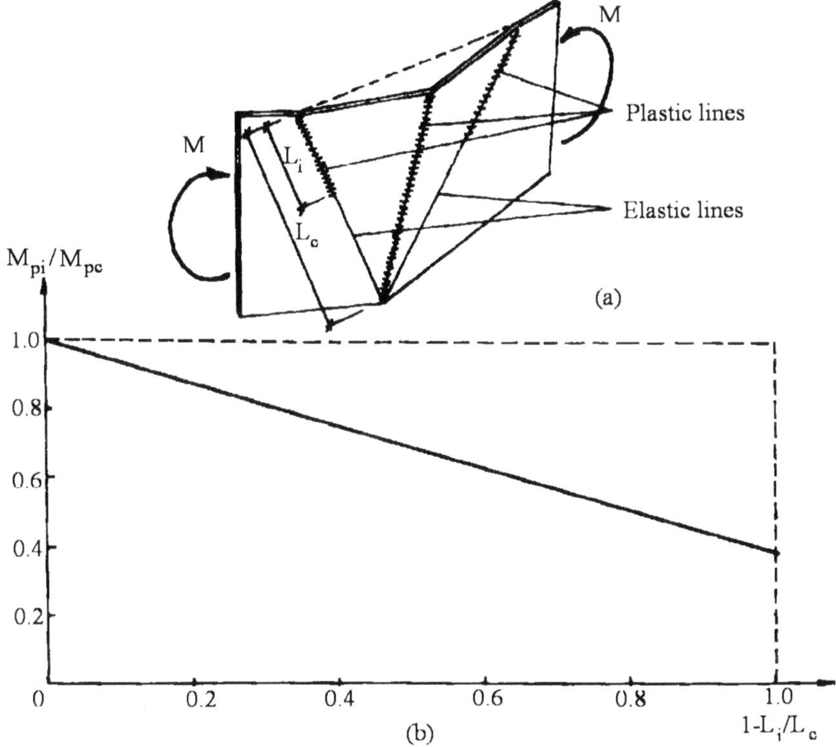

Figure 4.69: Influence of incomplete mechanisms

One can see that the influence of strain-rate may be defined as follows:

$$\frac{\theta_{fsr}}{\theta_f} = \frac{\left(\dfrac{1}{\rho_y} - 1\right)^2_{\text{strain rate}}}{\left(\dfrac{1}{\rho_y} - 1\right)^2_{\text{static}}} \tag{4.90}$$

This relation is presented in Fig. 4.70 for the strain-rate law proposed by Soroushian and Choi (1987). It is very interesting to observe that a very important reduction occurs when the strain-rate increases, even in the range considered as static loading. So, it can be argued that the effect of strain-rate on ductility should be taken into consideration even in static analysis, when plastic buckling may take place.

(iv) *Influence of cyclic loading.* For the analysis of the cyclic loading influence one must recall that the behaviour of a single plate is conditioned by the overall mechanism of the member section, which is composed by an ensemble of more plates. In addition, it is characterized by an asymmetry, the compression strains being more important than the tensile strains (see section 4.2.6). As a first step, the analysis of cyclic behaviour may be performed by neglecting the tensile strains.

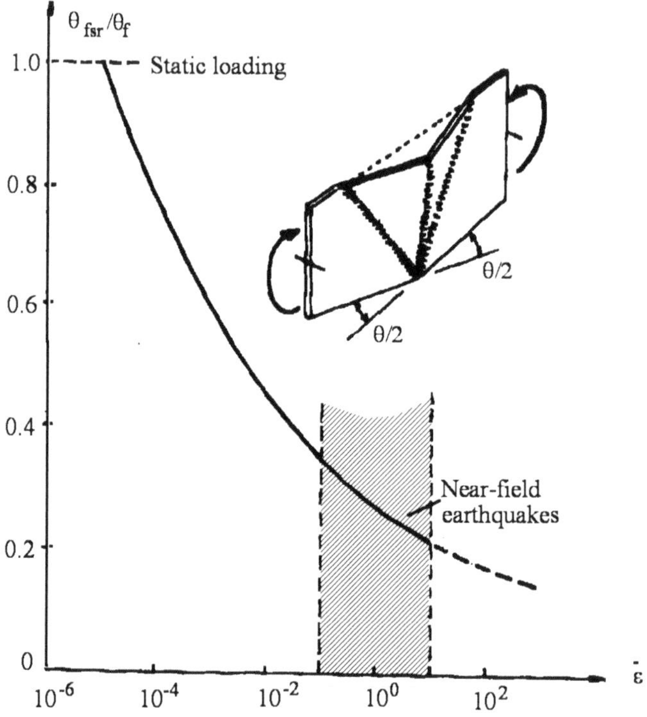

Figure 4.70: Influence of strain-rate

With this simplification, after each plastic deformation, a permanent deformation remains, which works as an initial geometrical imperfection for the next cycle. In Fig. 4.71 the evaluation of rotation limit of a bended plate can be done by means of four steps, each being composed of loading, unloading and reloading of a rigid-ideal plastic material. After the first two steps, the curve corresponding to an initial imperfection of half rotation limit is reached and a decreasing of bending moment occurs. In the following steps, a new decreasing is produced, due to the increasing of initial deformation. So, when reaching the rotation limit determined for monotonic loads, a reduced moment capacity is obtained, due to the cyclic loading. One can see that for this mechanism, the rotation limit for the cyclic loading is reduced to about half of the one corresponding to monotonic loads.

4.7 Ductility Classes for Elements

4.7.1. Behaviour Classes

In order to design sections and members able to provide sufficient ductility, the buckling of the constitutive elements have to be controlled. In particular, the occurrence of buckling in the elastic range has to be avoided

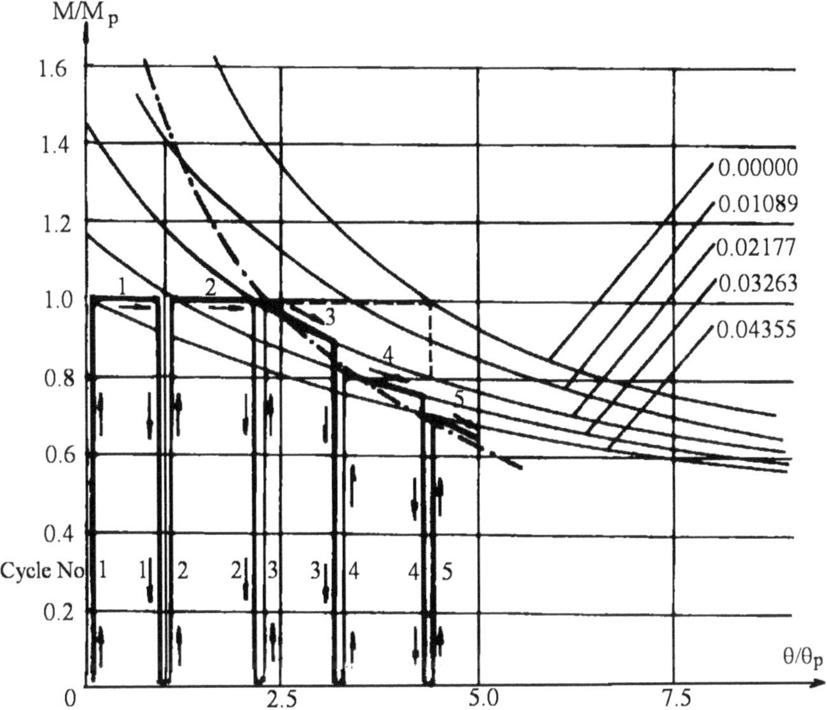

Figure 4.71: Influence of cyclic loading

for important structures. Therefore, the geometrical properties of a plate, described by the width-to-thickness ratio, b/t, have to guarantee the attainment of buckling in the plastic range. For this reasons, the design practice provides a classification of elements into four behavioural classes (Fig. 4.72):

-*plastic elements*, characterized by plastic buckling with high plastic deformations at collapse;

-*compact elements*, which provide plastic buckling with moderate plastic deformation at collapse;

-*semi-compact elements*, for which yield stress may be attained, but the buckling is mainly elastic;

-*slender elements*, for which the collapse occurs for elastic buckling.

4.7.2. Width-to-Thickness Ratio

The classification based on the above mentioned behavioural classes depends on the occurrence of local buckling, which can produce the collapse of elements. One must mention that this local buckling can occur for all elements indifferently of the behavioural class, the difference being given only by the level of strain at which it commences. This difference is governed by the element slenderness, defined by the width-to-thickness ratio. Table 4.5

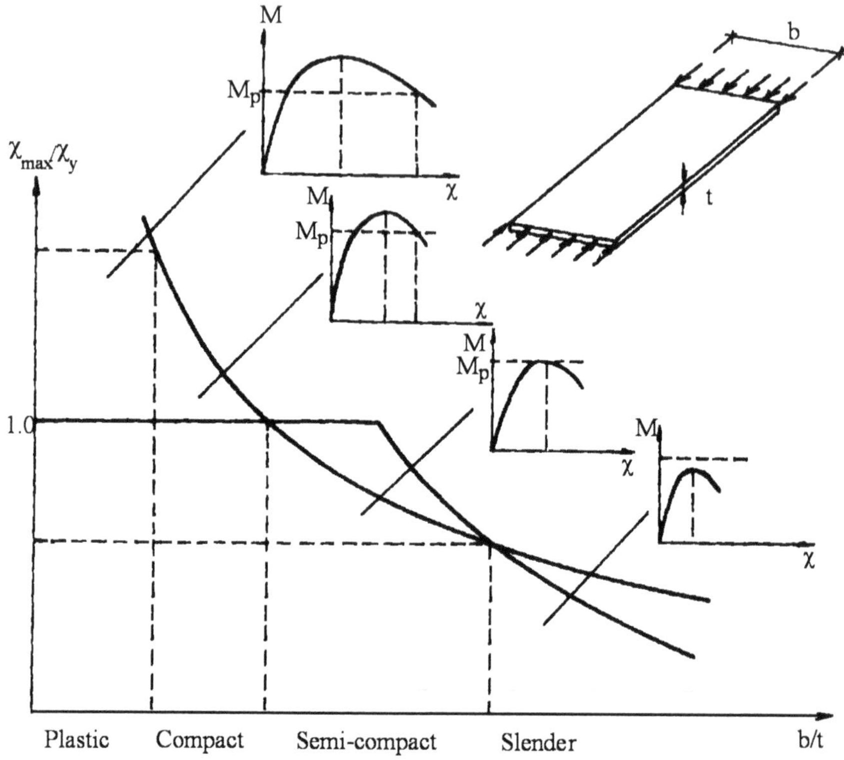

Figure 4.72: Behavioural classes

presents, after EC3, the proposal for classifying compressed elements into different behavioural classes, as a function of b/t ratio, plate type and stress distribution.

These behavioural classes related to b/t and d/t ratios have been defined as follows (Commision EC, 1988), (Fig. 4.73):

-for plastic cross-sections the limitation is obtained by the comparison between the required rotation of typical systems and the rotation capacity derived by tests;

-for compact sections the limitation is defined by the attainment of the plastic capacity;

-for semi-compact sections the limitation is defined by the attainment of the yielding capacity which is approximately 85 percent of the plastic capacity.

The rotation requirements depend on the structural system, loading conditions, used profiles and yield stress. Summarizing the results of static loaded structures concerning the required rotation capacity, a minimum value $\mu_r = 3$ should be considered.

Generally speaking, the development of rules for assessing the classification of plate elements into different behavioural classes did not achieve

Table 4.5: Maximum width-to-thickness ratios

Plate and loading	Plate type	Stress	Class plastic	Class compact	Class semi-compact
Plate in compression			$33\,\varepsilon$	$38\,\varepsilon$	$42\,\varepsilon$
			$9\,\varepsilon$	$10\,\varepsilon$	$14\,\varepsilon$
Plate in bending		Plastic Elastic	$72\,\varepsilon$	$83\,\varepsilon$	$124\,\varepsilon$
Plate in bending and compression		f_y, αd, Ψf_y Plastic Elastic	$\alpha > 0.5$ $\frac{39.6}{13\alpha-1}\varepsilon$ $\alpha < 0.5$ $\frac{36}{\alpha}\varepsilon$ $\varepsilon=\sqrt{235/f_y}$	$\alpha > 0.5$ $\frac{45.6}{13\alpha-1}\varepsilon$ $\alpha < 0.5$ $\frac{41.5}{\varepsilon}$	$\Psi > -1$ $\frac{42\,\varepsilon}{0.67+0.33\Psi}$ $\Psi \leqslant -1$ $\frac{62\varepsilon(1-\Psi)}{\sqrt{-\Psi}}$

a satisfactory level of maturity (Bild and Kulak, 1991). The classifying of buckling into elastic and plastic is not a subject of controversy, whereas the classification into plastic, compact and semi-compact elements is not a recognized standard in order to predict the available ductility of an element in relation with its behavioural class. This problem will be discussed in detail in the next chapter.

4.8 Conclusions

The main conclusions for the aspects presented in this chapter may be summarized as follows:

-it is a great mistake to judge the steel ductility according to the native properties, because severe erosion occurs due to loading systems, characterized by velocity and number of cycles;

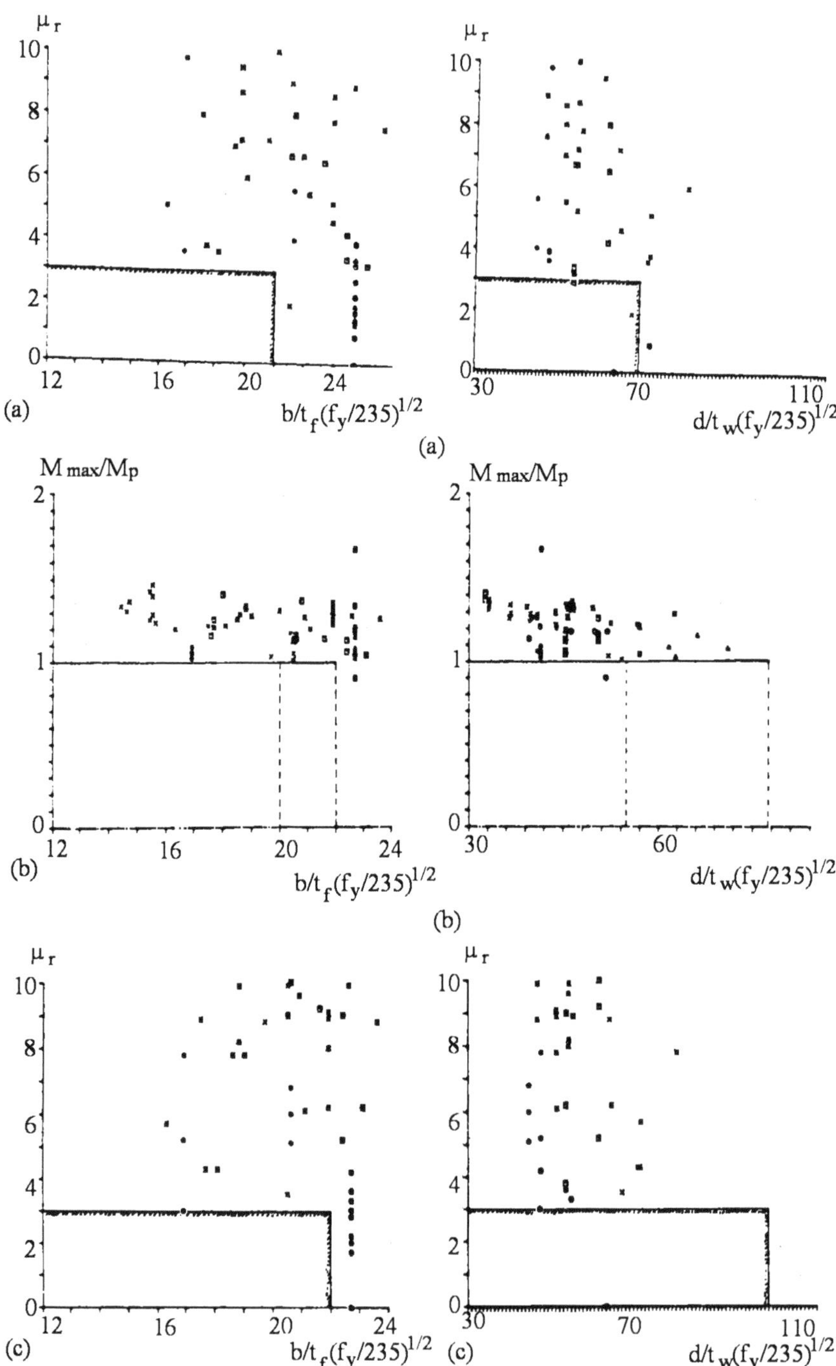

Figure 4.73: Width-to-thickness ratios: (a) Plastic sections; (b) Compact sections; (c) Semi-compact sections (after Commision EC, 1988)

-the steel fracture during an earthquake is due to the accumulation of plastic deformations and not to fatigue;

-the variability of yield stress has a great influence on the local ductility, due to the upper limit of yield stress which is not under control during the steel production;

-the element ductility is governed by the plastic buckling, for which a very effective method is the use of the plastic collapse mechanism theory;

-during the cyclic loading of an element, the cumulative plastic deformation at each cycle, for the next cycle plays the role of a initial geometrical imperfection;

-under monotonic or cyclic loading a fracture can occur for a given rotation, which depends on yield ratio, length of plastic mechanism and steel quality;

-for each element type there are more plastic mechanisms and the comparison between the post-crushing curves may decide which one is the most dangerous;

-the influence of strain-rate is very important, for high velocity transforming of the ductile rupture into a brittle fracture.

-the classification of elements in classes of ductility, used in codes, is a very disputable approach.

4.9 References

Agostoni, N., Ballio, G., Poggi, C. (1993): Indagine statistica sulle proprieta mechaniche degli acciai da costruzione. In XIV Congresso CTA, Viareggio, 24-27 Octobre 1993, Progettazione, Lavorazione, Montaggio, 22-34

Akiyama, H. (1999): Evaluation of fractural mode of failure in steel structures following Kobe lessons. In Stability and Ductility of Steel Structures, SDSS 99, Timisoara, 9-11 September 1999, Journal of Constructional Steel Research, Vol. 55, No.1-3, 211-227

Anastasiadis, A., Gioncu, V., Mazzolani, F.M. (2000): New trends in the evaluation of available ductility of steel members. In Behaviour of Steel Structures in Seismic Areas, STESSA 2000 (eds. F.M.Mazzolani, R. Tremblay), Montreal, 21-24 August 2000, Balkema, Rotterdam, 3-10

Ballio, G., Castiglioni, G.A. (1994a): Seismic behaviour of steel section. Journal of Constructional Steel Research, Vol. 29, 21-54

Ballio, G., Castiglioni, G.A. (1994b): An approach to seismic design of steel structures based on cumulative damage criteria. Earthquake Engineering and Structural Dynamics, Vol. 23, 969-986

Ballio, G., Castiglioni, G.A. (1995): An unified approach for the design of steel structures under low and/or high cycle fatigue. Journal of Constructional Steel Research, Vol. 34, 75-101

Barsom, J.M. (1992): Fracture design. In Constructional Steel Design. An International Guide, (eds. P.J. Dowling et al). Elsevier Applied Science, London, 611-625

Barsom, J.M., Rolfe, S.T. (1987): Fracture and Fatigue Control in Structures. Application of Fracture Mechanics. Prentice-Hall, New Jersey

Barsom, J.M., Korvink, S.A. (1998): Through-thickness properties of structural steels. Journal of structural Engineering, Vol. 124, No. 7, 727-735

Beg, D., Plumier, A., Remec, C., Sanchez, L. (2000): Influence of strain-rate. In Moment Resistant Connections of Steel Building Frames in Seismic Areas (ed. F.M. Mazzolani), E &FN Spon, London, 167-216

Bild, S., Kulak, G.L. (1991): Local buckling rules for structural steel members. Journal of Constructional Steel Research, Vol. 20, 1-52

Bjorhovde, R. (1998): Deformation considerations for the design of steel structures. Festschrift J. Lindner, Technische Universitat Berlin, 21-31

Boeraeve, Ph., Lognard, B., Janss, J., Gerardy, J.C., Schleich, J.B. (1993): Elasto-plastic behaviour of steel frame works. Journal of Constructional Steel Research, Vol. 27, 3-21

Bruneau, M., Uang, C.M., Whittaker, A. (1998): Ductile Design of Steel Structures. McGraw-Hill, New York

Calderoni, B., Mazzolani, F.M., Piluso, V. (1995): Quality control of material properties for seismic purposes. In Behaviour of Steel Structures in Seismic Areas, STESSA 94, (eds. F.M. Mazzolani, V.Gioncu), Timisoara, 26 June-1 July 1994, E&FN Spon, London, 111-120

Commision EC (1988): Background Documents for Eurocode 8, Vol. 2, Design rules

Cosenza, E., Manfredi, G. (1992): Low cycle fatigue: Characterisation of the plastic cycles due to earthquake ground motion. In Testing of Metals for Structures (ed. F.M Mazzolani), Napoli, 29-31 May 1990, E&FN Spon, London, 116-131

Croce, P., Cecconi, A., Salvatore, W. (1997a): Statistical distributions of the mechanical properties of structural steel elements. In XVI Congresso CTA, Ancona, 2-5 Octombrie 1997, 220-229

Croce, P., Cecconi, A., Salvatore, W. (1997b): Influence of production quality controls on partial safety factors of steel elements. In XVI Congresso CTA, Ancona, 2-5 Oct., 230-244

Dexter, R.J., Melendrez, M.I. (2000): Through-thickness properties of column flanges in welded moment connections. Journal of Structural Engineering, Vol. 126, No. 1, 24-31

Dodd, L.L., Restrepo-Posada, J.I. (1995): Model for predicting cyclic behaviour of reinforcing steel. Journal of Structural Engineering, Vol. 121, No. 3, 433-445

De Martino, A., Manfredi, G. (1994): Experimental testing procedures for the analysis of the cyclic behaviour of structural elements: activity of RILEM Technical committee 134 MJP. In Danneggiamento Ciclico e Prove Pseudo-dinamiche (ed. E. Cosenza), Napoli, 2-3 June 1994, Universita degli Studi Federico II, 3-20

EN 10020 (1988): Definition and classification of grades of steel

Feldmann, M. (1994): Zur Rotationskapazitat von I-Profilen statisch and dynamisch belastung Trager. Ph Thesis, RWTH Universitat Aachen

Fujimoto, M., Nanba, T., Nakagomi, T., Sasaki, S. (1988): Strength and deformation capacity of steel brace under high-speed loading. In 9th World Conference on Earthquake Engineering, Tokyo-Kyoto, 2-9 August 1988, Vol. IV, 139-144

Fukumoto, Y. (1994): New constructional steels and structural stability. In Link between Research and Practice, SSRC Meeting, Bethlehem, 21-22 June 1994, Lehigh University, 211-224

Fukumoto, Y. (2000): Reduction of structural ductility factor due to variability of steel properties. Engineering Structures, Vol. 22, 123-127

Galambos T.V. (1999): Recent research and design developments in steel and composite steel-concrete structures in USA. Keynote. In Stability and Ductility of Steel Structures, SDSS 99. Timisoara, 9-11 September 1999, Journal of Constructional Steel Research, Vol. 55, 289-303

Gioncu, V. (2000): Influence of strain-rate on the behaviour of steel members. In Behaviour of Steel Structures in Seismic Areas, STESSA 2000 (eds. F.M.Mazzolani, R. Tremblay), Montreal, 21-24 August 2000, Balkema, Rotterdam, 19-26

Gioncu, V., Mateescu, G., Petcu, D., Anastasiadis, A. (2000): Prediction of available ductility by means of local plastic mechanism method: DUCTROT computer program. In Moment Resistant Connections of Steel Building Frames in Seismic Areas, (ed. F.M.Mazzolani), E&FN Spon, London

Haaijer, G. (1957): Plate buckling in the strain-hardening range. Journal of Engineering Mechanics Division, Vol. 83, EM 2, 1212/1-1212/47

Hemerich, E., Milcic, V. (1992): High strength steel: Research on the basic phenomena concerning new design models in the probabilistic approach. In Testing of Metals for Structures (ed. F.M. Mazzolani), Napoli, 29-31 May 1990, E&FN Spon, London, 8-13

Inoue, T. (1994): Analysis of plastic buckling of rectangular steel plates supported along their four edges. International Journal of Solids and Structures, Vol. 31, No. 2, 219-230

Inoue, T., Kato, B. (1993): Analysis of plastic buckling of steel plates. International Journal of Solids and Structures, Vol. 30, No. 6, 835-856

Iwatsubo, K., Koganemam, T., Yamao, T., Sakimoto, T. (1997): Bending strength and ductility of H-section members made of high-strength steel with low-yield ratio. In Stability and Ductility of Steel Structures. SDSS 97 (ed. T. Usami), Nagoya, 29-31 July 1997, 981-988

Kaneko, H. (1997): Influence of strain-rate on yield ratio. In Kobe Earthquake Damage to Steel Moment Connections and Suggested Improvement. JSSC Technical Report No. 39

Kaneta, K., Kohzu, I., Fujimura, K. (1986): On the strength and ductility of steel structural joints subjected to high speed monotonic tensile loading. The 8th European Coference on Earthquake Engineering, Lisbon, Vol. IV

Kassar, M., Pan, C.L., Yu, W.W. (1992): Effect of strain-rate on cold-formed steel stub columns. Journal of Structural Engineering, Vol. 118, No. 11, 3151-3163

Kassar, M., Yu, W.W. (1992): Effect of strain rate on material properties of sheet steels. Journal of Structural Engineering, Vol. 118, No. 11, 3136-3150

Kato, B., Chen, W.F., Nakao, M. (1988): Effect of joint-panel shear deformation on frames. Journal of Constructional Steel Research, Vol. 10, 269-320

Kato, B., Aoki, H., Yamanouchi, H. (1990): Standardised mathematical expression for stress-strain relations of structural steel under monotonic and uniaxial tension loading. Materials and Structures, Vol. 23, 47-58

Kohzu, I., Suita, K. (1996): Single or few excursion failure of steel structural joints due to impulsive shocks in the 1995 Hyogoken Nanbu earthquake. 11th World Conference on Earthquake Engineering, Acapulco, 23-28 June 1996, CD-ROM, Paper No. 412

Korol, R.M., Sherbourne, A.N. (1972): Strength prediction of plates in uniaxial compression. Journal of the Structural Division, Vol. 98, ST 9, 1965-1986

Kotelko, M. (1996a): Ultimate load and postfailure behaviour of box-section beams under pure bending. Engineering Transactions, Vol. 44, 229-251

Kotelko, M. (1996b): Selected problems of collapse behaviour analysis of structural members built from strain-hardening material. Conference on Thin-walled Structures, 2-4 December 1996, Glasgow

Kotelko, M., Kolakowski, Z. (1995): Postbuckling and collapse behaviour of thin-walled beam-columns. In Lightweight Structures in Civil Engineering (ed. J.B. Obrebski), Warsaw, 25-29 September 1995, Vol. I, 384-387

Kragerup, J. (1982): Five notes on plate buckling. Technical University of Denmark, Department of Structural Engineering, Series R, No. 143

Kurobane, Y., Ogawa, K., Ueda, C. (1996): Kobe earthquake damage to high-rise Ashiyahama apartment buildings: Brittle tensile failure of box section columns. In Tubular Structures VII (eds I. Farkas and K. Jarmai), Miskolc, 28-30 August 1996, Balkema, Rotterdam, 277-284

Kuwamura, H. (1988): Effect of yield ratio on the ductility of high-strength steels under seismic loading. In 1988 Annual Technical Session of SSRC, Minneapolis, Minnesota, 26-27 April 1988, 201-210

Kuwamura, H. (1996): Fracture of steel welded joints under severe earthquake motion. In 11th World Conference on Earthquake Engineering, Acapulco, 23-28 June 1996, CD-ROM 466

Kuwamura, H. (1997a): Transition between fatigue and ductile fracture in steel. Journal of Structural Engineering, Vol. 123, No. 7, 864-870

Kuwamura, H. (1997b): Steel properties governing structural seismic behaviour. General report. In Behaviour of Steel Structures in Seismic Areas. STESSA 97, (eds F.M. Mazzolani and H. Akiyama), Kyoto, 3-8 August 1997, IO/I7, Salerno, 119-129

Kuwamura, H. (1997c): Ductility of steel members susceptible to brittle fracture. In Stability and Ductility of Steel Structures. SDSS 97, (ed. T.Usami), Nagoya, 29-31 July 1997, 925-932

Kuwamura, H., Akiyama, H. (1994): Brittle fracture under repeated high stress. Journal of Constructional Steel Research, Vol. 29, 5-19

Kuwamura, H., Yamamoto, K. (1997): Ductile crack as trigger of brittle fracture in steel. Journal of Structural Engineering, Vol. 123, No. 6, 729-735

Lay, G.M. (1965a): Yielding of uniformly loaded steel members. Journal of the Structural Division, Vol. 91, ST 6, 49-66

Lay, M.G. (1965b): Flange local buckling in wide-flange shapes. Journal of the Structural Division, Vol. 91, ST 6, 95-116

Leblois, C. (1972): Influence de la limite d'elasticite superieure sur la comportement en flexion et tension de l'acier doux. Ph Thesis, Liege

Mahendran, M. (1997): Local plastic mechanisms in thin steel plates in uniaxial compression. Journal of Constructional Steel Research, Vol. 27, No. 3, 245-261

Mahendran, M., Murray, N.W. (1991): Effect of initial imperfections on local plastic mechanisms in thin steel plates with in-plane compression. In Steel Structures. Recent Research and Development. ISCAS 91, (eds. S.L.Lee and N.E. Shanmugam), Singapore, Elsevier, London, 491-500

Mahin, A.S., Bertero, V.V., Atalay, M.B., Rea, D. (1972): Rate of loading effects on uncracked and repaired reinforced concrete members. Report EERC 72-9, University of California, Berkeley

Manjoine, M.J. (1944): Influence of rate of strain and temperature on yield stress of mild steel. Journal of Applied Mechanics, No. 11, 211-218

Mateescu, G., Gioncu, V. (2000): Member response to strong pulse seismic loading. In Steel Structures in Seismic Areas, STESSA 2000, (eds. F.M.Mazzolani, R. Tremblay) Montreal, 21-24 August 2000, Balkema, Rotterdam, 53-62

Mazzolani, F.M. (1974): Influence de l'effect Bauschinger sur le comportement des barres metalliques soumises a des cycles d'allongement. ECCS-Committee T16, Doc. 16-74-4

Mazzolani, F.M., Piluso, V. (1995): ECCS Manual on Design of Steel Structures in Seismic Zones. TC 13 Seismic Design Report

Mazzolani, F.M., Piluso, V. (1996): Theory and Design of Seismic Resistant Steel Frames. E&FN Spon, London

Mazzolani, F.M., Mele, E., Piluso, V. (1990): Statistical features of mechanical properties of structural steels. ECCS Document TC13.26.90

Mazzolani, F.M., Mele, E., Piluso, V. (1993): Statistical characterization of constructional steels for structural ductility control. Costruzioni Metalliche, No. 2, 89-101

McDermott, J.F. (1969): Local plastic buckling of A514 steel members. Journal of the Structural Division, Vol. 95, ST 9, 1837-1850

Moller, M., Johansson, B., Collin, P. (1997): A new analytical model of inelastic local flange buckling. Journal of Constructional Steel Research, Vol. 23, No. 1-3, 43-63

Morrison, J.L. (1932): The influence of rate of strain in tension tests. Engineer, 158, 183

Moy, S.S.J. (1985): Plastic Methods for Steel and Concrete structures. McMillan, London

Murray, N.W. (1973): Das aufnehmbare moment in einem zur Richtung der Normalkraft Schragliegenden plastischen Gelenk. Die Bautechnik, Vol. 50, No. 2, 57-58

Murray, N.W. (1986): Recent research into behaviour of thin-walled steel structures. In Steel Structures. Recent Advances and their Application to Design (ed. M.N. Pavlovici), Budva, Elsevier, 171-191

Murray, N.M. (1995): Some effects arising from impact loading of thin-walled structures. In Lightweight Structures in Civil Engineering (ed. J.B. Obrebski), Warsaw, 25-29 September 1995, Vol. I, 389-394

Nakagomi, T., Tsuchihashi, H. (1988): Fracture and deformation capacity of a welded T-shape joint under dynamic loading. In 9th World Conference on Earthquake Engineering, Tokyo-Kyoto, 2-9 August 1988, Vol. IV, 157-162

Nakamura,K., Mizuno, J., Matsuo, I., Suzuki, A., Tsubota, H. (1999): Effects of loading rate on reinforced concrete shear walls: Part 1, Dynamic properties of large-size rebars. In Earthquake Resistant Engineering Structures (eds G. Oliveto, C.A. Brebbia), Catania, June 1999, WIT Press, Southamton, 43-52

Nica, A. (1981): Mechanics of Aerospace Materials. Editura Academiei Bucharest, Elsevier, London

Obata, M., Goto, Y., Matsuura, S., Fujiwara, H. (1996): Ultimate behaviour of tie plates at high-speed tension. Journal of Structural Engineering, Vol. 122, No. 4, 416-422

Park, M.S., Lee, B.C. (1996): Prediction of bending collapse behaviour of thin-walled open section beams. Thin-Walled Structures, Vol. 25, No. 3, 185-206

Park, Y.S., Iwai, S., Kameda, H., Nonaka, T., Kang, S.H. (1997): Quantitative assessment of failure process to steel members under strong earthquake loading. In Stability and Ductility of Steel Structures, SDSS 97, (ed. T.Usami), Nagoya, 29-31 July 1997, 809-816

Piluso, V. (1992): Il comportamento inelastico dei telai seismo-resistenti in acciaio. Ph Thesis, Universita degli Studi di Napoli Federico II

Quinney, H. (1934): Time effect in testing metals. Engineer, 157, 332

Rao, N.R.N., Lohramann, M., Tall, L. (1966): Effect of strain-rate on the yield stress of structural steels. ASTM Journal of Materials, Vol. 1, No. 1

Restrepo-Posada, J.T., Dodd, L.L., Park, R., Cooke, N. (1994): Variable affecting cyclic behaviour of reinforced steel. Journal of Structural Engineering, Vol. 120, No. 11, 3178-3196

Rondal, J. (1998): Buckling and interactive buckling of metal columns, optimum design under stability constraints and code aspects. In Coupled Instabilities in Metal Structures. Theoretical and Design Aspects (ed. J.Rondal), Springer, Wien, CISM Lecture 379, Part VII, 345-372

Rondal, J., Maquoi, R. (1985): Stub-column strength of thin-walled square and rectangular hollow sections. Thin-Walled Structures, No. 3, 15-34

Sherbourne, A.N., Korol, R.M. (1972): Post-buckling of axially compressed plates. Journal of the Structural Division, Vol. 98, ST 10, 2223-2234

Soroushian, P., Choi, K.B. (1987): Steel mechanical properties at different strain rate. Journal of Structural Engineering, Vol. 113, No. 4, 863-872

SSRC (1987): Standard Methods and Definitions for Tests for Static Yield Stress. Technical Memorandum No 8

Suita, K., Nakashima, M., Morisako, K. (1998): Tests of welded beam-column subassemblies. II. Detailed behaviour. Journal of Structural Engineering, Vol. 124, No. 11, 1245-1252

Szabo, G., Ivanyi, M. (1995): The influence of Luders-Hartmann lines on stability of steel members. In Stability of Steel Structures (ed. M.Ivanyi), Budapest, 21-23 September 1995, Akademiai Kiado, Budapest, 1057-1064

Uang, C.M., Bondad, D.M. (1996): Dynamic testing of full-scale steel moment connections. In 11th World Conference on Earthquake Engineering, Acapulco, 23-28 June 1996, CD-ROM 407

Vayas, I. (1997): Investigation on the cyclic behaviour of steel beams by application of low-cycle fatigue criteria. In Behaviour of Steel Structures in Seismic Areas, STESSA 97, (eds. F.M. Mazzolani and H. Akiyama),Kyoto, 3-8 August 1997, IO/I7, Salerno, 350-357

Wallace, B.J., Krawinkler, H. (1989): Small-scale model tests of structural steel assemblies. Journal of Structural Engineering, Vol. 115, No. 8, 1999-2015

Wright, R.N., Hall, W.J. (1964): Loading rate effects in structural steel design. Journal of the Structural Division, Vol. 90, ST 5, 11-37

Yamada, M. (1992): Low fatigue fracture limits of structural materials and structural elements. In Testing of Metals for Structures (ed. F.M.Mazzolani), Napoli, 29-31 May 1990, E&FN Spon, London, 184-192

Zhao, X.L., Hancock, G.J. (1993a): A theoretical analysis of the plastic moment capacity of an inclined line under axial force. Thin-Walled Structures, Vol. 15, 185-207

Zhao, X.L., Hancock, G.J. (1993b): Experimental verification of the theory of plastic moment capacity of an inclined yield line under axial force. Thin-Walled Structures, Vol. 15, 209-233

Zhang, Y.C., Dong, Y.T., Luo, P.L., Ju, X.H. (1997): Buckling behaviour of steel plate under cyclic loading. In Stability and Ductility of Steel Structures, SDSS 97, (ed. T. Usami), Nagoya, 29-31 July 1997, 119-126

5

Section and Stub Ductilities

5.1 Ductility Erosion due to Section Behaviour

The design of steel seismic resistant structures took a dramatic turn after the last Californian and Japanese events. The heavy damage observed as a results of these earthquakes was never before recorded in the history of building design. These events gave rise to a general effort all over the world to improve the seismic resistance of steel structures. A comprehensive program started to evaluate both the design specifications and detail rules. In this perspective, clear attention has been paid to the evaluation of local ductility erosion at the level of sections.

The native steel properties and the potential power of seismic input energy dissipation are eroded by loading conditions at the material level and by random variability of strength and local buckling at the element level (see Chapter 4). New factors increase this erosion at the section level, especially due to the complex phenomena of plastic buckling (Fig. 5.1). Local plastic buckling produces a loss of strength and a severe degrading of the hysteretic behaviour associated with a poor structural response. To determine the section ductility, compression tests on stubs of steel members are used in practice. Control of all buckling forms is a very important design condition, if a good seismic performance is to be achieved. If these adverse effects are not controlled a proper way, a bad performance may affect the structure behaviour and the legendary good performance of steel structures during a severe earthquake remains only a dogma.

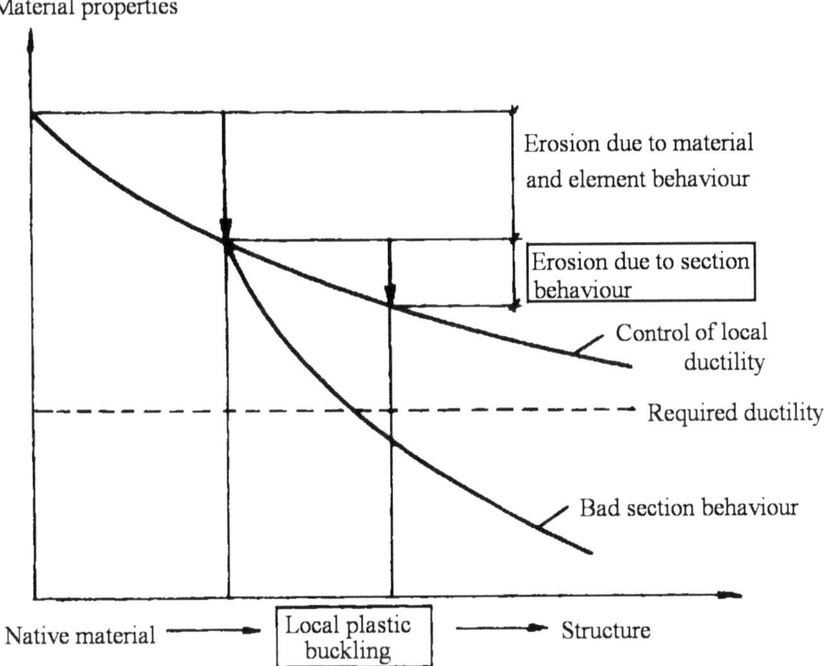

Figure 5.1: Erosion due to section behaviour

5.2 Section Ductility

5.2.1. Main Factors Influencing the Section Ductility

The collapse of a structure is produced, in the majority of cases, by local buckling and/or fracture of an element. The deformation is localized in a small part of the affected member and the ductility of the full member is mainly influenced by the behaviour of this critical zone. Due to this fact, a lot of research work is devoted to the ductility of member stubs, where the section properties play a leading role.

The main factors influencing the section ductility, considered in this Chapter, are:

(i) *Cross-section types* (Fig. 5.2a). For steel structures the most usual cross-sections are:

-I-sections;
-box (rectangular hollow) sections;
-concrete-filled box sections.

(ii) *Fabrication* (Fig. 5.2b). The profiles can be manufactured by:

-hot-rolling;
-welding;
-cold-forming.

(iii) *Buckling type* (Fig. 5.2c). The local buckling may be characterized

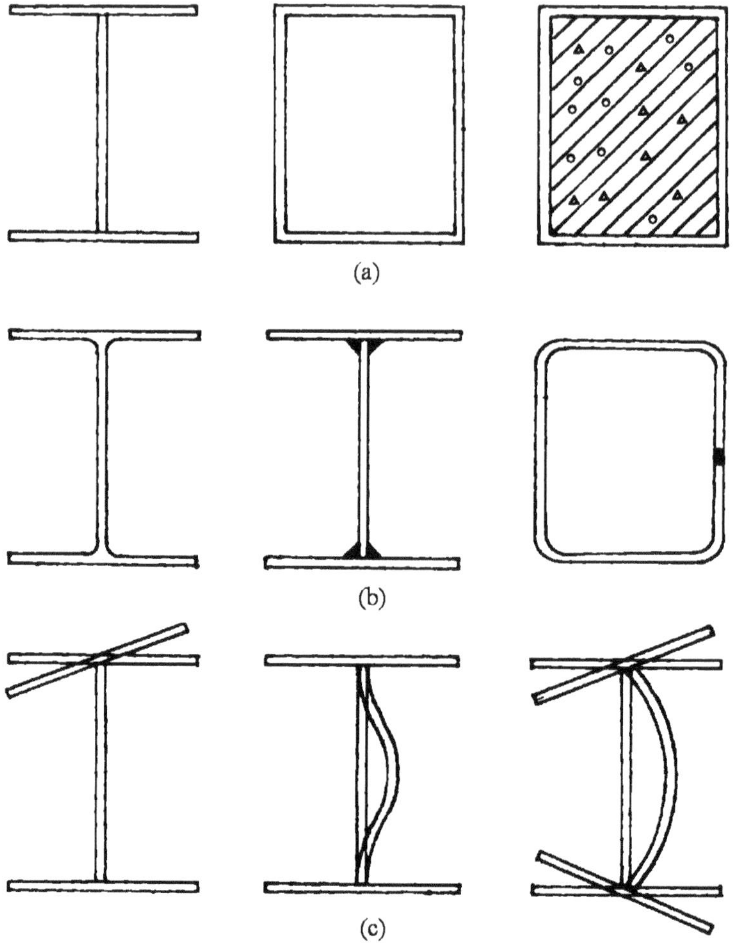

Figure 5.2: Factors influencing the section ductility: (a) Section type; (b) Fabrication; (c) Buckling type

by:

-buckling of flanges;
-buckling of web;
-interaction between the two buckling modes.

5.2.2. Plastic Behaviour of Sections

(i) *Pure flexural yielding.* The study of the behaviour of a bended section is based on the following hypothesis (Moy, 1985):

-section has at least one axis of symmetry parallel to the loading direction (Fig. 5.3a);

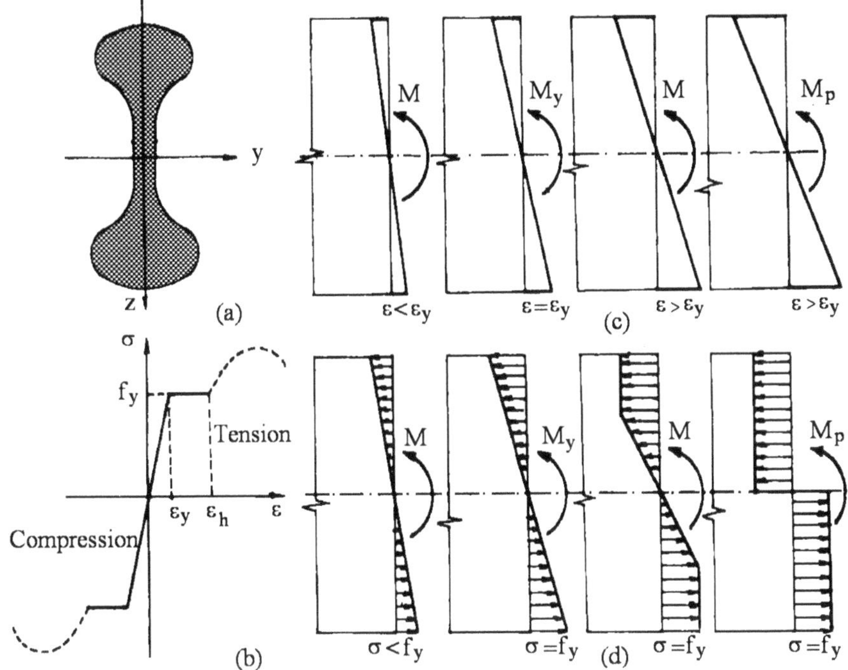

Figure 5.3: Pure bending: (a) Cross-section; (b) Simplified stress-strain relationship; (c) Strain distribution; (d) Stress distribution

-section fibres have the same behaviour as the native behaviour of steel (Fig. 5.3b);

-for pure flexural yielding only the elastic and yielding plateau is considered;

-plane sections before deformation remain plain also after deformation, even in plastic range (Fig. 5.3c);

-stress distribution represents a part of the stress-strain curve of native steel specimens (Fig. 5.3d), as a result of the last two hypotheses;

-local buckling is avoided;

-section is not subjected to axial, shear or torsional forces.

The last two hypotheses will be reconsidered in the next Sections.

If there is no yield in the material, there are straight line relationships for stress and strain over the whole depth of the section. There will be an elastic behaviour until the maximum stress reaches yield point. At this stage only, material fibres at the outside edges of the section are yielded and the corresponding moment is:

$$M_y = Zf_y \tag{5.1}$$

where Z is the elastic section modulus.

As the bending moment is increased, yielding spreads towards the axis of zero strain. The stress distribution shows two constant regions where

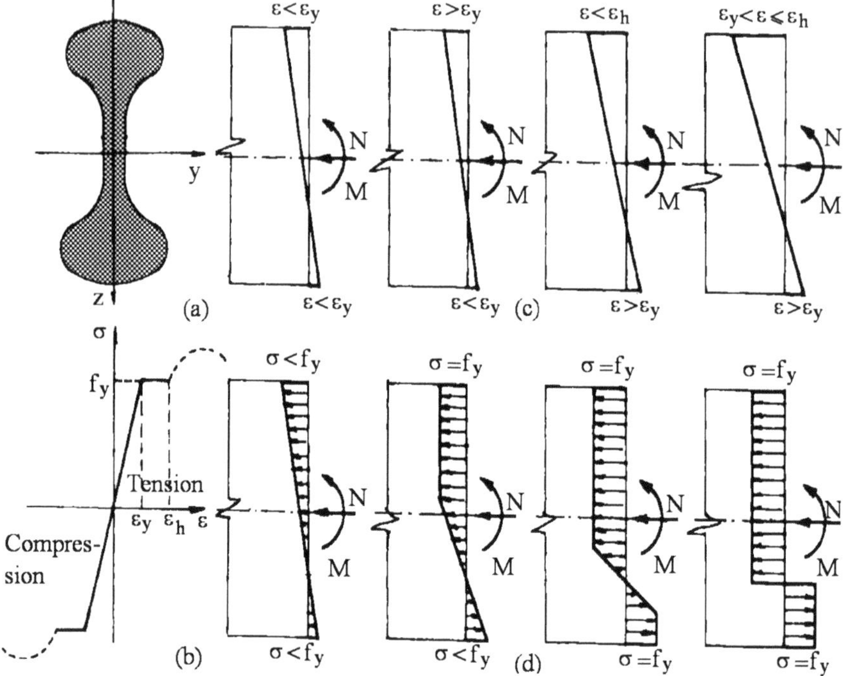

Figure 5.4: Influence of axial force: (a) Cross-section; (b) Simplified stress-strain relationship; (c) Strain distribution; (d) Stress distribution

yield occurred, jointed by an elastic stress distribution. For pure flexural yielding, the stresses are limited to yield stress, but the strains can increase by plastic yielding until the beginning of the strain-hardening. As limit situation, stresses are constant until the axis of zero strain. For this stress distribution the section behaves like a hinge, because the strain may increase everywhere along the section, without any change in stress. So, a *plastic hinge* is formed and the bending moment is the *full plastic moment*, M_p:

$$M_p = Z_p f_y \tag{5.2}$$

where Z_p is the plastic section modulus, a geometrical section property, like the elastic section modulus Z.

(ii) *Influence of axial forces.* Sections may have to carry significant axial forces in addition to the bending moment. The presence of axial force moves the axis of zero strain towards the tension zone (Fig. 5.4). The maximum axial force which the section can carry, ignoring the strain-hardening zone and local buckling, is the plastic axial load:

$$N_p = A f_y \tag{5.3}$$

where A is the total section area.

The expression for reduced plastic moment due to the presence of axial force can be obtained in a normalized format:

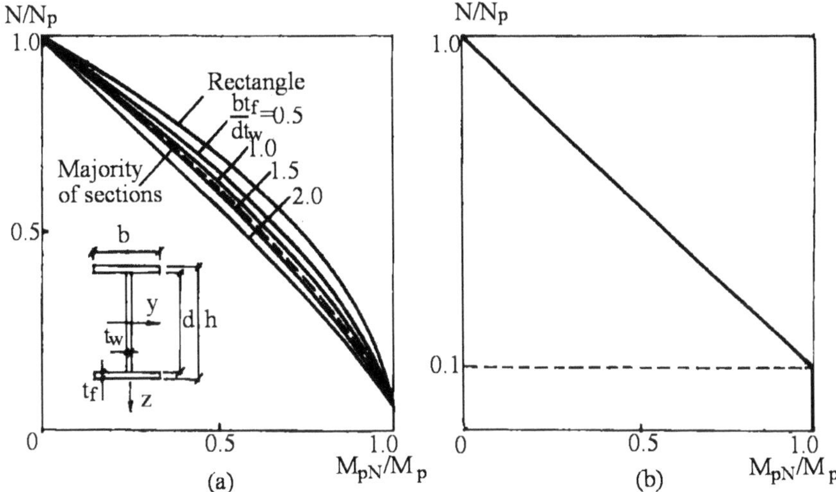

Figure 5.5: Moment-axial force interaction for strong axis: (a) M-N diagram for I-section; (b) Simplified design interaction diagram (after Bruneau et al, 1998)

- *I-section*, strong axis bending. The M–N interaction curves are presented in Fig. 5.5a (Bruneau et al, 1998). It can be observed that these curves are close to the one corresponding to rectangular sections when the I-sections have the lowest ratio of flange-to-web areas and close to a straight line when this ratio is the highest. The usual I-sections frame in the last category. For practical purposes, the EUROCODE 3 (1992) proposes the following relations (Fig. 5.5b):

$$\frac{M_{pN}}{M_p} = 1 \quad for \quad \frac{N}{N_p} \leq 0.10 \tag{5.4a}$$

$$\frac{M_{pN}}{M_p} = 1.11\left(1 - \frac{N}{N_p}\right) \quad for \quad \frac{N}{N_p} > 0.10 \tag{5.4b}$$

For weak axis bending, the M–N interaction curves are shown in Fig. 5.6a (Bruneau et al, 1998), being functions of the ratio between flange to web areas. For this bending type, the case of the smallest ratio is the most distant from the rectangular section. The most used I-sections show that for the practical design a simple relationship can be used (EUROCODE 3, 1992):

$$\frac{M_{pN}}{M_p} = 1 \quad for \quad \frac{N}{N_p} \leq 0.20 \tag{5.5a}$$

$$\frac{M_{pN}}{M_p} = 1.56\left(1 - \frac{N}{N_p}\right)\left(\frac{N}{N_p} + 0.6\right) \quad for \quad \frac{N}{N_p} > 0.20 \tag{5.5b}$$

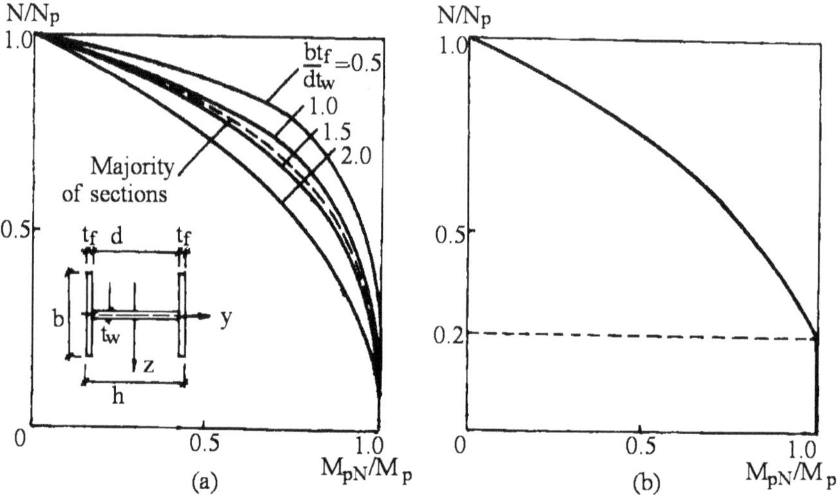

Figure 5.6: Moment-axial force interaction for weak axis: (a) M-N diagram for I-section; (b) Simplified interaction diagram (after Bruneau et al, 1998)

-*box-section*, strong axis bending. In the same way as for I-sections, simplified relationships are proposed in EC 3:

$$\frac{M_{pN}}{M_p} = \left(1 - \frac{N}{N_p}\right)(1 - 0.5a_w) \tag{5.6a}$$

where

$$a_w = \frac{A - 2bt_f}{A} \leq 0.5 \tag{5.6b}$$

In the case of uniform thickness, a simplified relationship can be used for square standardized sections:

$$\frac{M_{pN}}{M_p} = 1 \quad \text{for} \quad \frac{N}{N_p} \leq 0.20 \tag{5.7a}$$

$$\frac{M_{pN}}{M_p} = 1.26\left(1 - \frac{N}{N_p}\right) \quad \text{for} \quad \frac{N}{N_p} > 0.20 \tag{5.7b}$$

For rectangular sections:

$$\frac{M_{pN}}{M_p} = 1 \quad \text{for} \quad \frac{N}{N_p} \leq 0.25 \tag{5.8a}$$

$$\frac{M_{pN}}{M_p} = 1.33\left(1 - \frac{N}{N_p}\right) \quad \text{for} \quad \frac{N}{N_p} > 0.25 \tag{5.8b}$$

In the case of weak axis bending (EC3):

$$\frac{M_{pN}}{M_p} = \left(1 - \frac{N}{N_p}\right)(1 - 0.5a_f) \tag{5.9a}$$

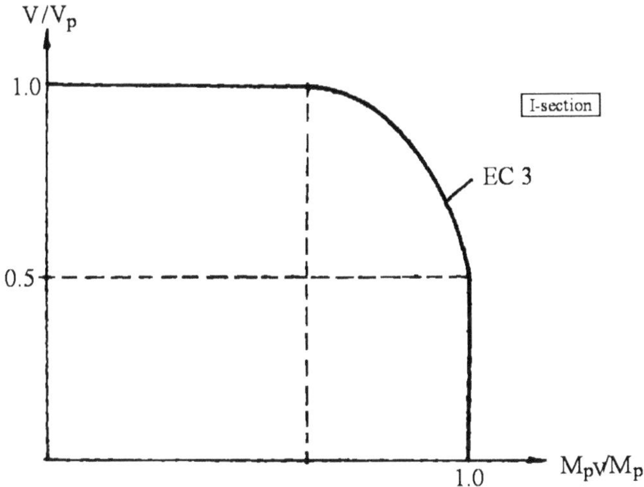

Figure 5.7: Moment-shear force interaction

where

$$a_f = \frac{A - 2ht_w}{A} \leq 0.5 \tag{5.9b}$$

A simplified relationship can be used for rectangular sections of uniform thickness:

$$\frac{M_{pN}}{M_p} = \left(1 - \frac{N}{N_p}\right)\left(0.5 + \frac{ht}{A}\right) \leq 1 \tag{5.10}$$

(iii) *Influence of shear forces.* Except in the zone of constant moment, all sections must carry a bending moment and a shear force. This means that there is a combination of axial stress σ due to moment and shear stress τ due to shear. In this circumstance it is necessary to use a criterion to determine the start of yield. The Tresca and Van Mises criteria are the most common for ductile materials and require that:

$$\left(\frac{\sigma}{f_y}\right)^2 + \left(\frac{\tau}{\tau_y}\right)^2 = 1 \tag{5.11a}$$

where

$$\tau_y = f_y/3^{1/2} = 0.577 f_y \tag{5.11b}$$

is the yielding shear stress. From this criterion, when the axial stresses reach yield, no shear stresses occur. Thus, when yield occurs in bending, the remaining elastic zones only undertake the shear stresses.

For I-sections, EC3 provides a simplified relationship (Fig. 5.7):

$$\frac{M_{pV}}{M_p} = 1 \quad ; \quad \frac{V}{V_p} \leq 0.5 \tag{5.12a}$$

$$\frac{M_{pV}}{M_p} = 1 - \frac{d^2 t_w f_{yw}}{M_p}\left(\frac{V}{V_p} - 0.5\right)^2 \quad ; \quad \frac{V}{V_p} > 0.5 \tag{5.12b}$$

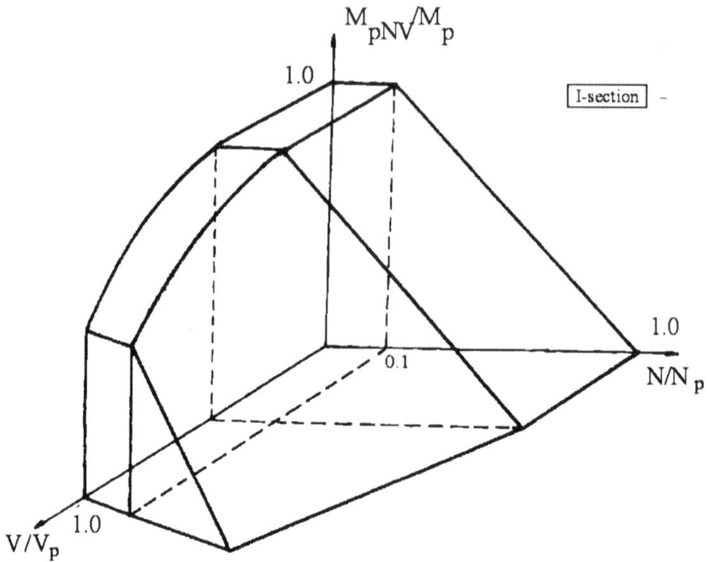

Figure 5.8: Moment-axial force-shear force interaction

where

$$V_p = dt_w\sigma_y \tag{5.12c}$$

(iv) *Combination of flexural, axial and shear forces.* In the case of combination of flexural, axial and shear forces the same procedures used for the individual effects can be considered. The interaction surface is presented in Fig. 5.8. For $N/N_p < 0.1$ and $V/V_p < 0.5$ no interaction exists, the full plastic moment being unaffected by the axial and shear forces.

(v) *Influence of the strain-hardening range.* The model of pure flexural yielding neglects the effect of strain-hardening. But, as it is shown in the previous Chapter, if the stress is controlled during the loading process, the steel sections are stressed in the strain-hardening range. So, keeping the hypotheses of plane sections also for strain exceeding the yield plateau, the strain and stress distributions over the section are presented in Fig. 5.9. At the ultimate stage, the influence of the yielding zone and of the first portion of strain-hardening is very small and the ultimate bending moment can be approximately determined considering that the ultimate stress is uniformly distributed over all the section:

$$M_u = \frac{f_u}{f_y}M_p = \frac{1}{\rho_y}M_p \tag{5.13}$$

where ρ_y is the yield ratio (see Section 4.2.3).

For usual steel grades, the equation (5.13) gives $M_u = (1.4... \ 1.55)M_p$. Lay and Galambos (1967) consider that it is unreasonable to expect that a plate section in bending reaches the large strains corresponding to the attainment of yield stress in a tension test and it is more rational to take

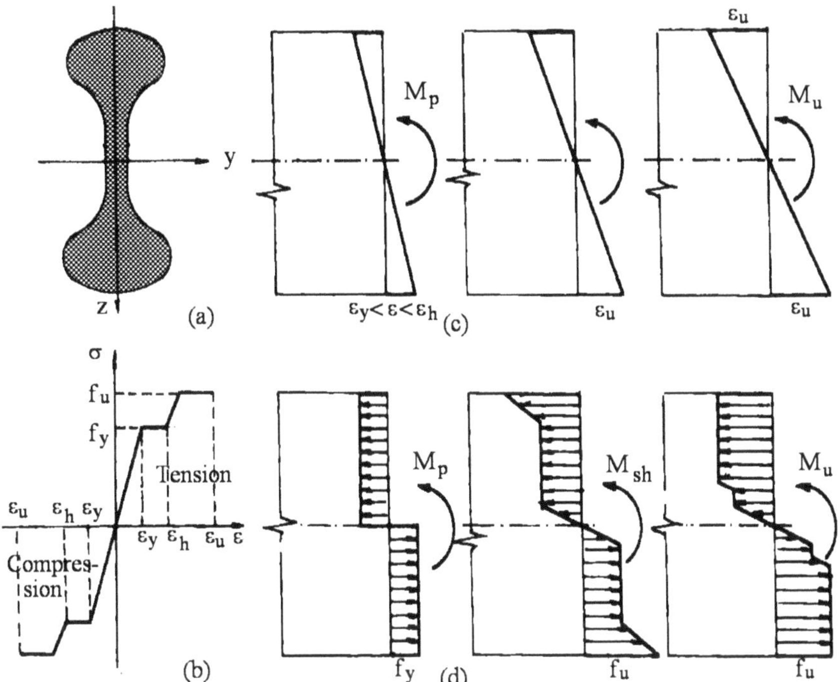

Figure 5.9: Pure bending in strain-hardening range: (a) Cross-section; (b) Simplified stress-strain curve; (c) Strain distribution; (d) Stress distribution

for bending a reduced ultimate tensile strength equal to $(f_u + f_y)/2$, giving:

$$M_u = \frac{1}{2}\left(1 + \frac{f_u}{f_y}\right) M_p = \frac{1}{2}\left(1 + \frac{1}{\rho_y}\right) M_p \tag{5.14}$$

from which $M_u = (1.22\text{–}1.27)M_p$. However, examining the experimental results, it seems that this proposal is too severe. For instance, in the tests performed by Sawyer (1961) for beams in which no plastic buckling occurs, the M_u exceeded M_p by 28–53%, where M_p is determined using the actual yield stresses, the differences being caused only by strain-hardening effects. So, it is more realistic to use a reduced ultimate tensile stress equal to $(f_y + 3f_u)/4$ (Gioncu and Petcu, 1997), the ultimate moment resulting as:

$$M_u = \frac{1}{4}\left(1 + 3\frac{f_u}{f_y}\right) M_p = \frac{1}{4}\left(1 + \frac{3}{\rho_y}\right) M_p \tag{5.15}$$

from which $M_u = (1.33 \div 1.40)M_p$. These values seem more appropriate than the ones resulting from the Lay and Galambos proposal. So, for design purposes, either (5.13) or (5.15) relationships can be used for determining the ultimate bending moment, depending on designer option.

The ultimate axial force, so-called crushing force, may be determined as

$$N_u = f_u A = \frac{1}{\rho_y} N_p \tag{5.16}$$

(vi) *Influence of local buckling.* Generally, the ultimate moment cannot be reached due to the fact that plastic buckling occurs in the strain-hardening range. Indeed, the experimental data for maximum bending moments show that there is an overstrength above the full plastic moment due to the local buckling in the strain-hardening range, even in the presence of axial forces (Nakashima, 1992). The value of this critical stress is very difficult to be determined by means of theoretical methods. Therefore, the use of empirical relationships based upon experimental data is justified. By examining the results of a great number of stub column tests, Kato (1989, 1990) has proposed to determine the maximum moment corresponding to plastic local buckling by this empirical formula:

$$M_{max} = m_b M_p \tag{5.17}$$

where m_b is a numerical coefficient, obtained by means of multiple regression analysis, valuable for ductile and compact sections. Therefore, this coefficient must be greater than 1, because for these sections, buckling phenomena are attained in the hardening range.

The values of this coefficient may be determined from the following relationships:

-I-sections:

$$\frac{1}{m_b} = 0.6003 + \frac{1.600}{\alpha_f} + \frac{0.1535}{\alpha_w} \tag{5.18a}$$

-welded box-sections:

$$\frac{1}{m_b} = 0.710 + \frac{0.167}{\alpha} \tag{5.18b}$$

-cold-formed box sections:

$$\frac{1}{m_b} = 0.778 + \frac{0.13}{\alpha} \tag{5.18c}$$

where the following notations have been used:

$$\alpha_f = \frac{E}{f_y}\left(\frac{t_f}{b/2}\right)^2 \quad ; \quad \alpha_w = \frac{E}{f_y}\left(\frac{t_w}{d_e}\right)^2 \tag{5.19a, b}$$

$$\alpha = \frac{E}{f_y}\left(\frac{t}{b_e}\right) \quad \text{for square box sections} \tag{5.19c}$$

$$d_e = \frac{1}{2}\left(1 + \frac{A}{A_w}\frac{N}{N_p}\right)d \quad ; \quad b_e = \frac{4}{3}\left(1 + \frac{N}{N_p}\right)b \tag{5.20a, b}$$

d_e and b_e being the effective depths, taking into account that the critical loads are experimentally determined on stub columns where the webs are uniformly compressed, while the webs in beams and beam-columns have a stress gradient.

More accurate relationships for m_b coefficients are given by Kato, (1990), for I-sections with the steel grades used in Japan:

-SM 41 ($f_y = 300N/mm^2$):

$$\frac{1}{m_b} = 0.689 + \frac{0.651}{\alpha_f} + \frac{0.0533}{\alpha_w} \pm 0.0303 \tag{5.21a}$$

-SM 50 ($f_y = 377N/mm^2$):

$$\frac{1}{m_b} = 0.689 + \frac{0.586}{\alpha_f} + \frac{0.0711}{\alpha_w} \pm 0.0538 \tag{5.21b}$$

-SM 58L ($f_y = 460N/mm^2$):

$$\frac{1}{m_b} = 0.716 + \frac{0.518}{\alpha_f} + \frac{0.0389}{\alpha_w} \pm 0.0325 \tag{5.21c}$$

-SM 58H ($f_y = 525N/mm^2$):

$$\frac{1}{m_b} = 0.881 + \frac{0.270}{\alpha_f} + \frac{0.0365}{\alpha_w} \tag{5.21d}$$

For high strength steel with yield stresses around $800N/mm^2$, a similar relationship is established by Bed and Hladnik, (1996):

$$\frac{1}{m_b} = 0.7353 + \frac{0.6439}{\alpha_f} + \frac{0.0072}{\alpha_w} \pm 0.0303 \tag{5.22}$$

5.2.3. Moment-Curvature Relationship

Let us assume that a short length δ_x of a beam initially straight is bent into an arch as in Fig. 5.10. This assumption is true for constant moment, but the error for gradient moment is very small. According to the hypothesis that plane sections before bending remain plane after bending, the length of top arc is $\delta_x(1-\epsilon_c)$ and the length of bottom arc is $\delta_x(1+\epsilon_t)$. From Fig. 5.10 it results:

$$R\theta = \delta_x \tag{5.23a}$$

$$(R + h_t)\theta = \delta_x(1 + \varepsilon_t) \quad ; \quad (R - h_c)\theta = \delta_x(1 - \varepsilon_c) \tag{5.23b,c}$$

From these equations the curvature χ can be evaluated as the inverse of radius of curvature:

$$\chi = \frac{1}{R} = \frac{\varepsilon_t + \varepsilon_c}{h} \tag{5.24}$$

As the axial strain is the measure of axial deformations, the curvature is a simple measure of section bending deformation. For double-symmetrical sections $\varepsilon_t = \varepsilon_c = \varepsilon_{max}$ and:

$$\chi = \frac{\varepsilon_{max}}{(h/2)} \tag{5.25}$$

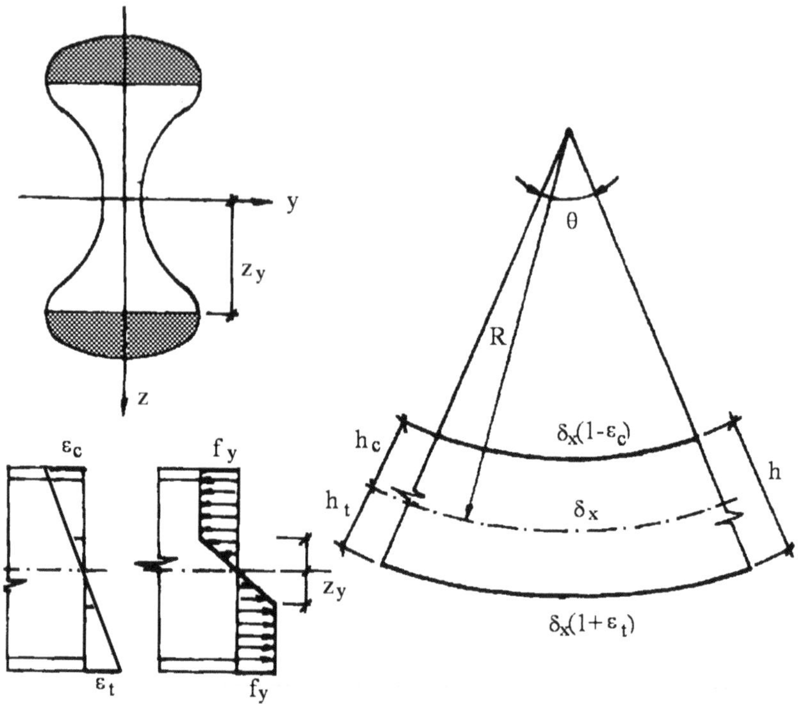

Figure 5.10: Definition of curvature: (a) Strain-stress distributions

where ε_{max} is the maximum strain. If the $\varepsilon_{max} = \varepsilon_y$ is located at the edges of section $h/2$:

$$\chi_y = \frac{\varepsilon_y}{(h/2)} \tag{5.26}$$

If the yield strain is located at distance z_y from the axis of zero strain, it results:

$$\chi_y = \frac{h}{2}\frac{\varepsilon_y}{\varepsilon_{max}} \tag{5.27}$$

From (5.24) and (5.25) one can see that the curvature directly depends on the variation of strain at the section edges. Taking into account the Navier's relationship, the elastic curvature results:

$$\chi = \frac{M}{EI} \tag{5.28}$$

where I is the second inertia moment of the section. For the yield strain results:

$$\chi_y = \frac{M_y}{EI} \tag{5.29}$$

where

$$M_y = Zf_y \tag{5.30}$$

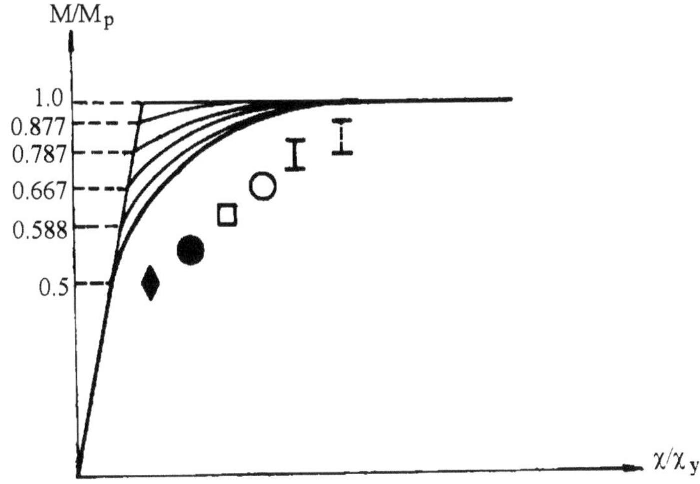

Figure 5.11: Moment-curvature relationship for different cross-sections

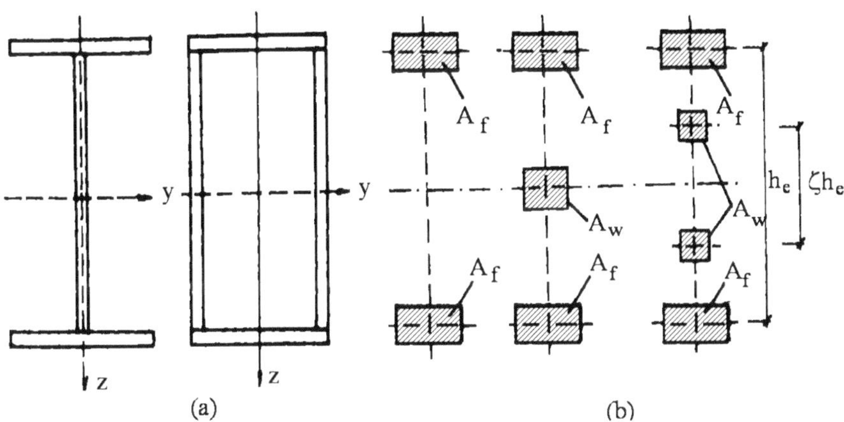

Figure 5.12: Idealized cross-sections: (a) Actual cross-section; (b) Different equivalent cross-sections

The moment-curvature relationship depends on the values of strain related to yield and strain-hardening. The $M-\chi$ relationships, in case that only the yield plateau is considered, are presented in Fig. 5.11 for different section types. For the rectangular sections, the yield moment M_y is 0.667 times the plastic moment M_p, while for I-sections, it is about 0.9 times.

But, as it is shown in Chapter 4, the section works in the yield plateau only if the rotation is controlled. For controlled loads the section works in the strain-hardening range. For determining the behaviour in this range an idealized cross-section is used (Fig. 5.12), corresponding to:

-two equivalent areas, A_f, corresponding to two-flanges section;

-three equivalent areas, two for flanges, A_f, and one for web, A_w, the last being concentrated in the section centroid:

-four equivalent areas, two for flanges, A_f, and two for web, A_w.

The equivalent areas are determined to have the same area and plastic moment as the actual section. The first model has been used by Kato (1988, 1989, 1990) and Mazzolani and Piluso (1992, 1993, 1996). The second and third models have been developed by Yamada et al (1970) and Yamada and Shirakawa (1991).

The moments and axial forces are:

-bending moment at the first yielding:

$$M_y = (A_f + \xi^2 A_w) h_e f_y \tag{5.31a}$$

-bending moment in a fully plastic range:

$$M_p = (A_f + \xi A_w) h_e f_y \tag{5.31b}$$

-plastic axial load:

$$N_p = (A_f + A_w) f_y \tag{5.31c}$$

For the first model $A_w = 0$ and for the second model $\xi = 0$.

For beams, the moment-curvature relationship is presented in Fig. 5.13. The curvature related to the yielding of flanges results from equation (5.29):

$$\chi_y = \frac{\varepsilon_y}{(h/2)} = \frac{M_y}{EI_e} \tag{5.32}$$

and for a fully plastic section:

$$\chi_p = \frac{\varepsilon_{max}}{(h_c/2)} = \frac{M_p}{EI_e} \tag{5.33}$$

where I_e is the second inertia moment for the equivalent section. The curvature corresponding to the beginning of strain-hardening is:

$$\chi_h = \frac{\varepsilon_h}{(h_e/2)} \tag{5.34}$$

The curvature at the ultimate strain is:

$$\chi_u = \frac{\varepsilon_u}{(h_e/2)} \tag{5.35}$$

Between these characteristic points the variation of the moment-curvature relationship is represented by straight lines.

If plastic buckling occurs in the strain-hardening range, the maximum moment results from (5.17) and the corresponding curvature is:

$$\chi_{max} = \chi_h + \frac{(m_b - 1) M_p}{E_h I_e} \tag{5.36}$$

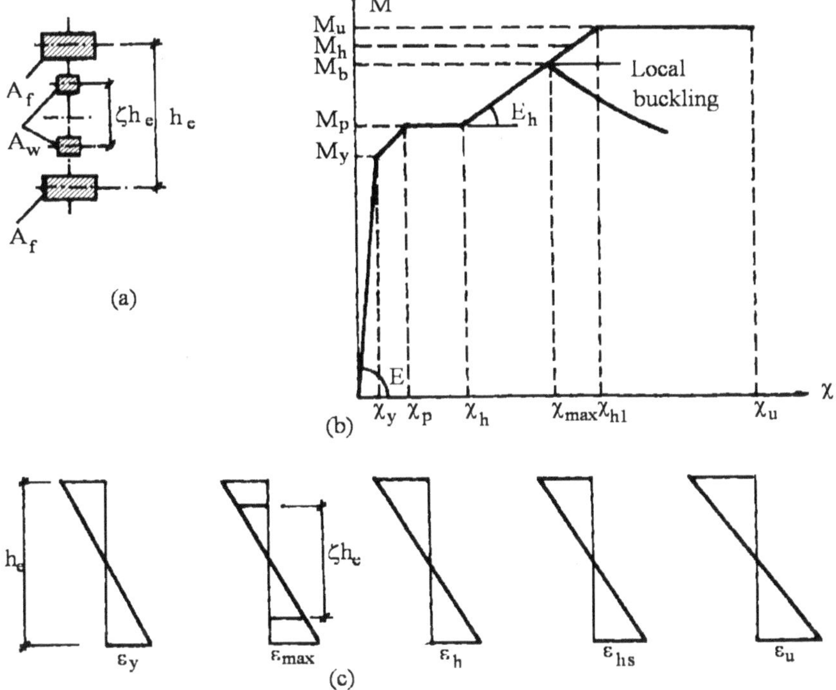

Figure 5.13: Moment-curvature relationship for beams: (a) Equivalent cross-section; (b) Moment-curvature diagram; (c) Strain distribution

In the case of beam-columns, the behaviour is not symmetrically related to the section centroid (Fig. 5.14a). Due to this fact, it can be shown that two behaviour types can occur:

-the maximum bending moment $M_k = m_h M_p$ is obtained when the lower tensile flange is still in elastic range. This situation occurs if:

$$m_h \leq 2n_p + 1 \tag{5.37a}$$

-the maximum bending moment is obtained when the lower tensile flange is in plastic or hardening range, if:

$$m_h > 2n_p + 1 \tag{5.37b}$$

The two cases are studied by Kato (1988, 1989, 1990) and Mazzolani and Piluso (1992, 1993, 1996) and are represented in Fig. 5.14b,c, being composed by a succession of straight lines.

The maximum moment corresponding to the plastic buckling is obtained for:

$$\chi_{max} = \chi_h + (1 - 2n_p)\chi_y + \frac{(m_b - 1)M_p}{E_r I_e} \quad ; \quad m_b \leq 2n_p + 1 \tag{5.38a}$$

$$\chi_{max} = \chi_h + \frac{(m_b - n_p - 1)M_p}{E_h I_e} \quad ; \quad m_b > 2n_p + 1 \tag{5.38b}$$

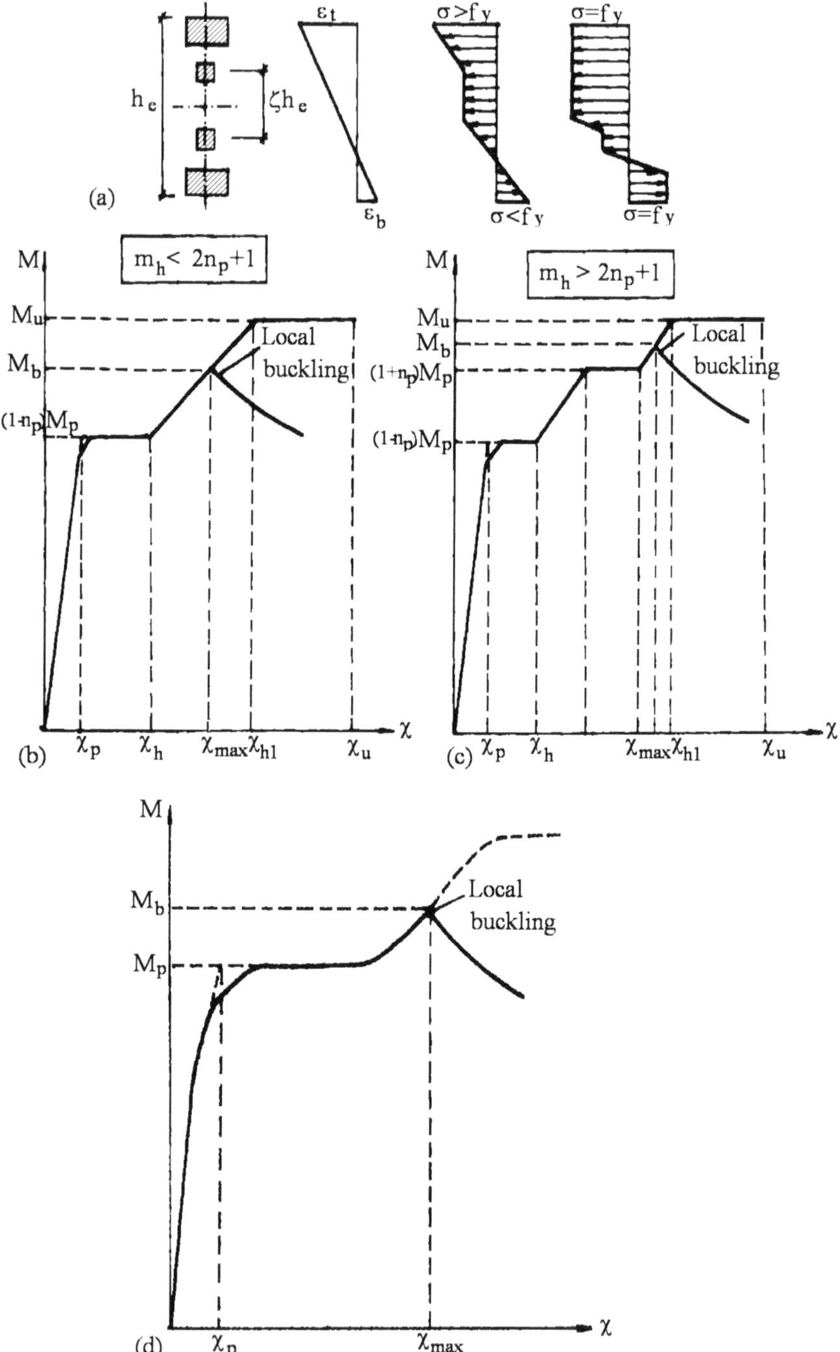

Figure 5.14: Moment-curvature relationship for beam-column: (a) Equivalent cross-section; (b) The lower tensile flange is still in elastic range; (c) The lower tensile flange is in plastic range; (d) Actual behaviour

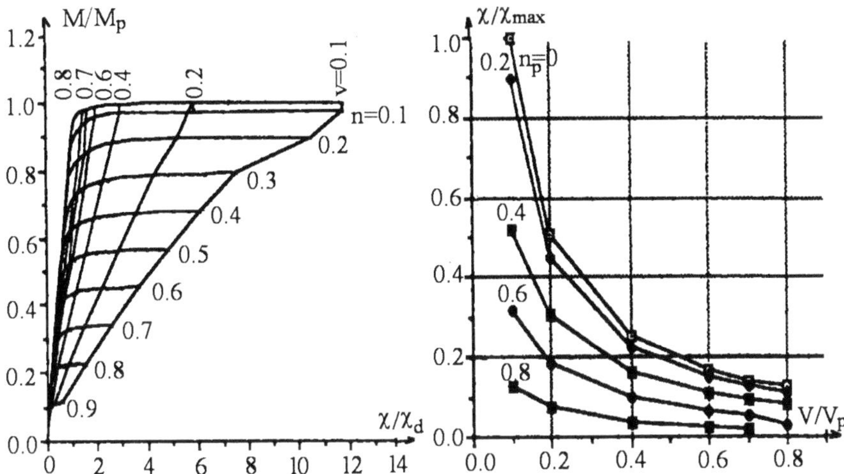

Figure 5.15: Axial and shear forces influences on moment-curvature relationship: (a) Influence of axial forces; (b) Influence of shear force (after Aiello and Ombres, 1995a)

were E_r is the reduced modulus of plasticity:

$$E_r = \frac{2EE_h}{E + E_h} \tag{5.39}$$

By examining the moment-curvature curves presented in Figs. 5.13, 5.14 and the actual curvatures, one can observe some differences. Due to the fact that the seismic action is of controlled loading type, jumps in the yield plateau occur at different levels of sections when the strains reach the yield value. So, instead of a succession of straight lines, the actual behaviour is characterized by a continuous increasing moment-curvature curve (Fig. 5.14d) until the attainment of the maximum bending moment for which a local plastic buckling occurs.

The influence of axial and shear forces on the moment-curvature and the cross-section ductility is studied by Aiello et al (1994) and Aiello and Ombres (1995a,b). In Fig. 5.15 the diagrams obtained for HE profiles are shown. One can see that the presence of axial and shear forces sensibly reduces the resistance and curvature of the cross-section. Particularly, when the level of shear force exceeds 0.3 and 0.4 for axial force, the resistance and ductility are remarkably reduced.

5.2.4. Ductility Classes of Sections

(i) *Section ductility.* The available ductility of a steel section is defined as the ratio between the curvature corresponding to the attainment of plastic buckling which leads to the maximum plastic moment and the curvature corresponding to yielding (Fig. 5.14):

$$\mu_\chi = \frac{\chi_{max}}{\chi_p} \tag{5.40}$$

The χ_{max} can be determined from the equation (5.36) for beams and (5.38) for beam-columns. The yield curvature corresponds to the formation of a plastic hinge, neglecting the difference between the elastic limit and the fully plastic moments.

The section ductility can be used for the classification of cross-sections in different behavioural classes. The classification in behavioural classes proposed in Section 4.7 considers plastic, compact, semi-compact and slender elements. The differences among these classes are governed by the element slenderness, defined by the width-to-thickness ratios.

For sections, the behaviour is governed by the buckling of flange and web plates, for which independent limitations are proposed by EC 3. This assumption is unreasonable because, obviously, the flange is restrained by the web and the web is restrained by the flanges. So, the interaction between the two buckling modes must be considered.

It is recognized that the plastic and compact sections buckle in the strain-hardening range. So, if $m_b = 1$ in equations (5.18) and (5.21), the resulting equations give the limit between slender and semi-compact sections, taking into account the interaction between the flange and the web. If the Kato's relationships are written in the form:

$$\frac{1}{m_b} = a_1 + \frac{a_2}{\alpha_f} + \frac{a_3}{\alpha_w} \tag{5.41}$$

the limit between slender and semi-compact sections results as:

$$\frac{\left(\frac{b/2}{t_f}\right)^2}{A} + \frac{\left(\frac{d}{t_w}\right)^2}{B} = 1 \tag{5.42}$$

where

$$A = \frac{1 - a_1}{a_2}\frac{E}{f_y} \quad ; \quad B = \frac{1 - a_1}{a_3}\frac{E}{f_y} \tag{5.43a, b}$$

The relationship (5.42) is plotted in Fig. 5.16b representing an interaction curve between flange and web buckling.

Another concept for classification is presented by Vayas and Psycharis (1995). Using the same criteria set up in EC 8 for reinforced concrete structures, the following classification is proposed, using the curvature ductility:

-class A (ductile) sections shall have a minimum curvature ductility of $\mu_\chi = 15$;

-class B (compact) sections shall have a minimum curvature ductility of $\mu_\chi = 6$;

-class C (semi-compact) sections shall have a minimum curvature ductility of $\mu_\chi = 3$.

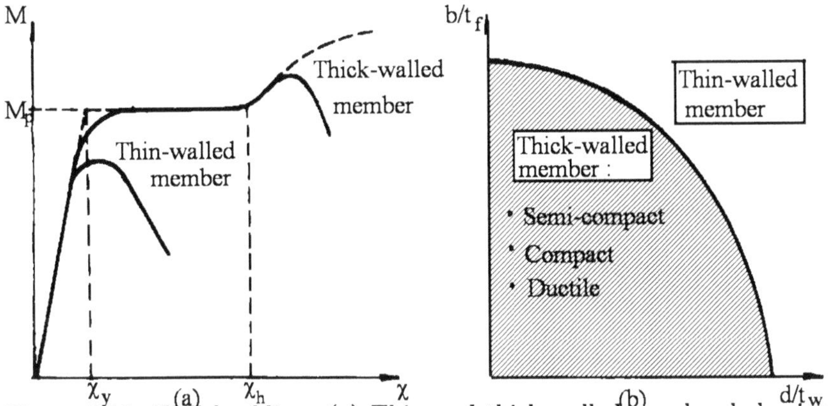

Figure 5.16: Wall buckling: (a) Thin and thick-walled member behaviour; (b) Interaction between flange and web buckling

Table 5.1: Width-to-thickness limitations for sections (f_y, MP_a)

	Ductility class		
	I	II	III
Beam			
H-section	$\frac{b}{2t_f}+0.09\frac{d}{t_w}\leq\frac{248}{\sqrt{f_y}}$	$\leq\frac{297}{\sqrt{f_y}}$	$\leq\frac{333}{\sqrt{f_y}}$
	applicable for $d\geq b$,	$\frac{b}{t_f}\leq\frac{249}{\sqrt{f_y}}$,	$\frac{d}{t_w}\leq\frac{1143}{\sqrt{f_y}}$
Column			
H-section	$\frac{b}{2t_f}+0.16\frac{d}{t_w}\leq\frac{248}{\sqrt{f_y}}$	$\leq\frac{297}{\sqrt{f_y}}$	$\leq\frac{333}{\sqrt{f_y}}$
	applicable for $d\geq b$,	$\frac{b}{t_f}\leq\frac{249}{\sqrt{f_y}}$,	$\frac{d}{t_w}\leq\frac{769}{\sqrt{f_y}}$
Box-section	$\frac{B}{t}\leq\frac{554}{\sqrt{f_y}}$	$\frac{B}{t}\leq\frac{624}{\sqrt{f_y}}$	$\frac{B}{t}\leq\frac{802}{\sqrt{f_y}}$

(ii) *Sectional classification.* For practical purposes, a more detailed classification of different section types is required. Based on the Kato's studies (1988, 1989, 1990, 1995), in Table 5.1 and Fig. 5.17a the interaction between the two buckling modes is considered for I-sections and square box sections (AIJ, 1990).

For I-sections, used as beams, another relationship for interaction flange-web buckling is proposed by Yabuki et al (1995):

-plastic sections:

$$\left(\frac{\overline{\lambda}_f}{0.4}\right)^2+\left(\frac{\overline{\lambda}_w}{0.5}\right)^2\leq 1 \qquad (5.44a)$$

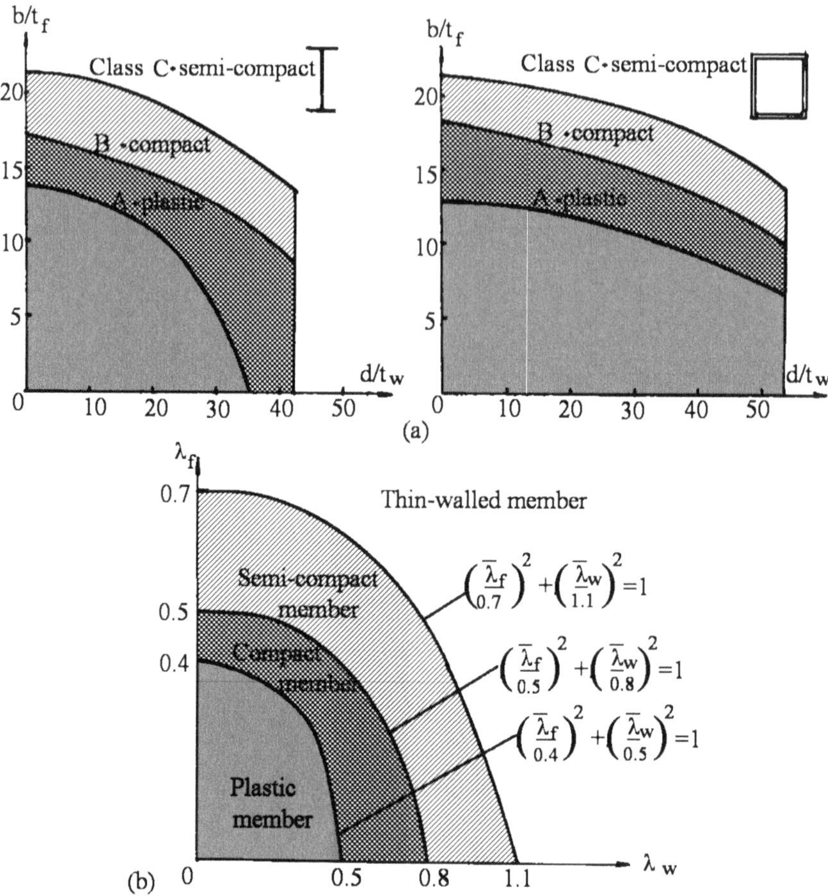

Figure 5.17: Sectional classification: (a) Based on depth-to-thickness ratios; (b) Based on slenderness ratios

-compact sections:

$$\left(\frac{\overline{\lambda}_f}{0.5}\right)^2 + \left(\frac{\overline{\lambda}_w}{0.8}\right)^2 \leq 1 \quad ; \quad \left(\frac{\overline{\lambda}_f}{0.4}\right)^2 + \left(\frac{\overline{\lambda}_w}{0.5}\right)^2 > 1 \tag{5.44b}$$

-semi-compact sections:

$$\left(\frac{\overline{\lambda}_f}{0.7}\right)^2 + \left(\frac{\overline{\lambda}_w}{1.1}\right)^2 \leq 1 \quad ; \quad \left(\frac{\overline{\lambda}_f}{0.5}\right)^2 + \left(\frac{\overline{\lambda}_w}{0.8}\right)^2 > 1 \tag{5.44c}$$

-thin-walled sections:

$$\left(\frac{\overline{\lambda}_f}{0.7}\right)^2 + \left(\frac{\overline{\lambda}_w}{1.1}\right)^2 > 1 \tag{5.44d}$$

where the following width-to-thickness ratios are defined as:
-for flange:

$$\overline{\lambda}_f = \left(\frac{f_y}{\sigma_{crf}}\right)^{1/2} \quad ; \quad \sigma_{crf} = 0.425\frac{\pi^2 E}{12(1-\mu^2)}\left(\frac{t_f}{b/2}\right)^2 \qquad (5.45a, b)$$

-for web:

$$\overline{\lambda}_w = \left(\frac{f_y}{\sigma_{crw}}\right)^{1/2} \quad ; \quad \sigma_{crf} = 23.9\frac{\pi^2 E}{12(1-\mu^2)}\left(\frac{t_w}{d}\right)^2 \qquad (5.46a, b)$$

The interaction of flange and web buckling is presented in Fig. 5.17b. The sectional slenderness parameter is defined as:

$$\overline{\lambda}_{sect} = (\overline{\lambda}_f^2 + \overline{\lambda}_w^2)^{1/2} \qquad (5.47)$$

The framing of section in a ductility class is performed using this sectional slenderness.

A comprehensive review of the local buckling and section ductility classes of a number of specifications is presented by Bild and Kulak (1991). The analysis was performed for Canada, USA, Germany, Switzerland, UK, Australia specifications and international codes such as ISO and Eurocode 3, for I-shapes, box sections, rectangular and circular hollow structural shapes, tee, channel and angle sections. Distinction between hot-rolled and welded sections is noticed.

5.3 Stub Ductility

5.3.1. Determination of Section Ductility Using the Stub Behaviour

For material ductility the conventional tensile test is used. For *section ductility*, compression tests on short length of steel members, the *stub*, is established to be a source of new information concerning the local buckling and ductility of members. The stub test is used to determine the effective squash load, influenced by local wall buckling, residual stresses and section geometrical imperfections, etc (Nethercot, 1992). The advantage of the stub use in determining the local behaviour is related to the fact that there is no stress gradient along its length. Therefore, in order to use steels with various characteristics for seismic-resistant structures, it is required to have a complete understanding of material properties in a more complex specimen than the tension one. It is indispensable to clarify the relation between material and plastic behaviour of members, especially concerning the local buckling.

The length of stub must be chosen to be sufficiently short to avoid the overall buckling effects and sufficiently long to allow local buckling. Studies concerning this aspect have shown that a ratio equal to 3 between length and depth is sufficient to eliminate stub length effect on the local buckling of walls.

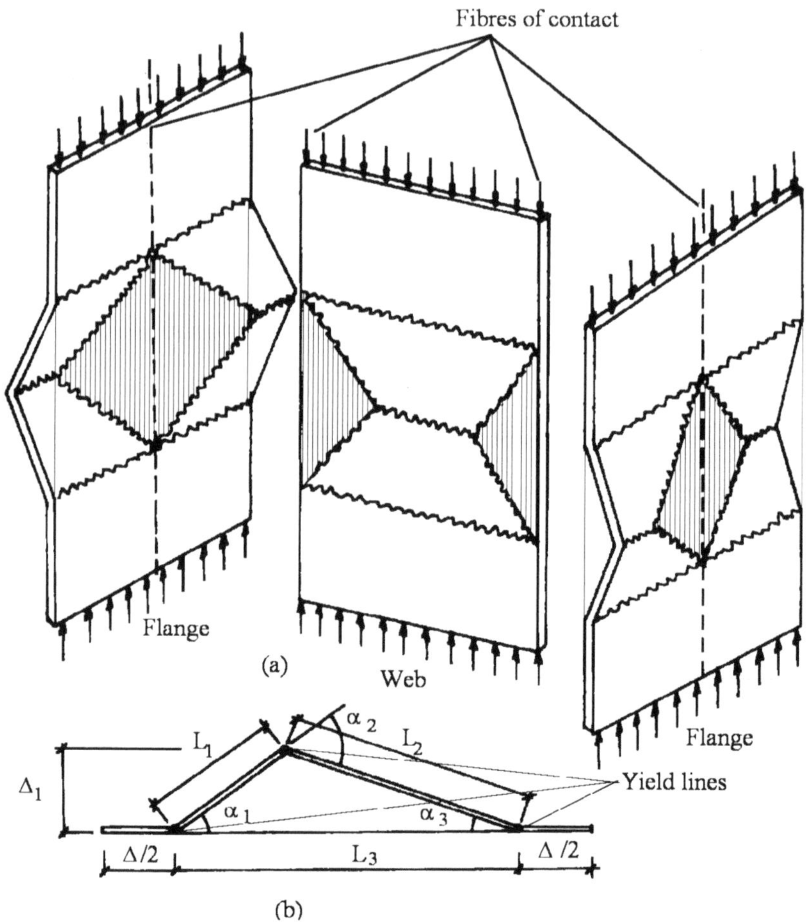

Figure 5.18: Assembling of wall plastic mechanisms: (a) Local mechanism; (b) Deformation compatibility

5.3.2. Local Plastic Mechanism Problems

In Chapter 4 it has been shown that the method of a plastic collapse mechanism is the most effective one for determining the ductility capacity of individual elements. In the case of section ductility it is necessary to assemble some different plastic mechanisms corresponding to the component elements, flanges and webs (Fig. 5.18a), by complying with the conditions of equal deformations (displacements and rotations) along the contact fibres.

The problems of using the plastic mechanisms for determining the local ductility are as follows:

-carrying out of a suitable plastic mechanism based on the examination of experimentally obtained buckled shapes, in which the interaction flange-web plays a very important role. This mechanism is obtained on the basis of the analysis of a lot of possible mechanisms, the minimum carrying capacity giving the criterion to choice the adequate one;

-checking of the validity of linear analysis for plastic rotation of mechanisms. Taking into account that the maximum rotations obtained to define the local ductility are generally found around 0.1 rad, the linear analysis can be used. In this case one can consider that $\sin\alpha \simeq \alpha$ and $\cos\alpha \simeq 1$;

-including the increased stresses due to the strain-hardening effect in the mechanism behaviour. This problem is connected to the definition of limit capacity. If it refers to fracture, it is crucial to include in analysis this stress increasing. Contrary to this, if the rotation capacity is involved as a result of the local buckling, the influence of strain-hardening stresses is much more reduced due to the lowering in moment capacity;

-considering the analysis of the effect of axial stresses producing a reduction of yield line rotation capacity. This effect decreases as the rotation increases. Due to the fact that the effects of strain-hardening and axial stresses are opposite, both influences can be neglected.

The displacement of a mechanism composed by two plates (Fig. 5.18b) is given by the equations:

$$\Delta_1 = \left[\frac{(L_1 + L_2 + L_3)(L_1 - L_2 + L_3)(L_1 + L_2 - L_3)(-L_1 + L_2 + L_3)}{2L_3}\right]^{1/2} \tag{5.48}$$

in which:

$$L_3 \approx L_1 + L_2 - \Delta \tag{5.49}$$

Δ being the horizontal displacement of mechanism. Taking into account that this displacement is small in comparison with the plate dimensions, it results from (5.48):

$$\Delta_1 \approx \left(\frac{2L_1L_2}{L_1 + L_2}\Delta\right)^{1/2} \tag{5.50}$$

In cases of equal plates $L_1 = L_2 = a$:

$$\Delta_1 = \sqrt{a\Delta} \tag{5.51}$$

The rotation angles of plastic mechanisms are:

$$\sin(\alpha_1) \approx \alpha_1 = \frac{\Delta_1}{L_1}; \quad \sin\alpha_3 \approx \alpha_3 = \frac{\Delta_1}{L_2} \tag{5.52a, b}$$

$$\sin\alpha_2 \approx \alpha_2 = \alpha_1 + \alpha_2 = \left(\frac{1}{L_1} + \frac{1}{L_2}\right)\Delta_1 \tag{5.52c}$$

5.4 Ductility of I-Section Stubs

5.4.1. Main Factors Influencing the Stub Ductility

The main factors are:

-*fabrication*, the section being produced by hot-rolling or by welding, with fillet welds, partial or full penetration welds (Fig. 5.19);

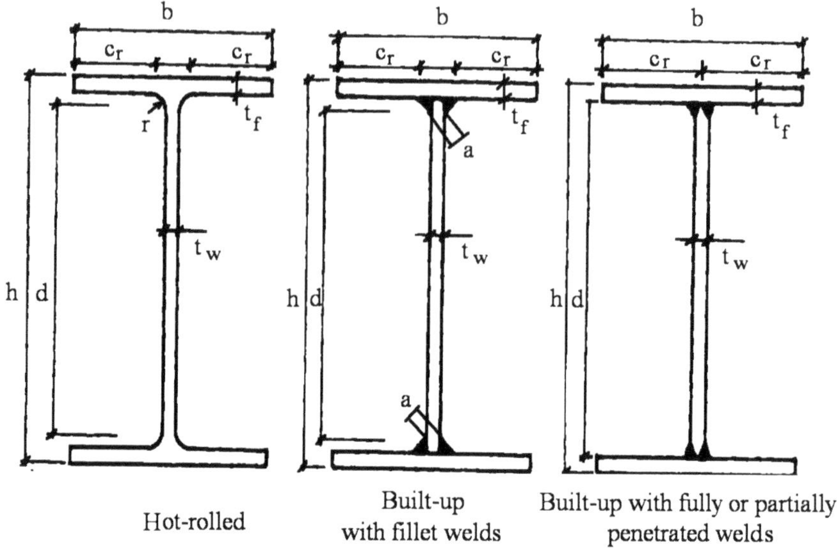

Figure 5.19: Fabrication processes for I-sections

-*steel properties* as yield stress, tensile strength, yield ratio, uniform strain, etc.;

-*flanges and web slendernesses* as the width-to-thickness ratio;

-*local buckling mode*, governed by flange or web instability and the interaction of these buckling modes;

-*geometrical imperfections*, associated to the sectional shape (Fig. 5.20a) (Moldovan, 1990);

-*residual stresses*, function of the fabrication process (Fig. 5.20b) (Qiung, 1989, Mazzolani, 1992, Rondal, 1992);

-*loading type*, as static or dynamic monotonic or pulsatory (Fig. 5.21).

5.4.2. Experimental Results

Extensive experimental research works have been performed by Ivanyi (1979, 1985), Yamao et al (1995), Ono et al (1996), Iwata et al (1997), Sivakumaran et al (1998). All specimens experienced local buckling failure of flanges with interaction with web buckling. Fig. 5.22 shows the shape of a buckled compression stub (Ivanyi, 1979). Plastic deformations are produced in a limited zone only, the remaining part working in the elastic range. In this plastic zone large deformations are concentrated, amplified by the plastic buckling.

Fig. 5.23 shows the compression stress-strain curves for two Japanese steel grades, SN 490B with yield plateau and HT 590 which does not show a clear yield plateau (Ono et al, 1996, 1997, Iwata et al, 1997). Three different width-to-thickness ratios for flanges and webs were plotted corresponding to the three ranks of the AIJ (1990) design standard. The dotted lines indicate the stress-strain curves for tension tests and the full lines show the

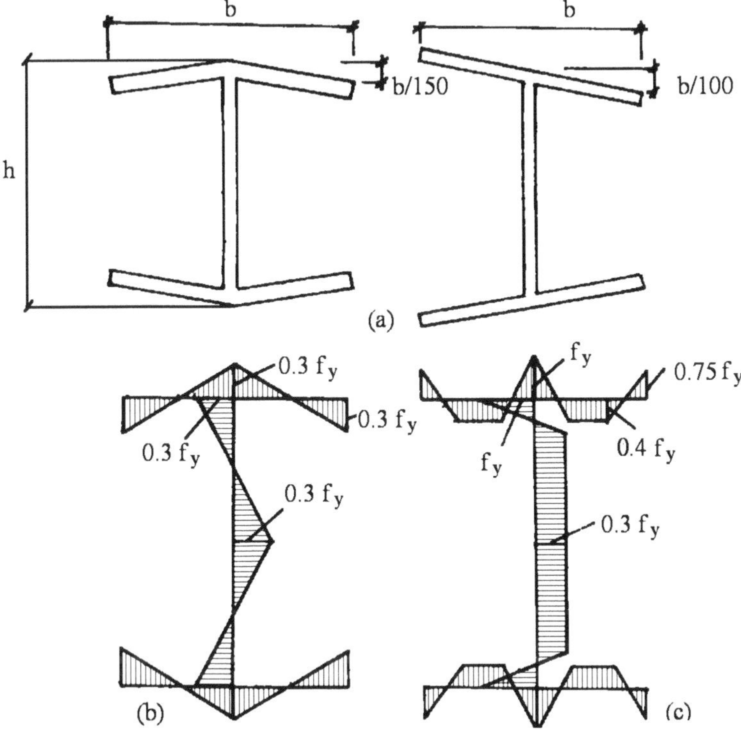

Figure 5.20: Imperfections due to fabrication: (a) Geometrical imperfections; (b) Residual stresses

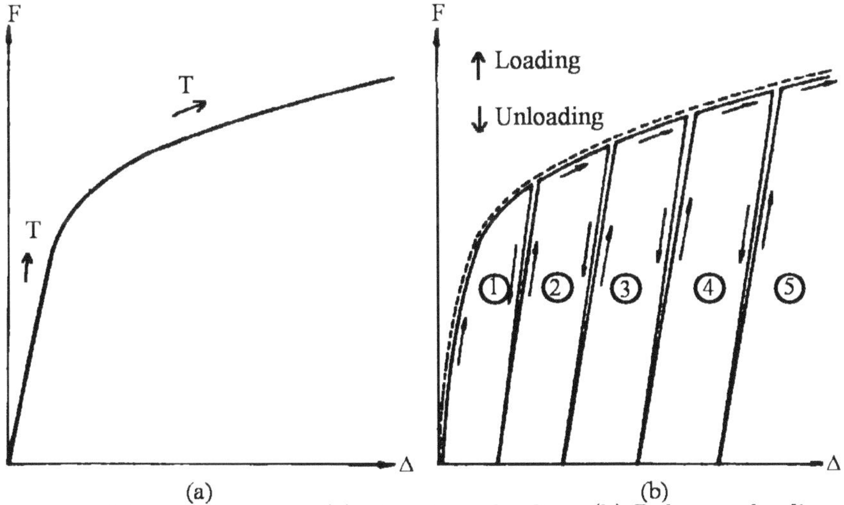

Figure 5.21: Loading type: (a) Monotonic loading; (b) Pulsatory loading

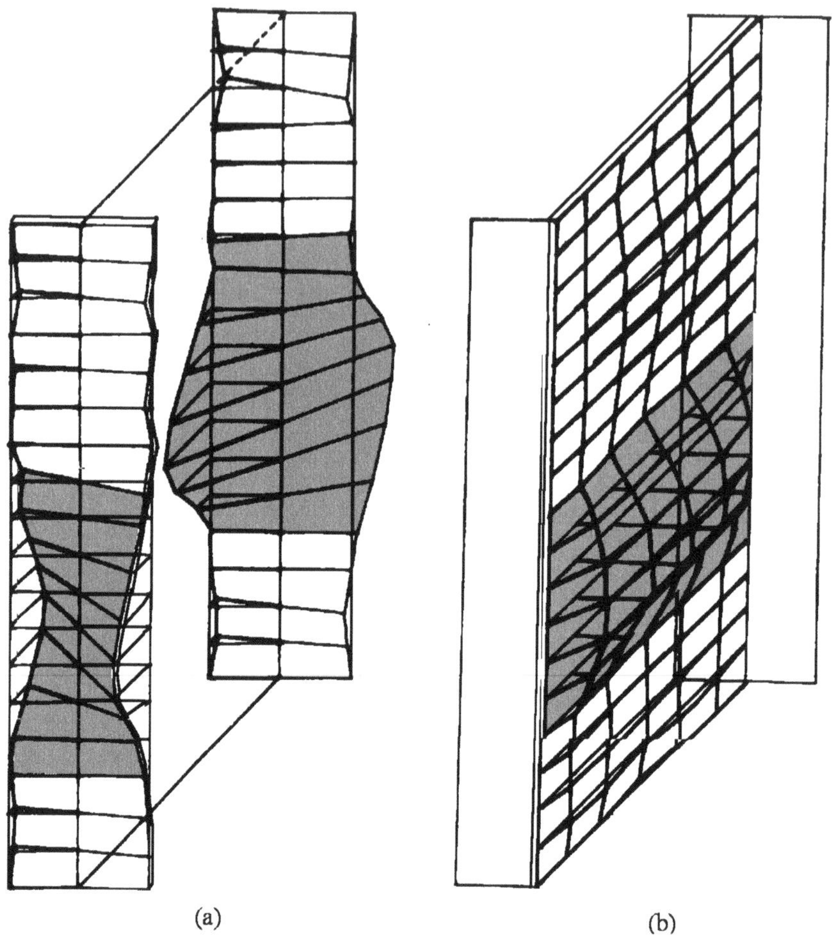

Figure 5.22: Plastic buckling of compression stub: (a) Flange buckling; (b) Web buckling (after Ivanyi, 1979)

compression tests for the stub specimens. The strength of the specimens rises to maximum values, when local buckling occurs, along the same line as for the tension tests. One can see that the local buckling and the gradient of deterioration are highly influenced by width-to-thickness ratios. The local ductility decreases with the increasing of flanges and web slenderness.

The local buckling is not only governed by width-to-thickness ratios, but also largely depends upon the material properties. Fig. 5.24a shows the experimental results corresponding to different steel grades (Sivakumaran and Yuan, 1998). The local buckling arises at the peak stress level. One can see that the ductility dramatically decreases with the increasing of steel grade and the stub available compression ductility is considerably lower than the corresponding material strain ductility. This erosion is due to the

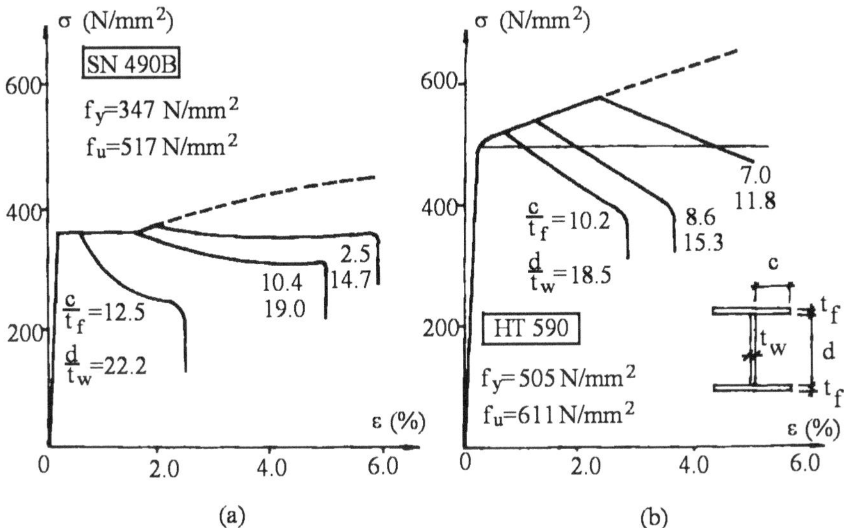

Figure 5.23: Influence of steel yield plateau: (a) Steel with yield plateau (after Iwata et al, 1997); (b) Steel without yield plateau (after Ono et al, 1996)

effects of local buckling of stub component walls. For each steel grade three identical specimens are tested. The experimental results exhibit somewhat similar behaviour up to peak stress. However, the post-ultimate stress-strain experimental curves differ from each other (Fig. 5.24b), even though they had the same geometrical dimensions and material properties. Such differences may be attributed to different initial geometrical and mechanical imperfections, which cannot be ignored in the ductility analysis.

5.4.3. Numerical Tests

The FEM is used to numerically calculate the ductility of a compressed stub (Fig. 5.25a) (Sivakumaran and Yuan, 1998). Geometrical imperfections and residual stresses are incorporated in the finite element mesh. Four b/t ratios and two steel grades were considered as parameters for the study (Fig. 5.25b). For 350W steel grade a good ductility can be noticed, which decreases as the slenderness increases. Contrary to this, the high steel grade 700 Q shows a very small strain ductility. Sections with b/t = 6.0–6.4 work in elastic range, without any ductility.

During the experimental tests it can be observed that the deformations are concentrated only in a limited plastic zone with a well defined length $L_p = b$, which can be characterized as a plastic mechanism (Fig. 5.26). This is a quasi-mechanism, being composed by yield lines and plastic zones (Ivanyi, 1979a, Ono et al, 1996), involving an interaction between flanges and web deformations. The stub ductility can be determined using this plastic mechanism (see Section 4.4.2). The total potential energy is given

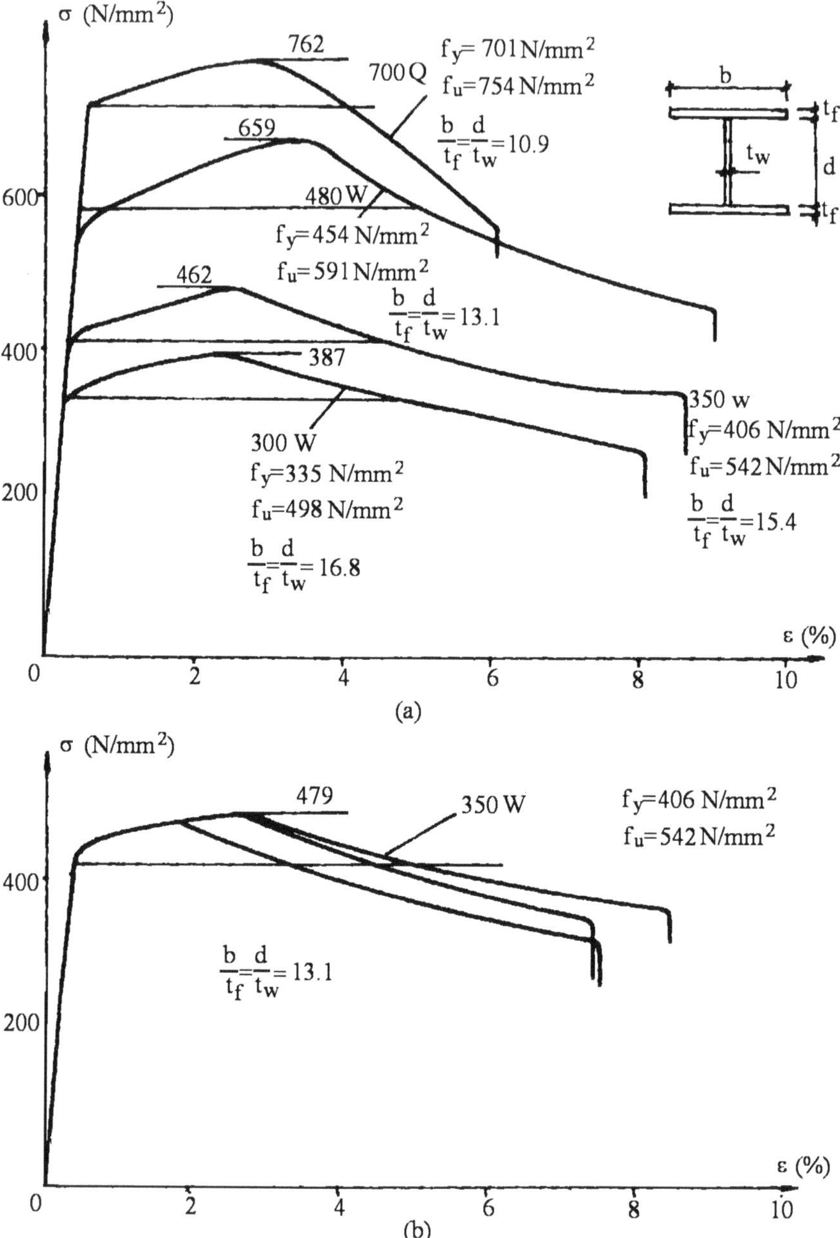

Figure 5.24: Influence of steel grade: (a) Ductility decreasing with increasing of steel grade; (b) Influence of stub imperfections (after Sivakumaran and Yuan, 1998)

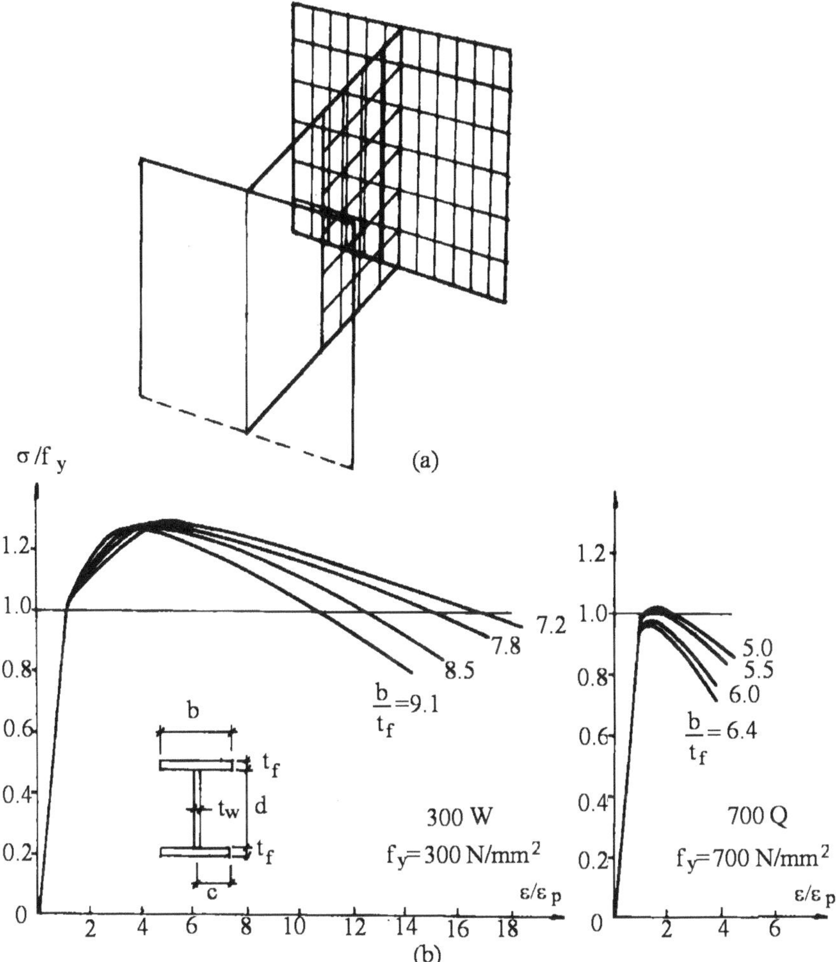

Figure 5.25: Numerical tests: (a) FE mesh for the stub; (b) Influence of steel grade and flange slenderness (after Sivakumaran and Yuan, 1998)

in Table 5.2, being composed by the potentials of plastic zones U_z and yield lines U_l. The axial force results from equation (4.38):

$$N = \frac{dU_z}{d\delta_r} + \frac{dU_l}{d\delta_r} \tag{5.53}$$

δ_r being the stub shortening, concentrated in the plastic zone. Using the values from Table 5.2 it results:

$$N = b_f(2t_f f_{yf} + t_w f_{yw}) + \frac{1}{2}\sqrt{c}\left[6t_f^2 f_{yf} + \left(1 + \frac{d}{c}\right) t_w^2 f_{yw}\right] \frac{1}{\sqrt{\delta}} \tag{5.54}$$

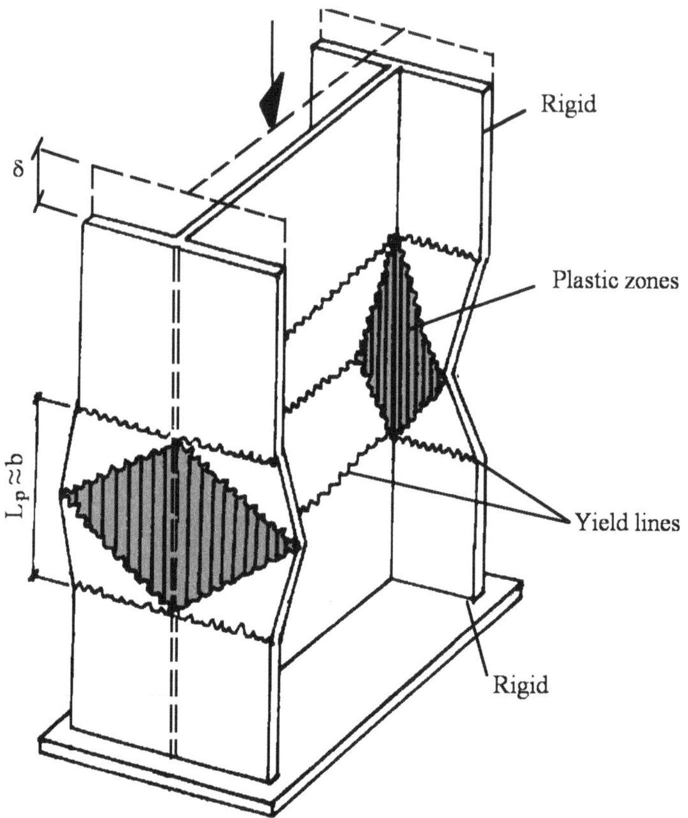

Figure 5.26: Plastic collapse mechanism

and, in a normalized form:

$$\frac{N}{N_p} = \alpha_{1N} + \alpha_{2N}\frac{1}{\sqrt{\delta_r/\delta_p}} \tag{5.55}$$

where:

$$N_p = 4ct_f f_{yf}\left(1 + \frac{1}{4}\frac{d}{c}\frac{t_w}{t_f}\frac{f_{yw}}{f_{yt}}\right); \qquad \delta_p = \frac{f_{yf}}{E}2c \tag{5.56a, b}$$

$$\alpha_{1N} = \frac{1}{2}\frac{1 + \dfrac{1}{2}\dfrac{t_w}{t_f}\dfrac{f_{yw}}{f_{yf}}}{1 + \dfrac{1}{4}\dfrac{d}{c}\dfrac{t_w}{t_f}\dfrac{f_{yw}}{f_{yf}}} \tag{5.57a}$$

$$\alpha_{2N} = \frac{3}{4}\left(\frac{E}{2f_{yf}}\right)^{\frac{1}{2}}\frac{t_f}{c}\frac{1 + \dfrac{1}{6}\left(1 + \dfrac{d}{c}\right)\left(\dfrac{t_w}{t_f}\right)^2\dfrac{f_{yw}}{f_{yf}}}{1 + \dfrac{1}{4}\dfrac{d}{c}\dfrac{t_w}{t_f}\dfrac{f_{yw}}{f_{yf}}} \tag{5.57b}$$

Table 5.2: Compression stub.Potential energy of plastic mechanism

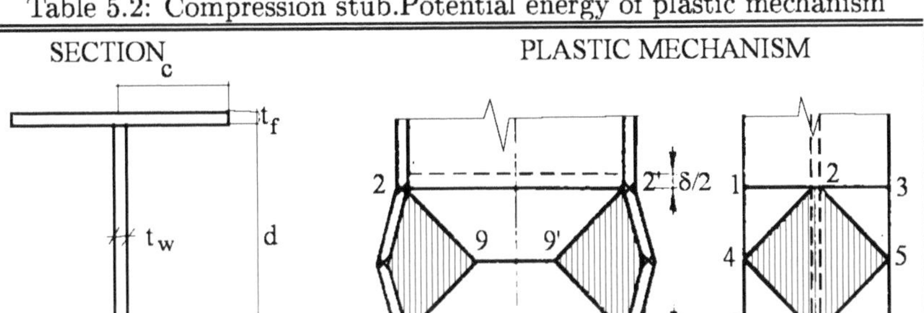

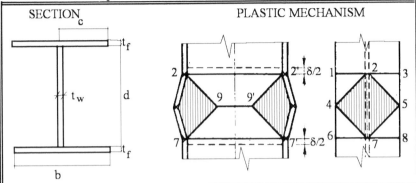

PLASTIC ZONES

Zones	No.	Volume	Strain	Stress	Potential energy per zones	Potential energy total
2574	2	$2c^2 t_f$	$\delta_r/2c$	f_{yf}	$c t_f f_{yf} \delta_r$	$2 c t_f f_{yf} \delta_r$
297	2	$c^2 t_f$	$\delta_r/2c$	f_{yw}	$\frac{1}{2} c t_w f_{yw} \delta_r$	$c t_w f_{yw} \delta_r$
TOTAL $U_z = c(2 t_f f_{yf} + t_w f_{yw})\delta_r$						

YIELD LINES

Lines	No.	Length	Rotation	Moment	Potential energy per zones	Potential energy total
123	4	2c	$(\delta_r/c)^{1/2}$	$\frac{c t_f^2}{4} f_{yf}$	$\frac{c^{1/2} t_f^2}{2} f_{yf} \delta_r^{1/2}$	$2 c^{1/2} t_f^2 f_{yf} \delta_r^{1/2}$
24	8	$2^{1/2} c$	$(2\delta_r/c)^{1/2}$	$2^{1/2} \frac{c t_f^2}{4} f_{yf}$	$\frac{c^{1/2} t_f^2}{2} f_{yf} \delta_r^{1/2}$	$4 c^{1/2} t_f^2 f_{yf} \delta_r^{1/2}$
22'	2	d	$(\delta_r/c)^{1/2}$	$\frac{d t_w^2}{4} f_{yw}$	$\frac{d t_w^2}{4(c)^{1/2}} f_{yw} \delta_r^{1/2}$	$\frac{d t_w^2}{2(c)^{1/2}} f_{yw} \delta_r^{1/2}$
99'	1	d-2c	$2(\delta_r/c)^{1/2}$	$\frac{(d-2c) t_w^2}{4} f_{yw}$	$\frac{(d-2c) t_w^2}{2(c)^{1/2}} f_{yw} \delta_r^{1/2}$	$\frac{(d-2c) t_w^2}{2(c)^{1/2}} f_{yw} \delta_r^{1/2}$
29	4	$2^{1/2} c$	$(2\delta_r/c)^{1/2}$	$2^{1/2} \frac{c t_w^2}{4} f_{yw}$	$\frac{c^{1/2} t_w^2}{2} f_{yw} \delta_r^{1/2}$	$2 c^{1/2} t_w^2 f_{yw} \delta_r^{1/2}$
TOTAL $U_l = \left(6 c^{1/2} t_f^2 f_{yf} + c^{1/2} t_w^2 f_{yw} + \frac{d}{(c)^{1/2}} t_w^2 f_{yw}\right)\delta_r^{1/2}$						

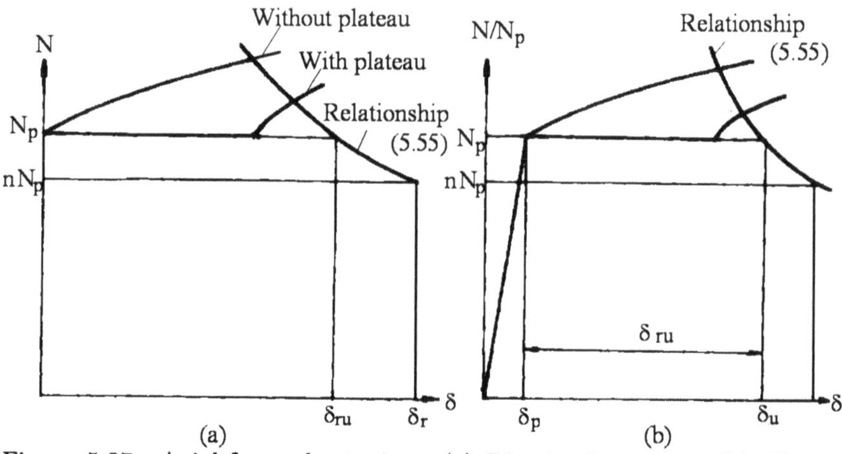

Figure 5.27: Axial-force shortening: (a) Plastic shortening; (b) Elasto-plastic shortening

The first term α_{1N} contains the strain energy corresponding to the strain energy of plastic zones, while the second one, α_{2N} refers to the strain energy of yield lines. One can see that the erosion of loading capacity due to local buckling is produced only by the second term.

The equation (5.55) represents a hyperbole corresponding to rigid-plastic behaviour, which intersects the primary curve (Fig. 5.27a). The ultimate capacity can be obtained at the intersection between the mechanism hyperbole (5.55) and the horizontal line corresponding to a fully plastic axial force. There are cases in which the ultimate capacity is determined for a fraction of the plastic axial force, nN_p $(n < 1)$. For this deformation results the axial ductility:

$$\mu_\delta = \frac{\delta_{ru}}{\delta_p} = \left(\frac{\alpha_{2N}}{n - \alpha_{1N}}\right)^2 \tag{5.58}$$

5.4.4. Comparison of Theoretical and Experimental Results

The results obtained from equation (5.55) are compared in Fig. 5.28 with the experimental tests performed by Sivakumaran and Yuan (1998). The comparison confirms the theoretical assumptions. The same good correspondence is obtained in the studies of Ivanyi (1979, 1985) (Fig. 5.29). One can see that for slender flanges (b/t higher than 10) the buckling occurs in the elastic field, without attainment of full axial plastic capacity. The results of the measured buckled shapes confirm the assumption used in the definition of plastic mechanisms that the mechanism lengths are equal to the flange widths (Fig. 29b). These comparisons between experimental and theoretical results show that the methodology of local plastic mechanisms may be used for determining the stub ductility and for the study of parameter influence.

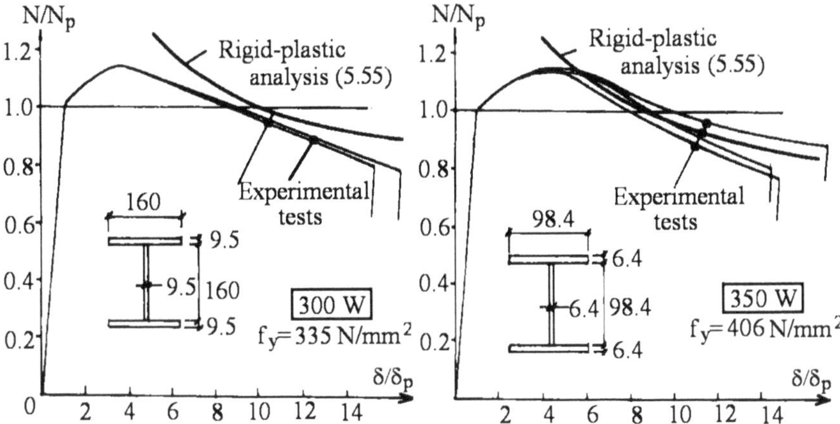

Figure 5.28: Comparison between theoretical and experimental results

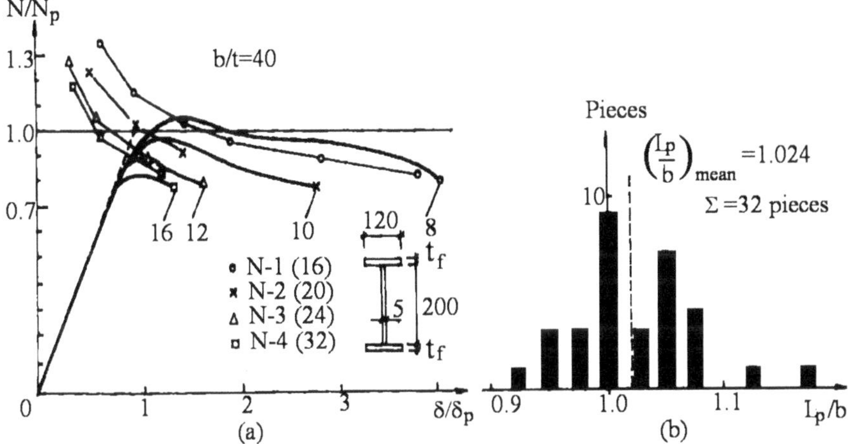

Figure 5.29: Experimental and theoretical curves: (a) Elastic and plastic buckling; (b) Length of buckled shapes (after Ivanyi, 1979)

5.4.5. Influence of Stub Parameters

(i) *Ductility criterium.* The stub ductility can be defined at different levels of axial forces nN_p (Fig. 5.30).

One can see that important increasing of ductility can be obtained if lower n values are used. For $n = 0.9$ an increase of 48 percent is obtained, while for $n = 0.8$, the ductility increase is about 140 percent. A reduced growth is obtained if the post-buckling degradation is important. Therefore, the ductility criterion for design practice is a very important decision and must be established by codes. The Background of EC 8 considers only $n = 1$, but this proposal must be revised, considering the favorable plastic

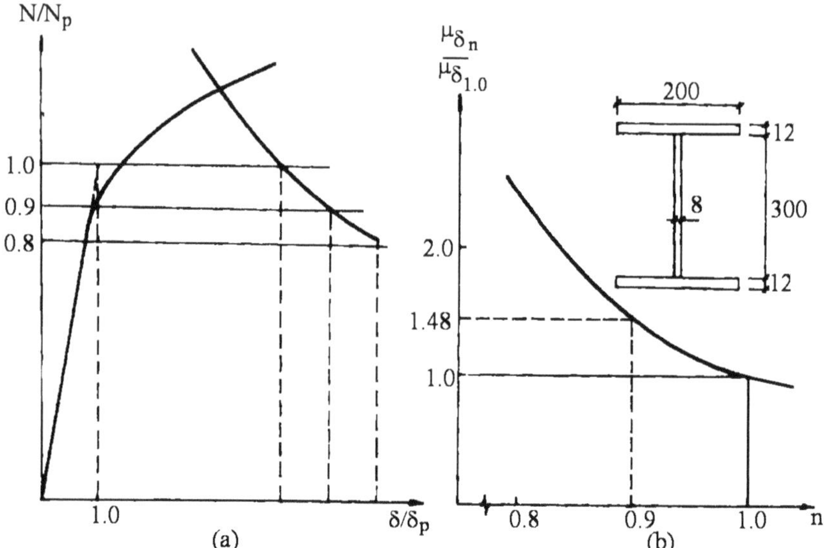

Figure 5.30: Influence of ductility criterium: (a) Influence of axial force level; (b) Variation of axial ductility with ductility criteriom

behaviour of some sections. It seems that a value $n = 0.9$ should be suitable for design purposes.

(ii) *Geometrical dimensions.* The equation (5.55) can be used to define the geometric factors influencing the axial ductility. Examining the coefficients α_{1N} and α_{2N} given by equations (5.57), one can notice that these factors are:

-*Flange slenderness.* The influence of flange slenderness on the axial ductility is presented in Fig. 5.31. The primary normalized curve of axial force versus shortening is intersected with the curves of the rigid-plastic behaviour, determining the axial ductility of compression stubs. These curves are plotted for different flange slenderness. One can see that only for the biggest flange thickness the stub works in a strain-hardening range. For $n = 1$, the increase of flange slenderness from 6.25 to 12.50 means the reduction of axial ductility from 16.20 to 5.36. A minimum ductility increase of 23 percent is obtained, if $n = 0.9$ is considered.

-*Depth-to-width ratio.* Fig. 5.32 shows the axial ductility of compression stubs, function of flange slenderness and depth-to-width ratios. One can see that both parameters have an important influence: high ductility is obtained for low d/c ratios, the increasing of depth producing a reduction of axial ductility. It is very important to notice this effect, because the codes neglect this factor as determinant for evaluating the local ductility.

(iii) *Fabrication process.* One problem for practical design is the definition of flange width as a function of the fabrication process. For I-section built-up by means of penetrated welds (Fig. 5.33a) with thin web, it is possible to consider the flange width c starting from the vertical symmetry axis. For hot-rolled and filled weld I-sections, the flange width must be

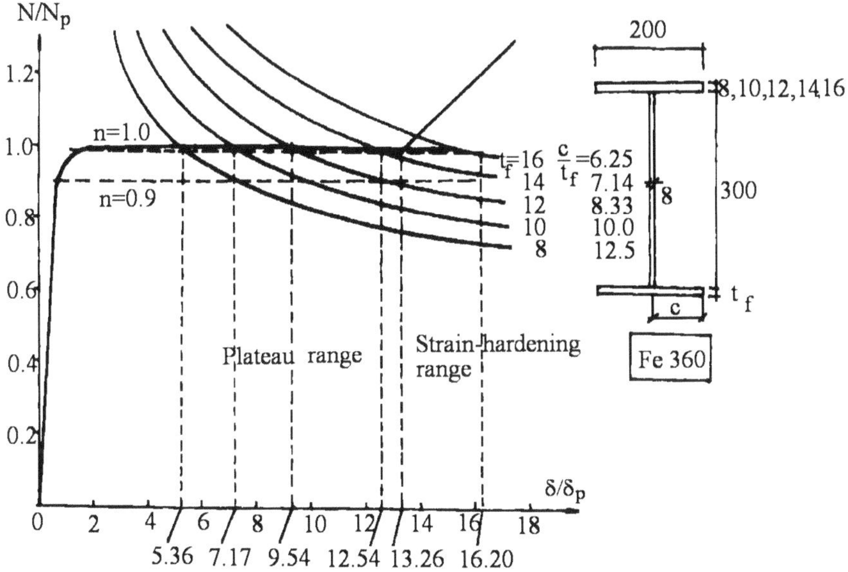

Figure 5.31: Influence of flange slenderness on stub ductility

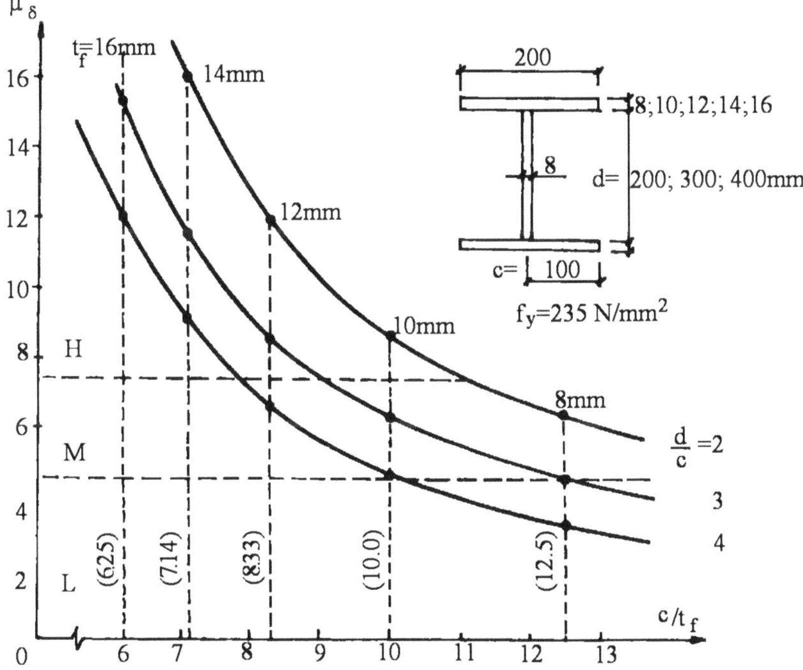

Figure 5.32: Influence of section shape

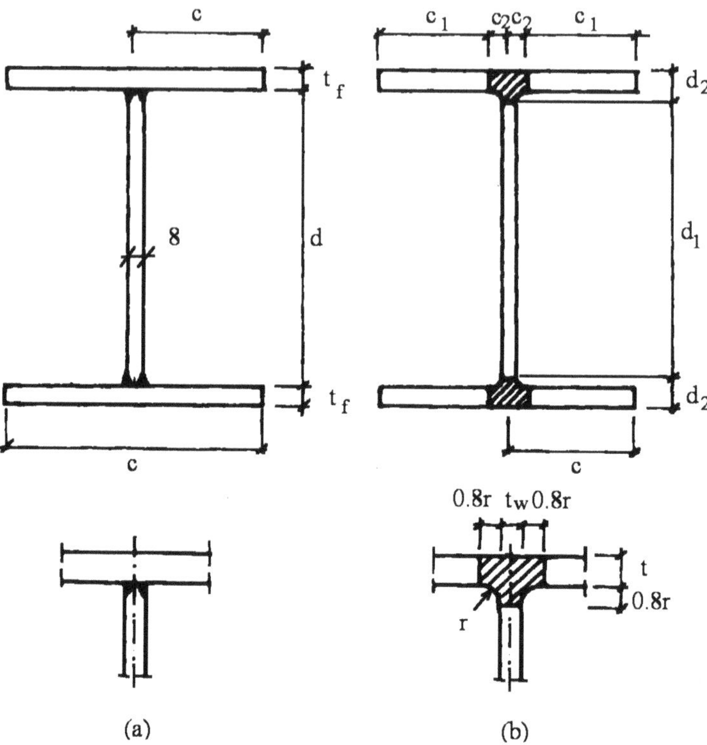

Figure 5.33: Influence of fabrication process: (a) Penetrated weld section; (b) Hot-rolled or fillet-welded section

determined as in Fig. 5.33b, considering a reduced width for flanges. Fig. 5.34 shows the shortening curves for I-sections with penetrated welds and hot-rolled or filled welds. One can see that the axial ductility is strongly influenced by the definition of flange width; the ductility is higher for hot-rolled I-sections. An approximate solution consists in the evaluation of the ductility capacity starting from the one of unreduced flange width and reducing this value by means of a correction factor (Anastasiadis, 1999, Anastasiadis and Gioncu, 1999):

$$c_r = \left(\frac{c}{c - t_w/2 - 0.8r} \right)^2 \tag{5.59}$$

(iv) *Geometrical imperfections.* For considering the geometrical imperfections one can adopt the assumption that the shape is related to the plastic buckled shape (Fig. 5.35a). Using the same algebra as in Section 4.6, it results

$$\delta + \delta_i = \frac{(\Delta + \Delta_i)^2}{c}; \qquad \delta_i = \frac{\Delta_i^2}{c} \qquad (5.60a, b)$$

where δ_i is the initial shortening and Δ_i, the initial deflection of flanges.

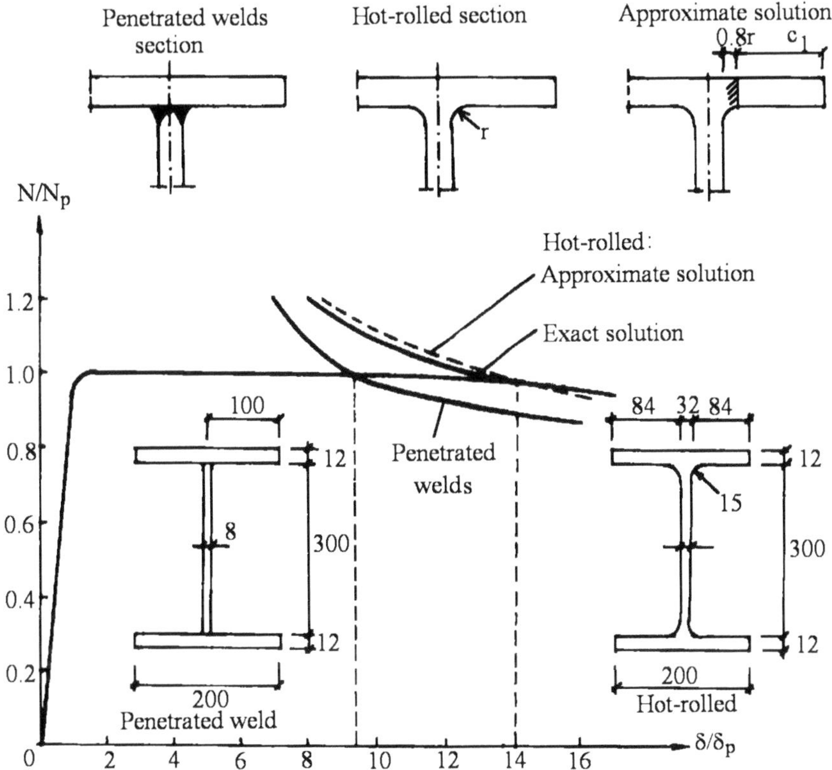

Figure 5.34: Ductilities of penetrated weld and hot-rolled sections

Geometrical imperfections affect only the part of plastic mechanism composed by yield lines. So, in the value of yield line potential from Table 5.2, the value of $\delta_r^{1/2}$ must be replaced by $(\delta_r + \delta_i)^{1/2}$ - $\delta_i^{1/2}$. It results:

$$N = c(2t_f f_{yf} + t_w f_{yw}) + \frac{1}{2}\sqrt{c}\left[6 + t_f^2 f_{yf} + \left(1 + \frac{d}{c}\right) t_w^2 f_{yw}\right] \frac{1}{\sqrt{\delta_r + \delta_i}} \tag{5.61}$$

and in normalized form:

$$\frac{N}{N_p} = \alpha_{1N} + \alpha_{2N} \frac{1}{[(\delta_r + \delta_i)/\delta_p]^{1/2}}; \qquad \left(\frac{N}{N_p}\right) = \alpha_{1N} \tag{5.62a, b}$$

where α_{1N} and α_{2N} are given by equations (5.57). One can see that the erosion of the carrying capacity due to geometrical imperfections is produced only by the term corresponding to yield lines and this capacity cannot fall bellow the value given by (5.62b). The ductility of actual stubs is given in the same way as for the ideal stubs:

$$\mu_\delta = \frac{\delta_{ru}}{\delta_p} = \left(\frac{\alpha_{2N}}{1 - \alpha_{1N}}\right)^2 - \frac{\delta_i}{\delta_p}; \qquad \frac{\delta_i}{\delta_p} = \frac{1}{2}\frac{E}{f_{yf}}\left(\frac{\Delta_i}{b_f}\right)^2 \tag{5.63a, b}$$

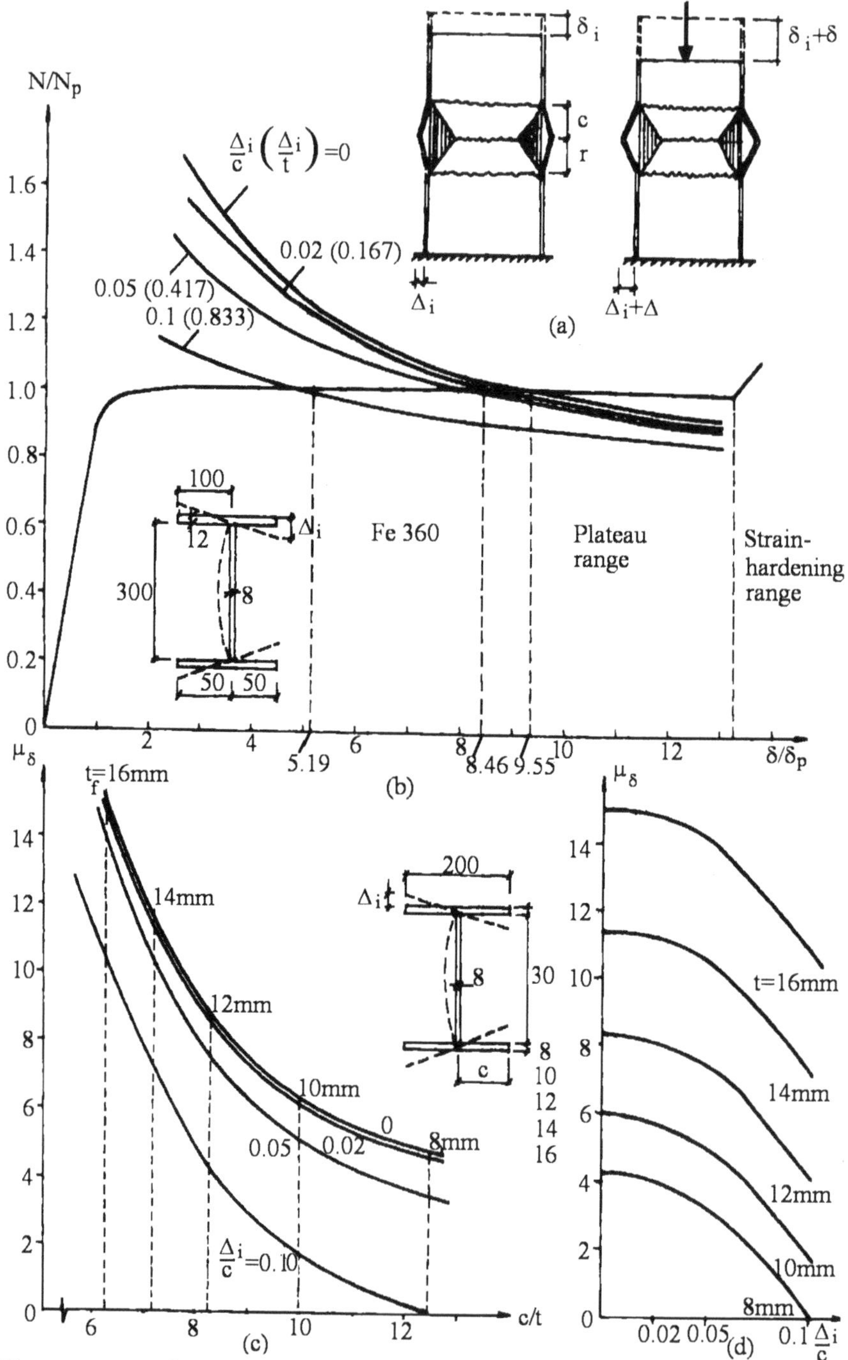

Figure 5.35: Influence of geometrical imperfections: (a) Stub deformations; (b) Ductility for ideal and actual stubs; (c), (d) Influence of geometrical imperfections and flange slenderness

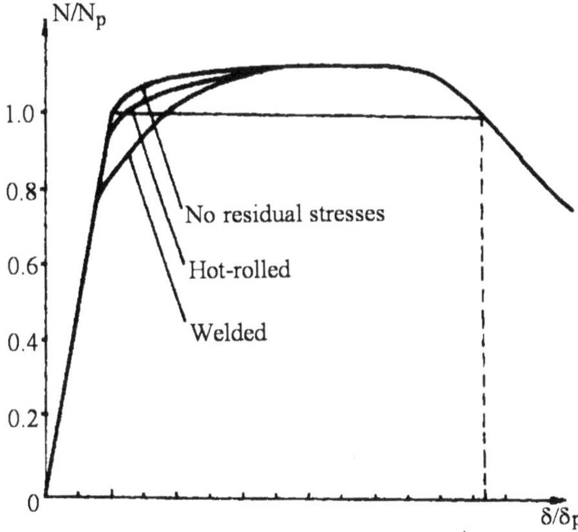

Figure 5.36: Influence of residual stresses

The influence of geometrical imperfections on the axial ductility is presented in Fig. 5.35b, for initial deformations $\Delta_i/c = 0$; 1/50; 1/20; 1/10. The influence of small imperfections is negligible. Contrary to this, the imperfections having the same magnitude of the flange thickness may produce an important decrease in axial ductility. For $\Delta_i/c = 1/10$ and $c/t = 12.5$ ($\Delta_i = 1.25$t) the ductility becomes zero (Fig. 5.35c,d).

(v) *Residual stresses.* Physically it is expected that there is no influence of residual stresses on the ductility capacity. In fact, the presence of residual stresses produces a premature starting of the plastic behaviour (Fig.5.36) and the reduction of the elastic limit is greater for welded sections, where the intensity of residual stresses is higher (Kuranishi, 1995). After the elastic limit, residual stresses can influence the extension of the inelastic range, but, when the section is completely yielded, their influence disappears. Therefore, after full plasticization, the section behaves in the same way as a section without residual stresses.

5.4.6. Influence of Steel Quality

(i) *Steel grade.* In Fig. 5.37a the influence of steel grade is presented, showing a very important degradation of local ductility as far as the steel grade increases. Regarding the coefficients α_{1N} and α_{2N}, one can see that only the last one is influenced by yield stress and that the reduction, related to Fe360 steel grade, is proportional to $235/f_y$. Section 4.3.3 looked at the fact that the specifications guarantee the minimum value only for yield stress, and, therefore, actual values can be higher. The variation of yield stress is given by equations (4.19)–(4.21). Taking into account these values, Fig. 5.37b shows the reduction of axial ductility due to the variation of yield

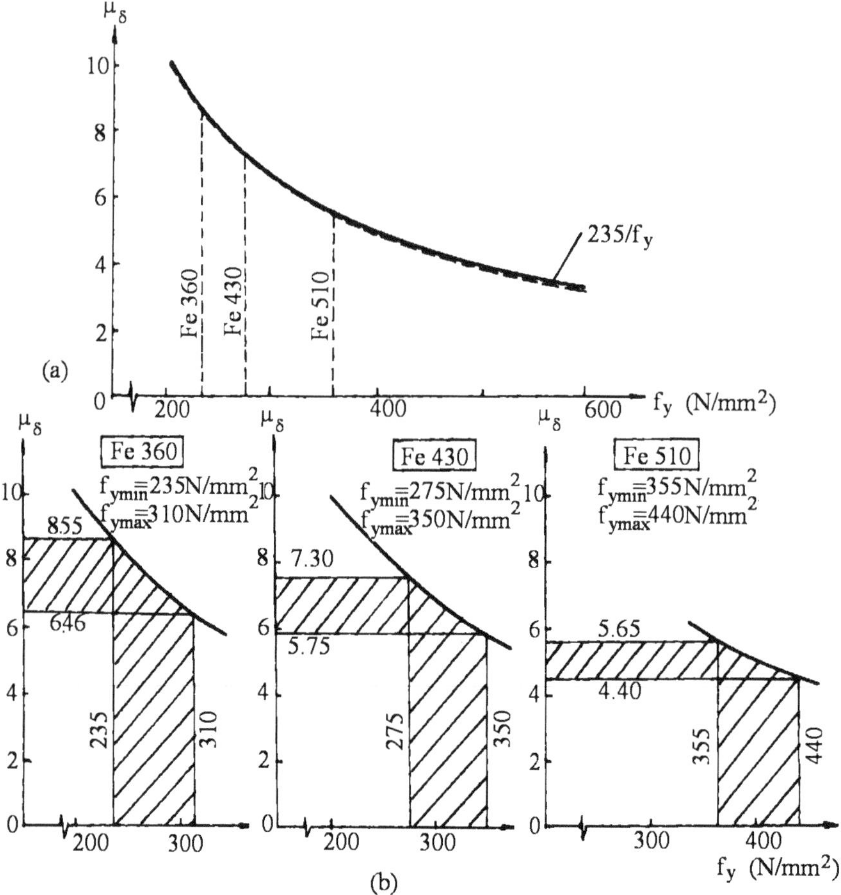

Figure 5.37: Influence of steel grade: (a) Decreasing of axial ductility with increasing of nominal steel grade; (b) Influence of variation of yield stress

stress. For the values given in the figure, the reduction is 25 percent for Fe360, 21 percent for Fe460 and 22 percent for Fe510 steel grades. Thus, the reduction in ductility due to the random variability of yield stress cannot be ignored in the design process.

(ii) *Yield ratio.* The axial ductility for Fe360 steel grade is characterised by a very good yield ratio. Due to the advances in fabrication technology, the recent trend is to use high strength steels, having yield stress of 700 MPa (see Section 4.2.3.). For these steel grades an increasing of yield ratios must be noted, if no special measures in fabrication are taken. The flange instability produces an important gradient moment in buckled zones (Fig. 5.38), and, therefore, a danger of premature fracture exists (see Section 4.5). Taking into account equations (4.59), (5.51) and (5.52), with $\Delta_i = 0$, the normalized axial shortening which produces the flange fracture, which

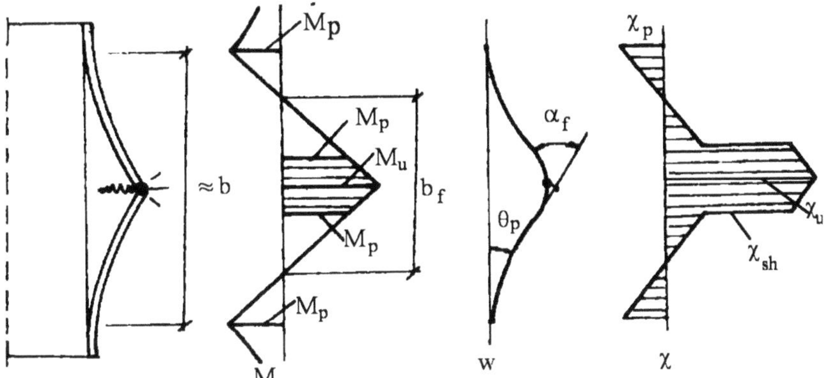

Figure 5.38: Gradient moment in buckled zones

represents the *axial fracture ductility* given by:

$$\mu_{\delta p} = \frac{\delta_{rf}}{\delta_p} = \frac{1}{2}\frac{E}{f_y}\zeta^2 \frac{(1/\rho_y - 1)^2}{(1+\rho_y)^3}\frac{c^3}{t_f^2}\varepsilon_u^2 \leq \frac{\varepsilon_u}{\varepsilon_y} \tag{5.64}$$

where ζ and ρ_y are defined by the equations (4.59a) and (4.3), respectively. Fig. 5.39 shows a very important effect of yield ratio on the flange fracture, the slope of curves having a dramatic reduction of fracture ductility with the increasing of the yield ratio. The fracture occurs earlier for thick flanges and for steel with high yield ratio. Fig. 5.40 presents the experimental results of a 300W steel grade stub, given by Sivakumaran and Yuan (1998) and the fracure strain obtained from relation (5.64), showing a good correspondence. But, honestly, one can mention that this relation must be considered only as an indicator for the yield ratio influence. Due to the behaviour at the ultimate load which tends to be very unstable at the formation of fracture lines, it is very difficult to control the deformations during this stage of experimental tests. All papers including tests until the ultimate behaviour in very advanced plastic fields do not mention the test stopping criterion. In many cases this was related to the test equipment capacity to follow the specimen deformation and the test was stopped before the stub fracture. Therefore, it was impossible to verify the relationship (5.64) with an eloquent number of experimental values found in the technical literature.

5.4.7. Influence of Loading Type

All the above influences have been established for static and monotonic loads. But during the earthquakes the loads are dynamic and cyclic and some important changing in stub behaviour occurs. The main influences are induced by the high strain-rate, the pulsatory loading and the combination of these two factors.

(i) *High strain-rate.* In Section 4.2.2 it has been shown that the field within 10^{-1} to 10^{1} corresponds to the strong near-field seismic actions. In

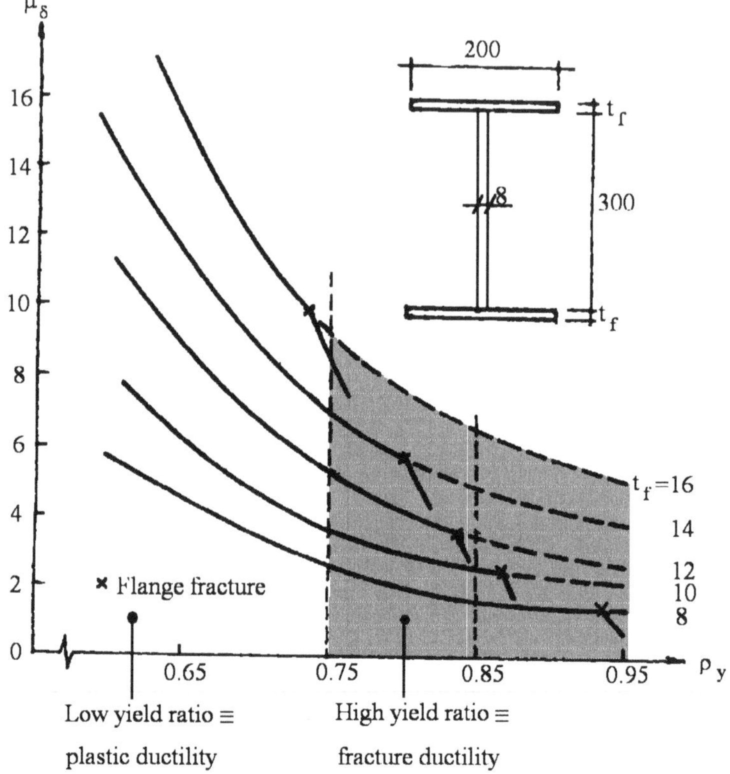

Figure 5.39: Influence of yield ratio on ductility type

this interval, using the equation (4.12) one can obtain that the increase in yield ratio is within 1.25 to 1.30. So, for the cross-section of Fig. 5.39 and for Fe360 steel grade with a variation of yield ratio within 0.6 to 0.73 due to the randomness of yield stresses, the increase of yield ratio is within 0.75 to 0.95 (Fig. 5.41). Therefore, it is very clear that the main effect of the strain-rate is to move the domain of plastic ductility into the one of fracture ductility, for which brittle fracture of buckled flanges can occur (Gioncu, 2000, Anastasiadis et al, 2000). The most affected are the thickest flanges. This observation is in concordance with the recorded failure during the last strong earthquakes, where a great amount of thick sections have been cracked.

(ii) *Pulsatory loads.* In the case of compression stubs subjected to repeated loads, only the pulsatory loads are interesting in practice. There are two pulsatory types, the first studies cumulative plastic behaviour (low cycles fatigue) and the second obtains information about the specimen collapse type. In the first case the displacements are constant for all the cycles, while in the second case the displacements increase after some constant cycles. In both cases the test is stopped when the cracks produce an important reduction in carrying capacity (for instance, half of monotonic loading).

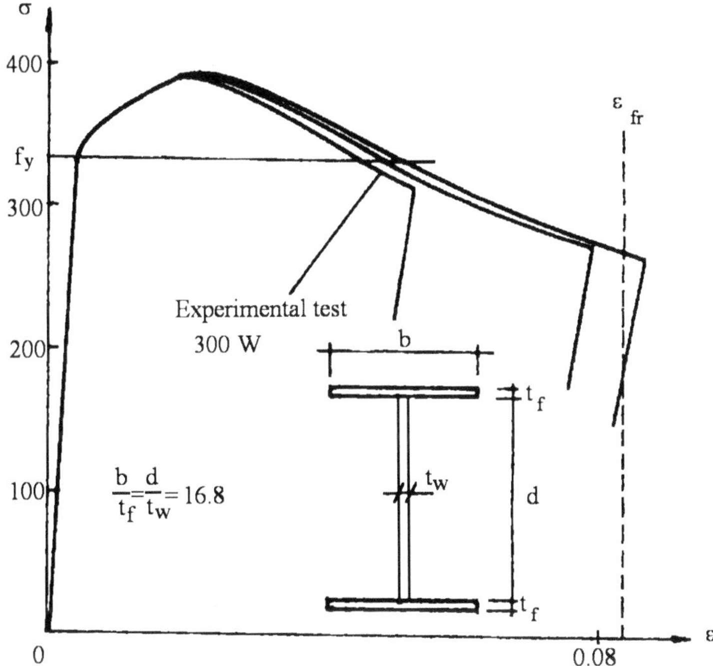

Figure 5.40: Theoretical and experimental comparison for flange fracture

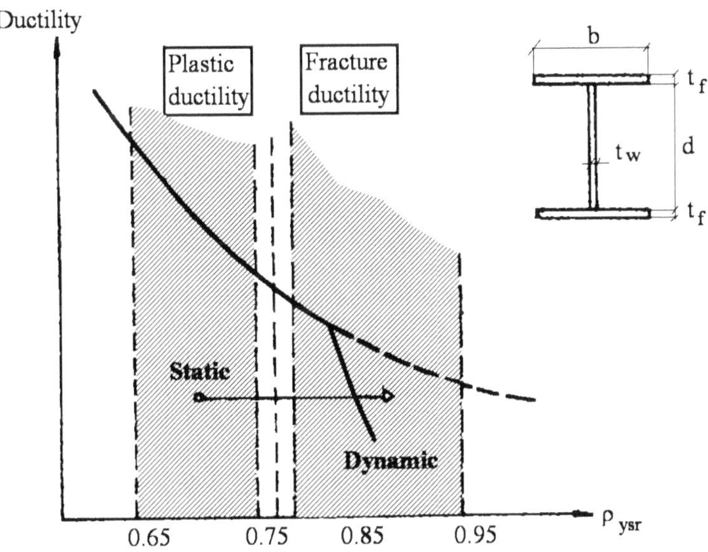

Figure 5.41: Influence of high strain-rate

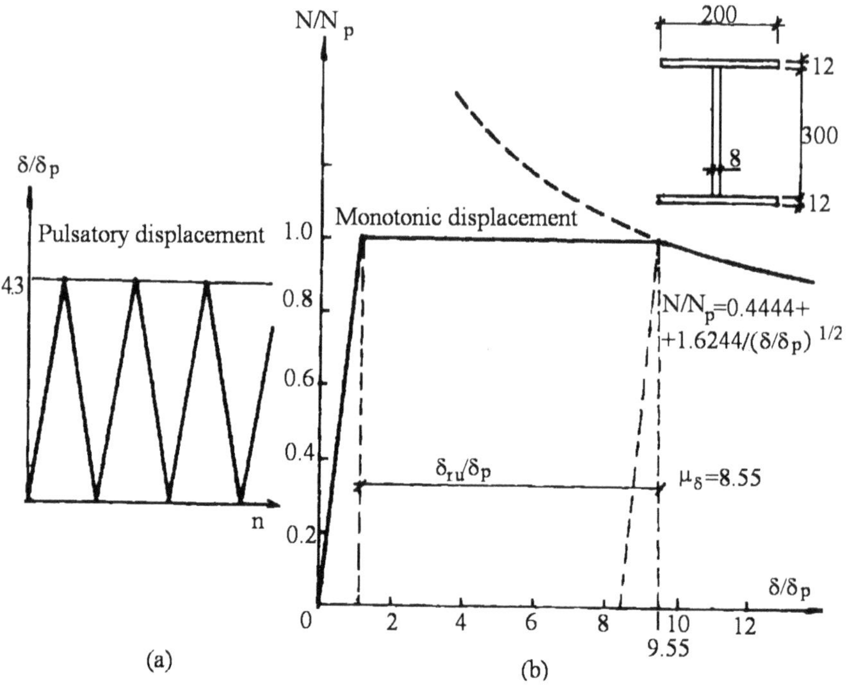

Figure 5.42: Pulsatory constant displacement: (a) Fixed imposed displacement; (b) Determination of monotonic ultimate displacement

-*Constant displacement* (Fig. 5.42a). The fixed imposed value for displacement is the main parameter of this test. Di Martino and Manfredi (1994) proposed using a constant displacement equal to 1/2 of the monotonic ultimate deformation. The effect of pulsatory constant displacement is studied for the stub profile presented in Fig. 5.42b for which the monotonic ultimate deformation is determined as $8.55\delta_p$. The imposed deformation for each pulse is equal to $4.28\delta_p$ (Fig. 5.42a). After each pulse a residual deformation $4.28\delta_p$ remains and plays for the next pulse the role of a geometrical imperfection. In this manner it is possible to determine the load-displacement curve (Fig. 5.43a) for the history function of total displacement (residual displacement + displacement for the corresponding pulse) and in Fig. 5.43b only the pulse deformation is considered. One can see that the reduction in loading after the third pulse is due to the accumulation of plastic deformation during the previous pulses. The plastic buckling of a stub occurs only at the third pulse. The increase in amplitude of a local buckle is presented in Fig. 5.44. One can see an important increase for the first pulses. After, the rate of growth decreases with the number of pulses. The same tendency is noted in the experimental tests of Krawinkler and Zohrer (1983). From Fig. 5.43b a stabilization tendency of the stub carrying capacity arises when the number of pulses increases. Indeed, if the residual deformation is introduced in the equation (5.62a) as a geometrical

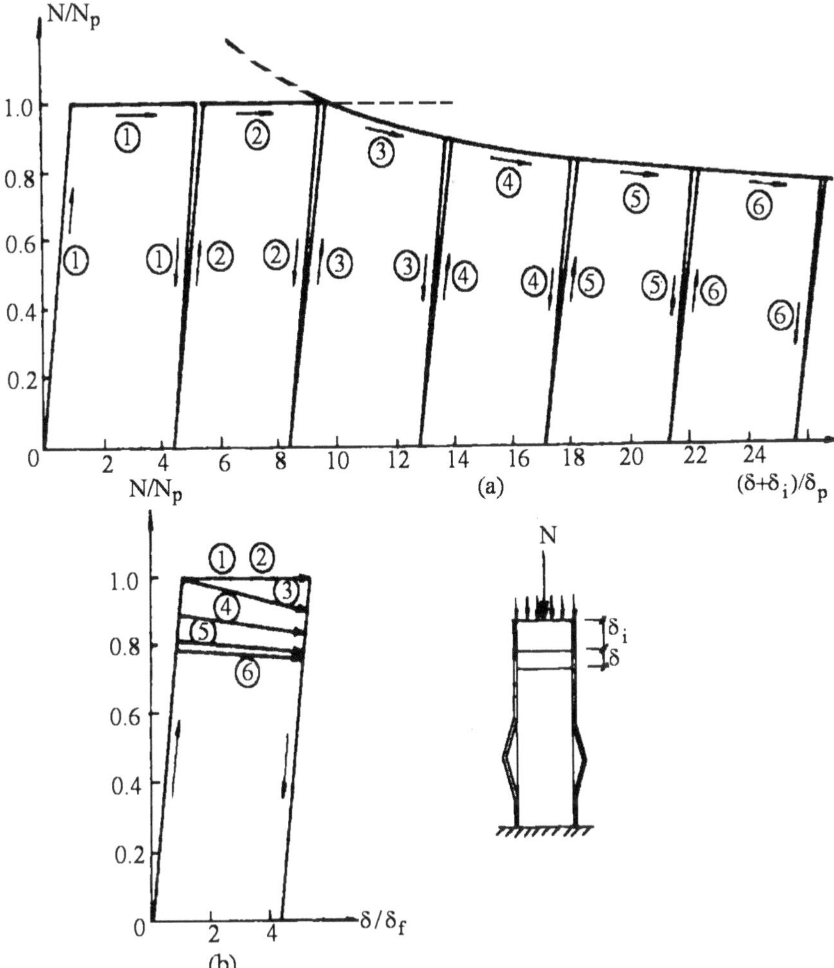

Figure 5.43: Axial force-axial shortening relationship for constant displacements: (a) Total displacements; (b) Pulse displacements

imperfection for the n pulse results:

$$\left(\frac{N}{N_p}\right)_{pulse} = \alpha_{1N} + \alpha_{2N} \frac{1}{\left[(n-1)\frac{1}{2}\frac{\delta_{ru}}{\delta_p} + \frac{\delta}{\delta_p}\right]^{1/2}} \tag{5.65}$$

One can see that only the last term corresponding to the plastic rotation of yield lines is eroded by the pulse action. So, the equation (5. 65) has an asymptotic value:

$$\left(\frac{N}{N_p}\right)_{pulse,minim} = \alpha_{1N} \tag{5.66}$$

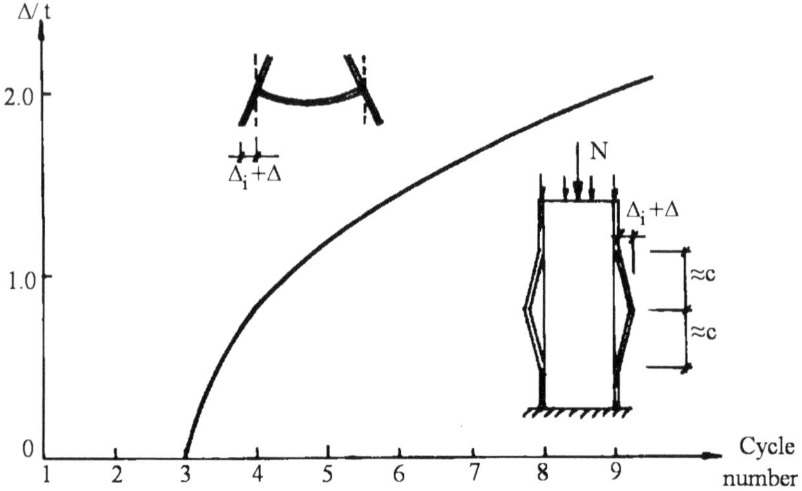

Figure 5.44: Increasing of amplitude of local buckled flange

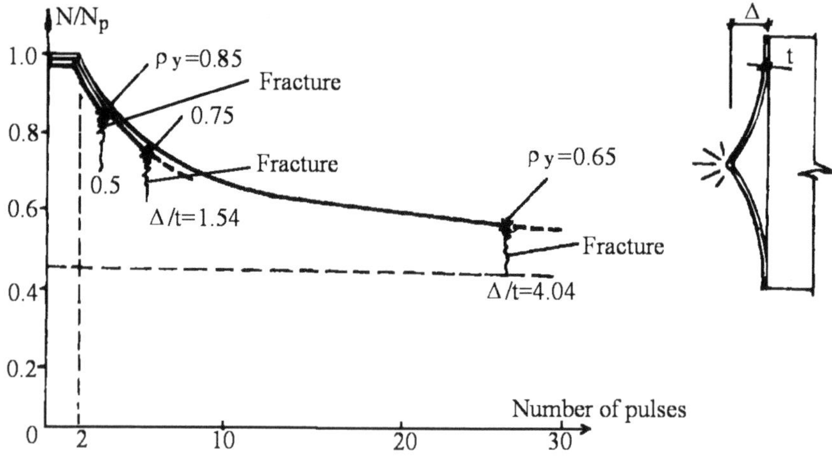

Figure 5.45: Influence of yield ratio on fracture pulse member

This value corresponds to the behaviour of a plastic zone, which is not eroded during the pulsatory loading. The equation (5.65) is plotted in Fig. 5.45. For the first two pulses no degradation is noticed, but for the next 5–6 pulses an important deterioration in stub carrying capacity can be observed, followed by a stabilization, where the rate of deterioration diminishes considerably.

The equation (5.65) gives the possibility to determine the number of cycles which produces the fracture of flange. This number results from

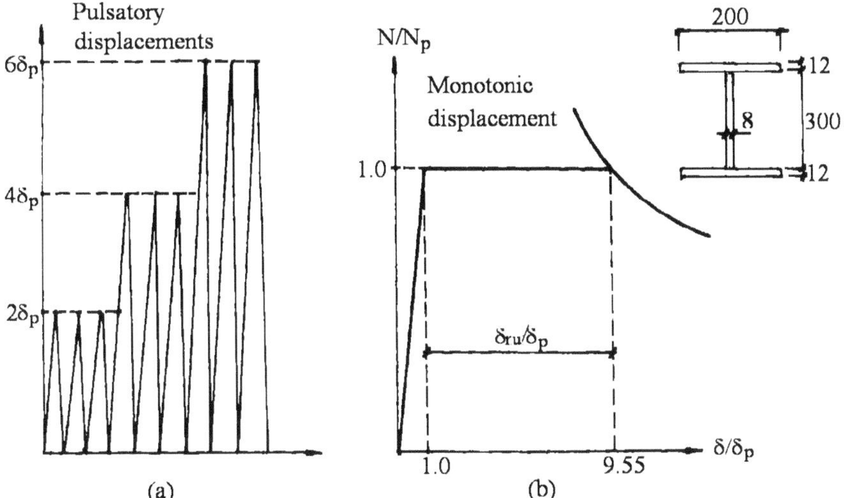

Figure 5.46: Pulsatory increasing displacement: (a) ECCS testing procedure; (b) Determination of monotonic ultimate displacement

equation (5.64):

$$n = \frac{\delta_{rf}}{\delta_p} \frac{1}{\frac{1}{2}\frac{\delta_{ru}}{\delta_p}} + 2 \tag{5.67}$$

Two pulses must be added, because the first two do not produce any flange buckling. For the stub presented in Fig. 5.42 it results:

-$\rho_y = 0.65$: $n = 104.3/4.28 + 2 \simeq 26.5$ pulses;
-$\rho_y = 0.75$: $n = 14.86/4.28 + 2 \simeq 5.5$ pulses:
-$\rho_y = 0.85$: $n = 1.66/4.28 + 2 \simeq 2.5$ pulses.

In Fig. 5.45 also the amplitudes of flange buckled shapes producing fracture are plotted. Examining these values one can note the drastic decreasing of fracture pulse number when the yield ratio increases, showing the great importance of this factor for pulsatory or cyclic loadings.

(ii)*Increasing displacement*, for which the ECCS testing procedure - (ECCS, 1986), presented in Fig. 5.46b, is used. In this methodology the yield displacement δ_p is the step of the increasing displacement. As this value is defined by the monotonic loading (Fig. 5.45b), three pulses in each interval $(2 + 2n)\delta_p$ with n = 0,1,2,3,.... are imposed (Fig. 5.46a). In the same manner as for constant displacement, the force-displacement curve is plotted in Fig. 5.47a, including the residual displacements, and in Fig. 5.47b the same curve without residual displacements. One can see that the stub buckles during the fourth pulse and a continuous decreasing of carrying capacity occurs after this buckling. In the same way it is possible to determine the number of pulses until the flange fracture. For ρ_y = 0.65, the fracture occurs during the 11th pulse, for ρ_y = 0.75, during the 5th pulse and for ρ_y = 0.85, during the 4th pulse.

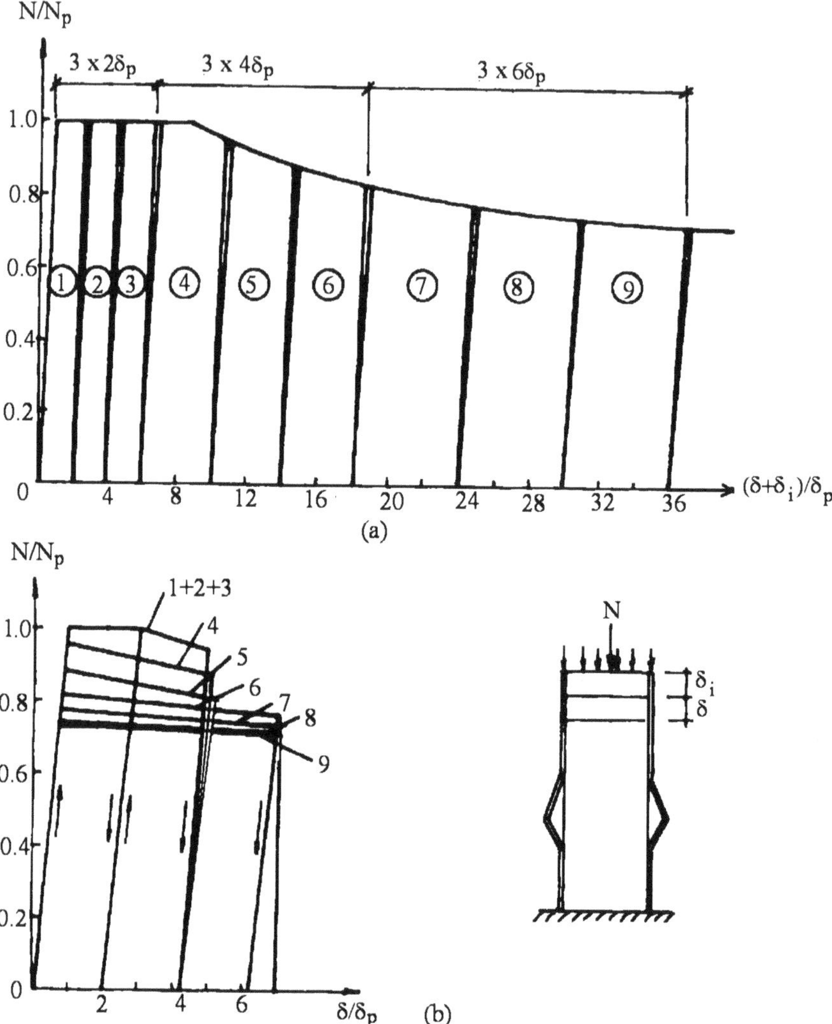

Figure 5.47: Axial force-shortening relationship for increasing displacements: (a) Total displacements; (b) Pulse displacements

(iii) *High velocity pulsatory loads.* The influence of pulsatory loads is, generally, examined by quasi-static loads, because dynamic analysis is a very complex problem. In many cases some members which behave very well under quasi-static loads in laboratory tests have shown a poor behaviour in the case of seismic actions. This is due to the fact that earthquakes act on the structures with high velocity cyclic loads, producing a premature local fracture of members. Therefore, the combined effects of pulsatory loads and strain-rates must be considered in design.

The strain-rate effect gives rise to a premature fracture of the buckled surface due to the increasing of yield stress and yield ratio. The pulsatory

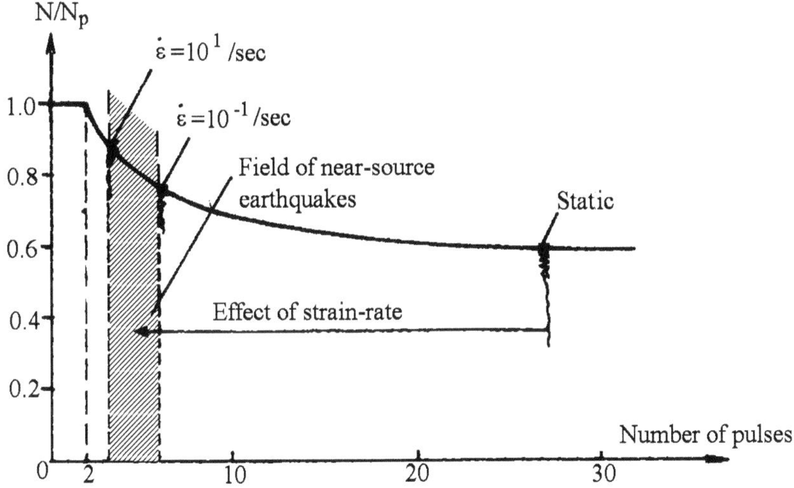

Figure 5.48: Influence of high strain-rate

loads also induce a fracture, by the accumulation of plastic displacement in buckled surfaces. Under the simultaneous action of strain-rate and pulsatory loads, a very early fracture is expected. In fact, for the stub presented in Fig. 5.44 the fracture under static pulsatory loads occurs for 27 cycles. For high velocity loading, characterized by high strain-rate, the number of cycles producing fracture decreases dramatically. Using the relation (4.12) between static and with strain-rate yield ratio, it results (for Fe360):

-static loading: $\rho_y = 0.65$; $\delta_{rf}/\delta_p = 104.3$; $n_f = 27$ pulses;
-strain-rate 10^{-1}/sec: $\rho_y = 0.808$; $\delta_{rf}/\delta_p = 13.7$; $n_f = 5.5$ pulses;
-strain-rate 10^{1}/sec: $\rho_y = 0.841$; $\delta_{rf}/\delta_p = 7.4$; $n_f = 4$ pulses.

Therefore, the strain-rate drastically reduces the number of pulses producing fracture (Fig.5.48), even in the case of a low strength steel like Fe 360. High strain-rate is generated in structures by near-field earthquakes, which conversely are characterized by a reduced number of high amplitude cycles.

5.4.8. Ductility Classes for I-Section Stubs

As it is shown in Section 5.2.4, the concept of section ductility classes depends only on the width-to-thickness ratios of flanges and webs. This procedure contains many shortcomings (Gioncu and Mazzolani, 1995), the most important being the neglecting of the interaction between flanges and webs during plastic buckling. The use of the local plastic mechanism allows us to consider this interaction, a relation between ductility and the main geometrical and mechanical characteristics of stub being possible to be established

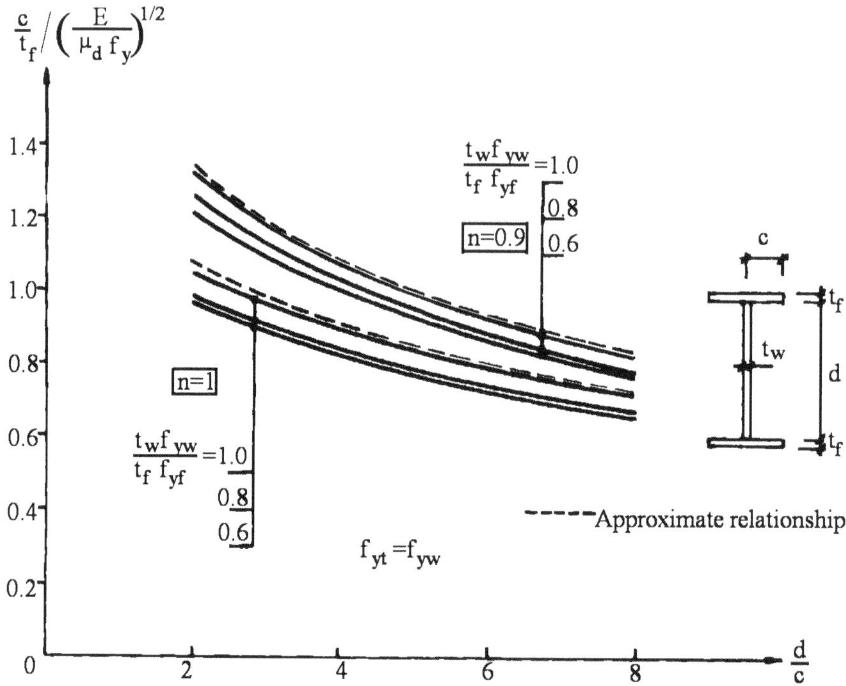

Figure 5.49: Flange-web slenderness interaction for two ductility criteria

using the equation (5.57) and (5.58):

$$\frac{c}{t_f} = \frac{3}{4}\left(\frac{E}{2f_{yf}}\right)^{1/2} \frac{1+\frac{1}{6}\left(1+\frac{d}{c}\right)\left(\frac{t_w}{t_f}\right)^2 \frac{f_{yw}}{f_{yf}}}{n\left(1+\frac{1}{4}\frac{d}{c}\frac{t_w}{t_f}\frac{f_{yw}}{f_{yf}}\right)-\frac{1}{2}\left(1+\frac{1}{2}\frac{t_w}{t_f}\frac{f_{yw}}{f_{yf}}\right)} \frac{1}{(\mu_\delta)^{\frac{1}{2}}} \tag{5.68a}$$

where n is defined by the ductility criterion ($n = 1.0$ or 0.9, see Section 5.4.5). The equation (5.68a) is presented in Fig. 5.49. One can see that the influence of ratio $t_w/t_f \times f_{yw}/f_{yf}$ is not very important, but the ratio d/c plays a very important role, showing the influence of flange restrain in webs. This restrain is more important for short webs than in the case of slender webs. From Fig. 5.49 one can note the importance of the ductility criterium, the choice of $n = 0.9$ leading to a benefit of 20–25 percent for flange slenderness with respect to $n = 1.0$.

Instead of the equation (5.68a), an approximate relationship can be used:

$$\frac{c}{t} = \frac{3}{4}\left(\frac{E}{2\mu_d f_{yf}}\right)^{1/2} \frac{1+\frac{1}{6}\left(1+\frac{d}{c}\right)}{n\left(1+\frac{1}{4}\frac{d}{c}\right)-\frac{3}{4}} \tag{5.68b}$$

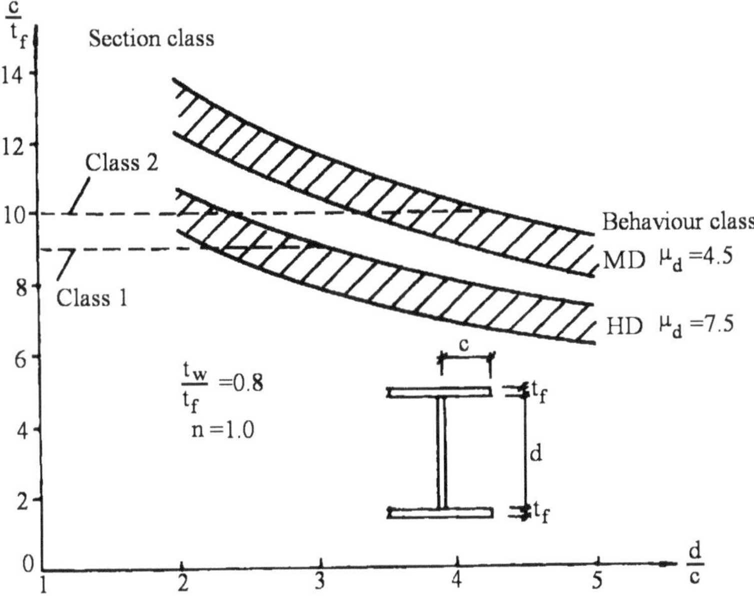

Figure 5.50: Correspondence between EC3 classes and behavioural classes

The equations (5.68a,b) could allow us to verify the provisions of Eurocode 8 concerning the section classes. But this is not possible directly, because there are no specifications for the values of ductility corresponding to the cross-section classes. Thus, in the case of calculation of this stub ductility it is not possible to identify the class in which the analysed cross-section can be framed.

In these conditions a new classification in behavioural classes is required. In the literature there are some proposed classification criteria (Mazzolani and Piluso, 1993, Vayas et al, 1995, Gioncu, 1995), among which no significant differences exist. So, for classification into different behaviour classes, Mazzolani and Piluso proposals were adopted:

-high ductility, HD: $\mu_d \geq 7.5$
-medium ductility, MD: $4.5 \leq \mu_d < 7.5$ (5.69)
-low ductility, LD: $\mu_d < 4.5$

Due to the fact that the calculated value for μ_d contains all principal factors influencing the ductility, this classification is more adequate to be used in practice than the one proposed in codes.

For the first and second classes Fig. 5.50 shows the values obtained from equation (5.68) considering the scattering of yield stress of about 30 percent and the limit values given in EC3. One can see a certain correspondence, the differences being introduced by the factor d/c neglected in code provisions.

The main conclusion of this analysis is that the behavioural classes can be used in design practice when the local ductility is determined by an adequate procedure.

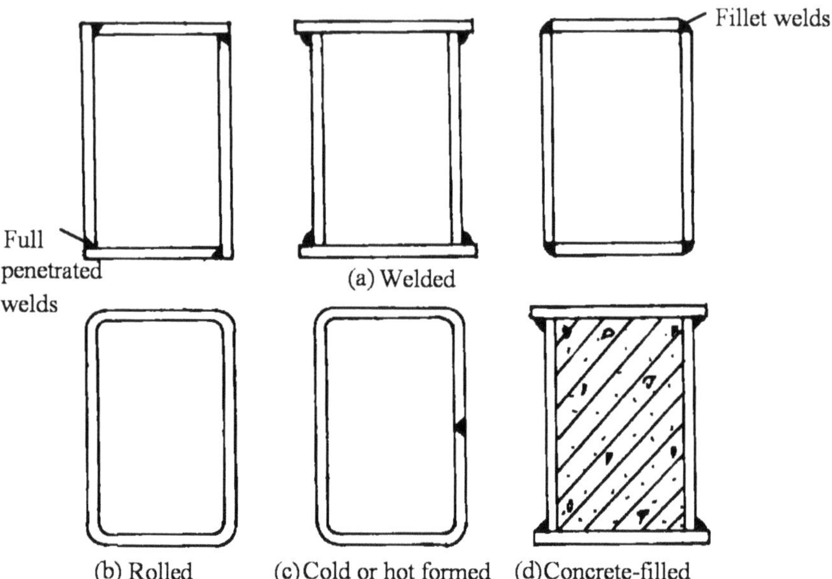

Figure 5.51: Fabrication processes of box-sections: (a) Welding: (b) Rolling; (c) Cold or hot-forming; (d) Concrete-filling

5.5 Ductility of Box-Section Stubs

5.5.1. Main Factors Influencing the Stub Ductility

Box-section columns are more and more widely used in high-rise buildings. This considerable interest is due to the fact that they are not subjected to lateral-torsional buckling which affects I-sections and other open sections bent around their strong axis (Dwyer and Galambos, 1965). It is very important to bear in mind that the use of box-sections requires a special knowledge, more than for the open profiles in conventional structural engineering (Rondal et al, 1992). The peculiarities of box-sections with respect to the open ones are presented in the following.

One must mention that the box-sections are also named in technical literature as rectangular and square hollow sections or tubes.

Generally, the main factors influencing the stub ductility are the same as for I-section stubs, but there are some differences concerning:

-*fabrication*, the section being produced with different processes: rolled, cold or hot press-formed, fillet or butt welded; in all cases, the section can be concrete-filled (Fig. 5.51);

-*geometrical imperfections* along the cross-section with special patterns (Fig. 5.52a);

-*residual stresses* due to pressing or welding (Fig. 5.52b). One must mention the bending residual stresses produced across wall thickness in cases of cold formed profiles;

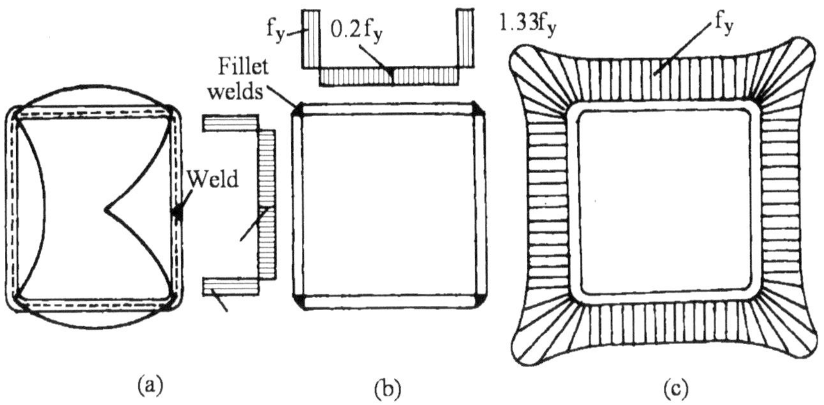

Figure 5.52: Profile characteristics: (a) Geometrical imperfections; (b) Residual stresses for welded sections; (c) Increasing of yield stresses due to cold-forming process (after Stranghoner, 1995)

-*variation of yield stress* along the cross-section and especially in corners due to the cold-forming process (Abdel-Rahman and Sivakumaran, 1997, Stranghoner, 1995) (Fig. 5.52c).

Recently, the cold-formed square box-section columns made of thick plates are commonly used for the design of super high-rise buildings in earthquake-prone Countries, especially in Japan (Akiyama and Yamada, 1997, Kato et al, 1997). These section types present some detrimental factors which increase the possibility of brittle fracture in tall building structures (Kuwamura and Akiyama, 1994).

5.5.2. Theoretical and Experimental Results

There are a lot of theoretical and experimental tests on box-sections, but the great majority refers to thin-walled sections. The results of roll-formed tubular sections are presented in Fig. 5.53 as a function of the normalized width-thickness ratio B (Aoki et al, 1996):

$$B = \frac{b}{t}\sqrt{\frac{\sigma_{0.2}}{E}\frac{3(1-\nu^2)}{\pi^2}} \tag{5.70}$$

where $\sigma_{0.2}$ is the 0.2 percent offset stress, which exceeds the yield stress due to the fabrication process. This increased yield stress is used also in Hancock's (1998) studies concerning the plastic buckling of box-sections. During the Aoki's experimental tests an increase of 80 percent is obtained for $\sigma_{0.2}$ in comparison with the nominal stress. From Fig. 5.53 one can see that the maximum stress decreases proportionally as the B coefficient increases. The following approximation is obtained for the tested data:

$$\frac{\sigma_{max}}{\sigma_{0.2}} = 1.63 - 0.74B \tag{5.71}$$

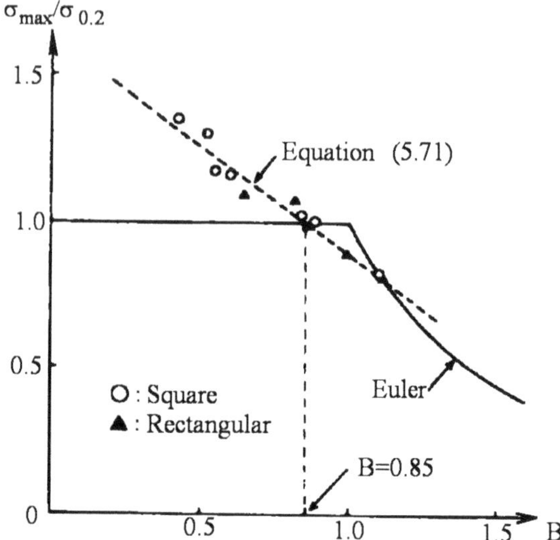

Figure 5.53: Experimental plastic buckling of box-section (after Aoki et al, 1996)

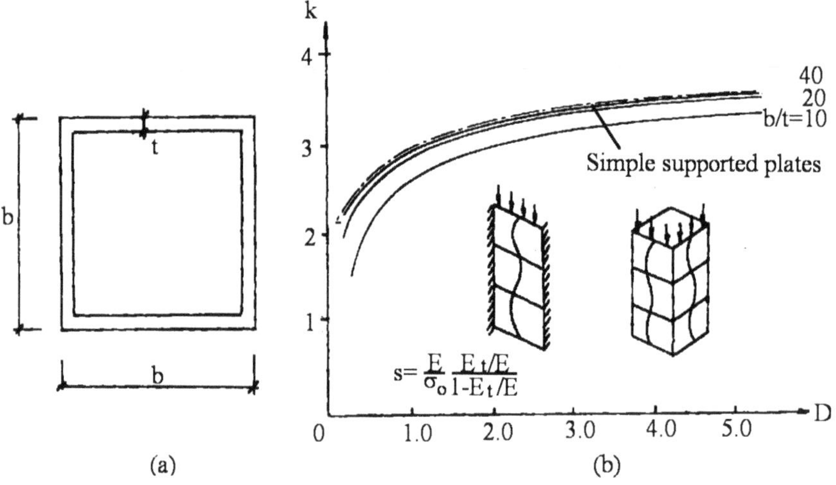

Figure 5.54: Theoretical plastic buckling of square box-section: (a) Geometrical dimensions; (b) Plate and box-section plastic buckling (after Li and Reid, 1992)

The plastic buckling analysis for compressed square box-sections is theoretically studied by Li and Reid (1992), in the case when the plastic deformations occur before buckling takes place. For the square box-section of Fig. 5.54a, the plastic buckling load is given by the equation:

$$N_b = k\frac{4\pi^2}{9}\frac{E_s t^3}{b} \tag{5.72}$$

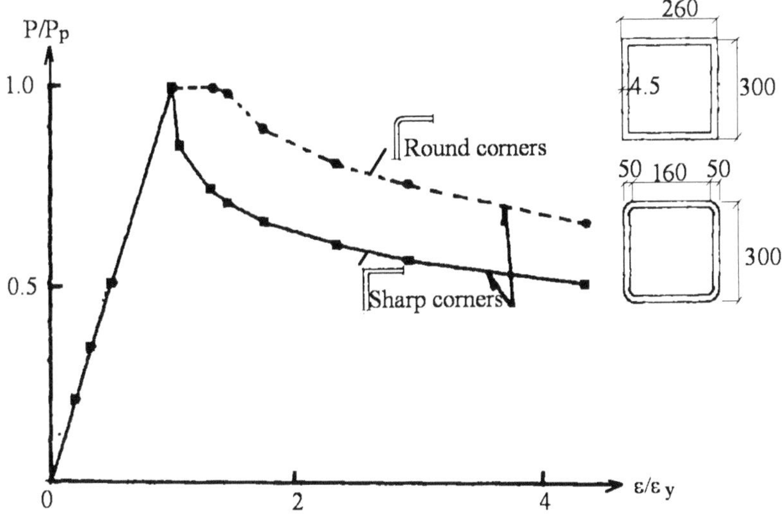

Figure 5.55: Influence of corner type (after Watanabe et al, 1990)

where N_b is the axial buckling load, k, a numerical coefficient, E_s, secant modulus. The coefficient k, varying from 3 to 4, is plotted in Fig. 5.54b as a function of wall slenderness and tangent modulus E_t. In the Figure the plastic component of plates is plotted also. One can see that no equivalence exists between box-sections and component plates when plastic buckling is concerned, the values for box-sections (full line) being lower than the plate ones (dashed and dotted line). The differences are due to the fact that the section corners deflect during the plastic buckling, while for plates the edges are considered undeformable. Therefore, the corner effect on plastic buckling of box-sections must be considered in analysis. Watanabe et al (1990) have studied this effect, considering two cases: sharp and round corners. Fig. 5.55 shows the relationship of axial compressive force versus axial shortening for both corner types. One can see that the sharp peak is not observed for round corners. On the other hand the sudden drop of bending moment is obtained for sharp corners.

Referring to the corner behaviour, another very important aspect must be noted. Fig. 5.56 shows the load-shortening curve for a box section (Rondal and Maquoi, 1985). One can see that, after the maximum load is reached, a local collapse mechanism is developed, with the corners working in a plastic range, but remaining straight. If the stub deformation continues in the advanced post-critical range, the corners buckle too, forming a spatial plastic mechanism. The different local inextensional buckling deformations, considering the conditions of kinematic continuity including the corner deformations, have been studied by Tarnai (1995) and presented in Fig. 5.57a. There are three possible buckling forms: with four bumps, with three bumps and one dimple and with two bumps and two dimples. The extensional deformations of a box-section has been studied by Wierzbicki

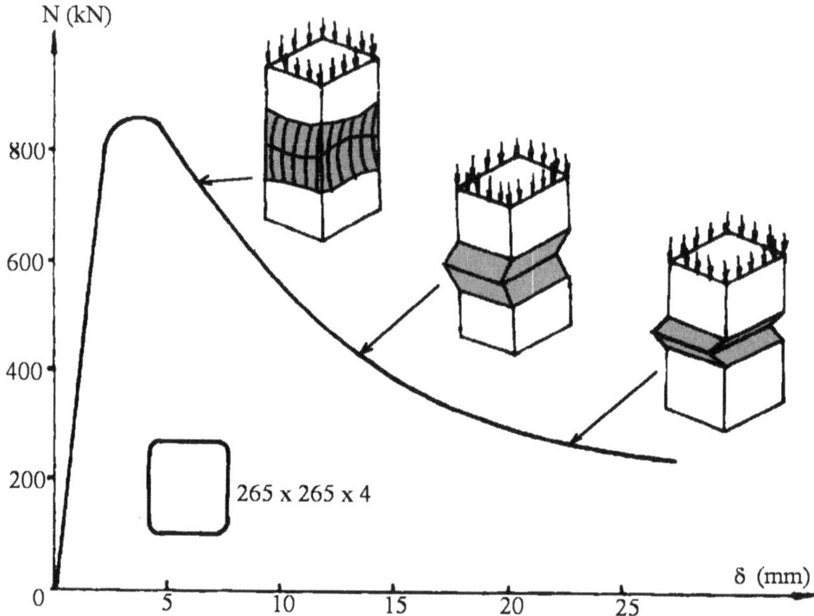

Figure 5.56: Development of local collapse mechanism (after Rondal and Maquoi, 1985)

(1983). Fig. 5.57b shows the difference between inextensional deformation which concentrates all plastic rotation along the yield line and the actual plastic deformation.

Very interesting experimental results have been obtained by Inoue (1994) (Fig. 5.58) for two series of stubs ($L/b = 1.5;\ 3.0$) with sharp corners. The specimens were welded with a partial penetration weld and annealed to eliminate the residual stresses. The main parameter of the study was the wall slenderness: $b/t = 10$ to 30. So, according to the EC 3 classification, all the cross-sections belong to the first class. The tests have been stopped when the fracture of welds occured. One can see that for slendernesses less than 20, the plastic buckling develops in the strain-hardening range, while for slendernesses higher than 22, the post-critical behaviour curves start from the yield plateau, showing a very strong deterioration of rigidity.

Experimental and theoretical tests on round corner stubs have been performed by Ono et al (1997) for cross-sections corresponding to first and second classes. The specimens were produced by cold roll forming and welding along the contact ends. Fig. 5.59a shows the stress-strain curves for stubs corresponding to class 1, but having different lengths. The strength of stubs rises to a maximum along the line corresponding to the tension test, but due to local buckling the strength strongly deteriorates, the longest specimen having the most important degradation. One can see that the ductility of this specimen is very reduced. Fig. 5.59b shows the influence of wall slenderness. The post-buckling deterioration is less marked for reduced slenderness

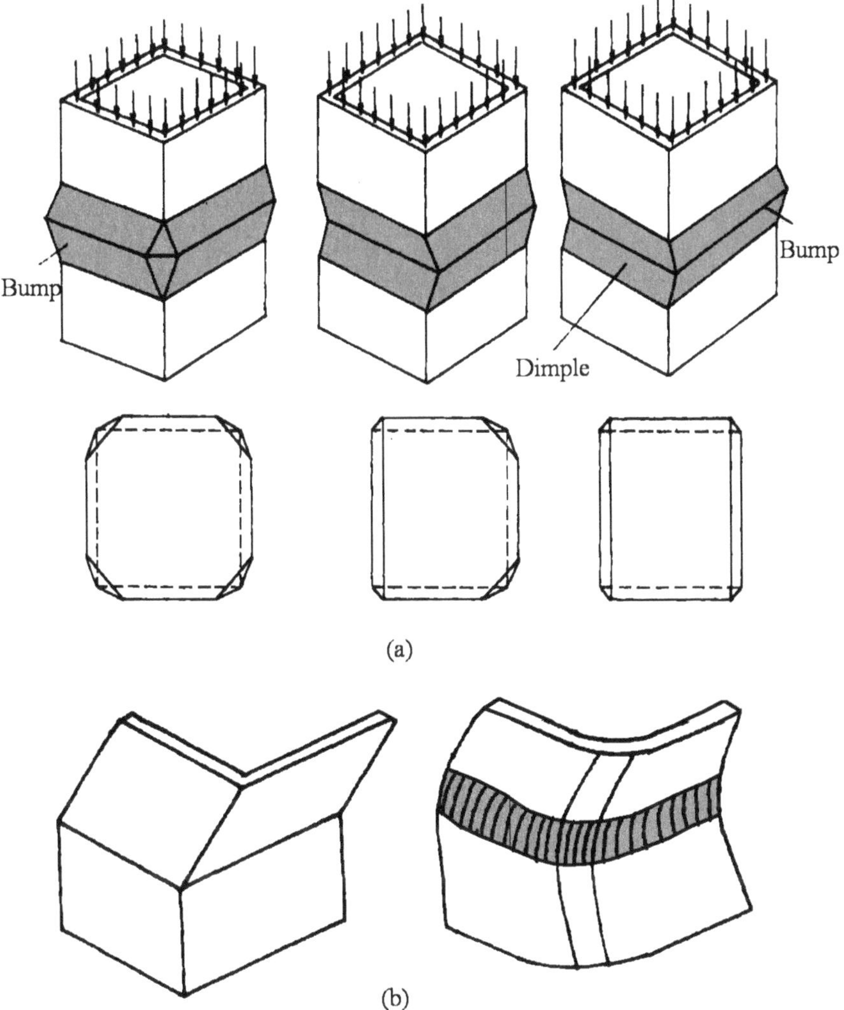

Figure 5.57: Local plastic mechanisms: (a) Mechanism types (after Tarnai, 1995); (b) Inextensional and extensional deformations along the yield line (after Wierzbicki, 1983)

and very strong for high slenderness. In the first case an important ductility is marked, while in the last a small ductility is obtained.

5.5.3. Rigid-Plastic Analysis

By examining the experimental collapse mode of box-section stubs it is clear that the ultimate condition is produced by the formation of a local plastic mechanism (Fig. 5.60). Therefore, the above developed methodology can be used for determining the stub ductility.

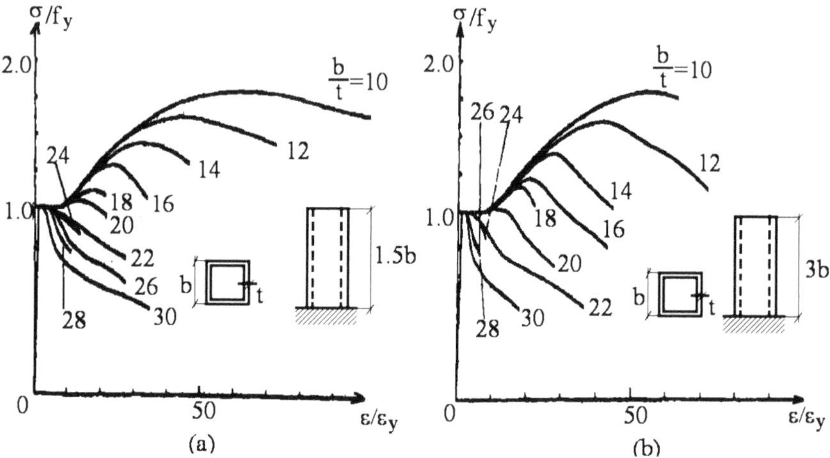

Figure 5.58: Stress-strain relationships for different wall slendernesses: (a) Short stubs; (b) Long stubs (Inoue, 1994)

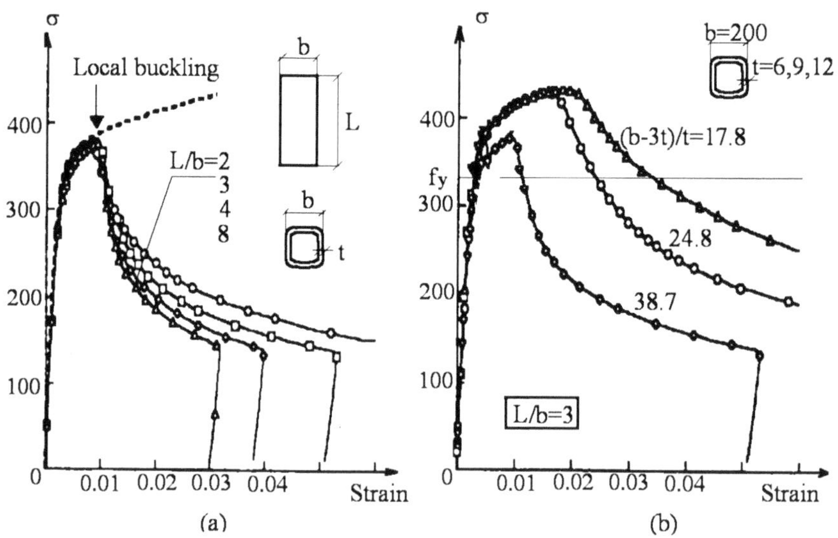

Figure 5.59: Stress-strain relationship: (a) Influence of stub length; (b) Influence of wall slenderness (after Ono et al, 1997)

A special attention has been payed by Ono et al (1997) to study the local buckling pattern and the length of buckling zone (Fig. 5.60a). One can see that the local buckling mode is a one-wave mode, composed by two half-waves. The theoretical length of a wave is L_b = 1.45b and a half-wave is L_p = 0.73b. The experimental obtained values have been 1.21b and 0.55b, respectively (Fig. 5.61). One can see that this length is independent of the wall slenderness. Even if there are two half-waves during

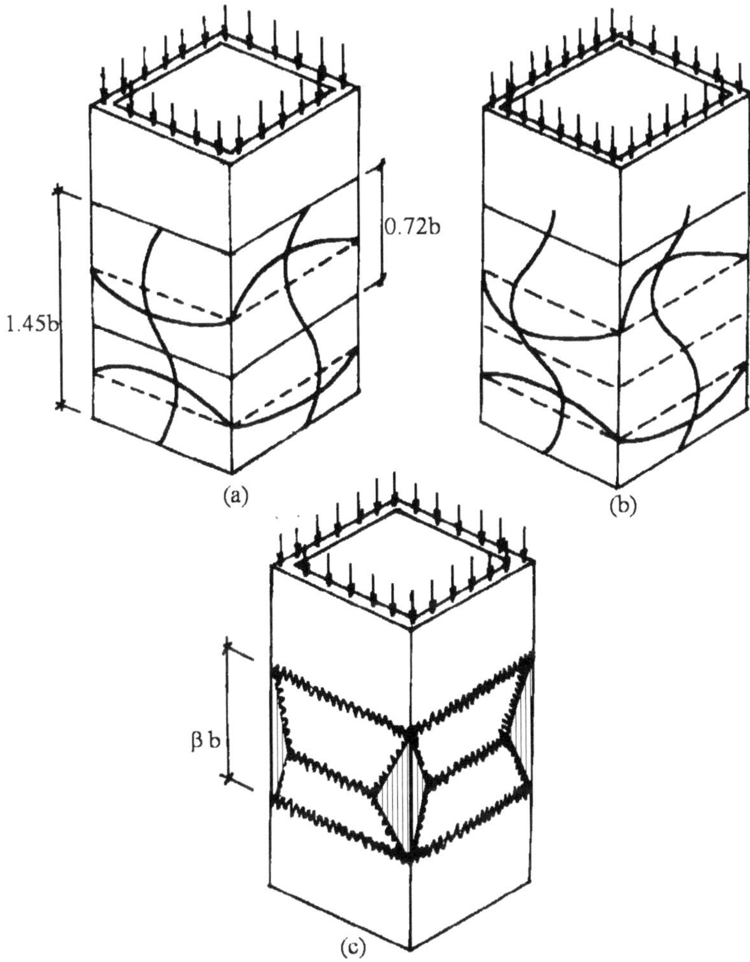

Figure 5.60: Local plastic mechanism: (a) First stage of plastic buckling; (b) Advanced stage of plastic buckling; (c) Theoretical local plastic mechanism

the first stage of buckling, one can note that only one is developed in the advanced stage of post-buckling behaviour (Fig. 5.60b). So, the local plastic mechanism proposed by Ono is composed by yield lines and plastic zones (stub corners). Fig. 5.62 shows an example of using this local plastic mechanism methodology. The intersection of the compression stress-strain curve with the rigid-plastic mechanism curve is supposed to be the local buckling point. The comparison with the experimental results shows a very good correspondence for the buckling load, but an important disagreement for the post-buckling curve. This difference is due to the fact that the corner lateral deformations, which produce a dramatic erosion of stub carrying capacity, have been neglected in the analysis (Sully and Hancock, 1995) (Fig. 5.62b).

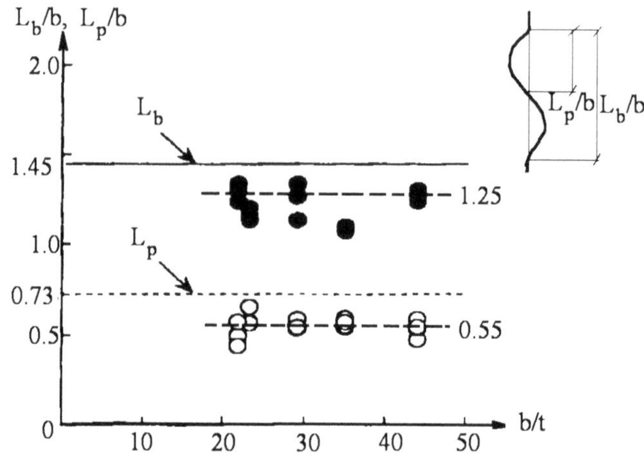

Figure 5.61: Length of local buckled shape (after Ono et al, 1997)

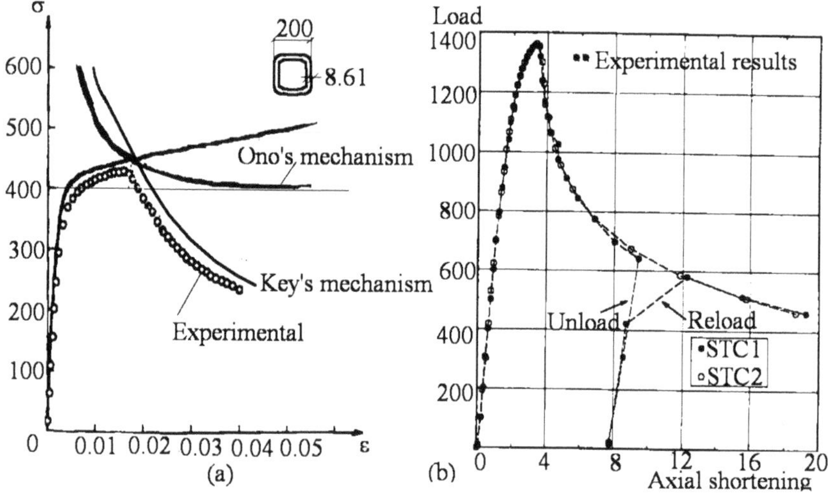

Figure 5.62: Experimental and plastic mechanism stress-strain curves: (a) Ono's (1997) and Key's (1988) results; (b) Sully and Hancock (1995) results

A spatial plastic mechanism which considers the corner deformations has been developed by Key et al (1988). This model fits the experimental results in the post-buckling range very well, contrary to Ono's model (see Fig. 5.62a). The Key's mechanism (Fig. 5.63a) considers that two opposite faces fold inwards and the remaining two fold outwards (Fig. 5.63b). Due to the compatibility of these deformations, the corner element is compressed and twisted (Fig. 5.63c). Therefore, the carrying capacity of stub can be determined as composed by three components:

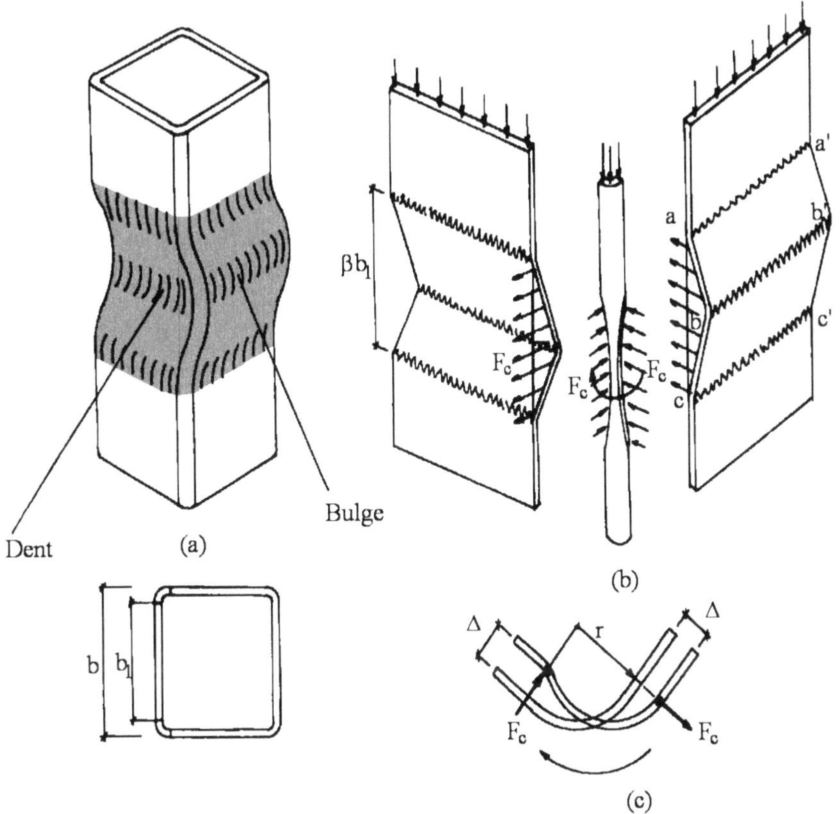

Figure 5.63: Spatial plastic mechanism: (a) Failure mode; (b) Face and corner plastic mechanisms; (c) Corner deformations

-the face local mechanism component is composed by three yield lines and rigid plates. The compressive force can be related to the out-of-plane displacement of mechanism :

$$N_f = A_f t b_1 \left\{ \left[\left(\frac{\Delta}{t} \right)^2 + 1 \right]^{1/2} - \frac{\Delta}{t} \right\} \sigma_f \qquad (5.73)$$

where σ_f is the face yield stress considering the strain-hardening range, A_f is the area of the flat part of the wall;

-the corner axial component. The deformations of corners being equal with the wall deformations. It results:

$$N_{ca} = A_c \sigma_c \qquad (5.74)$$

where A_c is the corner area and σ_c the corner yield stress, which is higher than the face yield stress due to the cold-forming;

-*the corner torsional component*; for the compatibility of deformations between the two folded plates. The corner elements are loaded by restraining forces F_c which generate a local torsional moment (Fig. 5.63c). The additional local component corresponding to this effect is given by:

$$N_{ct} = \frac{1}{4}\frac{t_f^2}{\Delta}\frac{p^2}{r}\sigma_c \tag{5.75}$$

where L_p is the length of plastic mechanism and r is the corner radius.

The total load is the sum of these three components:

$$N_{tot} = N_f + N_{ca} + N_{ct} \tag{5.76}$$

The total axial deformation is the sum of the axial deformation due to geometric changes in the spatial plastic mechanism and the elastic deformation due to the applied load:

$$\Delta L = \frac{2\Delta^2}{L_F} + \frac{N_{tot}L_p}{EA} \tag{5.77}$$

where L is the stub length and A is the total section area. In the normalized form it results:

$$\frac{N}{N_p} = \frac{A_f}{A}\left[\left(\frac{\Delta^2}{t^2}+1\right)^{1/2} - \frac{\Delta}{t}\right]\frac{\sigma_f}{f_y} + \frac{A_c}{A}\frac{\sigma_c}{f_y} + \frac{1}{4}\frac{L_p^2}{A_p}\frac{\sigma_c}{f_y}\frac{1}{\Delta/t} \tag{5.78a}$$

$$\frac{\delta}{\delta_p} = 1 + 2\frac{E}{f_y}\frac{1}{(L_p/t)^2}\frac{\Delta^2}{t^2} \tag{5.78b}$$

Using these equations the results for a 152x4.9 square box-section stub are presented in Fig. 5.64 (Key et al, 1988), in comparison with the experimental results. A good correspondence may be noticed. One can see that the component for the corners has the main contribution to the stub carrying capacity. But an important erosion of this component occurs when the deformation of buckled area increases. The results obtained using this spatial local mechanism for Ono's tested stub are presented in Fig. 5.62a. One can see an excellent correspondence between these results and the experimental ones.

The main conclusions of these analyses emphasize the important role of the corners on the behaviour and ductility of box-section stubs. If these corners remain straight during the plastic buckling, it is possible to use the local plastic mechanism presented in Fig. 5.60c. Generally this is the case of welded T corners (Fig. 5.65a). But for the round or sharp corners (Fig. 5.65b,c) are involved in the local plastic mechanism deformations and the spatial plastic mechanism, presented in Fig. 5.63a, must be used.

5.5.4. Pulsatory Axial Loading

An experimental study is performed by Kusama and Fukumoto (1983) on welded box-sections using a pulsatory axial loading. The result on a spec-

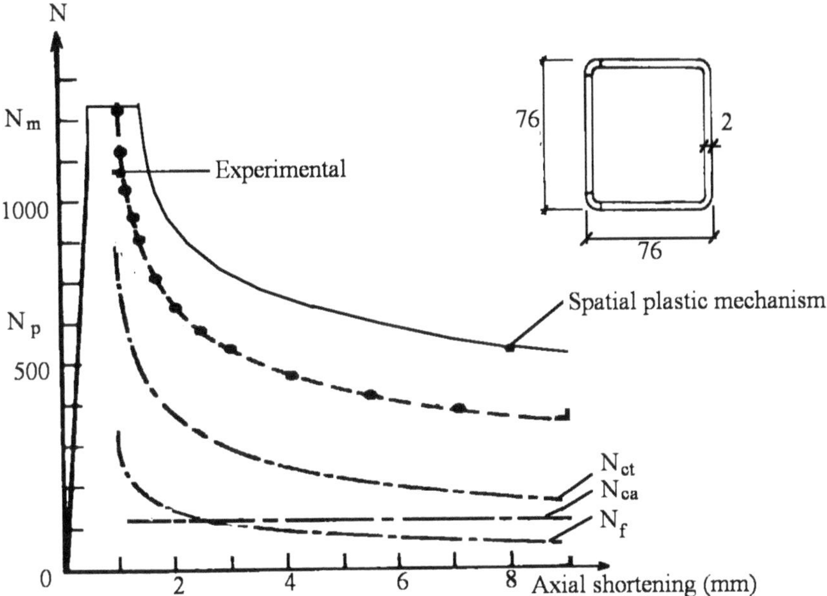

Figure 5.64: Experimental and spatial plastic mechanism stub column load shortening (after Key et al, 1988)

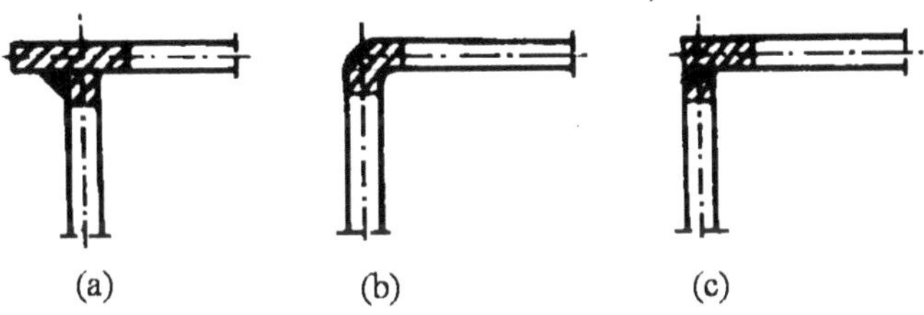

Figure 5.65: Corner types

imen is presented in Fig. 5.66, in comparison with the monotonic curve. The compressive peak stress versus axial deformation at the nth cycle is nearly equal to the one of stress and deformation at the initiation of unloading at the $(n-1)$th cycle. The envelope curve which passes across the points of peak stress at each cycle may coincide with the monotonic load-deformation curve. This result confirms the methodology elaborated in Section 5.4 for I-section stubs, which is based on the plastic deformation accumulation (compare Figs. 5.43 and 5.66).

5.5.5. Ductility Classes for Box-Section Stubs

Using the equations (5.78) the axial force-displacement curves are plotted in Fig. 5.67a. These diagrams allow to determine the axial ductility for box-

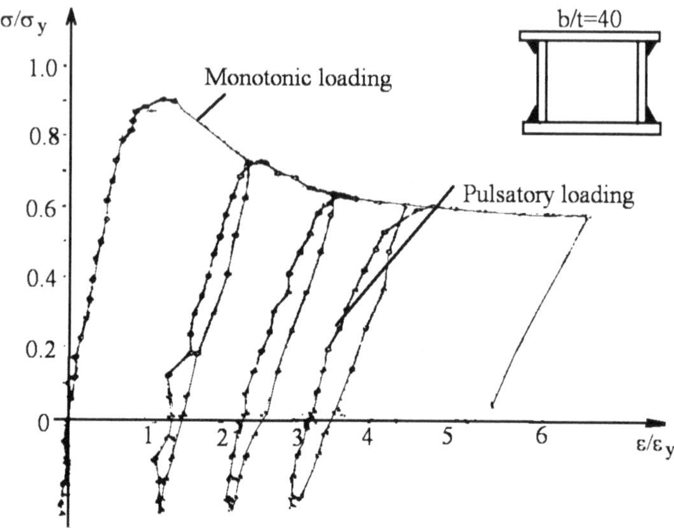

Figure 5.66: Pulsatory loading (after Kusama and Fukumoto, 1983)

sections and, to compare the results with the cross-section classes and the new proposal of behavioural classes (Fig. 5.67b). One can see an important variation of ductility in the function of wall slenderness. The range of high ductility corresponds very well with the EC 3 class one, but the range of class two is too restrictive.

5.5.6. Ductility of Concrete-Filled Box-Sections

The use of concrete-filled box-sections for high-rise buildings in seismic areas becomes very popular in the recent years as they provide several advantages with respect to steel or reinforced concrete: elimination of formwork and reinforcement, higher strength and stiffness by combining high strength steel box-sections with high strength concrete (Zhang and Shahrovz, 1999).

The main effect of concrete-filled box-sections is that the wall local buckling may be eliminated or at least delayed. The local buckling behaviour of concrete-filled box-sections is enhanced by the pressure of concrete infill when compared to hollow box-sections. The difference in local behaviour and local buckling is illustrated in Fig. 5.68 (Uy, 1998). While for hollow sections the walls are hinged along the corners and the local buckling may involve the corners too. For concrete-filled sections, the walls are fixed along the corners and they remain straight. This effect allows to use an increased wall slenderness and higher steel quality. In addition, the presence of the steel box-section produces an increase of concrete strength due to the confinement effect (Hajjar and Gourley, 1996, Roeder et al, 1998, Wakanishi et al, 1999, Tsuda et al, 1996).

Fig. 5.69 shows the experimental response of an axially loaded concrete-filled box-section stub (Varma et al, 1998). The yield strain of steel was

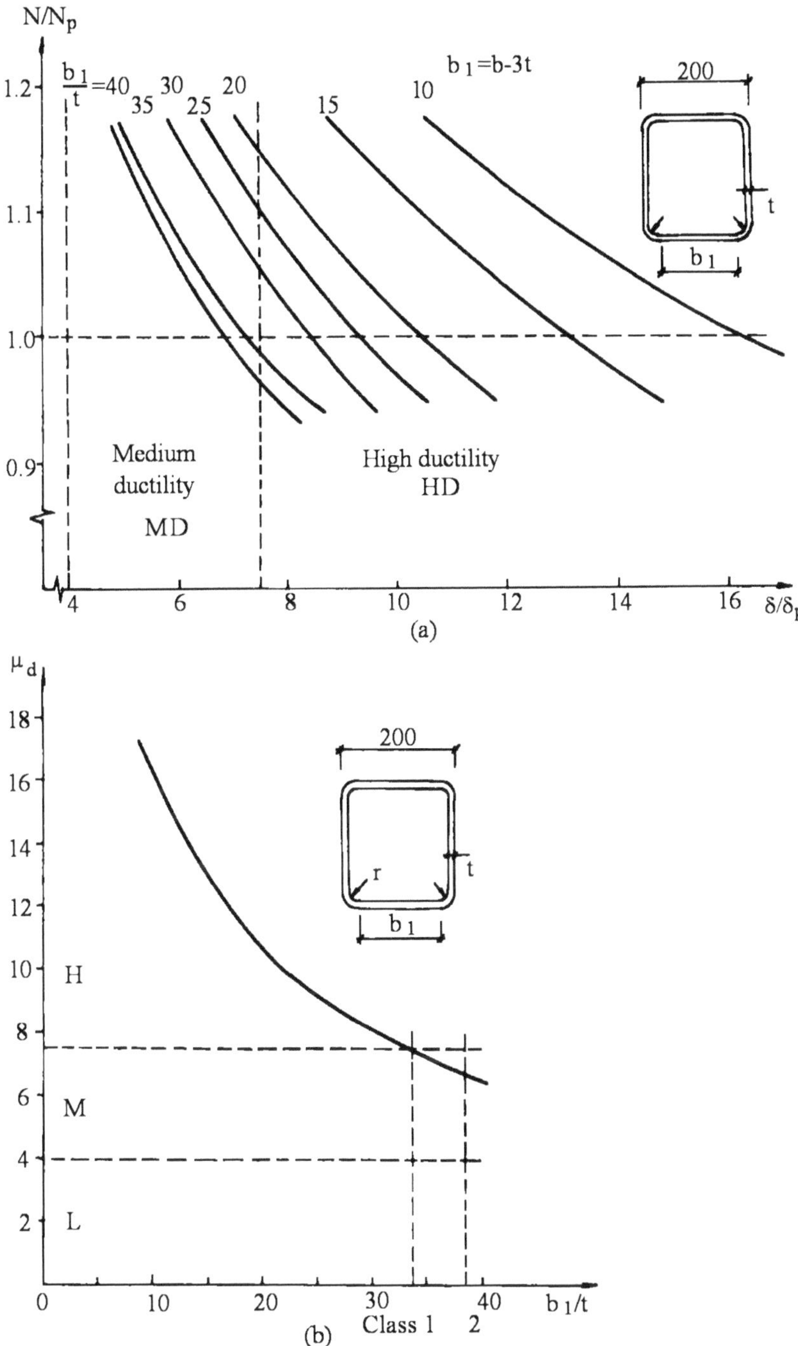

Figure 5.67: Ductility of box-section: (a) Axial force-axial shortening for different wall slendernesses; (b) EC3 and behavioural classes

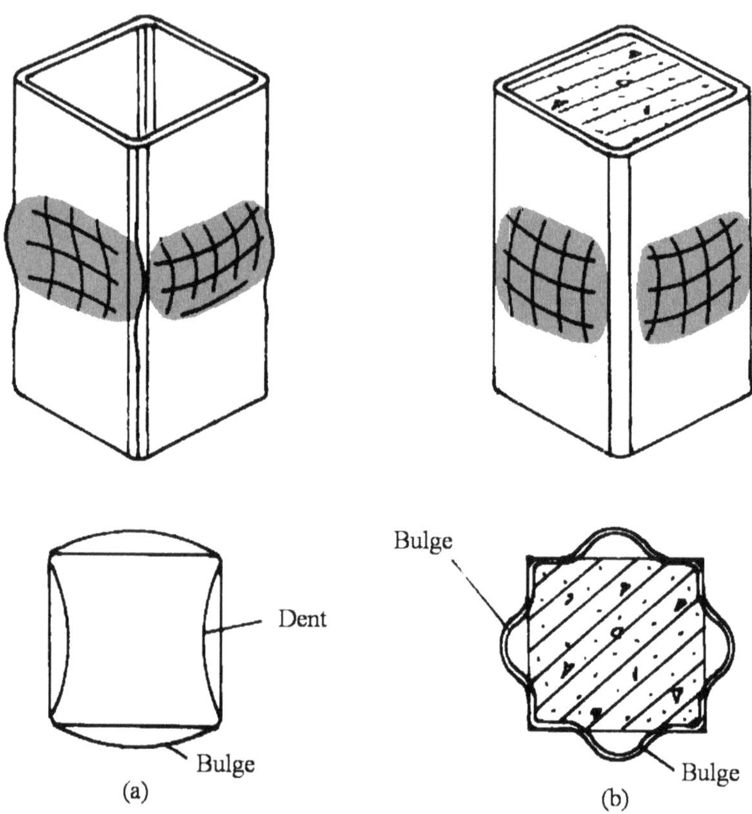

Figure 5.68: Concrete-filled effect in box-sections: (a) Buckling of hollow box-section; (b) Buckling of concrete-filled box-section

only 3 per cent higher than the concrete cylinder crushing strain. The stub response was linear up to the failure, where failure is defined as the unloading of the stub due to the local buckling of the steel tube and, crushing was heard at 90 per cent of the peak load. After the maximum load, the steel tube yielded in compression and the specimen started unloading due to concrete crushing and local buckling of tube walls. The stub unloaded slowly to 92 per cent of its peak capacity and then unloaded in a brittle and explosive manner to 70 per cent of its peak capacity. The test was continued in displacement control and the stub maintained its residual load capacity as it developed inelastic deformations. Similar results have been obtained by Tsuda et al (1996) and Uy (1998). Fig. 5.70 shows the load-axial shortening of three specimen types tested by Uy, in which a concrete-filled stub behaviour is compared with the hollow ones with the same geometry. One can see that the ultimate capacity for composite stubs is different from that of steel stubs due to additional contribution of concrete. But after the peak load an important loss in carrying capacity may be noted. In both Varma's and Uy's experimental results, this particular behaviour in the

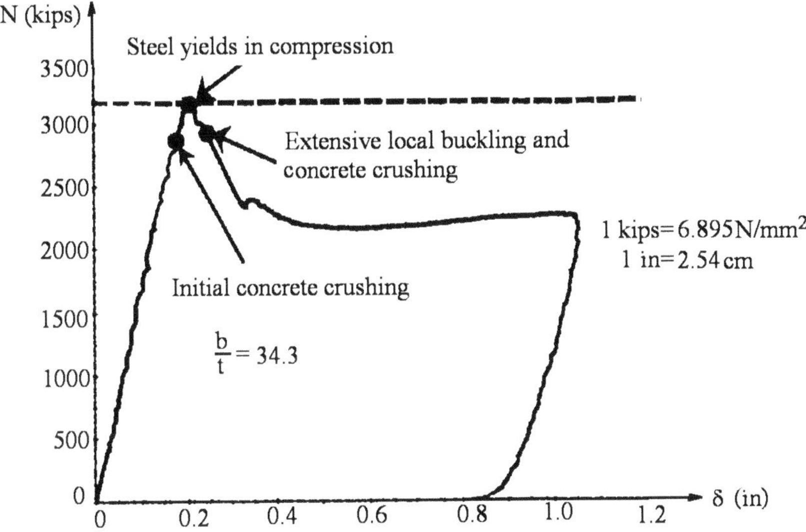

Figure 5.69: Experimental behaviour of concrete-filled box-section (after Varma et al, 1998)

unloading range is due to the sudden disappearing of concrete confinement after the local buckling of steel walls, when the interior concrete works as a plain concrete. In addition, during the experimental tests of Ge and Usami (1996) for welded box-sections with small wall slenderness, cracks in the welds were observed at 88 per cent of maximum load (Fig. 5.71), especially for cyclic loading. Accordingly, special attention is necessary to be payed on the corner welding. Concerning the ductility, the effect of concrete-filled box-section is to increase the plastic deformations due to the impeding of dimple buckling of walls and the deformations of corners. But the effect of concrete confinement must be used cautiously due to sudden looseness of this effect after the wall local buckling.

5.6 Conclusions

The main conclusions for the aspects presented in this Chapter may be summarized as follows:

-an important erosion of ductility occurs at the level of section, the main factor being the plastic local buckling of compression walls of the section;

-the plastic local buckling must be analyzed considering the flange to web interaction;

-theoretical and experimental studies on compression ductility must be performed on different section stubs;

-the experimental tests examination have shown that the behaviour of compression stubs can be studied by using a local plastic mechanism formed by yield lines and plastic zones;

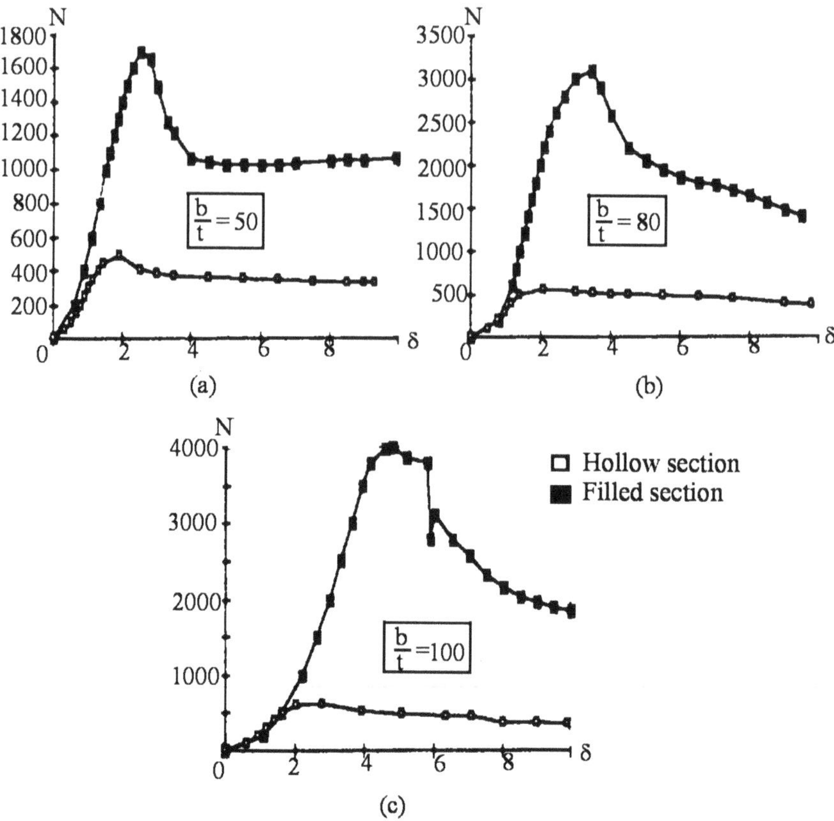

Figure 5.70: Axial-force shortening of concrete-filled box-sections for different wall slendernesses (after Uy, 1998)

-the main factors influencing the ductility of I-section stubs are the geometrical characteristics, geometrical imperfections, steel quality, etc.;

-the high velocity load produces an important strain-rate which introduces an early fracture of buckled flanges;

-the pulsatory loading in the plastic range can be analyzed as an accumulation of residual deformations after each step;

-in order to judge on the stub ductility, it seems more adequate to use the axial ductility classification into high, medium and low ductility classes instead of the EC 3 classification, in which the flange to web interaction is not considered;

-the study of box-section stub ductility has shown the importance to include the corner behaviour in the analysis of local ductility;

-the concrete-filled box-section stubs present an increased ductility due to the favorable effect of concrete presence on the local buckling of steel walls.

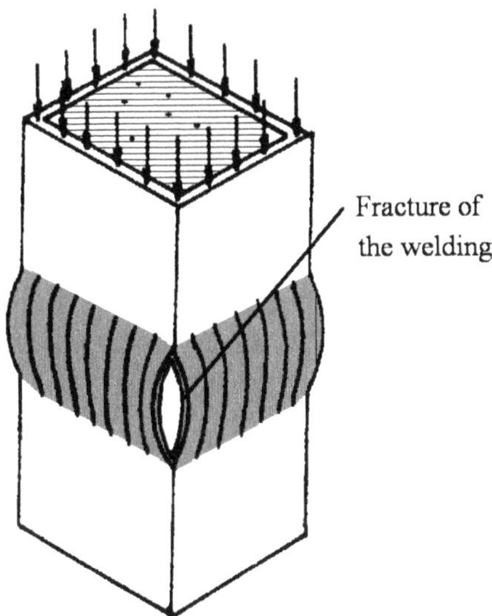

Figure 5.71: Fracture of welds

5.7 References

Abdel-Rahman, N., Sivakumaran, K.S. (1997): Material properties models for analysis of cold-formed steel members. Journal of Structural Engineering, Vol. 123, No. 9, 1135-1143

Aboutaha, R.S., Machado, P.I. (1999): Seismic resistance of steel-tubed high-strength reinforced-concrete columns. Journal of Structural Engineering, Vol. 125, No. 5, 485-494

Aiello, M.A., La Tegola, A., Ombres, L. (1994): Coupled instability of thin-walled members under combined bending moment, axial and shear force. Thin-Walled Structures, Vol. 19, No. 2-4, 285-297

Aiello, M.A., Ombres, L. (1995a): Influence of cyclic actions on the local ductility of steel members. In Behaviour of Steel Structures in Seismic Areas, STESSA 94, (eds. F.M. Mazzolani and V. Gioncu), Timisoara, 26 June-1 July 1994, E&FN Spon, London, 191-200

Aiello, M.A., Ombres, L. (1995b): Rotational capacity and local ductility of steel members. In 9th International Conference on Metal Structures, (ed. J. Murzewski), Krakow, 26-30 June 1995), Vol. 2, 271-282

AIJ, Architectural Institute of Japan (1990): Limit State Design of Steel Structures. Draft

Akiyama, H., Yamada, S. (1997): Seismic input and damage of steel moment frames. In Behaviour of Steel Structures in Seismic Areas, STESSA 97, (eds. F.M. Mazzolani and H. Akiyama), Kyoto, 3-8 August 1997, 10/17 Salerno, 789-800

Anastasiadis, A. (1999): Ductility problems of steel moment-resisting frames (in Romanian). Ph D Thesis, Polytechnica University Timisoara

Anastasiadis, A., Gioncu, V. (1999): Ductility of IPE and HEA beam-columns.

In Stability and Ductility of Steel Structures, SDSS 99, (eds. D.Dubina and M. Ivanyi), Timisoara, 9-11 September 1999, Elsevier, Amsterdam, 249-257

Anastasiadis, A., Gioncu, V., Mazzolani, F.M. (2000): New trends in the evaluation of available ductility of steel members. In Behaviour of Steel Structures in Seismic Areas, STESSA 2000, (eds. F.M. Mazzolani and R. Tremblay), Montreal, 21-24 August 2000, Balkema, Rotterdam, 3-10

Aoki, T., Migita, Y., Fukumoto, Y. (1990): Local buckling strength of closed polygon folded section columns. In Stability of Steel Structures, (ed. M.Ivanyi), Budapest, 25-27 April 1990, Akademiai Kiado, Budapest, Vol. 2, 1039-1046

Aoki, T., Hasegawa, K. (1996): Local and overall buckling strength of centrally loaded cold-formed square and rectangular section steel columns. In Coupled Instability in Metal Structures. CIMS 96, (ed. J. Rondal), Liege, 5-7 September 1996, Imperial College Press, London, 181-188

Beg, D., Hladnik, L. (1996): Slenderness limit of class 3 I cross-section made of high strength steel. Journal of Constructional Steel Research, Vol. 38, No. 3, 201-217

Bild, S., Kulak, G.L. (1991): Local buckling rules for structural steel members. Journal of Constructional Steel Research, Vol. 20, 1-52

Bruneau, M., Uang, C.M., Whittaker, A. (1998): Ductility Design of Steel Structures. McGraw-Hill, New York

Commission of EC (1992): Background Documents for Eurocode 8. Part 1, Vol. 2, Design rules

De Martino, A., Manfredi, G. (1994): Experimental testing procedures for the analysis of the cyclic behaviour of structural elements: Activity of RILEM Technical Committee 134 MJP. In Danneggiamento Ciclico e Prove Pseudodinamiche, (ed. E. Cosenza), Napoli, 2-3 July 1994, 1-20

Dwyer, T.J., Galambos, T.V. (1965): Plastic behaviour of tubular beam columns. Journal of Structural Division, Vol. 91, ST4, 153-168

ECCS, TWG1.3 (1986): Recommended testing procedure for assessing the behaviour of structural steel elements under cyclic loads. Doc. N.45/86

Eurocode 3, EC3 (1992): Design of Steel Structures, Part 1.1. General Rules and Rules for Buildings, ENV 1993 1-1, February 1992

Eurocode 8, EC8 (1994): Design Provisions for Earthquake Resistance of Structures. Part 1.3. Specific Rules for Various Material and Elements. ENV 1998, 1-3 November

Ge, H., Usami, T. (1996): Cyclic tests on concrete-filled steel box sections. Journal of Structural Engineering, Vol. 122, No. 10, 1169-1177

Gioncu, V. (1995): Some consideration on structural member classes. In 9th International Conference on Metal Structures, (ed. J. Murzewski), Krakow, 26-30June 1995, Vol. 2, 311-320

Gioncu, V. (2000): Influence of strain-rate on the behaviour of steel members. In Behaviour of Steel Structures in Seismic Areas, STESSA 2000, (eds. F.M. Mazzolani and R. Tremblay), Montreal, 21-24 August 2000, Balkema, Rotterdam, 19-26

Gioncu, V., Mazzolani, F.M. (1995): Alternative methods for assessing local ductility. In Behaviour of Steel Structures in Seismic Areas. STESSA 94, (eds. F.M. Mazzolani and V. Gioncu), Timisoara, 26 June-1 July 1994, E&FN Spon, London, 182-190

Gioncu, V., Petcu, D.(1997): Available rotation capacity of wide-flange beams and beam-columns. Part 1. Theoretical approaches. Part 2. Experimental and numerical tests. Journal of Constructional Steel Research, Vol. 43, No. 1-3, 161-217, 219-244

Hajjar, J.F., Gourley, B.C. (1996): Representation on concrete-filled steel tube cross-section strength. Journal of Structural Engineering, Vol. 122, No. 11, 1327-1223

Han, L.H., Zhao, X.L., Tao, Z. (2001): Tests and mechanics model for concrete filled SHS stub columns, columns and beam-columns. Steel and Composite Structures, Vol. 1, No. 1, 51-74

Hancock, G.J. (1998): Finite strip buckling and nonlinear analyses and distorsional buckling analysis of thin-walled structural members. In Coupled Instabilities in Metal Structures, (ed. J. Rondal), CISM Course, Udine, Springer Verlag, Wien, 225-289

Inoue, T.(1994): Analysis of plastic buckling of rectangular steel plates supported along their four edges. International Journal of Solids Structures, Vol. 31, No. 2, 219-230

Ivanyi, M. (1979): Yield mechanism curves for local buckling of axially compressed members. Periodica Polytechnica, Civil Engineering, Vol. 23, No. 3-4, 203-216

Ivanyi, M. (1985): The model for interactive plastic hinge. Periodica Polytechnica, Civil Engineering, Vol 29, No 3-4, 121-146

Iwata, M., Hayashi, K., Ono, T., Yoshida, F. (1997): Evaluation of deformation capacity of various steels. In Behaviour of Steel Structures in Seismic Areas, STESSA 97, (eds. F.M. Mazzolani and H. Akiyama), Kyoto, 3-8 August 1997, 10/17 Salerno, 168-175

Kato, B. (1988): Rotation capacity of steel members subjected to local buckling. In 9the World Conference on Earthquake Engineering, Tokyo-Kyoto, 2-8 August 1988, Vol. IV, 115-120

Kato, B.(1989): Rotation capacity oh H-section members as determined by local buckling. Journal of Constructional Steel Research, Vol. 13, 95-109

Kato, B. (1990): Deformation capacity of steel structures. Journal of Constructional Steel Research, Vol. 17, 33-94

Kato, B. (1995): Development and design of seismic-resistant steel structures in Japan. In Behaviour of Steel Structures in seismic Areas, STESSA 94, (eds. F.M. Mazzolani and V. Gioncu), Timisoara, 26 June-1 July 1994, E&FN Spon, London, 28-42

Kato, B., Morita, K., Maruoka, Y., Sugimoto, H., Teraoka, M.(1997): Seismic damage of steel beam-to-column rigid connections in the 1995 Hyogoken-Nanbu earthquake. Fabrication. In Behaviour of Steel Structures in Seismic Areas, STESSA 97, (eds. F.M. Mazzolani and H. Akiyama), Kyoto, 3-8 August 1997, 10/17 Salerno, 811-820

Key, P.W., Hasan, S.W., Hancock, G. (1988): Column behaviour of cold-formed hollow sections. Journal of Structural Engineering, Vol. 114, No. 2, 390-407

Krawinkler, H., Zohrei, M. (1983): Cumulative damage in steel structures subjected to earthquake ground motions. Computers and Structures, Vol. 16, No. 1-4, 531-541

Kuranishi, S. (1995): The shortening characteristics of column of strain hardening material. In Stability of Steel Structures, (ed. M. Ivanyi), 21-23 September 1995, Budapest, Akademiai Kiado, Budapest, 79-86

Kusama, H., Fukumoto, Y. (1983): Cyclic behaviour of thin-walled box stub-columns and beams. In Stability of Metal Structures, Paris, 16-17 November 1983, 211-218

Kuwamura, H., Akiyama, H. (1994): Brittle fracture under repeated high stresses. Journal of Constructional Steel Research, Vol. 29, No. 1-3, 5-19

Lay, M.G., Galambos, T.V. (1967): Inelastic beams under moment gradient. Journal of the Structural Division, Vol. 93, ST 1, 381-399

Li, S., Reid, S.R. (1992): The plastic buckling of axially compressed square tubes. Journal of Applied Mechanics, Vol. 59, June, 276-282

Mazzolani, F.M. (1992): Mechanical imperfections: Tests and models. In Testing of Metals for Structures, (ed. F.M. Mazzolani), Napoli, 29-31 May 1990, E&FN Spon, London, 364-378

Mazzolani, F.M., Piluso, V.(1992): Evaluation of the rotation capacity of steel beams and beam-columns. In 1st COST C1 Workshop, Strasburg

Mazzolani, F.M., Piluso, V. (1993): Member behavioural classes of steel beams and beam-columns. In XIV Congresso CTA, Viareggio, 24-27 Octobre 1993, Ricerca Teorica e Sperimentale, 405-416

Mazzolani, F.M., Piluso, V. (1996): Theory and Design of Seismic Resistant Steel Frames. E&FN Spon, London

Moldovan, M. (1990): Tolerances de fabrication. Comparison de quelques normes et recommandations. Construction Metallique, No. 2, 50-53

Moy, S.S.J. (1985): Plastic Methods for Steel and Concrete Structures. McMillan, London

Nakanishi, K., Kitada, T., Nakai, H. (1999): Experimental study on ultimate strength and ductility of concrete filled steel columns under strong earthquake. Journal of Constructional Steel Research, Vol. 51, No. 3, 297-319

Nakashima, M. (1992): Variation and prediction of deformation capacity of steel beam-columns. In 10th World Conference on Earthquake Engineering, Madrid, 19-24 July 1992, Blakema, Rotterdam, 4501-4507

Nethercot, D.A. (1992): Stub column test procedure. In Testing of Metals for Structures (ed. F.M. Mazzolani), Napoli, 29-31 May 1990, E&FN Spon, London, 379-380

Ono, T., Yoshida, F., Iwata, M., Hayashi, K. (1996): Effects of material properties on local buckling behaviour of metal members. In Coupled Instabilities in Metal Structures. CISM 96, (eds. J.Rondal, D.Dubina, V.Gioncu), Liege, 5-7 Sepember 1996, Imperial College Press, London, 539-546

Ono, T., Yoshida, F., Ishida, K. (1997): An experimental study on local buckling zone of box-section stub-column subjected to axial compression. In Stability and Ductility of Steel Structures, SDSS 97, (ed. T. Usami), Nagoya, 29-31 July 1997, 165-172

Qiang, G., Shaofan C., Ji, C. (1989): The in-plane ultimate load of I-section beam-columns with slender web. In Stability of Metal Structures, Beijing, 10-12 October 1989, International Academic Publishers, 341-348

Roeder, Ch. W., Cameron, B., Brown, C.B. (1999): Composite action in concrete-filled tubes. Journal of Structural Engineering, Vol. 125, No. 5, 477-484

Rondal, J. (1992): Residual stresses: Measurements, theoretical predictions and their use in the design of steel structures. In Testing of Metals for Structures, (ed. F. M. Mazzolani), Napoli, 29-31 May 1990, E&FN Spon, London, 381-391

Rondal, J., Maquoi, R. (1985): Stub-column strength of thin-walled square and rectangular hollow-sections. Thin-Walled Structures, Vol. 3, 15-34

Rondal, J., Wurker, K.G., Dutta, D., Wardenier, J., Yeomans, N. (1992): Structural stability of hollow sections. TUV Rheinland Verlag

Sawyer, H.A. (1961): Post-elastic behaviour of wide-flange steel beams. Journal of Structural Division, Vol. 87, ST 9, 43-71

Schneider, S.P. (1998): Axially loaded concrete-filled steel tubes. Journal of Structural Engineering, Vol. 124, No. 10

Sivakumaran, K.S., Yuan, B. (1998): Slenderness limit and ductility of high strength steel sections. In 2nd World Conference on Steel in Construction. San-Sebastian, 11-13 May 1998, CD-ROM 273

Stranghoner, N. (1995): Undersuchungen zum Rotationverhalten von Tragen aus Hohlprofilen. Ph Thesis, Technischen Hochschule Aachen

Sully, R.M., Hancock, G.J. (1995): Behaviour of cold-formed slender SHS beam-columns. University of Sydney, Research Report No. R707

Tarnai, T. (1995): Inextensional local buckling form of column with thin-walled closed polygon sections. In Stability of Steel Structures, (ed. M. Ivanyi), Budapest, 21-23 September 1995, Akademiai Kiado, Budapest, Vol. 2, 799-806

Tsuda, K., Matsui, C., Mino, E. (1996): Strength and behaviour of slender concrete filled steel tubular columns. In S tability Problems in Designing, Construction and Rehabilitation of Metal Structures, (eds. R.C. Batista and E. M. Batista), Rio de Janeiro, 5-7 August 1996, 489-500

Yabuki, T., Arizumi, Vinnakota, S. (1995): Mutual influence of cross-sectional and member classification on stability of I-beams. In Structural Stability and Design, (eds. S. Kitipornchai et al), Sydney, 30 October- 1 November 1995, Balkema, Rotterdam, 125-134

Yamada, M., Sakae, K., Shirakawa, K. (1970): Elasto-plastische Biegeformanderungen von Stahlstutzen mit I-Querschnitt. Teil I. Einseitige Biegung under konstanter Normalkrafteinwirkung. Der Stahlbau, Heft 12, 353-364

Yamada, M., Shirakawa, K. (1971): Elasto-plastische Biegeformanderung von Stahlstutzen mit I-Querschnitt. Teil II. Wechselseitig wiederholte Biegung under konstanter Normalkrafteinwirkung. Der Stahlbau, Heft 3, 5, 65-74, 143-151

Yamao, T., Sakimoto, T., Iwatubo, K. (1995): Strength and behaviour of stub-columns made of high-strength steel with low-yield ratio. In Stability of Steel Structures (ed. M. Ivanyi), Budapest, 5-7 September 1995, Akademiai Kiado, Budapest, 111-117

Zhang, W., Shahorovz, B.M. (1999): Comparison between ACI and AISC for concrete-filled tubular columns. Journal of Structural Engineering, Vol. 125, No. 11, 1213-1223

Uy, B.(1998): Local and post-local buckling of concrete filled steel welded box columns. Journal of Constructional Steel Research, Vol. 47, 47-72

Varna, A.H., Hull, B.K., Ricles, J.M., Sause, R., Lu, L. W. (1998): High strength square CFT columns: An experimental perspective. SSRC Annual Session,

Atlanta, 21-23 September 1998, 205-218

Vayas, I., Psycharis, I. (1995): Local cyclic behaviour of steel members. In Behaviour of Steel Structures in Seismic Areas, STESSA 94, (eds. F.M. Mazzolani and V. Gioncu), Timisoara, 26 June -1 July 1994, E&FN Spon, London, 231-241

Vayas, I., Syrmakezis, C., Sophocleous, A. (1995): A method for evaluation of the behaviour factor for steel regular and irregular buildings. In Behaviour of Steel Structures in Seismic Areas, STESSA 94, (eds. F.M. Mazzolani and V. Gioncu), Timisoara, 26 June- 1 July 1994, E&FN Spon, London, 344-354

Watanabe, E., Sugiura, K., Kanon, M., Takao, M., Emi, S. (1990): Hysteretic behaviour of thin tubular beam-column with round corners. In Stability of Steel Structures, (ed. M. Ivanyi), Budapest, 25-27 April 1990, Akademiai Kiado, Budapest, Vol. 2, 911-918

Wierzbicki, T. (1983): On the formation and growth of folding modes. In Collapse: The Buckling of Structures in Theory and Practice, (eds. J.M.T. Thompson and G.W. Hunt), IUTAM Symposium, London, 31 August-3 September 1982, Cambridge University Press, Cambridge, 19-30

6

Member Ductility Evaluation

6.1 Ductility Erosion Due to Member Behaviour

Current seismic design procedures implicitly permit the inelastic structural deformations under strong ground motions for economical reasons. The local ductility is used as a parameter for evaluating the available inelastic performances of structures. It is widely accepted that the material ductility cannot be used alone in evaluating the capacity of structure to dissipate seismic energy input. Due to this fact the code provisions take into account the ductility at the level of section, considering that the loss in ductility can be correctly predicted in this case. The Eurocode 3 classification in plastic, compact, semi-compact and slender sections just on the basis of sectional properties and static behaviour, provides a great flexibility for the designer, allowing some improvement of structure ductility. Unfortunately, the use of this classification leads to many questions and uncertainties about the erosion of ductility at the level of member, mainly caused by the dynamic characteristics of seismic actions. Indeed, the actual deformation of a structure is influenced by the global structure response to very complex seismic actions. The multiple buckling modes and the interaction between these modes, the continuous changing in moment gradient during the earthquake action, the strain-rate produced by high velocity of actions, the important accumulation of plastic deformations due to hysteretic behaviour under strong ground motions cannot be ignored in design. If these factors are out of control during the structure design, the erosion in ductility can by very high and the structure damage very important (Fig. 6.1).

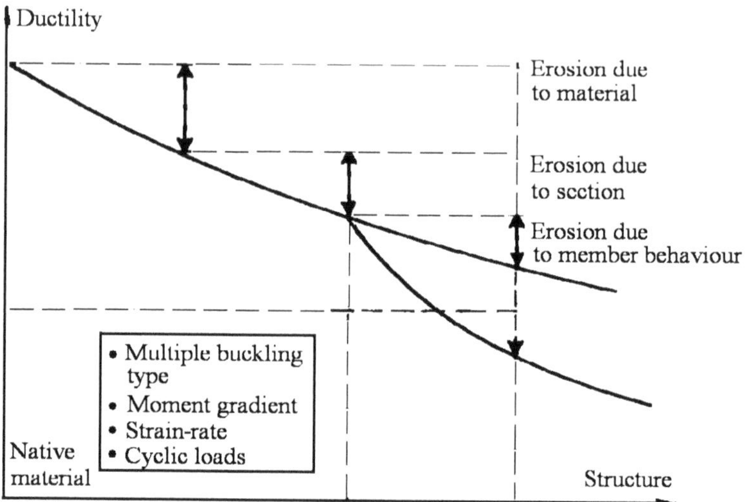

Figure 6.1: Erosion due to member behaviour

6.2 Main Factors Influencing the Member Ductility

Many factors influencing the member ductility are the same as already examined for stubs: cross-section types, fabrication, local buckling type, steel properties, geometrical and mechanical imperfections, loading types, etc. But there are some additional factors strictly related to the member features, like:

(i) *Member type* (Fig. 6.2):

-beam;

-beam-column.

(ii) *Member behaviour*:

-multiple buckling modes interaction. The in-plane buckling mode being associated to the out-of-plane local or global buckling modes, flange induced buckling, etc, which are influenced by local and overall member slendernesses, depending also on the lateral bracing system (Fig. 6.3);

-moment gradient, the main effect being the increasing of plastic moment as a result of the working in the strain-hardening range of a part of plastic zone (Fig. 6.4);

-strong change of moment and axial forces distribution during the seismic loading and, consequently, the permanent change in configuration due to formation of plastic zones and plastic hinges during an earthquake;

-presence of axial force producing an important decreasing of rotation capacity (Fig. 6.5);

-influence of member slenderness on local ductility.

(iii) *Influence of joints* (Fig. 6.6):

-type of joint, the main types being welded, bolted or mixed;

-supplementary rotations due to the deformations of joint components;

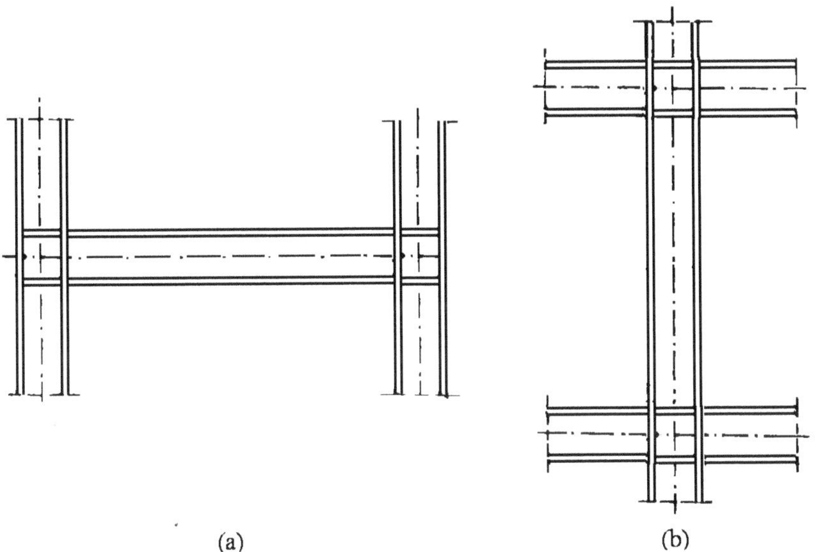

Figure 6.2: Member type: (a) Beam; (b) Beam-column

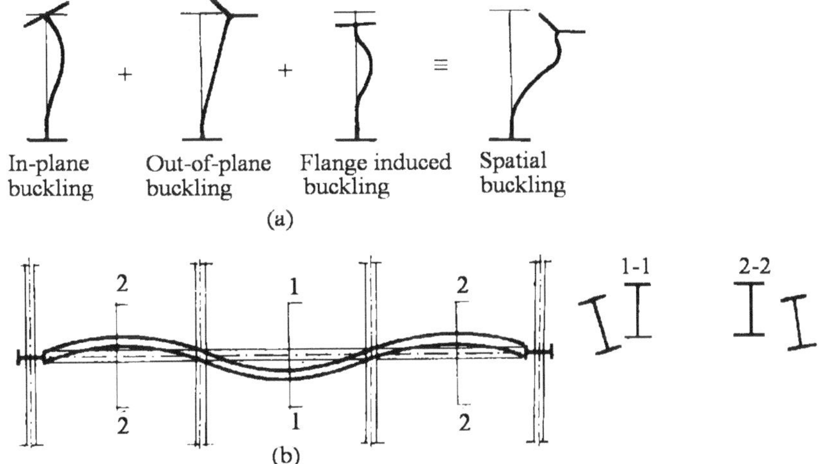

Figure 6.3: Buckling modes: (a) Local buckling; (b) Global buckling

-collapse of connections by fracture of welds, bolts or plates, produced before the formation of plastic hinges in connected members.

(iv) *Interaction with slabs* (Fig. 6.7):

-type of slab, the main types being reinforced concrete or composite ones;

-position of slab, in compression zone for positive (sagging) moment or negative (hogging) moment.

(v) *Earthquake type* (Fig. 6.8):

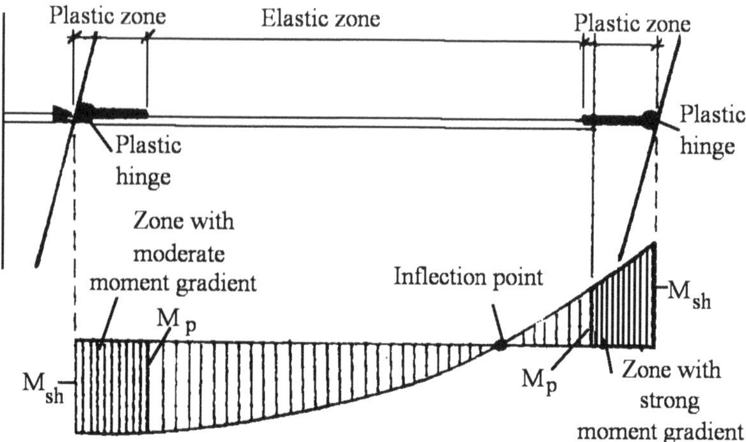

Figure 6.4: Influence of moment gradient

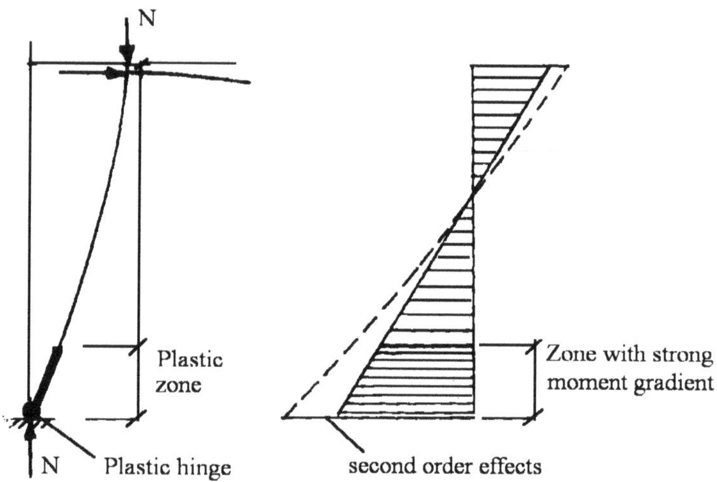

Figure 6.5: Influence of axial forces

-near-field earthquake, characterized by velocity pulse, producing high strain-rate in structure;

-far-field earthquake, characterized by cyclic ground motions, which give rise to important plastic deformation accumulation in the plastic hinges.

(vi) *Change of plastic hinges configuration* (Fig. 6.9):

-direction of rotation due to the earthquake's change of direction;

-change of the position due to the modification of earthquake in direction and value;

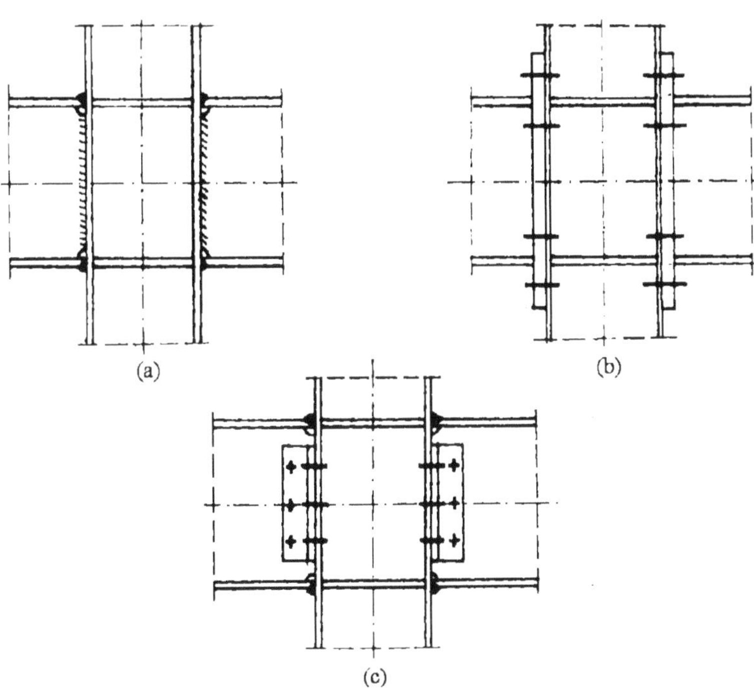

Figure 6.6: Joint types: (a) Welded; (b) Bolted; (c) Mixed

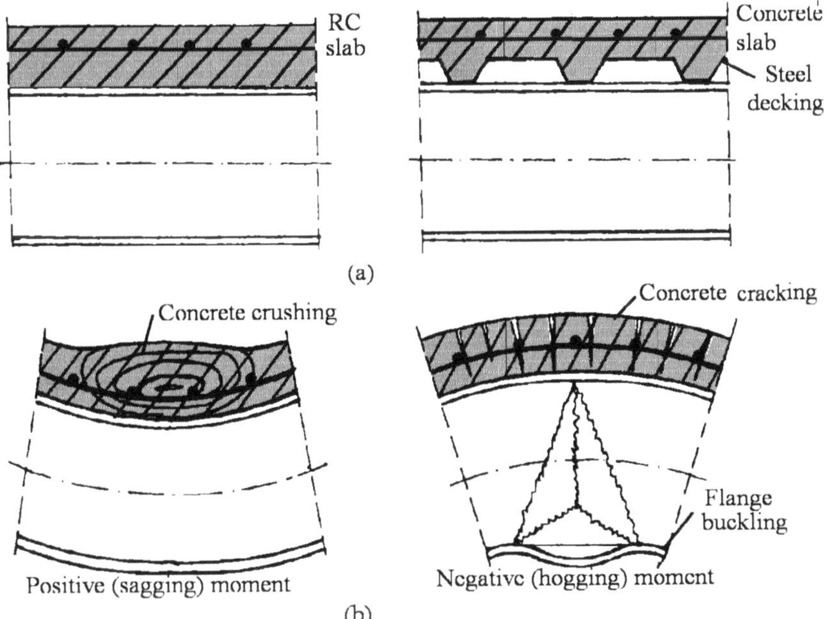

Figure 6.7: Interaction beam-slab: (a) Slab type; (b) Local failure

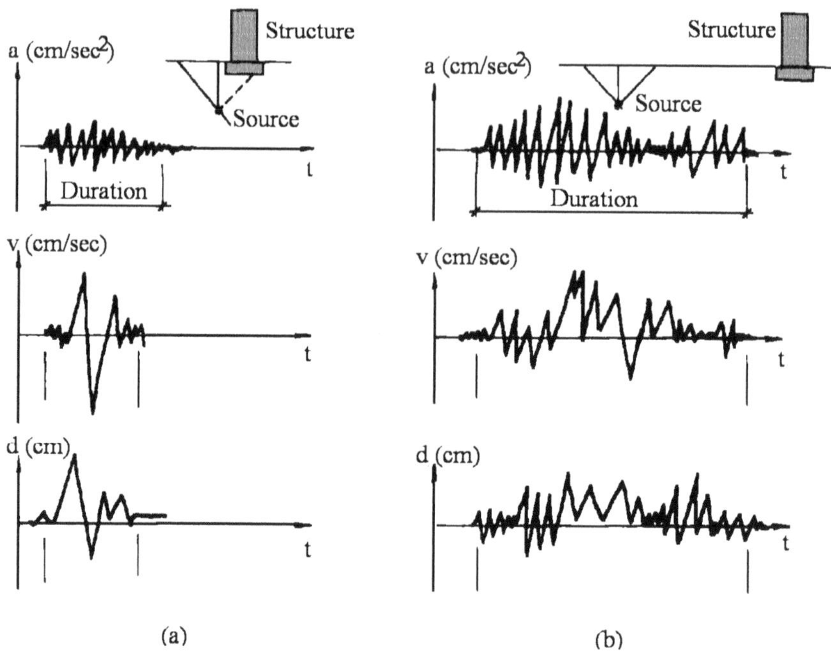

Figure 6.8: Earthquake type: (a) Near-field earthquake; (b) Far-field earthquake

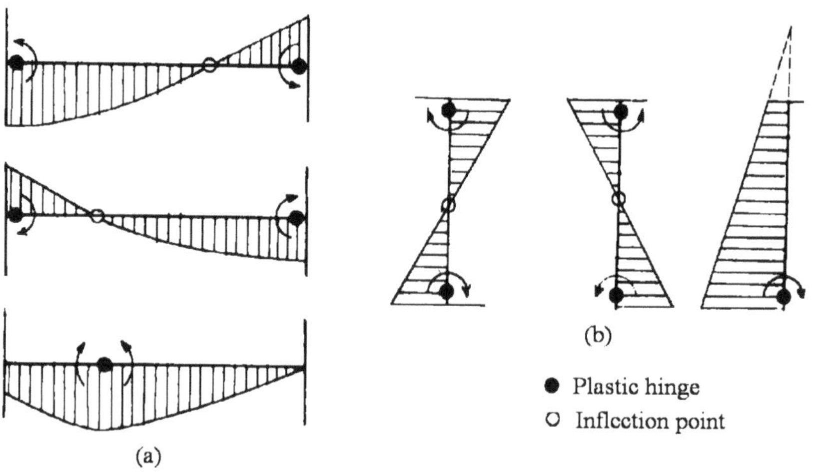

Figure 6.9: Change of plastic hinges configuration: (a) Beams; (b) Columns

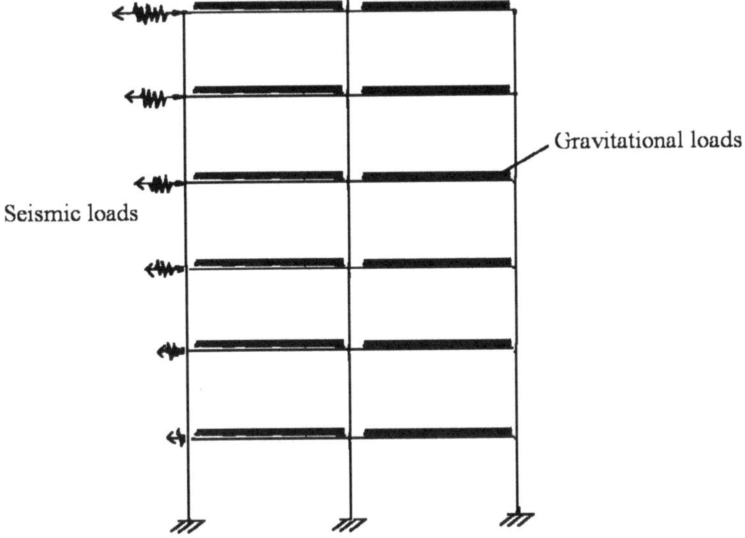

Figure 6.10: Structure actions

-change in member configuration, passing from double curvature moment to single curvature moment, or from a configuration with two plastic hinges to a configuration with only one.

All these very important factors have a great influence on member ductility and cannot be ignored. This is the reason why in many proposals for code improving (Mazzolani and Piluso, 1992, 1993, Gioncu, 1995a,b, Gioncu and Mazzolani, 1995, Gioncu and Petcu, 1997, Gioncu et al, 1995a, 2000a) the member ductility is proposed to be used for determining the available rotation capacity, instead of the section ductility commonly present now in code provisions.

6.3 Assessment of the Member Ductility in a Structure

6.3.1. Modelling the Member Behaviour

The available ductility of a member must be determined taking into account that the member belongs to a structure with a complex behaviour, being influenced by both gravitational loads and seismic actions (Fig. 6.10).

(i) *Beam behaviour.* An one-bay beam is extracted from the structure in order to study the member behaviour (Fig. 6.11a). The gravitational loads produce time constant moment diagrams, but the ones corresponding to seismic actions are time variable both in magnitude and in distribution. In current plastic methodology the interpretation of member behaviour is greatly simplified. The moment-rotation relationship is usually represented

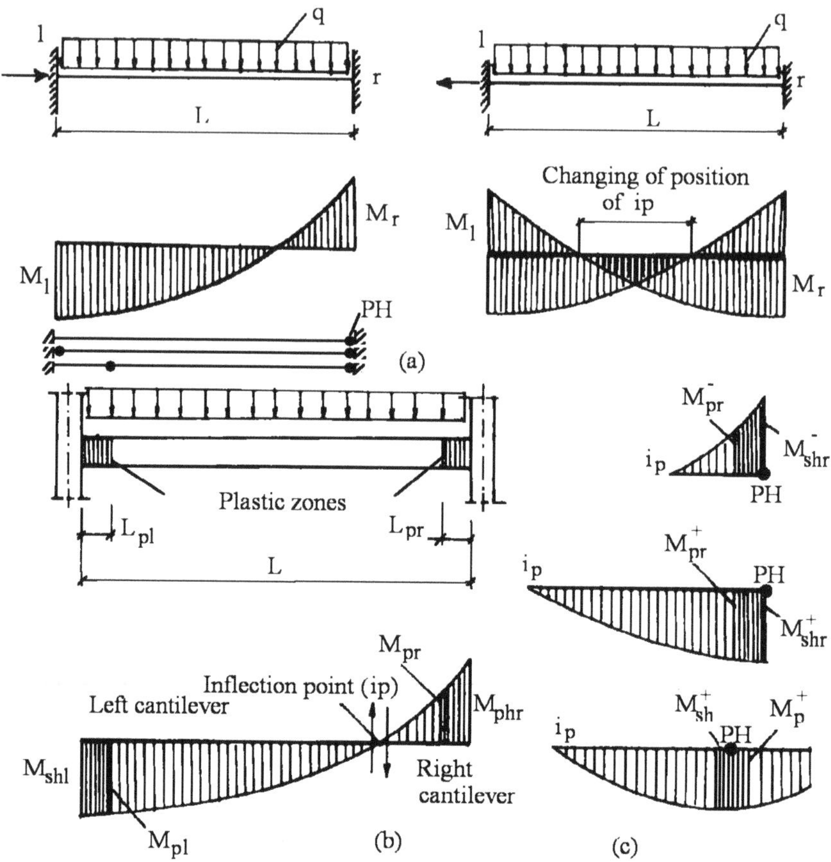

Figure 6.11: Beam behaviour in structure: (a) Plastic hinge (PH) positions; (b) Inflection point (ip) position; (c) Changing of moment diagrams

by a bilinear model for perfect elasto-plastic material or with a constant hardening after the elastic limit. This representation does not consider the presence of the hardening range after the yield plateau and does not allow to model the deterioration due to local instability. When local ductility arises, a more refined model should be used.

The traditional plastic analysis considers that the beam is a linear element with points of inelastic rotation at the ends, which are the well-known ***plastic hinges***, *PH*. Three different situations can be found out during the load increasing: PH only at right beam end, PHs at both beam ends and the left PH is formed at some distance from the end (Fig. 6.11a). But in reality the plastic deformations are not concentrated in a point, being extended on a given length at the beam ends. This length depends on the beam span, moment gradient and on the gravitational to seismic loads ratio. One can see that the beam can be divided in two cantilevers, separated by the inflection point (Fig. 6.11b). For the left cantilever, the gravitational and seismic loads act in opposite directions, so a reduction of bending moment occurs, together with a reduction of moment gradient. On the contrary, for the right cantilever, the two loads act in the same direction increasing both the moment value and the moment gradient. Therefore, the two plastic zones work in different conditions related to moment gradient. Due to this fact, for the right cantilever, some amount of strain-hardening is required for the plastic zone to grow along the beam length. Without this ability, the plastic zone cannot be formed at the beam end. So, the length of plastic zone is directly related to the length of cantilever (McMullin and Astaneh-Asl, 1997) and not to the depth of beam, as it is suggested in many papers. The ultimate seismic load can be achieved when the plastic hinge formed in the plastic zone loses its rotation capacity due to the local buckling of compression flanges or when a fracture occurs in the tension flange due to the reaching of ultimate strain.

(ii) *Beam-column behaviour.* The presence of significant axial forces leads to a much more complex behaviour of beam-column in comparison with simple beams. The current methodology, which considers the plastic zones as point hinges, becomes a very poor approximation for ductility determination. The plastic deformation capacity depends on many parameters and has a great variability (Nakashima, 1994, Krawinkler and Gupta, 1998). This is one of the reasons why in design practice it is recommended to size the structure in such a way that significant column plasticizations are avoided. Plastic hinges are accepted only at the column bases at the first level in order to form a global mechanism. But this recommendation is very difficult to be respected due to the uncertainties in the ground motions. Therefore, beam-columns must be designed in order to form plastic zones at their ends.

The moment diagrams in columns can present a great variation of patterns (Fig. 6.12a). When the rigidities of beams and columns are not very different and the shear deformations are dominant, the moment diagrams in columns have a double curvature. In the case of a high influence of second order vibration mode, the changing in moment diagrams can be very important, the double curvature diagrams change in single curvature ones, especially in the middle and top part of the frame. In the case of a strong column-weak beams, when cantilever deformations are dominant, the moment diagrams have a single curvature. In the function of the values of these moments, the plastic zones can form at the lower column end, upper end or at both ends (Fig. 6.12b).

Some moment diagrams for a column are plotted in Fig. 6.12c. One can see that the length of the plastic zone depends on the moment-gradient and moment diagram patterns. In all cases, the plastic zones work in strain-hardening range due to the variation of moment along the column height. The shortest plastic zone occurs when two plasticizations form at both column ends and the longest for single moment curvature. One can also see that for columns it is possible to use the same cantilever beam type as for the beam, its span being determined by the inflection point. In the case of a single moment curvature, the cantilever span exceeds the column height.

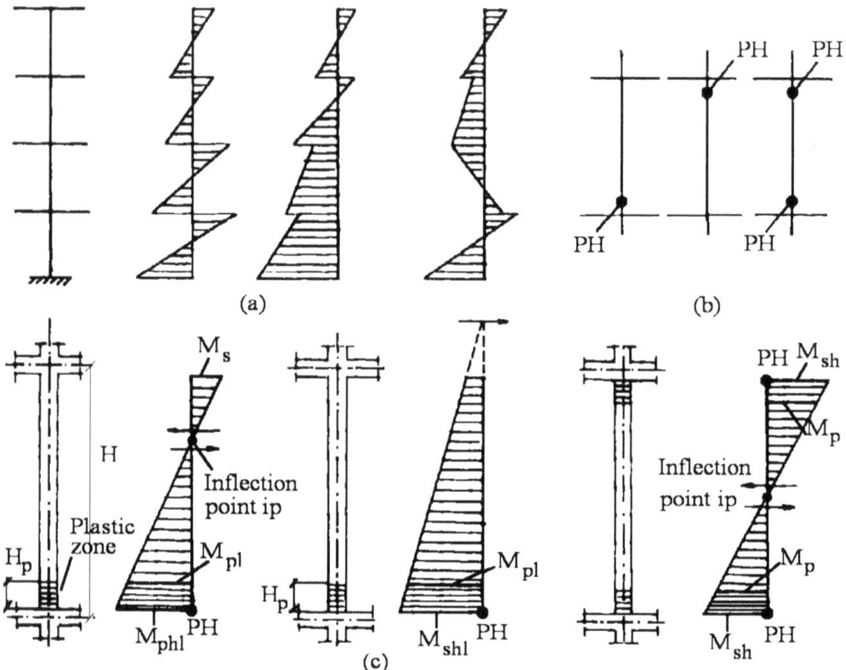

Figure 6.12: Beam-column behaviour in structure: (a) Variation of moment diagrams; (b) Plastic hinge (PH) positions; (c) Inflection point (ip) positions

(iii) *Methodologies for determination of member behaviour.* Three methodologies for structure analysis are considered:

-*Equivalent lateral force analysis*, in which lateral forces are statically applied and the structure is elastically analysed for reduced forces, taking into account the energy dissipation capacity of the structure. On the base of the obtained moment diagram the positions of inflection points are determined. The main hypothesis of this method considers that both load and moment distributions on the structure remain unchanged until collapse. In reality this is not true due to the dynamic nature of loads and plastic behaviour of structure. But this methodology gives a satisfactory representation concerning the position of inflection points.

-*Push-over method*, in which the analysis is performed in elasto-plastic range for static incremental lateral forces. The maximum load is determined for the formation of collapse mechanism. During the lateral increasing loads the positions of inflection points shift in function of the formation of plastic hinges. Therefore, the positions of these inflection points are more correctly determined as in the previous methodology. The only objections regarding the use of this method are related to the distribution of lateral loads along the structure height and the static characteristics of loading.

-*Time-history method*, for which the loads and moment diagrams change their pattern during the seismic actions and successively both beams-ends work in strain-hardening range. The inflection points change their positions continuously during the structure movements. For instance, for a beam, the right plastic zone can work in the following conditions (Fig. 6.11c):

• as a short cantilever with a very high moment gradient and negative bending moment corresponding to strain-hardening range;

• as a long cantilever with a reduced moment gradient and positive moment corresponding to yielding plateau;

• the plastic zone can move towards the beam centre due to the decreasing of non-equilibrated bending moments in the columns.

For beam-columns, it must be noted a continuous changing of inflection point position and, consequently, a continuous modification of cantilever span and moment gradient. At the same time the direction of plastic rotation changes.

6.3.2. Simplified Method using the Standard Beam Approach

Considering all the above mentioned aspects, it seems impossible that the complexity of structure behaviour could be analyzed during the design and a simplified methodology for practical purposes represents a pressing need.

In order to setup a method to be used in the design practice, two steps can be followed:

-determination of local ductility using a static approach; and

-correction of obtained values considering the dynamic effects of seismic loadings (changing in standard beam configuration, effect of strain-rate, cyclic loading, etc.)

It is important to simplify the analysis by substituting the actual member with a simple member having a very similar behaviour. The substitutive member is the so-called *standard beam* proposed by Spangemacher and Sedlacek (1992b) and Gioncu and Petcu (1995, 1997). One can see that in a complex structure the inflection point divides the member into two portions with positive and negative bending moments and each actual member in a structure can be replaced by a combination of two standard beam types (Fig. 6.13):

-SB 1, a central concentrated load beam for the case of members under linear moment gradient;

-SB 2, a distributed load beam for the case of members under weak moment gradient. In some cases, this beam can be replaced by two concentrated load beams.

In the following it is assumed that the seismic actions act on the structure from left to right.

Working with standard beams means to subdivide the structure into single span beams, by identifying the position of inflection points.

(i) *Asymmetrical bending moment.* The beam is loaded by *two asymmetric end bending moments* (Fig. 6.14). Generally, the two moments are different, due to the fact that the two beam-ends work in different conditions:

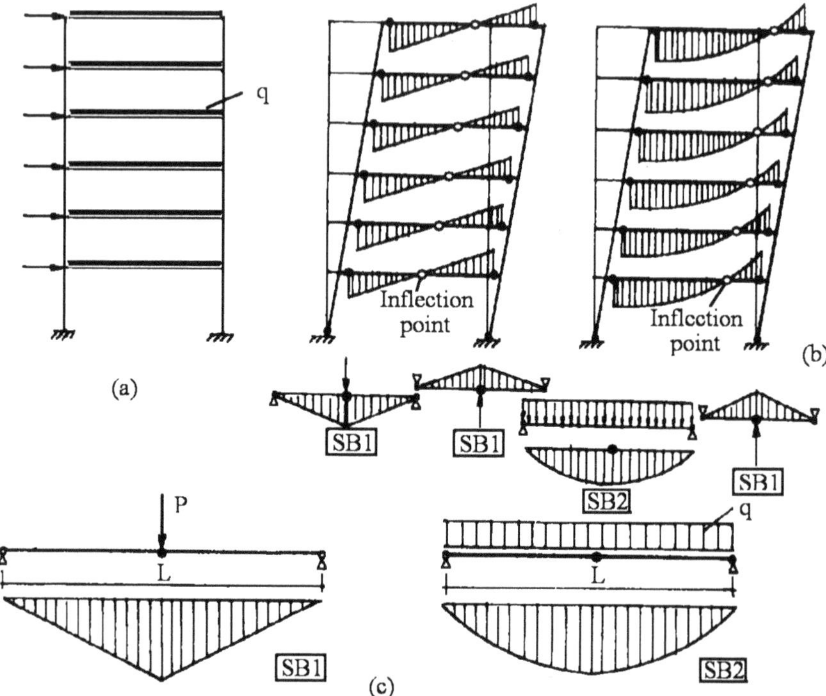

Figure 6.13: Using standard beam SB to determine rotation capacity: (a) Structure scheme; (b) Decomposition of structure into standard beams; (c) Standard beam types

-left bending moment:

$$M_A = m_A M_p \tag{6.1a}$$

the increasing over the full plastic moment. M_p being the result of the flange to slab interaction in the compression zone;

-right bending moment:

$$M_B = m_B M_p \tag{6.1b}$$

the increasing being produced by the strain-hardening and the effect of slab in the tension part.

In the following, the assumption that, after reaching the maximum value, the end moment that remains constant is used.

The inflection point is given by:

$$\frac{x_i}{L} = \frac{m_A}{m_A + m_B} \tag{6.2}$$

The behaviour of the beam can be studied using two standard beams of SB 1 type. The spans of these beams are:

$$L_l = 2x_i \quad ; \quad L_r = 2(L - x_i) \tag{6.3a, b}$$

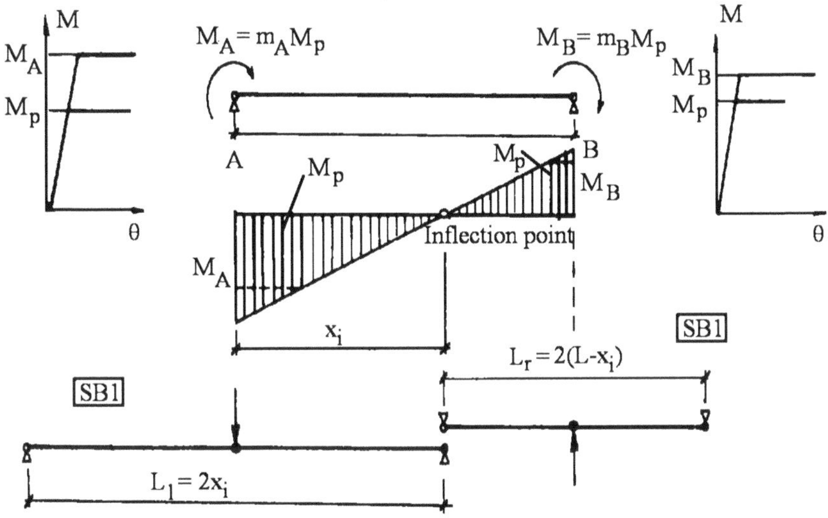

Figure 6.14: Standard beams SB under seismic action

The effect of the *distributed gravitational load q* is to increase the moment gradient at the right-end and to reduce it at the left-end. The inflection point is shifted to right side (Fig. 6.15a). The position of the maximum moment is given by:

$$x_m = \left(1 - \frac{M_A + M_B}{4M_0}\right)\frac{L}{2} \tag{6.4}$$

while the inflection point can be determined from:

$$x_i = \left\{1 - \frac{M_A + M_B}{4M_0} \pm \left[\left(1 - \frac{M_A + M_B}{4M_0}\right)^2 + \frac{M_A}{M_0}\right]^{1/2}\right\}\frac{L}{2} \tag{6.5}$$

where M_0 is the bending moment of a simply supported beam:

$$M_0 = \frac{qL^2}{8} \tag{6.6}$$

According to the hypothesis that the seismic forces act on the structure from the left to the right, the plastic moments at the two ends are different, due to the beam-slab interaction at the left-end and to the moment gradient at the right-end. Because usually the interaction can be high and the moment gradient is weak, the first plastic hinge always occurs at the right-end. For the left plastic hinge there are two cases:

-if

$$x_m < 0 \quad ; \quad \frac{m_A + m_B}{4}M_p > M_0 \tag{6.7}$$

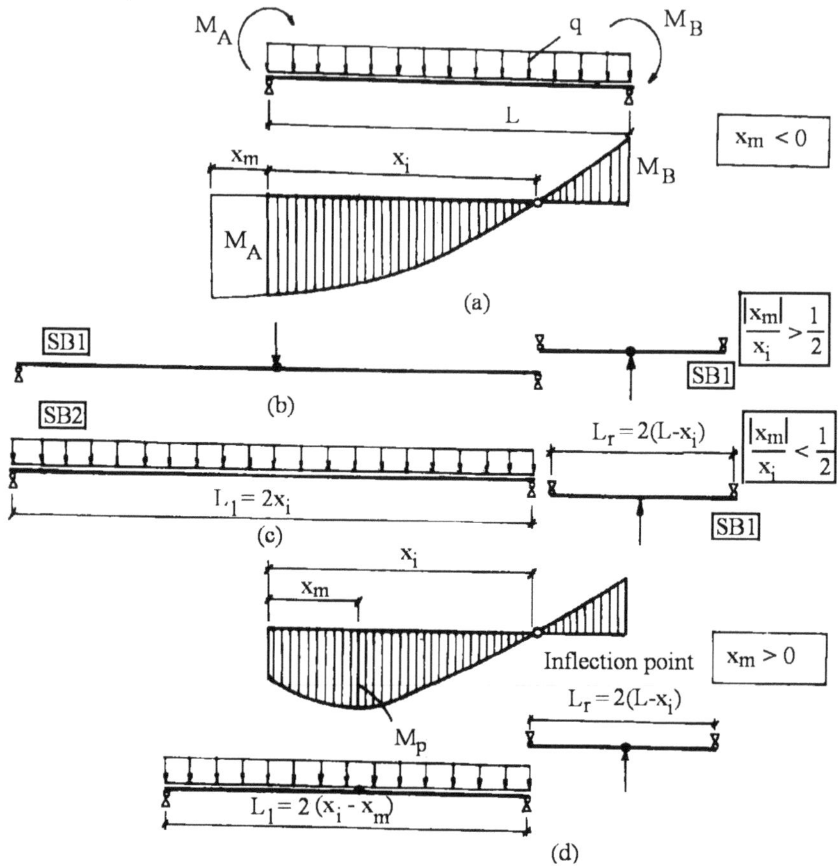

Figure 6.15: Standard beams SB under seismic and gravitational actions: (a) Position of inflection point; (b) Moment diagram with high left gradient; (c) Moment diagram with low left gradient; (d) Moment diagram with plastic hinge far from the beam end

the plastic hinge occurs at the beam end. The inflection point is given by:

$$x_i = \left\{ 1 - \frac{m_A + m_B}{4} \frac{M_p}{M_0} + \left[\left(1 - \frac{m_A + m_B}{4} \frac{M_p}{M_0} \right)^2 + m_A \frac{M_p}{M_0} \right]^{1/2} \right\} \frac{L}{2} \tag{6.8}$$

The actual beam can be replaced by two standard beams in the following manner:

- for:

$$\frac{|x_m|}{x_i} > \frac{1}{2} \tag{6.9a}$$

with two standard beams SB 1 type (Fig. 6.15b):

• for:

$$\frac{|x_m|}{x_i} < \frac{1}{2} \tag{6.9b}$$

with a standard beam SB 2 type at the left part and SB 1 type at the right part (Fig. 6.8c). The spans for the two standard beams are given by (6.15a,b);

-if

$$x_m > 0 \quad ; \quad \frac{m_A + m_B}{4} M_p < M_0 \tag{6.10}$$

the left plastic hinge occurs at a distance x_m from the beam end (Fig. 6.15d) and the inflection point is given by

$$x_i = x_m + (L - x_m)\left(\frac{1}{1 + m_B/m_M}\right)^{1/2} \tag{6.11}$$

where:

$$M_{max} = m_M M_p \tag{6.12}$$

is the maximum moment.

(ii) *Influence of beam-slab interaction.* The two beam ends are in different conditions regarding to the beam-slab interaction. At the right-end the slab is in tension zone, while at the left one it is in compression zone. There are the following situations.

-*Interaction must be considered* (Fig. 6.16) when the constructional details assure a good co-operation between the two elements. In this case the plastic behaviour at the right-end is dominated by the buckling of the lower compression flange, while the left-end behaviour is controlled by the crushing of the reinforced concrete slab. This situation creates a great difference in the plastic moments at the two ends, the smallest being the right plastic moment; an important shifting of the inflection point occurs. As a function of the ratio between the two plastic moments:

$$\nu_m = \frac{M_{pl}}{M_{pr}} \tag{6.13}$$

where M_{pr} is the plastic moment at the right-end in which the buckling of lower compression flange is considered and M_{pl} the plastic moment at the left-end, in which the interaction beam-slab is taken into account, two different cases concerning the position of the plastic hinge can be recognized:

-if:

$$\frac{\nu_m + 1}{4}\frac{M_B}{M_0} \geq 1 \tag{6.14}$$

the plastic hinge occurs at the beam's left-end ($M_A = M_{pl}$) and the position of the inflection point is given by:

$$x_i = \left\{1 - \frac{\nu_m + 1}{4}\frac{M_B}{M_0} + \left[\left(1 - \frac{\nu_m + 1}{4}\frac{M_B}{M_0}\right)^2 + m_A\frac{M_p}{M_0}\right]^{1/2}\right\}\frac{L}{2} \tag{6.15}$$

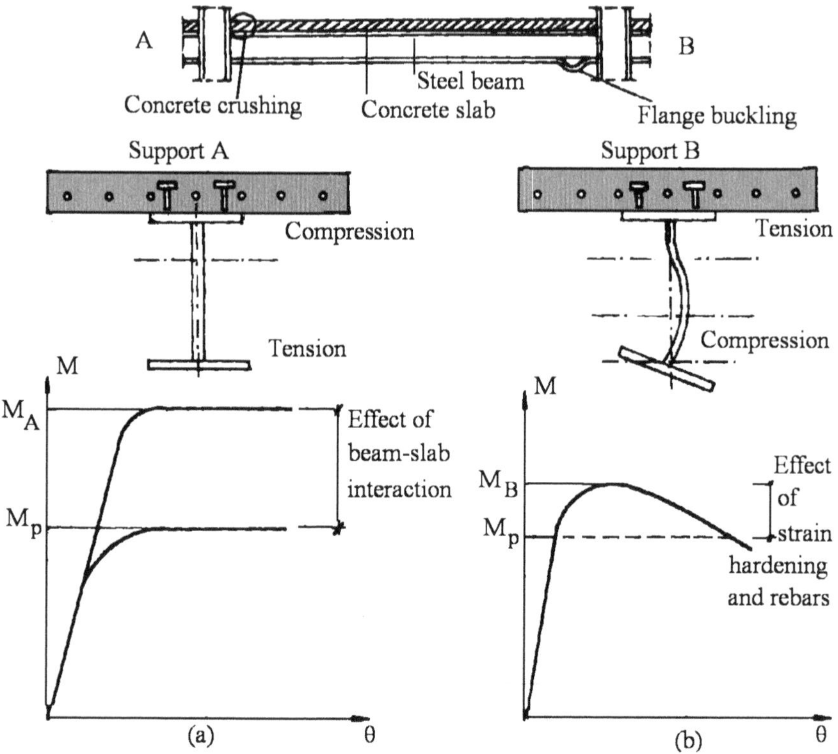

Figure 6.16: Effects of beam-slab interaction: (a) Left beam-end; (b) Right beam-end

-if:

$$\frac{\nu_m + 1}{4} \frac{M_B}{M_0} < 1 \tag{6.16}$$

the plastic hinge occurs at a given distance from the beam's left-end ($M_A \neq M_{pl}$, $M_{max} = M_{pl}$) and the position of the maximum moment may be determined by:

$$x_m = \left[1 - \left(\frac{\nu_m + 1}{4} \frac{M_B}{M_0} \right)^{1/2} \right] L \tag{6.17}$$

and the position of the inflection point:

$$x_i = x_m + \left(\nu_m \frac{M_B}{M_0} \right)^{1/2} \frac{L}{2} \tag{6.18}$$

The distance between the point of maximum moment and the inflection point is given by:

$$\Delta_i = x_i - x_m = \left(\nu_m \frac{M_B}{M_0} \right)^{1/2} \frac{L}{2} \tag{6.19}$$

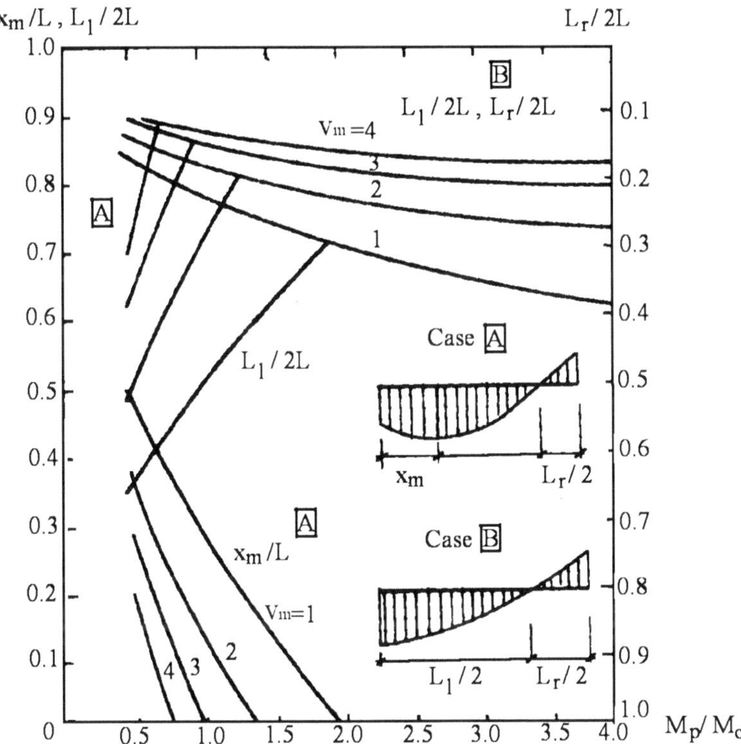

Figure 6.17: Determination of standard beam SB spans in case of beam-slab interaction

The variation of the characteristic distances is presented in Fig. 6.17. The A type behaviour corresponds to the plastic hinge produced at some distance from the edge, while the B type behaviour corresponds to the case when the plastic hinge occurs at the beam edge. One can see that the interaction beam-slab has a great influence on the plastic behaviour of the beam.

-*Interaction is avoided* by some special details aiming to eliminate the incertitude of the level of beam-slab interaction and to have a very clear plastic behaviour (Fig. 6.18a). There are some cases in which the presence of slab in compression part of beam does not assure the elimination of flange buckling (Fig. 6.18b). In this case the inequality of the two end bending moments is due to the difference of the moment gradient in the moment-rotation curve (Fig. 6.18c). At the right-end (support B) the moment-gradient is higher and the plastic moment is increased due to the strain-hardening $M_{pB} = m_h M_p$ and $m_B = m_h$. The left plastic hinge (support A) works with a quasi-constant moment $M_{pA} = M_p$ and $m_A = 1$.

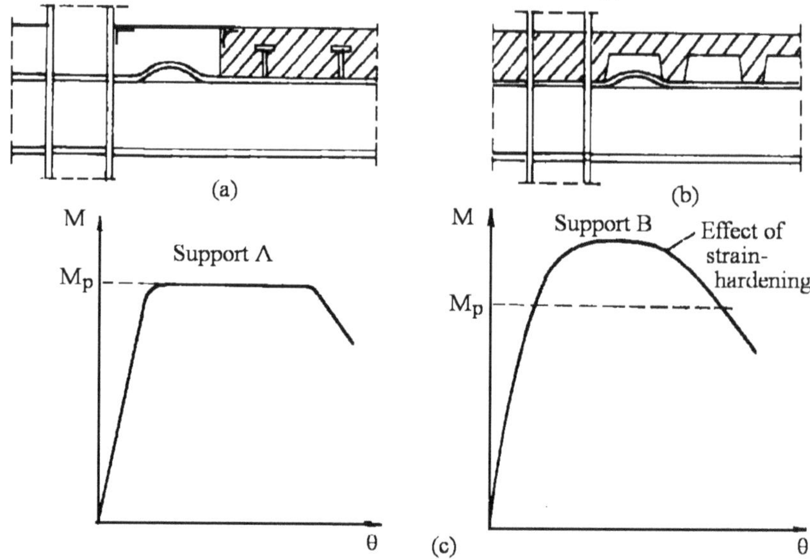

Figure 6.18: Elimination of beam-slab interaction: (a) Using special details; (b) Buckling of flange is not avoided by the slab presence; (c) Moment-rotation curves for left and right ends

The two behaviour cases are given by the conditions:
-if:

$$\frac{1+m_h}{4}\frac{M_p}{M_0} \geq 1 \tag{6.20}$$

the plastic hinge occurs at the beam left-end and the position of the inflection point results from:

$$x_i = \left\{1 - \frac{1+m_h}{4}\frac{M_p}{M_0} + \left[\left(1 - \frac{1+m_h}{4}\frac{M_p}{M_0}\right)^2 + \frac{M_p}{M_0}\right]^{1/2}\right\}\frac{L}{2} \tag{6.21}$$

-if:

$$\frac{1+m_h}{4}\frac{M_p}{M_0} < 1 \tag{6.22}$$

the plastic hinge occurs at some distance from the beam left-end and the position may be determined by:

$$x_m = \left[1 - \left(\frac{1+m_h}{4}\frac{M_p}{M_0}\right)^{1/2}\right] L \tag{6.23}$$

and the position of the inflection point:

$$x_i = x_m + \left(\frac{M_p}{M_0}\right)^{1/2}\frac{L}{2} \tag{6.24}$$

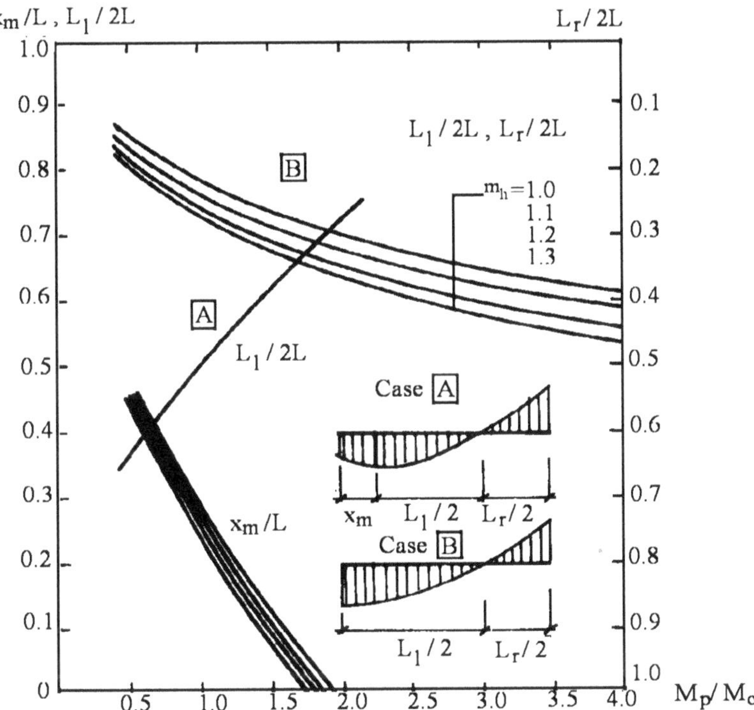

Figure 6.19: Determination of standard beam SB spans in case that interaction beam-span is avoided

the distance between the two positions being:

$$\Delta_i = x_i - x_m = \left(\frac{M_p}{M_0}\right)^{1/2} \frac{L}{2} \tag{6.25}$$

The distances between the characteristic points are presented in Fig. 6.19. It is very clear that when the interaction is avoided, the differences between the plastic moment values at the two ends have not a great influence on the plastic behaviour of the beam.

(iii) *Beam-columns.* For beam-columns, examining the collapse mechanisms (Fig.6.20), one can see that basically there are two different situations:

-in case of global mechanism (Fig. 6.20a) one plastic hinge occurs at the column base. If M_s is the elastic bending moment at the upper end of the column, the position of inflection point is given by:

$$\frac{H_s}{H} = \frac{1}{1 \pm \dfrac{M_s}{M_p}} \tag{6.26}$$

In (6.26) the plus sign is for the case of double curvature moment (Fig. 6.20a), while the minus sign corresponds to the single curvature moment

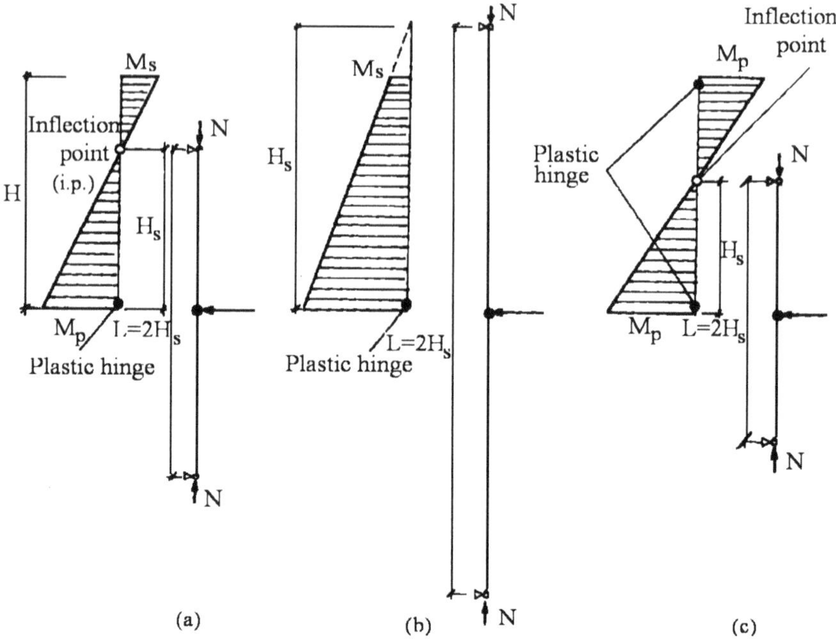

Figure 6.20: Standard Beam (SB) for beam-columns: (a) For double curvature moment diagram; (b) For single curvature moment diagram; (c) For storey mechanism

(Fig. 6.20b). The first case occurs if the column is weak related to the beam, while the second case corresponds to strong column;

-in case of storey mechanism, two plastic hinges occurs at both column ends (Fig. 6.20c). The inflection point is localized at the middle of the column.

In both cases, a standard beam SB1 type can be used, with a span corresponding to two times the distance between bottom and inflection point. The standard beam must be loaded also by the axial forces evaluated from the structure analysis.

6.3.3. Plastic Behaviour of Standard Beams

(i) *Standard beam SB 1 type.* The purpose of this Section is to describe the inelastic behaviour of a simply supported beam loaded by a concentrated force in the midspan, which produces a linearly varying moment diagram. The study is restricted to the case in which the force is applied in the web plane (Fig. 6.21a). Fig. 6.21b shows the behaviour of a fiber in the steel member. It has been shown in the previous chapter that, in case of controlled loading, the actual stress-strain curve for any fiber is discontinuous. So, the strain jumps from the yield strain to the strain corresponding to the start of the hardening range. Taking into account the relationship (5.25),

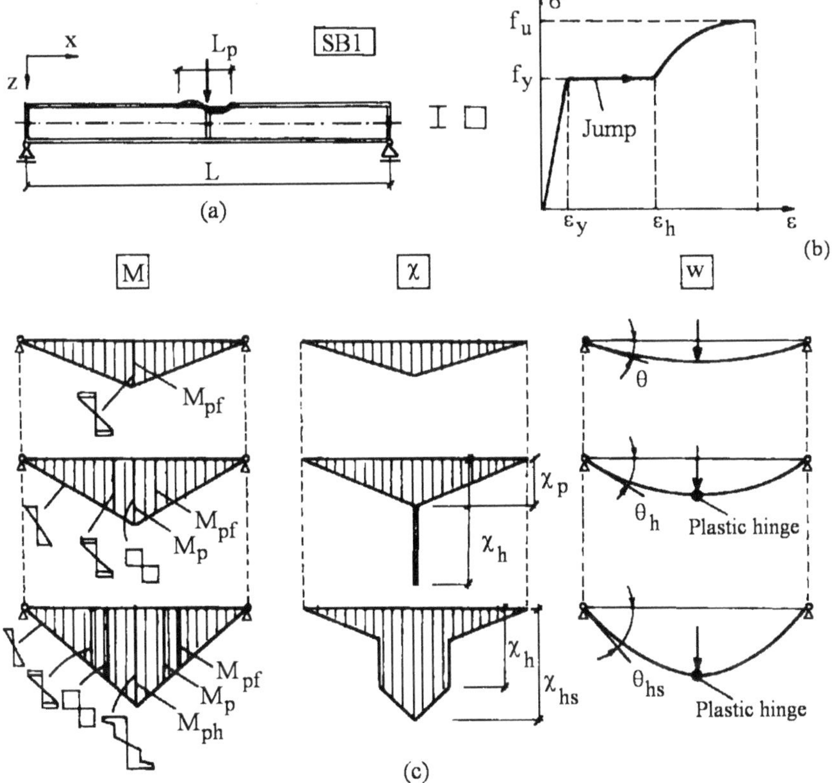

Figure 6.21: Behaviour of standard beam SB1: (a) Beam scheme; (b) Material behaviour; (c) Steps in beam behaviour

the variation of curvature corresponds to the variation of strain. So, also a jump in curvature from the elastic to the strain-hardening behaviour occurs.

The response of the beam under moment gradient can be divided in three steps (Fig. 6.21c):

-the initial behaviour is linear and elastic up to the moment M_y which corresponds to the first yielding. The moment increase beyond this value produces the full plasticization of the flanges at the moment M_{pf}. Because the flange thickness is small with respect to the member depth, it is reasonable to assume that the yielding occurs simultaneously at all points across the thickness: $M_y = M_{pf}$;

-in the second step, the full-plasticization of the cross-section occurs for the plastic moment M_p. The plastic hinge is theoretically punctual, but in reality its length along the flange is finite. At this level, due to the load control, a jump in curvature occurs from the elastic to the strain hardening range. The rotation corresponding to this step is θ_p;

-in the third step, there are three zones in the moment diagram. The first works in the elastic range, the second is plastic and the cross-section

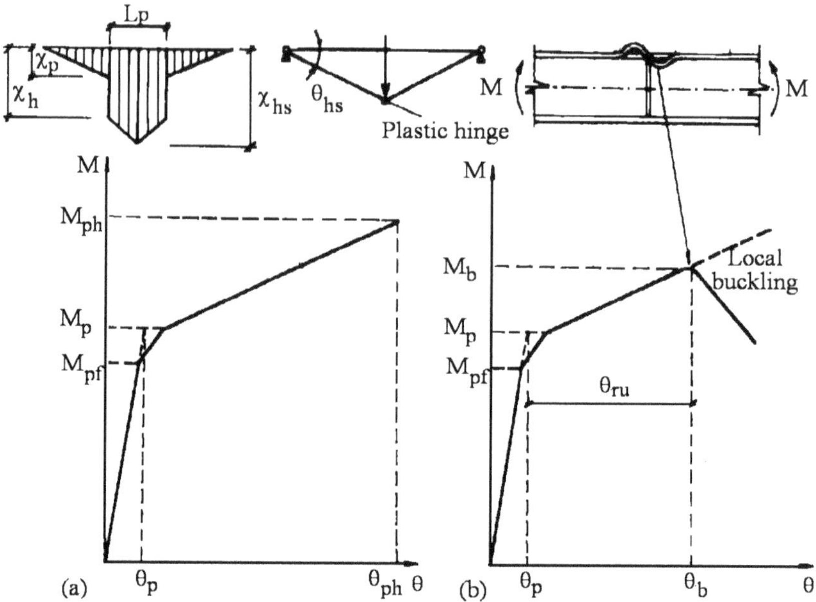

Figure 6.22: Determination of beam rotation for SB1 beam: (a) Moment-rotation curve; (b) Influence of compression flange buckling

of the third is fully plastic with some parts in the hardening range. The necessity of working in the hardening zone is clearly due to the gradient moment. The separation between the elastic range and the elasto-plastic range arises at the value M_{pf}. The full plastic moment, M_p, separates the elasto-plastic zone from the plastic hardening zone. The maximum moment M_{ph} occurs at the middle of the beam, this is the hardening zone where there is maximum stress.

The length of the fully plastic zone is given by:

$$\frac{L_p}{L} = 1 - \frac{M_p}{M_{ph}} \tag{6.27a}$$

and for flange plastic zone:

$$\frac{L_{pf}}{L} = 1 - \frac{M_{pf}}{M_{ph}} \tag{6.27b}$$

It is, therefore, evident that the length of plastic zone depends on the moment gradient.

The beam end rotation can be determined by the integration of the curvature diagram between support and mid-span (Fig. 6.22a). The rotation corresponding to the plasticization of flanges results:

$$\theta_{pf} = \int_0^{L/2} \chi dx = \frac{M_{pf} L}{4EI} \tag{6.28}$$

which corresponds to the elastic beam end rotation. For practical purposes, the differences between M_{pf} and M_p can be neglected and the rotation for the fully plastic condition is assumed equal to:

$$\theta_p = \frac{M_p L}{4EI} \tag{6.29}$$

which can be considered the limiting rotation of the elastic behaviour.

In the strain-hardening range, the bending moment and the corresponding curvatures:

$$M_{ph} = m_h M_p \tag{6.30a}$$

$$\chi_{ph} = s_h \chi_p \tag{6.30b}$$

Between s_h and m_h the following relationship exists:

$$s_h = s + (m_h - 1)e_h \quad ; \quad e_h = \frac{E_h}{E} \tag{6.31a, b}$$

where s is a factor measuring the length of the yielding plateau. The rotation in the strain-hardening range is given by:

$$\theta_{ph} = \int_0^{\frac{L}{2}-\frac{L_p}{2}} \chi dx + \int_{\frac{L}{2}-\frac{L_p}{2}}^{\frac{L}{2}} \chi dx =$$

$$= \frac{\chi_p}{2}\frac{L - L_p}{2} + \frac{\chi_p + \chi_{ph}}{2}\frac{L_p}{2} = \left[1 + (s + s_h - 1)\left(1 - \frac{1}{m_h}\right)\right]\frac{M_p L}{4EI} \tag{6.32}$$

The plastic rotation is:

$$\theta_r = \theta_{ph} - \theta_p = (s + s_h - 1)\left(1 - \frac{1}{m_h}\right)\frac{M_p L}{4EI} \tag{6.33}$$

If the plastic buckling of the compression flange and web occurs at the moment M_b (Fig. 6.22b), it results $m_h = m_b$. The corresponding plastic rotation is:

$$\theta_{ru} = (s + s_b - 1)\left(1 - \frac{1}{m_b}\right)\frac{M_p L}{4EI} \tag{6.34}$$

where

$$s_b = s + \frac{m_b - 1}{\varepsilon_y e_h} \tag{6.35}$$

The rotation ductility corresponding to the plastic buckling results:

$$\mu_\theta = \frac{\theta_b}{\theta_p} = (s + s_b - 1)\frac{L_p}{L} = (s + s_b - 1)\left(1 - \frac{1}{m_b}\right) \tag{6.36}$$

It is very clear that the rotation ductility depends on:

-the length of the plastic zone, L_p, which is directly related to the moment gradient;

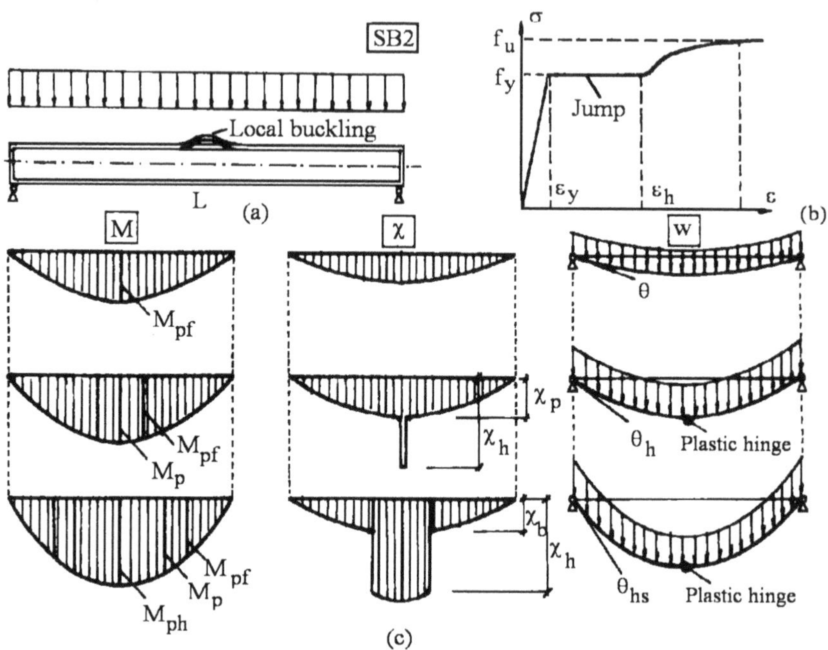

Figure 6.23: Behaviour of standard beam SB2: (a) Beam scheme; (b) Material behaviour; (c) Steps in beam behaviour

-the length of the yielding plateau, characterized by the coefficient s, which is a material feature;

-the level of incursion in the strain-hardening range, characterized by the coefficient s_b, depending on the critical load of the flange plastic buckling.

(ii) *Standard beam SB 2 type*. The beam with uniformly distributed load is used to study the ductility of beams under weak gradient moment. The behaviour of the beam is presented in Fig. 6.23 and it respects the general feature of the concentrated load beam, but the length of the plastic zone is larger:

$$\frac{L_p}{L} = \left(1 - \frac{M_p}{M_{ph}}\right)^{1/2} = \left(1 - \frac{1}{m_h}\right)^{1/2} \tag{6.37}$$

The plastic rotation is obtained by integrating the curvature diagram. The rotation corresponding to the plasticization of flanges results (Fig. 6.24a):

$$\theta_{pf} = \int_0^{\frac{L}{2}} \chi dx = \frac{M_{pf}L}{3EI} \tag{6.38}$$

and for the full plasticization of the cross-section:

$$\theta_p = \frac{M_pL}{3EI} \tag{6.39}$$

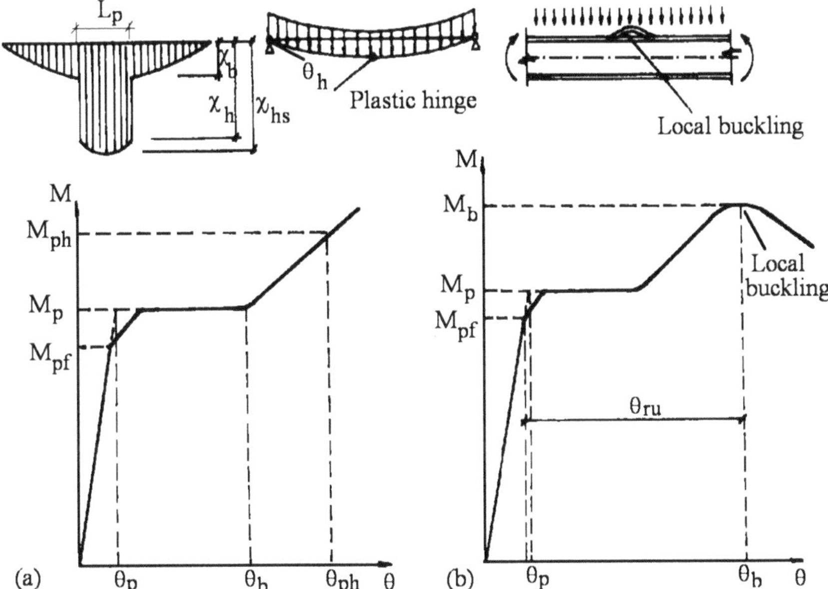

Figure 6.24: Determination of beam rotation for SB2 beam; (a) Moment-rotation curve; (b) Influence of compression flange buckling

The rotation in the strain-hardening range is:

$$\theta_{ph} = \int_0^{\frac{L}{2}-\frac{L_p}{2}} \chi dx + \int_{\frac{L}{2}-\frac{L_p}{2}}^{\frac{L}{2}} \chi dx =$$

$$= \frac{2}{3}\chi_p \frac{L-L_p}{2} + \frac{\chi_p + 2\chi_{ph}}{3}\frac{L_p}{2} = \left[1 + \left(\frac{s+2s_h}{2} - 1\right)\left(1 - \frac{1}{m_h}\right)^{1/2}\right]\frac{M_pL}{3EI} \tag{6.40}$$

where s, s_h, m_h are previously defined.

The plastic rotation is:

$$\theta_{ru} = \theta_{ph} - \theta_p = \left(\frac{s+2s_h}{2} - 1\right)\left(1 - \frac{1}{m_h}\right)^{1/2}\frac{M_pL}{3EI} \tag{6.41}$$

If the plastic buckling of compression flange and web occurs at moment M_b (Fig. 6.24b), $m_h = m_b$ and $\theta_r = \theta_b$ must be substituted in (6.41). The corresponding rotation ductility is given by:

$$\mu_\theta = \frac{\theta_b}{\theta_p} = \left(\frac{s+2s_b}{2} - 1\right)\left(1 - \frac{1}{m_b}\right)^{1/2} \tag{6.42}$$

(iii) *Comparison between the ductility of standard beam types.* A comparison between the behaviour of the two standard beam types is presented

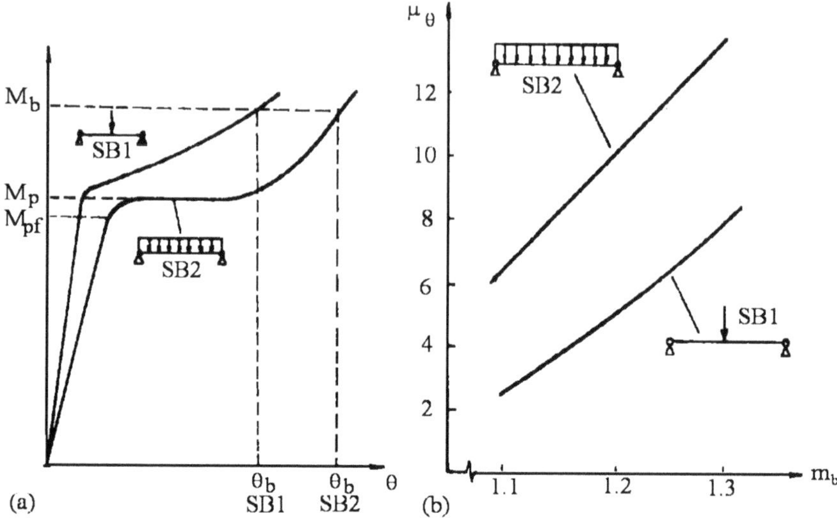

Figure 6.25: Comparison between SB1 and SB2 beams behaviour: (a) Moment-rotation curves; (b) Rotation capacity

in Fig. 6.25. It is clear that, for a given moment M_b corresponding to the plastic buckling of flange and web, the largest rotation ductility is obtained for distributed load and the lowest value results for concentrated load. This observation shows the great influence of the moment gradient on the rotation ductility, which is usually neglected in code specifications.

(iv) *Ductility of beam-columns.* Due to the presence of the axial force, the axis of zero strain moves towards the tension zone and an asymmetry in the strain between the two flanges occurs (Fig. 6.26a). If the (5.37a) condition is satisfied, the tension flange works in the elastic range, while for the (5.37b) condition, the tension flange is in the plastic or hardening range.

The rotation can be evaluated by integrating the moment versus curvature diagram (Mazzolani and Piluso, 1996) (Fig. 6.26b):

-for $m_p < 2n_p + 1$:

$$\theta_{ph} = \frac{1}{(m_h - n_p)}\Bigg\{(1 - n_p)^2 +$$

$$+(m_h - 1)\left[1 - 2n_p + s + (m_h - 1)\frac{E}{E_r}\right]\Bigg\}\frac{M_pL}{EI_e} \tag{6.43}$$

-for $m_p > 2n_p + 1$

$$\theta_{ph} = \frac{1}{(m_h - n_p)}\Bigg\{1 + n_p^2 - 2n_p(m_p - 1) + 2s(m_p - n_p - 1) +$$

$$+4n_p(m_p - n_p - 1)\frac{E}{E_r} + (m_p - 2n_p - 1)^2\frac{E}{E_h}\Bigg\}\frac{M_pL}{EI_e} \tag{6.44}$$

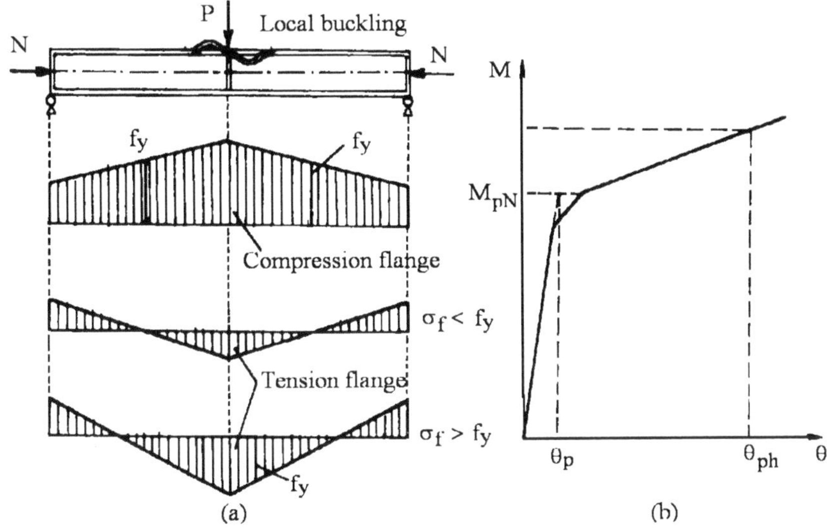

Figure 6.26: Beam-column behaviour: (a) Beam-flanges behaviour; (b) Moment-rotation curve

Taking into account that the rotation, corresponding to the first yielding, is:

$$\theta_p = (1 - n_p)\frac{M_p L}{4EI_e} \tag{6.45}$$

the rotation ductility is given:

-for $m_p < 2n_p + 1$

$$\mu_\theta = \frac{1}{m_h - n_p}\left\{(1 - n_p) + \frac{m_h - 1}{1 - n_p}\left[1 - 2n_p + s + (m_h - 1)\frac{E}{E_r}\right]\right\} - 1 \tag{6.46a}$$

-for $m_p > 2n_p + 1$

$$\mu_\theta = \frac{1}{(m_h - n_p)(1 - n_p)}\left\{1 + n_p^2 - 2n_p(m_p - 1) + 2s(m_p - n_p - 1) + \right.$$

$$\left. + 4n_p(m_p - n_p - 1)\frac{E}{E_r} + (m_p - 2n_p - 1)^2\frac{E}{E_h}\right\} \tag{6.46b}$$

The available rotation capacity in function of m_p and n_p is presented in Fig. 6.27. One can see that for each value of m_p a minimum rotation ductility occurs.

6.3.4. Actual Moment-Rotation Curve

There are two typical moment-rotation curves presented in Fig. 6.28.

(i) *Moment gradient* (Fig. 6.28a). In this case there are some characteristic points in the actual moment-rotation curve which mark some important

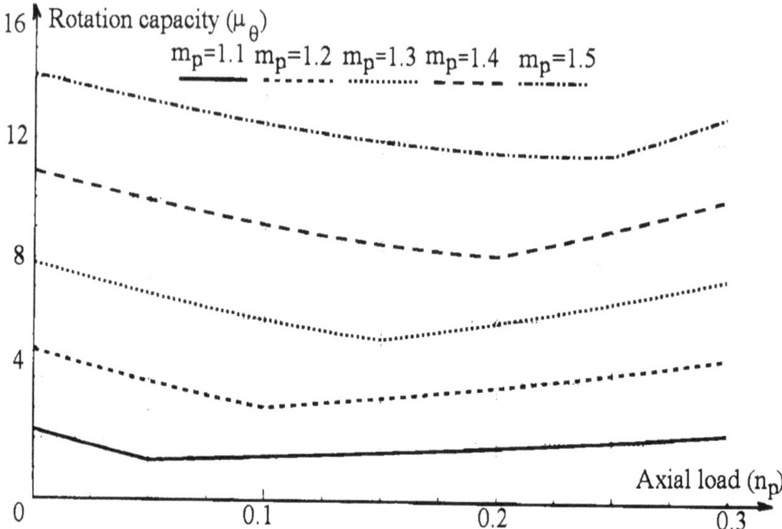

Figure 6.27: Influence of axial load on rotation capacity (after Mazzolani and Piluso, 1996)

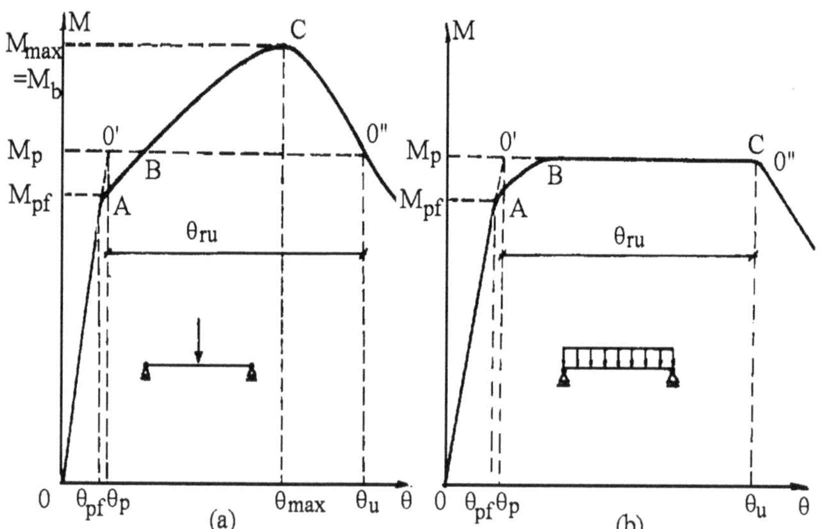

Figure 6.28: Actual moment-rotation curves: (a) Moment gradient; (b) Constant or quasi-constant moment

changes in behaviour. The first point A refers to the reaching of the flange yielding, corresponding to the bending moment M_{pf} and the rotation θ_{pf}; the member behaviour is elastic until this point. The second point B is defined by the occurrence of fully plastic moment M_p and the first plastic

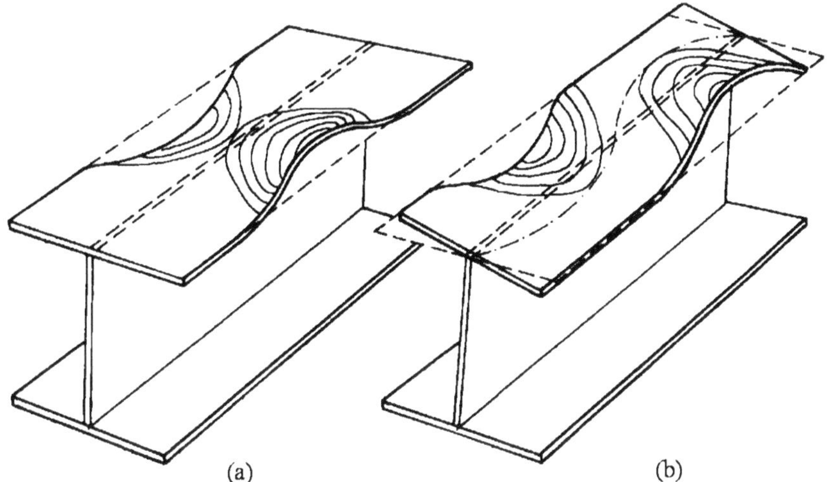

Figure 6.29: Buckling modes: (a) In-plane local buckling; (a) Out-of plane lateral-torsional buckling

rotation θ_p. To develop the plastic hinge, the rotation must increase and, as it has been shown in previous sections, the behaviour in strain-hardening range occurs. This fact explains the increasing of bending moment over the theoretical values of fully plastic moment. The maximum value M_{max} for and plastic rotation θ_{max} is reached at the point C, when plastic buckling occurs in the yielding zone of the compression flange. After point C the bending moment begins to decrease with the increasing of rotation and the equilibrium of the beam becomes unstable. One can see that there are some important differences between the actual behaviour and the bilinear relationship O-O'-O'' assumed in the simple plastic theory. Because in the first part of curve the difference between actual and simplified behaviours is not so important, for practical purposes it is reasonable to neglect the difference between flange yielding and full yielding $M_{pf} = M_p$.

(ii) *Constant or quasi-constant moment* (Fig. 6.28b). In this case, the first part of the curve is the same as for moment gradient. The differences can be observed in the post-yielding range. Due to the fact that in this case the plastic hinge can occur without or with a weak incursion into the hardening range, a plateau in the moment-rotation curve is noticed, contrary to the moment gradient, where this plateau is missing.

(iii) *Buckling modes.* In both the above described cases, the observed decreasing of the moment-rotation curve is due to:

-local buckling of flange (Fig. 6.29a) and/or web;

-lateral-torsion buckling (6.29b);

-interaction of these two different buckling modes.

Details concerning the influence of buckling modes on the rotation capacity will be presented in the next Chapters devoted to different section types.

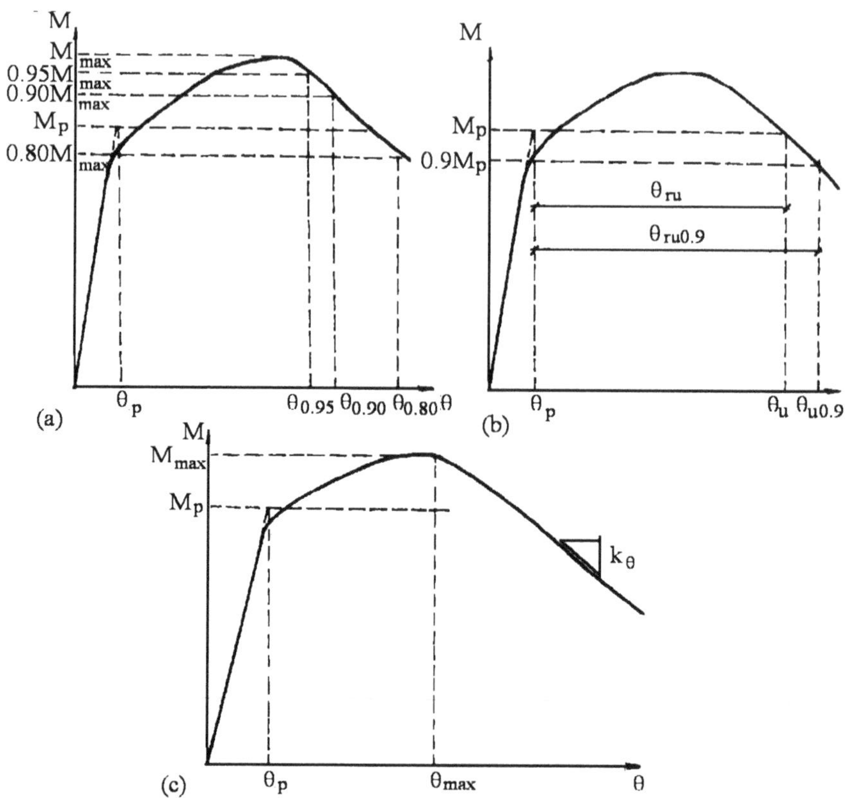

Figure 6.30: Definition of ultimate rotation capacity: (a) Related to the maximum moment; (b) Related to the fully plastic moment; (c) Related to post-buckling slope

6.3.5. Definitions of Ultimate Rotation Capacity for Monotonic Statically Loads

One must recognize that there is no standard definition of the ultimate rotation capacity, which is universally accepted by all the specialists. The analysis, of the research works presented in the literature, shows that there are three different approaches to evaluate this rotation:

(i) *Definition related to the maximum moment* (Fig. 6.30a). The rotation capacity definition is related to the maximum moment. This approach is proposed by Kato (1989, 1990), Mazzolani and Piluso (1992, 1993, 1995, 1996), Aiello and Ombres (1995), Suzuki et al (1994), Georgescu and Mazzolani (1996), Kemp (1985, 1996), Kemp and Dekker (1991), Dekker (1998). Gioncu and Mazzolani (1995) have recognized that the values of rotation capacity corresponding to maximum moment are very low and this approach can be used when a bilinear curve with hardening characterizes the structural analysis. In case the post-yielding curve has a quasi-horizontal plateau, this definition is too conservative. For this reason, by using the maximum moment as a reference value, Nakashima (1992, 1994) considers a fraction of this value (0.95; 0.90; 0.80) for determining the rotation capacity. This procedure gives a more rational value of the rotation capacity, but there are no rational criteria for choosing the moment level corresponding to the ultimate situation.

(ii) *Definition related to the fully plastic moment* (Fig. 6.30b). The rotation capacity is determined in the lowering post-buckling curve at the intersection with the theoretical fully plastic moment. This procedure is used by Kuhlmann (1986,1989), Roik and Kuhlmann(1987), Spangemacher (1991), Spangemacher and Sedlacek (1992a,b), Boerhave et al (1993), Climenhaga and Johnson (1972), Ivanyi (1979a,b, 1995, 1993), Piluso (1995), Gioncu et al (1989, 1995, 1996a). This definition is given also in the Background Document of EC 3. This method is very useful for practical purposes, because in many computer programs for structural analysis under seismic actions, a bilinear moment-rotation curve, with horizontal post-yielding behaviour, is used. Dekker (1998) has shown that it is very difficult to measure the rotation accurately, because the moment-rotation curve is nearly horizontal in that particular region. This observation can be justified by the large scatter of experimental values, when they are compared with the theoretical results. By examining the experimental data Dekker proposes to evaluate for practical purposes the ultimate rotation related to fully plastic moment as the double of the rotation at maximum moment, which is easier to calculate. There are some cases in which this definition is useless, especially for quasi-constant moments, when the experimental moment does not reach the theoretical value of the fully plastic moment.

(iii) *Definition related to post-buckling slope* (Fig. 6.30c). The rotation capacity is considered in a more general way. In addition to the rotation corresponding to the maximum moment, the slope k_θ of the decreasing branch of the moment-rotation curve is determined. This definition used by Axhag (1995) and Espiga (1998), is very promising, because it is very important to know the member behaviour after the local buckling. But this procedure is very difficult to be used in practice for structural analysis, because usually just the bilinear moment-rotation curve can be considered in the common computer program. Only very sophisticated computer programs can use a three-linear curve, but they normally operate for high performance research works.

Under these conditions, the approach related to the fully plastic moment seems to be the most reliable. The formula to calculate the ultimate rotation capacity is given by:

$$\mu_{\theta r} = \frac{\theta_{ru}}{\theta_p} = \frac{\theta_u}{\theta_p} - 1 \tag{6.47}$$

or

$$\mu_{\theta r0.9} = \frac{\theta_{ru0.9}}{\theta_p} = \frac{\theta_{u0.9}}{\theta_p} - 1 \tag{6.48}$$

where the index 0.9 refers to the case when the rotation capacity is determined for $0.9M_p$, θ_u is the ultimate rotation. θ_p is the rotation cor-

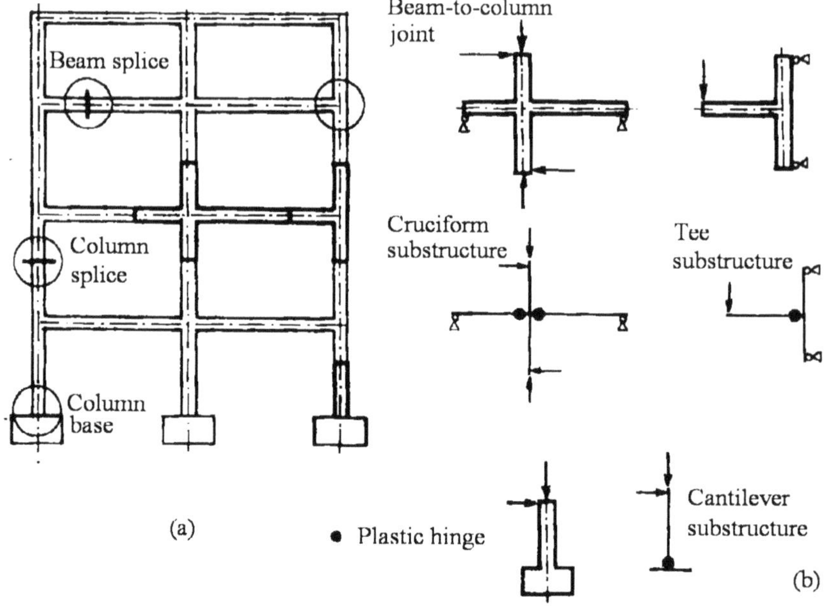

Figure 6.31: Joint types: (a) Structure scheme; (b) Substructure types

responding to the first plastic hinge, θ_{ru} is the plastic ultimate rotation. The alternative to use M_p or $0.9M_p$ for evaluating the rotation capacity remains a decision of the designer, in function of the beam behaviour. For gradient moment it is possible to choose both procedures, but in case of quasi-constant moment the reduced moment use is recommended.

Due to the above mentioned scattered experimental values, the available ultimate rotation must be determined from the relation:

$$\mu_{\theta a} = \frac{1}{\gamma_m} \left\{ \begin{array}{c} \mu_{\theta r} \\ \mu_{\theta r 0.9} \end{array} \right\} \qquad (6.49)$$

where γ_m is a partial safety factor which covers the uncertainties in determining the available rotation capacity. Values of $\gamma_m = 1.3...1.5$ have been proposed by Gioncu and Petcu (1997) and Spangemacher (1991).

6.4 Effect of Joints on the Member Ductility

6.4.1 Representation of Joints in the Structure Analysis

Conventional joints (beam-to-column, column base, beam and column splices) (Fig. 6.31) have been treated either as nominally pinned without any strength or stiffness (simple joints) or as fully rigid with full strength (continuous joints) (Huber et al, 1998). In addition, the traditional design approach assumes that the joint, despite its finite dimensions, is concentrated in one point at the intersection of the member centrelines. This approach has been used in the past in structure analysis due to incomplete knowledge on joint moment-rotation behaviour and due to lack of modelling. The use of these assumptions can lead to an non-economical joint detailing as well as wrong assessment of the actual structural behaviour in terms of rigidity, strength and ductility. The Northridge and Kobe earthquakes dramatically demonstrated that the above hypothesis has been, in many cases, erroneous, because the premature fracture of joints have impeded the development of plastic hinges in members, contrary to the base approach of the analysis (Mazzolani, 1998).

A modern approach now is basically oriented to develop efficient joint types firstly and then to take their actual behaviour into consideration in frame analysis. An accurate modelling of the joint behaviour forms the basis for the correct and safe design, where the available joint properties have the leader role (Huber et al, 1998). So, the configuration of standard beams (studied in the previous sections) must be modified, by including the effect of joints. So, Fig. 6.31 shows the manner in which the behaviour of a complex structure can be substituted by the sum of the behaviour of some simple substructures, composed by beams, columns and joints. These substructures can be defined in function on the inflection point positions in the structure. There are three substructure types (Fig. 6.31b):

-*cruciform substructure* for inner columns, where two beams are connected to the column;

-*tee substructure* for outer columns, where a single beam is connected to the column;

-*cantilever substructure* for the column base, where the connection of column to foundation is modelled.

6.4.2. Behaviour of Joints

(i) *Definitions.* Fig. 6.32 gives an example of a joint. Some definitions are necessary for the terms used in the literature (Gioncu et al, 2000a, Faella et al, 2000):

-panel zone is the portion of column web corresponding to the connection dimensions;

-connection is the location at which two members are connected and the means of this interconnection (welds, bolts, plates, etc);

-joint is the assembly of basic components (panel zone and connections) which enables members to connect together in such a way that the relevant internal forces and moments can be transferred among them;

-node covers the joint and the adjacent beam and column-ends where plastic deformations may occur.

(ii) *Joint types.* There are several technological systems for connecting members. Non-linear behaviours ranging from the quasi-perfect rigid to the flexible one are possible (Cosenza et al, 1989, Mazzolani and Piluso, 1996, Faella et al, 2000). The intermediate positions correspond to the range of semi-rigidity. Some common types of joints, listed according to the lowering

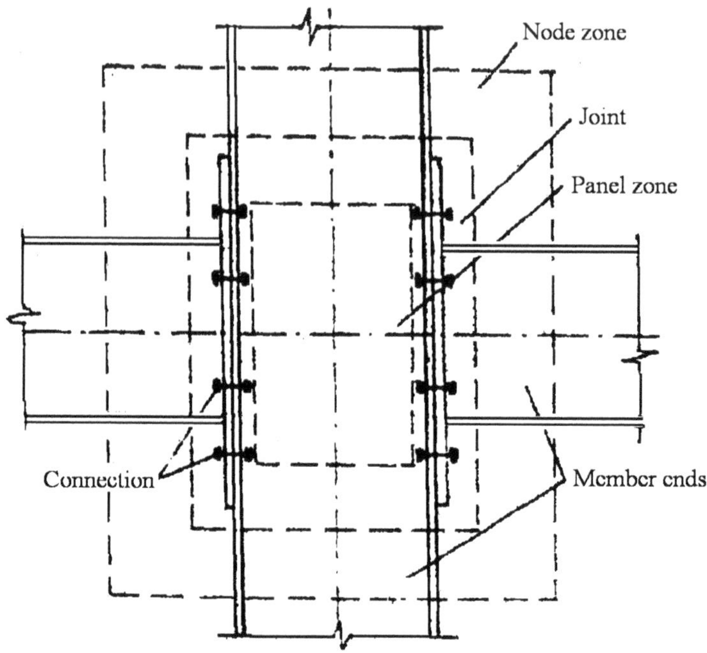

Figure 6.32: Panel, joint and node definitions (after Gioncu et al, 2000a)

of rigidity, are (Fig.6.33):

-fully welded;

-extended end plate;

-tee stubs;

-flush end plate;

-flange and web angles;

-flange angles;

-web angles.

The fully welded and extended end-plate joints may be considered as quasi-perfectly rigid and the standard beam may be used for determining the member ductility. The web angle joint is practically a pinned one and cannot develop any ductility. For the semi-rigid joints, the *generalized standard beam* (see Section 6.4.4) must be used for determining the member ductility.

(iii) *Joint collapse.* In addition to the influence on the global frame rigidity, the presence of joint may introduce a new source of collapse (Fig. 6.34). The collapse modes of joints are (Gioncu et al, 2000):

-for welded joints, buckling or crushing of panel zone or fracture in tension zone;

-for bolted joints, fracture of bolts, end-plates or weld between plate and beam.

(iv) *Component method for joint ductility.* As joints are composed by many parts, the determination of this complex microstructure is a very difficult task. Alternatively, a macroscopic view of the whole joint, by sub-

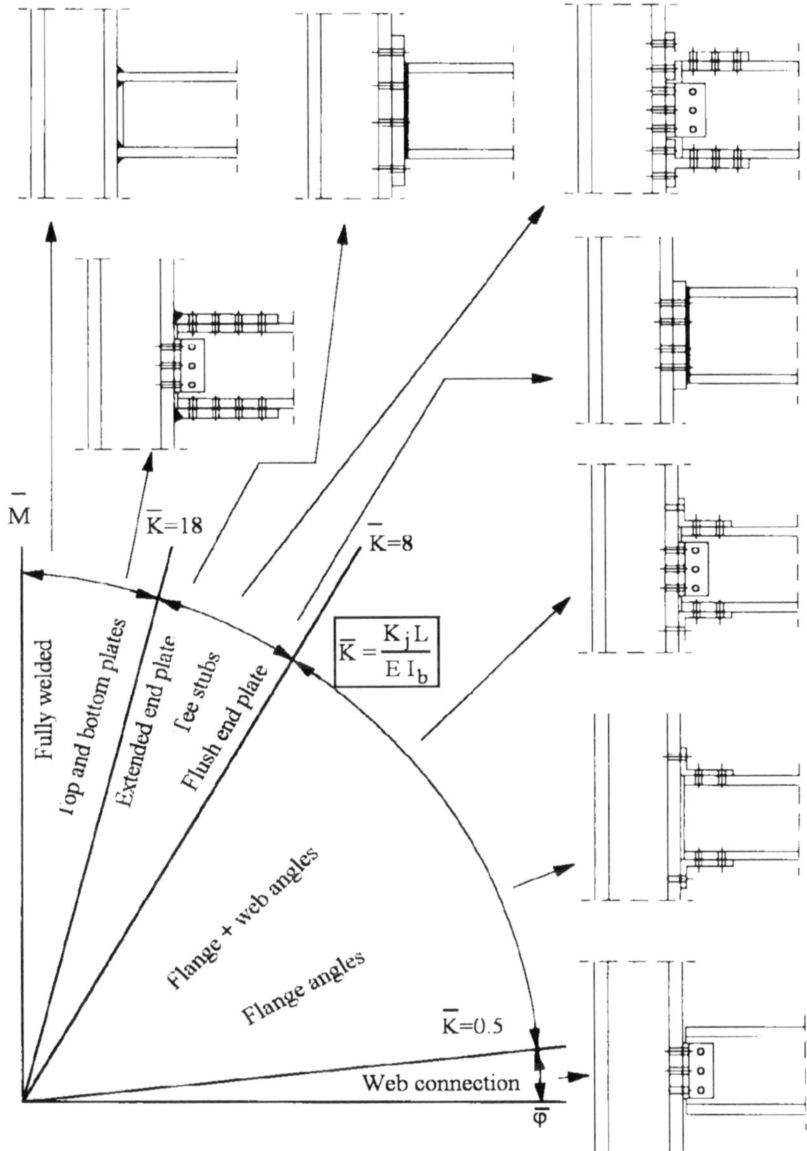

Figure 6.33: Typical values of the non-dimensional stiffness of different connections (after Cosenza et al, 1989, Mazzolani and Piluso, 1996)

dividing it into individual basic components, has proved to give good results. This procedure can be developed through three distinct steps (Jaspart et al, 1998):

-identification of the joint components;

-determination of the properties of these components;

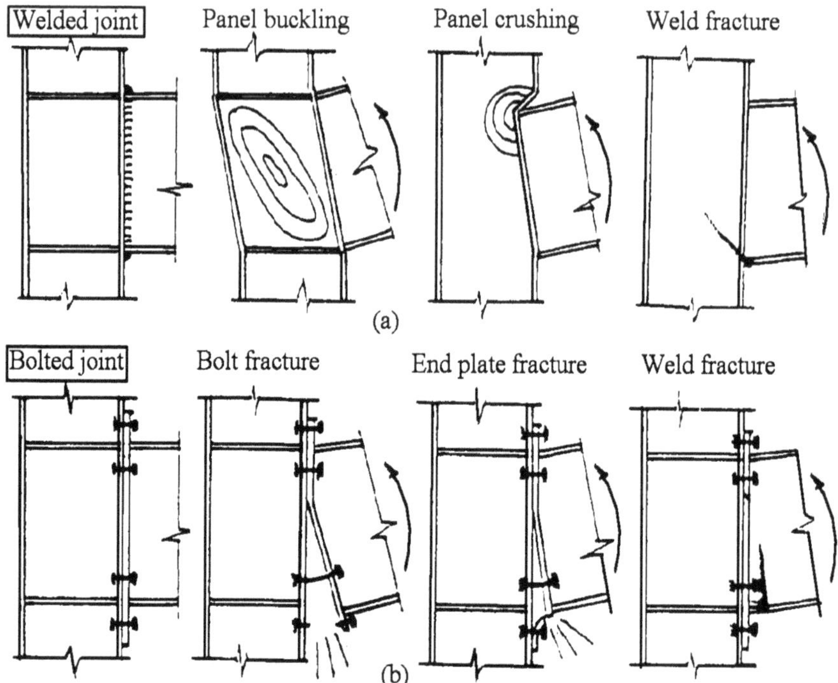

Figure 6.34: Joint collapse types: (a) Welded joints; (b) Bolted joints (after Gioncu, 2000a)

-assembly of the component behavioural curves.

From the ductility point of view, the components may be classified as (Fig. 6.35):

-high ductile component, with an almost unlimited increasing of the deformation capacity. When the ultimate strain is reached, a ductile fracture causes collapse.

-*limited ductile component* is when the moment-rotation curve presents a moderate decreasing in capacity, after attending the maximum values, and the collapse is produced by local plastic buckling;

-*reduced ductile component*, with brittle fracture. Especially high strength bolts and some welds present this behaviour type.

The main hypothesis of the component method states that the overall behaviour of joints is dictated by the behaviour of the weakest component. The assembly of two components is presented in Fig. 6.36, the components being connected in series. The feature of the two components is different: one is high ductile, while the other has limited or reduced ductility. One can see that the ductility of the assembly depends only on the weaker strength component. It is absolutely irrelevant if the stronger element is ductile or brittle.

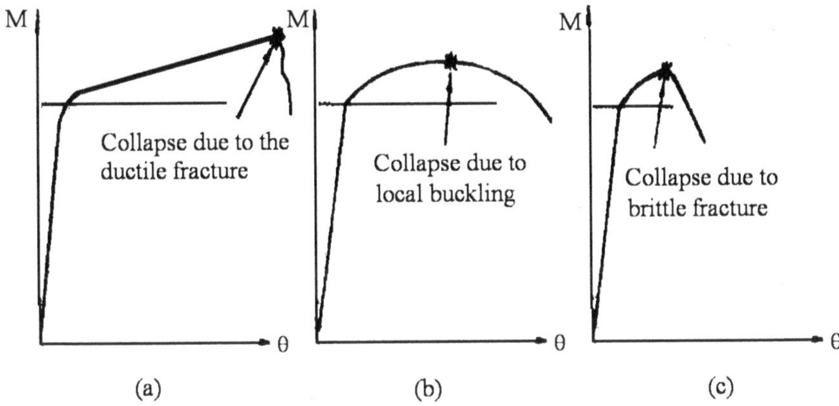

Figure 6.35: Component ductility types: (a) High ductility; (b) Limited ductility; (c) Reduced ductility (after Gioncu et al, 2000a)

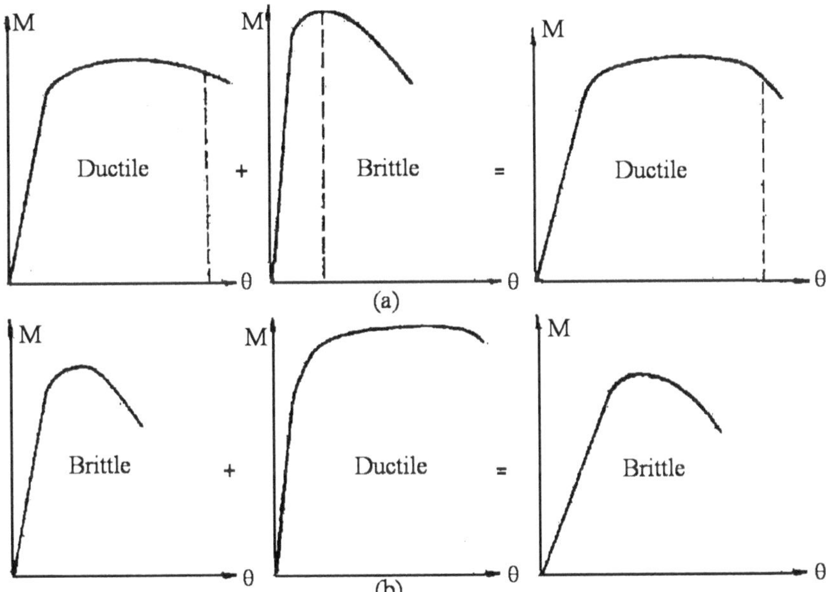

Figure 6.36: Component assembly types: (a) Ductile behaviour; (b) Britle fracture (after Gioncu et al, 2000a)

A special case occurs when the ductile component is dominant. However, the strongest brittle component has the carrying capacity close to the first component and an interaction between these two components may occur. It is possible that, due to randomness of material properties, the behaviour of the joint should be determined by the second component with brittle behaviour. So, in this case, a statistical analysis has to be considered in the estimation of the joint behaviour.

6.4.3. New Joint Design Philosophy

(i) *Overview of steel structure damage.* As a consequence of the damage observed in steel frames, built in the US and Japan after the Northridge and Kobe earthquakes respectively, major research and development programs have been undertaken. In order to understand the behaviour of steel structures in condition of near-source strong earthquakes, a very careful comparative examination of damage, in both Countries, has been performed by Nakashima et al (2000). There are similarities and differences which have played a great importance in the development of the new design philosophy of steel structures. Some notable similarities are summarized as follows:

-the Northridge and Kobe earthquakes have shown for the first time the potential significant damage in welded steel structures;

-the old and new buildings were damaged at the same level, showing that damage was not associated with old technologies and design practices;

-even if many joint fractures were disclosed, no steel building collapsed, indicating the great adaptability of these structures under difficult conditions;

-many welded beam-to-column connections failed and fractured, indicating that welded connections were one of the weakest locations in steel structures;

-no damages in old riveted connections were depicted.

At the same time there are some differences:

-the majority of fractures in Northridge steel structures have not shown plastification in either beams or columns. Contrary, in Kobe steel structures, at beam-to-column fractured joints there are evident plastic deformations in connected members, showing that the beams dissipated some energy before fracture;

-the material properties controls are different in the two Countries;

-the welding processes and the erection procedures are also different: the Japanese welding was often done in the shop, whereas the American welding was commonly done in the field;

-connection details are different, the Japanese column being made of box-sections, whereas I-sections are used in US. This basic difference is accompanied by many differences in local connection details (see Chapter 2);

-the Japanese steel structures were built with rigid connections everywhere, whereas in US rigid connections were concentrated only in the perimetral frames.

(ii) *Post-Northridge and post-Kobe attitudes.* The main conclusion resulting after these post-earthquake analyses is that the joint fracture before the development of plastic hinges in connected members must be avoided, due to the brittle characteristic of these fractures. Keeping this scope in mind, the joints have to possess a sufficient overstrength, with respect to connected members, to allow for complete development of the plastic deformation capacity of these members.

But, due to the differences between the type of fracture in steel structures in Japan and the US, two different attitudes concerning the measures to achieve this demand have been followed in these Countries (Nakashima et al, 2000). In Japan it is a dominant a point of view that a sufficient ductility capacity can be ensured only by modifying connection details with good welding. In the US the attitude has pursued three courses:

-firstly, to move the plastic hinge away from the connection, by reinforcing the beam end or by reducing the beam section, considering that in this way the welds are protected;

-secondly, by substituting the welded connections with bolted ones, solutions are found to eliminate the welding disadvantages (e.g. residual stresses, stress concentration, brittle fracture, modification of parent material properties in the heat affected zones, etc);

-thirdly, the improving of local details for conventional non-reinforced joints.

(iii) *Improving of conventional joints.* Following this approach, the typologies of joints remain the same as before Northridge and Kobe earthquakes. The improving being referred only to a better consideration of joint material properties in condition of a strong action, joint behaviour and welding procedures, in order to obtain a sufficient *joint overstrength* with respect to connected members.

According to Eurocode 8 (1994), the overstrength requirement is automatically fulfilled in the case of fully welded connections with butt or fully penetrated welds. On the contrary, in the case of fillet weld connections or bolted connections it is deemed that the moment resistance of the joint is at least 1.2 times the design flexural resistance of the connected beam. A severe degree of overstrength is required by the Japanese code, where the value 1.3 is suggested. Mazzolani et al (1998) have shown that these design criteria are not sufficient to allow for the complete development of the plastic deformation capacity of connected members due the following reasons:

-a stringent limitation of the width-to-thickness ratios of the connected member section is required by EC 3 (ductile or compact sections) can completely prevent the occurrence of local buckling in plastic range, so that failure of a fully welded connection can occur near to the column flange in the heat affected zone prone to fracture;

-the actual overstrength of plastic member capacities due to the strain-hardening range is, in many cases, greater than the required value. Therefore, there are many member sections for which the EC 8 demanded overstrength which is not sufficient for developing the whole plastic deformation capacity of the connected members;

-the random material variability of connected components and fastening elements must be considered. The results of Mazzolani et al (1998) show that the overstrength level suggested by EC 8 is not sufficient if the variabil-

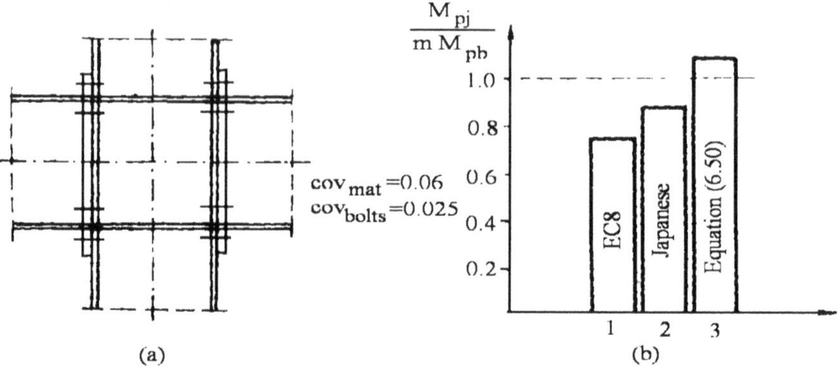

Figure 6.37: Joint overstrength: (a) Joint type; (b) Obtained main overstrenght after different requirements (after Mazzolani et al, 1998)

ity of material properties is considered. A slight improvement is obtained when the overstrength level suggested by Japanese code is applied.

A good approach is obtained if

$$\frac{M_{pj}}{mM_{pb}} \geq \frac{f_{yt}}{f_y} \frac{1}{1 - 1.64COV} \tag{6.50}$$

where M_{pj} is the fully plastic moment of joint, M_{pb} is the fully plastic moment of the connected beam, m is the numerical coefficient which considers the strain-hardening effect, f_{yt} is the yield stress value depending on the thickness of the beam flanges, and COV is the coefficient of variation of beam yield stress. So, for instance, if m = 1.29, f_{yt}/f_y = 1.32 and COV = 0.06, an overstrength value of 1.89 results, this is much greater than the one required by the Japanese code. Only for this high overstrength it is a surety, with a 95% probability, that the plastic hinge is formed in members and not in joint (Fig. 6.37).

Another direction of joint improving is related to the *welding process*, by using electrodes having large toughness, control of deposition rate and a more severe in-site inspection practice.

Regarding the *connection details*, many efforts have been made to improve them by:

-elimination of fracture sources, produced by backing bars. The vertical unfused interface between the backing bar and column flange act resulting in open fine cracks (Fig. 6.38a). In order to eliminate this effect, there are two possibilities: to place a continuous fillet weld along the backing bar contact to column flange (Fig. 6.38b) or to remove the backing bar and then apply a fillet weld (Fig. 6.38c);

-improving the geometry of web copes, used to weld across the full width of flange without interruption. The presence of this cope produces a sudden change in geometry at the toe and causes a stress concentration at the flange-web junction of the beam, being a source of cracking of the beam. The effect of modification of size and shape of access hole, aiming to reduce

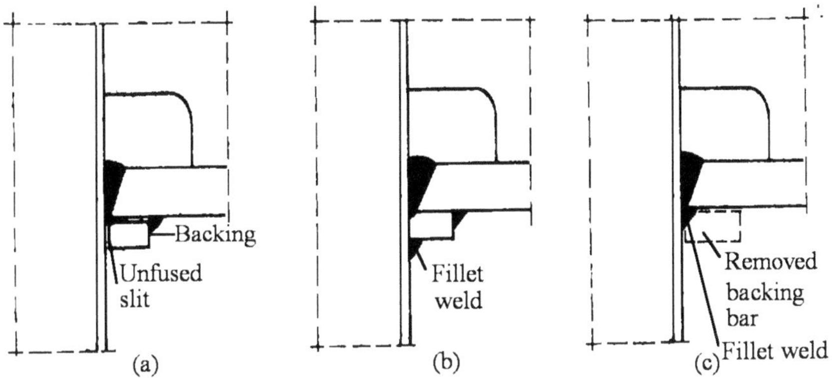

Figure 6.38: Elimination of fracture source: (a) Unfused slit; (b) Continuous fillet weld; (c) Removing the backing bar

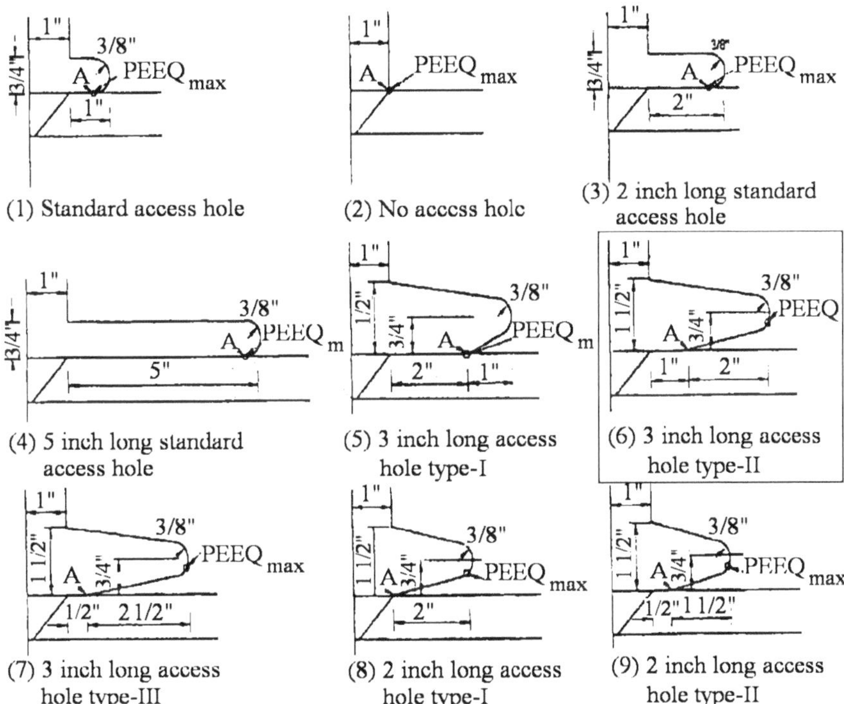

Figure 6.39: Improving the geometry of web cope, with PEEQ, the effective plastic strain (after Lu et al, 2000)

the stress concentration is discussed by Nakashima et al (1998) and Lu et al (2000). Different weld access hole configurations have been examined (Fig. 6.39), and the configuration (6) appears as the best, being recommended to be used for seismic-resistant structures. An alternative to this solution is the

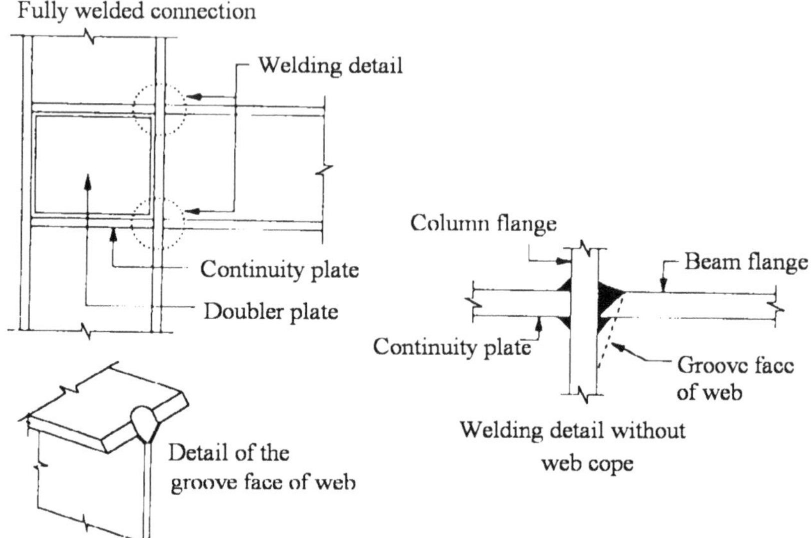

Figure 6.40: Welding without web cope (after Matsui and Sakai, 1992, Mazzolani and Piluso, 1995)

welding without cope (Matsui and Sakai, 1992, Mazzolani and Piluso, 1995, Nakashima et al, 2000, Suita et al, 2000) (Fig. 6.40). The experimental tests have shown that no early fracture occurred at the vicinity of welds in this connection type.

(iv) *Strengthening of beam ends* solution (Fig. 6.41) is used in order to move the plastic hinge away from the column face. A wide variety of strengthening solutions have been proposed since the Northridge and Kobe earthquakes. Some samples are illustrated in Fig. 6.42 (Chen et al, 1997, Engelhardt and Sabol, 1997, Anastasiadis and Gioncu, 1998, Bruneau et al, 1998):

-cover plates (Fig. 6.42a), in which the demand to welds at the column flange is reduced. The bottom cover plate is rectangular and sized wider than the beam flange to allow fillet welding between the two plates;

-upstanding ribs (Fig. 6.42b), in which one or two vertical ribs can be used for upper and bottom beam flanges;

-haunches (Fig. 6.42c), usually at the top and bottom sides or at bottom side only, are used especially to repair damaged joints;

-lateral reinforcing plates (Fig. 6.42d).

All experimental tests on these reinforced beam-ends demonstrated that plastic hinges always develop away from the welded connections. But this solution has the disadvantage of increasing the beam-end size; in order to maintain the strong column-weak beam design conception, it produces the increasing of the column size too.

(v) *Weakening of beam section*, closed to the connections ('dogbone' concept) (Fig. 6.43), in order to move the plastic hinge away from the con-

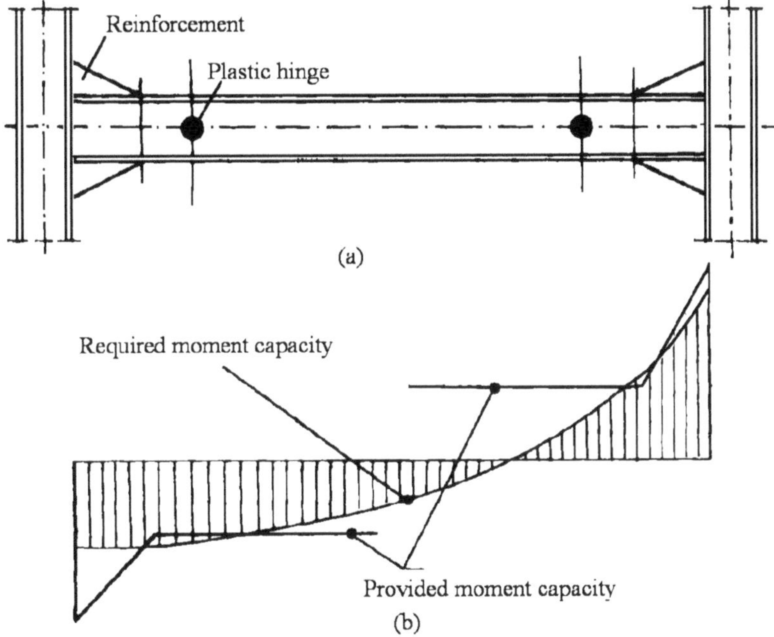

Figure 6.41: Strengthening of beam-ends: (a) Beam configuration; (b) Required and provided moment capacity

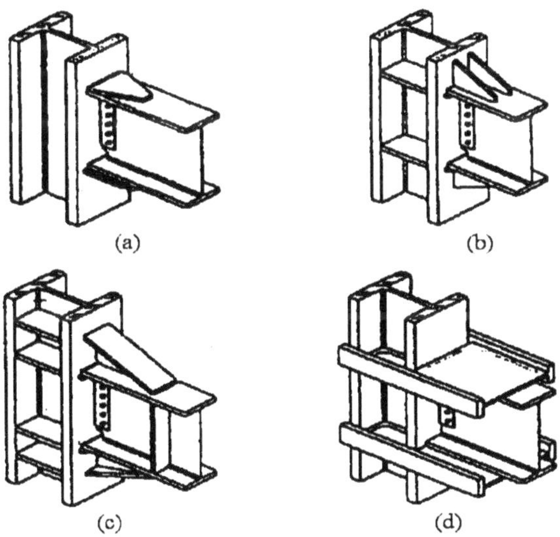

Figure 6.42: Strengthening types: (a) Cover plates; (b) Upstanding ribs; (c) Haunches; (d) Side plates (after Mazzolani, 2000)

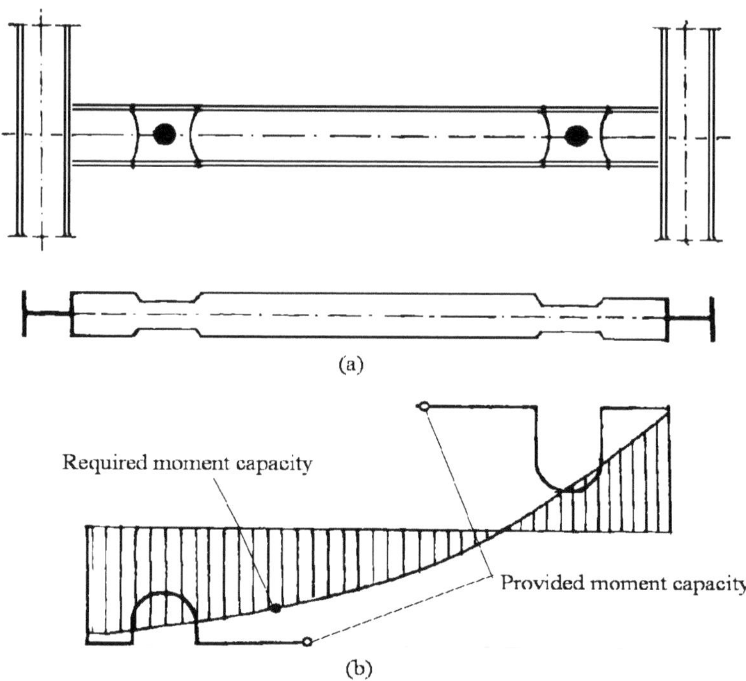

Figure 6.43: Weakening of beam sections: (a) Beam configuration; (b) Required and provided moment capacity

nections at a selected location. The idea was developed in principle during the 80's by Plumier (1996a,b) and patented by ARBED. The weakening can be obtained in different ways (Plumier, 1996a,b, Chen et al, 1996, 1997, Bruneau et al, 1998, Carter and Iwankiw, 1998, Anastasiadis and Gioncu, 1998)(Fig. 6.44):

-drilled holes (Fig. 6.44a) with constant or tapered diameters;

-polygonal cuts (Fig. 6.44b) with straight or smoother connecting lines between the straight lines;

-fully curved cut (Fig. 6.44c)

-tapered cut (Fig. 6.44d), in which the plastic resistance of reduced section is adjusted to the shape of the bending moment diagram.

In all cases the experimental tests have shown that no failure of connections occurs and plastic hinge is formed in the reduced section.

The advantage of this solution is, contrary to the reinforcing of beam-ends, that no increasing of column cross-section is required. In addition, the ductility of the beam is increased due to the reducing of flange slenderness in the cutted section. The disadvantage of this solution is the increased beam sensitivity to lateral-torsional buckling (Bruneau et al, 1998).

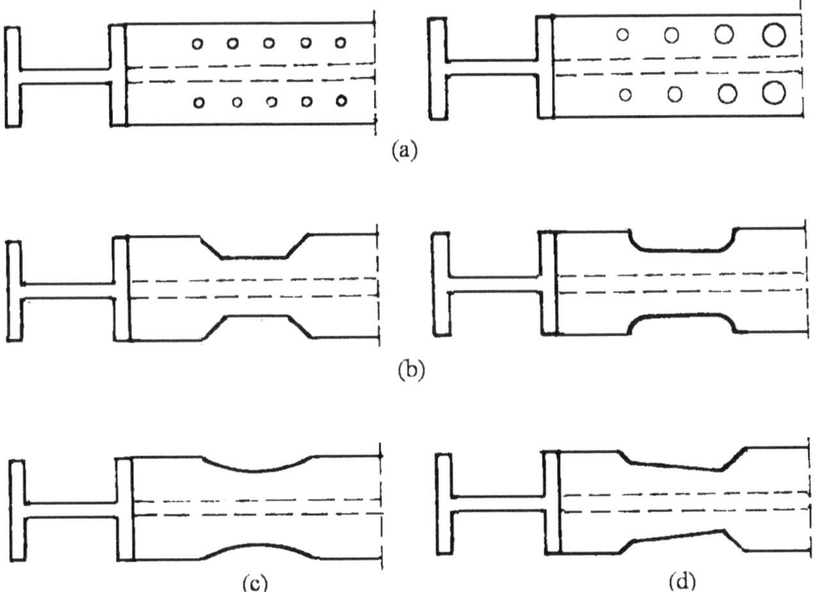

Figure 6.44: Weakening types: (a) With holes; (b) Polygonal cuts; (c) Curved cuts; (d) Tapered cuts (after Mazzolani, 2000)

6.4.4. Generalized Standard Beam

(i) *Joint influence on standard beam behaviour.* In the previous section (see 6.3.2), the behaviour of the standard beam is studied without considering the presence of the joint, assuming that it does not influence its behaviour. But, in some cases, this hypothesis must be modified, because the premature fracture of joints may impede the development of plastic hinges in members. Therefore, the configuration of the standard beam must be modified by including the joint presence.

A typical moment resisting frame, in which the joints are identified, is illustrated in Fig. 6.45a. The maximum earthquake moments are primarily produced at the column face (Fig. 6.45b). Fig. 6.45c shows the standard beam models which produce the same moment gradient as in the frame under earthquake loading. Contrary to the standard beams presented in Fig. 6.13, they contain a portion of the column, allowing for the possibility to include the connection effects. These substructures are the *generalized standard beams*, which can be used to study the influence of joints on the member ductility. As for the standard beams, there are two types (Fig. 6.45c):

-GSB 1, central concentrated load beam for the case of member under linear moment gradient;

-GSB 2, distributed load beam for the case of member under weak moment gradient.

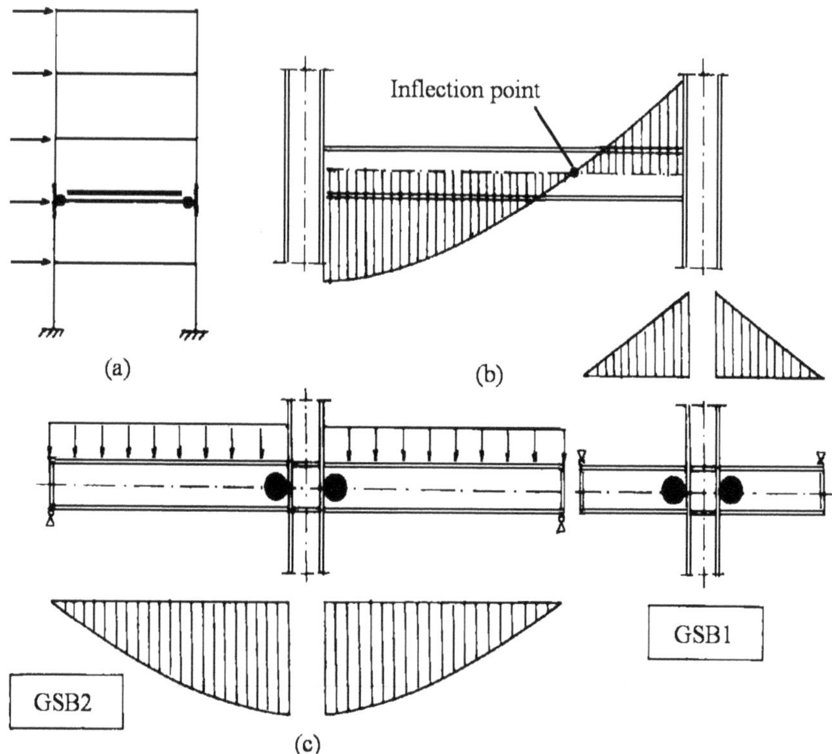

Figure 6.45: Generalized standard beams GSB: (a) Structure scheme; (b) Substructure configuration; (c) Generalized standard beam definition

(ii) *Classification of joints.* The joints may be classified according to their capability to restore the behavioural properties (e.g. rigidity, strength and ductility) of connected members, giving rise to different classification proposals (Gomes et al, 1998). With respect to the global behaviour of the connected members, two classes are defined (EC 3, 1998) (Fig. 6.46):

-fully restoring joints, which are designed in such a way to have behavioural properties always equal to, or higher than, those of connected members. The moment-rotation curve of the joint always lies above the one of the connected member. The existence of this joint may be ignored in the structural analysis;

-partially restoring joints, for which the behavioural properties of the connection do not reach those of the connected member, due to the lack of capability to restore either elastic rigidity, ultimate strength or ductility of the connected member. The moment-rotation curve of the joint could fail in some part below the connected member. The structural analysis has to consider the presence of such a joint.

With respect to a single behavioural property of the connected member, joints may be classified in (Fig. 6.47):

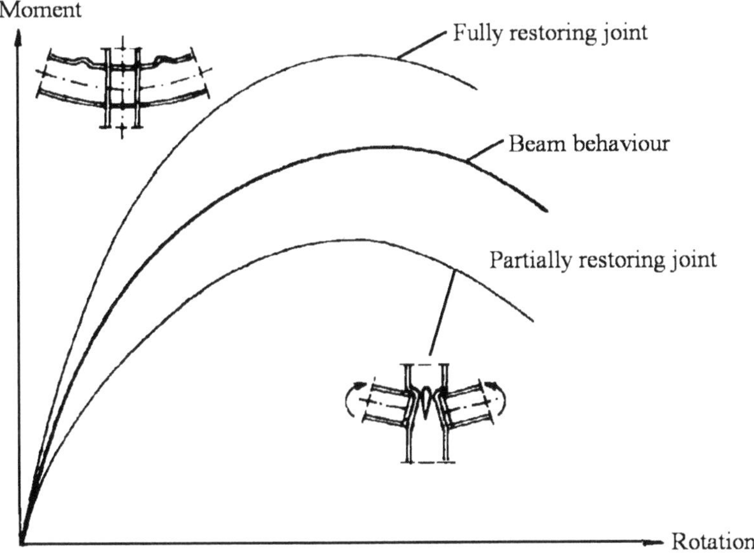

Figure 6.46: Joint types

-full strength (strength restoring) joints, or
-partial strength (strength non-restoring) joints,
with respect to strength of connections;
-rigid (rigidity restoring) joints, or
-semi-rigid (rigidity non-restoring) joints,
with respect to rigidity of connections;
-ductile (ductility restoring) joints, or
-semi-ductile (ductily non-restoring) joints,
-brittle joints (ductility less than the elastic rotation of connected members),
with respect to ductility of connections.

The joint should exhibit ductile or semi-ductile behaviour throughout the loading history. Joints which exhibit brittle behaviour prior to reaching the required plastic rotation must be avoided.

A general classification considering the three contemporary types of restoring (rigidity, strength and ductility) has been introduced in Eurocode 9 for Aluminium Structures (Mazzolani et al, 1996).

6.4.5. Improved Standard Beam

In order to consider the new philosophy of joint design, in which the location of plastic hinge is moved away from the column face, an *improved standard beam* must be considered.

(i) *Strengthened standard beam*, SSB, (Fig. 6.48a), for which one of the solutions presented in Fig. 6.42 is used for strengthening the member ends. The location of plastic hinges away from the column face should be deter-

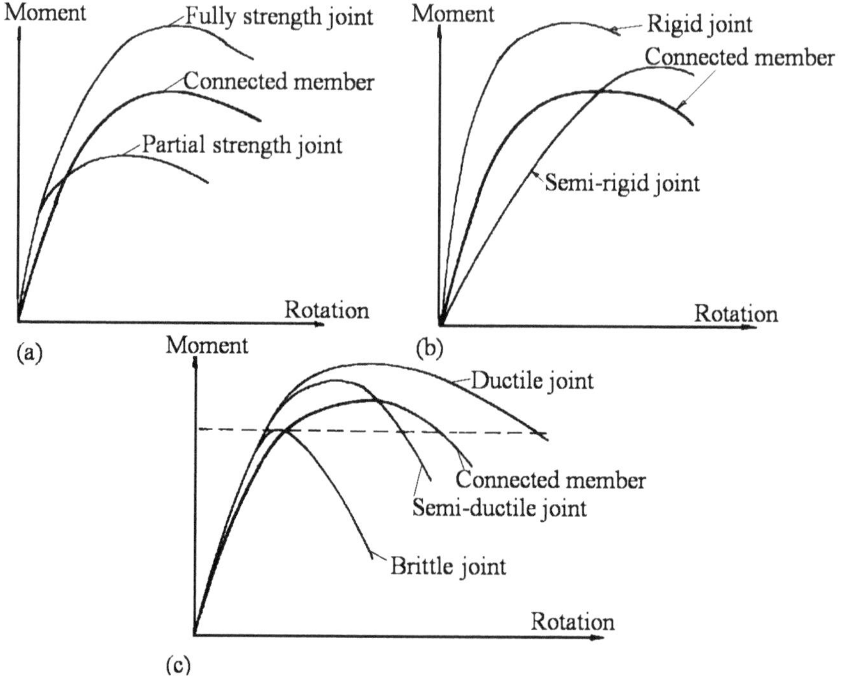

Figure 6.47: Joint classification: (a) Related to strength; (b) Related to rigidity; (c) Related to ductility

mined as a basic parameter for the calculation. For beams where gravity loads represent a small portion of the total flexural demand, the location of the plastic hinge may be assumed to occur at (Fig. 6.48b) (FEMA, 1997):

-$h/4$ beyond end of cover plate;

-$h/3$ beyond toe of haunch;

-$h/3$ beyond toe of vertical ribs.

The plastic hinge must form at a sufficient distance from the column flange face. But in the case of too long distance, the location of plastic hinge can be placed in a zone with reduced moment and the danger of plastic hinge occurring at the column face remains.

In order to avoid this incovenience and to complete the design of joint it is necessary to evaluate the shear and flexural strength demand at the column face. They may be determined by taking off the portion of beam-end located between the column face and plastic hinge (Fig. 6.48c):

$$M_f = M_p + V_p L_p - \frac{q L_p^2}{2} \tag{6.51}$$

where M_f is the moment at column face, M_p and V_p are the fully plastic moment and shear force of connected beam, L_p is the distance of plastic hinge from the column face and q is the gravity load.

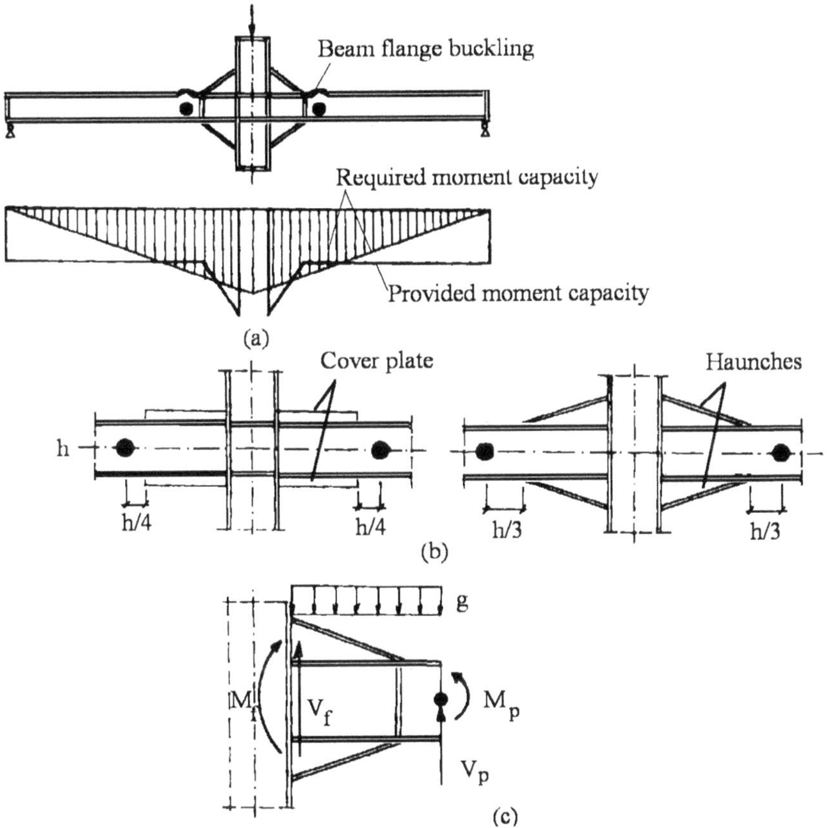

Figure 6.48: Strengthened standard beam SSB: (a) SSB scheme; (b) Position of plastic hings; (c) Column face forces

(ii) *Weakened standard beam*, WSB, (Fig. 6.49a), for which one of the solutions presented in Fig. 6.44, is used for weakening the beam at some distance from the column face. The reduced beam section should be located at a sufficient distance from the connections to avoid significant inelastic behaviour of the beam-to-column flange joint. Based on testing evidence, it appears that a value $L_1 = h/4$ is sufficient. The total length of the reduced section of beam flange, L_w, should be on the order of $3h/4$ to h. The location of plastic hinge, L_h, may be taken as (FEMA, 1997):

-straight reduced section, $L_w/2$;

-circular reduced section, $L_w/2$;

-tapered reduced section, $L_w/4$ towards the column from the location of the minimum section.

The plastic section modulus, Z_r, may be calculated from the equation (Fig. 6.49c):

$$Z_{pr} = Z_p - b_r t_f \left(d + \frac{t_f}{2}\right) \tag{6.52}$$

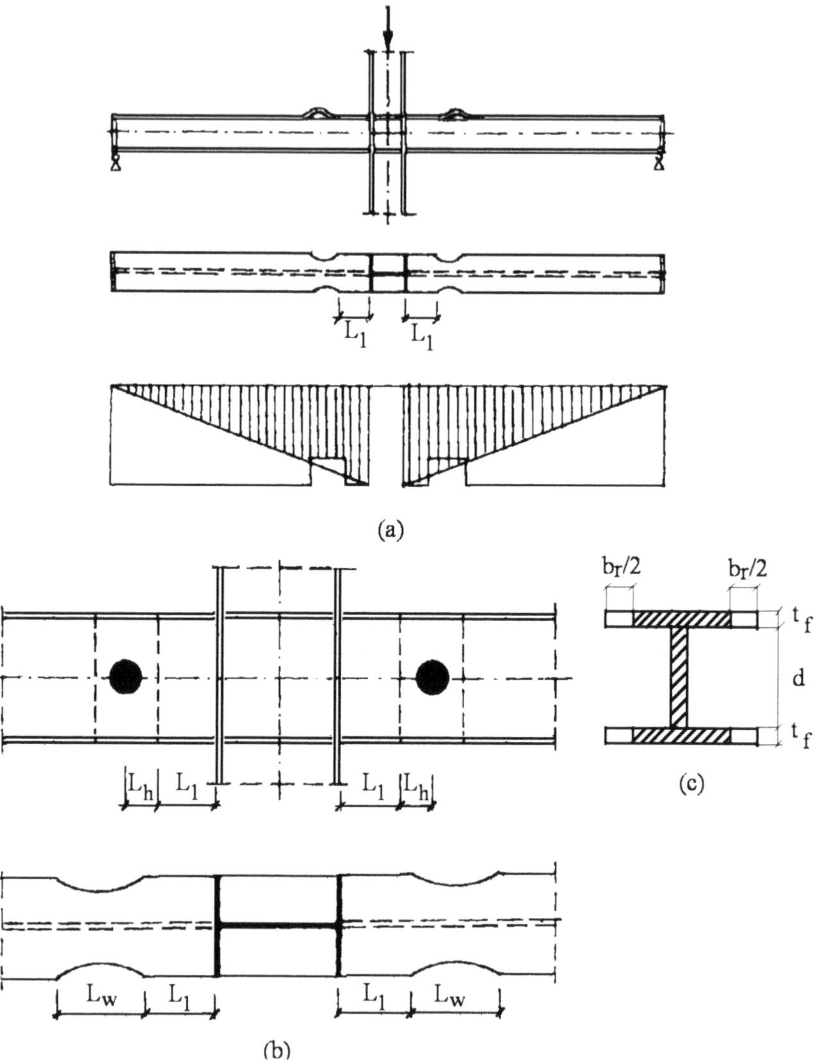

Figure 6.49: Weakened standard beam WSB: (a) WSB scheme; (b) Position of plastic hinges; (c) Reduced section

where Z_p is the plastic modulus of unreduced section, b_r is the total width of material removed from the beam flange. The material removed section must be determined by reducing the plastic moment 5–10% less than the bending moment determined in the location of the reduced section, in order to be sure that the plastic hinge occurs in the predetermined location.

6.5 Effects of Seismic Loading on the Member Ductility

6.5.1. Influence of Seismic Loading Type

Seismic-resistant structures are usually designed relying on their ability to sustain high plastic deformations. The earthquake input energy is dissipated through the hysteretic behaviour of the plastic hinges due to a number of cycles of seismic loading. The plastic rotation must show a stable hysteretic behaviour with a sufficient ductility to allow for dissipating this input energy. According to this design philosophy, the structure may be designed for lower forces than those it has to resist. This seismic design philosophy has proved to be valuable in many great earthquakes and it corresponds to the methodology included in the modern codes.

But the Northridge and Kobe earthquakes, both near-source events, produced widespread and unexpected brittle fracture of joints with little or without evidence of plasticization and local buckling of members. As a consequence, in the fractured structures the amount of seismic input energy dissipated before the joint failure was not clearly apparent. From these circumstances, it has been questioned whether the current philosophy, based on the dissipation of seismic input energy by means of plastic rotations, is effective also in the case of near-source earthquakes. Among the causes assumed to explain this bad behaviour, the effect of high velocity of the pulse seismic loading, producing important structure strain-rate, is considered by the specialists to be the main factor.

Therefore, there are two different types of member behaviour influenced by the position of structure related to the epicentre:

-for pulse seismic loadings, characteristic for near-source earthquakes, the great velocity induces very high strain-rate with the consequence of increasing the yield stress, reducing the ductility and producing the brittle fracture of member or joint at the first or second cycle (Fig. 6.50a);

-for cyclic seismic loadings, characteristic for far-source earthquakes and soft soils, an accumulation of plastic deformation occurs, producing a degradation in behaviour and leading to fracture after some cycles (Fig. 6.50b). The number of cycles depends on the velocity of loading, a high velocity drastically reducing this number.

6.5.2. Dynamic Behaviour of the Standard Beam

In case of static forces the standard beam is loaded by an increasing force until the collapse mechanism is produced. When the response to dynamic forces is considered (Fig. 6.51a), the prediction of standard beam collapse mode is more intricate. First, the formation of plastic mechanism is not significant, the beam being able to carry the load due to the very quick changing of the load direction. Second, the decrease in stiffness of beam, as a result of presence of plastic hinge, produces a modification of dynamic forces, depending on the acceleration spectrum. Thus, the amplification of lateral

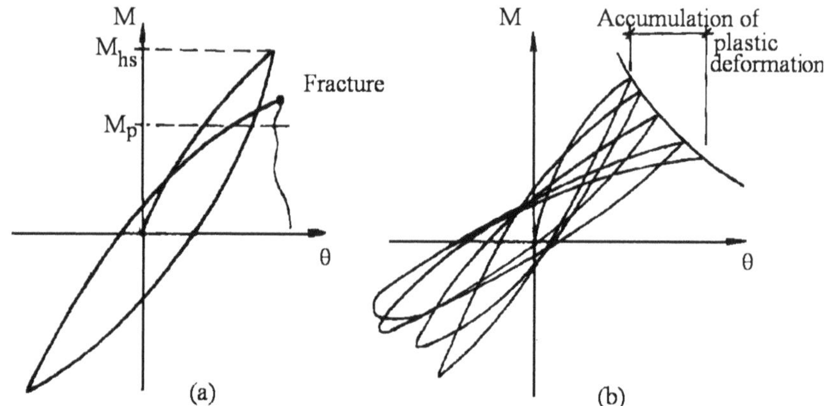

Figure 6.50: Member behaviour under seismic actions: (a) Pulse seismic loading; (b) Cyclic seismic loading

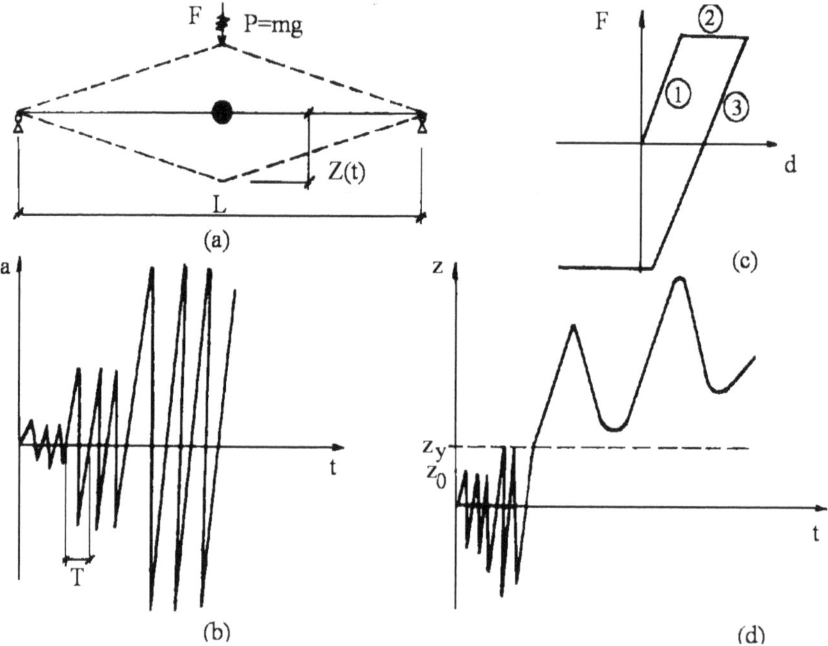

Figure 6.51: Dynamic behaviour of the standard beam: (a) Standard beam scheme; (b) Sequential accelerogram; (c) Force-displacement curve; (d) Standard beam response

forces, used in the static push-over method, must be replaced in dynamic analysis by the amplification of acceleration intensity, corresponding to some basic accelerogram:

$$a_g = a_0 \sin \phi t = a_0 \sin \frac{2\pi t}{T_g} \tag{6.53}$$

where a_0 is the amplitude of acceleration, ϕ and T_g, the natural circular frequency and natural period, respectively. By amplification of this accelerogram with a numerical multiplier λ, one can find the value λ_y for which the first plastic hinge is produced:

$$a_y = \lambda_y a_0 \tag{6.54}$$

The analysis continues with the determination of the standard beam behaviour for sequential accelerogram (Fig. 6.51b):

$$a_n = n a_y \tag{6.55}$$

until the collapse of the beam, generally produced by fracture of beam flange or joint, is attained:

$$a_u = n_u a_y \tag{6.56}$$

For an elasto-plastic force-displacement curve presented in Fig. 6.51c, the equations of motions are expressed by:

$$m\ddot{z}(t) + c\ddot{z}(t) + \left\{ \begin{array}{ll} \text{path 1:} & kz(t) \\ \text{path 2:} & kz_y \\ \text{path 3:} & k[z(t) - z_{y0}] \end{array} \right\} = -ma_g(t) \tag{6.57}$$

where the first term in the left expression corresponds to the inertial force, the second to damping and the third to the return force. The right term corresponds to the force induced by dynamic action. One can see that the motion is divided in three distinct fields, corresponding to elastic, plastic and unloading behaviour. The solutions for the equations (6.57) are (Gioncu, 1995):

(i) *Elastic behaviour* (path 1), for which the beam displacement is:

$$z(t) = \frac{a_0}{\phi^2 - \omega^2}\left[\frac{\phi}{\omega}e^{-\nu\omega t}\sin\omega t - \sin\varphi t + \frac{2\nu\omega\phi}{\phi^2 - \omega^2}(e^{-\nu\omega t}\cos\omega t - \cos\phi t)\right] \tag{6.58}$$

with

$$\omega = \left(\frac{k}{m}\right)^{1/2} = \frac{2\pi}{T_s}; \quad \nu = \frac{c}{c_{cr}} = \frac{c}{2k/\omega} \tag{6.59a, b}$$

where ω is the natural circular frequency, T_s is the natural vibration period of structure, and ν is the fraction from the critical damping constant.

In order to find the acceleration which produces the plastic hinge, the acceleration is increased. After some tedious algebra, one can obtain the value

$$\lambda_y = \alpha\frac{M_p}{PL}\frac{g}{a_0} \tag{6.60}$$

g being the gravity acceleration and

$$\alpha = \frac{T_s/T_g - 1}{\sin\dfrac{2k\pi}{1 + T_s/T_g}}; \quad k = \left\{ \begin{array}{ll} 2 & \text{for } T_s/T_g \leq 0.2 \\ 1 & \text{for } T_s/T_g > 0.2 \end{array} \right. \tag{6.61}$$

The effect of damping is neglected due to very reduced influence of this factor for steel structures.

(ii) *Plastic behaviour* (path 2), when the plastic deformation is obtained from equation (6.57). After some algebra:

$$z_p(t) = z(t) - z_y = \frac{a_0}{2\pi} T_g \left(\bar{t} - \frac{\sin(2\pi\bar{t}/T_g)}{2\pi/T_g} \right) + z_0 \qquad (6.62)$$

where $\bar{t}$ is the time measured from the one corresponding to the formation of the first plastic hinge and z_o, the displacement of system at $\bar{t} = 0$. The effect of damping is neglected, due to its very weak influence on motion in the plastic range. One can see that the rotation is composed by an oscillatory motion, due to the force oscillation mainly depending on its natural vibration, and a continuous motion, corresponding to the acceleration and velocity reached at the time corresponding to the first plastic hinge occurrence.

(iii) *Unloading behaviour* (path 3) is an elastic one, the same of the first path, but with a modified acceleration.

The history of standard beam displacement in function of acceleration history is presented in Fig. 6.51d. For elastic behaviour, an oscillatory motion is produced, characterized by the interaction between dynamic action and beam. In the plastic range for acceleration a_y the character of motion is changed. First, the beam-action interaction is missing, the beam motion being governed only by action characteristic and natural period T_g. Secondly, after each oscillation, a new plastic deformation is accumulated and the oscillations are produced around a new position. If the natural period of action is long, the plastic deformations are increasing very much, so the return motion is not able to pass in the opposite position. If the level of maximum acceleration is not diminished, the collapse of structure is produced by accumulation of plastic deformations until the beam fracture occurs. For short action period, or for short duration of cyclic motion, the danger of such collapse is reduced.

6.5.3. Behaviour of the Standard Beam Under Pulse Loading

The main characteristics of near-source earthquakes (Gioncu et al, 2000b) are the very strong pulse loadings produced with high velocity of the ground motion. The pulse loadings may arise from acceleration, velocity or displacement of the ground motion. Examining a great number of earthquakes recorded in epicentral areas (Gioncu et al, 2000b, Hall et al, 1995, Sasani and Bertero, 2000, Alavi and Krawinkler, 2000) the conclusions were that the pulse characteristic of ground motion is related to velocity. Sometimes this velocity pulse is referred to as a fling source. The main characteristics of this ground motion type are presented in Section 3.3.

(i) *Spectra for pulse loading.* An artificial generated time-history representation of pulse loading is presented in Fig. 3.40 for asymmetric velocity,

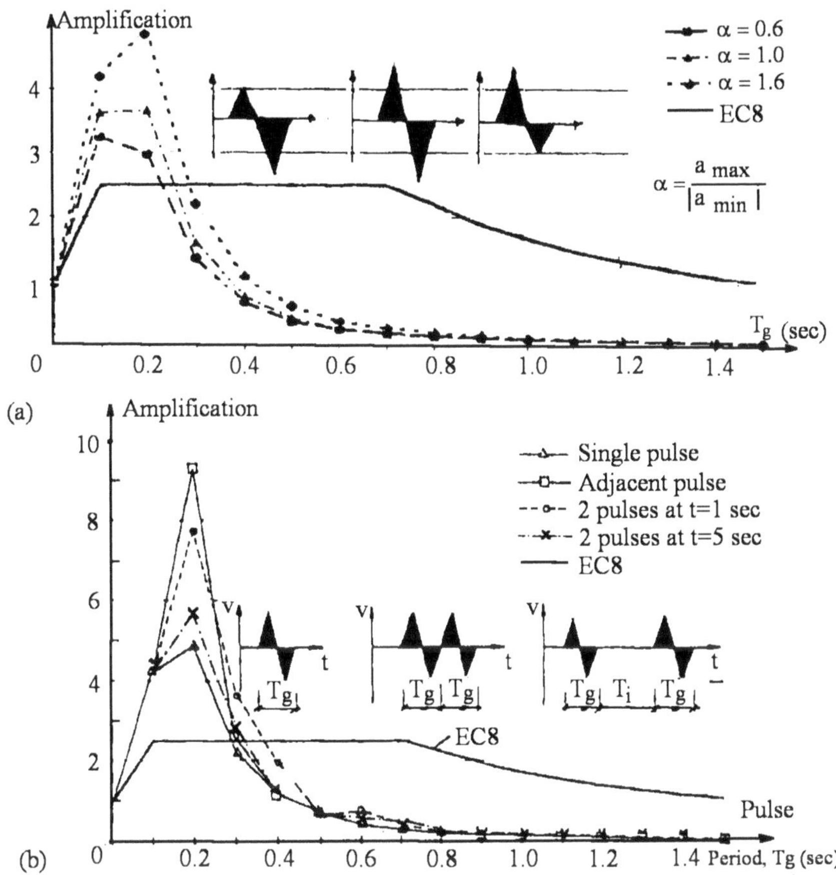

Figure 6.52: Elastic spectra for pulse loading: (a) Single pulse; (b) Multiple pulses (after Gioncu et al, 2000b)

adjacent pulses, and distinguished distant pulses. Fig. 6.52a shows the elastic spectra for different pulse periods and patterns and Fig. 6.52b the spectra for multiple adjacent pulses. The spectra shows a very high amplification of beam acceleration in the field of very short periods and an abrupt reduction for larger periods. The comparison with the EC 8 spectrum shows that, in the range of reduced periods, the amplification is higher than the one considered in the code. Contrary for larger periods, the code design spectrum appears to be too conservative. The presence of multiple adjacent pulses produces an important increasing of amplification factor, but only in the field of short periods.

(ii) *Strain-rate for standard beam.* The main characteristic of pulse loading is the velocity, while at the level of structure, it is the strain-rate for

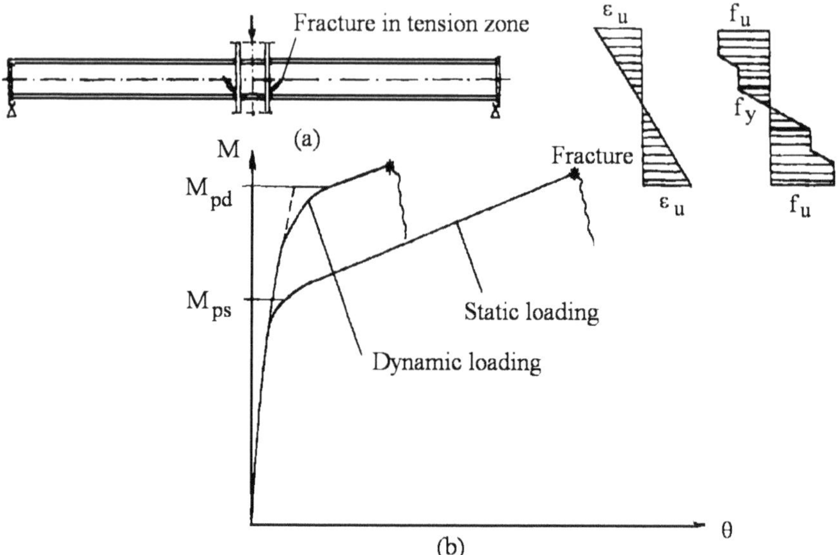

Figure 6.53: Strain-rate effects for stocky flange standard beams: (a) Beam behaviour; (b) Moment-rotation curves for static and dynamic loadings

the plastic hinge. A relationship between these values must be established. From the equation (6.62) it results:

$$z_p(t) = v_0 \left(\bar{t} - \frac{\sin(2\pi\bar{t}/T_g)}{2\pi/T_g} \right) + z_0 \tag{6.63}$$

The rotation of the plastic hinge is:

$$\theta_p(t) = \frac{2v_0}{L} \left(\bar{t} - \frac{\sin(2\pi\bar{t}/T_g)}{2\pi/T_g} \right) + \frac{2z_0}{L} \tag{6.64}$$

and the velocity of the rotation is:

$$\dot{\theta}_p(t) = \frac{2v_0}{L} \left(1 - \cos \frac{2\pi\bar{t}}{T_g} \right) \tag{6.65}$$

The strain-rate of the plastic hinge may be determined in function of the position of the rotation centre:

-in case of stocky flanges, when the plastic buckling of compressed flange is delayed, the rotation of the section occurs around a point situated at the middle of the beam (Fig. 6.53a). Considering the plastic hinge length equal to the beam depth, the strain-rate results:

$$\dot{\varepsilon} = \frac{2}{L} \left(1 - \cos \frac{2\pi\bar{t}}{T_g} \right) v_0 \tag{6.66}$$

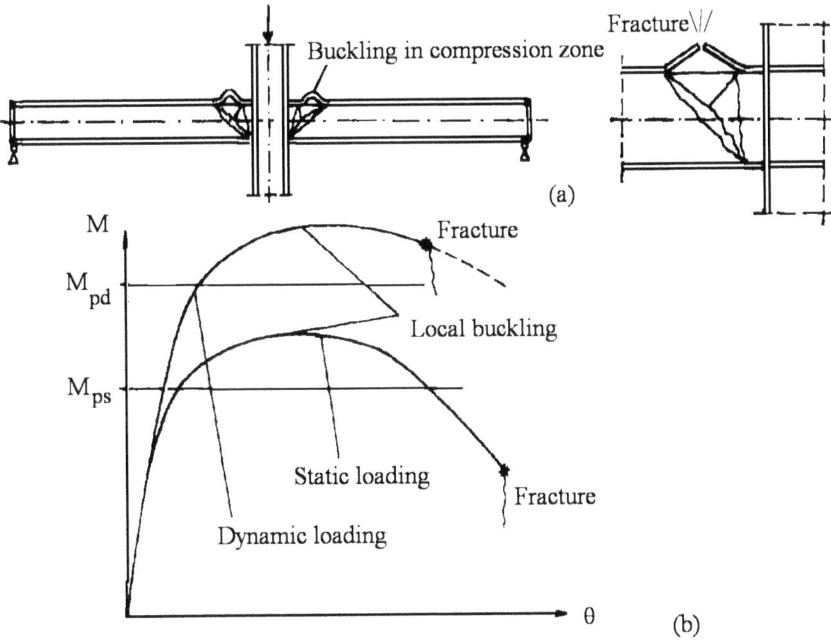

Figure 6.54: Strain-rate effects for plastic-section standard beams: (a) Beam behaviour; (b) Moment-rotation curves for static and dynamic loadings

with the maximum value for $\bar{t} = T_g$:

$$\dot{\varepsilon}_{max} = \frac{4}{L} v_0 \tag{6.67}$$

The moment-rotation curve is presented in Fig. 6.53b; the main effect of the strain-rate is the increasing of the yield stress (see Chapter 4) and, consequently, the increasing of the fully plastic moment. The collapse of the beam occurs because of the fracture of the tension flange in the beam or column;

-in case of buckling of compressed flanges, the rotation of section occurs around the point situated at the tension flange and the length of plastic mechanism is equal to the flange width (Gioncu, 2000b), (Fig. 6.54a). So, the strain-rate is:

$$\dot{\varepsilon} = \frac{2}{L} \frac{h}{b} \left(1 - \cos \frac{2\pi \bar{t}}{T_g} \right) v_0 \tag{6.68}$$

with the maximum value

$$\dot{\varepsilon} = \frac{4}{L} \frac{h}{b} v_0 \tag{6.69}$$

Fig. 6.54b shows the moment-rotation curve in case of plastic buckling of compression flange. One can see that, due to plastic buckling, the cross-section does not experience a high degree of strain hardening as it would be in the case of the fully plastic section. Local buckling of flange precludes extensive strain hardening. So, contrary to the opinion of many designers that the buckling of flanges must be avoided, this buckling operates as a filter against large strains in the tension flange, reducing the danger of cracking due to the strain-rate effect (Gioncu and Petcu, 1997).

6.5.4. Behaviour of the Standard Beam Under Cyclic Loading

This type of earthquake is characteristic of far-source ground motion. During an earthquake the structure resists hundreds of loading cycles, but only few cycles cause high plastic deformations. In case of an earthquake with a short duration, less than 5 cycles with large plastic rotations occur, while for an earthquake with a long duration, less than 20 cycles with plastic excursions are induced. Many research works have classified the failure of structural members during the seismic actions as low-cycle fatigue, but this classification is questionable due to the reduced number of cycles. An accumulation of plastic deformations along some yield lines induces cracks or ruptures of component plates, rather than a reduction of the material strength, which produces failure under high cycle fatigue.

The determination of the effects of seismic loads are a very intricate problem due to the chaotic movement history, depending on a great number of factors. The deformation induced by seismic actions varies from one member to another and also from one earthquake to another. Therefore, it is extremely difficult to select a particular deformation time-history to generate the earthquake induced deformations. Considering these uncertainties, one must adopt a cautious approach to determine the member rotation capacity under cyclic loading, this one being the severe loading system. For cyclic loading, the following aspects must be underlined:

(i) *Accumulation of plastic deformations.* The behaviour of the I-section is illustrative for the feature of cyclic loading (Fig. 6.55a). The first cycles produce only plastic deformation and the envelope of cyclic curves corresponds to the monotonic loading curve (Fig. 6.55b). At the semi-cycle, which produces the buckling of compression flange, a section rotates round a point located in, or near, the opposite tension flange. As a result, the tension forces are very small. For the reversal semi-cycle, the new compression flange buckles too. The most important observation is that the opposite flange remains unchanged, because of the small tension force, it being incapable to straighten the buckled flange (Fig.6.55c)(Gioncu et al, 2000a). Therefore, during the next cycle, the section works having initial deformations resulted from the previous cycle. So, after each cycle, a new rotation is superposed over the previous one and an accumulation of plastic rotation occurs.

Examining the moment-rotation curve (Fig. 6.55b) one can see that, after the buckling of compressed flanges, the envelope diverges from the monotonical one. This is due to the fact that the plastic mechanism consists

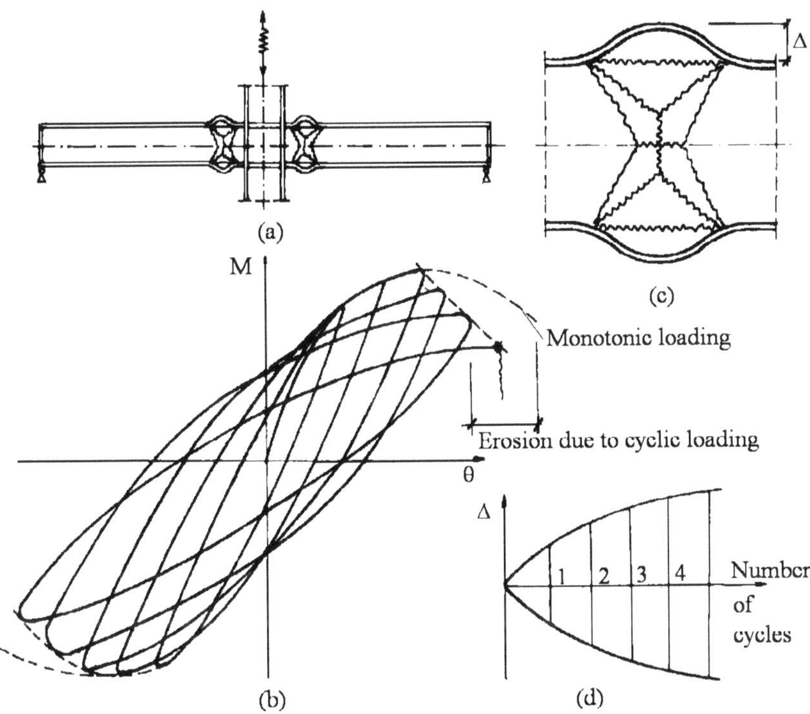

Figure 6.55: Behaviour of standard beam under cyclic loading: (a) Standard beam configuration; (b) Erosion of ductility due to accumulation of plastic deformation; (c) Local plastic mechanism under cyclic loading; (d) Accumulation of plastic flange displacement under cyclic loading

of two superimposed deformed shapes, each one for the corresponding flange and some portion of the yield lines becomes inactive. Due to this change in the plastic mechanism pattern, a reduction in the ultimate rotation capacity occurs. The main characteristic of the cyclic loading is the number of cycles which produces the fracture of member due to the accumulation of plastic rotations.

(ii) *Influence of cyclic loading with high velocity.* The effect of high velocity cyclic loading is to increase the yield ratio in function of the strain-rate and, consequently, to reduce the number of cycles until fracture. Fig. 6.56 presents the number of fracture cycles in function of strain-rate (Gioncu et al, 2000a). So, the coupling of these two erosion effects, cyclic loading and strain-rate, can produce a premature fracture. One can see that, in the case of very high strain-rate values, corresponding to strong near source earthquakes, the fracture may occur during the early cycles.

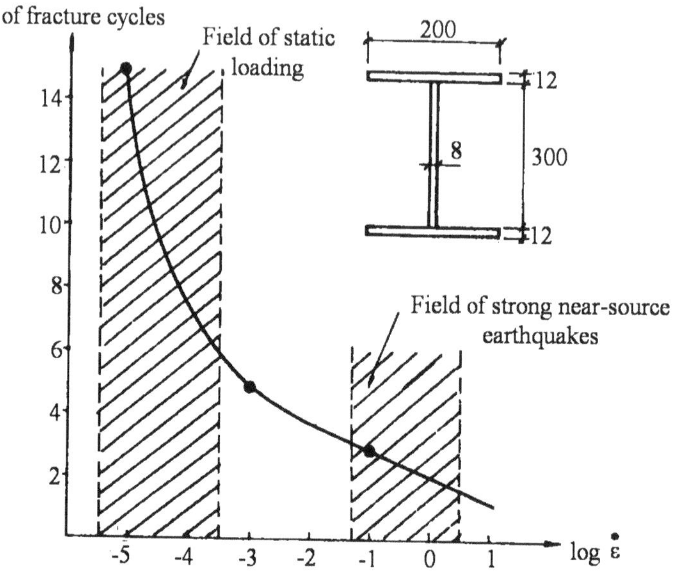

Figure 6.56: Influence of cyclic loading with high velocity (after Gioncu et al, 2000a)

6.6 Member Behavioural Classes

Ductility classes have been defined in the previous chapters for elements (Section 4.7), sections (Section 5.2.4), stubs (Sections 5.4.6 and 5.5.5). At the level of member, there are some additional factors influencing the ductility (i.e. the beam span, flange to web width ratio, member slenderness, moment gradient, level and eccentricity of axial load, etc). As a consequence of these factors, it seems that the above behavioural classes should be substituted by the concept of member behavioural classes (Mazzolani and Piluso, 1993, Gioncu and Mazzolani, 1995, Gioncu, 1995a).

There are in literature some proposed classification criteria for members according to ductility classes. Between them, the Mazzolani and Piluso (1993) proposal seems to be the most rational one:

-HD, high ductility, $\mu_{\theta r} \geq 7.5$

-MD, medium ductility, $4.5 \leq \mu_{\theta r} < 7.5$

-LD, low ductility, $1.5 \leq \mu_{\theta r} < 4.5$

A correlation between member behavioural classes and cross-sectional classes after EC 3 is presented in Fig.6.57 (Gioncu et al, 2000a). One can observe great differences between the two classifications and so it is clear that the member classification is more adequate for checking the structure ductility than the cross-sectional classes.

Slenderness requirements for classification of beams with I-sections are proposed by Vayas (2000) and Vayas et al (1999), taking into account the interactive effects between local buckling (including flange and web interaction) and lateral torsional buckling (Fig. 6.58).

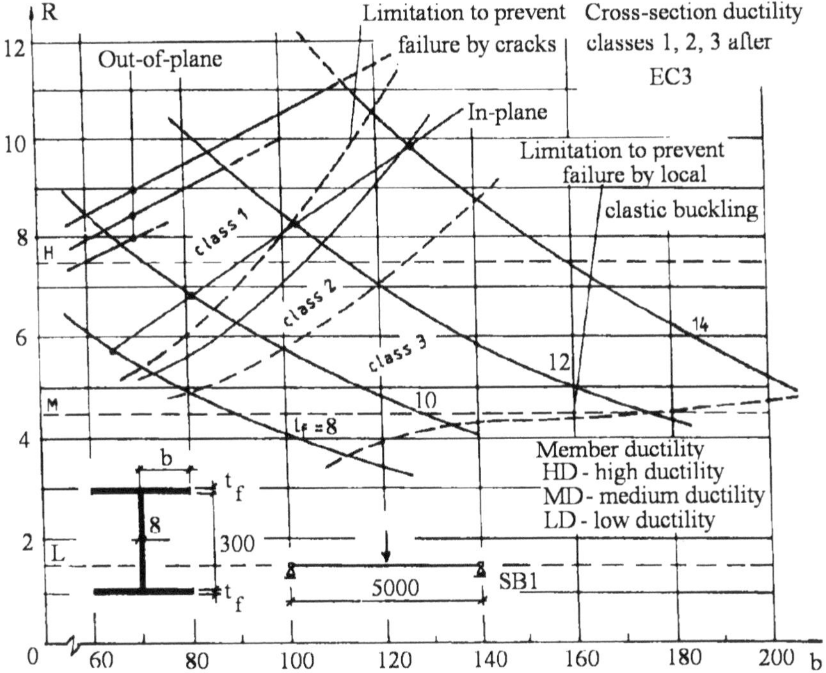

Figure 6.57: Correlation between member and cross-sectional classes (after Gioncu et al, 2000a)

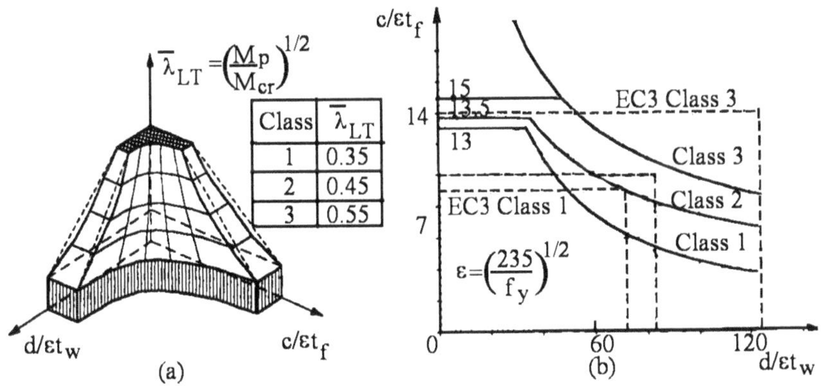

Figure 6.58: Member classification: (a) Limitation for beam classes; (b) Limitation for cross section (after Vayas et al, 1999)

6.7 Conclusions

The main conclusions for the aspects presented in this chapter may be summarized as follows:

-the ductility of a complex structure may be determined using the so-called standard beam;

-two standard beam types must be used, one for the case of important moment gradient and one for the quasi-constant moment;

-the most adequate definition for the ultimate rotation capacity is the one determined in the lowering post-buckling curve at the intersection with the theoretical plastic moment;

-the effect of joints on the member ductility may be evaluated using the so-called generalized standard beam;

-an adequate design methodology must eliminate the possibility of a joint collapse before the flange buckling of connected member ends;

-two approaches are proposed for this requirement, firstly, by improving the design of conventional joint configuration or by moving the member plastic hinges away from the connections, secondly, by strengthening or weakening the member-ends;

-the effects of seismic loadings are very different for the cases of near-source or far-source structure location;

-the pulse loading can produces very high strain-rate and premature fracture of tensioned or buckled compression flanges;

-the cyclic loadings produce an accumulation of plastic deformations, reducing the member ductility by altering the local plastic mechanism.

6.8 References

Aiello M.A., Ombres L. (1995a): Rotation capacity and local ductility of steel members. In 9th International Conference on Metal Structures (ed. J. Murjewski), 26-30 June, Krakow, Vol. 2, 271-282

Alavi,B., Krawinkler, H. (2000): Consideration of near-fault ground motion effects in seismic design. In 12th World Conference on Earthquake Engineering, Auckland, 30 January-4 February 2000, CD-ROM 2665

Anastasiadis, A. (1999): Ductility problems of steel moment resisting frames. Ph D Thesis, Politechnica University Timisoara

Anastasiadis, A., Gioncu, V. (1998): Influence of joint details on the local ductility of steel MR frames. In 3rd National Conference on Steel Structures. Thessaloniki, 30-31 October 1998, 311-319

Anastasiadis, A., Gioncu, V. (1999): Ductility of IPE and HEA beams and beam-columns. In Stability and Ductility of Steel Structures, SDSS 99,(eds. D.Dubina and M. Ivanyi), Timisoara, 9-11 September 1999, Elsevier, Amserdam, 249-257

Anastasiadis, A., Gioncu, V., Mazzolani, F.M. (2000): New trends in the evaluation of available ductility of steel members. In Behaviour of Steel Structures in Seismic Areas, STESSA 2000, (eds. F.M Mazzolani and R. Tremblay),

Montreal, 21-24 August 2000, Balkema, Rotterdam, 3-10

Axhag, F. (1995): Plastic design of composite bridges allowing for local buckling. Technical Report, Lulea University of Technology, 09T

Boerave, Ph., Lognard, B., Janss, J., Gerardy, J.C., Schleich, J.B. (1993): Elasto-plastic behaviour of steel frameworks. Journal of Constructional Steel Research, Vol. 27, 3-21

Bruneau, M., Uang, C.M., Whittaker, A. (1998): Ductility Design of Steel Structures. McGraw-Hill, New York

Charter, C.J., Iwankiw, N.R. (1998): Improved ductility in seismic steel moment frames with dogbone connections. In 11th European Conference on Earthquake Engineering, Paris, 6-11 September 1998, CD-ROM 253

Chen, C.C., Lu, L.W. (1991): Cyclic tests of a box-column and truss-girder asemblage. In Inelastic Behaviour and Design of Frames, SSRC Annual Session, Chicago, 15-17 April 1991, 303-312

Chen, S.J., Yeh, C.H., Chu, J.M. (1996): Ductile steel beams-to-column connections for sismic resistance. Journal of Structural Engineering, Vol. 122, No. 11, 1292-1299

Chen, S.J., Chu, J.M., Chou, Z.L. (1997): Dynamic behaviour of steel frames with beam flanges shaved around connection. Journal of Constructional Steel Research, Vol. 42, No. 1, 49-70

Climenhaga, J.J., Johnson, R.P. (1972a): Moment-rotation curves for locally buckling beams. Journal of the Structural Division, Vol. 98, ST 6, 1239-1254

Climenhaga, J.J., Johnson, R.P. (1972b): Local buckling in continuous composite beams. The Structural Engineer, Vol. 50, No. 9, 367-374

Cosenza, E., De Luca, A., Faella, C. (1989): Inelastic buckling of semirigid sway frames. In Structural Connection: Stability and Strength (ed. R. Narayanan), Elsevier, Chapter 9

Dekker, N.W. (1998): Interactive buckling model for the prediction of the rotation capacity of steel beams. In 2nd World Conference on Steel in Construction. San Sebastian, 11-13 May, 1998, CD-ROM 298

Earls, C.J. (1999b): Factors influencing ductility in high performance steel I-shaped beams. In Stability and Ductility of Steel Structures, SDSS 99, (eds. D. Dubina and M. Ivanyi), Timisoara, 9-11 September 1999, Elsevier, Amsterdam, 269-278

Engelhardt, M.D., Sabol, T.A. (1997): Seismic steel moment connections: Development since the 1994 Northridge earthquake. Construction Research Communications Limited, 68-77

Espiga, F. (1998): Ductility assessment methods and safety evaluation based on rotation capacity approach. In 2nd World Conference on Steel in Construction, San Sebastian, 11-13 May, CD-ROM 22

EUROCODE 3, EC 3 (1992): Design of Steel Structures, Part 1.1. General rules and rules for buildings. ENV 1933 1-1, February 1992

EUROCODE 8, EC 8 (1994): Design provisions for earthquake resistance of structures. Part 1.3. Specific rules for various materials and elements. ENV 1998 1-3

Faella, C., Piluso, V., Rizzano, G. (2000): Structural Steel Semirigid Connections. Theory, Design and Software. CRC Press LLC,USA

FEMA-267A (1997): Program to reduce the earthquake hazards of steel moment frame structures. Interim guid lines advisory No. 1. Report SAC-96-03

Georgescu, D., Mazzolani, F.M. (1996): The influence of unsymmetrical post-elastic properties of the structural steel on the rotation capacity of the beam. In Coupled Instabilities in Metal Structures, CIMS 96 (eds. J. Rondal, D. Dubina, V. Gioncu), Liege, 5-7 September 1996, Imperial College Press, London, 292-300

Gioncu, V. (1995a): Some considerations on structural member classes. In 9th International Conference on Metal Structures (ed. J. Murjewski), Krakow, 26-30 June 1995, Vol. 2, 311-320

Gioncu, V. (1995b): Local and global ductility in seismic design of MR frames. In Steel Structures, Eurosteel 95 (ed. A. Kounadis), Athens, 18-20 May 1995, Balkema, Rotterdam, 469-477

Gioncu, V. (2000a): Framed structures. Ductility and seismic response. General report. In Stability and Ductility of Steel Structures, SDSS 99, Timisoara, 9-11 September 1999, Journal of Constructional Steel Research, Vol. 55, No. 1-3, 125-154

Gioncu, V. (2000b): Influence of strain-rate on the behaviuor of steel members. In Behaviour of Steel Structures in Seismic Areas, STESSA 2000, (eds F.M. Mazzolani and R. Tremblay), Montreal, 21-24 August 2000, Balkema, Rotterdam, 19-26

Gioncu, V., Mateescu, G., Orasteanu, S. (1989): Theoretical and experimental research regarding the ductility of welded I-sections subjected to bending. In Stability of Metal Structures, SSRC Conference, Beijing, 10-12 October 1989, 289-298

Gioncu, V., Mateescu, G., Iuhas, A. (1995a): Contributions to the study of plastic rotational capacity of I-steel sections. In Behaviour of Steel Structures in Seismic Areas, STESSA 94 (eds. F.M Mazzolani and V. Gioncu), Timisoara, 26 June-1 July 1994, E&FN Spon, London, 169-181

Gioncu, V., Mateescu, G., Petcu, D., Anastasiadis, A. (2000a): Prediction of available ductility by means of plstic mechanism method: DUCTROT computer program. In Moment Resistant Connections of Steel Frames in Seismic Areas. Design and Reliability, RECOS, (ed. F.M. Mazzolani), E&FN Spon, London, 97-146

Gioncu, V., Mateescu, G., Tirca, L., Anastasiadis, A. (2000b): Influence of the type of seismic ground motions. In Moment Resistant Connections of Steel Frames in Seismic Areas. Design and Reability, RECOS, (ed. F.M. Mazzolani), E&FN Spon, London, 57-92

Gioncu, V., Mazzolani, F.M. (1995): Alternative methods for assessing local ductility. In Behaviour of Steel Structures in Seismic Areas, STESSA 94 (eds. F.M. Mazzolani and V. Gioncu), Timisoara, 26 June- 1 July 1994, E&FN Spon, London, 182-190

Gioncu, V., Petcu, D. (1995): Numerical investigations on the rotation capacity of beams and beam-columns. In Stability of Steel Structures, (ed. M. Ivanyi), SSRC Colloquium, Budapest, 21-23 September 1995, Akamemiai Kiado, Budapest, Vol. 1, 163-174

Gioncu, V., Petcu, D. (1997): Available rotation capacity of wide-flange beams and beams-columns. Part 1. Theoretical approaches. Part 2. Experimental and numerical tests. Journal of Constructional Steel Research, Vol. 43, No.

1-3. 161-217, 219-244

Gioncu, V. Tirca, L, Petcu, D. (1996a): Interaction between in-plane and out-of-plane plastic buckling of wide-flange section members. In Coupled Instabilities in Metal Structures, CIMS 96, (eds. J. Rondal, D. Dubina, V. Gioncu) Liege, 5-7 September 1996, Imperial College Press, 273-281

Gioncu, V., Tirca, L., Petcu, D. (1996b): Rotation capacity of rectangular hollow section beams. in Seven International Symposium on Tubular Structures (eds. I. Farkas and K. Jarmai), 28-30 August 1996, Miskolc, Balkema, Rotterdam, 387-395

Gomez, F.C.T., Kuhlmann, U., De Matteis, G., Mandara, A. (1998): Recent developments on classification of joints. In Control of Semi-Rigid Behaviour of Civil Engineering Structural Connections, COST 1, Liege, 17-19 September 1998, 187-198

Grzebieta, R., Zhao, X.L., Purza, F. (1997): Multiple low cycle fatigue of SHS tubes subjected to gross pure bending deformation. In Stability and Ductility of Steel Structures, SDSS 97 (ed. T.Usami), Nagoya, 29-31 July 1997, 847-854

Guruparan, N.I., Walpole, W.R. (1990): The effect of lateral slenderness on the cyclic strength of steel beams. In 4th US National Conference on Earthquake Engineering, Palm Springs, 20-24 May 1990, Vol. 2, 585-594

Haaijer, G, Thurlimann, B. (1958): On inelastic bucling in steel. Journal of Engineering Mechanics Division, Vol. 83, EM 2

Hall, J.F., Heaton, T.H., Halling, M.W., Wald, D.J. (1995): Near-source ground motion and its effects on flexible buildings. Earthquake Spectra, Vol. 11, No. 4, 569-605

Hoglund, T., Nylander, H. (1970): Maximiforhallande B/t for tryckt flans hos valsad I-balk vid dimensionering med granslastruetod. Technical University Stockholm, Report 83/70

Huang, P.Ch., Deierlein, G.G. (1996): Investigation of beam rotation capacity using finite element analysis. In 5th SSRC International Conference on Stability of Metal Structures. Future Direction in Stability Research and Design. Chicago, 15-18 April 1996, 383-392

Huber, G., Kronenberger, H.J., Weynand, K. (1998): Representation of joints in the analysis of structural systems. in Control of the Semi-Rigid Behaviour of Civil Engineering Structural Connections. Liege, 17-19 September 1998, 105-114

Ivanyi, M. (1979a): Yield mechanism curves for local buckling of axially compressed members. Periodica Polytechnica, Civil Engineering, Vol. 23, No. 3/4, 203-216

Ivanyi, M. (1979b): Moment rotation characteristics of locally buckling beams. Periodica Polytechnica, Civil Engineering, Vol. 23, No. 3/4, 217-230

Ivanyi, M. (1985): The model of interactive plastic hinge. Periodica Polytechnica, Vol. 29, No. 3/4, 121-146

Ivanyi, M. (1993): Interactive plastic hinge based method for analysis of steel frames. Proceedings of 1993 Annual SSRC Technical Session, Milwaukee, 5-6 April 1993, 153-177

Kato, B. (1965): Buckling strength of plates in the plastic range. Publication of International Association of Bridge and Structural Engineering, Vol. 25,

127-141

Kato, B. (1988): Rotation capacity of steel members subjected to local buckling. In 9th World Conference on Earthquake Engineering, Tokyo-Kyoto, 2-9 August 1988, Vol. IV, 115-120

Kato, B. (1989): Rotation capacity of H-section members as determined by local buckling. Journal of Constructional Steel Research, Vol. 13, 95-109

Kato, B. (1990): Deformation capacity of steel structures. Journal of Constructional Steel Research, Vol. 17, 33-94

Kato, B., Akiyama, H. (1981): Ductility of members and frames subjected to buckling. ASCE Convention, 11-15 May 1981, New York, Preprint 81-100

Kawaguchi, J., Morino, S., Machida, T. (1997): Energy dissipation capacity of CFT beam column failing in local buckling. In Behaviour of Steel Structurs in Seismic Areas (eds. F.M. Mazzolani and H.Akiyama), Kyoto, 3-8 August 1997, 10/17 Salerno, 311-318

Kecman, D. (1983): Bending collapse of rectangular and square section tubes. International Journal of Mechanical Science, Vol. 25, No. 9-10, 623-636

Kemp, A.R. (1985): Interaction of plastic local and lateral buckling. Journal of Structural Engineering, Vol. 111, ST 10, 2181-2196

Kemp, A.R. (1986): Factors affecting the rotation capacity of plastically designed members. The Structural Engineer, Vol. 64B, No. 2, 28-35

Kemp, A.R. (1996): Inelastic local and lateral buckling in design codes. Journal of Structural Engineering, Vol. 122, No. 4, 374-382

Kemp, A.R. (1999): A limit states criterion for ductility of class 1 and 2 composite and steel beams. In Stability and Ductility of Steel Structures, SDSS 99 (eds. D. Dubina and M. Ivanyi), Timisoara, 9-11 September 1999, Elsevier, Amserdam, 291-298

Kemp, A.R., Dekker, N.W. (1991): Available rotation capacity in steel and composite beams. The Structural Engineer, Vol. 69, No. 5, 88-97

Kemp, A.R., Dekker, N.W., Trinchero, P. (1995): Differences in inelastic properties of steel and composite beams. Journal of Constructional Steel Research, Vol. 34, 187-206

Kollar, L., Dulacska, E. (1984): Buckling of Shells for Engineers. Akademiai Kiado, Budapest

Korol, R.M., Hudoba, J. (1972): Plastic behaviour of hollow structural sections. Journal of Structural Division, Vol. 98, ST 5, 1007-1023

Kotelko, M. (1996a): Selected problems of collapse behaviour analysis of structural members built from strain hardening material. Bi-Centenar Conference on Thin-Walled Structures. Glasgow, 2-4 December, 1996

Kotelko, M. (1996b): Ultimate load and postfailure behaviour of box-section beams under pure bending. Engineering Transactions, Vol. 44, No. 2, 229-251

Krawinkler, H., Zohrei, M. (1983): Cumulative damage in steel structures subjected to earthquake ground motions. Computers and Structures, Vol. 16, 531-541

Krawinkler, H., Gupta, A. (1998): Modelling issues in evaluating nonlinear response for steel moment. In 11th European Conference on Earthquake Engineering, Paris, 6-11 Seprember 1998, CD-ROM 274

Kubo, M., Kitahori, H. (1997): Buckling strength and rotation capacity of monosymmetric I-beams. In Stability and Ductility of Steel Structures, SDSS 97, (ed. T. Usami), Nagoya, 29-31 July, 1997, 523-530

Kuhlmann, U. (1986):Rotations kapazitat biegebeanspruchter I-Profile under Berucksichtigung des plastischen Beulens. Technical Report, Bochum University, Mitteilung Nr. 86-5

Kuhlmann, U. (1989): Definition of flange slenderness limits on the basis of rotation capacity values. Journal of Constructional Steel Research, Vol. 14, 21-40

Kuhlmann, U. (1999): On the rotational capacity of structural connections. In Semirigidity in Connections of Structural Steelworks: Theory, Analysis and Design. CIMS course, Udine 20-24 September 1999

Kusawa, H., Fukumoto, Y. (1983): Cyclic behaviour of thin-walled box stubcolumns and beams. In Stability of Metal structures, SSRC Conference, Paris, 16-17 November 1983, 211-218

Kuwamura, H. (1988): Effect of yield ratio on the ductility of high-strength steels under seismic loading. Annual SSRC Technical Session Computer Technology to Structural Stability. Minneapolis, 26-27 April 1988, 201-210

Kuwamura, H., Akiyama, H. (1994): Brittle fracture under high stress. Journal of Constructional Steel Research, Vol. 29, 5-19

Lay, M.G. (1965): Flange local buckling in wide-flange shapes. Journal of the Structural Division, Vol. 91, ST 6, 95-116

Lay, M.G., Galambos, T.V. (1965): Inelastic steel beams under uniform moment. Journal of the Structural Division, Vol. 91, ST 6, 67-93

Lay, M. G., Galambos T.V. (1967): Inelastic beams under moment gradient. Journal of the Structural Division, Vol. 93, ST 1, 381-399

Lee, G.C., Lee, E.T. (1994): Local buckling of steel sections under cyclic loading. Journal of Constructional Steel Research, Vol. 29, 55-70

Lu, L.W., Ricles, J.M., Mao, C., Fisher, J.W. (2000): Critical issues in achieving ductile behaviour of welded moment connections. In Stability and Ductility of Steel Structures, SDSS 99, Timisoara, 9-11 September 1999, Journal of Constructional Steel Research, Vol. 55, No. 1-3, 325-341

Lukey, A.F., Adams, P.F. (1969): Rotation capacity of beams under moment gradient. Journal of the Structural Division, Vol. 95, ST 6, 1173-1188

Mamaghani, I.H.P., Usami, T., Mizuno, E., Kajikawa, Y. (1997): Cyclic inelastic behaviour of compact steel box columns. In Stability and Ductility of Steel Structures, SDSS 97 (ed. T. Usami), Nagoya, 29-31 July 1997, 259-266

Mateescu, G., Gioncu, V.(2000): Member response to strong pulse seismic loading. In Behaviour of Steel structures in Seismic Areas, STESSA 2000 (eds. F.M. Mazzolani and R. Tremblay), Montreal, 21-24 August 2000, Balkema, Rotterdam, 55-62

Matsui, C., Mitani, I (1977): Inelastic behaviour of high strength steel frames subjected to constant vertical and alternating horizontal loads. In 6th World Conference on Earthquake Engineering, New Delhi, 1977, Vol. 3, 3169-3174

Matsui, C., Sakai, J. (1992): Effect of collapse modes on ductility of steel frames. In 10th World Conference on Earthquake Engineering, Madrid, 19-24 July 1992, 2949-2954

Mazzolani, F.M. (1998): Design of steel structures in seismic regions: The paramount influence of connections. In Control of Semi-Rigid Behaviour of Civil

Engineering Structural Connections, COST 1, Liege, 17-19 September 1998, 371-384

Mazzolani, F.M. (2000): Design of moment resisting frames. In Seismic Resistant Steel Structures (eds F.M. Mazzolani and V. Gioncu), CISM courses, Udine, 18-22 October 1999, Springer, Wien, 159-240

Mazzolani, F.M., De Matteis, G., Mandara, A. (1996): Classification system for aluminium alloy connections. In IABSE Colloquium on Semi-Rigid Structural Connections, Istambul, 83-94

Mazzolani, F.M., Piluso, V. (1992): Evaluation of the rotation capacity of steel beams and beams-columns. In First State of the Art Workshop COST C1, Strassburg

Mazzolani, F.M., Piluso, V. (1993): Member behavioural classes of steel beams and beam-columns. In XIV Congresso CTA, Viareggio, 24-27 October 1993, Ricerca Teorica e Sperimentale, 405-416

Mazzolani, F.M., Piluso, V. (1995): Design of Steel Structures in Seismic Zones. ECCS Manual, Doc. No. 76

Mazzolani, F.M., Piluso, V. (1996): Theory and Design of Seismic Resistant Steel Frames. E&FN Spon, London

Mazzolani, F.M., Pluso, V., Rizzano, G. (1998): Design of full-strength extended end-plate joints according for random material variability. In Control of the Semi-Rigid Behaviour of Civil Engineering Structural Connections, COST 1, Liege, 17-19 September 1998, 405-416

McDermott, J.F. (1969): Plastic bending of A514 steel beams. Journal of Structural Division, Vol. 95, ST 6, 1173-1188

McMullin, K.M., Astaneh-Asl, A. (1997): Analytical study of hinge formation in steel moment-resisting frames. In Behaviour of Steel Structures in Seismic Areas, STESSA 97, (eds. F.M.Mazzolani and H. Akiyama), Kyoto, 3-8 August 1997, 10/17 Salerno, 319-325

Migiakis, C.E., Elghazouli, A.Y. (1998): Slab effects in composite floor systems subjected to earthquake loading. In 3th national Conference on Steel Structures (eds. K.T. Thomopoulous, C.C. Baniolopoulos, A.V. Avdelas) Thessaloniki, 30-31 October 1998, 328-336

Mitani, I., Makino, M., Matsui, C. (1977): Influence of local buckling on cyclic behaviour of steel beam-column. In 6th World Conference on Eartquake Engineering, New Delhi, 1977, Vol. 3, 3175-3180

Mitani, I., Makino, M. (1980): Post-local buckling behaviour and plastic rotation capacity of steel beam-columns. In 7th World Conference on Earthquake Engineering, Istambul

Moen, L.A., De Matteis, G., Hopperstad, O.S., Langseth, M., Landolfo R., Mazzolani, F.M. (1999): Rotational capacity of aluminium beams under moment gradient.Numerical simulations. Journal of Structural Engineering, Vol. 125, No. 8, 921-929

Moller, M., Johansson, B., Collin, P. (1997): A new analytical model of inelastic local flange buckling. Journal of Constructional Steel Research, Vol. 43, 43-63

Morino, S., Kawaguchi, J., Fukao, H., Tawa, H. (1995): Analysis of elasto-plastic behaviour of beam-column failure in local buckling. In Structural Stability and Design (eds. S.Kitiporuchai, G. Hancock, and J. Bradford), Sydney, 30

October - 1 November 1995, Balbema, Rotterdam, 181-186

Murray, N. W. (1986): Stability and ductility of steel elements. Journal of Constructional Steel Research, Vol. 44, No. 1-2, 23-50

Nakai, H., Yoshikawa, O., Terada, H. (1986): An experimental study on ultimate strength of composite columns for compression or bending. Structural Engineering and Earthquake Engineering, Vol. 3, No. 2, 235-245

Nakamura, T. (1988): Strength and deformability of H-shaped steel beams and lateral requirements. Journal of Constructional Steel Research, Vol. 9, 217-228

Nakashima, M. (1992): Variation and prediction of deformation capacity of steel beam-columns. In 10th World Conference on Earthquake Engineering, Madrid, 19-24 July 1992, Balkema, Rotterdam, 4501-4507

Nakashima, M. (1994): Variation of ductility capacity of steel beam-column. Journal of Structural Engineering, Vol. 120, No. 7, 1941-1960

Nakashima, M. (1997): Uncertainties associated with ductility performance of steel building structures. In Seismic Design Methodologies for the Next Generation of Codes (eds. P.Fajfar and H.Krawinkler), Bled, 24-27 June 1997, 111-118

Nakashima, M., Suita, K., Morisako, K., Maruoka, Y. (1998): Tests on welded beam-column subassemblies. I. Global behaviour. Journal of Structural Engineering, Vol. 124, No. 11, 1236-1244

Nakashima, Roeder, C.W., Maruoka, Y. (2000): Steel moment frames for earthquakes in United States and Japan. Journal of Structural Engineering, Vol. 126, No. 8, 861-868

Piluso, V. (1992): Il comportamento inelastico dei telai seismo-resistenti in acciaio. Ph Thesis, Universita di Napoli

Piluso, V. (1995): Post-local behaviour of rolled steel beams subjected to nonuniform bending. XV Congresso CTA, Riva del Garda, 15-18 October 1995, 542-562

Plumier, A. (1996a): Problems and options in seismic design of composite frames. SPEC Internal Report

Plumier, A. (1996b): Reduced beam section: A safety concept for structures in seismic zones. Buletinul Stiintific al Universitatii Politehnica Timisoara, Transactions on Civil Engineering, Vol. 41(55), No. 2, 46-60

Plumier, A. (1999): Seismic resistant composite structures. In Seismic Resistant Steel Structures. Progress and Challenges (eds. F.M. Mazzolani and V. Gioncu), CISM course, Udine 18-22 October 1999, Springer, Wien

Plumier, C., Doneux, C. (1998): Dynamic tests on the ductility of composite steel-concrete beams. In 11th European Conference on Earthquake Engineering, Paris, 6-11 September 1998, CD-ROM 205

Plumier, A., Doneux, C. (1999): Design of composite structures in seismic regions. In Stability and Ductility of Steel Structures (eds D. Dubina and M. Ivanyi), Timisoara, 9-11 September 1999, Elsevier, Amsterdam, 27-34

Plumier, A., Doneux, C., Bouwkamp, J.G., Plumier, C. (1998): Slab design in connection zone of composite frames. In 11th European Coference on Earthquake Engineering, Paris, 6-11 September 1998, CD-ROM 30

PROFIL-FEM-3D (1991): Ein elektronischer simulator fur Bauteile aus Stahl. D. Bohman, Lehrstuhl fur Stahlbau. RWTH Aachen

Roik, K., Kuhlmann, U. (1987): Rechnerische Ermittlung der Rotationkapazitat biegebeanspruchter I-Profile. Stahlbau, No. 11, 321-327

Rondal, J. et al (1994): Rotation capacity of hollow beam sections. Final Report/CIDECT Research Project No 2P

Sasani, M., Bertero, V.V. (2000): Importance of severe pulse-type ground motions in performance-based engineering. Historical and critical review. In 12th World Conference on Earthquake Engineering, Auckland, 30 January-4 February 2000, CD-ROM 1302

Sawyer, H.A. (1961): Post-elastic behaviour of wide-flange steel beams. Journal of the Structural Division, Vol. 87, ST 9, 43-71

Schilling, C.G. (1988): Moment rotation tests of steel bridge girders. Journal of structural Engineering, Vol. 114, No. 1, 134-141

Sedlacek, G., Stutki, Ch., Bilt, St. (1987): A contribution of the background of the B/T ratios controlling the applicability of analysis models in Eurocode 3. In Stability of Plate and Shell Structures (eds P. Dubas and D. Vandepitte), Ghent, 6-8 April 1987, 33-40

Sedlacek, G. et al (1995): Investigation of the rotation behaviour of hollow section beams. RWTH Aachen and MSM Liege Report

Spangemacher, R. (1991): Zum Rotationsnachweis von Stahlkonstruktion, die nach dem Traglastverfahren berechnet werden. Ph Thesis, Aachen University

Spangemacher, R., Sedlacek, G. (1992a): On the development of a computer simulation for tests of steel structures. In Constructional Steel Design. World Developments (ed. P.S. Dowling), Acapulco, 6-9 December, 1992, Elsevier Applied Science, London, 593-611

Spangemacher, R., Sedlacek, G. (1992b): Zum Nachweis ausreichender Rotationsfahigheit von Fliessgelenhken bei der Anwendung des Fliessgelenkverfahrens. Stahlbau, Vol. 61, Heft 11, 329-339

Stranghoner, N. (1995): Undersuchungen zum Rotationsverhalten von Tragern aus Hohlprofilen. Ph Thesis, Aachen University

Stranghoner, N., Sedlacek, G. (1997): Zum Tragverhalten von kalt und warmgefertigten quadratischen Hohlprofiltragern. Stahlbau, Heft 4, 198-204

Suita, K., Nakashima, M., Morisako, K. (1998): Tests of welded beam-column subassemblies. II: Detailed behaviour. Journal of Structural Engineering, Vol. 124, No. 11, 1245-1252

Sully, R.M., Hancock, G.J. (1996): Behaviour of cold-formed SHS beam-column. Journal of Structural Engineering, Vol. 122, No. 3, 326-336

Sully, R.M., Hancock, G.J. (1999): Stability of cold-formed tubular beam-columns. In Light-Weight Steel and Aluminium Structures (eds. P. Makelainen and P. Hassinen), Espoo, 20-23 June 1999, Elsevier, Amsterdam, 141-154

Suzuki, T., Ono, T. (1977): An experimental study on inelastic behaviour of steel members subjected to repeated loading. In 6th World Conference on Earthquake Engineering, New Delhi, 1977, Vol. 3, 3163-3168

Suzuki, T., Ogawa, T., Ikaraski, K. (1994): A study on local buckling behaviour of hybrid beams. Thin-Walled Structures, Vol. 19, 337-351

Suzuki, T., Ogawa, T., Ikarashi, K. (1996): A study on plastic deformation capacity of H-shaped hybrid beam-column. In Coupled Instabilities in Metal Structures, CIMS 96, (eds. J. Rondal, D. Dubina, V. Gioncu), Liege, 5-7

September 1996, Imperial College Press, London, 301-308

Suzuki, T., Ogawa, T., Ikarashi, K. (1997): Evaluation of the plastic deformation capacity modified by the effect of ductile fracture. In Behaviour of Steel Structures in Seismic Areas, STESSA 97, (eds. F.M.Mazzolani and H. Akiyama), Kyoto, 3-8 August 1997, 10/17 Salerno, 326-333

Tehani, M. (1997): Local buckling in class 2 continuous composite beams. Journal of Constructional Steel Research, Vol. 43, No. 1-3, 141-159

Vayas, I. (1997): Stability and ductility of steel elements. Journal of Constructional Steel Research, Vol. 44, Nos 1-2, 23-50

Vayas, I. (2000): Interaction between local and global properties. In Moment Resistant Connections of Steel Frames in Seismic Areas. Design and Reability. RECOS, (ed. F.M. Mazzolani), E&FN Spon, London, 409-458

Vayas, I., Psycharis, I. (1990): Behaviour of thin-walled steel elemnts under monotonic and cyclic loading. In Structural Dynamic (ed. W.B. Kratzig), Balkema, Rotterdam, 579-583

Vayas, I., Psycharis, I. (1992): Dehnungsorientierte Formulierung der Methode der wirksamen Breite. Stahlbau, Vol 61, 275-283

Vayas, I., Psycharis, I. (1993): Ein dehnungsorientiertes Verfahren zur Ermittlung der Duktilitat von Tragern aus I-Profilen. Stahlbau, Vol. 63, 333-341

Vayas, I., Psycharis, I. (1995): Local cyclic behaviour of steel members. In Behaviour of Steel Structures in Seismic Areas, STESSA 94, (eds. F.M. Mazzolani and V. Gioncu), Timisoara, 26 June-1 July, 1994, E&FN Spon, London, 231-241

Vayas, I., Syrmakezis, C., Sophocleous, A. (1995): A method for evaluation of the behaviour factor for steel regular and irregular buildings. In Behaviour of Steel Structures in Seismic Areas, STESSA 94, (eds. F.M. Mazzolani and V. Gioncu), Timisoara, 26 June-1 July, 1994, E&FN Spon, London, 344-354

Vayas, I., Rangelov, N., Georgiev, T. (1999): Schlankheitsanforderungen zur Klassifizierung von Tragern aus I-Querschnitten. Stahlbau, Vol. 68, Heft 9, 713-724

Watanabe, E., Sugiura, K., Harimoto, S., Hasegawa, T. (1992): Ductile cross sections for bridge piers. In Constructional Steel Design. World Development (eds. P.J. Dowling et al), Acapulco, 6-9 December 1992, Elsevier Applied Science, London, 707-710

Watanabe, E., Emi, S., Sugiura, K., Takao., M., Higuchi, Y.(1991): Ductility evaluation of thintubular beam-column under cyclic loadings. In Steel Structures. Recent Research and Development (eds. S.L. Lee and N.E. Shanmugam), Singapore, 22-24 May 1991, Elsevier, London, 847-856

Weston, G., Nethercot, D.A. (1987):Continuous composite bridge beams. Stability of the steel compression flange in hogging bending. In Stability of Plate and Shell Structures (eds. P. Dubas and D. Vandepitte), Ghent, 6-8 April 1987, 47-52

White ,D.W., Barth, K.E. (1998): Strength and ductility of compact flange I-girders in negative bending. Journal of Constructional Steel Research, Vol. 45, No. 3, 241-280

Wilkinson, T., Hancock, G.J. (1998): Tests to examine compact web slenderness of cold-formed RHS. Journal of Structural Engineering, Vol. 124, No. 10, 1166-1174

Yamada, M., Kawamura, H., Tani, A., Iwanaga, K., Sakai, Y., Nishikawa, H., Masui, A., Yamada, M. (1988): Fracture ductility of structural elements and of structures. In 9th World Conference On Earthquake Engineering, Tokyo-Kyoto, 2-9 August 1988, Vol. 4, 219-241

Yamada, M. (1992): Low fatigue fracture limits of structural materials and structural elements. In Testing of Metals for Structures (ed. F.M.Mazzolani), Napoli, 29-31 May 1990, E&FN Spon, London, 184-192

Yoshizumi, T., Matsui, C. (1988): Cyclic behaviour of steel frames influenced by local and lateral buckling. In 9th World Conference on Earthquake Engineering, Tokyo-Kyoto, 2-9 August 1988, Vol. IV, 225-230

Yu, Q.S.K., Uang C.M. (2001): Effects of near-fault loading and lateral bracing on the behaviour of RBS moment connections. Steel and Composite Structures, Vol. 1, No. 1, 145-158

Yura, A. Galambos, T.G., Ravindra, M.K. (1978): The bending resistance of steel beams. Journal of the Structural Division, Vol. 104, ST 9, 1355-1370

Zambrano, A., Mulas, M.G., Castiglioni, C. (1999): Steel members under cyclic loads: A constitutive law accounting for damage. In XVII Congresso CTA, Napoli, 3-7 October 1999, 323-334

7

Advances in Member Ductility

7.1 Recent Development

Although the plastic deformations of members are considered the main source of seismic input energy dissipation, the member ductility has not been completely quantified as well as many other characteristics of their structural response. This remark is valid even for the I-section members, which are used mostly in seismic resistant structures, in spite of quite a large amount of experimental tests and theoretical results. It is probably due to the complexity of analysing the inelastic behaviour, when plastic-local buckling, lateral-torsional buckling, fractures due to high velocity or accumulation of plastic deformations play an important role. Only recently, the problem of evaluating the ductility became a primary interest, as testified by numerous published papers dealing with different aspects. As it is mentioned in Section 6.3.1, the monotonic and static behaviour is studied as a first step and the effects of dynamic behaviour are analysed in a second step.

The calculation of the *ductility of a monotonic statically loaded* member involves the following problems:

-evaluation of the moment–rotation curve, considering the elastic, plastic and strain-hardening deformations;

-determination of rotation capacity, the maximum rotation being limited by the effect of local buckling or by flange fracture;

-providing the member with lateral rigidity and bracings to prevent the lateral-torsional buckling in order to assure an energy dissipation only by bending;

-determination of rotation capacity of high-performance steel members, knowing that the increasing of tensile strength produces a substantial reduction of ductility and an increasing of premature fracture danger.

The *influence of seismic actions* on the rotation capacity involves the following aspects:

-high velocity action producing a high–strain rates, with the consequence of increasing of brittle fracture danger;

Table 7.1: Methods for determining the ultimate rotation capacity

Methods	Approach	Scheme
Experimental	Moment gradient	
	Constant moment	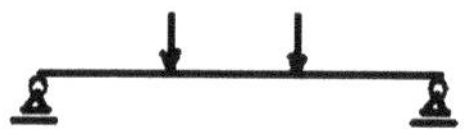
Theoretical	Analytical	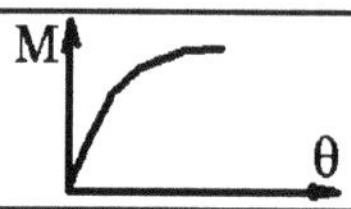 M θ
	Numerical	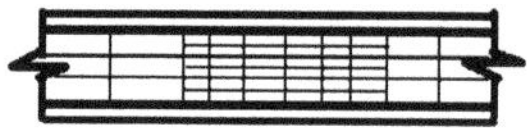
	Plastic collapse mechanim	
Empirical	Based on experimental tests	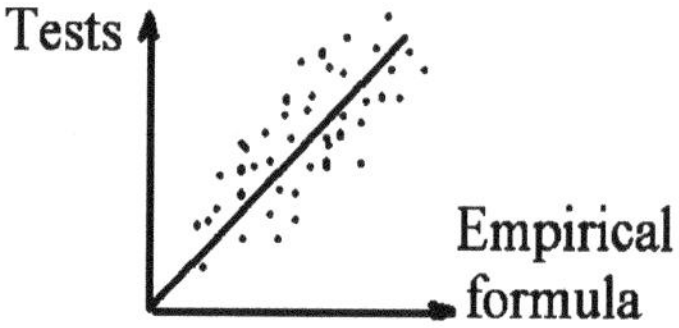 Tests Empirical formula
	Based on numerical tests	

-cyclic loading producing an accumulation of plastic rotation which gives rise to a ductile fracture.

The methods to obtain answers to these problems may be classified into experimental, theoretical (analytical, numerical and plastic collapse mechanism) and empirical (based on experimental or numerical tests) (Table 7.1). The main results are obtained using these methods, and published in the literature which will be presented in the following Sections.

7.2 Ductility of I-Section Members

7.2.1. Monotonic Experimental Tests

(i) *Examination of experimental results.* The experimental research selected from the technical literature for monotonic loading are presented in Table 7.2. The large field of typologies and parameters assure a comprehensive covering of all aspects. But at the same time a very large spread of values are obtained, so the designer should be interested in the degree of confidence in these results.

Table 7.2: I-section; Experimental tests

No	Year	Author(s)	Profile type	Spec. no.	Moment variation	Slenderness ratio flange	Slenderness ratio web	Member type
1	1957	Haaijer	RS	6	CM	6.6..9.2	24.8..40.9	BM
2	1961	Sawyer	RS	21	MG	5.4..9.3	21.1..55.9	BM
3	1969	McDermott	RS	9	MG,CM	3.2..12.3	18.6..31.8	BM
4	1969	Luckey & Adams	RS	12	MG	7.0..9.7	32.7..54.5	BM
5	1970	Höglund & Nylander	RS	3	CM	3.0..10.7	19.0..44.0	B
6	1972	Climenhaga & Johnson	WS	14	MG	4.5..8.5	23.0..62..0	BM
7	1985	Kemp	WS	12	MG	7.5..8.4	31.0..46.8	BM,BC
8	1986	Kuhlmann	WS	24	MG	8.8..11.9	35.0..43.0	BM
9	1988	Schilling	WS	3	MG	9.1..9.9	113..172	BM
10	1989	Gioncu et al	WS	3	CM	9..13.8	60..71	BM
11	1989 1990 1991	De Martino et al	CS	10	CM	10..100	40..150	BM
12	1991	Spange-maher	RS+WS	40	MG	6.7..11.2	18.8..35.9	BM
13	1991	Wargsjo	WS	10	MG	6.4	80..120	BM
14	1993	Boeraeve et al	RS	5	MG	6.7..7.1	10.7..11.0	BM
15	1994	Susuki et al	WS	9	MG	8.3	22.0	BM
16	1994	Nakashima	RS+WS	224	MG,CM	-	-	BM,BC
17	1995	Axhag	WS	15	MG	4.3..6.9	35..114	
18	1996	Susuki et al	WS	9	MG	8.33	12.5..2.5	BM
19	1997	Kubo & Kitahori	WS	7	MG	3.5..7.0	75.0	BM
20	1997	Susuki et al	WS	9	MG	3.1..7.5	17..25	BM
21	1999	Kuhlmann	RS	6	MG	7.1..13.3	20.6..31.7	BM

RS rolled section; WS welded section; CS cold-formed section
MG moment gradient; CM constant moment
BM beam; BC beam-column

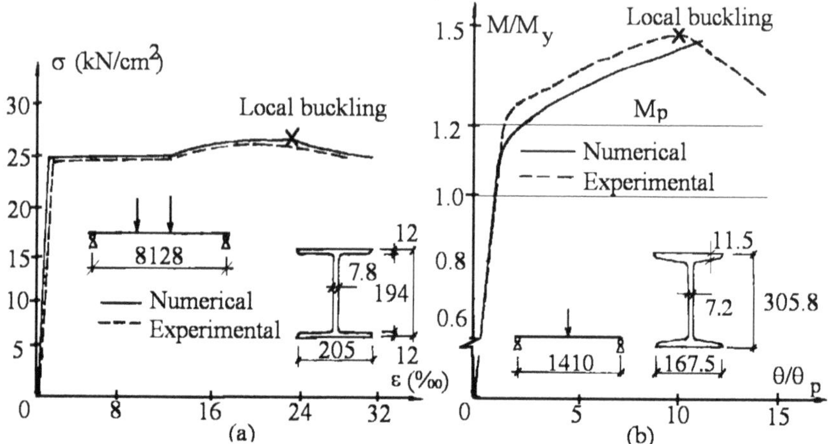

Figure 7.1: Differences in beam behaviour: (a) Constant moment (after Haaijer, 1957); (b) Moment gradient (after Sawyer, 1961)

The first research on the *rotation capacity* of I-section beams was performed by Haaijer (1957) and Haaijer and Thurlimann (1958), for SB 2 standard beams with a constant moment (Fig. 7.1a). The flange buckling occurs in the strain-hardening range, after the yield plateau is crossed. After the local buckling the post-critical curve becomes unstable, showing the degradation in flange rigidity.

A wide range of experimental research work has been performed by Sawyer (1961) for SB 1 standard beams, in order to establish the *influence of moment gradient, strain-hardening and instability phenomena*. Some experimental results are plotted in Fig. 7.1b. One can see that, due to the moment gradient, the moment–rotation curves are different from the ones obtained for SB 2 beams with constant moments. The moment plateau disappears in the plastic range as the result of stresses in strain-hardening range. A continuous increasing of the moment–rotation curve is obtained. One must mention the great influence of beam span on the rotation capacity, the parameters are ignored by the present design codes.

The influence of *cross-section dimensions* and *fabrication types* has been studied by McDermott (1969) (Fig.7.2). The beams 3, 4 and 5 collapse due to local in-plane buckling, while at beams 6 and 7, an interaction between local and lateral buckling occurs. Due to this interaction, an important reduction in the rotation capacity may be noted.

The SB 1 standard-beam types with different flange slendernesses have been tested by Lukey and Adams (1969). This experimental program is the most commonly known among the researchers, many numerical and analytical methods are verified using these experimental results. The typical moment–rotation curve is plotted in Fig. 7.3. After the first yield the curve departs from the idealized curve and continues to rise, exceeding the plastic moment due to the strain-hardening of the outer fibres. Local plas-

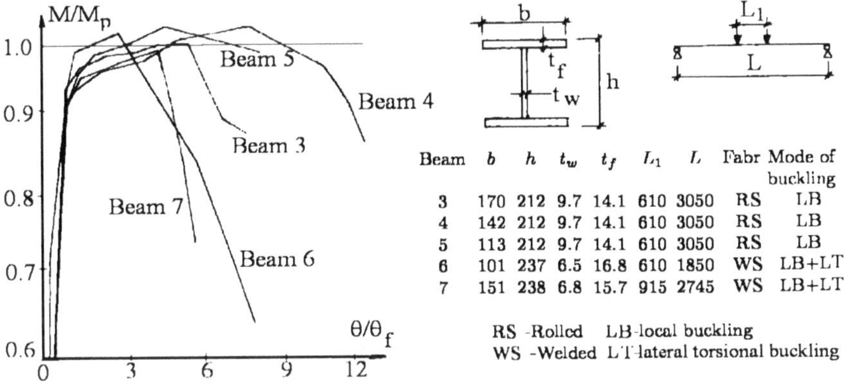

Beam	b	h	t_w	t_f	L_1	L	Fabr	Mode of buckling
3	170	212	9.7	14.1	610	3050	RS	LB
4	142	212	9.7	14.1	610	3050	RS	LB
5	113	212	9.7	14.1	610	3050	RS	LB
6	101	237	6.5	16.8	610	1850	WS	LB+LT
7	151	238	6.8	15.7	915	2745	WS	LB+LT

RS -Rolled LB-local buckling
WS -Welded LT-lateral torsional buckling

Figure 7.2: Influence of cross-section dimensions and fabrication type (after McDermott, 1969)

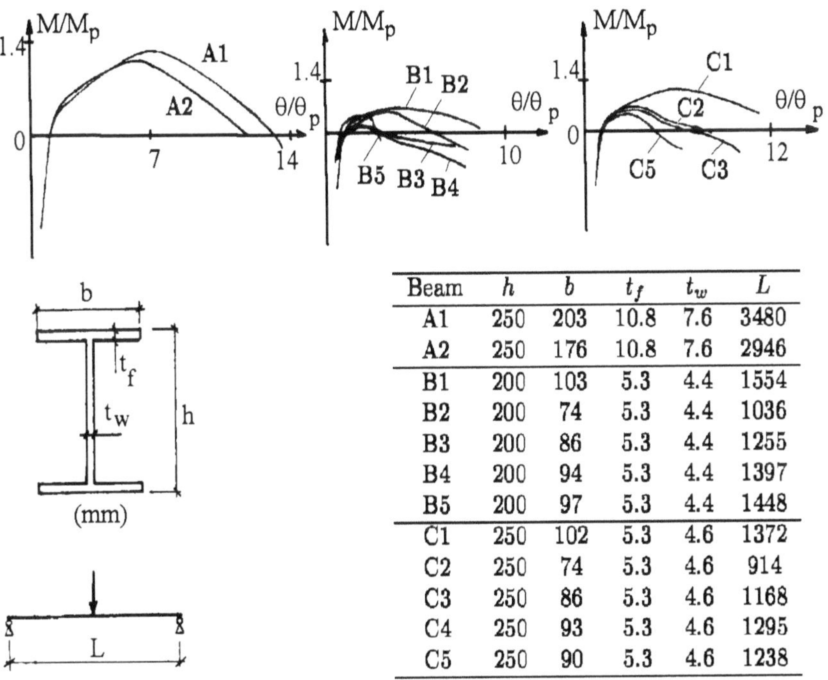

Beam	h	b	t_f	t_w	L
A1	250	203	10.8	7.6	3480
A2	250	176	10.8	7.6	2946
B1	200	103	5.3	4.4	1554
B2	200	74	5.3	4.4	1036
B3	200	86	5.3	4.4	1255
B4	200	94	5.3	4.4	1397
B5	200	97	5.3	4.4	1448
C1	250	102	5.3	4.6	1372
C2	250	74	5.3	4.6	914
C3	250	86	5.3	4.6	1168
C4	250	93	5.3	4.6	1295
C5	250	90	5.3	4.6	1238

Figure 7.3: Influence of flange slenderness (after Lukey and Adams, 1969)

tic buckling occurs within the yielded portion of the compression flange and precipitates a drop in moment capacity. This behaviour is typical for beams subjected to loads, which produce moments which vary along the length of the member, the so-called moment gradient variation. This is the case of the

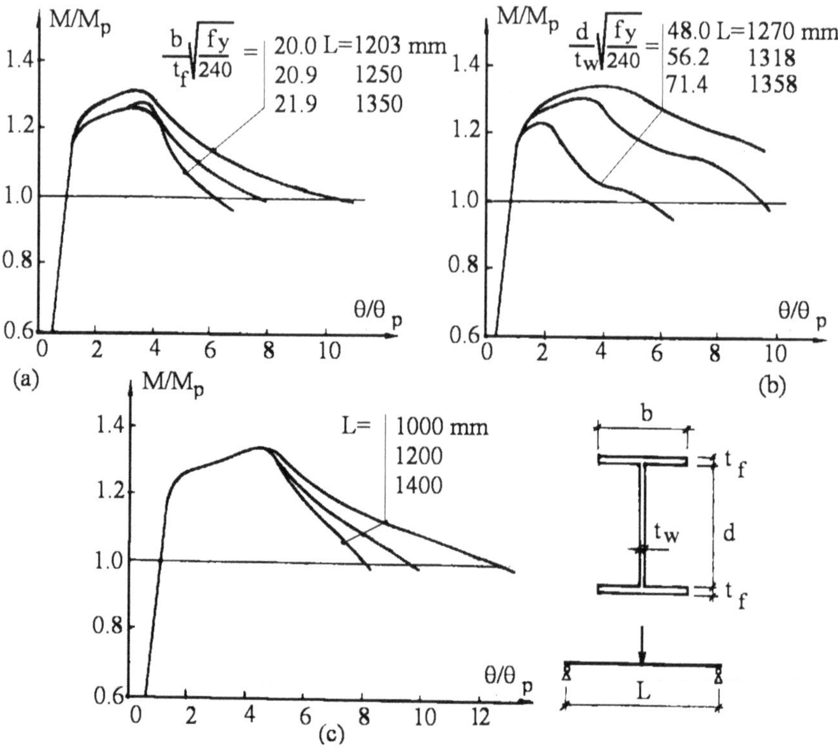

Figure 7.4: Influence of geometrical dimensions (after Kuhlmann, 1986, 1989)

majority of members working in a structure subjected to seismic actions. The experiments have emphasized that the local buckling can occur both in-plane or out-of-plane (S-mode), as a function of geometrical parameters. The possibility to determine the ultimate rotation, as the one at which the moment, drops below plastic moment, has been shown for the first time by these experimental tests.

Experimental tests of Höglund and Nylander (1970) on SB 2 standard beams made of hot-rolled HEA, HEB and IPE profiles were devoted to determine the *range of plastic buckling* of flanges and webs. The influence of steel quality has been studied. For steel with $f_y = 225$ MPa a very good ductility has been obtained, local buckling occuring with very large plastic rotation. Contrary to this, for steel with $f_y = 320$ MPa the ultimate rotation is reduced, showing the great influence of steel grade on the rotation capacity of beams.

The first systematic experimental studies on the *influence of geometrical dimensions* on the rotation capacity of I-section beams have been performed by Kuhlmann (1986, 1989) and Roik and Kuhlmann (1987). Fig. 7.4 shows the experimental moment–rotation curves as a function of three geometrical

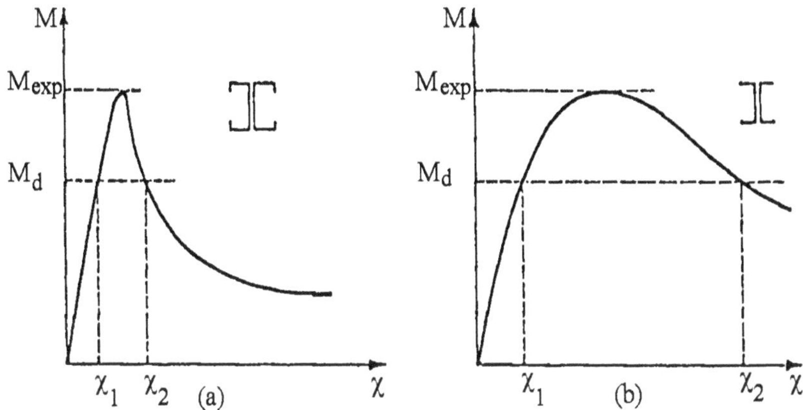

Figure 7.5: Flexural behaviour of cold-formed open cross-sections: (a) Stiffened cross-section; (b) Unstiffened cross-section (after De Martino et al, 1989, 1990, 1991)

parametres: flange and web slendernesses and beam span (moment gradient). Examining these experimental tests clearly results that both slendernesses have a leading role in the beam ductility. It has been observed that the region where local buckling occurs is limited to a length of about $1.2b$ and this value is independent of the beam span. As a consequence, the deformation caused by local buckling has greater importance for the rotation of a short-span beam than for a long span one.

Similar experimental results were obtained by Kemp (1985), Gioncu et al (1989), and Boeraheve et al (1993). A very low ductility has been obtained during the experimental tests performed by Schilling (1988) due to very slender webs (class 4), even the flange slenderness belongs to class 1.

The *flexural behaviour of cold-formed open cross-sections*, including ductility, has been widely examined at the University of Naples (De Martino et al, 1989a,b, 1990a,b, 1991). The beams are composed by two back-to-back channels, covering all the 4 cross-sectional classes of EC 3. The monotonic loading system corresponds to the SB2 beam. The collapse mechanism is strictly connected to the type of section. Basically two different behaviours were observed, according to whether the section is stiffened or not. The stiffened cross-sections reach the maximum load with an almost linear load versus displacement relationships. Then the response curve suddenly lowers, because of the local buckling of compressed edge stiffeners. So the ductility of these cross-section types is reduced (Fig. 7.5a). Contrary to this, the unstiffened cross-sections show a smoother response, with a progressive decrease of strength before reaching the maximum load and a slower decreasing of loading capability in the softening branch (Fig. 7.5b), showing that these cross-section types have some amount of ductility. These experimental tests have shown that some cold-formed thin-gauge sections can be used in dissipative zones of seismic-resistant structures under some circumstances (what is now always forbidden by the present seismic codes).

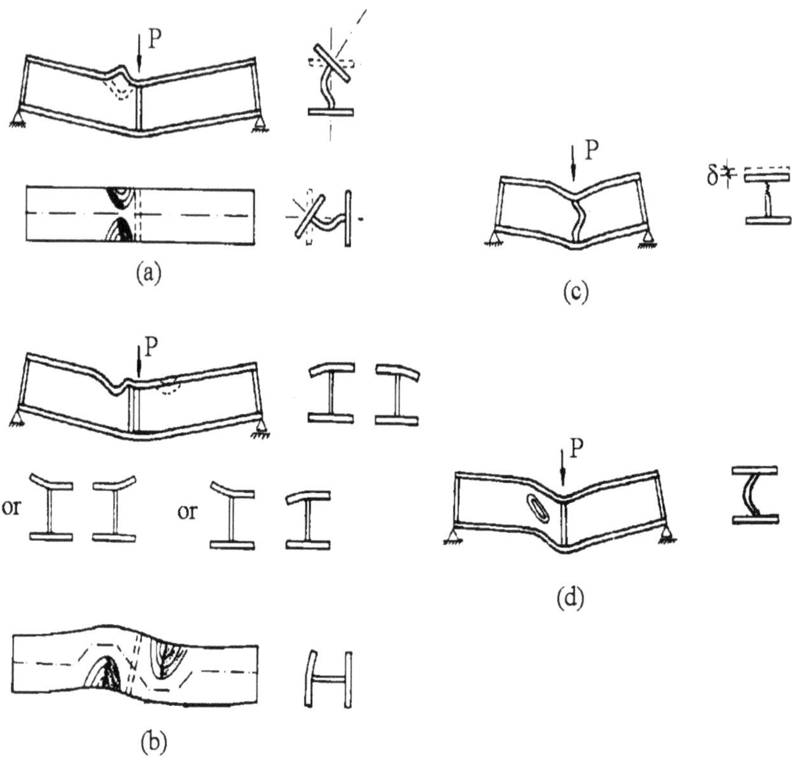

Figure 7.6: Possible collapse modes: (a) In-plane buckling; (b) Out-of-plane buckling; (c) Flange induced buckling; (d) Web shear buckling (after Spangemaher, 1991)

In order to obtain information concerning the *influence of flange and web slendernesses* and *steel grade*, a very wide experimental program has been performed by Spangemaher (1991) and Spangemaher and Sedlacek (1992a,b). An identification of the possible collapse modes is shown in Fig. 7.6: in-plane, out-of-plane, flange-induced and shear-collapse buckling modes.

In the experimantal tests of Axhag (1995) the interest was concentrated to obtain the *maximum plastic moment* and the *slope of the post-critical curve.* A statistical study referring to the maximum moment and the slope of decreasing moment–rotation branch has been performed.

Susuki et al (1992) have studied the *ductility of hybrid beams* with high-strength steel webs and mild steel flanges. The hybrid beams behave better in plastic than the homogeneous beams. A very interesting experimental study concerning the use of *high-strength steel* has been performed by Suzuki et al (1997). For 780 MPa nominal tensile strength steel a lot of beams with different flange and web slendernesses and beam spans were tested. For high slendernesses of flange and web the collapse is due to the local buckling of compression flanges, while for smaller slendernesses or short span the collapse is produced by fractures of tensile flanges.

A detailed analysis of tests performed in Japan concerning the ***ductility of beam-column members*** has been performed by Nakashima (1992, 1994). The experimental results obtained from 224 specimens were grouped according to different criteria: definition of ductility, type of loading, level of axial force, member slenderness. The analysis reveals that neither residual stresses nor initial geometrical imperfections are major factors affecting the ductility, but strain-hardening characteristics are mainly responsible for the large variation in experimental data. The main conclusion of this study is that, for beam-column members, the normalized slenderness ratio $n_p\lambda^2$ is a good indicator for estimating the member ductility.

(ii) *Conclusions on experimental tests.* After the examination of the experimental pictures showing the collapse of specimens, the experimental diagrams and some partial conclusions, the following observations can be drawn in order to obtain some general information which can be used for developing a consistent methodology:

-the majority of experimental tests have been performed on SB 1 standard beams, characterized by the effect of moment gradients. This is a very good situation, because it represents the most specific case for the seismic loaded structures;

-the plastic deformations are only produced in a limited zone, the remaining part of the member works in the elastic field. In this plastic zone large rotations are concentrated and sometimes amplified by the buckling of flange and web;

-during the plastic deformation, crumpling of flange and web can be observed, forming a clear local plastic mechanism which remains even after the specimen unloading;

-for SB 1 standard beams two symmetric buckling zones are expected to develop (at the left and right parts of vertical stiffener), but during the experimental test only one is completely formed due to the well known phenomenon of plastic concentration, the second one remaining undeveloped;

-a part of the plastic mechanism works with stress levels corresponding to yielding, but there are some zones in which the stresses are in strain-hardening range;

-the plastic mechanism may be composed by yield lines, fracture lines and plastic zones in function of the plastic rotation extension;

-the buckling for usual specimens starts from flange crumpling which induces the web deformations. Thus a very important interaction between flange buckling and web deformation exists;

-examining the aspects of beam failures shown in Fig. 7.6, it appears that there are many forms of plastic buckling, depending on the geometrical proportions of specimen: in-plane and out-of-plane buckling, produced by bending moments or shear forces;

-an interaction between these local mechanisms is observed during the experimental tests. In the majority of cases, the plastic buckling starts with in-plane buckling, but due to the weakening in lateral rigidity caused by the plastic buckling, the lateral buckling occurs as a second step. So, the lowering path of the moment–rotation curve is dominated by the interaction between the two buckling modes, producing a more pronounced degradation;

-for high-strength steel the collapse of members may be produced by the fracture of buckled flange along some lines working in the strain-hardening range;

-a good ductility is obtained from the plastic behaviour of webs: for very slender webs, the ductility is consequently reduced;

-the main factors influencing the local ductility of a beams are the flange and web slendernesses, lateral rigidity, beam span, moment gradient, steel grade of flanges and web, yield ratio of steel and strain-hardening characteristics;

-for beam columns, in addition to these factors one must mention the column slenderness and the level of axial force;

-the cross-section classification based on the wall slenderness, according to EC 3, must be revised on the evidence of new experimental results;

-the interpretation of some experimental results on cold-formed sections, even if they belong to the slender class, are encouraging in view of their application in dissipative zones of seismic resistant structures. It is a very important issue as the cold-forming technology is becoming more and more popular and economically attractive.

7.2.2 Analytical Approaches

Table 7.3 gives the references devoted to the main aspects of analytical approach, which are the definition of the moment–rotation curve and the determination of the rotation capacity.

(i) *The moment–rotation curve.* The studies of Sawyer (1961) are devoted to determine the moment-rotation curve until the maximum moment. The post-critical path is ignored. The moment–rotation curve is obtained by the integration of the moment-curvature equation, resulting the following expressions:

- for $M_y < M < M_p$:

$$\theta = \frac{L}{2d}\left\{\varepsilon_y + \frac{1}{2M}\left[-M_y\varepsilon_y + (M-M_y)^2\frac{\varepsilon_{sh}-\varepsilon_y}{M_p-M_y}\right]\right\} \tag{7.1a}$$

- for $M_p < M < M_{max}$:

$$\theta = \frac{L}{2d}\left\{\varepsilon_{sh} + \frac{1}{2M}\left[M_p\varepsilon_y - (M_p+M_y)\varepsilon_{sh} + (M-M_p)^2\frac{\varepsilon_{max}-\varepsilon_{sh}}{M_{max}-M_y}\right]\right\} \tag{7.1b}$$

where M_y is the elastic limit bending moment, M_p is the fully plastic bending moment, M_{max} is determined from the stress–strain curve for a strain

Table 7.3: I-section; Analytical approaches

Moment–rotation curve	Rotation capacity	
	in-plane buckling	out-of-plane buckling
• Sawyer (1961) • Lay & Galambos (1967) • McDermott (1969) • Mazzolani & Piluso (1992, 1993) • Vayas & Psycharis (1992, 1993, 1995) • Morino et al (1995)	• Lay (1965) • Lay & Galambos (1967) • Höglund & Nylander (1970) • Kemp (1985, 1986, 1996, 1999) • Spangemaher (1999) • Mazzolani & Piluso (1992, 1993, 1996) • Daali & Korol (1995) • Georgescu & Mazzolani (1996) • Dekker (1998) • Cryssanthopoulos et al (1999)	• Lay & Galambos (1965, 1967) • Yura et al (1978) • Kemp (1985, 1986, 1991) • Nakamura (1988) • Daali & Korol (1994) • Kemp (1996) • Dekker (1998) • Vayas (1997)

ϵ_{max} given by:

$$\varepsilon_{max} = 5\varepsilon_{avg} = \frac{1280}{\left(\frac{b}{t_f}\right)^2 \cdot \left(\frac{L}{2b}\right)^{3/4} \frac{d}{t_w}} \tag{7.2}$$

being the relationship assessed from the experimental evidence.

Lay and Galambos (1967) proposed the equation:

$$\frac{\theta}{\theta_p} = \frac{M}{M_p} + \left(1 - \frac{M_p}{M}\right)\left[2s + \frac{1}{e_h}\left(\frac{M}{M_p} - 1\right)\right] \tag{7.3}$$

with the notations given by (6.30) and (6.31). This equation indicates the necessity to consider the strain hardening in the evaluation of the rotation capacity. In order to obtain the maximum rotation capacity, the characteristic coefficients for the strain-hardening range s and e_h should be as large as possible. The dominant term in equation (7.3) is the coefficient $2s$ which marks the jump from the yield point to the outset of strain hardening. A similar equation was developed by McDermott (1969). Introducing the value of M_{max} obtained for plastic buckling of the flange, one can determine the rotation capacity corresponding to maximum moment value.

The studies of Kato (1989, 1990) and Mazzolani and Piluso (1992, 1993) use the same integration methodology of moment curvature, but considers the differences in stress levels for compression and tension flanges.

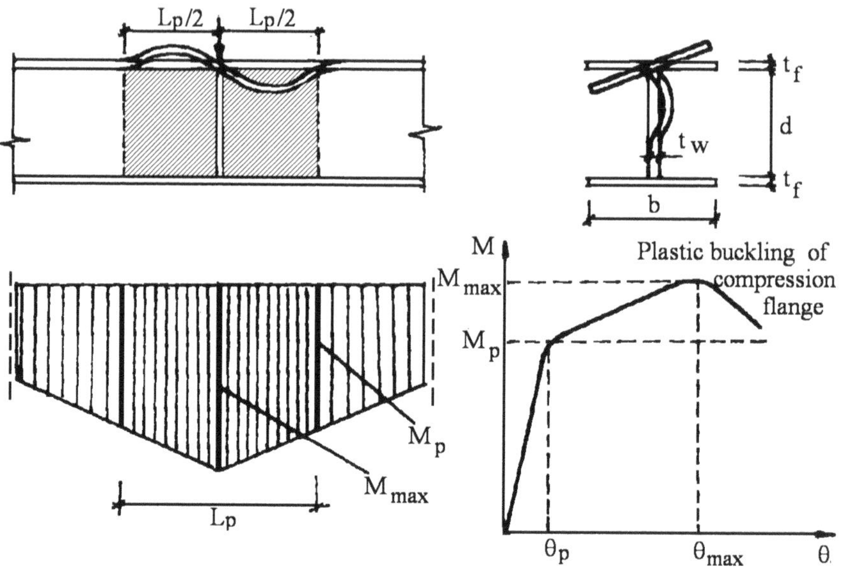

Figure 7.7: Length of plastic zone (after Lay, 1965, Lay and Galambos, 1967)

Vayas and Psycharis (1990, 1992, 1993, 1995) and Vayas (1997) developed a methodology for plotting the moment–rotation curve in its complete form, including the post-critical path. The method of *effective width* is used in order to obtain the moment–curvature diagram for the cross-section and the rotation capacity for the beam in the lowering branch.

The moment–rotation curve is determined also by Morino et al (1995) by means of two approaches. First the beam is divided into segments and for each segment a different stress–strain law is used. In the second method, which corresponds to the rigid-plastic analysis, the plastic deformation is concentrated only in the first element. Some differences exist between these two approaches. The rigid-plastic analysis estimates that the plastic rotation is a little larger than the one of the first more exact approach. But being very simple and applicable to plastic analysis, it is recommended to be used for beam and beam-column failing in local plastic buckling mode.

(ii) ***Rotation capacity for in-plane buckling.*** The first consistent studies have been performed by Lay (1965) and Lay and Galambos (1967). The proposed approach considers that the maximum value of bending moment occurs when the length of the plastic zone reaches the length of one wave of plastic buckling of compression flange (Fig. 7.7). It is assumed that the flange is restrained against buckling by a torsional spring that represents the effect of web. In this case, taking account of Haaijder's observation about the similarity between local and torsional buckling (see Section 4.3.4), the strain corresponding to buckling may be expressed in the following form:

$$\varepsilon_b = \frac{12}{E_h b^3 t_f}\left[\frac{G_h b t_f^3}{3} + E_h I_w \left(\frac{2\pi}{L_p}\right)^2 + K\left(\frac{L_p}{2\pi}\right)^2\right] \tag{7.4}$$

where I_w is the warping constant and K the torsional stiffness of the web. The first term in the bracket corresponds to free torsion, the second to warping torsion and the last to the web effect. For these values, Lay considers that:

$$\frac{G_h}{G} = \frac{2}{1 + \dfrac{1}{4(1+\nu)}\dfrac{1}{e_h}} \tag{7.5}$$

$$I_w = \frac{7}{16}\frac{b^3 t_f^3}{144}; \qquad K = \frac{G_h t_w^3}{3d} \tag{7.6a, b}$$

where ν is Poisson's ratio ($\nu = 0.5$ for the plastic range). The minimum of equation (7.4) is obtained when the two last terms are equal, resulting:

$$L_p = 2\pi\left(\frac{E_h I_w}{K}\right)^{1/4} \tag{7.7}$$

Using equation (7.7) with some approximations, the plastic zone length becomes:

$$L_p = \frac{\pi}{4}\left\{\frac{7}{3}\left[\frac{1}{4} + \frac{1}{e_h}(1+\nu)\right]\right\}^{1/4}\left(\frac{A_w}{A_f}\right)^{1/4}\frac{t_f}{t_w}b \tag{7.8}$$

A_w and A_f being the areas of web and flange, respectively. After eliminating material constants some new approximations, result:

$$L_p \approx 0.713\frac{t_f}{t_w}\left(\frac{A_w}{A_f}\right)^{1/4} b \tag{7.9a}$$

or in terms of section dimensions:

$$L_p \approx 0.713\left(\frac{d}{b}\right)^{1/4}\left(\frac{t_f}{t_w}\right)^{3/4} b \tag{7.9b}$$

One can see that this buckling length is independent of the beam span and moment gradient, being a function of the cross-section characteristics.

The first requirement for local plastic buckling is that it is able to form the full plastic pattern due to the conditions which one likely to occur in structural schemes. These cases are illustrated in Fig. 7.8. In the case of SB 1 standard beams, buckling waves can be asymmetrical with respect to the middle beam span, while the beam end is restrained, the full wave length must be in the plastic zone. But, according to Section 7.2.1, the experimental tests have shown that only one half wave is developed due to the phenomenon of plastic deformation concentration. One of the two half

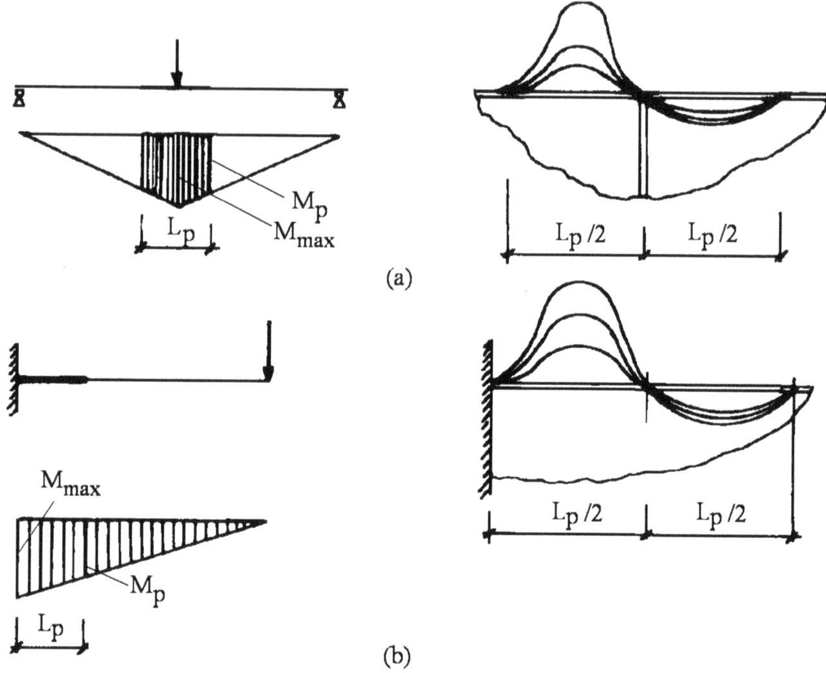

Figure 7.8: Development of buckled waves: (a) For simply supported beams; (b) For cantilever beams

waves is inevitably weaker than the other one in resistance (in case of the standard beam) or in the bending moment (in the case of cantilevers), and, due to the decreased bending moment after local buckling, the second half wave cannot develop.

The comparison between the length of the plastic zone is determined by equation (7.9) and the measured length from the experimental works of Haaijer (1957) and Lukey and Adams (1969) are presented in Fig. 7.9a. One can see a good correspondence between experimental and theoretical values, the average value being 0.999 with a coefficient of variation 11.7%. Kuhlmann (1989) has obtained an average value $L_p/b \simeq 1.2$, with a very good correspondence with a theoretical average result given by equation (7.9), which is equal to 1.176 (Fig. 7.9b). One must mention that this correspondence between theoretical and experimental values is excellent, considering the difficulties in measuring the actual buckling lengths.

Lay (1965) has determined the flange slenderness for which no plastic buckling occurs:

$$\frac{b}{t} \le \frac{3.6}{\left[\left(3 + \frac{f_u}{f_y}\right)(1 + 0.192 e_h)\right]^{1/2}} \left(\frac{E}{f_y}\right)^{1/2} \tag{7.10}$$

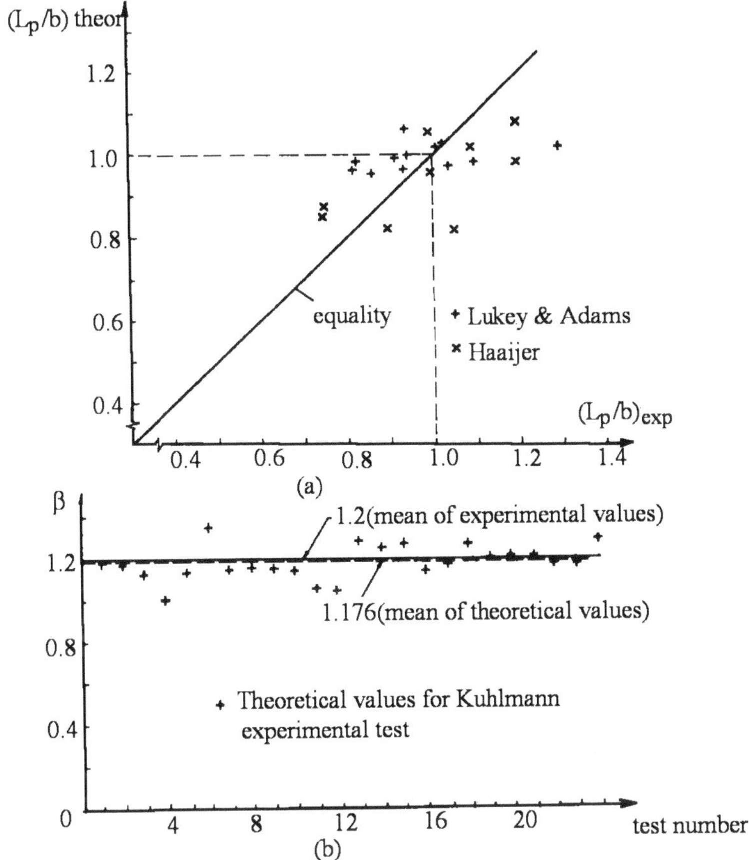

Figure 7.9: Theoretical-experimental correlation for length of buckled waves: (a) Haaijar and Lukey & Adams experimental results; (b) Kuhlmann results (after Gioncu and Petcu, 1997)

An extensive theoretical and experimental study of Hoglund and Nylander (1970) has confirmed the validity of this condition.

Based on the equation (7.9) it is possible to determine the value of rotation corresponding to the plastic buckling of flange. The moment producing the buckling is given by:

$$M_b = \frac{M_p}{1 - \dfrac{L_p}{L}} \tag{7.11a}$$

and therefore:

$$m_b = \frac{1}{1 - \dfrac{L_p}{L}} \tag{7.11b}$$

With this coefficient the rotation ductility corresponding to the plastic buckling results:

$$\mu_\theta = \left[2s + \frac{1}{e_h}\left(\frac{1}{1 - L_p/L} - 1\right)\right]\frac{L_p}{L} = \left(2s + \frac{1}{e_h}\frac{L_p/L}{1 - L_p/L}\right)\frac{L_p}{L} \quad (7.12)$$

One can see that, contrary to this, the plastic rotation, the rotation ductility corresponding to the plastic buckling depends on the beam span.

Based on the Lay and Galambos proposals, Kemp (1985, 1986, 1996), Kemp and Dekker (1991), Kemp et al (1995), has rearranged the expressions and modified some hypothesis, giving rise to the following equations:

-*local buckling of flange* occurs for the following slenderness:

$$\left(\frac{b}{t_f}\right)_b = \left[\frac{2}{1.5\varepsilon_b - c_f(\pi t_f/L_p)^2}\right]^{1/2} \quad (7.13a)$$

where L_p is given by (7.9) and

$$\varepsilon_b = \varepsilon_{yf}\left(s + \frac{1}{2e_h}\frac{L_p/L}{1 - L_p/L}\right) \quad (7.13b)$$

c_f considers the web restraint (0.5 for no warping restraint and 1.0 for web warping restraint);

-*local buckling of web* occurs for the following slenderness:

$$\left(\frac{d_{fc}}{t_w}\right)_b = 4\left(\frac{0.97}{\varepsilon_{yw}}\right)^{1/2}(e_h)^{-1/4}\frac{d_{fc}}{L_w} = \frac{2}{c_w}\left(\frac{0.97}{\varepsilon_{yw}}\right)^{1/2} e_h^{-1/4} \quad (7.14a)$$

where d_{fc} is the distance between the plastic neutral axis and the centre of the compression flange (Fig. 7.10a), L_w, the length of the buckling zone of web, resulting from the equation:

$$L_w = 2c_w d_{fc} \quad (7.14b)$$

In this equation c_w is the ratio of the effective depth of web in compression to the actual depth of web in compression d_{fc} and is given by (Fig. 7.10b):

-if compression flange is not buckled:

$$c_w = 0.75 - 0.25n_p \quad (7.15a)$$

-if compression flange is already buckled:

$$c_w = 1.0 - 0.25n_p \quad (7.15b)$$

n_p being the ratio between the axial force and the fully yielding force of the web depth. For bending $d_{fc} = 0.5d$, $c_w = 0.75$ and 1.0, respectively, resulting in $L_w = d$ or $0.75d$, in function of web restraint in flange.

Relations for determining the ductility at the maximum moment capacity are proposed by Mazzolani and Piluso (1992, 1993, 1996), in the same

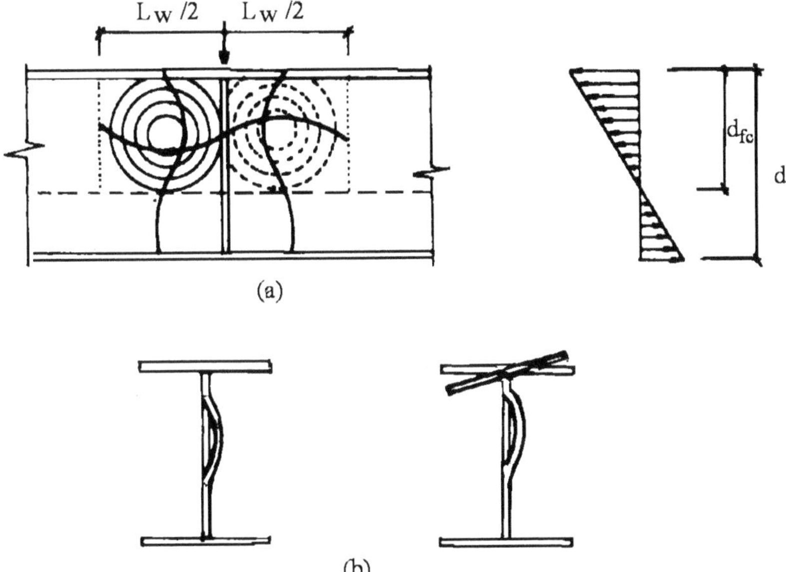

Figure 7.10: Local buckling of web; (a) Web buckled shape; (b) Influence of compression flange deformation (after Kemp and Dekker, 1991, Kemp et al, 1995)

form as the equations (6.43–6.46). Daali and Korol (1995) have established an equation to evaluate the rotation capacity, which is the same as the equation (7.12).

The different behaviour of compression and tension flanges has been considered by Georgescu and Mazzolani (1996). The rotation capacity related to maximum moment is obtained.

Based on Kemp's studies, Dekker (1998) has proposed an equation for evaluating the buckling strain:

$$\varepsilon_b = 7.895\left(1 - 0.017\frac{d_c}{t_w}\right)\left(\frac{t_f}{b}\right)^2 \qquad (7.16)$$

Considering the idealized stress–strain variation, it is possible to calculate the value of compression buckling stress.

Chyssanthopoulos et al (1999) has examined the effect of material variability (yield stress and strain hardening) on the deflection ductility of a cantilever beam type (equivalent to the SB 1 standard beam type). The material properties are susceptible to random variability as a result of making, forming, rolling or welding. The most important effect in the reduction of ductility is due to the variability of yield stress; the effect of strain-hardening variability reduces as far as the level of axial force increases. From the designer's point of view this study can be used to quantify acceptable limits for material variability in order to ensure a characteristic ductility value.

(iii) *Rotation capacity for out-of-plane buckling.* In case of a uniform moment the relationship between the lateral slenderness factor λ_y and the rotation capacity μ_θ is given by (Lay and Galambos, 1965):

$$\lambda_y = \frac{\pi}{\left(1 + 0.7\dfrac{\mu_\theta}{s-1}\dfrac{1}{e_h}\right)^{1/2}} \frac{1}{\varepsilon_y^{1/2}} \tag{7.17}$$

with

$$\lambda_y = k\frac{L}{r_y} \tag{7.18}$$

where k is the effective length factor, L, the beam length between two braces, r_y, the weak axis radius of gyration. In the majority of cases the value of the effective length factor is $k = 0.54$. However, if the adjacent spans are under full plastic moment, $k = 0.8$. The equation (7.17) allows to determine the maximum support spacing needed to assure that a given rotation capacity is obtained. Alternatively, the rotation capacity supplied by a given support spacing may be determined. A good agreement between relation (7.17) and experimental results is obtained.

For a beam with a moment gradient, Lay and Galambos (1967) have proposed the equation:

$$\lambda_{y\,max} = 1.87\frac{t_w}{t_f}\left(\frac{A_f}{A_w}\right)^{1/4}\frac{b}{L}\varepsilon_y^{1/2} \tag{7.19}$$

For $\lambda < \lambda_{ymax}$ the local plastic buckling occurs, and for $\lambda > \lambda_{ymax}$, lateral-torsional buckling is produced. But an interaction between these two buckling modes exists: lateral buckling occurring after the local buckling produces a reduction of ductility capacity.

The problem of lateral bracing of a beam has been studied also by Yura et al (1978).

Kemp (1985,1986) and Dekker (1998) have proposed, for the lateral-torsional buckling of a beam with moment gradient, an equation to evaluate the critical lateral slenderness ratio for which plastic lateral buckling occurs:

$$\frac{L_b}{r_y} = C\frac{1}{\varepsilon^{1/2}} \tag{7.20}$$

with

$$C = \frac{(1.08 - 0.04\beta)(m - \beta)}{(1-\beta)(0.6 - 0.4\beta) + (m-1)[e(0.6m + 0.4)]^{1/2}/c_t} \tag{7.21}$$

where $m = M_{max}/M_p$ is the ratio of maximum moment to fully plastic moment, β, ratio of moment at lateral restraint to fully plastic moment, $c_t = 1.0$ if the full flange resists buckling in plastic range or $c_t = 0.667$ if only two-thirds of width is effective in plastic range due to the presence of flange local buckling. The interaction between the flange, web and lateral buckling has been also studied.

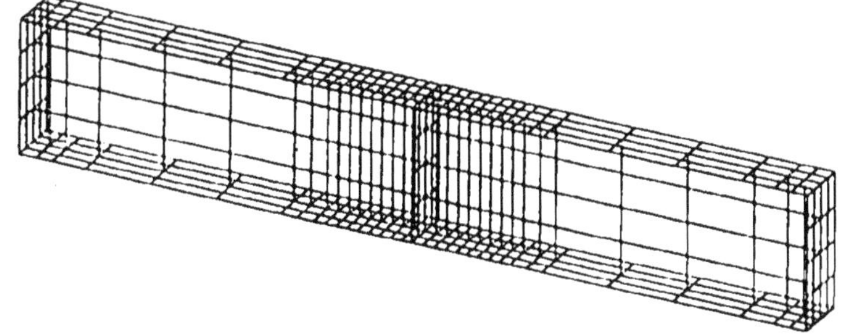

Figure 7.11: Mesh of finite elements

(iv) *Conclusions on analytical approaches.* After the examination of analytical results, the following observations may be noted:

-the analytical approaches consider just the first part of the moment–rotation curve, until reaching the maximum moment and the rotation capacity is determined for this value only;

-for out-of-plane buckling, the studies are concentrated on determining the distance between braces in order to avoid the lateral buckling;

-the ultimate rotation capacity is strongly influenced by the interaction between local plastic buckling and lateral-torsional buckling. So, it is very important to assure sufficient lateral rigidity or some constructional measures to impede the lateral deformations;

-the length of the plastic buckled zone can be determined using the equation (7.9) proposed by Lay, which is confirmed by the experimental results;

-the local ductility is strongly influenced by the variability of steel properties, especially by the increasing of yield stress.

7.2.3. FEM Numerical Tests

(i) *Examination of numerical tests.* Several simulations of bended beams have been performed with the material and geometrical non-linear finite element method, FEM. The most commonly known computer programs are ABAQUS (1996), FINELG (1996), PROFIL (1991), these being used in the majority of numerical tests. The element subdivision along the beam length is usually characterized by a more refined area in the region of possible local plastic buckling, and by a coarser mesh in the less loaded regions (Fig. 7.11). In the refined modeling, the flange–web junction of rolled profiles is introduced, while for welded profiles, the reduced strength of heat affected zone, HAZ, must be considered.

A list of the main studies in which FEM numerical tests have been performed for determining the rotation capacity of I-section members is presented in Table 7.4, with the principal characteristics of each research work and the used program.

The first study belongs to Kuhlmann (1986, 1989) and Roik and Kuhlmann (1987). A special software has been developed to calculate the ro-

Table 7.4: I-section; Numerical tests

No	Year	Author(s)	Specimen no.	Test type	Steel grade	Objective	Used program
1	1986 1989	Kuhlmann	12	MG	ms	PS,EC	special software
2	1990	Greschik et al	4	MG	ms	PS,EC	STARS
3	1991	Spangemaher	69	MG	ms	PS,EC	ABACUS PROFIL
4	1993	Boerave et al	14	MG	ms	PS,EC	FINELG
5	1994	Suzuki et al	10	MG	hs+ms	PS,EC	-
6	1995	Axhag	20	MG	ms	PS,EC	ABAQUS
7	1996	Espiga & Anza	77	MG	ms	PS,ES	ABAQUS
8	1996	Huang & Deierlein	-	MG	ms	PS,ES	ABAQUS
9	1997	Suzuki et al	4	MG	hs	PS,EC	-
10	1998	Earls & Galambos	14	MG,CM	hs	PS,EC	ABAQUS
11	1999	Moen et al	4	MG	al	EC	ABAQUS
12	2000	Earls	56	BG	ms+hs	PS	ABAQUS

MG moment gradient — CM constant moment
EC experimental calibration — PS parametrical study
ms mild steel — hs high strength steel
al aluminium alloy

tation capacity. The aim of this research was the *numerical evaluation of the rotation capacity* in function of some governing parameters, as flange and web slendernesses and beam spans. A good correspondence between numerical results and experimental values obtained by Lukey and Adams (see Fig. 7.3) exists (Fig. 7.12).

Greschik et al (1990) have investigated the *use of FEM for predicting the rotation capacity*. Some differences between the experimental and numerical results are explained by the presence of residual stresses, geometrical imperfections and the idealization of the stress–strain curve.

A wide numerical testing program for I-rolled and welded sections has been carried-out by Spangemaher (1991) and Spangemaher and Sedlacek (1992a,b). The research was divided in two parts: the first is devoted to the *verification of numerical results* by comparison with the experimental values, the second being concentrated on the *parametrical study* in order to elaborate a simplified methodology. The main parameters under examination were the moment gradient, flange and web thicknesses and steel quality. Fig. 7.13 shows the influence of some parameters on the rotation capacity.

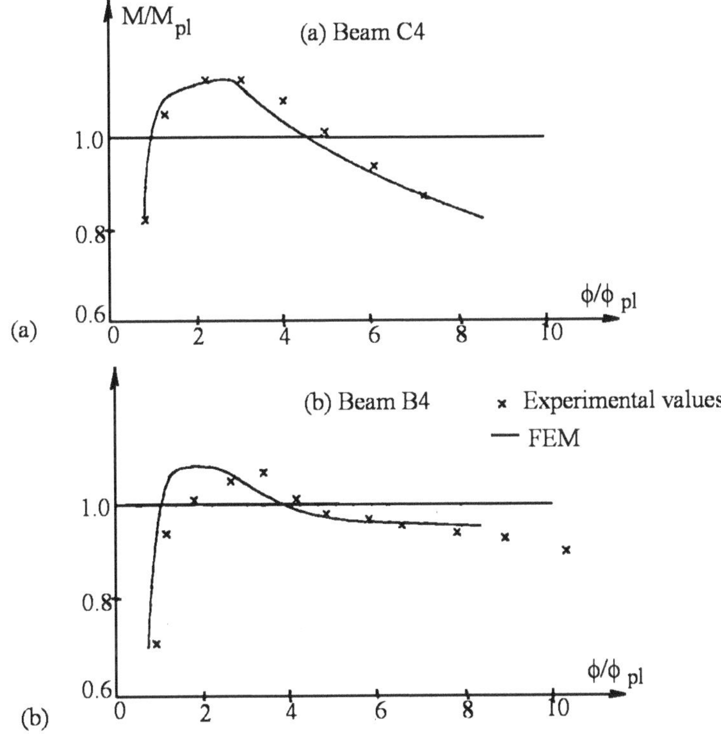

Figure 7.12: Correspondence between experimental values (Lukey and Adams, see Fig. 7.3) and FEM analysis (after Kuhlmann, 1986, 1989)

The main reason of numerical tests performed by Boeraeve et al (1993) has been the verification of the *potential of the FEM to determine the rotation capacity* by comparison of numerical simulation with the experimental results obtained on five specimens. The good correspondence between these values has stimulated the analysis of the influence of yield stress and yield ratio.

The numerical study of Susuki et al (1994) is devoted to the *rotation capacity of hybrid beams* with high-strength steel webs and mild steel flanges. Numerical tests have shown that the rotation capacity of these hybrid beams is larger when compared with homogeneous beams, confirming that the post-critical behaviour is governed by web strength.

In order to obtain the *maximum moment values* and the *slope of post-critical curves*, Axhag (1995) has numerically tested different members having geometrical imperfections and residual stresses. Because in the majority of numerical tests the slenderness of web belongs to class 4, the results are not very interesting for the seismic applications.

The *simulation of flange buckling* and the unstable post-critical behaviour of I-section beams is studied by Espiga and Anza (1996) and Espiga

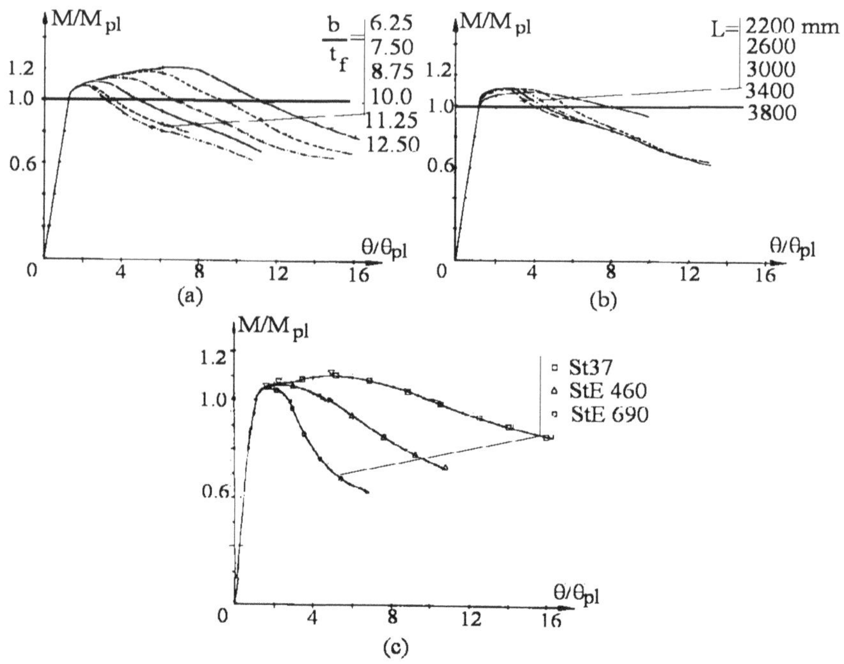

Figure 7.13: Numerical simulation for different parameter influences: (a) Flange slenderness; (b) Beam span; (c) Steel quality (after Spangemaher, 1991)

(1997a, b, c, 1998). Some FEM simulated deformations of beams with buckled flanges are presented in Fig. 7.14a. An important number of simulations have been carried-out to quantify the influence of some important factors: position of load, concentrated and uniformly distributed loads, lateral restrains, moment gradient, steel grade, profile type (HEA and IPE). The influence of beam span on rotation capacity results from Fig. 7.14a,b. One can see that, for the same beam span, the IPE profiles have a more reduced rotation ductility than the HEA profiles. Therefore, a question arises concerning the frequent utilization of IPE sections for beams of moment-resisting frames, designed to work in a global plastic mechanism. The second very important observation is related to the influence of moment–gradient (beam span), which cannot be neglected in design, as in EC 3. The influence of geometrical imperfections is not very important.

A very interesting study performed by Huang and Deierlein (1996), considers the inelastic behaviour of SB 1 standard beams tested by Lukey and Adams (Fig. 7.15a). In Fig. 7.15b the curvature along the left half span of the beam at the corresponding load steps is plotted. One can see the *concentration of plastic deformations* is near to the beam midspan. The plastic buckling occurs at step 14, but this does not affect the beam behaviour.

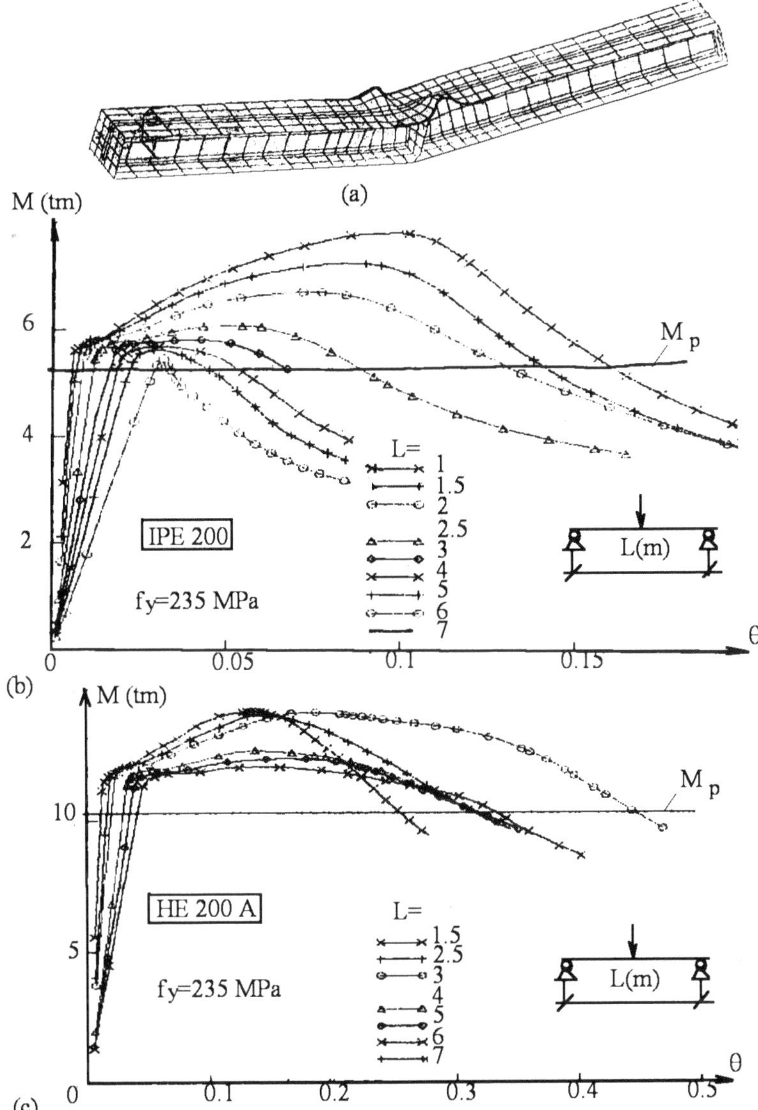

Figure 7.14: Numerical simulation for beam span influence: (a) Beam behaviour; (b) IPE 200 profile; (c) HE 200 A profile (after Espiga, 1997)

Only at step 18 the beam unloads due to a large increase in curvature near point C produced by local plastic buckling. Between steps 18 and 29 the curvature peak increases slightly, while the zone of large curvatures spreads along the plastified length. The lateral restraints (beam framing into stocky columns or the effect of braces) impeding the out-of-plane rotation has an important impact on the rotation capacity (Fig. 7.15c).

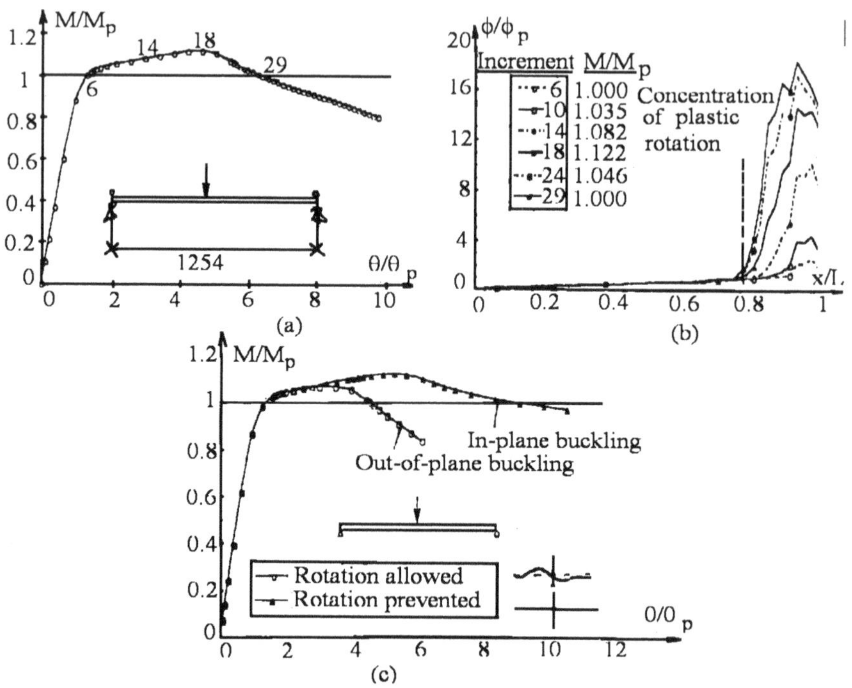

Figure 7.15: Numerical simulation of B3 specimen (Lukey and Adams, see Fig. 7.3): (a) Moment–rotation curve; (b) Concetration of plastic rotation; (c) Influence of lateral restrains (after Huang and Deierlein, 1996)

The *ductile fracture* of beams made from high-strength steel (f_y = 780 MPa) is studied by Susuki et al (1997). The fracture is defined by the displacement at the time when the strain of tensile flange reaches the strain of tensile strength on specimen tests. The ductility of beams limited by the ductile fracture; for small flange and web slendernesses the plastic ductility turns into the ductile fracture.

The studies of Earls and Galambos (1998) and Earls (1999a,b) considers the behaviour of beams made of *high-performance steel* (HSLA 80). In-plane and out-of-plane plastic buckling modes, constant and moment gradients, cross-sectional imperfections, and lateral bracing are all considered parameters. The moment–rotation curves for different geometrical imperfection types are plotted in Fig. 7.16. One can see that the influence of these imperfections on rotation capacity depends on the lateral rigidity, i.e. on the plastic buckling mode. The influence of geometrical factors and material properties (yield strength, strain hardening, yield plateau) is studied by Earls (2000a,b). All these factors play fundamental roles in influencing which of the two modes, in-plane or out-of-plane, govern the beam failure and its ductility.

Numerical research on *aluminium beams* have been performed by Moen et al (1999). The comparison between experimental and numerical results

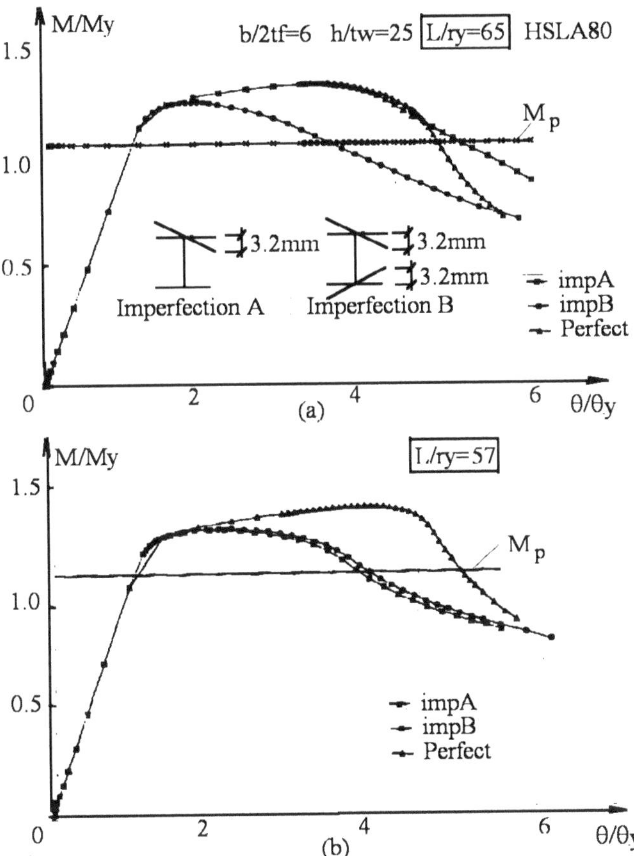

Figure 7.16: Behaviour of high steel beams (HSLA 80) with geometrical imperfections: (a) $L/ry = 65$; (b) $L/ry = 57$ (after Earls, 1999)

shows that the FEM is capable to predict the beam behaviour and rotation capacity even when the constitutive stress–strain law is strongly nonlinear.

(ii) *Conclusions on numerical tests.* Examining the above mentioned numerical results, the following remarks can be drawn:

-in order to obtain good results a dense element mesh in the plastic zone must be used. Therefore, the computer time and cost for use of FEM analysis are very high. So the method can be used in research only in order to calibrate with some more simple formulations;

-due to the possibility offered by this very refined analysis, some more detailed aspects concerning the inelastic behaviour can be obtained;

-a good correlation with experimental results in the field of post-critical path has been reached;

-the main experimental conclusions concerning the rotation capacity of steel members are confirmed by the numerical tests.

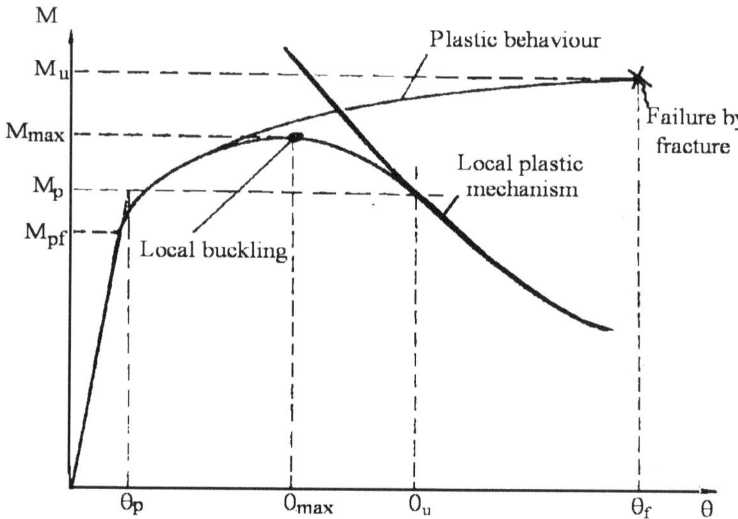

Figure 7.17: Post-critical curve of local plastic mechanism

7.2.4. Plastic Collapse Mechanism Method

Generally, the analytical approaches are limited in plotting the moment–rotation curve just until the maximum moment value, due to the difficulties for describing the post-critical behaviour. The FEM analysis can provide the complete curve, but with very high costs which are not suitable for practical design. Contrary to this, the plastic collapse mechanism method is only involved to plot the post-critical curve (Fig. 7.17). Therefore, this method is very useful to determine the ultimate plastic rotation and the rotation capacity.

This method has been used for the first time to analyze the collapse of cylindrical shells. The so-called Yoshimura–pattern is well known as an inextensible mapping of cylindrical surfaces (Kollar and Dulacska, 1984). Table 7.5 presents the main studies which use the plastic collapse mechanism method for determining the member rotation capacity in combination with an analytical method.

(i) *In-plane plastic collapse mechanisms.* The first application of the plastic mechanism method for I-section members is due to Kato (1965), who proposed that *flange crumpling* was produced by yielding. The main deficiency of the Kato's method is that the buckling of plates forming the I-section develops in a way which is geometrically independent of each other.

This deficiency has been corrected by Climenhaga and Johnson (1972a) by means of *geometrically compatible yield line models* which have been adapted to the form of actual experimental buckling shapes (Fig. 7.18). Some important hypothesis were used concerning this mechanism pattern: the length of local buckling is considered equal to the compression flange width, and the centre of rotation is located in the tension flange.

Table 7.5: I-section; Plastic collapse mechanism method

In-plane	Out-of-plane
• Kato (1965) • Climenhaga & Johnson (1972) • Ivanyi (1979, 1985, 1993) • Kuhlmann (1986, 1989) • Gioncu et al (1989, 1995a,b) • Feldmann (1994) • Piluso (1995) • Gioncu & Petcu (1995, 1997) • Tehami (1997) • Moller et al (1997) • Anastasiadis (1999) • Anastasiadis & Gioncu (1999) • Gioncu et al (2000)	• Climenhaga & Johnson (1972) • Ivanyi (1985, 1993) • Gioncu et al (1995a,b, 1996a) • Gioncu & Petcu (1997) • Gioncu et al (2000)

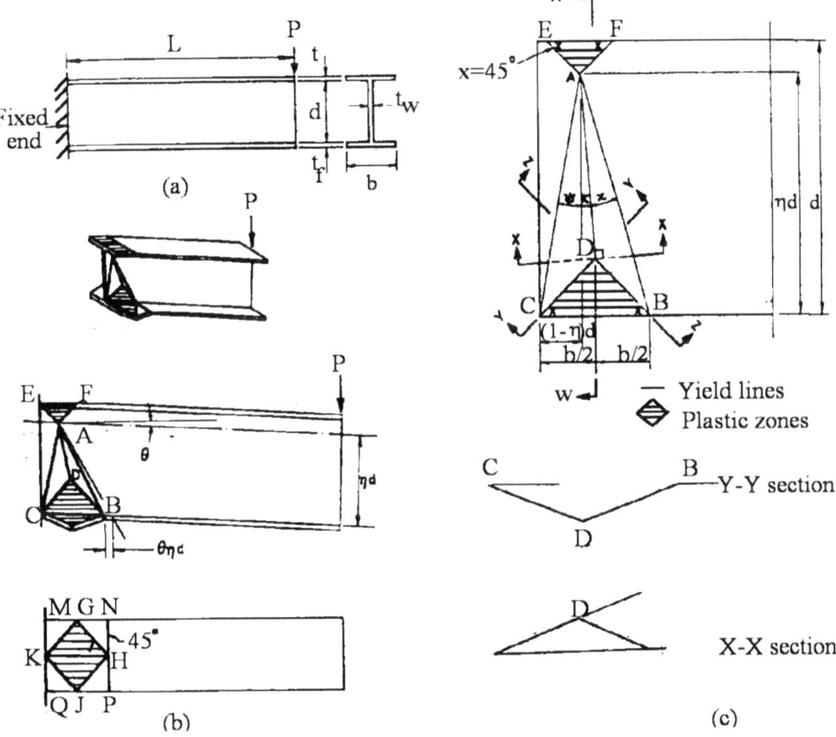

Figure 7.18: In-plane collapse mechanism for cantilever beam: (a) Beam configuration; (b) Collapse mechanism shape; (c) Collapse mechanism deformation (after Climenhaga and Johnson, 1972a)

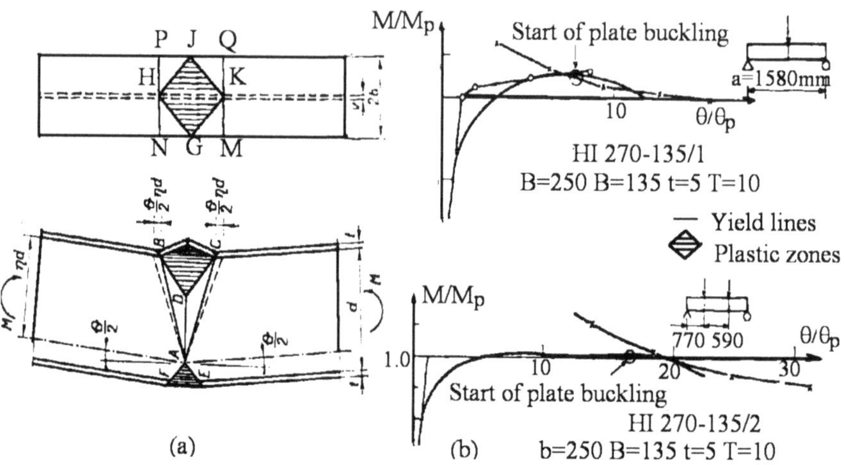

Figure 7.19: Plastic hinge model: (a) Collapse mechanism shape; (b) Correspondence with experimental results (after Ivanyi, 1979b, 1985)

An important progress in using this methodology was performed by Ivanyi (1979a,b, 1985, 1993), in the so-called *interactive plastic hinge model.* A methodology based on the principle of virtual work was used, and the plastic collapse mechanism curve can be determined as illustrated in Fig. 7.19. It can be observed that these curves coincide very well with the ones obtained from the experimental results.

The use of the *semi-pyramidal shape plastic mechanism*, in which an interaction with the web deformation is considered, allowed Kuhlmann (1986, 1989) to develop a special software to calculate the post-buckling behaviour and rotation capacity. The same mechanism type has been used by Feldmann (1994) and Moller et al (1997).

On the basis of Climenhaga, Johnson and Ivanyi plastic mechanisms, Gioncu et al (1989, 1995a,b) and Gioncu and Petcu (1995, 1997) have considered a lot of *suitable plastic collapse mechanisms* for in-plane and out-of-plane local buckling. A succession of computer programs (POSTEL, DUCTROT 93, DUCTROT 96, DUCTROT M) have been elaborated on the basis of these plastic collapse mechanisms in order to determine the rotation capacity of beams and beam columns with different cross-sectional types. This methodology produced satisfactory results when compared with experimental data coming from the technical literature (Fig. 7.20).

Piluso (1995) used the same plastic mechanism to determine the *rotation capacity of the rolled I-section* in which the effect of flange–web junction is considered. The same effect has been analyzed by Anastasiadis (1999) and Anastasiadis and Gioncu (1999). An important increasing of rotation capacity is obtained following this way.

Tehami (1997) has proposed a *modified local plastic mechanism* in which the length of the buckled flange is related to the beam depth (not to the

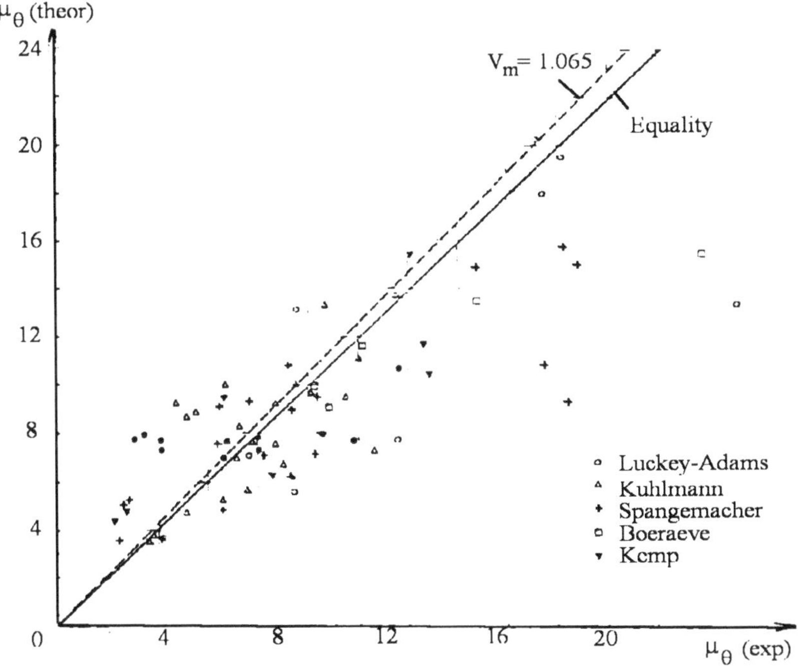

Figure 7.20: Suitability of plastic collapse mechanism method (after Gioncu and Petcu, 1997)

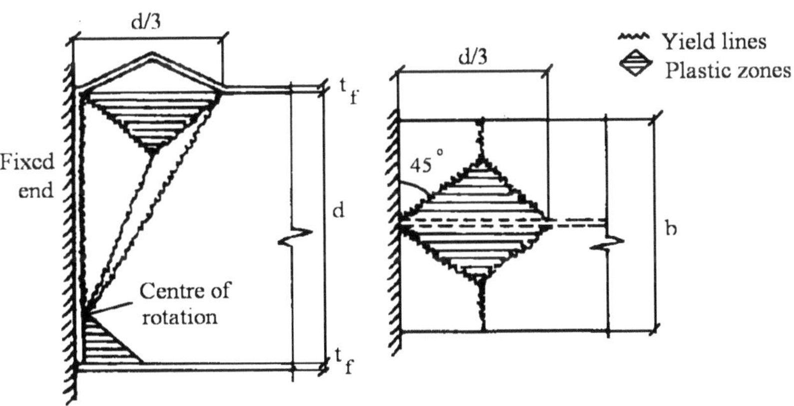

Figure 7.21: Modified plastic mechanism (after Tehami, 1997)

flange width as in the other studies), and the mechanism rotates around a point eccentrically situated with respect to the buckled flange (Fig. 7.21).

The state of the art concerning the use of a local plastic mechanism for determining the available ductility is presented by Gioncu et al (2000).

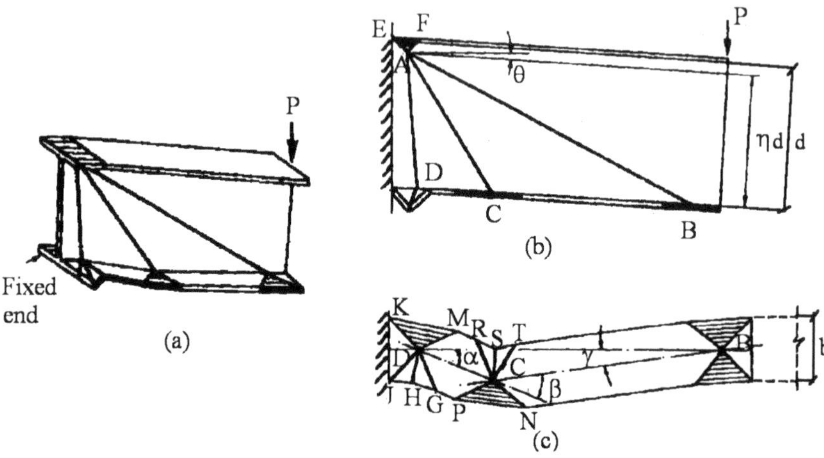

Figure 7.22: Out-of-plane collapse mechanism: (a) Buckling mechanism; (b) Side elevation; (c) Buckling of compression flange (after Climenhaga and Johnson, 1972)

(ii) *Out-of-plane plastic collapse mechanisms.* The first study concerning this mechanism type belongs to Climenhaga and Johnson (1972), studying the so-called *S- mechanism type* (Fig. 7.22). The research activity has been continued by Ivanyi (1985, 1993), Gioncu et al (1995a,b, 1996a) and Gioncu and Petcu (1995, 1997) who have studied the interaction between in-plane and out-of-plane buckling modes, produced in the post-critical range. Examining the experimental tests one can find that even if the buckling is produced by in-plane deformations, an out-of-plane mechanism occurs in the advanced post-critical range, due to the considerably weakened flange rigidity caused by plastic deformations. Therefore, a reduction of rotation capacity may be noted due to the interaction of the two buckling modes.

(iii) *Conclusions on this methodology.* After examining the use of the plastic collapse mechanism method to determine the rotation capacity, the following conclusions result:

-this method can only be used to plot the post-critical path, but this is very convenient to determine the ultimate plastic rotation and the rotation capacity;

- a combination with an analytical solution for plotting the first part of the moment–rotation curve is recommended in order to have the complete curve over all significant ranges;

-it is very important to choose a proper pattern for the plastic mechanism by examining the experimental and numerical simulation of buckled shapes;

-there are many forms of plastic mechanisms, and a study is required to evaluate which one gives the minimum value for moment and ultimate plastic rotation;

Table 7.6: I-section; Empirical methods

In-plane buckling	Out-of-plane buckling
• Susuki & Ono (1977) • Mitani et al (1977) • Mitani & Makino (1980) • Kato & Akiyama (1981) • Spangemaker & Sedlacek (1991, 1992) • Axhag (1995) • Gioncu & Mazzolani (1997) • White & Barth (1998, 2000) • Anastasiadis (1999) • Anastasiadis et al (2000)	• Nakamura (1988) • Kemp & Dekker (1991) • Daali & Korol (1994) • Daali (1995) • Dekker (1998)

-the use of this method allows the determination of the available rotation capacity with an accuracy which is confirmed by experimental and numerical results;

-the method can also be used for the elaboration of performance programs which are very useful for design practice.

7.2.5. Empirical Methods

The main research in which empirical methods have been developed are presented in Table 7.6 for in-plane and out-of-plane modes.

(i) *Rotation capacity for in-plane buckling.* For beams, Suzuki and Ono (1977) have established the following equations:

-for uniform moment:

$$\mu_r = 10^4 \frac{1}{(k\lambda_y)^2}\left(\frac{275}{f_y}\right)^2 \left(\frac{b}{d}\right)^{1/2} \tag{7.22a}$$

-for moment gradient:

$$\mu_r = 10^4 \frac{1}{\lambda_y^2} m \left(\frac{275}{f_y}\right) \frac{t_f}{b} \tag{7.22b}$$

where m is the end moment ratio.

The influence of web buckling in the presence of axial forces is studied by Mitani et al (1977). The empirical expressions deduced from experimental results are the following:

$$\mu_r = 9.9 - 0.00157 \frac{d}{t_f}\sqrt{f_y} \qquad \text{for } \frac{N}{N_p} = 0 \tag{7.23a}$$

$$\mu_r = 10.8 - 0.001 \frac{d}{t_f}\sqrt{f_y} \qquad \text{for } \frac{N}{N_p} = 0.3 \tag{7.23b}$$

Mitani and Makino (1980) proposed the following formulae for the rotation capacity of beam columns based on experimental results:

$$\mu_r = \alpha[80(\lambda_f - 0.65)^2 - 4.0\lambda_w + 6] \qquad \text{if } \frac{\sigma_N}{f_y} < \frac{A_w}{2A} \tag{7.24a}$$

$$\mu_r = \alpha[50(\lambda_f - 0.65)^2 - 5.5\lambda_w + 7] \qquad \text{if } \frac{\sigma_N}{f_y} > \frac{A_w}{2A} \tag{7.24b}$$

where

$$\alpha = \left(\frac{500}{k\dfrac{L_\lambda L_b}{i_x i_y}}\right)^{1/2} \varepsilon; \qquad \varepsilon = (235/f_y)^{1/2} \tag{7.25a, b}$$

L_b is the distance between lateral bracings, L_x, the distance between the plastic hinge and inflection point:

$$L_x = \frac{L_b}{1 + \dfrac{M_2}{M_1}} \tag{7.25c}$$

i_x, i_y, the gyration radius around the strong and weak axes, respectively, k, the numerical coefficient depending on moment distribution defining the effective length for lateral-torsional buckling:

$$k = 0.7\left(\frac{1.75}{c_b}\right)^{1/2}; \qquad c_b = 1.75 + 1.05\frac{M_2}{M_1} + 0.3\left(\frac{M_2}{M_1}\right)^2 \leq 2.3 \tag{7.25d, e}$$

A simplified moment–rotation curve for a member subjected to anti-symmetrical bending has been established by Kato and Akiyama (1981). The actual curve is idealized by three linear segments, as it is shown in Fig. 7.23. For:

$$\frac{b}{t_f} \leq \frac{104}{\varepsilon^{1/2}}; \qquad \frac{d}{t_w} = \frac{2.4}{\varepsilon^{1/2}} \tag{7.26a, b}$$

it results:

-maximum moment index:

$$s = \frac{M_{max}}{M_p} = \max\begin{cases} 1 + \left[\left(0.043 - 0.0372\dfrac{b}{t_f}\varepsilon_y^{1/2}\right)^2 - \right. \\ \left. \quad - \left(0.00024\dfrac{d}{t_w}\varepsilon^{1/2} - 0.00025\right)\right]\dfrac{1}{\varepsilon_y} \\ 1.46 - \left[0.315\dfrac{b}{t_f} + 0.053\dfrac{d}{t_w} + 0.02(\lambda_y - 50)\right]\varepsilon^{1/2} \end{cases} \tag{7.27a, b}$$

-stiffness ratio for strain hardening:

$$k_p = 0.03 \tag{7.28}$$

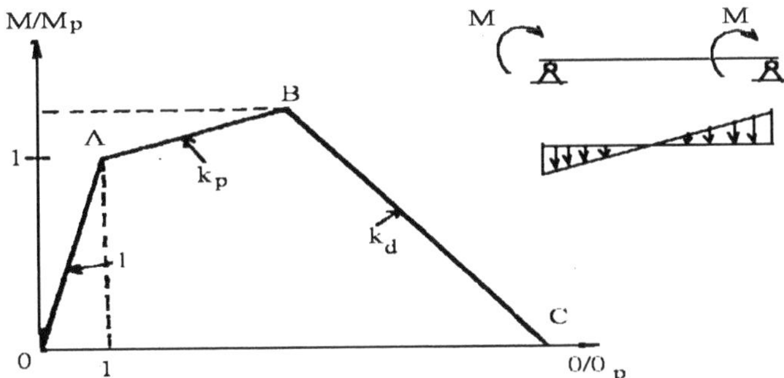

Figure 7.23: Simplified moment–rotation curve (after Kato and Akiyama, 1981)

-stiffness ratio for negative slope region:

$$k_d = \min \begin{cases} -0.355\dfrac{d}{t_w}\varepsilon_y \\ -\left[-1.33 + \left(5.3\dfrac{b}{t_f}\varepsilon^{1/2} - 2\right)\left(0.63 + 0.33\dfrac{d}{t_w}\varepsilon_y^{1/2}\right)\right]\varepsilon^{1/2} \end{cases} \tag{7.29a, b}$$

If $\lambda_y > 100$ the term d/t_w should be factored by 1.5. The rotation capacity becomes:

$$\mu_r = (s-1)\left(\frac{1}{k_p} + \frac{1}{|k_d|}\right) \tag{7.30}$$

Spangemaher (1991) and Spangemaher and Sedlacek (1992a,b) have performed a large number of numerical simulations, using FEM analysis, in order to obtain an approximate equation. The simulation has been divided into different series, in which a single parameter affecting the rotation capacity has been varied, keeping all the others constant. The considered parameters include moment gradient, flange and web slendernesses and steel quality. The analysis of results has led to the empirical equation:

$$\mu_r = \mu(\rho) + \mu(t_f) + \mu\left(\frac{b}{t_f}\right) + \mu\left(\frac{L}{b}\right) - \mu(K_0) \tag{7.31}$$

where

-$\mu(\rho)$ considers the influence of yield ratio:

$$\mu_\rho = \frac{0.75}{\rho_y^{6.5}} \tag{7.32a}$$

-$\mu(t_f)$ considers the influence of flange thickness:

$$\mu(t_f) = \alpha\left(\frac{2.53}{\rho_y} - 2.63\right)(15 - t_f) \tag{7.32b}$$

$$\alpha = \begin{cases} 1.0 & \text{for } t_f < 15 \text{ mm} \\ 0.5 & \text{for } t_f > 15 \text{ mm} \end{cases}$$

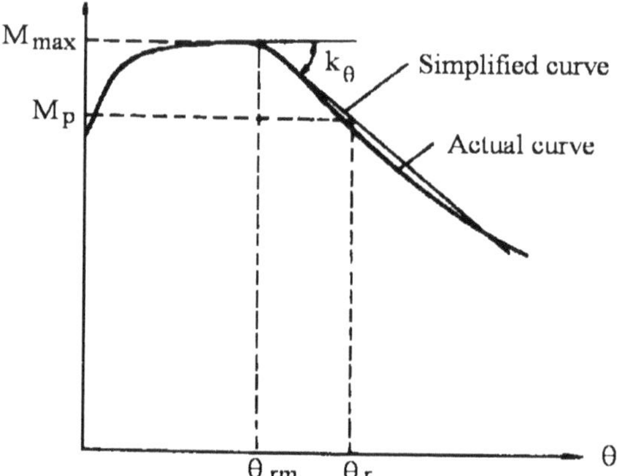

Figure 7.24: Simplified method considering the rotation for the maximum moment and slope of the decreasing curve (after Axhag, 1995)

-$\mu(b/t_f)$ considers the influence of flange slenderness:

$$\mu\left(\frac{b}{t_f}\right) = \left(\frac{2.81}{\rho_y} - 2.74\right)\left(20 - \frac{b}{t_f}\right) \tag{7.32c}$$

-$\mu(L/b)$ considers the influence of moment gradient:

$$\mu\left(\frac{L}{b}\right) = 5.40\left(\frac{1}{\rho_y} - 1\right)\left(2.5 - \frac{L}{b}\right) \tag{7.32d}$$

-$\mu(K_\theta)$ considers the effect of flange–web interaction:

$$\mu(K_\theta) = S_k \Delta K$$

$$\Delta K = 9.31 - \frac{0.035}{\rho_y^{6.5}} - \frac{G_{hw}}{3}\frac{t_w^3}{d} \tag{7.32e}$$

$$S_k = \begin{cases} 0.35/{\rho_y}^4 & \text{if } \Delta K > 0 \\ 0 & \text{if } \Delta K < 0 \end{cases}$$

In equations (7.32) ρ_y is the yield ratio given by equation (4.3) and G_{hw} is the strain-hardening tangential modulus of web, evaluated according to Lay's equation (4.31). This approximate formulation can be applied only to beams.

Based on the analysis of experimental results, Axhag (1995) has proposed the following equations (Fig. 7.24):

-plastic rotation corresponding to the start of decreasing moment capacity:

$$\theta_{rm} = 10^{2.401}\left(\frac{2d}{t_w}\right)^{-1.04}\left(\frac{b}{t_f}\right)^{-1.707}\left(\frac{E}{f_y}\right)^{0.773}\left(\frac{M_{max}}{M_p}\right)^{1.684} \tag{7.33a}$$

-the descending slope of the moment–rotation curve:

$$\frac{k_\theta}{M_{max}} = 10^{-0.027}\left(\frac{V}{V_{cd}}\right)^{0.529}\left(\frac{2d}{t_w}\right)^{-0.867}\left(\frac{b_f}{t_f}\right)^{2.060}\left(\frac{E}{f_y}\right)^{-0.583} \cdot \left(\frac{M_{max}}{M_p}\right)^{-3.365} \tag{7.33b}$$

where:

$$V_{cd} = \omega_v d t_w f_y \tag{7.34a}$$

$$\omega_v = \frac{0.79}{0.35\frac{d}{t_w}\left(\frac{f_y}{E}\right)^{1/2} + 0.70} \tag{7.34b}$$

Using these equations one can determine the ultimate rotation:

$$\theta_r = \theta_{rm} + (M_{max} - M_p)\frac{k_\theta}{M_{max}} \tag{7.35}$$

A simplified equation based on a large number of numerical tests using DUCTROT 96 computer program, is proposed by Gioncu and Mazzolani (1997)

$$\mu_r = 6 \times 10^4 \frac{t_f^2}{bL}\varepsilon^2\left(0.8 + 0.2\frac{f_{yw}}{f_{yt}}\right); \qquad \varepsilon = \left(\frac{235}{f_{yf}}\right)^{1/2} \tag{7.36a, b}$$

The comparison obtained with the theoretical results has shown a maximum scatter of about ±10%.

For compact flanges and noncompact webs, White and Barth (1998) and Barth et al (2000) have developed a model for moment–plastic rotation (Fig. 7.25) which gives:

-maximum moment:

$$\frac{M_{max}}{M_p} = 1 + \frac{3.6}{\left(\frac{2d}{t_w}\right)^{1/2}} + \frac{0.1}{2\frac{dt_w}{bt_f}} - 0.4\frac{M_p}{M_y} \tag{7.37}$$

where M_y corresponds to the yield moment capacity;

-the decreasing start of plastic rotation:

$$\theta_{rm} = 0.128 - 0.0119\frac{b}{2t_f} - 0.0216\frac{d}{b} + 0.002\frac{b}{2t_f}\frac{d}{b} \tag{7.38}$$

-the lowering path of the moment–rotation curve:

$$\frac{M}{M_{max}} = 1 - 16(\theta_p - \theta_{rm}) + 100(\theta_p - \theta_m)^2 \tag{7.39}$$

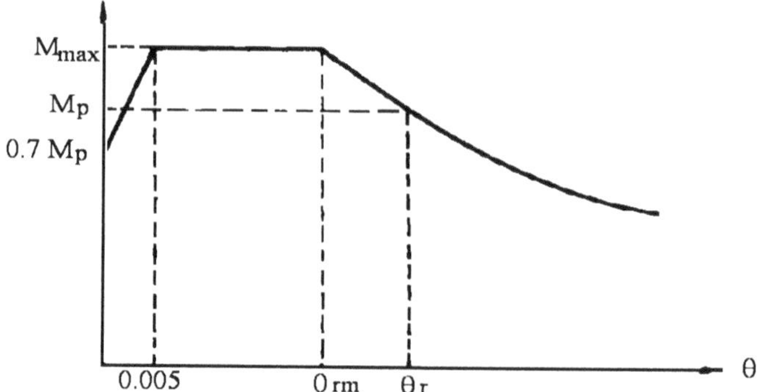

Figure 7.25: Simplified moment–rotation curve for compact flanges and noncompact webs (after White and Barth, 1998, Barth et al, 2000)

These equations are available if:

$$\frac{V}{V_n} < 0.6; \qquad \frac{L_b}{r_y} \leq \left(0.124 - 0.0759\frac{M_{min}}{M_{max}}\right)\frac{E}{f_y} \qquad (7.40a, b)$$

where V_n is the nominal web shear capacity , L_b, the lateral unsupported length between brace points, M_{max} and M_{min} , the larger and smaller moments at the ends of unsupported length. The White and Barth proposal specifically refers to the case when the behaviour of a beam is governed by the web buckling. Its use for the other cases is not verified.

For beam columns, Anastasiadis (1999) and Anastasiadis et al (2000) have proposed the equation:

$$\mu_r = r_1\left(\overline{\lambda}\frac{b}{2t_f}f_y^{1/2}\right)^{r_2} \qquad (7.41)$$

where, for $0.1 < n_p < 0.4$

$$r_1 = 275(1 + 44.2n_p); \qquad r_2 = -1.25(1 + 0.72n_p) \qquad (7.42a, b)$$

$$\overline{\lambda} = \left(\frac{N}{N_{cr}}\right)^{1/2}; \qquad N_{cr} = \frac{\pi^2 EI}{L_b^2} \qquad (7.42c, d)$$

L_b being the buckling length.

(ii) ***Rotation capacity for out-of-plane buckling.*** Based on 121 experimental tests, Nakamura (1988) has evaluated the rotation capacity as a function of lateral slenderness. Fig. 7.26a shows the results as a function of moment distribution and generalized slenderness ratio:

$$\overline{\lambda} = (M_p/M_e)^{1/2} \qquad (7.43)$$

where M_e is the elastic lateral buckling moment. It results :

$$\frac{\theta_u}{\theta_p} = B\frac{1}{\overline{\lambda}^2} \qquad (7.44)$$

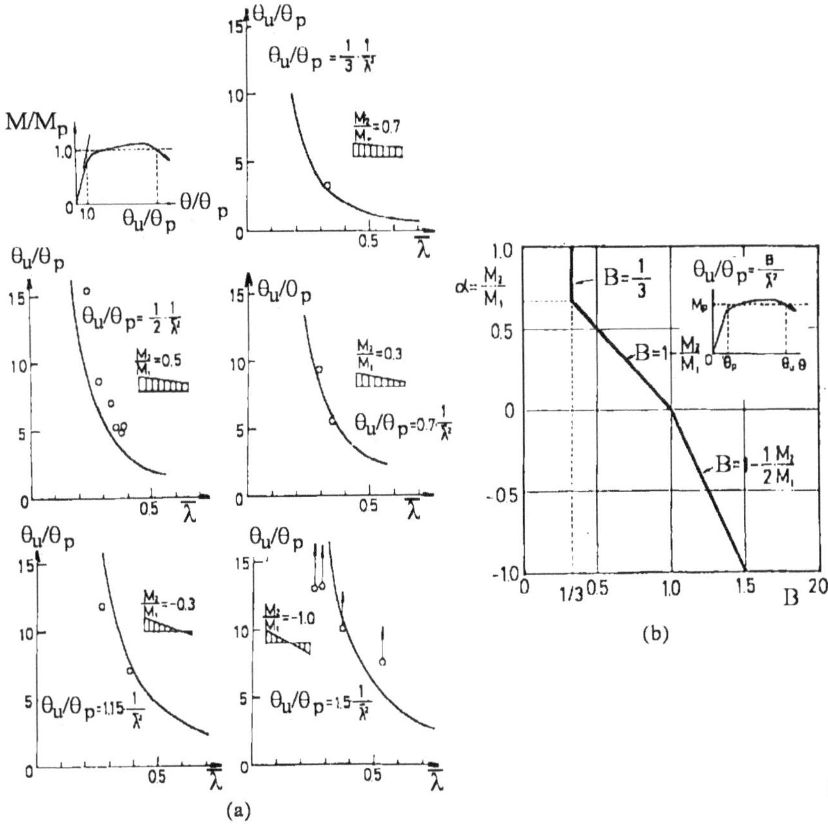

Figure 7.26: Rotation capacity for out-of-plane buckling: (a) Influence of moment distribution; (b) Variation of B coefficient (after Nakamura, 1988)

The variation of the coefficient B as a function of moment gradient is given in Fig. 7.26b. One can see that this coefficient varies between 0.33 and 1.5:

$$B = 0.33 \qquad \text{for } \frac{M_{min}}{M_{max}} > \frac{2}{3}$$

$$B = 1 - \frac{M_{min}}{M_{max}} \qquad \text{for } 0 < \frac{M_{min}}{M_{max}} < \frac{2}{3} \tag{7.45a – c}$$

$$B = 1 - \frac{1}{2}\frac{M_{min}}{M_{max}} \qquad \text{for } -1 < \frac{M_{min}}{M_{max}} < 0$$

An empirical expression has been proposed by Kemp and Dekker (1991) showing the relationship between available rotation capacity and effective lateral slenderness:

$$\mu_r = 3\left(\frac{60}{\lambda_e}\right)^{1.5}\frac{1}{2\alpha} \tag{7.46}$$

where $\alpha = d_c/d$ is the compression portion of member depth (for beams $\alpha = 0.5$) and:

$$\lambda_e = k_f k_w \frac{L_i}{r_y \varepsilon} \tag{7.47}$$

where:

$$k_f = \frac{1}{20}\frac{b}{t_f \varepsilon} \tag{7.48a}$$

$$k_w = \begin{cases} \dfrac{1}{400}\left(\dfrac{\alpha d}{33 t_w \varepsilon}\right)^{1/2}\left(460 - \dfrac{L_i}{r_y \varepsilon}\right) & \text{for class 1: } \frac{\alpha d}{t_w \varepsilon} \leq 33 \\ \dfrac{1}{33}\dfrac{\alpha d}{t_w \varepsilon} & \text{for class 2: } 33 < \frac{\alpha d}{t_w \varepsilon} \leq 40 \end{cases} \tag{7.48b}$$

For $L/r_y > 60$ the lateral-torsional buckling is the dominant effect. If $L/r_y < 60$, the local flange buckling occurs. The equation (7.46) represents the lower bound of the experimental scatter.

Using the Kemp and Dekker empirical equation, Daali and Korol (1994) have proposed a revised expression of (7.46):

$$\mu_r = (6.831 - 0.43\alpha_l)\left(\frac{60}{\lambda_e}\right)^{1.5} \tag{7.49}$$

where

$$\lambda_e = \alpha_f \alpha_w^{1/2} \alpha_l (0.7224 - 0.000667\alpha_l) \tag{7.50a}$$

$$\alpha_f = \frac{b}{2t_f \varepsilon}; \quad \alpha_w = \frac{d}{t_w \varepsilon}; \quad \alpha_l = \frac{L}{2b\varepsilon}; \quad \varepsilon = \left(\frac{300}{f_y}\right)^{1/2} \tag{7.50b - e}$$

Another empirical expression is proposed by Daali (1995):

$$\mu_r = 1 - 9.2\alpha_e + 10.71\alpha_l^{-0.293} \tag{7.51}$$

where

$$\alpha_l = \frac{\alpha_f \alpha_w \alpha_l}{30072} \tag{7.52}$$

the numerical coefficients being given by equation (7.50).

A survey of the empirical approaches is presented by Dekker (1998).

(iii) *Conclusions on empirical methods.* Examining these empirical approaches, the following remarks can be outlined:

-there are a lot of empirical equations proposed by different authors, obtained on the base of experimental or numerical results, very different in shape and in range of validity. On the other hand, unfortunately, there are no studies involved in the calibration of these equations in order to provide the designer with the choice of the most reliable one;

-the great merit of these approximate equations is to emphasize the main factors influencing the rotation capacity: flange and web slendernesses, lateral rigidity, depth-to-width ratio, depth-to-span ratio, moment gradient and axial forces. Many of these factors are ignored in the code provisions concerning the ductility classes.

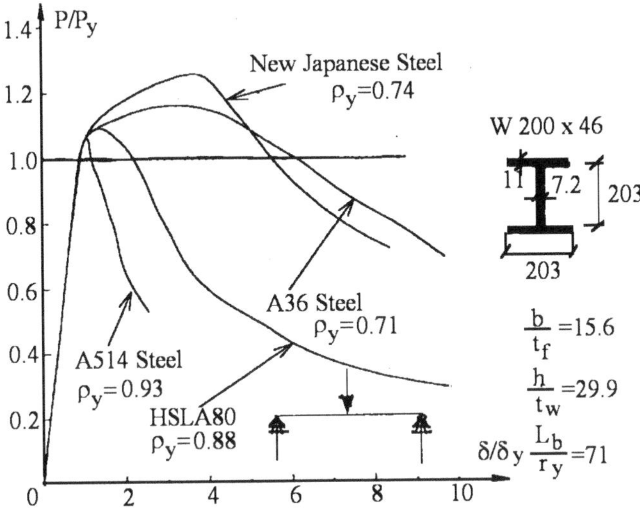

Figure 7.27: Influence yield ratio on rotation capacity (after Galambos, 1999)

7.2.6. Influence of Yield Ratio

Generally, for usual steel grade, the nominal values of the yield ratio may assure a good ductility, the fracture occurring only for very large plastic rotations. But, due to the fact that the actual fabrication of steel guarantees only the minimum value for yield stress (see Section 4.3.2), the increased yield ratio could produce a premature fracture, which prevents the development of a good ductility.

Beams and columns in a moment-resisting frame under earthquake loading are subjected to moment gradients along the member axes. Since, in general, the bending moment is maximum at the end of the member, the plasticity starts at the end and spreads to the middle of the member. So, the length of the plastic zone of a steel member under moment gradient is governed by the yield ratio of steel. There are two different behavioural types where the influence of yield ratio is very important.

(i) *Reduction of rotation capacity.* The load–deflection curves of Fig. 7.27 were obtained by using FEM analysis with the ABAQUS program (Galambos, 1999)for SB 1 standard beams with important moment gradient. One can see that the shape of these curves shows a tremendous difference on the inelastic rotation capacity of the structural member, due to the yield ratio ρ_y of steel.

(ii) *Increasing of fracture danger*, which can occur in two different situations:

-*Local buckling is prevented* and load carrying capacity is governed by the fracture of tension flange. In this case Fig. 7.28 shows the SB 1 standard beam, for which the length of plastic zones in the compression and tension

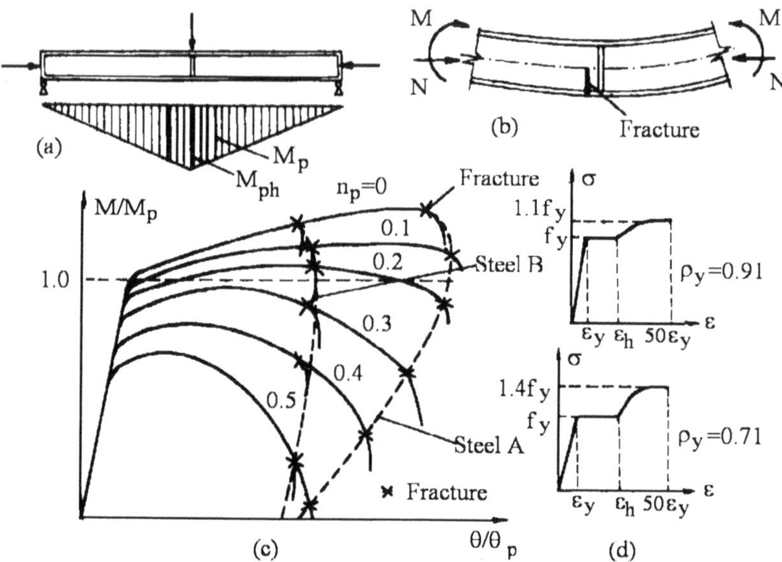

Figure 7.28: Influence of yield ratio on tension flange fracture: (a) Beam scheme; (b) Fracture of tension flange; (c) Moment–rotation curves for different axial force levels; (d) Steel properties (after Kuwamura, 1998)

flanges at the maximum moment, M_u, producing the member fracture given by:

$$L_p = (1 - \rho_y)L \tag{7.53}$$

where ρ_y is the yield ratio. Thus, this equation indicates that a lower yield ratio gives a wider plastic zone, with a larger plastic rotation and a higher ductility. The influence of yield ratio on the moment–rotation relationship of a beam column is very clearly presented in Fig. 7.37 (Kuwamura, 1988). Two steel qualities are considered, having the same yield stress, but with different ultimate strengths: $1.4f_y$ for steel A with a yield ratio of 0.71 and $1.1f_y$ for steel B with yield ratio of 0.91. The limit moment–rotation curve is obtained in both cases, when ϵ_u reaches the ultimate strain of the material, considered equal to $50\epsilon_y$. The figure shows that, for a given force ratio, steel A with a lower yield ratio has a ductility significantly higher than steel B with a very high yield ratio.

-***Local buckling is considered*** and the ultimate behaviour is governed by the crushing of the compression flange (Fig. 7.29a) (Gioncu et al, 2000) The relationship between the beam plastic rotation θ_{fb} and the maximum local rotation of the buckled flange θ_f is given by:

$$\theta_{fb} = \frac{1}{16}\frac{L_p}{h_0}\theta_f^2 \tag{7.54}$$

where L_p is the length of the buckled shape (defined by the inflection points), h_0 is the distance of flange to the centre of the plastic hinge rotation

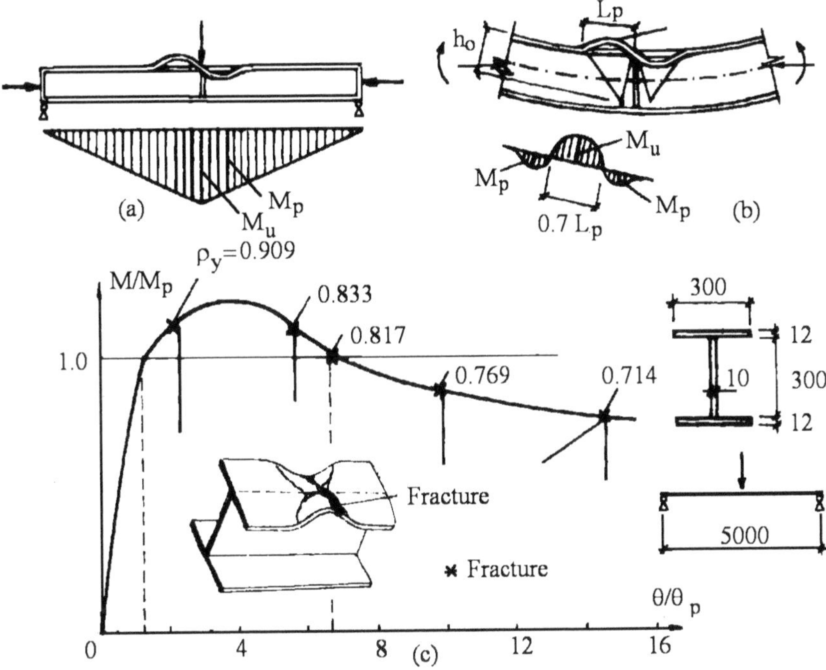

Figure 7.29: Influence of yield ratios on compression flange fracture: (a) Beam scheme; (b) Buckled shape of compression flange; (c) Fracture of compression flange for different yield ratios (after Gioncu et al, 2000)

(Fig. 7.29b). In equation (7.54) the fracture rotation is given by equation (4.59). Fig. 7.29c shows the influence of the yield ratio for the I-section. One may determine the yield ratio ρ_y for which the fracture of the buckled flange occurs. One can see a very important reduction in rotation capacity when the yield ratio increases. For values $\rho_y > 0.817$, the fracture can occur before reaching the ultimate rotation capacity, defined as the rotation where the bending moment drops below the full plastic moment on the unloading branch of the moment–rotation curve.

Fig. 7.30 shows the influence of yield ratio on beam rotation capacity (Gioncu, 2000). For a given yield ratio the plastic rotation capacity is changed in a fracture collapse. One must note that the thicker flanges are more sensitive to fracture than the thiner ones, in a very good accordance with the observations after the Kobe earthquake.

7.2.7. Influence of Strain-Rate

In case of a near-field earthquake, the failure of a steel member due to a strong seismic action may occur when the carrying capacity is exceeded by a single excursion in the plastic range. In this case, the strain rate, with the corresponding velocity, is very high (Gioncu, 2000). It has been shown

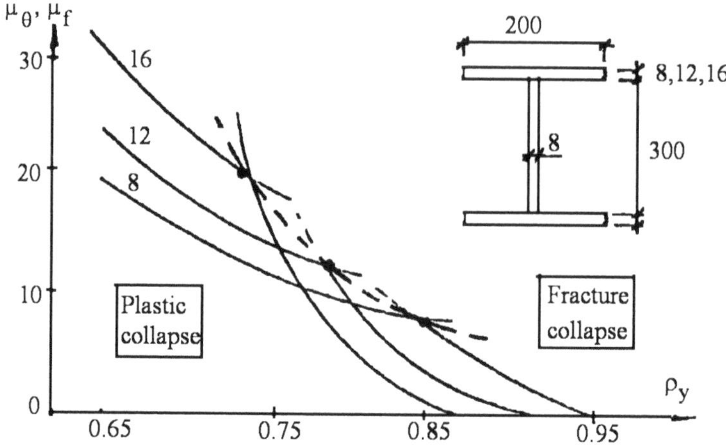

Figure 7.30: Plastic and fracture collapse related to yield ratio (after Gioncu et al, 2000)

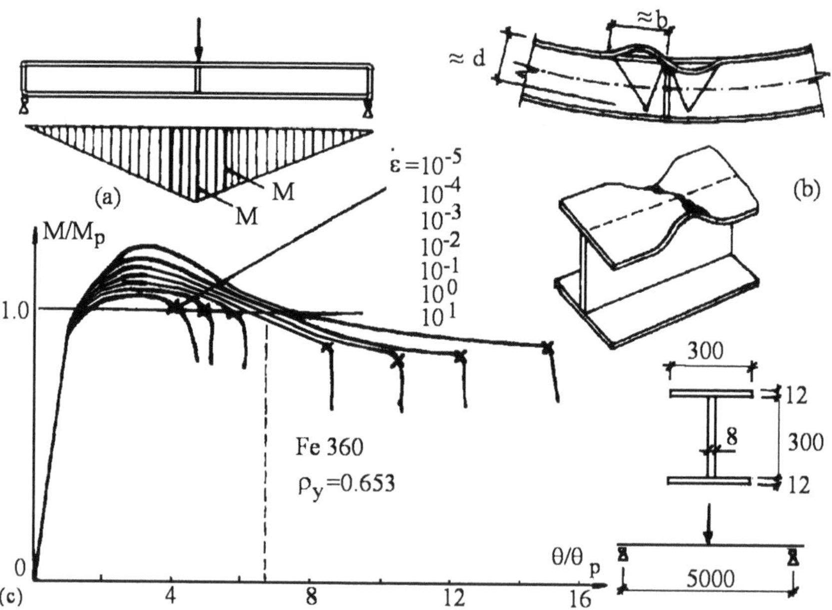

Figure 7.31: Influence of strain rate on fracture rotation: (a) Beam scheme; (b) Fracture of compression flange; (c) Moment–rotation curve for different strain rates (after Gioncu, 2000)

in Section 4.2.4 that the increase in strain rate has the effect to increase the yield ratio. For a simply supported beam with Fe 360 steel grade, Fig. 7.31 shows the fracture rotation as a function of strain rate. One can see that this parameter has a great influence on the fracture rotation, especially if

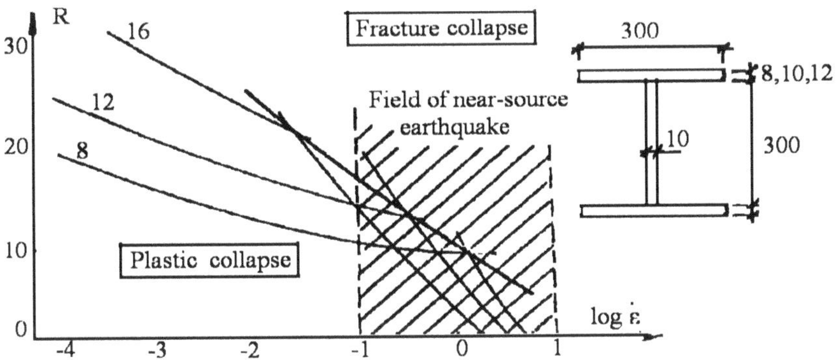

Figure 7.32: Plastic and fracture collapses related to strain rate (after Gioncu, 2000)

the yield ratio is increased because the actual yield stress is greater than the nominal one due to a lack of severe controls on the yield stress scatter. Due to high strain rate, a premature fracture can occur without reaching the ultimate rotation capacity, especially when the actual steel grade does not have a proper yield ratio. Even for steel with good yield ratio as with the Fe 360 steel grade, seismic actions producing strain rates higher than 10^{-1}/sec can produce fractures before reaching the plastic ductility. This fact can explain in some measure the unreliable behaviour of steel structures during the last great seismic events.

Fig. 7.32 shows the influence of strain rate on the rotation capacity as a function of the flange thickness. One can see that the influence of strain rate is higher for thick flanges and that the occurence of fracture collapse is possible for near-field earthquakes with high velocities (Gioncu, 2000).

7.2.8. Cyclic loading

In previous sections the plastic behaviour of I-sections has been examined under monotonic loading only. In seismic design this is just an indicator concerning the potential ductility, but a more complex behaviour analysis must include the cyclic feature of ground motion and its influence on the monotonic ductility.

(i) *Experimental tests.* The main experimental research for cyclic loading, selected from literature, are presented in Table 7.7 with their most important characteristics.

The experimental tests of Bertero and Popov (1965) revealed that the main parameters of cyclic loading are the *controlled deformations* of each cycle and the *number of cycles* until specimen fracture. The relationship between the number of cycles causing the beam fracture is shown in Fig. 7.33. The controlling strain, measured at the edges of the clamped end, varied from 0.01 to 0.025, with corresponding plastic rotations of 0.02 to 0.05 rad. For specimens tested under a controlled strain of 0.025 (0.05

Table 7.7: I-section; Cyclic experimental tests

No	Year	Author(s)	Pro-file type	Spec. no.	Test type	Mem-ber type	Loa-ding	Slenderness ratio	
								flan-ge	web
1	1965	Bertero & Popov	RS	11	MG	BM	cd	6.35	1.24
2	1977	Mitani et al	WS RS	90	MG	BM BC	cd id	6.2..16.0	158..640
3	1977	Suzuki & Ono	WS	-	CM MC	BM BC	-	-	-
4	1977	Matsumi& Mitami	WS	9	MG	BC	cd id	-	-
5	1986	Ballio & Calado	WS	4	MG	B	cd	10..25	38.3..60.5
6	1988	Yoshizumi& Matsui	WS	10	MG	BM BC	id	7.9..16.2	31.3..42.2
7	1987	Castiglioni	RS	14	MG	BM	id	7.0..10.0	19.8..33.2
8	1989	Gioncu et al	WS	6	CM	BM	id	9.0..13.0	60.0..71.0
9	1990	Guruparan& Walpole	WS RS	6	MG	BM	id	8.2..8.5	53.7..80.0
10	1993 1995	Ballio& Castiglioni	RS	45	MG	BM	cd rd	6.9..10.0	19.8..39.2
11	1994	Lee & Lee	RS	8	CM	BM BC	cd	9.6	30.4
12	1997	Green et al	WS	13	MG	BM	id	5.9..9.0	27.9..29.3
13	2000	Mateescu& Gioncu	WS	11	MG	BM	id pd	10.9..12.3	24.0..26.3

MG moment gradient BM beam cd constant displacement
CM constant moment BC beam column id increasing displacement
rd random displacement
pd pulse displacement

rad), which is higher than the strain at the onset of strain hardening of the material, the buckling of flanges was observed during the second half of the first cycle. The wrinkles were enlarged considerably after each new cycle and the first crack became visible during the 9th cycle. Failure occurred during the 16th cycle as the result of the enlargement of the cracks in the wrinkle. For controlled cyclic strains at 0.01 (0.02 rad), the initiation of local buckling was detected after 70 cycles and fracture after 650 cycles. So, the imposed deformation of a cycle, expressed in strain, displacement or rotation, plays a leading role in determining the number of cycles until fracture.

The influence of *flange and web slendernesses* and *axial forces* under

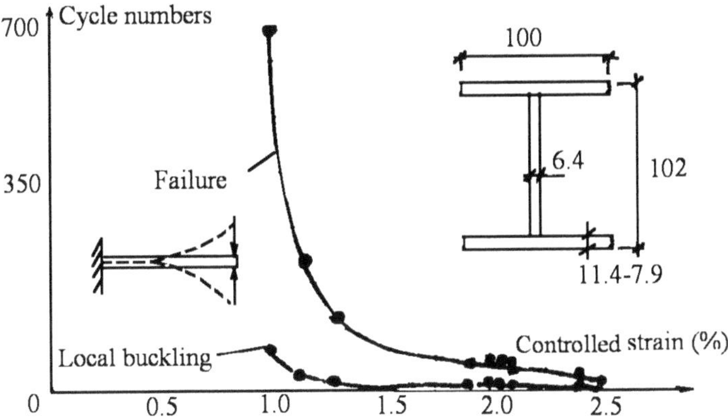

Figure 7.33: Number of cycles causing buckling and failure as a function on controlled deformations (after Bertero and Popov, 1965)

cyclic horizontal loads have been examined by Mitani et al (1977). The buckling of flanges produced a deterioration of the carrying capacity, very important in the case of very slender flanges. A very drastic deterioration has been observed when web buckling occurs contemporary to the flange buckling.

The differences in the ***behaviour of beams and beam columns***, subjected to monotonic and cyclic loadings, uniform moments and moment gradients are experimentally studied by Suzuki and Ono (1977). Cyclic loading in comparison with monotonic loading for a beam and a beam column is presented in Fig. 7.34. One can see that the carrying capacity for cyclic loading is a little higher than for monotonic, but the degradation after the local buckling is larger. In presence of constant axial loads, the decreasing rate of loops for the beam column is larger than the one of the beam.

The interaction between local and lateral buckling modes in cyclic loading was studied by Yoshizumi and Matsui (1988). The difference in behaviour of ***high strength beam-columns*** in comparison with the mild steel ones was experimentally observed by Matsui and Mitani (1977).

A very important experimental program has been performed at the Milano Polytechnic (Ballio and Calado, 1986a, b, Ballio et al, 1986, Ballio and Perotti, 1987) in order to investigate the influence of the ***flange slenderness ratio*** on the cyclic behaviour of bent sections. Welded profiles have been tested under monotonic and cyclic loadings. Different increasing displacement variation and the ECCS Recommendations (1985) for cyclic loading of laboratory tested specimens were used. The most striking indication of these results was that, once a flange has buckled, the maximum load cannot be reached in the following cycles. It was also observed that the deterioration range increased with the increase in the width-thickness ratio of flanges, as a result of the early occurrence of local buckling in the flange (Fig. 7.35). The interaction between local buckling and lateral-torsional buckling caused a reduction of the load carrying capacity with respect to the one developed

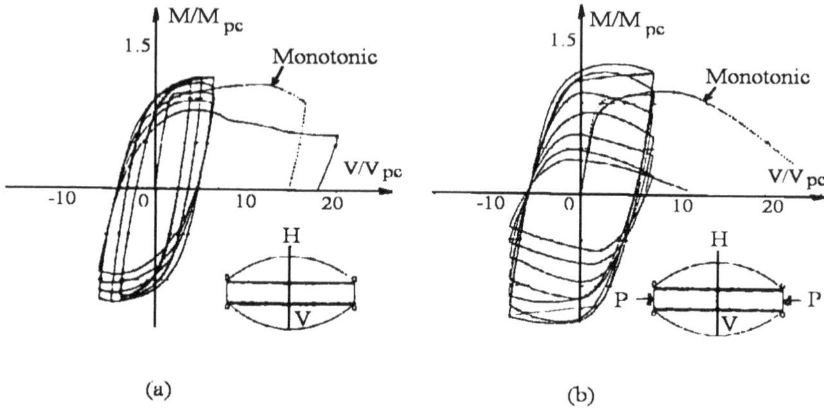

Figure 7.34: Monotonic and cyclic loads; (a) Beam behaviour; (b) Beam-column behaviour (after Susuki and Ono, 1977)

in pure flexural buckling. Local buckling took place after several cycles of loading and deterioration of load carrying capacity was not observed. But for the new increase in displacements, the local instability began to amplify and a significant decrease in load carrying capacity could be observed. In the zone of the flange–web welded junction some cracks were found, which have the tendency to extend up to the flange end. The maximum load reached in the monotonic tests was slightly smaller than the one which was found in the cyclic tests. This is the consequence of the deterioration which affects the cyclic tests.

A new series of experimental tests on European profiles have been performed at the Milano Polytechnic for HEA, HEB and IPE profiles (Castiglioni, 1987, Castiglioni and Di Palma, 1988, Castiglioni and Losa, 1991, 1992). Tests were performed for different loading histories in accordance with the procedure proposed by ECCS (1985). The collapse of specimens was caused by the complete fracture of the flange which was first compressed. IPE specimens, having the flange slenderness ratio comparable with that of HEB profiles, have shown a reduced ductility due to the increased web slenderness.

The main purpose of the experimental tests performed by Gioncu et al (1989) on beams subjected to monotonic and cyclic loadings was the ***study of buckled shape*** differences in these two loading conditions. They have shown that in the last case a superposition of two plastic mechanisms occurs, caused by the opposite bending moments.

The influence of ***lateral slenderness*** on the behaviour of I-section members (3 rolled and 3 welded) has been studied by Guruparan and Walpole (1990). It was found that a very good behaviour was obtained for a lateral slenderness of 37, with an acceptable strength degradation. When the lateral slenderness was increased to 57, a greater strength degradation was observed. For lateral slenderness of about 96, a very large amount of lateral movement and twisting was produced.

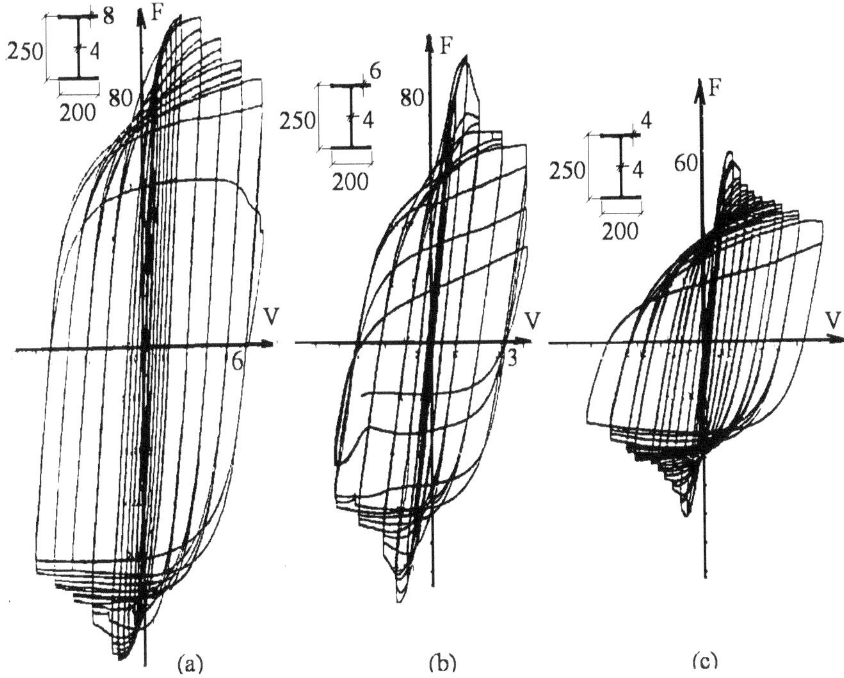

Figure 7.35: Influence of flange slenderness on the cyclic behaviour: (a) $b/t_f = 25$; (b) $b/t_f = 33.33$; (c) $b/t_f = 50$ (after Ballio and Calado, 1986a)

Low cycle fatigue due to cyclic loading has been studied by Ballio and Castiglioni (1993, 1994a,b, 1995a,b). Constant and variable amplitude tests were performed for rolled profiles (HE 220A, HE220B, IPE 300). Fig. 7.36 shows the deterioration of resistance and rigidity ratios as a function of the number of cycles. The constant amplitude $\Delta v/v_y$ corresponds to the beginning of strain hardening, where important plastic deformations are involved. One can see that the HE 220A profile has a flange slenderness ratio larger than that of HE 220A and IPE 300 profiles, while its web slenderness is intermediate between those of HE 220B and IPE 300. The best behaviour is presented by the HE 220B profile, having the most important strain hardening effect and the most reduced deterioration of load carrying capacity and rigidity. The fastest degradation is presented by the IPE 300 profile having the largest web slenderness.

Examining the Ballio and Castiglioni experimental results and the ones carried out by Krawinkler and Zohrei (1983) one can observe that, after the plastic buckling, there are *three ranges of deterioration* (Fig. 7.37):

-first range, deterioration proceeds at high rates in association with a continuous growth of flange buckles, coupled with lateral buckling;

-second range, deterioration proceeds at a slow and almost constant rate, showing a certain stabilization of behaviour;

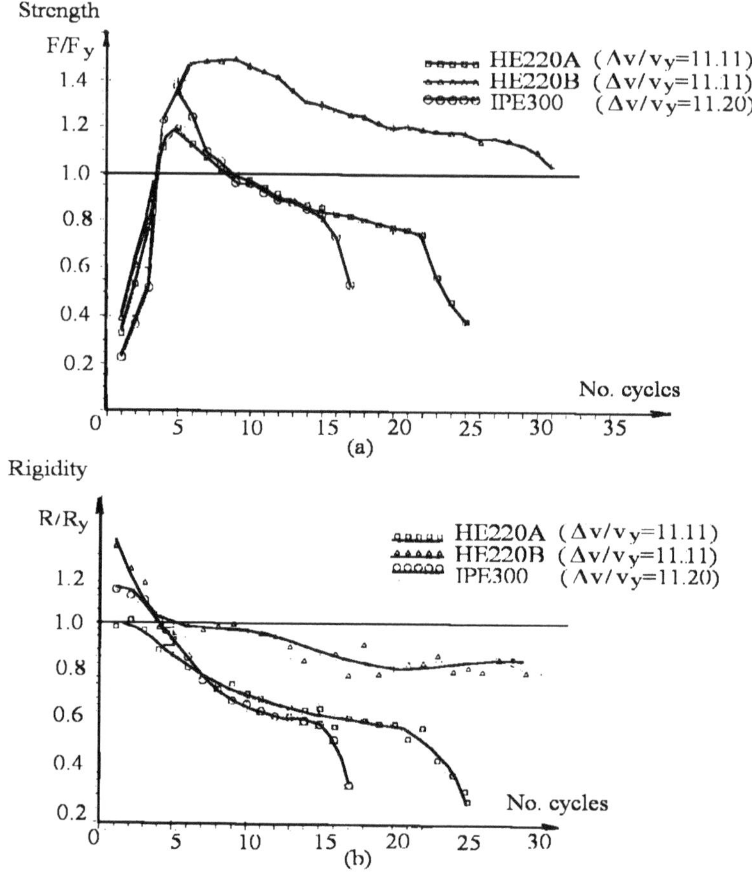

Figure 7.36: Low cycle fatigue influence: (a) Strength reduction; (b) Rigidity reduction (after Ballio and Castiglioni, 1993, 1994, 1995)

-third range, a rapid deterioration is produced due to the cracks of propagation. Small surface cracks formed early during the first cycles have no noticeable effect on strength and stiffness until they grow through the thickness of flange and propagate transversaly.

The cycle number corresponding to each range depends on the steel quality and the amplitude of imposed displacement.

A *unified approach* for structure design under low- and high-cycle fatigue has been proposed by Ballio and Castiglioni (1995b), the S–N curves being valid in both cases. The Miner's rule can be adopted to define an unified collapse criterium.

The *influence of axial forces* on the beam column cyclic ductility has been experimentally studied by Lee and Lee, (1994), (Fig. 7.38). For beams, the local buckling in pure cyclic bending happened faster (around 60%) than the one in monotonic pure bending. For beam columns, two cases have been studied: constant axial force and proportionally increasing force. In the case

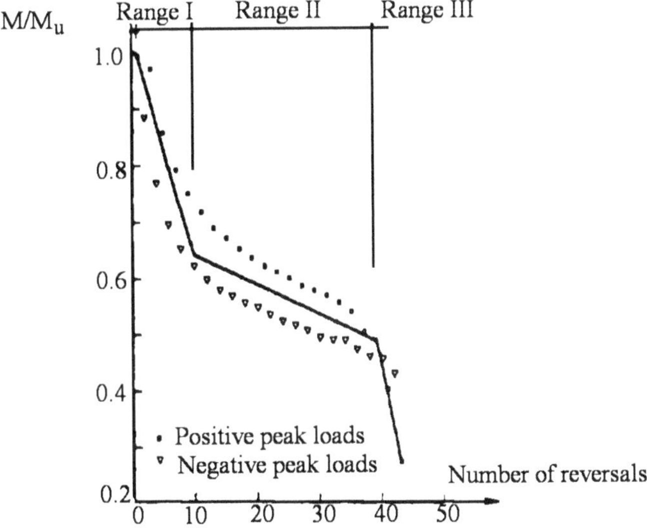

Figure 7.37: Three ranges of cyclic deterioration (after Krawinkler and Zohrei, 1983)

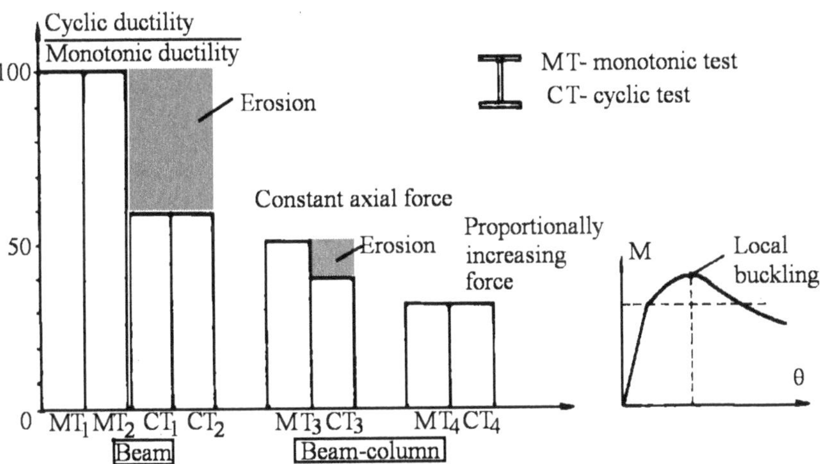

Figure 7.38: Influence of cyclic loading on beam and beam-column ductility (after Lee and Lee, 1994)

of monotonic bending with axial force, the ductility is 50% below the pure bending ductility, the higher reduction being caused by the proportional loadings. In the presence of the axial force, the reduction of local ductility due to cyclic loads is not at the same level as for pure bending. The high rate of degradation was observed during the first 6 cycles.

The experimental tests performed by Green et al (1997) on beams with *high-yield ratios* ($\rho_y = 0.89$) have very clearly shown that these members

Table 7.8: I-section; Theoretical studies on cyclic loading

Numerical tests	Constitutive law method
• Ballio & Calado (1986a,b) • Castiglioni & Di Palma (1988) • Butterworth & Beamish (1993) • Aiello & Ombres (1995)	• Castiglioni et al (1987, 1990) • Vayas (1997) • Zambrano et al (1999)
Low-cycle fatique approach	Plastic deformation accumulation method
• Yamada et al (1988, 1992) • Castiglioni (1994) • Ballio & Castiglioni (1995) • Castiglioni et al (1997)	• Krawinkler & Zohrei (1983) • Gioncu et al (1989) • Calado & Azevedo (1989) • Calado (1992) • Gioncu et al (2000) • Anastasiadis et al (2000)

have a more reduced cyclic rotation capacity than the similar steel members having a lower yield ratio ($\rho_y = 0.68$).

The influence of *three cyclic loading types* (increasing displacement, decreasing displacement and pulse displacement) has been studied by Mateescu and Gioncu (2000). For the first two cases the fracture modes were almost the same. Contrary to this, the impulsive types composed by one, two or three adjacent pulses with strain rates of about 0.04 to 0.1/sec produced a very early fracture, during the first or second cycle.

(ii) *Theoretical studies.* In the last few years several theoretical studies on the influence of cyclic loads have been carried out. The main contributions are presented in Table 7.8. Four principal methodologies have been developed:

-*Numerical tests.* The strip method, in which the cross-section of the member is divided in a finite number of strips, each strip being characterized by area, distance from the centroid, residual stress, yield stress and ultimate strain (Ballio and Calado, 1986a,b, Castiglioni and Di Palma ,1988 and Aiello and Ombres, 1995). To model the *member damage* due to local buckling (Ballio and Calado, 1986a,b), fracture or deterioration, it is assumed that the strip area is reduced according to a certain law depending on the type of structural damage. The application of this method is presented in Fig. 7.39. One can see that the loss of stiffness is higher when the flange width-thickness ratio increases. The significant loss of stiffness at the end of the simulation process was attributed to the tension failure in the area surrounding the web–flange junction. According to the damage model for fracture, cracking started when the flange reached a given deformation. For the profile from Fig. 7.39b, cracking started too early, and consequently, a considerable loss of stiffness was observed. Contrary to this, if cracking started late, the loss of stiffness was smaller.

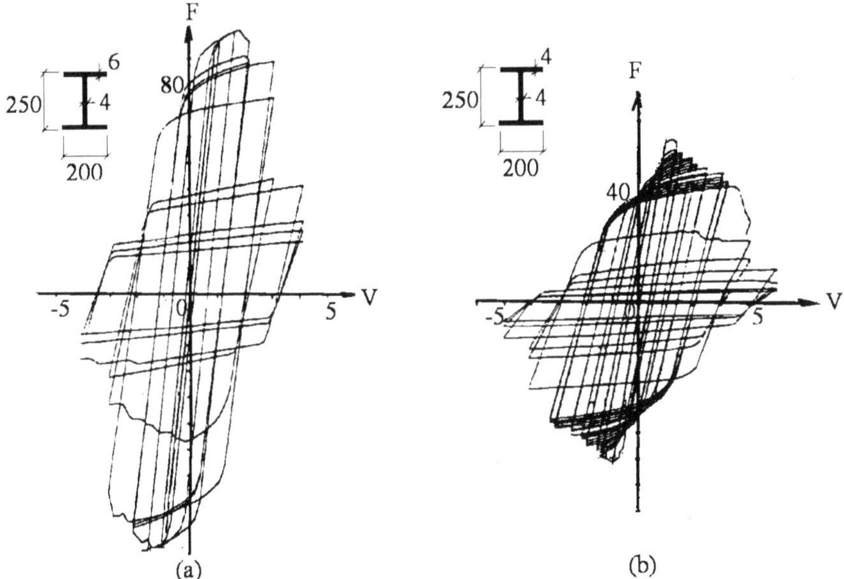

Figure 7.39: Numerical tests using the strip method (after Ballio and Calado, 1986a,b)

The FEM analysis was performed by Butterworth and Beamish (1993), including material with Bauschinger effect, residual stresses, initial deformations and eccentrical loading. The numerical tests provided results in the close agreement with experimental tests on members of different types (welded and rolled). Despite the satisfactory output, this methodology has proved to bee too heavy in terms of computing time for being used in design practice.

-*Constitutive law method* considers the possibility of problem formulation in terms of general behaviour rather than in terms of local variables. This approach can be achieved by formulating constitutive laws capable of reproducing the inelastic response of members in the presence of damage in terms of force-displacement parameters. Following this direction, Castiglioni (1987), Castiglioni et al (1990) have used a three-linear constitutive model (Fig. 7.40a), in which the effects of local buckling and fractures were simulated by modifying, cycle by cycle, the constitutive law of parameters. Fig. 7.40b shows that for the amplitude of a semi-cycle a reduced load-carrying capacity is observed due to local buckling. But after some cycles a stabilization occurs, the reduction of load-carrying capacity within each cycle becoming almost constant.

A method based on the material law, including the local buckling, has been worked out by Vayas (1997), who extended to cyclic loading the *effective width* methodology developed for monotonic loading. The strength and stiffness degradation of the section under cyclic loading is much more pronounced than under monotonic loading.

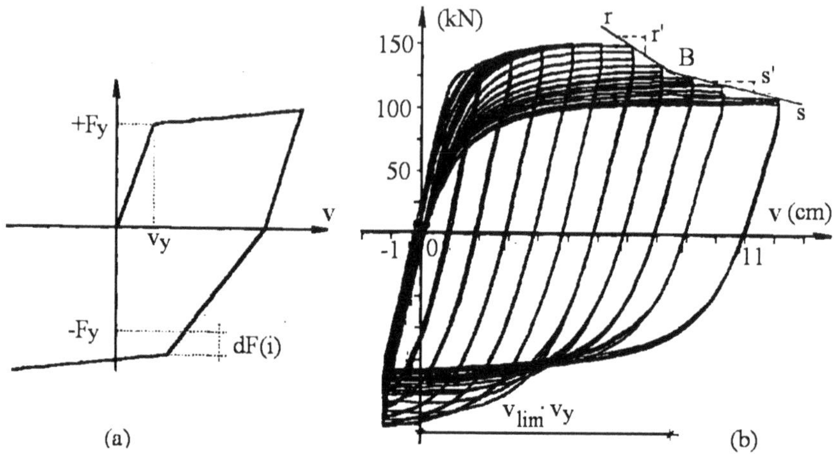

Figure 7.40: Numerical test using the constitutive law method (after Castiglioni et al, 1987, 1900)

A study concerning the cyclic behaviour of rolled HEA and IPE profiles and welded profiles has been performed by Zambrano et al (1999), by using a degrading constitutive law.

-*Low fatigue analysis.* The *fracture of buckled flanges* under cyclic loading has been studied by Yamada et al, 1988, Yamada, 1992. The fracture strains decrease when the number of cycles increase, due to low-cycle fatigue. A procedure to unify design and damage assessment methods for structures under high- and low-cycle fatigue has been proposed by Castiglioni (1994), Ballio and Castiglioni (1995), Castiglioni et al (1997). The methodology is based on the S–N curve approach, having a general validity, in both high- and low-cycle fatigue. The fatigue failure prediction function has the form:

$$N \cdot S^m = K \tag{7.55}$$

where N is the number of cycles until the failure at constant stress or strain range S, while the non-dimensional constant m and dimensional parameter K depend on both the typology and the mechanical properties of the considered member. A log–log equation represents a straight line with a slope equal to (-1/m) called the fatigue resistance line (Fig. 7.41).

-*The plastic deformation accumulation method.* There are basically two developed models considering the accumulation of plastic deformations:

• *The cumulative damage model* (Krawinkler and Zohrei, 1983, Calado and Azevedo, 1989, Calado, 1992). The rate of deterioration for a constant cycle amplitude can be accurately expressed as a function in the form:

$$\Delta d = a(\Delta\theta_p)^b \tag{7.56}$$

where a and b are parameters which depend on the properties of the members, and $\Delta\theta_p$ is the plastic rotation. Parameter b provides a measure of the

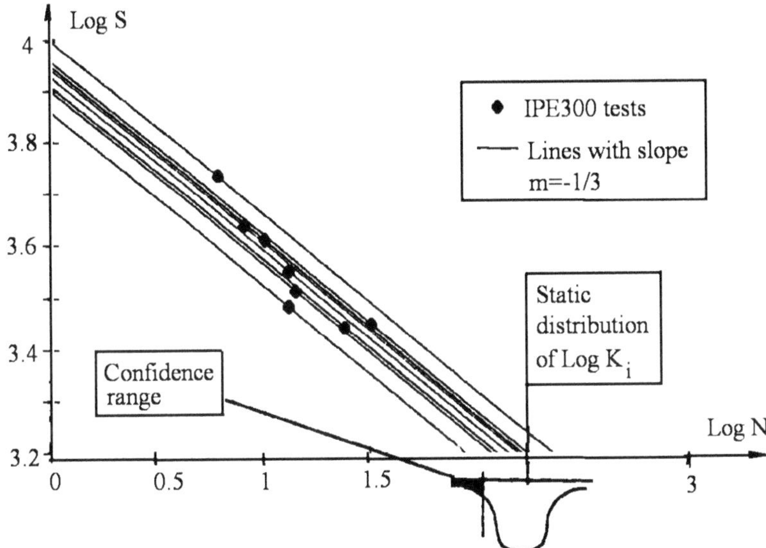

Figure 7.41: Log S – Log N lines for low fatigue analysis (after Castiglioni et al, 1997)

gradient of deterioration trend. If the hypothesis of linear damage accumulation is accepted for reversals with variable amplitudes, the accumulated deterioration can be expressed as:

$$d = \sum_{i=1}^{n} \Delta d_i = a \sum_{i=1}^{n} (\Delta\theta_p)^b \tag{7.57}$$

where n is the number of reversals or plastic excursions. Failure can be defined as the attainment of a limiting value of deterioration d_{lim}; therefore the number of reversals to failure for constant-amplitude cycles is given by:

$$N_f = \frac{d_{lim}}{a} (\Delta\theta_p)^{-b} \tag{7.58}$$

The total damage D under Miner's assumption of linear damage accumulation, can be expressed as:

$$D = \sum_{i=1}^{n} \frac{1}{N_{fi}} = \frac{a}{d_{lim}} \sum_{i=1}^{n} (\Delta\theta_p)^b \tag{7.59}$$

The value $D = 1$ represents the attainment of the collapse conditions. By considering the value $\Delta_p = \theta_u$, the ultimate plastic rotation leading to collapse $D = 1$ under monotonic loading ($n = 1$), results:

$$a = \frac{d_{lim}}{\theta_u} \tag{7.60}$$

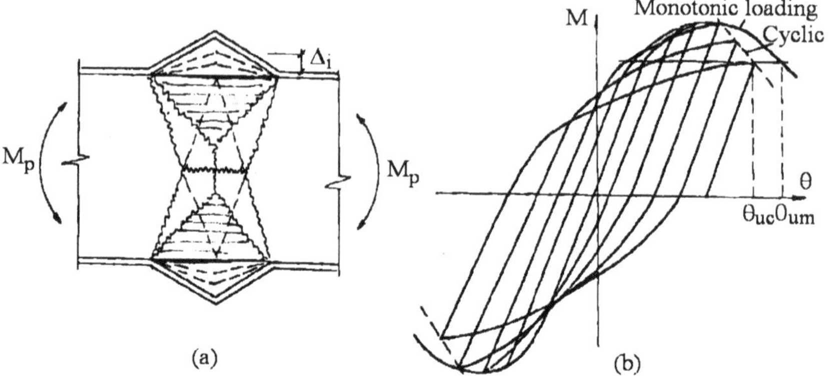

Figure 7.42: Plastic collapse mechanism method for cyclic loading: (a) Collapse mechanism shape; (b) Monotonic and cyclic moment–rotation curves (after Gioncu et al, 1989, 2000)

and

$$D = \frac{\sum(\Delta\theta_p)^b}{\theta_u} \tag{7.61}$$

• *Plastic collapse mechanism model* for cyclic loading is proposed for the first time by Gioncu et al (1989), by superimposing two plastic collapse mechanisms for two opposite bending moments (Fig 7.42a). The reliability of this methodology has been the object of the following research. (Gioncu, 2000, Gioncu et al, 2000, Anastasiadis et al, 2000). The behaviour of the I-section has a particular feature under cyclic loading. The first semi-cycle produces a buckling of the compression flange and the section rotates around a point located in or near to the opposite flange. As a result of this behaviour, the tensile forces are very small. During the reversal semi-cycle, the new compressed flange buckles too. The most important observation is that the opposite flange remains unchanged in the buckled shape because the small tension force is not able to straighten the buckled flange. Therefore, during the next cycle, the section works with an initial geometrical deformation resulted from the previous cycle. In this way after each cycle an additional rotation is superimposed on the previous one, according to the pattern of the plastic mechanism which is assumed to form in the flanges and web. Due to the fact that the superimposed plastic mechanism works with an incomplete web plasticization, an important reduction in plastic deformation occurs. So, as the result of the accumulation of residual displacements in flanges and incomplete formations of yield lines in webs, the moment–rotation curve in the softening range shows a degradation with respect to the monotonic loading (Fig. 7.42b).

(iii) *Conclusions on cyclic loading.* The examination of the experimental and theoretical results gives rise to the following conclusions:

-the main characteristic of cyclic loading is the number of cycles until the member fracture occurs;

-the imposed amplitude of deformation (strain, displacement, rotation) plays a leading role in determining the number of cycles until fracture;

-as for monotonic loading, a crumpling of flanges and webs can be observed, but both flanges are involved in this deformation. The local formed plastic mechanism consists of two superimposed plastic mechanisms, each corresponding to the buckling of the compressed flange, as it is obtained for the monotonic loading (with some small difference);

-the deformation is characterized by an accumulation of plastic deformation, the deformation from a new cycle being superimposed over the deformations obtained during the previous cycles;

-the maximum loading capacity is limited by the local buckling of flanges. Once a flange is buckled, the maximum load cannot be reached in the following cycles;

-a very drastic deterioration in carrying capacity has been observed when the web buckling occurs immediately after the flange buckling or when lateral-torsional buckling is produced;

-the web slenderness determines the extent of carrying capacity and ductility deterioration;

-an important deterioration after the local buckling occurs for the first cycles, followed by a stabilisation tendency until the fracture;

-due to the effect of cyclic loading, the danger of fracture collapse increases as a result of plastic-rotation accumulation;

-the impulse cyclic loading may produce a very early fracture during the first or second cycle.

7.3 Ductility of Box- and Hollow-Section Members

7.3.1. Design Aspects

Box- and hollow-sections are produced by means of a great variety of fabrication processes: rolling, welding, cold or hot forming, etc. Their inner space gives the possibility to increase strength and ductility by filling them of concrete with or without reinforcement. Due to their production, there are some important features in mechanical properties influencing the local ductility, as mentioned in Section 5.5.1.

In seismic resistant structures the box-section members are used mainly for beam columns, due to the excellent distribution of material which assures considerably higher buckling stability and torsional strength than those offered by the conventional open profiles, as well as, great bending resistance about both principal axes. But, due to the large variety of fabrication, their use in design practice requires extensive experimental tests to determine the manufacturing imperfections as residual stresses, non-uniform distributions of yield stress, section and member geometrical imperfections, etc.

From the ductility point of view the main characteristic is the presence of high axial forces, which can drastically reduce the rotation capacity of beam columns.

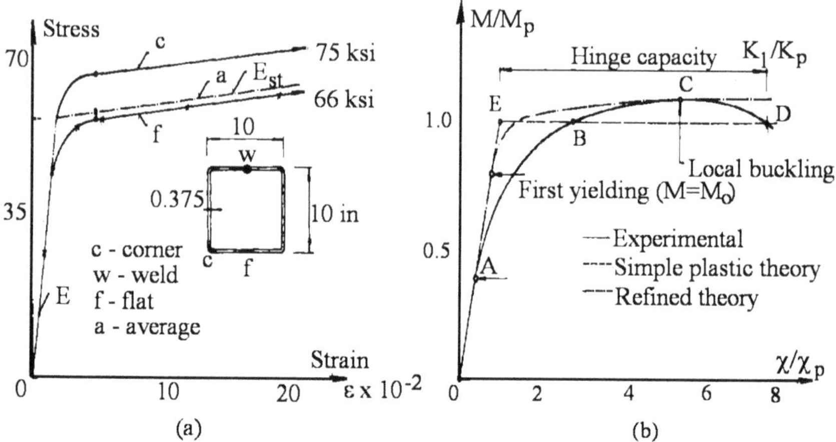

Figure 7.43: Hollow section ductility: (a) Increasing of yield stress; (b) Moment-curvature relationship (after Korol and Hudoba, 1972)

7.3.2. Experimental Tests

(i) *Survey on experimental tests.* Table 7.9 summarise the main features of the experimental activity on the ductility of box and hollow sections. The first systematic experimental research work on the *ductility of hollow sections* has been performed by Korol and Hudoba (1972). Due to the press forming process, an increase in yield stress is obtained, different for corner and flate portions of the profile (Fig. 7.43a). The ultimate curvature can be determined for parent yield stress or for increased yield stress. The obtained ultimate curvature is presented in Fig. 7.43b. Using these curves one can determine the flange slenderness as a function of the required ductility and the yield stress.

For high ductility, HD, it results;

$$\frac{b}{t} < 1.85 \left(\frac{f_y}{345}\right)^{1/2} \tag{7.62a}$$

and for moderate ductility, MD:

$$\frac{b}{t} < 20.5 \left(\frac{f_y}{345}\right)^{1/2} \tag{7.62b}$$

Kusawa and Fukumoto (1983) have tested welded box sections at *cyclic bending* for two values of slendernesses. The flexural rigidity of beams gradually decreases associated to the increase of bending cycle, the reducing being more pronounced for thiner walls.

A valuable experimental program was performed by Kecman (1983) for different sections having aspect ratios from 0.33 to 3.0 and slenderness ratios from 9.1 to 128. The main purpose of these experiments was to observe the *plastic collapse mechanisms* and to propose a simplified design model.

Table 7.9: Box and hollow sections; Experimental tests

No.	Year	Author(s)	Fabrication	Specimen number	Wall slenderness	Moment variations	Loading type	Member type
1	1972	Korol & Hudoba	HFS	26	13.7..37.0	CM	MG	BM
2	1983	Kusawa & Fukumoto	WS	6	40..80	CM	CM	BM
3	1983	Kecman	-	56	9.1..128	MG	MG	BM
4	1986	Nakai et al	CFS	8	457	CM	MG	BM
5	1986	Ballio & Calado	WS	4	33..72	MG	MG,CM	BM
6	1991	Watanabe et al	WS,CF	18	22.2..66.7	CM	MG,CM	BC
7	1992	Watanabe et al	WS	12	-	CM	MG,CM	BC
8	1994	Kuwamura & Akyama	CF,WS	3	18.4..27.3	MG	CM	BM
9	1994 1995	Rondal et al, Sedlacek et al, Stranghoner	CFS	30	15.4..27.8	CM	MG	BM
10	1996 1999	Sully & Hancock	CFS	7	20.4..21.2	CM	MG	BC
11	1997	Grzebieta et al	CRS	9	19.7..30.5	CM	MG,CM	BM
12	1997	Kawaguchi et al	CoFS	46	22.2..43.5	MG	CM	BC
13	1998	Wilkinson & Hancock	CFS	44	26.7..54.9	CM	MG	BM

CRS	cold rolled	CFS	cold formed	CM	constant momment
HFS	hot formed	CoFS	section filled	MG	moment gradient
MT	monotonic	BM	beam		
CT	cyclic	BC	beam column		

Ballio and Calado (1986a) have tested welded box sections for monotonic and cyclic loading. One can see an important degradation of load-carrying capacity in comparison with the monotonic loading (Fig. 7.44).

The behaviour of *concrete filled box sections* has been studied by Nakai et al (1986). The ultimate strength is influenced by the local buckling of flanges. The study of the influence of some constructional measures to prevent the slipping between interior surface of box plates and encased concrete has shown that the natural bond has the same behaviour. So, the restrictions of slips may not be a serious problem in designing composite members.

Watanabe et al (1991,1992) have shown that the *round of corners* can improve ductility and strength of box sections in both monotonic and cyclic loading tests.

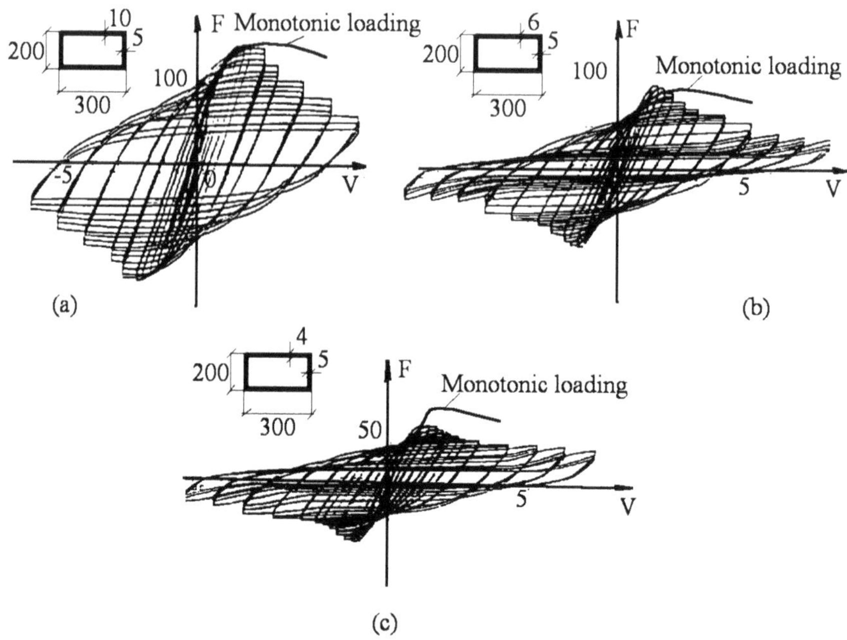

Figure 7.44: Cyclic loading of box sections: (a) $b/t = 30$; (b) $b/t = 50$; (c) $b/t = 75$ (after Ballio and Calado, 1986a)

The *brittle fracture* of cold press-bracked square tubes has been studied by Kuwamura and Akiyama (1994) for cyclic loading. High-strength heavy sections with a reduction in notch toughness associated to large heat-input welding procedures and cold forming of thick plates, have shown an increased possibility of brittle fractures. The ductility of these steel members is substantially impaired by the high value of the yield ratio. It is very important to observe that, if the b/t ratio is large enough to develop local buckling, no brittle fracture can occur. Contrary to this, if the slenderness is reduced, brittle fractures dramatically takes place associated to a sudden unloading and explosive cracks. These conclusions interprete the collapse of some members during the Kobe earthquake.

An extensive experimental test program was performed at the Aachen and Liege Universities (Rondal et al, 1994, Sedlacek et al, 1995, Stranghoner, 1995), for *square and rectangular hollow section beams*, subjected to monotonic loading. Some experimental moment–rotation curves are presented in Fig. 7.45. All specimens failed with one or two buckles when the maximum moment was reached. The effect of cold and hot forming was also considered. One can see that cold-formed sections reach their maximum bearable moment in a plateau, whereas the hot-formed sections reach a peak moment followed by an early important lowering. Due to the press forming, the increasing of the yield strength in flanges has an average value of 8%. The yield strength in corners increases with a mean value of 34%.

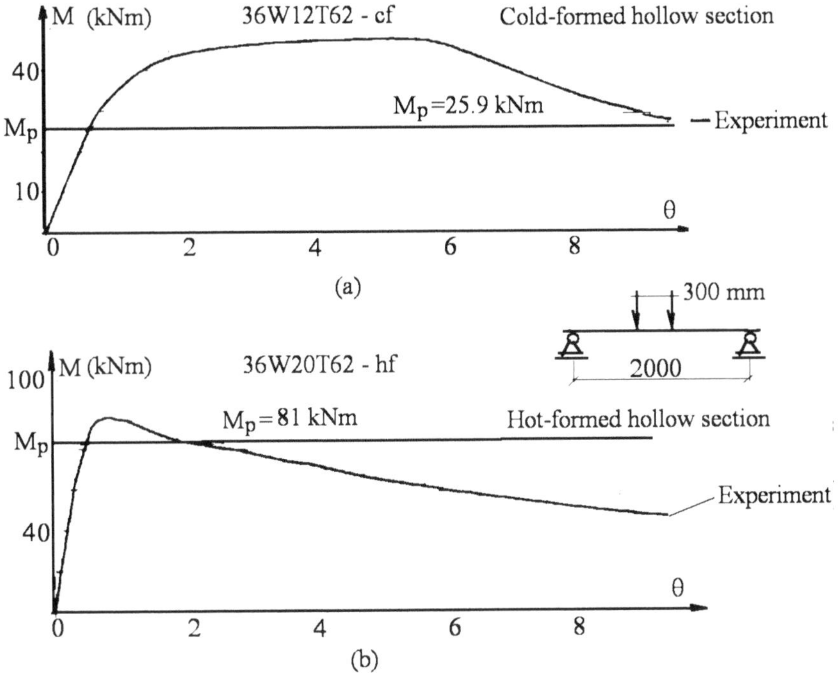

Figure 7.45: Effect of forming on hollow section beams: (a) Cold-forming; (b) Hot-forming (after Rondal et al, 1994, Sedlacek et al, 1995, Stranghoner, 1995)

The influence of axial forces and moment variation has been experimentally analysed by Sully and Hancock (1996, 1999).

The *ductility of slender, compact and plastic sections* for monotonic and cyclic loading were studied by Grzebieta et al (1997). Plastic sections which do not form a localized collapse mechanism during deformation have high ductility for both monotonic and cyclic loading. Contrary to this, for slender sections, if a localized mechanism is formed, the beam displays a rapid loss of strength after the first cycle.

The *ductility of concrete-filled tubular* beam columns subjected to cyclic bending has been studied by Kawaguchi et al (1997). The strength deterioration due to local buckling was very severe for specimens without filled-concrete. Contrary to this, in the case of concrete-filled specimens, the degradation was fairly small, showing a stable hysteresis loop. The energy dissipation of concrete-filled specimens was much greater compared with steel specimens, mainly due to the fact that the occurrence of local buckling was delayed by the presence of concrete.

The influence of *web slenderness* of rectangular hollow sections has been analyzed by Wilkinson and Hancock (1998). Fig 7.46a illustrates various

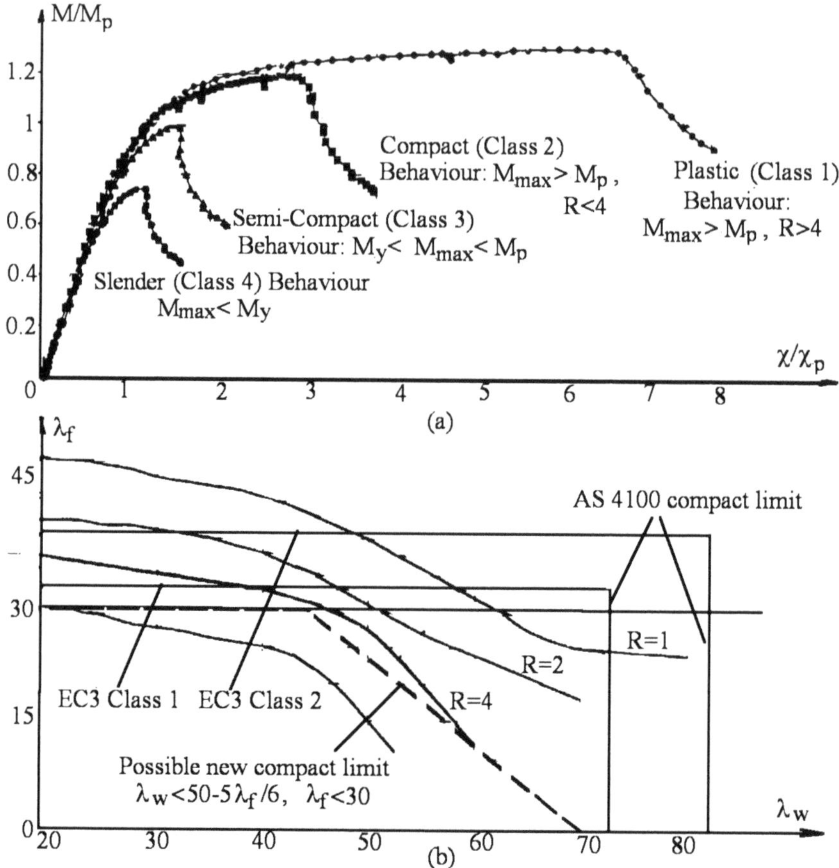

Figure 7.46: Influence of wall slenderness on ductility: (a) Moment-curvature relationships; (b) Flange-web slenderness interaction (after Wilkinson and Hancock, 1998)

types of behaviour for different section classes. Based on these experimental results an interaction between web and flange slendernesses is proposed (Fig. 7.46b), very different from the Eurocode 3 specifications.

(ii) *Conclusions on experimental tests.* Examining the experimental results, the following conclusions can be noted:

-the fabrication processes as rolling, welding hot or cold forming, concrete-filling, have a very great influence on the behaviour of this section type;

-in the case of cold-formed sections, the increase of yield stress in the corners has a great influence on the ductility;

-the round corners produce an improvement of local ductility;

-the danger of brittle fractures is higher for thick walls than for the thin ones, where local buckling is the dominant phenomenon;

-the concrete-filled members have an increased ductility due to the delaying of local buckling.

Table 7.10: Box and hollow sections; Theoretical results

Approximate solutions	FEM numerical results	Plastic collapse mechanism
• Kato & Akiyama (1981) • Ge & Usani (1995)	• Ballio & Calado (1992) • Watanabe et al (1992) • Rondal et al (1994) • Sedlacek et al (1995) • Stronghoner (1995) • Sully & Hancbok (1996, 1999) • Mamaghani et al (1997)	• Kecman (1983) • Kotelko and Krolak (1993) • Kotelko (1995, 1996a,b) • Stranghoner (1995) • Gioncu & Petcu (1995, 1996)

7.3.3. Theoretical Results

The principal results are presented in Table 7.10, being classified in approximate, numerical and plastic mechanism methods.

(i) *Approximate solutions.* For *square hollow-section beams*, Kato and Akiyama (1981) have established an approximate formulation similar to the one proposed for I-section members (see equations (7.26-7.30)):

-maximum moment:

$$s = \frac{M_{max}}{M_p} = \max \begin{cases} 0.98 + 0.0202 \left[\left(2.624 \frac{b}{t} \varepsilon_y^{1/2} - 5.79 \right)^2 - 1.368 \right] \cdot \\ \cdot (1.81 - \lambda \varepsilon_y^{1/2}) \\ 1/(0.526 + 0.341 \frac{b}{t} \varepsilon^{1/2}) \end{cases} \tag{7.63}$$

in which λ is the slenderness ratio with respect to principal axis;

-stiffness ratio for the strain-hardening range:

$$k_p = 0.03 \tag{7.64}$$

-stiffness ratio for the negative slope region:

$$k_d = -0.1377 \left[\left(2.674 \frac{b}{t} \varepsilon^{1/2} + 0.4 \right)^2 - 3.76 \right] (\lambda \varepsilon^{1/2} - 0.1906) \varepsilon_y^{1/2} \tag{7.65}$$

The rotation capacity can be determined using the equation (7.30). Another approximate solution is given by Ge and Usami (1995).

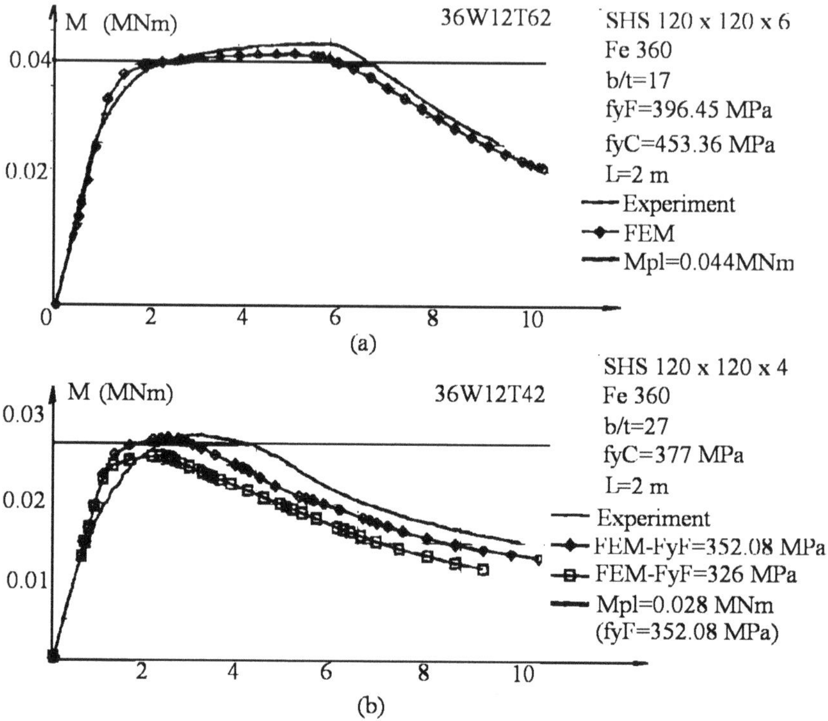

Figure 7.47: Numerical tests using FEM analysis and comparison with experimental tests: (a) $b/t = 17$; (b) $b/t = 27$ (after Rondal et al, 1994, Sedlacek et al, 1995, Stronghoner, 1995)

(ii) *FEM numerical results.* Ballio and Calado (1986) have used the strip method to study the cyclic behaviour of welded box sections with some degradations due to the cracks. The Watanabe et al (1992) FEM analysis has confirmed the important *effect of round corners* on improving the member ductility.

Very important results concerning the *ductility of square and rectangular hollow sections* have been obtained by Rondal et al (1994), Sedlacek et al (1995) and Stronghoner (1995), using the FE package ABAQUS (1988). Numerical simulations of some cold-formed sections compared with the experimental results are presented in Fig. 7.47a. Fig. 7.47b shows the same curves obtained for hot-formed sections. The differences in behaviour are due to the different stress–strain curves used for these two section types. A good correspondence with experimental results can be observed. The numerical results confirm the experimental conclusion that the ductility of cold-formed profiles is higher than hot-formed profiles.

The *effect of geometrical imperfections* of walls has been numerical studied by Sully and Hancock (1999) using the same ABAQUS computer program. The welded profiles are usually affected by very large imperfections.

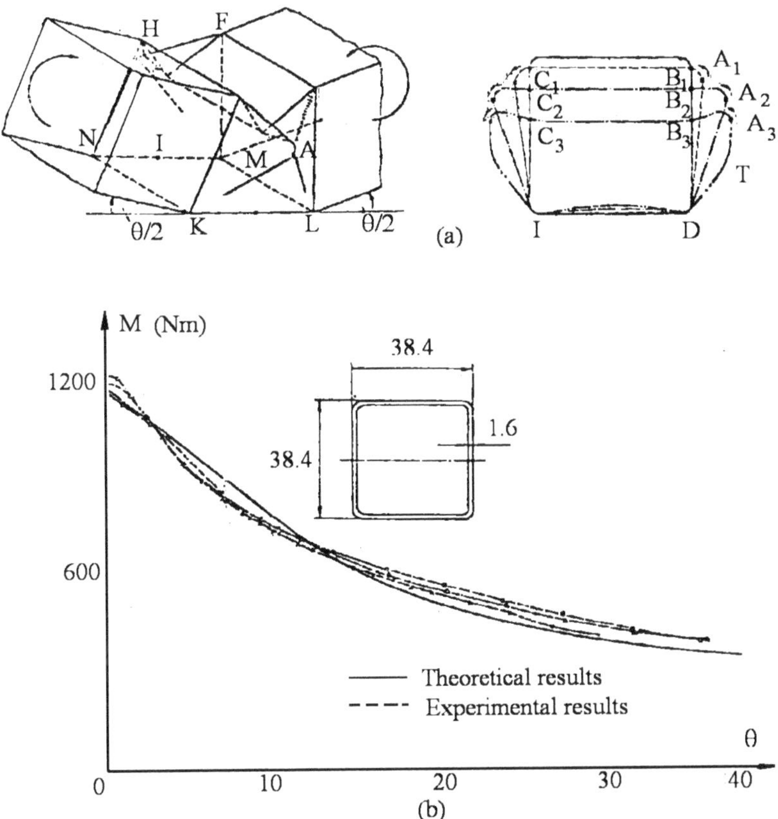

Figure 7.48: Plastic collapse mechanism method: (a) Collapse mechanism shape; (b) Theoretical and experimental results (after Kecman, 1983)

The cyclic behaviour of compact welded box sections has been numerically simulated by Mamaghani et al (1997), using elasto-plastic FE formulation. No local buckling effect has been considered, so the results have a limited application.

(iii) *Plastic collapse mechanism method.* The first theoretical model for the *true plastic mechanism* of rectangular hollow sections was evaluated by Kecman (1983) (Fig. 7.48a). The mechanism is composed by stationary yield lines and travelling yield lines. Using the principle of virtual work a theoretical value for post-critical behaviour is obtained. A comparison between experimental and theoretical results shows a very good concordance (Fig. 7.48b).

Using the method developed by Kecman, Kotelko (1995, 1996a, b) has studied *three plastic mechanism types* (Fig. 7.49a). A comparison between the post-critical behaviour of these plastic mechanisms and the experimental results (Fig. 7.49b) shows that the first one of Kecman is the most adequate model among those examined.

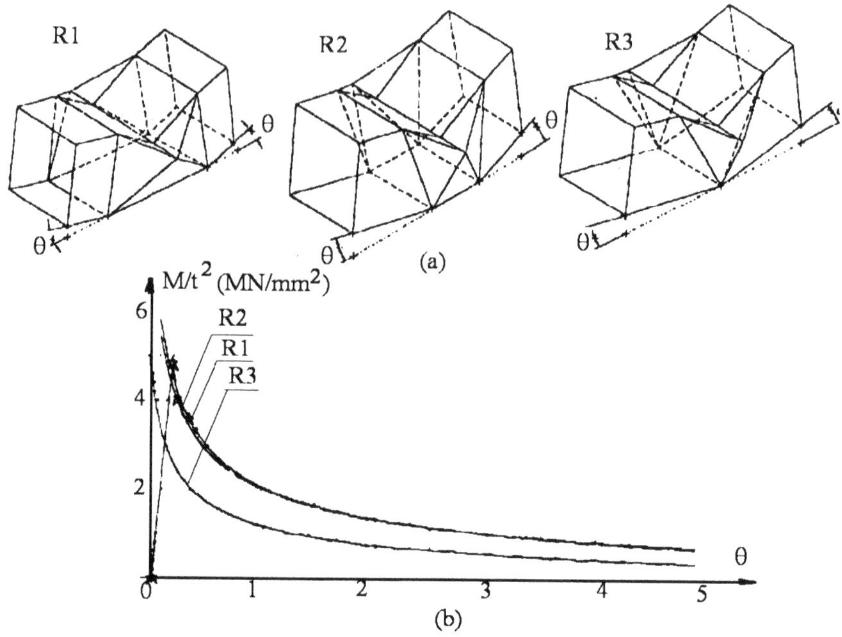

Figure 7.49: Influence of plastic collapse mechanism types: (a) Different mechanism types; (b) Moment–rotation curves (after Kotelko, 1996)

A complete study of the *plastic collapse mechanism* has been performed by Stranghoner (1995) and Stranghoner and Sedlacek (1997), (Fig. 7.50). A comparison between experimental results and FEM numerical simulation, presented in Fig. 7.50, shows that the plastic mechanism method is very valuable for practical purposes. For determining the rotation capacity for quasi-constant moment gradients, it is proposed to use the value 0.9 M_p, due to the fact that many experimental results do not reach the full plastic moment value (Fig.7.51).

The *quasi-mechanism type*, in which the corners are involved in plastic zones instead of plastic lines, has been studied by Gioncu and Petcu (1995) and Gioncu et al (1996). The comparison with the experimental results shows a good correspondence for cold-formed profiles tested by Rondal et al (1994) and some differences for hot-formed profiles. It seems that for higher wall slenderness, the true mechanism is more adequately used than the quasi-mechanism type.

(iv) *Conclusions on theoretical studies.* Examining the theoretical results the following conclusions can be noted:

-the influence of fabrication processes, variations of yield stress, residual stresses and geometrical imperfections plays a very important role in determining the member ductility;

-a very good solution to improve the ductility of hollow sections is to fill them with concrete;

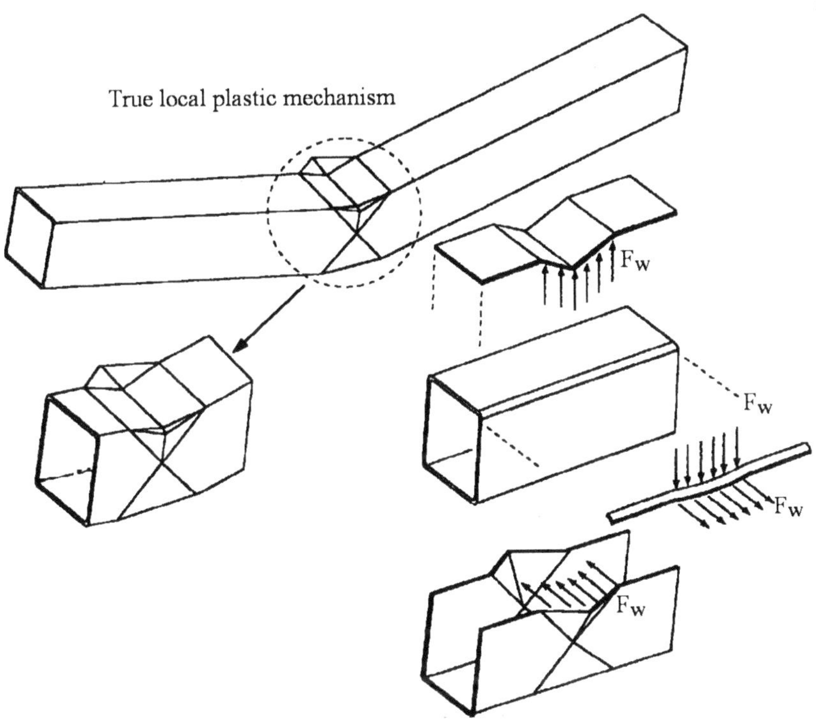

Figure 7.50: Plastic collapse mechanism types including corner deformations (after Stronghoner, 1995)

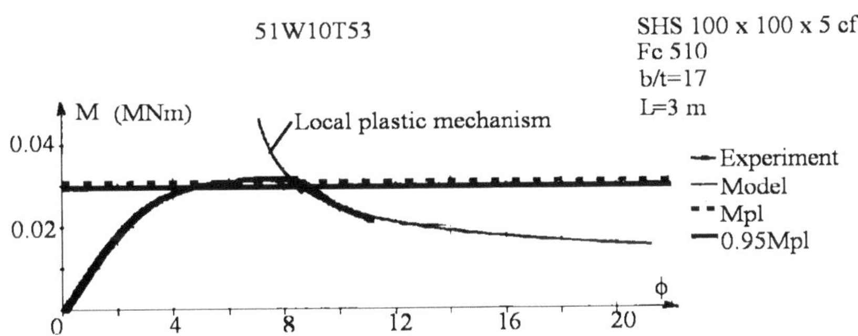

Figure 7.51: Correlation of experimental tests and theoretical results (after Stronghoner, 1995)

-the plastic collapse mechanism can be successfully used to evaluate the rotation capacity;

-as a function of the wall slenderness the true mechanism or the quasi mechanism must be used in determining the member ductility.

Table 7.11: Composite I-section

Effective width	Positive (sagging) moment	Negative (hogging) moment
• Plumier et al (1998) • Doneaux & Parung (1998) • Migiakis & Elghazouli (1998) • Plumier (1999) • Plumier & Doneaux (1999) • Amadio et al (1999)	• Ansourian (1982) • Aribert (1992) • Bruzzese & Ghersi (1992) • Doneaux (1998) • Plumier & Doneaux (1998, 1999) • Plumier (1999)	• Climenhaga & Johnson (1972 a,b) • Weston & Nethercot (1987) • Kemp & Dekker (1991) • Cosenza et al (1992) • Faella et al (1993) • Aribert (1994) • Consalvo et al (1994) • Anhag (1997) • Dekker et al (1995) • Kemp et al (1995) • Tehami (1997) • Cosenza et al (1997a,b) • Fabbrocino (1998) • Fabbrocino et al (1998) • Kemp (1999) • Bradford (2000)

7.4 Ductility of Composite I-Section Beams

7.4.1. Design Aspects

As it has been shown in Section 3.4.2, the modern buildings very often use trapezoidal sheeting integrated by casted concrete as floor deck. In this case the floor beams perform as a steel-concrete composite system, which provides a very efficient solution for enhancing the structure behaviour. The slab may be a significant source of overstrength for the floor beam. On the other hand, the interaction between the concrete deck and steel girder increases the complexity of models for the rotation capacity evaluation. The main contributions related to this interaction are presented in Table 7.11 with reference to the basic problems, which are also visualised in Fig. 7.52:

-correct estimation of effective width of the slab which interacts with the steel beam;

-determination of ductility for positive (sagging) moments, when the slab is in the compressed section part, the collapse being produced by the yielding of the steel beam and crushing of concrete slab;

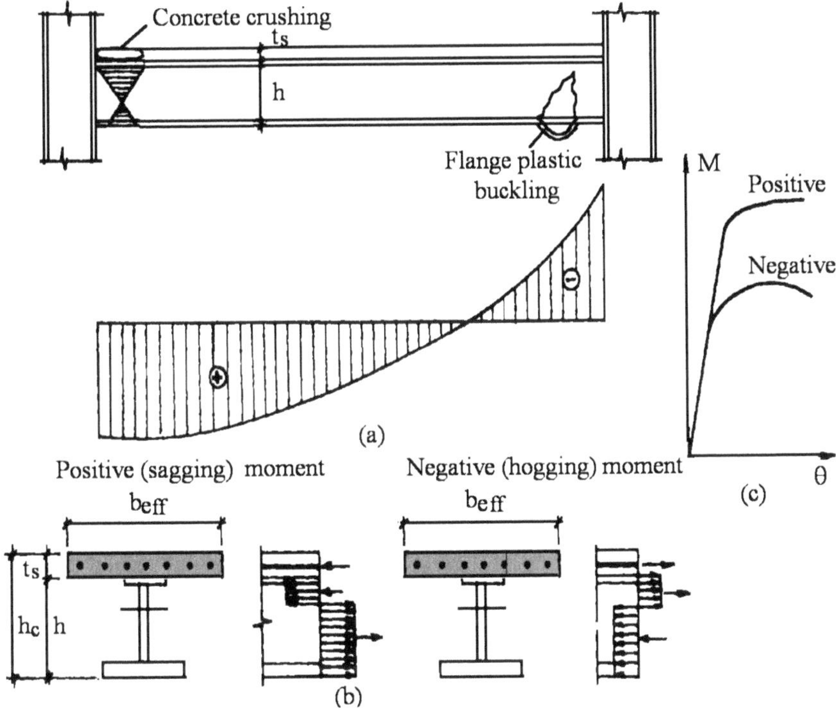

Figure 7.52: Composite I-section beam: (a) Beam scheme; (b) Positive and negative moments; (c) Moment–rotation curves

-determination of ductility for negative (hogging) moments, when the compression section part is in the steel beam, producing the local buckling of bottom flange, and tension is in the slab, causing cracks in concrete and yielding in reinforcement bars, so-called rebars;

-assuring a good behaviour of a composite beam under seismic actions, by avoiding an unfavorable local buckling in the steel section, and an early concrete crushing in the slab.

7.4.2. Effective Width

The effect of the slab on the steel beam may be quantified by introducing the concept of effective width. Accordingly, only a given portion of the slab width is assumed to participate in the resistance of the composite beam (Fig. 7.53). This effective width is a very important parameter for the calculation of moment capacity and member ductility. Lower values than the actual ones could be obtained by neglecting the composite action, but this would lead to a rather over conservative estimation of bending resistance. On the other hand, from the seismic point of view, the underevaluation of the structure stiffness consequently produces an underestimation of the earthquake actions applied to the structural system, what is unconservative.

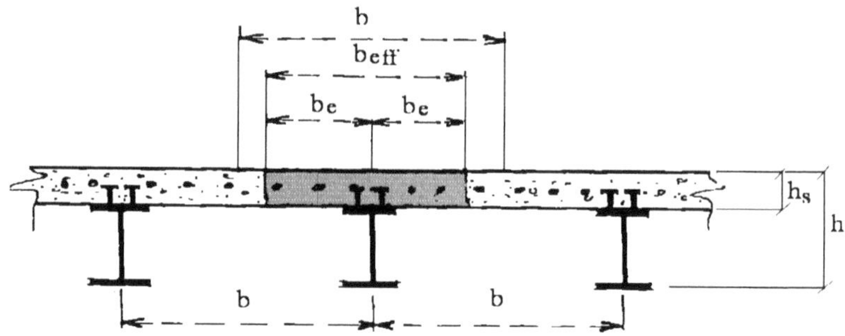

Figure 7.53: Effective width

In addition, the weak beam-strong column design procedure requires correct information on the plastic behaviour of beams, which depend on the definition of the actual resisting cross-section. For all these reasons, the effective width of a member in a seismic resistant structure cannot be evaluated for gravitational loads only, but requires a more specific characterization. A simplified formulation is given by Eurocode 4, only valid for vertical loads, where the effective width b_e is a portion of the flange on each side of the center line of steel webs:

$$b_e = \frac{L_0}{8} < b \tag{7.66}$$

The length L_0 is the distance between the inflection points of the bending moment diagram.

For earthquake loading the definition of the appropriate length L_0 is given in Fig. 7.54 (Plumier, 1999) and so a reduced effective width results:

-for positive (sagging) moments:

$$b_e^+ = \frac{L_0^+}{8} = 0.075L \tag{7.67a}$$

-for negative (hogging) moments:

$$b_e^- = \frac{L_0^-}{8} = 0.10L \tag{7.67b}$$

A more careful examination of the transfer mechanism from slabs to columns has emphasized the influence of transverse beams and rebars (Plumier, 1999, Plumier and Doneux, 1999, Plumier et al, 1998, Doneux and Parung, 1998), (Fig. 7.55a). The transfer of the horizontal force from slabs to columns by shear, bending and torsion can be achieved by mobilising the connectors placed on the transverse beams (Fig. 7.55b). The rebars refer to a supplementary slab reinforcement (Fig. 7.55c), called seismic rebars, in order to assure the slab forces transfer and to improve the ductility of plastic zones.

The effective width of slabs for composite I-sections used in the elastic analysis (in the case of serviceability limit states) and plastic analysis (for

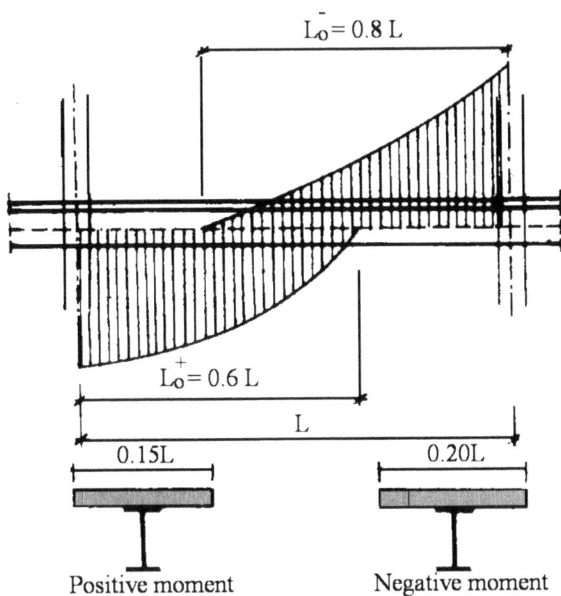

Figure 7.54: Effective widths for positive and negative moments (after Plumier, 1999)

damageability and survivability limit states) is defined in Table 7.12 and Fig. 7.56. It must be noticed that the contribution of the effective width of slab to the plastic moment of a beam is not proportional to effective width itself. The results for composite IPE 300 beams show that additional width over $b_e = 100$ cm does not produce important increment of plastic moments M_p (Plumier et al, 1998), (Fig. 7.57).

The study of Amadio et al (1999) carried out by numerical tests has pointed out that for vertical loads the effective width for positive moments can be determined according the EC 4 specifications, but for negative moments the effective width must be two times more than the one given by EC 4.

A very interesting study has been performed by Migiakis and Elghazouli (1998) in which the flexural stiffness of slabs related to the total beam stiffness ratio and the end beam rotation due to seismic moments are considered (Fig. 7.58). One can see an important variation of effective width along the beam and the great influence of loadings. These aspects make it more difficult for a proper determination of the effective width, and a realistic evaluation of beam rotation capacity. The main conclusions of this study is that for vertical loads the value of effective width is accurately approximated by EC 4. Contrary to this, for seismic loads, the effective widths are doubled compared to the design value suggested by EC 4, in good concordance with the results obtained by Amadio.

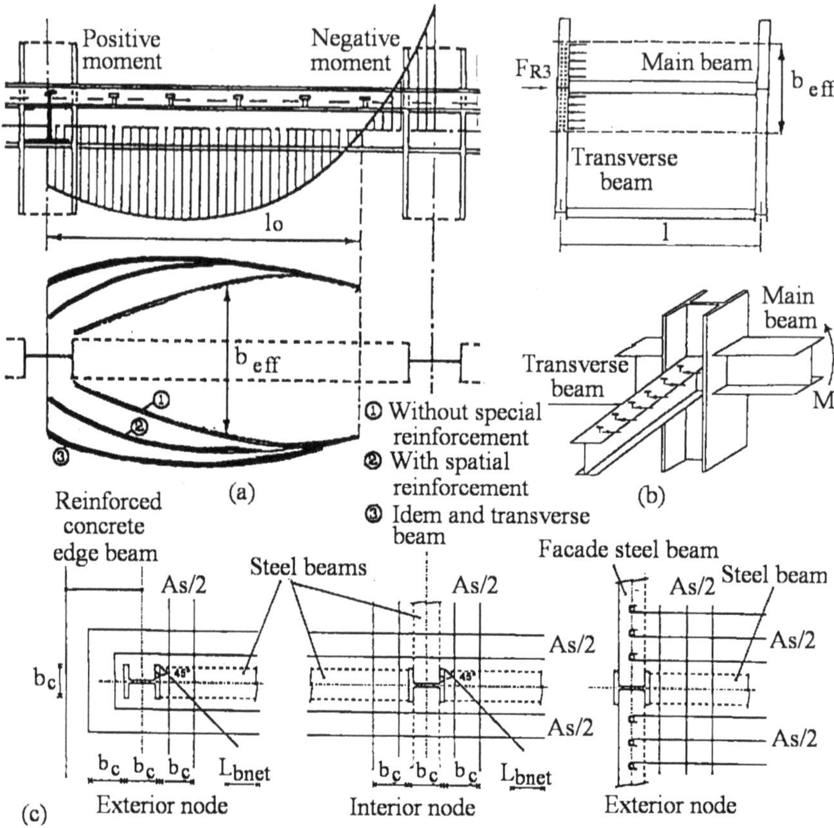

Figure 7.55: Transfer of horizontal forces from slab to column: (a) Effective widths; (b) Influence of transverse beam; (c) Special reinforcements (after Plumier, 1999)

7.4.3. Ductility of Beams under Positive (Sagging) Moments

In plastic design of composite beams under vertical loads no attention is paid to the positive bending zone. Only composite sections under negative bending are concerned because of the redistribution of problems. Contrary to this, in seismic design the required rotation capacities under positive bending moments are as important as under negative bending moments. A good overall mechanism type for moment-resistant frames is the global plastic mechanism which considers the occurrence of plastic hinges at both beam ends for positive and negative moments.

The first problem for reliable behaviour of a composite beam under a positive moment is to assure that no buckling of compression flange can arise due to too large spacing between connectors. Therefore, in the potential plastic zones, this spacing must be smaller than the flange width.

Table 7.12: Definition of effective width b_e of slab

Column	Transverse beam	M	Plastic analysis	Elastic analysis
Interior	Present, fixed to the column, with connectors for full shear	M^+ M^-	$0.1L_0$ $0.075L_0$	$0.05L_0$ $0.0375L_0$
Interior	Not present or present, but not fixed to the column, or not having connectors for full shear	no proposal		no proposal
Exterior	Present as an edge beam fixed to the column in the plane of the columns, with connectors for full shear and specific detailing for anchorage of rebars exterior to the column plane, with re-bars of the hair pin type	M^+ M^-	$0.1L_0$ $0.075L_0$	$0.05L_0$ $0.0375L_0$
Exterior	Not present or no re-bars anchored	M^+ M^-	0 $b/2$ or $h/2$	0 $b/4$ or $h/4$

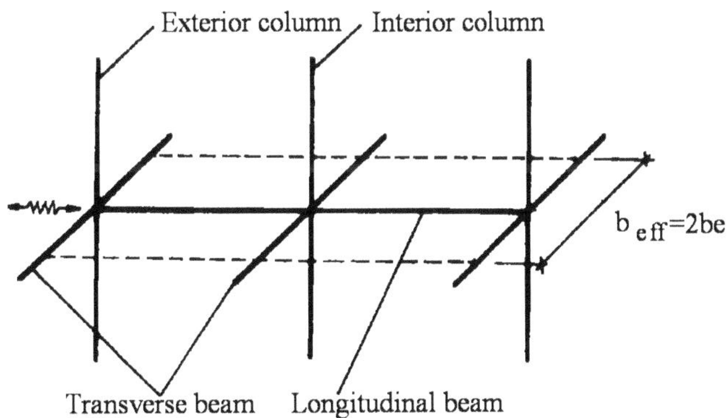

Figure 7.56: Determination of effective width (after Plumier, 1999)

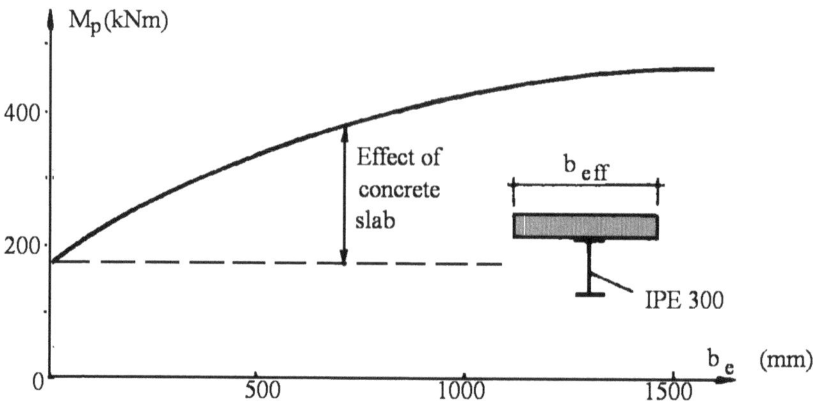

Figure 7.57: Effect of concrete slab on the plastic moment (Plumier et al, 1998)

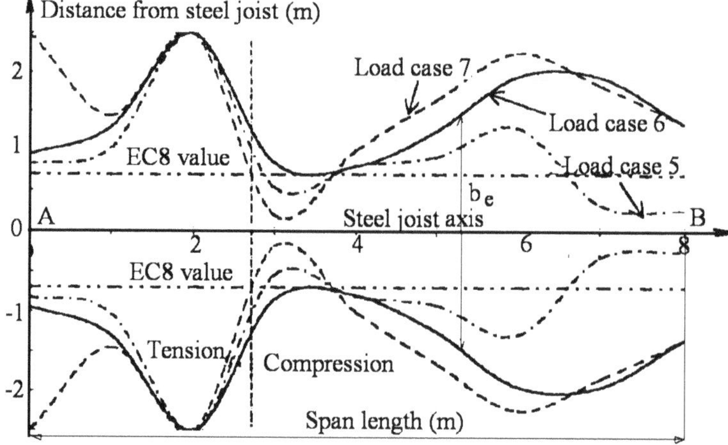

Figure 7.58: Variation of effective width along the beam (after Migiakis and Elghazouli, 1998)

The second problem is to obtain a plastic hinge where there is no crushing of the concrete slab, before complete plasticization in tension of the steel profile, limiting the rotation capacity of the composite beam. A detailed cross-sectional analysis of composite I-sections has shown that two distinct modes of behaviour arise under positive bending. On one hand, the strain hardening and fracture of the bottom flange of the steel beam occurs before the crushing failure of the slab. On the other hand, concrete crushing occurs before significant strain-hardening is developed in steel (Ansourian, 1982). The first type of collapse mode can be classified as ductile and that of the

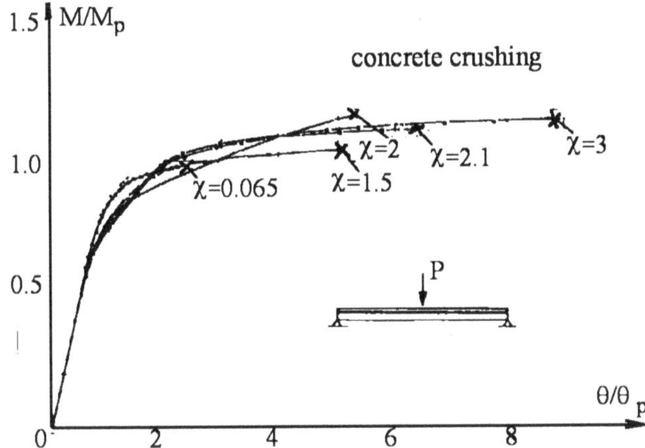

Figure 7.59: Ductility of beam for positive moments (after Ansourian, 1982)

second type as brittle. Introducing the ductility parameter:

$$\chi = \frac{0.72 f_c b_{eff} \varepsilon_{cu}(t_s + h)}{A_s f_y(\varepsilon_{cu} + \varepsilon_{sh})} \tag{7.68}$$

when f_c is the cylindrical strength of concrete, ϵ_{cu}, ultimate strain of concrete, A_s, cross-section area of steel beam, the remain notations being known. The value $\chi = 1$ corresponds to a beam in which the ϵ_{sh} and ϵ_{cu} values are reached simultaneously. So, for $\chi > 1$ the beam behaviour is ductile, while for $\chi < 1$, the beam collapse is brittle. The experimental tests of Ansounian (1982) have verified the equation:

$$\frac{\theta_u}{\theta_p} = 2.3\chi - 1.85 \tag{7.69}$$

where θ_u is the ultimate rotation corresponding to concrete crushing and θ_p the rotation corresponding to the plastic hinge. The influence of the ductility factor χ on the composite beam behaviour is shown in Fig. 7.59. High values of the χ parameter assure a good ductility, as it results from theoretical and experimental tests. If the condition $\chi > 1$ is satisfied and the corresponding connectors are used, the ductility of the beam for positive moments is high. This fact has been confirmed also by Aribert (1992a,b) and Bruzzese and Ghersi (1992).

In EC 4 (1992), for static actions, it is prescribed that the distance from the maximum concrete compression fibre to the plastic neutral axis shall not exceed:

$$\frac{x}{h_c} < 0.15 \tag{7.70}$$

where x is the distance from the neutral axis to the top of the concrete slab, and h_c is the total depth of the composite section.

The EC 8 is not explicit in defining the condition to obtain a good ductility in case of seismic actions. A more direct formulation linked to the physics of the problem has been proposed by Doneux (1998), Plumier (1999) and Plumier and Doneux (1998, 1999). The condition (7.70) can be translated in the form:

$$\frac{x}{h_c} < \frac{\varepsilon_{cu}}{\varepsilon_{cu} + \varepsilon_{apl}} \tag{7.71}$$

where ϵ_{cu} is the ultimate strain of concrete and ϵ_{apl}, the plastic strain of the steel bottom flange. The most disputable parameter of this equation is the concrete ultimate strain, the value considered by Plumier and Doneux (1999) being $\epsilon_{cu} = 2\text{x}10^{-3}$. Contrary to this, Doneux (1998) considers that the minimum value is 3.5×10^{-3}, but due to the favourable situation of concrete in beams, the concrete behaviour being more ductile than determined for specimens, the value $\epsilon_{cu} = 5 \times 10^{-3}$ better represents the reality. The experimental results of Plumier and Doneux (1998) have shown that for dynamic actions this value is reduced to $2.4\text{x}10^{-3}$. The yield strain ϵ_{apl} can be estimated on the basis of the Plumier proposal (1996) $\epsilon_{apl} = q\epsilon_y$, with a behaviour factor $q = 6$ for well designed earthquake resistant structures. Considering all these proposals, a rational value $\epsilon_{cu} = 2{,}5\text{x}10^{-3}$ can be assumed. So, the following limitation values are obtained:

$$\begin{aligned} &\text{Fe 360 } x/h_c \leq 0.267 \\ &\text{Fe 430 } x/h_c \leq 0.237 \\ &\text{Fe 510 } x/h_c \leq 0.194 \end{aligned} \tag{7.72}$$

For a beam with span of 5m and effective width 1m, the general curves for x/h_s versus slab thickness t_s are sketched in Fig. 7.60, for steel Fe 360 and for two profiles IPE and HEA (Doneux, 1998). On the basis of all these curves, one can determine the slab thickness that satisfied the condition (7.72). One can see that the EC 4 limiting condition (7.70) is too severe for seismic loading and the limits obtained from (7.72) seem to be more suitable. One can see that for Fe 360 IPE profiles there are no problems when obtaining a good ductility. But for Fe 510 steel profiles there are some limitations for obtaining ductile sections. For HEA profiles some problems arise in finding a ductile section, due to the greater thickness of their walls.

On the basis of these results it is clear that the use of soft steel and high strength concrete would be a good solution to obtain a ductile composite section.

7.4.4. Ductility of Beam under Negative (Hogging) Moments

The resistance of cross-section for negative moments is influenced by the magnitude of the tensile force in the concrete slab, the compression force in the bottom flange and the distance between these stress resultants. The cracking of a concrete slab normally occurs at a load level largely below the ultimate moment and in this cracked section it is commonly assumed that only the rebars work. In the compression part, considerable restraint for lateral-torsional deformation is provided by the concrete slab connection to

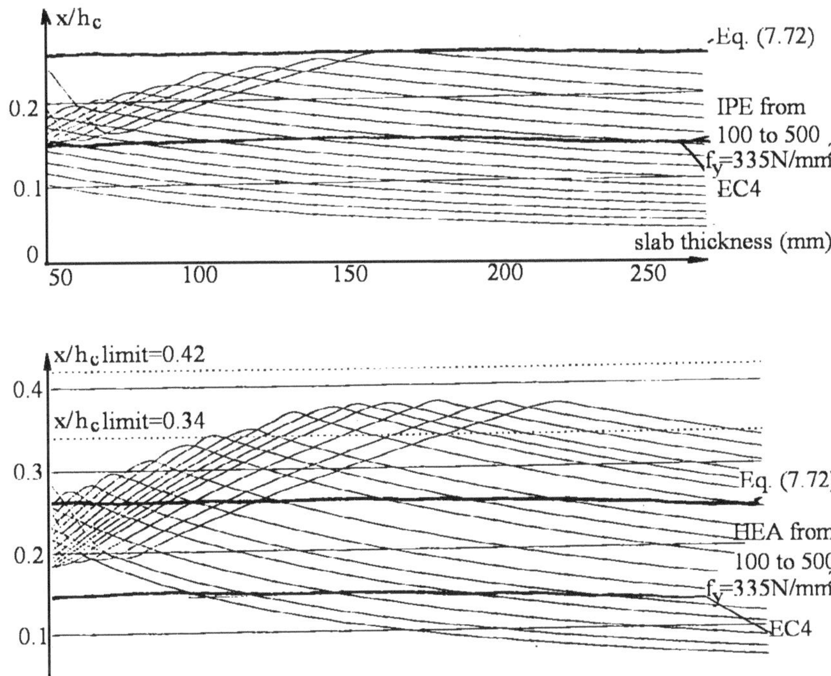

Figure 7.60: Conditions to avoid the crushing of a concrete slab: (a) IPE profiles; (b) HEA profiles (after Doneaux, 1998)

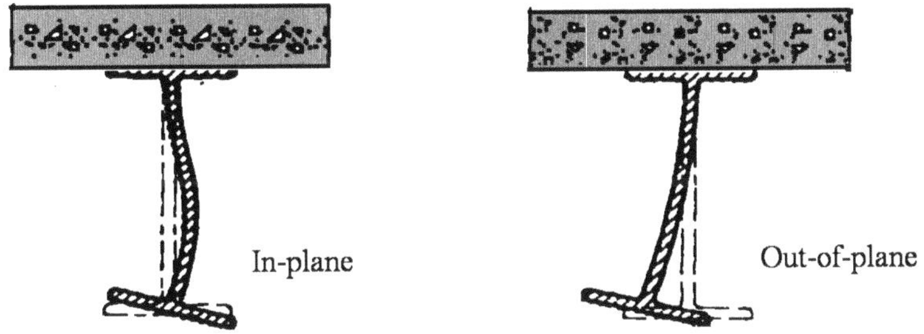

Figure 7.61: Buckling modes for negative moments: (a) In-plane buckling of compression flange; (b) Out-of-plane buckling of compression flange

the tension flange. The following buckling modes can occur (Fig. 7.61):
-local buckling of compression flange and web;
-lateral distorsional buckling of steel section;
-interaction between these two buckling modes.

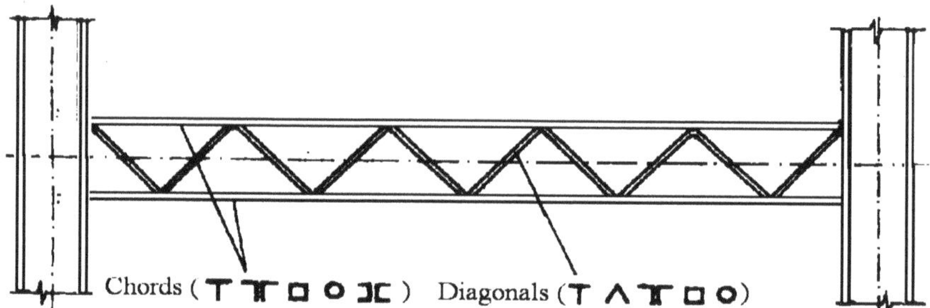

Figure 7.62: Open web members

The Table 7.11 presents the main theoretical and experimental contributions concerning the ductility of composite beams under negative moments. Generally, the problems are the same as for the steel sections, presented in Section 7.2. No major differences can be identified for steel and composite steel beams (Kemp and Dekker, 1991, Kemp et al, 1996), so, the results presented in Section 7.2 are available for composite sections under negative moments.

7.5 Ductility of Open-Web Members (Trusses)

7.5.1 Design Aspects

Open-web members (trusses) are often used for building structures when steel members with large cross-sections are requested, but are not available. This system has a lot of advantages, the most important one being economy, especially for long-span beams. The connection to a column requires relative simple detailing and piping and ductwork can be placed through web openings, resulting in a better utilization of span and reduced interstoreys in many cases (Goel and Itani, 1994a,b).

The basic elements of an open-web member are presented in Fig. 7.62, consisting in chords and diagonal web bars, which are usually made of T, angles, double angles in back-to-back configuration, rectangular or circular hollow sections.

In spite of the above advantages, the knowledge on their behaviour and ductility performance during severe ground motions is not satisfactory. Due to the poor behaviour of the Pino Suarez buildings during the 1985 Mexico City earthquake (see Section 2.2.3), engineers are now reluctant to use this system for beams in active seismic regions. As it is well known, the basic concept for frame design is to obtain a weak beam-strong column system, with very ductile beams, allowing for a global plastic mechanism with good seismic energy input dissipation. This design philosophy is difficult to be achieved by using open-web members, due to their reduced ductility. The variant of weak column-strong beam design philosophy is not considered as a very desirable solution for seismic resistant structures. So, due to this lack

Table 7.13: Open-web members

Conventional solutions	Special systems
• Chen & Lu (1991) • Cheng & al (1992) • Ger & Chen (1992, 1993) • Goel & Itani (1994a) • Kleiser & Uang (1999)	• Goel & Itani (1994b) • Goel (1995)

of information, the designer considers that this system belongs to structures for which the occurrence of a good collapse mechanism is not assured.

References related to the main problems of using this system for seismic resistant structures are listed in Table 7.13.

7.5.2. Ductility of Conventional Systems

After the collapse of the Pino Suarez buildings produced by the failure of open-web girders, some experimental and theoretical results were obtained on conventional type systems, without any special measure for improving their ductility.

Chen and Lu (1991), Cheng and al (1992), Ger and Chen (1992, 1993) have performed careful experimental tests and numerical simulations in order to explain the Pino Suarez collapse. Two open-web girder types have been tested under cyclic loading (Fig. 7.63). The failure of T1 girder was caused by the severe local buckling of top and bottom chord bars. A failure ductility of 1.71 and 1.72 has been obtained for the two tested beams. Analytical modelling (Cheng et al, 1992, Ger and Chen, 1993a) has confirmed the poor ductility of these systems due to the buckling of members without a good post-critical behaviour.

Experimental and theoretical studies have been carried out by Goel and Itani (1994a), (Fig. 7.64). The buckling of web bars due to axial compression produces an instant drop in the girder carrying capacity.

There are three main factors that contribute to the early collapse of the open-web girders:

-the ductility capacity of the beams is primary influenced by the web and the buckling of the diagonals leads to a sudden drop of carrying capacity;

-as a compression diagonal buckles, the adjacent tension diagonal creates an unbalanced force in the chord members. In absence of vertical web bars, this produces the increasing compression force in chord bars, and, consequently, the buckling of these bars and the squashing of the girder;

-slender sections are generally used for diagonal bars with very reduced ductility and this poor behaviour after buckling directly affects the girder ductility capacity.

The major conclusion from this experimental and theoretical study is that the conventional solutions for open-web girders show very poor ductility

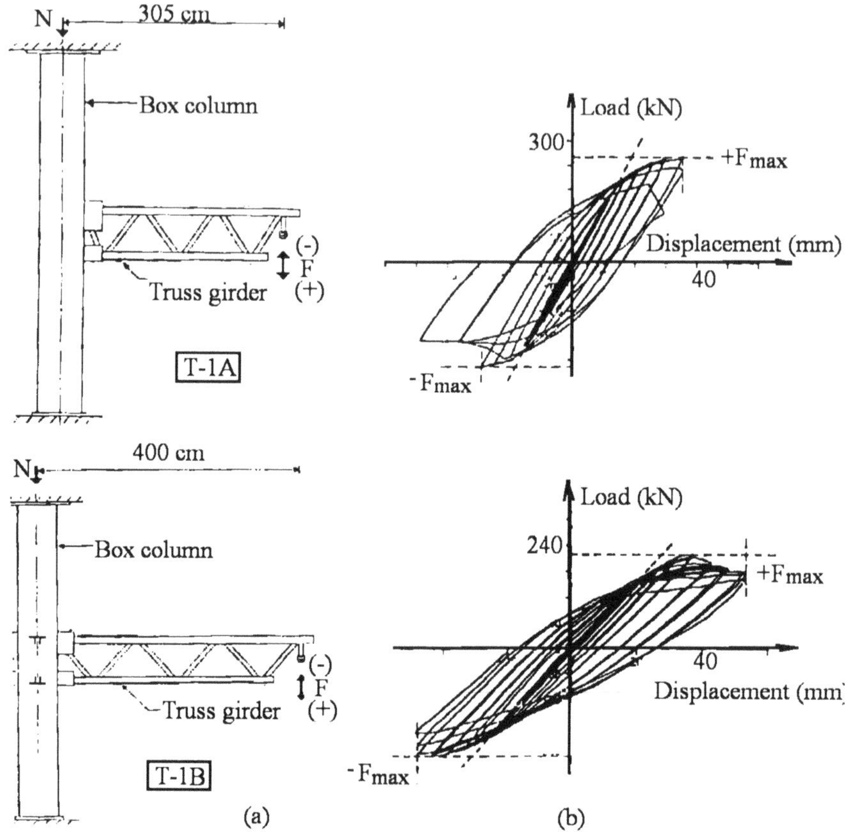

Figure 7.63: Experimental tests on open-web beams: (a) Test assemblages; (b) Hysteresis loops (after Chen and Lu, 1991)

due to the early fracture of girder bars. The same results are obtained for cyclic loading, showing that such systems poorly respond to severe ground motions with large storey drifts and excessive inelastic deformations.

7.5.3. Ductility of Special Systems

In order to improve the hysteretic behaviour of the open-web girders, a special solution has been worked out by Goel and Itani (1994b) and Goel (1995). Instead of a single diagonal, it is proposed to use a panel composed by X diagonals and vertical bars, so that the buckling of compression diagonals is counterbalanced by the tension diagonals and vertical bars. The diagonal bars should be made of compact sections in order to improve the girder ductility.

The most important aspect for using this improved system is the strategy to place these panels in a portion of a girder which assures the maximum ductility. There are two possibilities (Fig. 7.65):

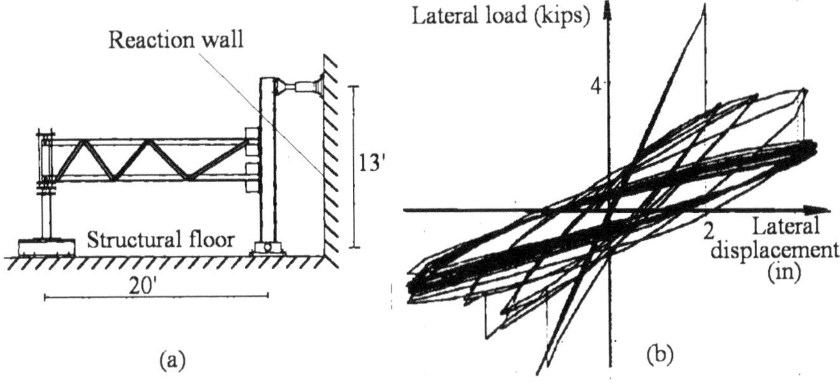

Figure 7.64: Experimental tests on conventional open-web beams: (a) Schematic of test set-up; (b) Hysteretic loops (after Goel and Itani, 1994a)

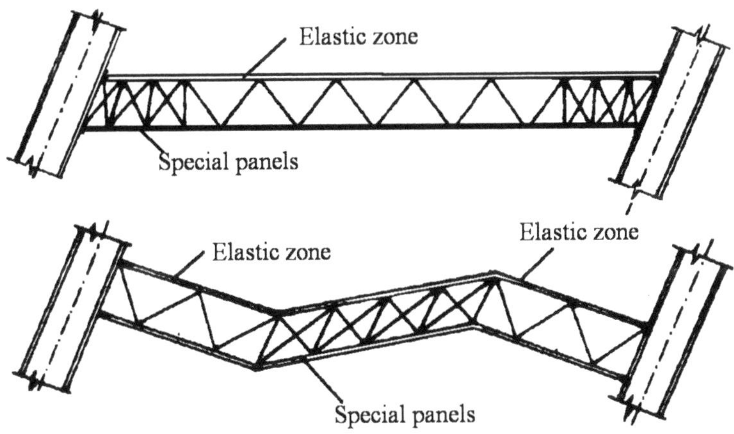

Figure 7.65: Special systems for open-web beams with ductile panels

-to place these panels at the ends of girders in the zone of maximum moment, assuring the formation of a global plastic mechanism by means of the plastic hinges formed in the panels;

-to place these panels near to the midspan of beams, the concept being similar to that used in ductile eccentric braced frames.

Goel and Itani (1994b) recommend the use of second solution. The experimental tests have shown a quite stable hysteretic loop (Fig. 7.66), contrary of that obtained for conventional open-web girders. Thus, this system can be an excellent and efficient component in seismic resistant framing systems for certain classes of building structures.

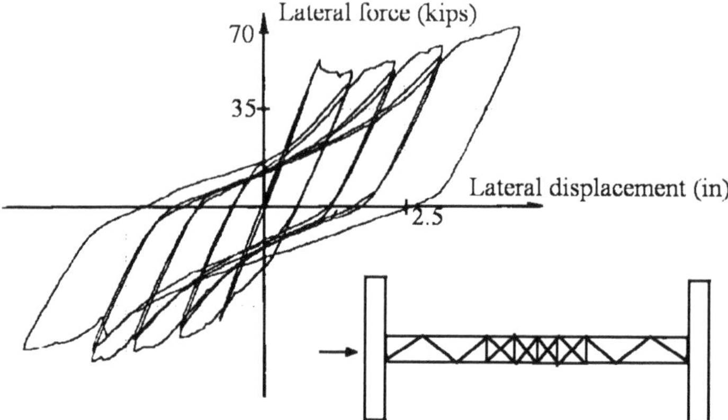

Figure 7.66: Hysteretic loops of special-web beams (after Goel and Itani, 1994b)

7.6 Conclusions

The main conclusions from the aspects presented in this chapter may be summarized as follows:

-the ductility of I-sections, box and hollow sections, composite I-sections and open-web members has been examined on the basis of research published in literature;

-the main results of experimental and theoretical research have been collected in order to elaborate a consistent methodology for computing the available ductility.

-the methodology based on the collapse plastic mechanism seems to be the most suitable one for evaluating the available ductility of members for design purposes;

-the experimental and theoretical research works have shown that there are some very significant factors influencing the local ductility of members:

- section type (I, box and hollow, composite, open web);
- flange and web slenderness;
- section shape;
- member span;
- moment gradient;
- axial force;
- yield stress and randomness;
- yield ratio;
- strain rate;
- cyclic loading;
- number of plastic excursions.

The available ductility must consider all these above mentioned factors. For this purpose, a devoted computer program must be elaborate in order to deliver a quantified value which must be compared with the required ductility determined from the global analysis of structure subjected to a given earthquake.

7.7 References

ABAQUS (1996): Standard theory manual. Hibbit, Karlsson and Sorensen Inc

Aiello M.A., Ombres L. (1995a): Rotation capacity and local ductility of steel members. In 9th International Conference on Metal Structures (ed. J. Murzewski), 26-30 June, Krakow, Vol. 2, 271-282

Amadio, C., Briganti, D., Fragiacomo, M.(1999): Effective width in steel-concrete composite beams fultimate analysis. In XVII Congresso CTA, Napoli, 3-7 October 1999, Vol. 2, 239-249

Anastasiadis, A. (1999): Ductility problems of steel moment resisting frames. Ph D Thesis, Politechnica University Timisoara

Anastasiadis, A., Gioncu, V. (1999): Ductility of IPE and HEA beams and beam-columns. In Stability and Ductility of Steel Structures, SDSS 99, (eds. D.Dubina and M. Ivanyi), Timisoara, 9-11 September 1999, Elsevier, Amserdam, 249-257

Anastasiadis, A., Gioncu, V., Mazzolani, F.M. (2000): New trends in the evaluation of available ductility of steel members. In Behaviour of Steel Structures in Seismic Areas, STESSA 2000, (eds. F.M Mazzolani and R. Tremblay), Montreal, 21-24 August 2000, Balkema, Rotterdam, 3-10

Ansourian, P. (1982): Plastic rotation of composite beams. Journal of the Structural Division, Vol. 108, ST 3, 643-659

Aribert, J.M. (1992a): Slip and uplift measurement along the steel and concrete interface of various types of composite beams. In Testing of Metals for Structures (ed. F.M. Mazzolani), Napoli, 29-31 May 1990, E&FN Spon, London, 395-427

Aribert, J.M. (1992b): Effect of the local plate buckling on the ultimate strength of continuous composite beams. Thin-Walled Structures, Vol. 20, No. 1-4, 279-300

Axhag, F. (1995): Plastic design of composite bridges allowing for local buckling. Technical Report, Lulea University of Technology, 09T

Ballio, G., Calado, L. (1986a): Steel bent sections under cyclic loads. Experimental and numerical approaches. Costruzioni Metalliche, No. 1, 1-23

Ballio, G., Calado, L. (1986b): Local buckling of steel sections under cyclic loads. In 8th European Conference on Earthquake Engineering, Lisbon, 1986, Vol. 4, 7.2/57-64

Ballio, G., Calado, L., Leoni, F., Perotti, F. (1986): Numerical simulation of the cyclic behaviour of steel structural systems. Costruzioni Metalliche No. 5, 269-294

Ballio, G., Perotti, F. (1987): Cyclic behaviour of axially loaded members: Numerical simulation and experimental verification. Journal of Constructional Steel Research, Vol. 7, 3-41

Ballio, G., Castiglioni, C.A. (1993): Le costruzioni metalliche in zona sismica: un criterio di progetto basato sull' accumulazione del danno. In XVI Congresso CTA, Viareggio, 24-27 October 1993, Ricerca Teorica e Sperimentale, 99-109

Ballio, G., Castiglioni, C.A. (1994a): Seismic behaviour of steel sections. Journal of Costructional Steel Research, Vol. 29, 21-54

Ballio, G., Castiglioni, C.A. (1994b): An approach to seismic design of steel structures based on cumulative damage criteria. Earthquake Engineering and Structural Dynamics, Vol. 23, 969-986

Ballio, G., Castiglioni, C.A. (1995a): Damage assessment in steel members under seismic loading. In Behaviour of Steel Structures in Seismic Areas, STESSA 94 (eds F.M. Mazzolani and V. Gioncu), Timisoara, 26 June-1 July 1994, E&FN Spon, London, 63-76

Ballio, G., Castiglioni, C.A. (1995b): An unified approach for the design of steel structures under low and/or high cycle fatigue. Journal of Constructional Steel Research, Vol. 34, 75-101

Ballio, G., Castiglioni, C.A., Perotti, F. (1988): On the assessment of structural design factors for steel structures. In 9th World Conference on Earthquake Engineering, Tokyo-Kyoto, 2-9 August 1988, 1167-1171

Barth, K.E., White, D.W., Bobb, B.M. (2000): Negative bending resistance of HPS70W girders. Journal of Constructional Steel Research, Vol. 53, No. 1, 1-31

Bertero, V.V., Popov, E. P. (1965): Effect of large alternating strains of steel beams. Journal of the Structural Division, Vol. 91, ST 1, 1-12

Boerave, Ph., Lognard, B., Janss, J., Gerardy, J.C., Schleich, J.B. (1993): Elasto-plastic behaviour of steel frameworks. Journal of Constructional Steel Research, Vol. 27, 3-21

Bradford, M.A. (2000): Strength of compact steel beams with partial restraint. Journal of Constructional Steel Research, Vol. 53, 183-200

Bruneau, M., Uang, C.M., Whittaker, A. (1998): Ductility Design of Steel Structures. McGraw-Hill, New York

Bruzzese, E., Ghersi, A. (1992): Metal-concrete composite structures: Tests and models. In Testing of Metals for Structures (ed. F.M. Mazzolani), Napoli, 29-31 May 1990, E&FN Spon, London, 411-427

Butterworth, J.M. (1995): Inelastic local buckling in grade 300 universal columns. In Structural Stability and Design (eds Kitipornchai et al), Sydney, 30 October-1 November 1995, Balkema, Rotterdam

Butterworth, J.M., Beamish, M.J. (1993): Inelastic buckling in three-dimensional steel structures. In Space Structures 4 (eds G.A.R. Parke and C.M. Howard), Guildford, 5-10 September 1993, Thomas Teford Service Ltd, London, 51-61

Calado, L.(1992): Cyclic behaviour of steel subassemblages. In Testing of Metals for structures (ed. F.M. Mazzolani), Napoli, 29-31 May 1990, E&FN Spon, 93-103

Calado, L, Azevedo, J. (1989): A model for predicting failure of structural steel elements. Journal of Constructional Steel Research, Vol. 14, 41-64

Castiglioni, C.A. (1987): Numerical simulation of steel shapes under cyclic bending effect of the constititive law of the material. Costruzioni Metalliche, No. 3, 154-175

Castiglioni, C.A. (1994): Damage assessment in structural steel members and welded joints under seismic loading. In Danneggiamento Ciclico e Prove Pseudodinamiche, (ed. E. Cosenza), Napoli, 2-3 July 1994, 21-35

Castiglioni, C.A., Di Palma, N. (1988): Steel members under cyclic loads: Numerical modelling and experimental verifications. Costruzioni Metalliche, No. 6, 288-312

Castiglioni, C.A., Di Palma, N., Morretta, E. (1990): A trilinear constitutive model for the seismic steel structures. Costruzioni Metalliche, No. 2, 80-96

Castiglioni, C.A., Losa, P.L.(1991): Validazione sperimentale dell'applicabilita di un modello di accumulazione lineare del danno nel caso di elementi inflessi in acciaio sottoposti ad azioni sismiche. In XIII Congreso CTA, Ricerca Teorica e Sperimentale, Abano Terme, 27-30 October 1991, 159-170

Castiglioni, C.A., Losa, P.L. (1992): Local buckling and structural damage in steel members under cyclic loading. In 10th World Conference on Earthquake Engineering, Madrid, 19-24 July 1992, 2891-2896

Castiglioni, C.A., Bernuzzi, C., Agatino, M.R. (1997): Low-cycle fatigue: A design approach. In XVI Congrsso CTA, Ancona, 2-5 October 1997, 167-175

Chen, C.C., Lu, L.W. (1991): Cyclic tests of a box-column and truss-girder asemblage. In Inelastic Behaviour and Design of Frames, SSRC Annual Session, Chicago, 15-17 April 1991, 303-312

Cheng, F.Y., Lu, L.W., Ger, J.F., Chen, C.C. (1992): Observations on behaviour of tall steel building under earthquake excitation. In Earthquake Stability Problems in Eastern North America, SSRC Annual Technical Session, Pittsburgh, 6-7 April 1992, 15-26

Chryssanthopoulos, M. K., Manzocchi, G.M.E., Elnashai, A.S. (1999): Probabilistic assessment of ductility for earthquake resistant design of steel members. Journal of Constructional Steel Research, Vol. 52, 47-68

Climenhaga, J.J., Johnson, R.P. (1972a): Moment-rotation curves for locally buckling beams. Journal of the Structural Division, Vol. 98, ST 6, 1239-1254

Climenhaga, J.J., Johnson, R.P. (1972b): Local buckling in continuous composite beams. The Structural Engineer, Vol. 50, No. 9, 367-374

Consalvo, V., Faella, C., Nigro, E. (1994): Influence of delayed effects on the required rotation capacity and design problems in continuous composite beams with moment redistribution. Costruzioni Metalliche, No. 6, 19-37

Cosenza, E., Mazzolani, S., Pecce, M. (1992): Ultimate limit state checking of continuous composite beams designed to Eurocode 4 recommendations. Costruzioni Metalliche, No. 5, 273-287

Cosenza, E., Fabbrocino, G., Manfredi, G. (1997a): Capacita' rotazionale di travi composite acciaio-calcestruzzo soggette a momento negativo. In XVI Congresso CTA, Ancona, 2-5 October 1997, Stato della Ricerca, 196-207

Cosenza, E., Fabbrocino, G., Manfredi, G. (1997b): The influence of rebar ductility in the rotational capacity of composite beams. In Stability and Ductility of Steel Structures, SDSS 97, (ed. T. Usami), Nagoya, 29-31 July 1997, 879-908

Daali, M.L. (1995): Damage assessment in steel structures. In 7th Canadian Conference on Earthquake Engineering, Montreal, 517-522

Daali, M.L., Korol, R.M. (1994): Local buckling rules for rotation capacity. Engineering Journal, AISC, Second Quarter, 41-47

Daali, M.L., Korol, R.M. (1995): Prediction of local buckling and rotation capacity at maximum moment. Journal of Constructional Steel Research, Vol. 32, 1-13

Dekker, N.W. (1998): Interactive buckling model for the prediction of the rotation capacity of steel beams. In 2nd World Conference on Steel in Construction. San Sebastian, 11-13 May, 1998, CD-ROM 298

Dekker, N.W., Kemp, A.R., Trinchero, P. (1995): Factors influencing the strength of continuous composite beams in negative bending. Journal of Constructional Steel Research, Vol. 34, 161-185

De Martino, A., Ghersi, A., Mazzolani, F.M., De Martino, F.P. (1989): Il comportamento flessionale di profili sottili sagomati a freddo: impostazione della ricerca. Acciaio, No 9, 415-422. Idem in XII Congresso CTA, Capri, 22-25 October 1989, 535-548

De Martino, A., Ghersi, A., Mazzolani, F.M. (1990): Bending behaviour of double-C thin walled beams. In Xth International Specialty Conference on Cold-Formed Steel Structures, St. Louis, Missouri, October 1990

De Martino, A., De Martino, F.P., Ghersi, A., Mazzolani, F.M. (1990): Il comportamento flessionale di profili sottili sagomati a freddo: indagine sperimentale. Acciaio, No 12, 577-588

De Martino, A., De Martino, F.P., Ghersi, A., Mazzolani, F.M. (1991): Bending behaviour of double-C thin-gauge beams: Experimental evidence versus codification. In 4th International Colloquium on Structural Stability, Istanbul, September 1991

Doneux, C. (1998): Ductility of composite sections under positive moments. Report

Doneux, C., Parung, H. (1998): A study on composite beam-column sub-assemblages. In 11th European Conference on Earthquake Engineering, Paris, 6-11 September 1998, CD-ROM 11

Earls, C.J. (1999a): On the inelastic failure of high strength steel I-shaped beams. Journal of Constructional Steel Research, Vol. 49, 1-24

Earls, C.J. (1999b): Factors influencing ductility in high performance steel I-shaped beams. In Stability and Ductility of Steel Structures, SDSS 99, (eds. D. Dubina and M. Ivanyi), Timisoara, 9-11 September 1999, Elsevier, Amsterdam, 269-278

Earls, C.J. (2000a): Geometric factors influencing structural ductility in compact I-shaped beams. Journal of Structural Engineering, Vol. 126, No. 7, 780-789

Earls, C.J. (2000b): Influence of material effects on structural ductility of compact I-shaped beams. Journal of Structural Engineering, Vol. 126, No. 11, 1268-1278

Earls, C.J., Galambos, T.V. (1998): Inelastic failure of high strength steel wide flange beams under moment gradient and constant moment loading. SSRC Annual Technical Session, Atlanta, 21-23 September 1998, 131-153

ECCS-TWG 1.3 (1985): Recommended testing procedure for assessing the behaviour of structural elements under cyclic loads. Doc. 45/86

Espiga, F. (1997a): Safety evaluation for rotation capacities based on plastic hinge stability considerations. Labein Report 1997, Bilbao

Espiga, F. (1997b): Moment gradient and LTB restraints influence on available rotation. Labein Report 1997, Bilbao

Espiga, F. (1997c): Numerical simulation of RWTH tests. Labein Report 1997, Bilbao

Espiga, F. (1998): Ductility assessment methods and safety evaluation based on rotation capacity approach. In 2nd World Conference on Steel in Construction, San Sebastian, 11-13 May, CD-ROM 223

Espiga, F., Anza, J.J. (1996): Numerical investigation on global and local inelastic buckling modes of I-beams. In Coupled Instabilities in Metal Structures, CIMS 96, (eds. J. Rondal, D. Dubina, V. Gioncu), Liege, 5-7 April, 1996, Imperial College Press, London, 85-92

EUROCODE 3, EC 3 (1992): Design of steel structures, Part 1.1. General rules and rules for buildings. ENV 1933 1-1

EUROCODE 4, EC 4 (1992): Design of composite steel and concrete structures, Part 1.1. General rules and rules for buildings, ENV 1994-1-1

EUROCODE 8, EC 8 (1994): Design provisions for earthquake resistance of structures. Part 1.3. Specific rules for various materials and elements. ENV 1998 1-3

Fabbrocino, G. (1998): Modellazione e comportamento sperimentale di travi continue composite acciaio-calcestruzzo. Ph D Thesis, Napoli University Federico II

Fabbrocino, G., Manfredi, G., Cosenza, E. (1998): Rotation capacity of steel-concrete composite beams: Influence of reinforcing steel properties. Costruzioni Metalliche, No 6, 43-50

Faella, C., Consalvo, V., Nigro, E. (1993): Moment redistribution and required rotational capacity in continuous composite beams with delayed effects. In XIV Congresso CTA, Viareggio, 24-27 October 1993, Ricerca Teorica e Sperimentale, 330-341

Feldmann, M. (1994): Zur Rotationskapazitat von I-Profilen statisch and dynamisch belastung Trager. Ph Thesis, RWTH Universitat, Aachen

FINELG (1996): Nonlinear finite element program. User's manual. Version 6.2, University of Liege

Galambos, T.V (1999): Recent research and design development in steel and composite steel-concrete structures in SUA. Keynote Lecture. Journal of Constructional Steel Research, Vol. 55, No. 1-3, 289-303

Ge, H., Usami, T. (1995): Moment-thrust-curvature curves for locally buckled steel box columns, In Stability of Steel Structures (ed. M. Ivanyi), Budapest, 21-23 September 1995, Akademiai Kiado, Budapest, 43-50

Georgescu, D., Mazzolani, F.M. (1996): The influence of unsymmetrical post-elastic properties of the structural steel on the rotation capacity of the beam. In Coupled Instabilities in Metal Structures, CIMS 96 (eds. J. Rondal, D. Dubina, V. Gioncu), Liege, 5-7 September 1996, Imperial College Press, London, 292-300

Ger, J.F., Chen, F.Y. (1992): Collapse assessment of tall steel building damaged by 1985 Mexico earthquake. In 10th World Conference on Earthquake Engineering. Madrid, 19-24 July 1992, Balkema, Rotterdam, 51-56

Ger, J.F., Chen, F.Y. (1993): Postbuckling and hysteresis models of open-web girders. Journal of Structural Engineering, Vol 119, No 3, 831-851

Gioncu, V. (1995): Some considerations on structural member classes. In 9th International Conference on Metal Structures (ed. J. Murjewski), Krakow, 26-30 June 1995, Vol. 2, 311-320

Gioncu, V. (2000): Influence of strain-rate on the behaviour of steel members. In Behaviour of Steel Structures in Seismic Areas, STESSA 2000, (eds. F.M. Mazzolani and R. Tremblay), Montreal, 21-24 August 2000

Gioncu, V., Mateescu, G., Orasteanu, S. (1989): Theoretical and experimental research regarding the ductility of welded I-sections subjected to bending. In Stability of Metal Structures, SSRC Conference, Beijing, 10-12 October 1989, 289-298

Gioncu, V., Mateescu, G., Iuhas, A. (1995a): Contributions to the study of plastic rotational capacity of I-steel sections. In Behaviour of Steel Structures in Seismic Areas, STESSA 94 (eds. F.M Mazzolani and V. Gioncu), Timisoara, 26 June-1 July 1994, E&FN Spon, London, 169-181

Gioncu, V., Mateescu, G., Petcu, D., Anastasiadis, A. (2000): Prediction of available ductility by means of local plastic mechanism method: DUCTROT computer program. In Moment Resistant Connections of Steel Frames in Seismic Areas. Design and Reliability (ed. F.M. Mazzolani), E&FN Spon, London, 95-146

Gioncu, V., Mazzolani, F.M. (1995): Alternative methods for assessing local ductility. In Behaviour of Steel Structures in Seismic Areas, STESSA 94 (eds. F.M. Mazzolani and V. Gioncu), Timisoara, 26 June-1 July 1994, E&FN Spon, London, 182-190

Gioncu, V., Mazzolani, F.M. (1997): Simplified approach for evaluating the rotation capacity of double T steel sections. In Behaviour of Steel Structures in Seismic Areas, STESSA 97, (eds. F.M. Mazzolani and H. Akiyama), Kyoto, 3-8 August 1997, 10/17, Salerno, 303-310

Gioncu, V., Petcu, D. (1995): Numerical investigations on the rotation capacity of beams and beam-columns. In Stability of Steel Structures, (ed. M. Ivanyi), SSRC Colloquium, Budapest, 21-23 September 1995, Akamemiai Kiado, Budapest, Vol. 1, 163-174

Gioncu, V., Petcu, D. (1997): Available rotation capacity of wide-flange beams and beams-columns. Part 1. Theoretical approaches. Part 2. Experimental and numerical tests. Journal of Constructional Steel Research, Vol. 43, No. 1-3. 161-217, 219-244

Gioncu, V., Tirca, L., Petcu, D. (1995b): Interaction between in-plane and out-of-plane plastic buckling of wide-flange section member. In Coupled Instabilities in Metal Structures, CIMS 96 (eds. J. Rondal, D. Dubina, V. Gioncu), Liege, 5-7 September 1996, Imperial College Press, London, 273-282

Gioncu, V. Tirca, L, Petcu, D. (1996a): Interaction between in-plane and out-of-plane plastic buckling of wide-flange section members. In Coupled Instabilities in Metal Structures, CIMS 96, (eds. J. Rondal, D. Dubina, V. Gioncu) Liege, 5-7 September 1996, Imperial College Press, 273-281

Gioncu, V., Tirca, L., Petcu, D. (1996b): Rotation capacity of rectangular hollow section beams. In Seven International Symposium on Tubular Structures (eds. I. Farkas and K. Jarmai), 28-30 August 1996, Miskolc, Balkema, Rotterdam, 387-395

Goel, S.C. (1995): Truss moment frames for ductile behaviour. In Leigthweight Structures in Civil Engineering (ed. J.B. Obrebski), Warsaw, 25-29 September 1995, Vol. 1, 464-470

Goel, S.C., Itani, A.M. (1994a): Seismic behaviour of open-web truss-moment frames. Journal of Structural Engineering, Vol. 120, No. 6, 1763-1780

Goel, S.C., Itani, A.M. (1994b): Seismic-resistant special truss-moment frames. Journal of Structural Engineering, Vol 120, No 6, 1781-1797

Green, P.S., Ricles, J.M., Sause, R. (1997): Response of high performance steel flexural menbers to inelastic cyclic loading. In Behaviour of Steel Structures in Seismic Areas, STESSA 97, (eds. F.M. Mazzolani and H. Akiyama), Kyoto, 3-8 August 1997, 10/17 Salerno, 160-167

Greschik, G., White, D.W., McGuire, W., Abel, J.F. (1990): Toward the prediction of flexural ductility of wide-flange beams for seismic design. In 4th US National Conference on Earthquake Engineering, Palm Springs, 20-24 May 1990, 107-113

Grzebieta, R., Zhao, X.L., Purza, F. (1997): Multiple low cycle fatigue of SHS tubes subjected to gross pure bending deformation. In Stability and Ductility of Steel Structures, SDSS 97 (ed. T.Usami), Nagoya, 29-31 July 1997, 847-854

Guruparan, N.I., Walpole, W.R. (1990): The effect of lateral slenderness on the cyclic strength of steel beams. In 4th US National Conference on Earthquake Engineering, Palm Springs, 20-24 May 1990, Vol. 2, 585-594

Haaijer, G. (1957): Plate buckling in the strain-hardening range. Journal of the Engineering Mechanics Division, Vol. 84, EM 2, 1212/1-1212/47

Haaijer, G., Thurlimann, B. (1958): On inelastic buckling in steel. Journal of Engineering Mechanics Division, Vol. 83, EM2

Hoglund, T., Nylander, H. (1970): Maximiforhallande B/t for tryckt flans hos valsad I-balk vid dimensionering med granslastruetod. Technical University Stockholm, Report 83/70

Huang, P.Ch., Deierlein, G.G. (1996): Investigation of beam rotation capacity using finite element analysis. In 5th SSRC International Conference on Stability of Metal Structures. Future Direction in Stability Research and Design. Chicago, 15-18 April 1996, 383-392

Ivanyi, M. (1979a): Yield mechanism curves for local buckling of axially compressed members. Periodica Polytechnica, Civil Engineering, Vol. 23, No. 3/4, 203-216

Ivanyi, M. (1979b): Moment rotation characteristics of locally buckling beams. Periodica Polytechnica, Civil Engineering, Vol. 23, No. 3/4, 217-230

Ivanyi, M. (1985): The model of interactive plastic hinge. Periodica Polytechnica, Vol. 29, Nos 3/4, 121-146

Ivanyi, M. (1993): Interactive plastic hinge based method for analysis of steel frames. Proceedings of 1993 Annual SSRC Technical Session, Milwaukee, 5-6 April 1993, 153-177

Kato, B. (1965): Buckling strength of plates in the plastic range. Publication of International Association of Bridge and Structural Engineering, Vol. 25, 127-141

Kato, B. (1988): Rotation capacity of steel members subjected to local buckling. In 9th World Conference on Earthquake Engineering, Tokyo-Kyoto, 2-9 August 1988, Vol. IV, 115-120

Kato, B. (1989): Rotation capacity of H-section members as determined by local buckling. Journal of Constructional Steel Research, Vol. 13, 95-109

Kato, B. (1990): Deformation capacity of steel structures. Journal of Constructional Steel Research, Vol. 17, 33-94

Kato, B., Akiyama, H. (1981): Ductility of members and frames subjected to buckling. ASCE Convention, 11-15 May 1981, New York, Preprint 81-100

Kawaguchi, J., Morino, S., Machida, T. (1997): Energy dissipation capacity of

CFT beam column failing in local buckling. In Behaviour of Steel Structurs in Seismic Areas (eds. F.M. Mazzolani and H.Akiyama), Kyoto, 3-8 August 1997, 10/17 Salerno, 311-318

Kecman, D. (1983): Bending collapse of rectangular and square section tubes. International Journal of Mechanical Science, Vol. 25, No. 9-10, 623-636

Kemp, A.R. (1985): Interaction of plastic local and lateral buckling. Journal of Structural Engineering, Vol. 111, ST 10, 2181-2196

Kemp, A.R. (1986): Factors affecting the rotation capacity of plastically designed members. The Structural Engineer, Vol. 64B, No. 2, 28-35

Kemp, A.R. (1996): Inelastic local and lateral buckling in design codes. Journal of Structural Engineering, Vol. 122, No. 4, 374-382

Kemp, A.R. (1999): A limit states criterion for ductility of class 1 and 2 composite and steel beams. In Stability and Ductility of Steel Structures, SDSS 99 (eds. D. Dubina and M. Ivanyi), Timisoara, 9-11 September 1999, Elsevier, Amserdam, 291-298

Kemp, A.R., Dekker, N.W. (1991): Available rotation capacity in steel and composite beams. The Structural Engineer, Vol. 69, No. 5, 88-97

Kemp, A.R., Dekker, N.W., Trinchero, P. (1995): Differences in inelastic properties of steel and composite beams. Journal of Constructional Steel Research, Vol. 34, 187-206

Kollar, L., Dulacska, E. (1984): Buckling of Shells for Engineers. Akademiai Kiado, Budapest

Korol, R.M., Hudoba, J. (1972): Plastic behaviour of hollow structural sections. Journal of Structural Division, Vol. 98, ST 5, 1007-1023

Kotelko, M. (1996a): Selected problems of collapse behaviour analysis of structural members built from strain hardening material. Bi-Centenar Conference on Thin-Walled Structures. Glasgow, 2-4 December, 1996

Kotelko, M. (1996b): Ultimate load and postfailure behaviour of box-section beams under pure bending. Engineering Transactions, Vol. 44, No. 2, 229-251

Kotelko, M., Kolakowski, Z. (1995): Postbuckling and collapse behaviour of thin-walled beam-columns. In Lightweight Structures in Civil Engineering (ed. J.B. Obrebski), Warsaw, 25-29 September 1995, Vol. 1, 384-387

Krawinkler, H., Zohrei, M. (1983): Cumulative damage in steel structures subjected to earthquake ground motions. Computers and Structures, Vol. 16, 531-541

Krawinkler, H., Gupta, A. (1998): Modelling issues in evaluating nonlinear response for steel moment. In 11th European Conference on Earthquake Engineering, Paris, 6-11 Seprember 1998, CD-ROM 274

Kubo, M., Kitahori, H. (1997): Buckling strength and rotation capacity of monosymmetric I-beams. In Stability and Ductility of Steel Structures, SDSS 97, (ed. T. Usami), Nagoya, 29-31 July, 1997, 523-530

Kuhlmann, U. (1986):Rotations kapazitat biegebeanspruchter I-Profile under Berucksichtigung des plastischen Beulens. Technical Report, Bochum University, Mitteilung Nr.86-5

Kuhlmann, U. (1989): Definition of flange slenderness limits on the basis of rotation capacity values. Journal of Constructional Steel Research, Vol. 14, 21-40

Kuhlmann, U. (1999): On the rotational capacity of structural connections. In Semirigidity in Connections of Structural Steelworks: Theory, Analysis and Design. CIMS course, Udine 20-24 September 1999

Kusawa, H., Fukumoto, Y. (1983): Cyclic behaviour of thin-walled box stub-columns and beams. In Stability of Metal Structures, SSRC Conference, Paris, 16-17 November 1983, 211-218

Kuwamura, H. (1988): Effect of yield ratio on the ductility of high-strength steels under seismic loading. Annual SSRC Technical Session Computer Technology to Structural Stability. Minneapolis, 26-27 April 1988, 201-210

Kuwamura, H., Akiyama, H. (1994): Brittle fracture under high stress. Journal of Constructional Steel Research, Vol. 29, 5-19

Lay, M.G. (1965): Flange local buckling in wide-flange shapes. Journal of the Structural Division, Vol. 91, ST 6, 95-116

Lay, M.G., Galambos, T.V. (1965): Inelastic steel beams under uniform moment. Journal of the Structural Division, Vol. 91, ST 6, 67-93

Lay, M. G., Galambos T.V. (1967): Inelastic beams under moment gradient. Journal of the Structural Division, Vol. 93, ST 1, 381-399

Lee, G.C., Lee, E.T. (1994): Local buckling of steel sections under cyclic loading. Journal of Constructional Steel Research, Vol. 29, 55-70

Lukey, A.F., Adams, P.F. (1969): Rotation capacity of beams under moment gradient. Journal of the Structural Division, Vol. 95, ST 6, 1173-1188

Mamaghani, I.H.P., Usami, T., Mizuno, E., Kajikawa, Y. (1997): Cyclic inelastic behaviour of compact steel box columns. In Stability and Ductility of Steel Structures, SDSS 97 (ed. T. Usami), Nagoya, 29-31 July 1997, 259-266

Mateescu, G., Gioncu, V.(2000): Member response to strong pulse seismic loading. In Behaviour of Steel structures in Seismic Areas, STESSA 2000 (eds. F.M. Mazzolani and R. Tremblay), Montreal, 21-24 August 2000, Balkema, Rotterdam, 55-62

Matsui, C., Mitani, I (1977): Inelastic behaviour of high strength steel frames subjected to constant vertical and alternating horizontal loads. In 6th World Conference on Earthquake Engineering, New Delhi, 1977, Vol. 3, 3169-3174

Mazzolani, F.M., Piluso, V. (1992): Evaluation of the rotation capacity of steel beams and beams-columns. In First State-of-the-Art Workshop COST C1, Strasburg

Mazzolani, F.M., Piluso, V. (1993): Member behavioural classes of steel beams and beam-columns. In XIV Congresso CTA, Viareggio, 24-27 October 1993, Ricerca Teorica e Sperimentale, 405-416

Mazzolani, F.M., Piluso, V. (1995): Design of Steel Structures in Seismic Zones. ECCS Manual, doc. No. 76

Mazzolani, F.M., Piluso, V. (1996): Theory and Design of Seismic Resistant Steel Frames. E&FN Spon, London

McDermott, J.F. (1969): Plastic bending of A514 steel beams. Journal of Structural Division, Vol 95, ST 6, 1173-1188

Migiakis, C.E., Elghazouli, A.Y. (1998): Slab effects in composite floor systems subjected to earthquake loading. In 3th national Conference on Steel Structures (eds. K.T. Thomopoulous, C.C. Baniolopoulos, A.V. Avdelas), Thessaloniki, 30-31 October 1998, 328-336

Mitani, I., Makino, M., Matsui, C. (1977): Influence of local buckling on cyclic behaviour of steel beam-column. In 6th World Conference on Eartquake Engineering, New Delhi, 1977, Vol 3, 3175-3180

Mitani, I., Makino, M. (1980): Post-local buckling behaviour and plastic rotation capacity of steel beam-columns. In 7th World Conference on Earthquake Engineering, Istanbul

Moen, L.A., De Matteis, G., Hopperstad, O.S., Langseth, M., Landolfo R., Mazzolani, F.M. (1999): Rotational capacity of aluminium beams under moment gradient. Numerical simulations. Journal of Structural Engineering, Vol. 125, No. 8, 921-929

Moller, M., Johansson, B., Collin, P. (1997): A new analytical model of inelastic local flange buckling. Journal of Constructional Steel Research, Vol. 43, 43-63

Morino, S., Kawaguchi, J., Fukao, H., Tawa, H. (1995): Analysis of elasto-plastic behaviour of beam-column failure in local buckling. In Structural Stability and Design (eds. S.Kitiporuchai, G. Hancock, and J. Bradford) Sydney, 30 October - 1 November 1995, Balkema, Rotterdam, 181-186

Murray, N. W. (1986): Stability and ductility of steel elements. Journal of Constructional Steel Research, Vol. 44, No. 1-2, 23-50

Nakai, H., Yoshikawa, O., Terada, H. (1986): An experimental study on ultimate strength of composite columns for compression or bending. Structural Engineering/Earthquake Engineering, Vol. 3, No. 2, 235-245

Nakamura, T. (1988): Strength and deformability of H-shaped steel beams and lateral requirements. Journal of Constructional Steel Research, Vol. 9, 217-228

Nakashima, M. (1992): Variation and prediction of deformation capacity of steel beam-columns. In 10th World Conference on Earthquake Engineering, Madrid, 19-24 July 1992, Balkema, Rotterdam, 4501-4507

Nakashima, M. (1994): Variation of ductility capacity of steel beam-column. Journal of Structural Engineering, Vol. 120, No. 7, 1941-1960

Nakashima, M. (1997): Uncertainties associated with ductility performance of steel building structures. In Seismic Design Methodologies for the Next Generation of Codes (eds P. Fajfar and H. Krawinkler), Bled, 24-27 June 1997, 111-118

Piluso, V. (1992): Il comportamento inelastico dei telai seismo-resistenti in acciaio. Ph Thesis, Universita di Napoli Federico II

Piluso, V. (1995): Post-local behaviour of rolled steel beams subjected to nonuniform bending. XV Congresso CTA, Riva del Garda, 15-18 October 1995, 542-562

Plumier, A. (1996): Problems and options in seismic design of composite frames. SPEC Internal Report

Plumier, A. (1999): Seismic resistant composite structures. In Seismic Resistant Steel Structures. Progress and Challenges (eds F.M. Mazzolani and V. Gioncu), CISM course, Udine, 18-22 October 1999, Springer, Wien, 289-347

Plumier, C., Doneux, C. (1998): Dynamic tests on the ductility of composite steel-concrete beams. In 11th European Conference on Earthquake Engineering, Paris, 6-11 September 1998, CD-ROM 205

Plumier, A., Doneux, C. (1999): Design of composite structures in seismic regions.

In Stability and Ductility of Steel Structures (eds D. Dubina and M. Ivanyi), Timisoara, 9-11 September 1999, Elsevier, Amsterdam, 27-34

Plumier, A., Doneux, C., Bouwkamp, J.G., Plumier, C. (1998): Slab design in connection zone of composite frames. In 11th European Coference on Earthquake Engineering, Paris, 6-11 September 1998, CD-ROM 30

PROFIL-FEM-3D (1991): Ein elektronischer simulator fur Bauteile aus Stahl. D. Bohman, Lehrstuhl fur Stahlbau. RWTH Aachen

Roik, K., Kuhlmann, U. (1987): Rechnerische Ermittlung der Rotationkapazitat biegebeanspruchter I-Profile. Stahlbau, No 11, 321-327

Rondal, J., Boeraeve, Ph., Sedlacek, G., Stranghoner, N., D'Hernoncourt, A. (1994): Rotation capacity of hollow beam sections. Final Report/CIDECT Research Project No. 2P

Sawyer, H.A. (1961): Post-elastic behaviour of wide-flange steel beams. Journal of the Structural Division, Vol. 87, ST 9, 43-71

Schilling, C.G. (1988): Moment rotation tests of steel bridge girders. Journal of Structural Engineering, Vol. 114, No. 1, 134-141

Sedlacek, G., Stutki, Ch., Bilt, St. (1987): A contribution of the background of the B/T ratios controlling the applicability of analysis models in Eurocode 3. In Stability of Plate and Shell Structures (eds P. Dubas and D. Vandepitte), Ghent, 6-8 April 1987, 33-40

Sedlacek, G., Dahl, W., Rondal, J., Boeraeve, Ph., Stranghoner, N., Kalinowski, B. (1995): Investigation of the rotation behaviour of hollow section beams. RWTH Aachen and MSM Liege Report

Spangemacher, R. (1991): Zum Rotationsnachweis von Stahlkonstruktion, die nach dem Traglastverfahren berechnet werden. Ph Thesis Aachen University

Spangemacher, R., Sedlacek, G. (1992a): On the development of a computer simulation for tests of steel structures. In Constructional Steel Design. World Developments (ed. P.S. Dowling), Acapulco, 6-9 December, 1992, Elsevier Applied Science, London, 593-611

Spangemacher, R., Sedlacek, G. (1992b): Zum Nachweis ausreichender Rotationsfahigheit von Fliessgelenhken bei der Anwendung des Fliessgelenkverfahrens. Stahlbau, Vol. 61, Heft 11, 329-339

Stranghoner, N. (1995): Undersuchungen zum Rotationsverhalten von Tragern aus Hohlprofilen. Ph Thesis, Aachen University

Stranghoner, N., Sedlacek, G. (1997): Zum Tragverhalten von kalt und warmgefertigten quadratischen Hohlprofiltragern. Stahlbau, Heft 4, 198-204

Sully, R.M., Hancock, G.J. (1996): Behaviour of cold-formed SHS beam-column. Journal of Structural Engineering, Vol. 122, No. 3, 326-336

Sully, R.M., Hancock, G.J. (1999): Stability of cold-formed tubular beam-columns. In Light-Weight Steel and Aluminium Structures (eds. P. Makelainen and P. Hassinen), Espoo, 20-23 June 1999, Elsevier, Amsterdam, 141-154

Suzuki, T., Ono, T. (1977): An experimental study on inelastic behaviour of steel members subjected to repeated loading. In 6th World Conference on Earthquake Engineering, New Delhi, 1977, Vol. 3, 3163-3168

Suzuki, T., Ogawa, T., Ikaraski, K. (1994): A study on local buckling behaviour of hybrid beams. Thin-Walled Structures, Vol. 19, 337-351

Suzuki, T., Ogawa, T., Ikarashi, K. (1996): A study on plastic deformation capacity of H-shaped hybrid beam-column. In Coupled Instabilities in Metal Structures, CIMS 96, (eds. J. Rondal, D. Dubina, V. Gioncu), Liege, 5-7 September 1996, Imperial College Press, London, 301-308

Suzuki, T., Ogawa, T., Ikarashi, K. (1997): Evaluation of the plastic deformation capacity modified by the effect of ductile fracture. In Behaviour of Steel Structures in Seismic Areas, STESSA 97, (eds. F.M.Mazzolani and H. Akiyama), Kyoto, 3-8 August 1997, 10/17 Salerno, 326-333

Tehami, M. (1997): Local buckling in class 2 continuous composite beams. Journal of Constructional Steel Research, Vol. 43, No. 1-3, 141-159

Vayas, I. (1997): Stability and ductility of steel elements. Journal of Constructional Steel Research, Vol. 44, No. 1-2, 23-50

Vayas, I., Psycharis, I. (1990): Behaviour of thin-walled steel elemnts under monotonic and cyclic loading. In Structural Dynamic (ed. W.B. Kratzig), Balkema, Rotterdam, 579-583

Vayas, I., Psycharis, I. (1992): Dehnungsorientierte Formulierung der Methode der wirksamen Breite. Stahlbau, Vol. 61, 275-283

Vayas, I., Psycharis, I. (1993): Ein dehnungsorientiertes Verfahren zur Ermittlung der Duktilitat von Tragern aus I-Profilen. Stahlbau, Vol. 63, 333-341

Vayas, I., Psycharis, I. (1995): Local cyclic behaviour of steel members. In Behaviour of Steel Structures in Seismic Areas, STESSA 94, (eds. F.M. Mazzolani and V. Gioncu), Timisoara, 26 June-1 July, 1994, E&FN Spon, London, 231-241

Vayas, I., Syrmakezis, C., Sophocleous, A. (1995): A method for evaluation of the behaviour factor for steel regular and irregular buildings. In Behaviour of Steel Structures in Seismic Areas, STESSA 94, (eds. F.M. Mazzolani and V. Gioncu), Timisoara, 26 June-1 July, 1994, E&FN Spon, London, 344-354

Wargsjo, A. (1991): Plastisk Rotationskapacitet hos Svetsade Stalbalker. Ph Thesis, Lulea University of Technology

Watanabe, E., Sugiura, K., Harimoto, S., Hasegawa, T. (1992): Ductile cross sections for bridge piers. In Constructional Steel Design. World Development (eds. P.J. Dowling et al), Acapulco, 6-9 December 1992, Elsevier Applied Science, London, 707-710

Watanabe, E., Emi, S., Sugiura, K., Takao., M., Higuchi, Y.(1991): Ductility evaluation of thin-tubular beam-column under cyclic loadings. In Steel Structures. Recent Research and Development (eds. S.L. Lee and N.E. Shanmugam), Singapore, 22-24 May 1991, Elsevier, London, 847-856

Weston, G., Nethercot, D.A. (1987): Continuous composite bridge beams. Stability of the steel compression flange in hogging bending. In Stability of Plate and Shell Structures (eds. P. Dubas and D. Vandepitte), Ghent, 6-8 April 1987, 47-52

White, D.W., Barth, K.E. (1998): Strength and ductility of compact flange I-girders in negative bending. Journal of Constructional Steel Research, Vol. 45, No. 3, 241-280

Wilkinson, T., Hancock, G.J. (1998): Tests to examine compact web slenderness of cold-formed RHS. Journal of Structural Engineering, Vol. 124, No. 10, 1166-1174

Yamada, M., Kawamura, H., Tani, A., Iwanaga, K., Sakai, Y., Nishikawa, H., Masui, A., Yamada, M. (1988): Fracture ductility of structural elements and of structures. In 9th World Conference on Earthquake Engineering, Tokyo-Kyoto, 2-9 August 1988, Vol. 4, 219-241

Yamada, M. (1992): Low fatigue fracture limits of structural materials and structural elements. In Testing of Metals for Structures (ed. F.M.Mazzolani), Napoli, 29-31 May 1990, E&FN Spon, London, 184-192

Yoshizumi, T., Matsui, C. (1988): Cyclic behaviour of steel frames influenced by local and lateral buckling. In 9th World Conference on Earthquake Engineering, Tokyo-Kyoto, 2-9 August 1988, Vol. IV, 225-230

Yura, A. Galambos, T.G., Ravindra, M.K. (1978): The bending resistance of steel beams. Journal of the Structural Division, Vol. 104, ST 9, 1355-1370

Zambrano, A., Mulas, M.G., Castiglioni, C. (1999): Steel members under cyclic loads: A constitutive law accounting for damage. In XVII Congresso CTA, Napoli, 3-7 October 1999, 323-334

8

Comprehensive methodology for ductility design

8.1 Basis of a Comprehensive Methodolog

8.1.1. Limits of the Current Seismic Design Practice

Basically the current seismic design practice agrees with the principl multi-level seismic design. According to this philosophy the building n resist minor earthquakes without significant damage, moderate earthqu with repairable damage and major earthquakes without collapse. Bu practice, the current seismic codes are involved with the minimal loss of only. The usual procedure for checking the fulfillment of the requireme that a seismic resistant structure has to posses, is based on elastic anal under equivalent static lateral forces. These forces are defined by redu the base shear forces by means of the so-called q factor, which takes i account the structural ductility and the seismic energy dissipation. safety against the ultimate state is considered automatically verified, p vided that the detailing rules and procedures, suggested by seismic co are satisfied.

This procedure leads to a great simplification in the design proc but its limits are evident: no direct checking of limiting states. A typ seismic response of a structure designed by this procedure is shown in I 8.1 (Leelataviwat et al, 1999). Point 1 corresponds to the response of equivalent elastic system. Point 2 represents the strength design level ultimate states and point 3 corresponds to an allowable stress design le of serviceability state, determined as a fraction of the strength design fo The curve 024 represents the structure design behaviour. But, due to fact that structures undergo significant inelastic deformations during sev ground motions, the actual responses differ from the design behaviour in unpredictable manner. Generally, the ultimate response can be evalua as follows:

-the structure is capable of developing a ductile mechanism (curve (but the degree of overstrength is not sufficient and excessive intersto

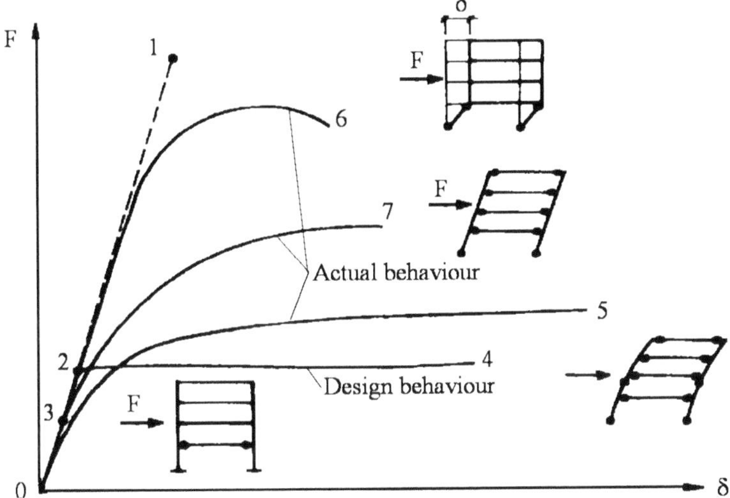

Figure 8.1: Typical seismic response of moment resisting frames (after Leelataviawat et al, 1999)

drifts occur. For serviceability limit states no control of structure damage is required (the control of nonstructural damage is performed only) so some plastic hinges in undesirable positions can occur;

-the structure can develop a high degree of overstrength but with a limited ductility due to unexpected plastic mechanisms (curve 06);

-the structure is capable of developing an adequate strength and ductile mechanism (curve 07). The result is a desirable structure.

Therefore, using the current seismic design practice, the results do not always lead to the foreseen failure mode and to the expected ductility. At the same time, for minor but frequent earthquakes, the control of structural damage is not performed with a rational methodology. The weakest point of the current seismic design practice is the lack of practical rules to obtain a structure with a predictable response for all limit states.

8.1.2. Design Methodology Based on Multi-Level Criteria

The main purpose of this Section is to establish a comprehensive methodology for ductility design. This is not possible without developing a general coherent methodology for the seismic design of structures, in which ductility plays an important role, but not the only one.

The performance based seismic design of earthquake-resistant structures has been identified as the best methodology for the next generation of codes (Fajfar and Krawinkler, 1997). This methodology basically identifies some performance levels which should be achieved for increasing intensity levels of earthquake actions. As it has been shown in Section 3.1.5, a coherent strategy must consider three verification levels (Gioncu, 2000):

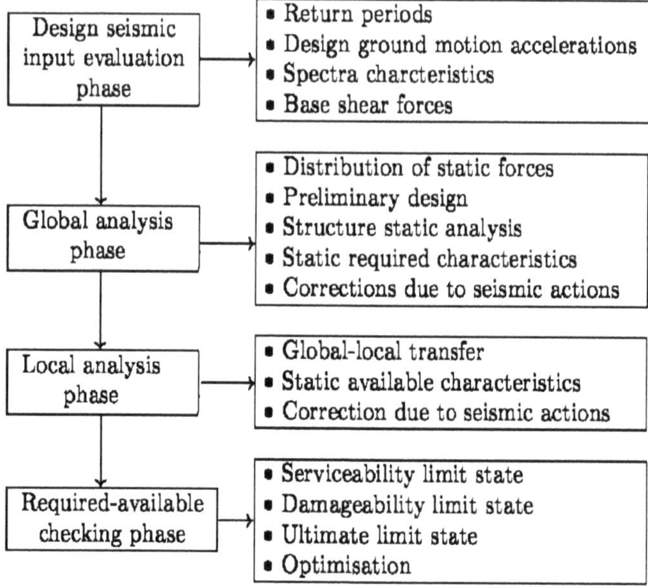

Figure 8.2: Principal phases in a comprehensive methodology

-*serviceability limit states*, for minor but frequent earthquakes, during which it is required that structural and nonstructural element damage is prevented;

-*damageability limit states*, for occasional moderate to strong seismic events, during which the reduced damage occurring in structures can be repaired without great technical difficulties. Important damage occurs in nonstructural elements which, in most cases, must be replaced by new ones;

-*ultimate limit states (survivability limit states)* for rare but very strong earthquakes, during which very important structural damage occurs, but the partial or total collapse of structure has to be prevented.

This coherent methodology consists of four phases presented in Fig. 8.2: seismic action evaluation, global analysis, local analysis and required-available checking phases.

(i) *Design seismic input evaluation phase* consists in the following steps:

-Assessment of *return period*, p_r, for the three limit states. This problem is discussed in detail in Section 3.1.6. For damageability and ultimate limit states there are no disputable values, the return period of 475 and 970 years, respectively, being accepted by the specialists. Contrary to this, for serviceability limit states, there are different proposals, comprising between 10 to 75 years. This dispute is caused by the fact that two levels, fully operational and functional, considered in the Vision 2000 Performance Based Seismic Design are included in one only, the serviceability limit state. A return period of 20 years seems to be a reasonable value.

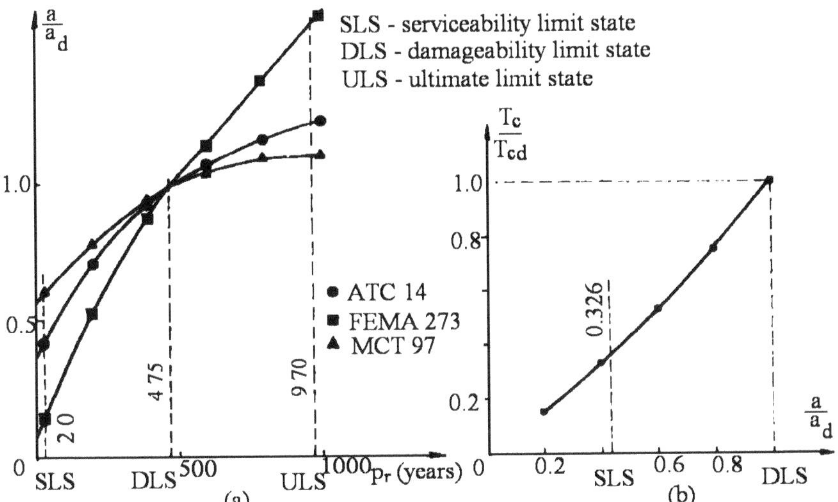

Figure 8.3: Characteristics of ground motions: (a) Acceleration-return period relationship (after Lee et al, 2000); (b) Corner period-acceleration relationship

-Choice of the ***ground motion acceleration*** for the three limit states can be determined as a function of the above accepted return periods. If the acceleration for damageability limit states, a_d, is considered as a basic value for ground motion acceleration in the structure site, for the other return periods the corresponding accelerations values can be determined from the equation (Lee et al, 2000) (Fig. 8.3a):

$$\frac{a}{a_d} = \left(\frac{p_r}{p_{rd}}\right)^{0.28} \tag{8.1}$$

based on ATC 14 (1987) proposals. So, for serviceability and ultimate limit states results:

$$a_s = 0.412 a_d \tag{8.2a}$$

$$a_u = 1.22 a_d \tag{8.2b}$$

There are also two proposals for the equation (8.1), given by FEMA (1997) and MTC (1997) (Lee et al, 2000), which can be used if, in some cases, this equation gives unsatisfactory results.

-Set up of ***spectra characteristics***, normalized to the acceleration due to gravity, g, for the three limit states (Fig. 8.4). The main characteristics of spectrum are the corner periods T_B, T_C, and the amplification factor, β. For the steel structures the corner period T_B is not significant. The corner period T_C depends on the level of the earthquake magnitude. A simplified relationship is proposed by Alavi and Krawinkler (2000):

$$\log T_c = -1.76 + 0.31M \tag{8.3}$$

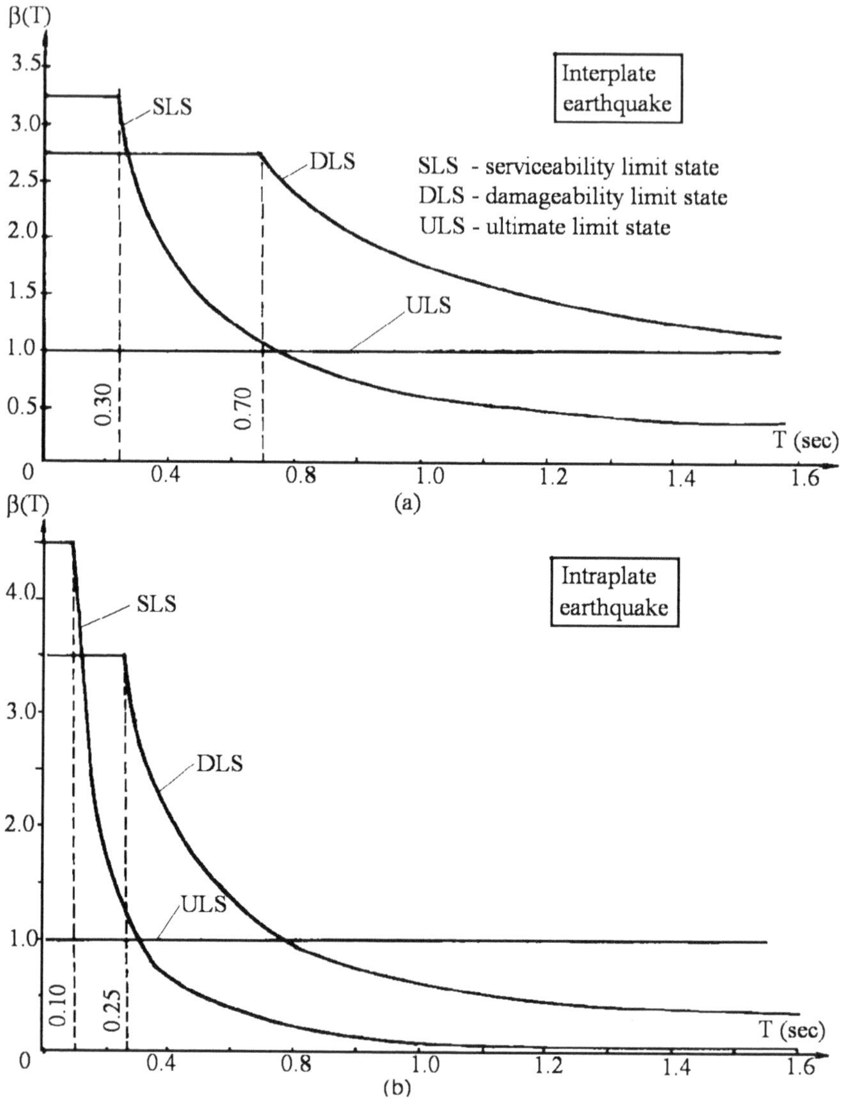

Figure 8.4: Design spectra: (a) Interplate earthquakes; (b) Intraplate earthquakes ($S = 1$, $Z = 1$)

If the corner period of damageability limit state is accepted as a basic value, after some algebra, using eqns (3.9) and (8.2), the corner periods for other acceleration levels are given by (Fig. 8.3b):

$$\frac{T_c}{T_{cd}} = \left(\frac{a}{a_d}\right)^{1.265} \tag{8.4}$$

Using the value obtained from equation (8.2a), it results:

$$T_{cs} = 0.326 T_d \tag{8.5}$$

One must mention that for ultimate limit states, when the structure works as a plastic mechanism, the notion of natural period is lost.

The laws for spectra are different for the three limit states and must be established as a function of earthquake type, propagation effects and local conditions (see Section 3.2). For instance, a typical variation for amplification factor β can result from the following equations:

-for interplate earthquakes (variation suggested by EC 8 provisions for strong and moderate earthquakes) (Fig. 8.4a):

- serviceability limit states (SLS):

$$\beta_s(T) = \frac{0.8\eta}{T^{1.1}}; \quad \beta \leq 3.25 \tag{8.5a}$$

- damageability limit states (DLS):

$$\beta_d(T) = \frac{2.0 S\eta}{T^{0.67}}; \quad \beta \leq 2.5; \tag{8.5b}$$

- ultimate limit states (ULS):

$$\beta_u(T) = 1.0 \tag{8.5c}$$

-for intraplate earthquakes (variation for DLS sugested by Lam et al, 1996) (Fig. 8.4b):

- serviceability limit states (SLS):

$$\beta_s(T) = \frac{0.18\eta}{T^{1.4}}; \quad \beta \leq 4.5 \tag{8.6a}$$

- damageability limit states (DLS):

$$\beta_d(T) = \frac{0.75 S\eta}{T^{1.1}}; \quad \beta \leq 3.5; \tag{8.6b}$$

- ultimate limit states (ULS):

$$\beta_u(T) = 1.0 \tag{8.6c}$$

In these equations T is the structure vibration period, S, coefficient considering the soil influence, η, damper correction factor. These typical spectra are presented in Fig. 8.4 as examples only, considering the main characteristics of earthquakes. If more adequate data are available, they can be using for setting up other spectra. The spectrum for serviceability limit states is characterized by high amplification for short periods, but with a steep reduction in amplification for vibration periods exceeding the corner period. For damageability limit states the amplification is not so high and the reduction more gentle. For ultimate limit states no amplification occurs, because the structure is a plastic mechanism which works as the ground motions impose, without any interaction.

The site influence is characterized by the coefficient S which is very reduced for serviceability limit states (therefore $S = 1$) and very important for damageability limit states. The values proposed in EC 8 may be considered (S =1.0 to 1.4, function of subsoil classes, see Section 3.2.4). Contrary to this, is the damper correction factor:

$$\eta = \left(\frac{10}{5+\xi}\right)^{1/2} \tag{8.7}$$

where ξ is the value of the viscous damping ratio of the structure, expressed in percent, which is more important for serviceability limit states, due to the fact that the viscous damping ratio is more reduced ($\xi_{max} = 2\%$) than for damageability limit states.

-Determination of ***base shear forces***, F_B, from spectra for different limit states:

• for serviceability limit states (SLS):

$$F_{bs} = \frac{a_s}{g}\beta_s(T)W \tag{8.8a}$$

where a_s is the design ground acceleration for SLS, β_s is determined from the corresponding spectrum for elastic structure vibration period, W, the vertical loads (including the dead and live loads);

• for damageability limit states (DLS):

$$F_{bd} = \frac{a_d}{g}\beta_d(T_d)W/q \tag{8.8b}$$

where a_d is the design acceleration for DLS, β_d is determined from the corresponding spectrum, but for an increased period of vibration, due to the occurrance of some plastic hinges, without forming a plastic mechanism (Priestly, 1997):

$$T_d = \mu_\delta^{1/2}T \simeq q^{1/2}T \tag{8.9}$$

where μ_δ is the displacement ductility, practically equal to q factor. The increase of vibration period as a function of structure degradation is presented in Fig. 8.5. One can see the evolution of plastic hinge patterns for different q values (Plumier and Boushaba, 1988): even for $q = 6$ the structure is not a plastic mechanism, showing that the use of a high q factor value does not assure the formation of an adequate plastic mechanism;

• for ultimate limit states (ULS):

$$F_{bu} = \frac{a_u}{g}W \tag{8.8c}$$

where a_u is the ultimate acceleration determined from the equation (8.2b). One can see that the response of the plastic mechanism is not sensible to the elastic vibration periods (Leelataviwat et al, 1999).

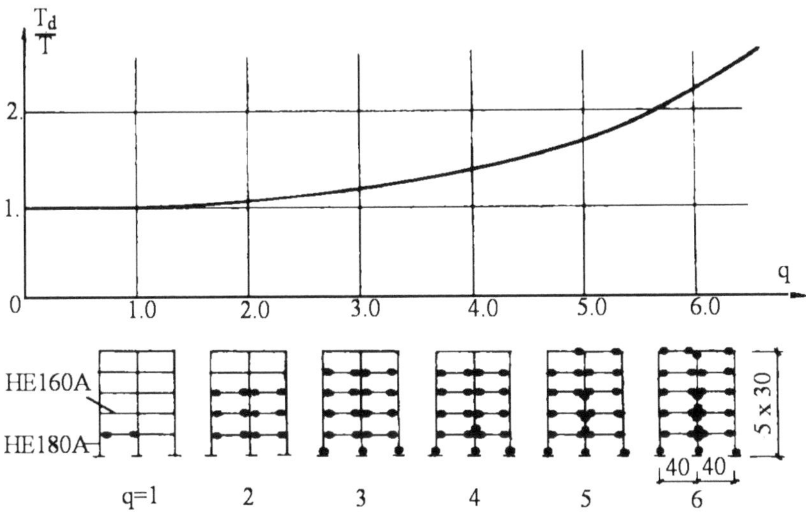

Figure 8.5: Influence of degradation on the structure period of vibration

The ratio between the shear forces corresponding to serviceability and damageability limit states results from the equations (8.8a,b):

$$\frac{F_{bs}}{F_{bd}} = \frac{a_s}{a_d}\frac{\beta_s(T)}{\beta_d(T_d)/q} \tag{8.10}$$

The ratio (8.10) is plotted in Figs. 8.6a,b. One can see that it depends on earthquake type, structure elastic period, reduction factor and soil conditions. For q factors greater than 4.5, the control of design is given by the serviceability limit state, showing that the benefit of large behaviour factors may sometimes be unprofitable. Fig. 8.6c shows that the use of the EC 8 provisions leads to an enlarging of the serviceability field.

(ii) *Global analysis phase* consists of the following steps:

-Evaluation of *shear force distribution* on the structure height, which depends on the evolution of plastic hinge patterns. For serviceability limit states, the lateral load distribution can be selected considering the elastic response, when the influence of higher vibration modes is very important. Therefore, it is reasonable to consider an inverse parabolic distribution (Fig. 8.7a). For damageability limit states, due to the increasing of structure vibration mode and the decreasing of superior mode influence, a parabolic shape distribution may be considered (Fig. 8.7b). For the ultimate limit state, when the structure is transformed into a plastic mechanism, the notion of vibration mode makes no sense at all. The load distribution can be evaluaued as an inverted triangular shape, due to the increase of accelerations from base to structure top (Fig. 8.7c).

-*Preliminary design* which is one of the most important issues during the seismic design process. A bad choice of structure configuration can easily result in a poor final design which is inefficiently adapted to comply with the

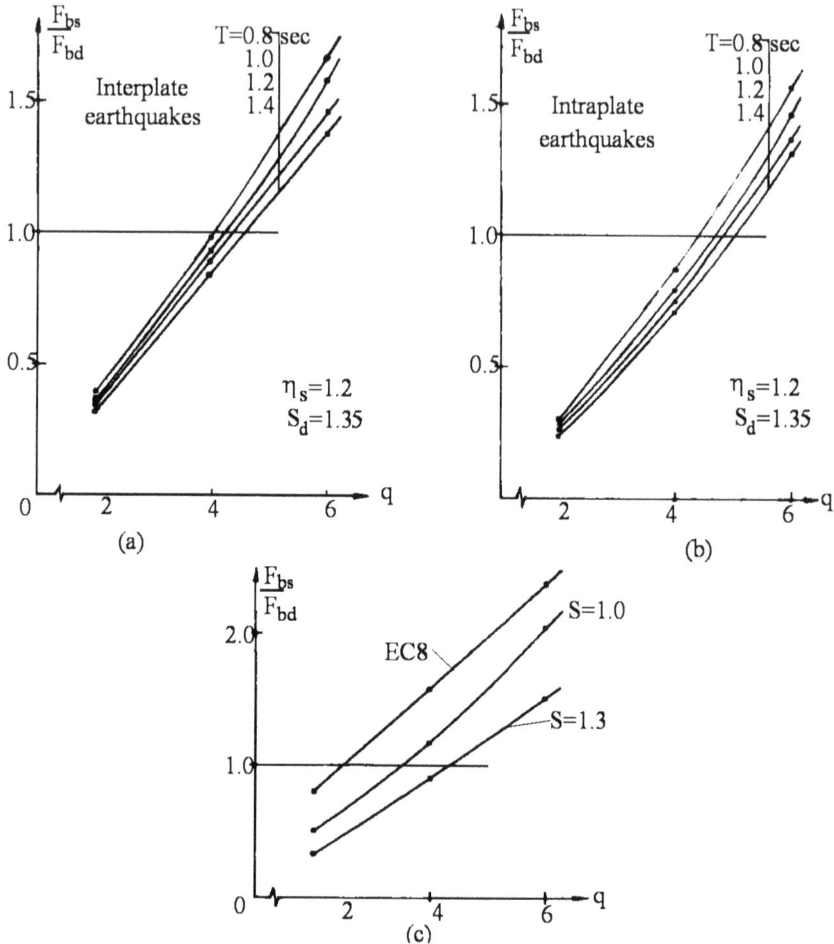

Figure 8.6: Shear forces of damageability and serviceability limit state ratios: (a) Interplate earthquakes; (b) Intraplate earthquakes; (c) Effect of soil conditions and comparison with EC8 provisions

three limit states (Teran-Guilmore, 1998). Consequently, the determination of a member strength hierarchy, failure mechanism and structure strength becomes the primary step in the preliminary design process. In particular, with reference to seismic resistant steel frames, the following types can be recognised (Mazzolani and Piluso, 1997a):

- ordinary moment resisting frame, designed without the limitations of interstorey drifts and verification of plastic mechanism type;
- special moment resisting frame, designed without verification of plastic mechanism type;
- global moment resisting frame, designed to be able to perform a global plastic mechanism.

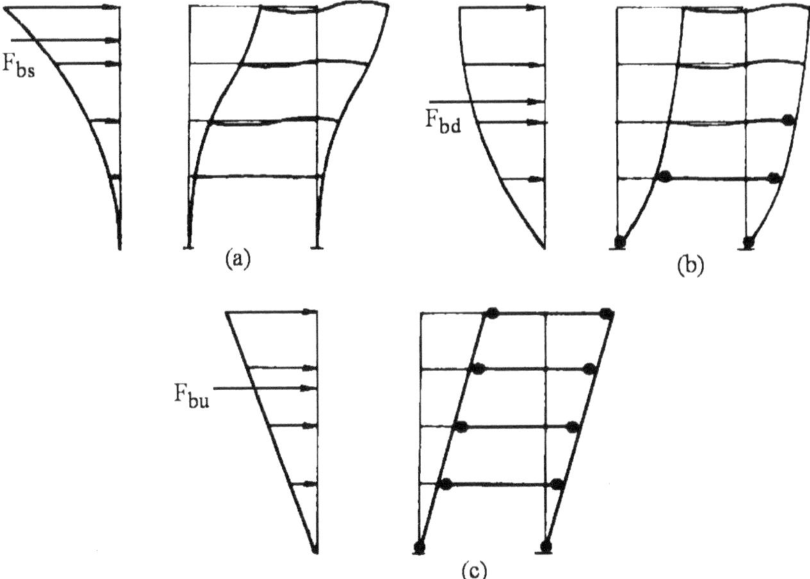

Figure 8.7: Shear force distributions on the structure height: (a) Serviceability limit state; (b) Damageability limit state; (c) Ultimate limit state

When achieving a good ductility is the major objective of structure design (areas with high seismicity), the global moment resisting frame type is the only solution which can offer a controlled behaviour. In this case the primary aim of the preliminary design is to choose the frame sizes so as to eliminate the possibility of formation of plastic hinges in the columns, by obtaining the so-named strong column-weak beam (SC-WB) frame type. Two main problems must be solved to assure the formation of an adequate global mechanism (Fig. 8.8a): determination of actual fully plastic moment capacity of beams and design of columns in order to be stronger than the joined beams. The beams must be sized for vertical and horizontal loads. The dimensions of cross-sections result from the nominal plastic moment:

$$M_{pb} = Z_p f_y = m_h M_w; \quad m_h = (V_i/V_{ns})^{1/2} \qquad (8.11a, b)$$

where M_w is the moment corresponding to vertical loads and m_h considers the influence of horizontal loads, V_i and V_{ns} being the storey shear forces corresponding to levels i and n_s (roof level) (Leelataviwat et al, 1999). The actual plastic moment may consider the difference between the actual yield stress, f_{ya}, and the nominal yield stress, f_y (see Section 4.3.3):

$$M'_{pb} = m_y M_{pb}; \quad m_y = \frac{f_{ya}}{f_y} \qquad (8.11c, d)$$

In order to obtain a strong column-weak beam frame there are three main procedures (Mazzolani, 2000): hierarchy criterion (EC 8), Mazzolani-Piluso

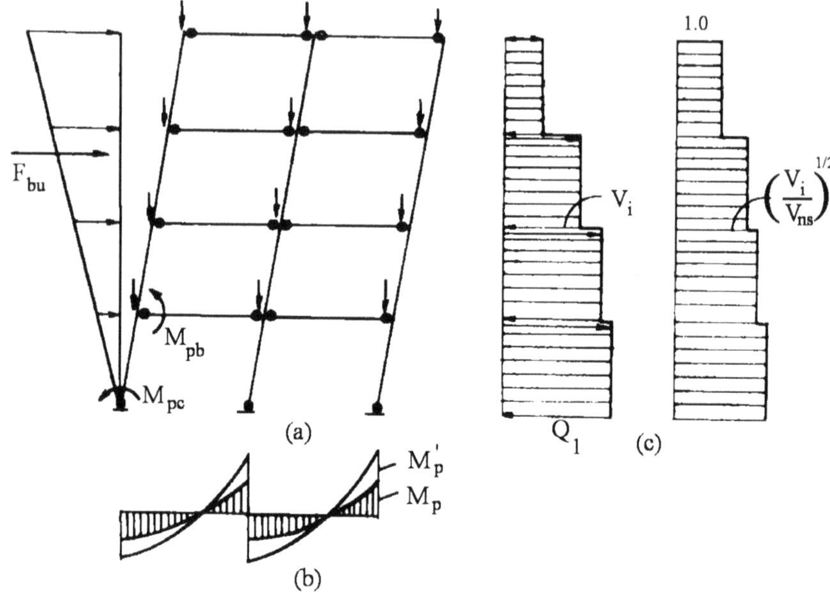

Figure 8.8: Characteristics of global mechanism: (a) Forces acting on structure; (b) Actual full plastic moments of beams; (c) Distribution of shear forces

(1995a,b, 1997a,b) global mechanism method and Ghersi-Neri (Ghersi et al, 1999) simplified method. The following characteristics of frame is adopted (Fig. 8.9):

-n_s, n_c, n_b, the number of storeys, columns and bays, respectively;

-k, i, j, the storey, column and bay index;

-$M_{pc,i1}$, $M_{pb,jk}$, the plastic moments of the i-th column of the first storey and the plastic moments of beam of the j-th bay and k-th storey, respectively.

By equating the energy dissipated by plastic hinges to the work of horizontal forces owed to a rigid displacement according to the global mechanism (Fig. 8.10a), the collapse multiplier α_c results (Ghersi et al, 1999):

$$\alpha_c = \frac{\sum_{i=1}^{n_c} M_{pc,i1} + \sum_{j=1}^{n_b}\sum_{k=1}^{n_s} 2M_{pb,jk}}{\sum_{k=1}^{n_s} F_k H_k} \tag{8.12a}$$

where F_k and H_k are the force and height of k storey, respectively. Knowing the multiplier α_c it is possible to evaluate the sum of bending moments $M_{c,ir}$ at the top or bottom of column at storey r, by imposing the rotation

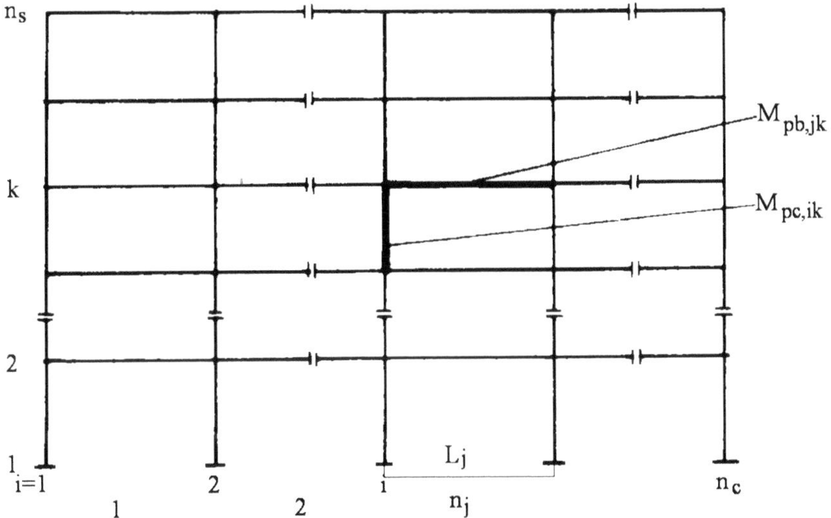

Figure 8.9: Characteristics of frames

equilibrium of the upper part of the frame (Fig. 8.10b):

$$\sum_{i=1}^{n_c} M_{c,ir} = \alpha_c \sum_{k=r}^{n_s} F_k(H_k - H) - \sum_{j=1}^{n_b} \sum_{k=r}^{n_s} 2M_{pb,jk} \tag{8.12b}$$

where $H = H_r$ for the top section and $H = H_{r-1}$ for the bottom section of the columns. Equilibrium conditions do not allow us to evaluate the bending moment at each column. Push-over analysis shows that when all columns at each storey have the same section, the bending moments in the ultimate limit state are quite similar for all columns. It is possible, therefore, to use an average value:

$$M_{ac,ir} = \frac{1}{n_c} \sum_{i=1}^{n_c} M_{c,ir} \tag{8.13a}$$

and the axial forces in columns are:

$$N_{rj} = N_W \pm \sum_{k=r}^{n_s} \frac{2M_{b,kj}}{L_j} \mp \sum_{k=r}^{n_s} \frac{2M_{k,kj+1}}{L_{j+1}} \tag{8.13b}$$

The variation of bending moments and axial forces along the frame height are presented in Fig. 8.10c. Using these diagrams one can size the columns in such a way as to remain in the elastic range with the exception of base moments at the first storey. The influence of the value of plastic moments at the base is studied by Ghersi et al (1999). When they are too small, the maximum moments will be attained at the middle storeys and for technological reasons require the use of sections larger than those strictly necessary in most columns. On the contrary, when they are too large, the

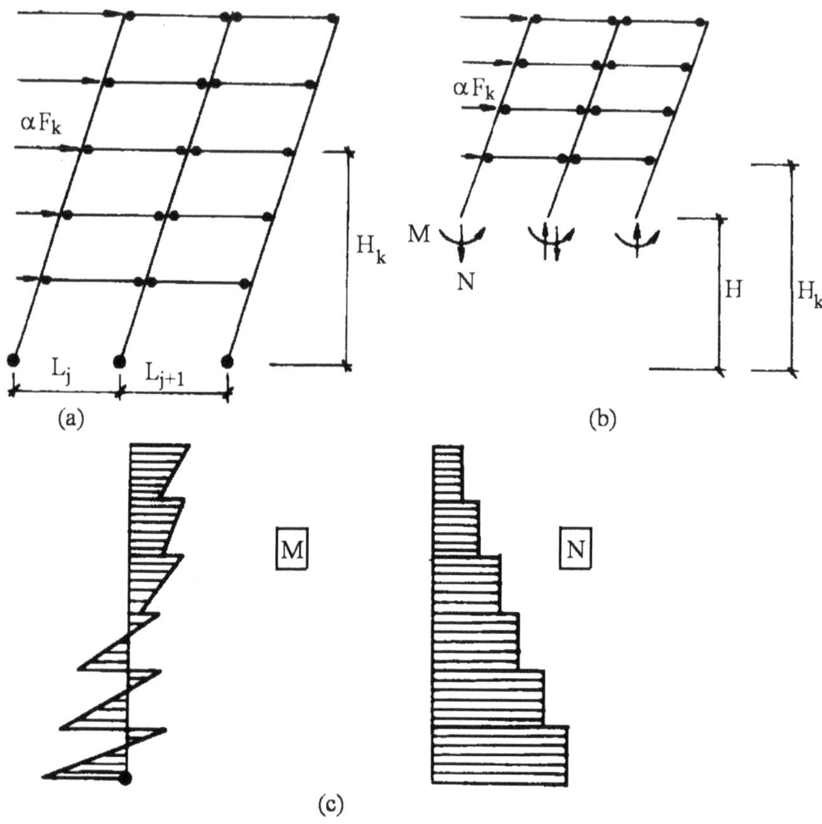

Figure 8.10: Global collapse mechanism: (a) Equilibrium of external and internal forces; (b) Equilibrium of upper part of the scheme; (c) Bending moments and axial forces distributions

columns will be severely stressed at the lower storeys. A better distribution is obtained by imposing that the maximum positive and negative moments along the frame height be coincident. After a set of numerical analysis, an approximate relationships is provided in order to obtain the optimal value for base plastic moment $M_{pc,i1}$:

$$\sum_{i=1}^{n_c} M_{pc,i1} = (0.7 + 0.15n_s) \sum_{j=1}^{n_b} 2M_{pb,j1} \tag{8.14}$$

where the right sum represents the sum of beam plastic moments of the first storey. In this design method all beams are assumed to reach their maximum overstrength due to the actual yield stress. This assumption may appear to be too conservative and, therefore, the use of a uniform reduced overstrength $m_y = 1.1$ for all beams along the entire structure height appears to be reasonable for calculating the design forces in columns.

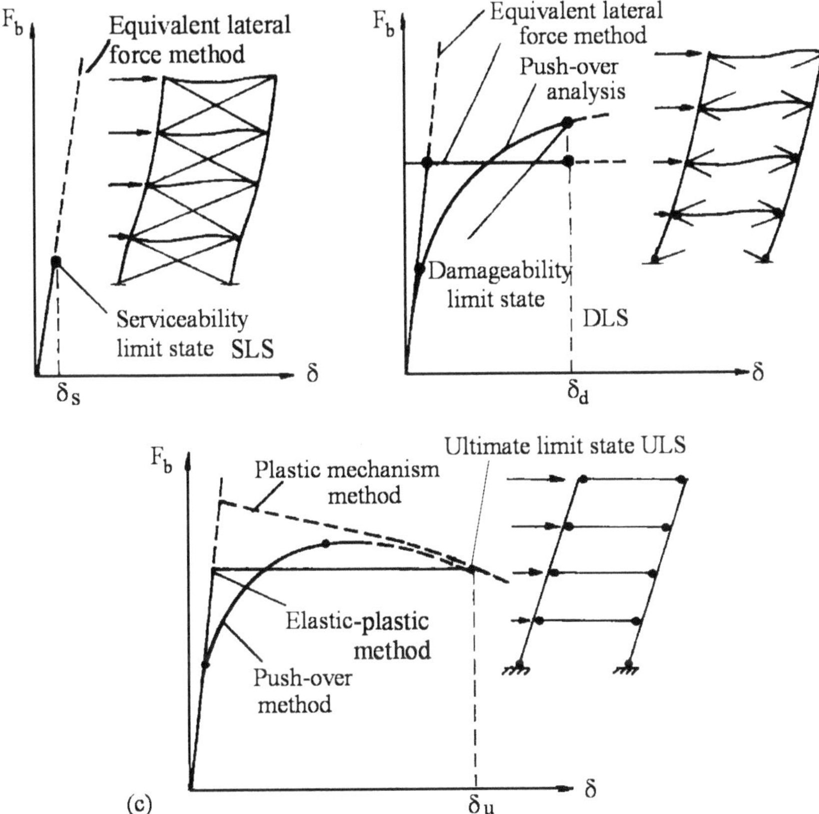

Figure 8.11: Static analysis of structure: (a) Serviceability limit state; (b) Damageability limit state; (c) Ultimate limit state

-Structure ***static analysis*** must be performed for all three limit states. The first analysis is for serviceability limit states. Because any damage in structure (no plastic deformations) and in nonstructural elements (no cracks in partition walls) is permitted, an elastic analysis of structure must be performed, by considering an interaction between structure-nonstructural elements (Fig. 8.11a). The second analysis refers to the damageability limit state, which is performed by using an elasto-plastic method. The usual procedure replaces the complex analysis with a more simple elastic one, the equivalent lateral force method (Fig. 8.11b), which takes into account the structural ductility and energy dissipation by means of the q factor. This simple design method, proposed by modern seismic codes for ordinary structures, does not always lead to frames failing in ductile mode. The number and location of the developed plastic hinges are out of the design control. For special structures or in the case of strong earthquakes, a push-over elasto-plastic method must be used for structure analysis. The effects of nonstructural elements must be neglected, given that they are totally or partially damaged. For the ultimate limit state a kinematic analysis, based on the developed plastic mechanisms, must be performed (Fig. 8.11c). This analysis may be performed using a push-over method or a global plastic method. In the first case the complete load-displacement can be obtained, whilst in the second case only an asymptotical elasto-plastic behaviour can be obtained.

-Determination of *static required values* for the three state limits. For each limit state principal and secondary values are determined. For the serviceability limit state the interstorey drifts are determined as principal values, in order to protect against damage to nonstructural elements. The secondary values are the bending moments, in order to verify the elastic behaviour of a structure (Fig. 8.12a). For damageability limit states the principal determined characteristics are the cross-section required strength capacities, in order to assure the formation of plastic hinges in desirable positions. The secondary characteristics are the interstorey drifts, in order to impede the catastrophic failure of nonstructural elements, and plastic rotation capacities, for a good redistribution of bending moments in plastic range. For ultimate limit state, the first characteristic is the local plastic rotation capacity of plastic hinges in order to allow an adequate plastic mechanism and to save the structure from a global collapse. The secondary characteristic is the verification of redistribution of bending moments.

-*Correction of static required characteristics* as a function of earthquake type. For serviceability limit states no correction of statical values, because the seismic actions are low. For damageability and ultimate limit states the influence of high velocity (for near-field earthquakes) and repeated cyclic actions (for far-field earthquakes) must be considered. The effects of seismic actions produce the increasing of all the required characteristics (Fig. 8.12b): the intersorey drifts and plastic rotation capacity, due to the accumulation of plastic deformations; the high velocity may affect the moment redistribution due to the increasing of plastic moments as a result of the strain-rate.

(iii) *Local analysis phase* consists in the following steps:

-*Global to local transfer*, in which, using the inflection points determined from static analysis, a complex structure can be divided into sum of standard beams. Each beam is subdivided into two parts, the left one with quasi-constant moment and the right one with an important gradient moment. The analysis of a beam is replaced by the sum of the analysis of two different standard beams, SB1 and SB2 types (Fig. 8.13a) (see Section 6.3). The beam-column behaviour is studied by means of SB1 standard beam type, due to the moment gradient (Fig. 8.13b).

-Determination of *static available characteristics* for the three limit states. For serviceability limit states the available interstorey drifts are determined as a function of nonstructural element performances. The damageability limit state determines the cross-section strength capacities as a function of geometrical and mechanical properties. The available interstorey drifts for this limit state must be higher than the one for the serviceability limit state (generally a value four times higher is admitted). Using the standard beam and a specific methodology (the local plastic mechanism is recommended

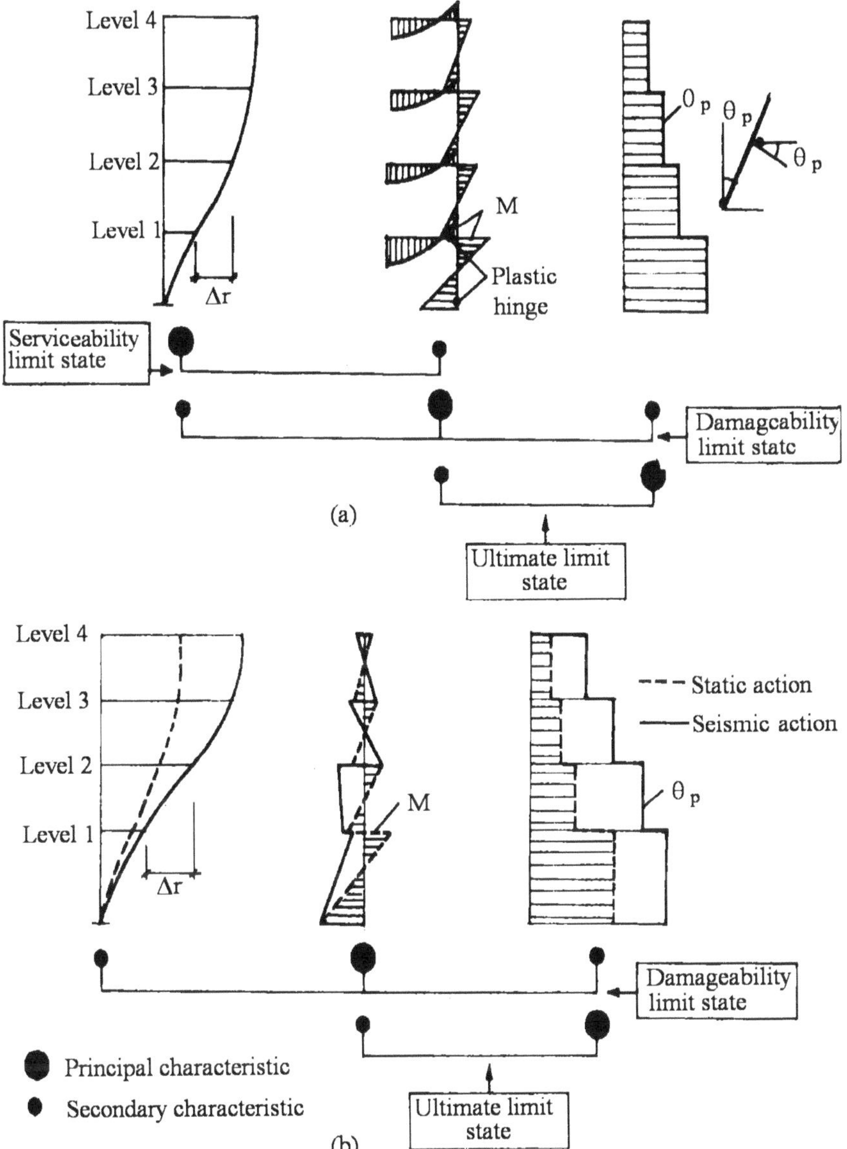

Figure 8.12: (a) Static required values; (b) Corrected values due to earthquakes

to be used), the available ductility for beams and beam-columns can be determined (Fig. 8.14a).

-*Correction of static available characteristics* as a function of earthquake type. For serviceability limit states no one correction is necessary. For dam-

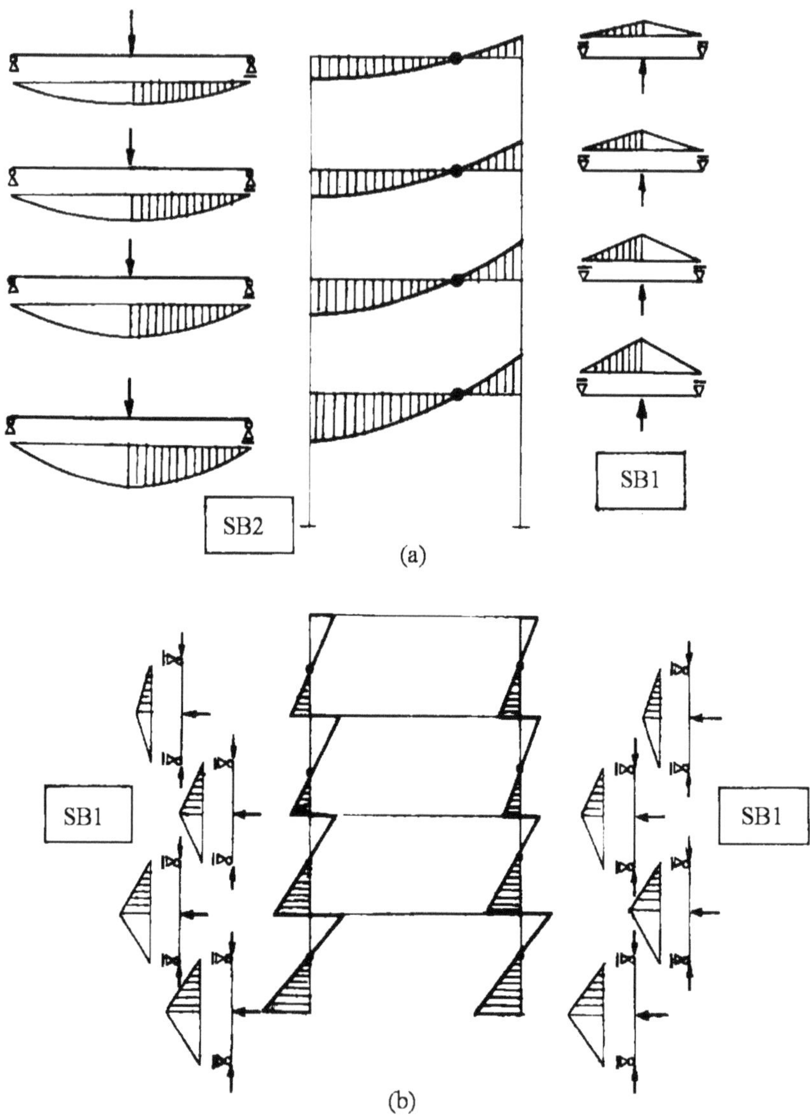

Figure 8.13: Standard beams for a moment resisting frame: (a) Standard beams for beams; (b) Standard beams for columns

ageability limit states, in case of near-field earthquakes, the correction refers to the increase of plastic moments due to the strain-rate effects. Contrary to this, a reduction of plastic moment capacity occurs for far-field earthquakes due an important number of plastic excursions in plastic range. For the plastic rotation capacity both types of earthquake produce a reduction, but in a different manner: in the case of near-field earthquakes the high

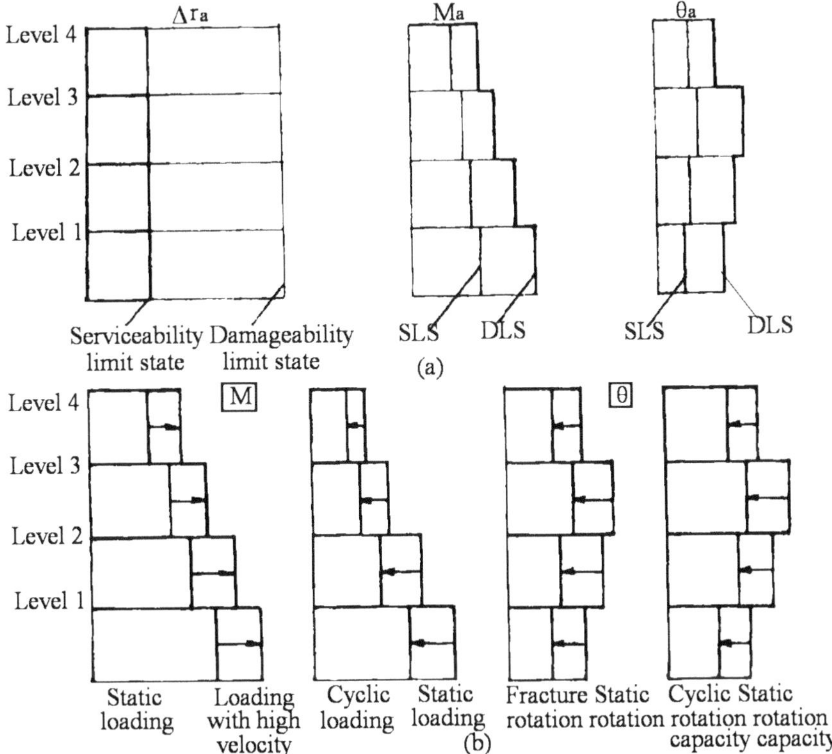

Figure 8.14: Available characteristic values: (a) Static available values; (b) Corrected values due to earthquakes

velocities can produce a section fracture, while in the far-field cases, the repeated cyclic actions generate an accumulation of plastic rotations (Fig. 8.14b).

(iv) *Required-available checking phase.* To achieve the design objective, the seismic problem is assessed through the required-available formulation:

Seismic required characteristics < *Seismic available characteristics*

Generally the required-available check must be considered for all three limit states and in all seismic areas. In any case, it would appear logical that in zones of low seismicity the design effort be more reduced than in zones with high seismicity (Kennedy and Medhekar, 1999). Table 8.1 presents a strategy for considering the seismicity of the site. The serviceability limit state must be considered for all seismic zones. For seismic areas with very low seismicity the checking of seviceability limit states is sufficient for a good structure behaviour. For seismic areas with low seismicity the checking must be performed for only both seviceability and damageability limit states, because the seismic actions are not so important to transform the structure in a plastic mechanism. In the case of moderate earthquakes only serviceability and ultimate limit states must be verified. All three limit

Table 8.1: Design strategy

Seismic area	Magnitude M	Frame type	Limit state			Analysis	
			SLS	DLS	ULS	LEF	PO
Very low seismicity	< 3.0	SMRF	x	-	-	x	-
Low seismicity	3.0–4.5	SMRF	x	x	-	x	-
Moderate seismicity	4.5–6.5	GMRF	x	–	x	x	x
High seismicity	> 6.5	GMRF	x	x	x	x	x

states must be considered only for areas with high seismicity. These considerations limit the design effort in the case of low seismic actions, the respect of constructional details required by codes assuring a good structure performance. Concerning the analysis procedure, the lateral equivalent force methodology must be used for all limit states, while the push-over method is required for seismic areas with moderate or high seismicity. Special moment-resisting frames may be used for very low and low seismicity areas, while for areas with moderate and high seismicity, global moment-resisting frames must be designed. Therefore, this design strategy considers that the ductility checking is necessary to be performed for important ground motions only.

8.2 Required Ductility

8.2.1. Reasons for a Simplified Method

The first problem in calculating an adequate required ductility is the determination of the expected ground motions on the structure site. As a result of the extensive monitoring of many areas with high seismic risk, today it is possible to dispose of a large number of records which can be used. The second problem is to determine the structure response to these seismic actions and to calculate the required ductility. Due to the large development of computer science, today it is not a problem to perform a nonlinear time history analysis. Therefore, from the scientific point of view, the evaluation of an adequate required ductility is a normal procedure. But, from the engineering point of view, this approach is very questionable. Each ground motion is basically unique, offering new surprises in the vulnerability of structures affected by earthquakes, showing the great complexity of the phenomenon. Due to the large variability of recorded ground motion features (peaks, periods, patterns, durations, etc) which influence the structural response, their characteristic values show a very large scattering.

Because of this randomness of results, a blind confidence on computed results can be a source of potential damage. Therefore, for seismic structural design, a simplified method which considers the main parameters of seismic actions and structure responses, as an envelope for all possible situations, can be a more efficient methodology than an advanced analysis based on unreliable actions.

8.2.2. Simplified Push-Over Method

The most advanced method for determining the required ductility is the time history analysis. Due to the complexity of the computation, this method remains a research procedure only, which can be used profitably to validate the other simplified methodologies.

A suitable procedure for structure static inelastic analysis is the push-over method. There is an increasing trend in the use of this method as a tool for evaluating resistance and safety of structures in the earthquake engineering field. This procedure is generally more realistic in gauging the vulnerability of buildings during earthquakes than the linear procedures commonly contained in current codes (Kim and D'Amore, 1999). Especially, when large structure deformations must be considered for checking the ductility at the ultimate limit state under strong earthquakes, the push-over analysis is indispensable. In this method (see Section 3.5.2) the lateral loads are increased to evaluate the inelastic capacity of structures. At each lateral load increment, the sequence of plastic hinge formation and the elasto-plastic behaviour are found.

Because the results are obtained without dynamic behaviour consideration, there are some differences in comparison with the results obtained using the time history method (Kim and D'Amore, 1999):

-there is an interaction between structure behaviour and seismic action, producing a continuous changing of lateral loads;

-the duration of the earthquake and the effect of cumulative plastic rotation are ignored;

-only two parameters are considered, base shear force and top drift, being very difficult to capture all structural variations with these two parameters only.

In spite of these deficiencies, the push-over method remains the only one able to consider the large displacements in the elasto-plastic range in a way which is affordable for practice. The main problem during the design process is to correct these deficiencies.

The determination of ductility at the ultimate limit state has some peculiarities which must be considered in a simplified method:

-the behaviour of a structure is dominated by the deformation of a plastic mechanism formed by the occurrence of some plastic hinges at the beam and column ends;

-the horizontal load distribution is determined by the mechanism pattern (Fig. 8.15a);

-the structure is sized to form a strong column-weak beam system and the collapse is dominated by a global plastic mechanism (Fig.8.15b);

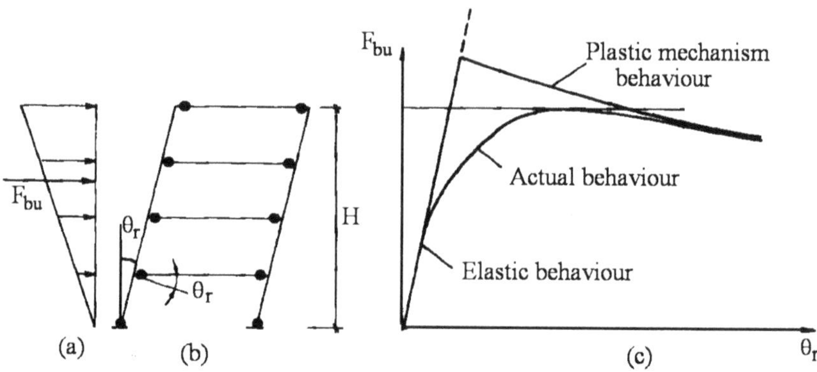

Figure 8.15: Simplified push-over method: (a) Load distribution; (b) Global plastic mechanism; (c) Idealized rigid-plastic mechanism behaviour

-the determination of the ultimate top drift corresponding to a plastic mechanism is independent of the actual first part of the force-displacement curve. In the field of high plastic rotations there are minor important differences between the actual behaviour and idealized elasto-plastic behaviour. Therefore, an idealized elasto-plastic behaviour may be considered (Fig. 8.15c).

Taking into account these aspects, a simplified push-over method for determining the required plastic rotation will be presented in the following. This procedure is based on the research results of Mazzolani and Piluso (1995a,b, 1996, 1997a,b) and it is intended to transform the complex push-over analysis in a more simple one, addressed to design practice.

8.2.3. Required Plastic Rotation Capacity: Kinematic Ductility

Within the framework of a comprehensive methodology, a simple approach for evaluating the required ductility of seismic-resistant steel frames is presented. The seismic forces are scaled at the level of ultimate base shear force, F_{bu}, given by equation (8.8c). A multiplier α is used to determine the ultimate horizontal forces (Fig. 8.16a). Assuming an inverted triangular force distribution along the height of the structure, the forces at the level k can be related to the base shear force by:

$$\alpha F_k = \frac{w_k H_k}{\sum_{k=1}^{n_s} w_k H_k} \alpha F_{bu} \tag{8.15a}$$

where w_k is the structure weight at the level k and H_k, the height of beam level k from the ground. The global vertical load, which remains constant

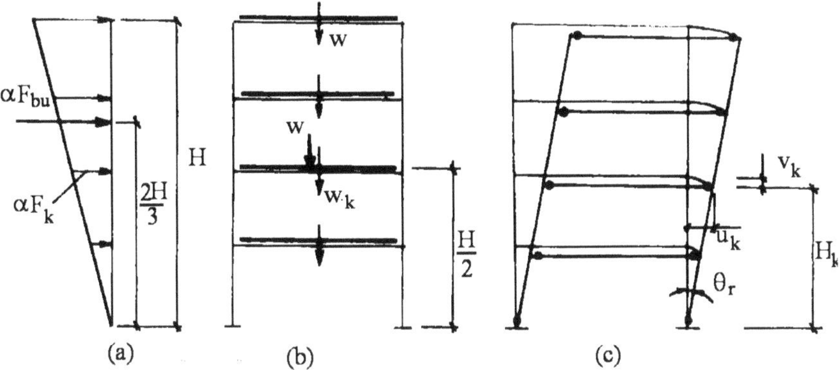

Figure 8.16: Global plastic mechanism behaviour: (a) Horizontal loads; (b) Vertical loads; (c) Plastic mechanism deformations

during the earthquake, is:

$$W = \sum_{k=1}^{n_s} w_k \tag{8.15b}$$

The horizontal and vertical resultants are situated at $2H/3$ and $H/2$ distances from the structure base, respectively (Fig. 8.16b). During the deformation, the plastic mechanism rotates around the base, the same rotation θ_r occurring for all plastic hinges in both beams and columns. The horizontal u_k and vertical v_k displacements of storey k are (Fig. 8.16c):

$$u_k = H_k \sin\theta_r; \quad v_k = H_k(1 - \cos\theta_r) \tag{8.16a, b}$$

and considering that the rotation is small, $\sin\theta_r \simeq \theta_r$ and $\cos\theta_r \simeq 1 - 0.5\theta_r^2$, it results:

$$u_k = H_k\theta_r; \quad v_k = H_k\frac{\theta_r^2}{2} \tag{8.17a, b}$$

The displacements of horizontal resultant force are:

$$u_F = \frac{1}{3}H\theta_r; \quad v_F = \frac{2}{3}H\frac{\theta_r^2}{2} \tag{8.18a, b}$$

and for vertical resultant force:

$$u_W = \frac{1}{2}H\theta_r; \quad v_W = \frac{1}{2}H\frac{\theta_r^2}{2} \tag{8.19a, b}$$

The work of a plastic mechanism implies that the largest amount of energy is absorbed by plastic hinges, so that the elastic deformations can be neglected. The approach is based on the principle that a minimum of the total potential energy, is composed of internal and external potential energy (see Section 4.2.3).

The internal potential energy is obtained from:

$$U = \left(\sum_{i=1}^{n_c} M_{pc,i1} + \sum_{k=1}^{n_s} \sum_{j=1}^{n_b} 2M_{pb,jk} \right) \theta_r \tag{8.20}$$

where the first sum refers to the plastic hinges formed in columns and the second to the plastic hinges in beams.

The external potential energy results as:

$$L_p = \frac{2}{3} \alpha F_{bu} H \theta_r + \frac{1}{2} W H \frac{\theta_r^2}{2} \tag{8.21}$$

The total potential energy is therefore:

$$V = \left(\sum_{i=1}^{n_c} M_{pc,i1} + \sum_{k=1}^{n_s} \sum_{j=1}^{n_b} 2M_{pb,sk} \right) \theta_r + \frac{2}{3} \alpha F_{bu} H \theta_r + \frac{1}{2} W H \frac{\theta_r^2}{2} \tag{8.22}$$

The equilibrium equation results from the variation of the potential energy (see equation (4.34)), giving:

$$\alpha = \alpha_1 - \alpha_2 \theta_r \tag{8.23}$$

where

$$\alpha_1 = \frac{3}{2} \frac{\sum_{i=1}^{n_c} M_{pc,i1} + \sum_{k=1}^{n_s} \sum_{j=1}^{n_b} 2M_{pb,jk}}{F_{bu} H} < 1.0 \tag{8.24a}$$

$$\alpha_2 = \frac{3}{4} \frac{W}{F_{bu}} \tag{8.24b}$$

The condition $\alpha_1 < 1$ is necessary to assure that for seismic actions corresponding to the ultimate limit state a plastic mechanism is formed.

The equation (8.23) represents a straight line in θ_r with a negative slope (Fig. 8.17a). The first term α_1 introduces the structure's plastic characteristics related to the seismic actions, while the second one, α_2, refers to the second order effect of gravitational loads. If the expression (8.9c) is used for F_{bu}, it results:

$$\alpha_2 = \frac{3}{4} \frac{1}{a_u/g} \tag{8.25}$$

This coefficient depends on the level of ultimate acceleration only.

For an ideal elasto-plastic structure the elastic rotation θ_e must be added to the plastic rotation θ_r:

$$\theta = \theta_e + \theta_r \tag{8.26}$$

If $\theta_1 = \delta_1 / H$ is the rotation corresponding to the design value of a seismic horizontal force F_{bu} for which the multiplier α has been scaled ($\alpha = 1$ for this force), it results:

$$\theta_e = \alpha_1 \theta_1 \tag{8.27}$$

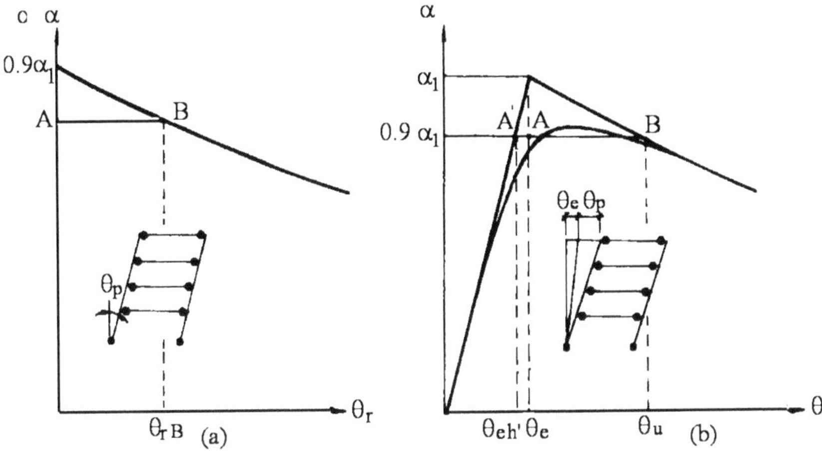

Figure 8.17: Determination of rotation capacity: (a) Ultimate plastic rotation; (b) Ultimate rotation

The value of θ_1 can be determined by computing the structure under the horizontal force F_{bu}, or using a simplified method, based on empirical formula for evaluating the fundamental vibration period T (sec):

$$\theta_1 \simeq 0.375\frac{T^2}{H}\frac{F_{bu}}{W} \qquad \text{(H,m)} \tag{8.28a}$$

An empirical formula for the fundamental vibration period T is proposed by Goel and Chopra (1997):

$$T = 0.116H^{0.8} \qquad \text{T, (sec), H, (m)} \tag{8.28b}$$

This relationship has been established on the examination of 42 steel frames. By equating (8.28a) and (8.28b), it results:

$$\theta_1 = 0.51 \times 10^{-2} H^{0.6}\frac{F_{bu}}{W} = 0.51 \times 10^{-2}\frac{a_u}{g}H^{0.6} \tag{8.28c}$$

The elastic rotation can be evaluated by substituting equation (8.28c) into (8.27).

The α-θ_r curve obtained for an ideal elasto-plastic behaviour considers that plastic hinges simultaneously occur in all plastic zones and the structure behaves elastically until a fully plastic mechanism is formed. But, in reality, the plastic hinges occur successively and the cusp formed by the perfect elastic and plastic behaviours is smoothed, the maximum load bearing capacity being reached for $\alpha_{max}F_{bu}$ (Fig. 8.17b). This value is easily determined using a push-over analysis, but this procedure is considered too complicated for practical purposes. In order to obtain a simple procedure to determine the required ultimate rotation, the actual continuous curve can be substituted by a three-linear curve OABC, where the plateau AB is obtained for the value $0.9\alpha_1$. The required ultimate rotation is given for the point B. So, one can obtain (Fig. 8.17b):

$$\theta_{rB} = 0.1\frac{\alpha_1}{\alpha_2}; \qquad \theta_{eA} = 0.9\theta_e \tag{8.29a, b}$$

and

$$\theta_{rr} = 0.1\left(\frac{\alpha_1}{\alpha_2} + \theta_e\right) \tag{8.29c}$$

which represents the required ultimate plastic rotation. The ratio:

$$\frac{\alpha_1}{\alpha_2} = 2\frac{\sum_{i=1}^{n_c} M_{pc,i1} + \sum_{k=1}^{n_s}\sum_{j=1}^{n_b} 2M_{pb,jk}}{WH} \tag{8.30}$$

is predominant in the required rotation capacity, the rotation θ_e being small.

The kinematic ductility is the ratio between plastic rotation and elastic rotation:

$$\mu_\theta = \frac{\theta_{rr}}{\theta_{eA}} = 0.111\left(1 + \frac{\alpha_1}{\alpha_2}\frac{1}{\theta_e}\right) \tag{8.31}$$

One can observe that in both equations (8.30) and (8.31) the characteristics of seismic actions are absent.

8.2.4. Influence of Seismic Actions: Hysteretic Ductility

(i) *Factors influencing the required ductility.* In the above static methodology for determining the kinematic ductility, the ground motion accelerations for different limit states are used as a single measure of the earthquake destructiveness. But it is very well known that the peak ground acceleration is not a very accurate means of classifying the severity of strong ground motions with regard to structure damage potential (Anderson and Bertero, 1987). There are two important factors, the duration of strong ground motion and the number of strong pulses which can have an important influence on earthquake damage effects (Sarma and Srbulov, 1998). The introduction of these additional factors, influencing the required ductility, can be done only the using a dynamic analysis. But the variability associated to the predictive values of these factors is very large and only an approximate deterministic design assessment may be considered.

The duration and the number of strong pulses are correlated factors, the short duration being characterized by very few inelastic pulses, while for long duration the number of these pulses may be very high. As it is presented in Section 3.2.3, the duration of earthquakes depends of the site-source distance:

-in the near-field areas and for rock or firm soil, the earthquake duration is short and contains only few fully reversed large amplitude pulses. For 170 recorded ground motions with duration under 20 sec, in 95% of cases

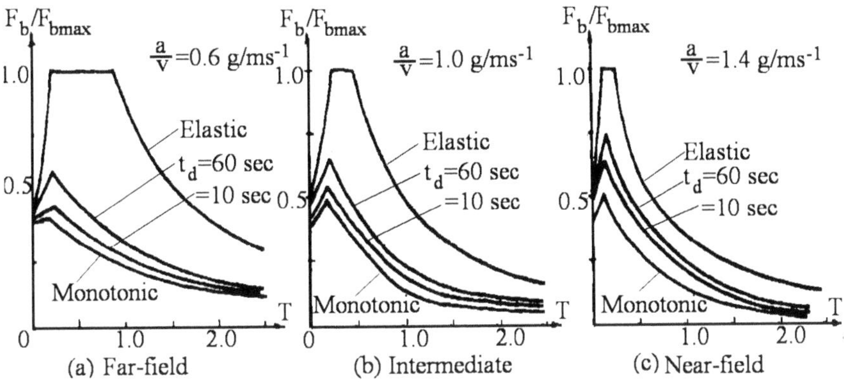

Figure 8.18: Influence of earthquake duration on the design spectra: (a) Far-field earthquake; (b) Intermediate-field earthquakes; (c) Near-field earthquakes (after Chai et al, 1998)

the important pulses were less than four (Driver et al, 2000). For these earthquakes the a/v acceleration-to-velocity ratio is high;

-in the far-field areas the earthquake duration is long and contains a larger number of important pulses (maximum 15 to 20). For these earthquake types the a/v acceleration-to-velocity ratio is low.

The field of high, normal and low acceleration-to-velocity ratio is presented in Section 3.2.4. In the case of long duration the hysteretic ductility, as a result of cyclic actions, plays a very important role.

Fig. 8.18 shows the influence of duration on the design spectra (Chai et al, 1998). One can observe that the corner periods T_c depends on the a/v ratio, being longer for the far-field earthquakes and shorter for the near-field ones. The base shear coefficients increase with the duration of ground motions, the monotonic static spectra having the smallest values. This increase is not uniform in the short period range. The highest base shear coefficient occurs at a period of $0.4T_c$, but the largest increasing percentage occurs at the corner period T_c. In the long period range the percentage of increased values due to duration is rather uniform. However, it should be noted that strong motion duration of 40 to 60 sec is unlikely to be associated with ground motions of high a/v ratios. The same observation is valuable for durations of 10 to 20 sec with low a/v ratios.

Fig.8.19 shows the effect of accumulation of plastic rotation on the required rotation capacity due to multiple pulses (De Matteis et al, 2000). The kinematic and hysteretic values have been obtained using the push-over and time-history methods, respectively. One can observe an important increase of required rotation capacity for the earthquakes characterized by multiple pulses.

The influence of seismic actions on required ductility for different q factors is presented in Fig. 8.20 (Zhu et al, 1988). One can observe that this influence depends on earthquake type, structure period and q factor. The influence of earthquake type is reduced for short period ranges and very

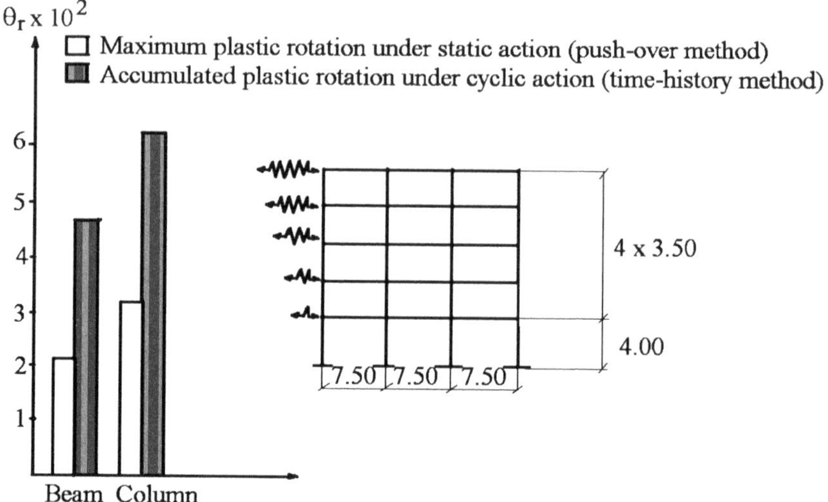

Figure 8.19: Maximum and accumulated plastic rotations (after De Matteis et al, 2000)

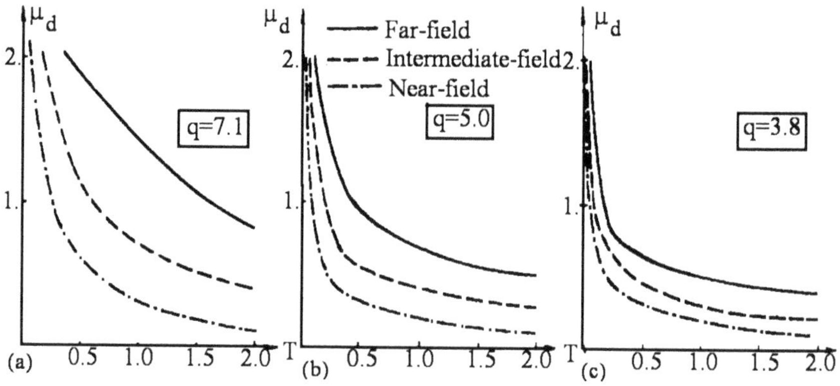

Figure 8.20: Influence of q-factor and earthquake types on design spectra: (a) High value of q-factor; (b) Medium value of q-factor; (c) Reduced value of q-factor (after Zhu et al, 1998)

important for long period ranges. This influence is much more pronounced for high q values.

Another aspect referring to the seismic actions is related to the distribution of ductility requirements on the structure height. If the curvature ductility has a regular variation on structure height, the maximum ductility being required at the first level, for hysteretic ductility (cumulative plastic rotation), especially for tall structures, a concentration of required ductility is found near the top of structures (Tso et al, 1993) (Fig. 8.21). This

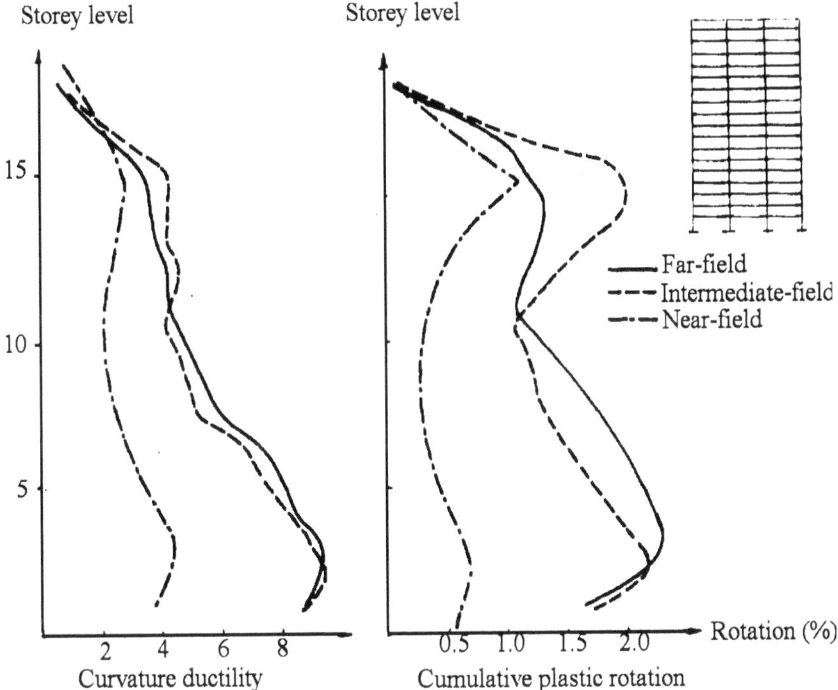

Figure 8.21: Distribution of ductility requirements on the structure height: (a) Curvature ductility; (b) Cumulative plastic rotation-hysteretic ductility (after Tso et al, 1993)

concentration is due to the effects of higher modal responses and flexible structures where the short ground motion periods accentuate the higher modal contributions.

The differences between the kinematic and hysteretic ductilities for different natural periods and ground motion durations are presented in Fig. 8.22 (Lam et al, 1996). The kinematic ductility appears to be insensitive to earthquake duration. Contrary to this, the duration significantly affects the hysteretic ductility due to accumulation of dissipated energy caused by cyclic actions. The amplification is higher for short natural periods and soft soils.

The main conclusions of these aspects are that the determination of required ductility based on a static monotonic action does not give a consistent control of the structural damage for all structures subjected to ground motions.

(ii)*Influence of strain-rate.* The strain-rate produces an increase of yield stress and, consequently, an increase of full plastic moment:

$$M_{psr} = \frac{f_{ysr}}{f_y} M_p \tag{8.32a}$$

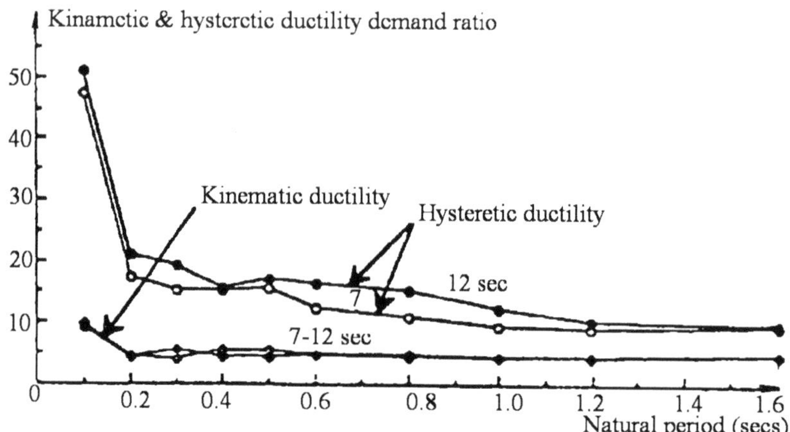

Figure 8.22: Differences between kinematic and hysteretic ductilities (after Lam et al, 1996)

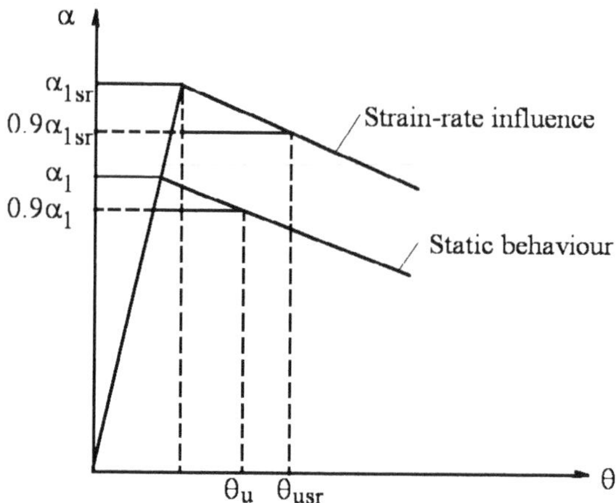

Figure 8.23: Influence of strain-rate on required ductility

where the ratio f_{ysr}/f_y is given by one of equations (4.8). This increase of beam and column plastic moments gives rise to an increase of collapse load characterized by multiplier α_{1sr}, given by:

$$\alpha_{1s1} = \frac{f_{ysr}}{f_y}\alpha_1 \qquad (8.32b)$$

where α_1 is determined from equation (8.24a).

The coefficient α_2 remains unaffected by the strain-rate. So, the required rotation θ_{rrsr}, in which the influence of strain-rate is considered, results from (Fig. 8.23):

$$\theta_{rrsr} = \frac{f_{ysr}}{f_y}\theta_{rr} \qquad (8.32c)$$

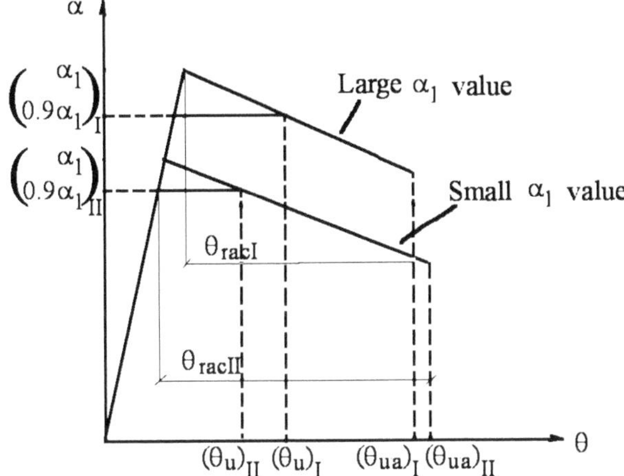

Figure 8.24: Influence of accumulated plastic rotations on required ductility

Table 8.2: Coefficients γ_a

α_1	γ_a	
	Beams	Columns
0.2	3.5	4.0
0.4	2.5	3.0
0.6	1.5	2.0

where θ_{rr} is determined from the equation (8.29c). Taking into account the increase of yield stress for near-field earthquakes, the increase of required rotation capacity is about 30 to 50 percent.

(iii) *Influence of plastic rotation accumulation.* The accumulation of plastic rotations produces an increase of required rotation capacity. As result of the above investigations, this influence depends on the collapse multiplier. For small values, the number of plastic rotations producing plastic accumulation is higher than for large values, where this number is smaller (Fig.8.24). Consequently, the required plastic rotation can be determined from the relationship:

$$\theta_{rac} = \gamma_a \theta_{rr} \tag{8.33}$$

where γ_a is a numerical coefficient which considers the influence of accumulated plastic rotation, presented in Table 8.2. There are differences for beams and columns, the latter being higher. These values are given as an example only, just showing the general trend. If other more adequate data are available, they can be used instead.

8.3 Available Ductility under Static and Monotonic Actions

8.3.1. Member Plastic Rotation Capacity

The main conclusion of the previous Chapters analyzing the possibility of determining the member ductility has been that the most suitable methodology for design practice is based on the local plastic mechanism. Therefore, for the determination of the available ductility under static and monotonic actions of different section types this method is used. The influence of the main common factors will be discussed in the following.

(i) *Random material variability.* As it has been shown in Section 4.3.2, the structural steel does not always have the mechanical characteristics assumed in design. Due to the fact that only the lower limit of yield stress is guaranteed by producer, the upper bounds being not specified, the plastic moment may be considerably higher than the one determined theoretically from the specified minimum yield stress. If the actual yield stress values are proportionally higher than the minimum ones in all members of the structure, the pattern of hinge formation would remain the same as the designed one. Contrary to this, when the degree of randomness in yield stress is very high, plastic hinges may form at totally unexpected positions and it may cause an unfavourable performance in the structure in the ultimate limit state (Kawamura and Sasaki, 1990). The random variability of a plastic hinge moment can be considered approximately equal to the one of yield stress and so, the investigation of yield stress randomness plays a very important role. A measure of this randomness is the coefficient of variation, COV, and the increase of yield stress with a 95% probability result (Mazzolani et al, 1998):

$$f_{yr} = \frac{1}{1 - 1.64\text{COV}} f_y \tag{8.34}$$

The COV varies in function of thickness, being high for thin plates and reduced for thick plates (Fig. 8.25). For the main properties of steel, Galambos and Ravindra (1978) propose the values and the COV given in Table 8.3. One can observe that the greatest COV value is provided for the strain-hardening modulus. The proposed increased values of flange and web yield stresses considers the differences in thicknesses. If equations (4.19) to (4.21) are used for the determination of the yield stress function of plate thickness, this increasing in value need not necessarily be included. For the ultimate strength, it is on the safe side to consider the minimum values corresponding to the nominal ones, but for ductility design the maximum values must be used.

(ii) *Collapse by plastic buckling or fracture?* Fig. 8.26a shows the stress distribution on the cross-section. After the elastic behaviour, yielding of the two flanges occurs and a full plasticization of cross-section is rapidly produced. Due to the strain dynamic jump (see Section 4.2.3) the maximum stresses correspond to the hardening range, both in tension and compression

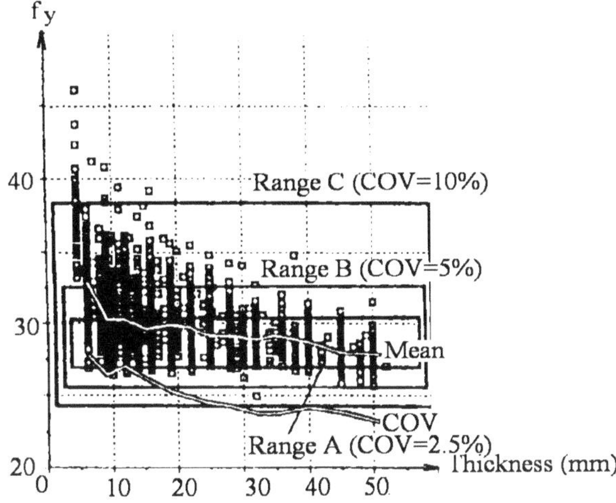

Figure 8.25: Plot of actual yield stresses (after Kuwamura and Sasaki, 1990)

Table 8.3: Steel properties

Properties	Mean values N/mm^2	COV
Elasticity modules		
• tension and compression	205000	0.06
• shear	77000	0.06
Strain-hardening modulus	4100	0.25
Yield stress		
– flanges	$1.05f_y$	0.10
– webs	$1.10f_y$	0.11
Shear yield stress	$0.64f_y$	0.10

flanges. In cases of very stocky flanges, where no buckling occurs, the symmetry is preserved until fracture of the tension flange at the ultimate strain ϵ_u and strength f_u. In the case of plastic and compact sections the symmetry in stresses is broken if the plastic buckling of compression flange occurs, followed by the decreased of compression stresses and the moving down of the neutral axis to equilibrate this compression stress reduction. The absence of local buckling increases the risk of member failure by fracture of the tension flange. The plastic buckling of compression flange operates as a filter against large strains in the tension flange, reducing the danger of brittle cracking (Gioncu and Petcu, 1997). Thus, contrary to what the majority of designers believe, the structure design must be faced to obtain plastic hinges where the ultimate situation is produced by plastic buckling of the compression flange.

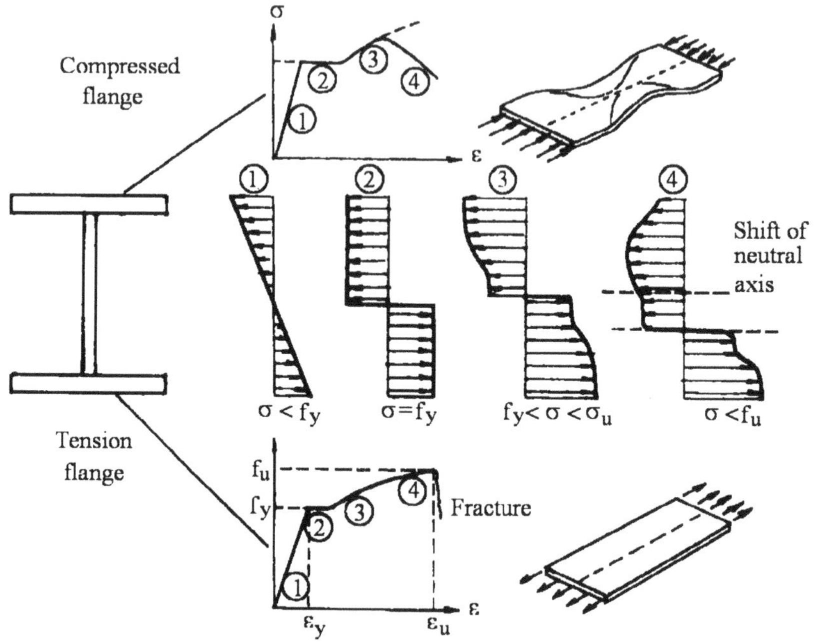

Figure 8.26: Stress distribution on the cross-section

(iii) *Moment-rotation curve.* The fully plastic moment is given by equation (5.2):

$$M_p = Z_p f_{ya} \tag{8.35a}$$

Z_p being the plastic section modulus, f_{ya}, the actual yield stress, which can be evaluated by means of equation (8.34). The ultimate moment can be determined from equation (5.13), using the nominal ultimate strength f_u:

$$M_u = Z_p f_u \tag{8.35b}$$

The moment-rotation curve directly depends on the moment variation along the member axis (Fig. 8.27). There are two cases:

-*quasi-constant moment* for which the moment-rotation curve is given by:

$$\text{for } \theta_p < \theta \le \theta_h, \quad M = M_p \tag{8.36a}$$

$$\text{for } \theta_h < \theta \le \theta_u, \quad M = M_p + (M_u - M_p)\frac{\theta - \theta_h}{\theta_u - \theta_h} \tag{8.36b}$$

-*moment gradient*, for which the relationship proposed by Lay and Galambos (1967) is used:

$$\frac{\theta}{\theta_p} = \frac{M}{M_p} + \left(1 - \frac{M_p}{M}\right)\left[2s + \frac{E}{E_h}\left(\frac{M}{M_p} - 1\right)\right] \tag{8.37}$$

the notations being defined in Section 7.2.2.

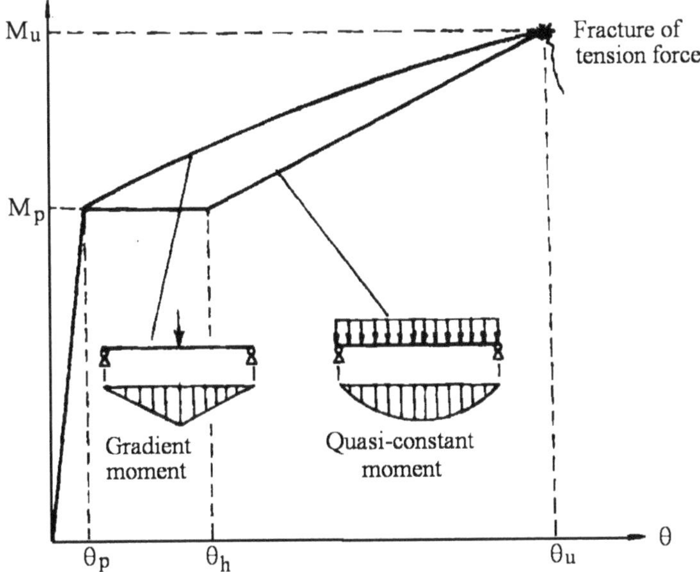

Figure 8.27: Moment-rotation curves

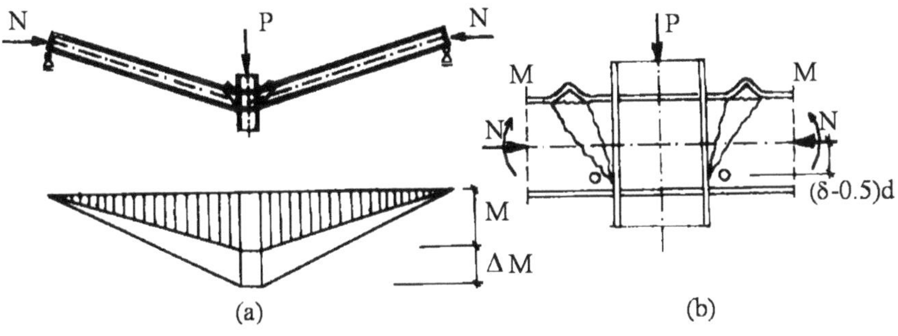

Figure 8.28: Axial load and member slenderness influences

(iv) *Axial load and member slenderness influences.* Fig. 8.28 shows a standard beam SB1 with axial force N at the two ends. The effect of this axial force is to increase the primary maximum moment at the value corresponding to second order effect (Allen and Bulson, 1980):

$$M_N = M \frac{1 + \frac{Nd}{M}(\delta - 0.5)}{1 - \frac{\pi^2}{12}\frac{N}{N_{cr}}} = M \frac{1 + n_p \frac{M_p}{M}\frac{1}{\overline{e}_0}(\delta - 0.5)}{1 - 0.822 n_p \overline{\lambda}^2} \tag{8.38}$$

where

$$n_p = \frac{N}{N_p}; \quad m_p = \frac{M}{M_p}; \quad \overline{\lambda} = \left(\frac{N_p}{N_{cr}}\right)^{1/2}; \quad \overline{e}_0 = \frac{M_p}{N_p d} \qquad (8.39a, b, c)$$

N_p being the plastic axial load and N_{cr} the elastic critical load of the strut:

$$N_{cr} = \frac{\pi^2 EI}{L^2} \qquad (8.39d)$$

$\bar{\lambda}$, normalized slenderness ratio, d, beam depth and for δ see Fig. 8.28. In these relationships the coefficient m_p must consider the interaction moment-axial forces (see equation (5.4)). In cases of standard beam SB2 the numerical coefficient 0.822 in the relationship (8.38), for concentrate forces, must be replaced by the value 1.0, for distributed forces. One can observe from the equation (8.38) that the influence of both axial force and slenderness is introduced by the factor $n_p\bar{\lambda}^2$, in agreement with the Nakashima (1994) conclusions which define it as a good indicator for estimating the axial force and member slenderness effects.

(v) *Local plastic mechanisms.* The patterns of local plastic mechanism are different for the two standard beams, which are characteristic of moment gradients, SB1, and for quasi-constant moments, SB2. In the first case, the plastic zone around the mid span is narrow and only one buckled shape can form at each side (Fig. 8.29a). Contrary to this, in a case of quasi-constant moments the plastic zone is larger and a minimum of two buckled shapes at each mid span side can form (Fig. 8.29b). Due to the increasing number of plastic mechanisms an increasing of ductility occurs. The pattern of local plastic mechanism is established as a function of section type and buckling mode, being examined from the experimental results. As it was shown in Section 4.4, the mechanisms are composed of plastic zones, yield or fracture lines, stationary and travelling yield lines, etc. The member collapse load is given by the minimum load obtained from the examination of all possible mechanism types.

(vi) *Local ductility definition.* Section 6.3.5 has shown the three existing approaches to evaluating local ductility. In the following the definition related to the fully plastic moment is used. The ductility is defined as a function of rotation capacity determined in the lowering post-buckling curve at the intersection with the theoretical fully plastic moment (Fig. 8.29c). In order to compensate the detrimental effects of seismic actions (strain-rate, cyclic actions, accumulation of plastic deformations, etc), the favourable effect of a moderate post-critical erosion can be considered by using the ultimate rotation determined for $0.9M_p$. Especially, this value is recommended for determining the rotation capacity in case of quasi-constant moment variation (Fig. 8.29d).

8.3.2. I-Section Members

The I-section member is one of the most used elements for seismic-resistant steel structures. Therefore, the detemination of the available ductility is

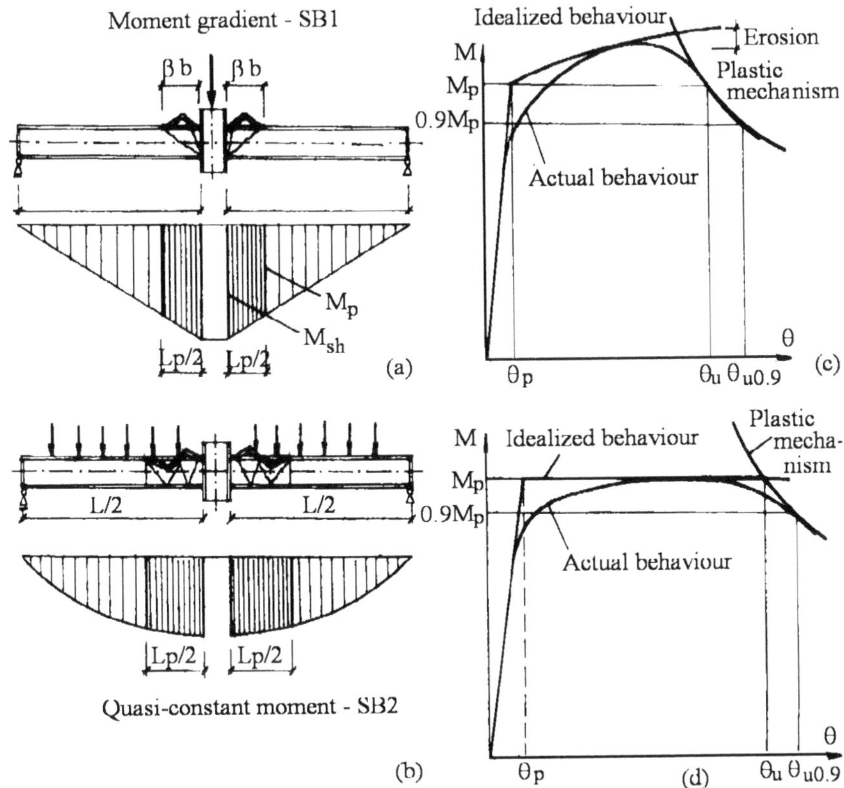

Figure 8.29: Local plastic mechanism patterns: (a) Moment gradient; (b) Quasi-constant moment

presented as an archetype procedure valid for other section shapes. Two I-section types are considered:

-built-up I-section made of welded plates with different sizes for compression and tension flanges (Fig. 8.30a);

-rolled I-section for which the junction between flanges and web plays a very important role, reducing the flange slendernesses and increasing the ductility (Fig. 8.30b).

The main collapse types are presented in Fig. 7.6, being identified by Spangemaher (1991) as the in-plan, out-of-plane, flange induced buckling, web shear buckling plastic mechanisms. The last two are specifically for beams with slender webs as in the case of bridges. In the case of steel building frames, where plastic or compact sections are used, only the first two mechanism types can be considered in design practice.

(i) *In-plane plastic mechanism.* The evolution of using different collapse mechanism types is presented in Fig. 8.31:

-Climenhaga and Johnson (1972)-Ivanyi (1979) model, characterized by a length of mechanism equal to the flange width. The shortcomings of this

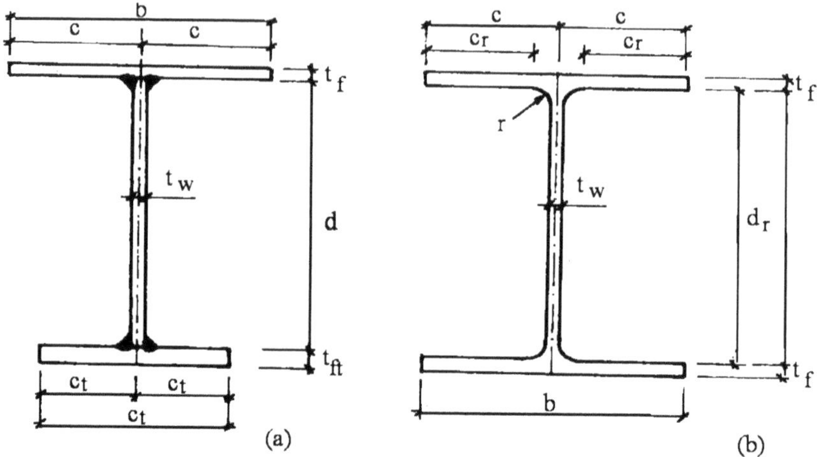

Figure 8.30: I-section members: (a) Welded section; (b) Rolled section

model are related to the experimental and theoretical results, which have shown that this length depends also on other section dimensions (see Section 7.2.2);

-Kuhlmann model (1986) characterized by a semi-pyramidal shape mechanism for buckled flange and a partial mechanism for compression zone of webs. An inconsistancy exists between the plastic deformations of flanges and webs;

-Feldmann model (1994), characterized by a semi-pyramidal shape mechanism for buckled flanges and elastic behaviour of webs. This model is more adequate for thin-walled members, where the true mechanisms (see Section 4.4.2) are quite similar;

-Gioncu and Petcu model (1995, 1997) characterized by using the Climenhaga-Johnson-Ivanyi model but with a mechanism length including all cross-section dimensions. The shortcoming of this model is related to the pattern of plastic mechanisms;

-Tehami model (1997), for which the length of mechanism is dependent on section depth only, without considering the flange width as an influencing parameter, contrary to the experimental evidence;

-Gioncu and Petcu improved model (2001), which may be considered as a synthesis of the above presented models, both flange width and web depth being considered as mechanism parameters.

In the followings the last Gioncu and Petcu plastic mechanism will be used in order to determine the ultimate rotation capacity of beams and beam-columns.

A complex local plastic mechanism is composed by a number of local plastic mechanisms, depending of moment variation:

-for *gradient moment*, the global mechanism is presented in Fig. 8.32a, being composed by two local asymmetrical plastic mechanisms. The mechanism rotates around the rotation center O with the angle θ, giving rise to

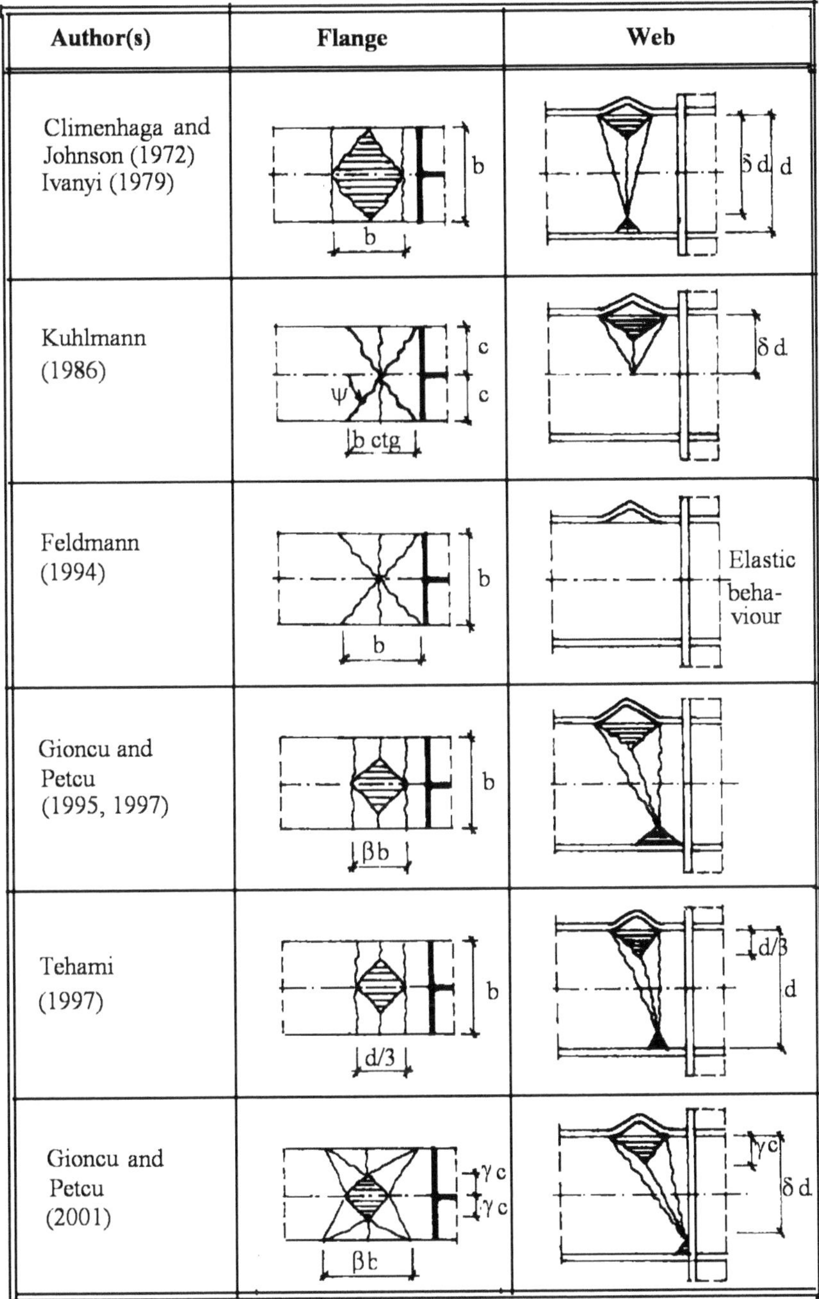

Figure 8.31: Different collapse types

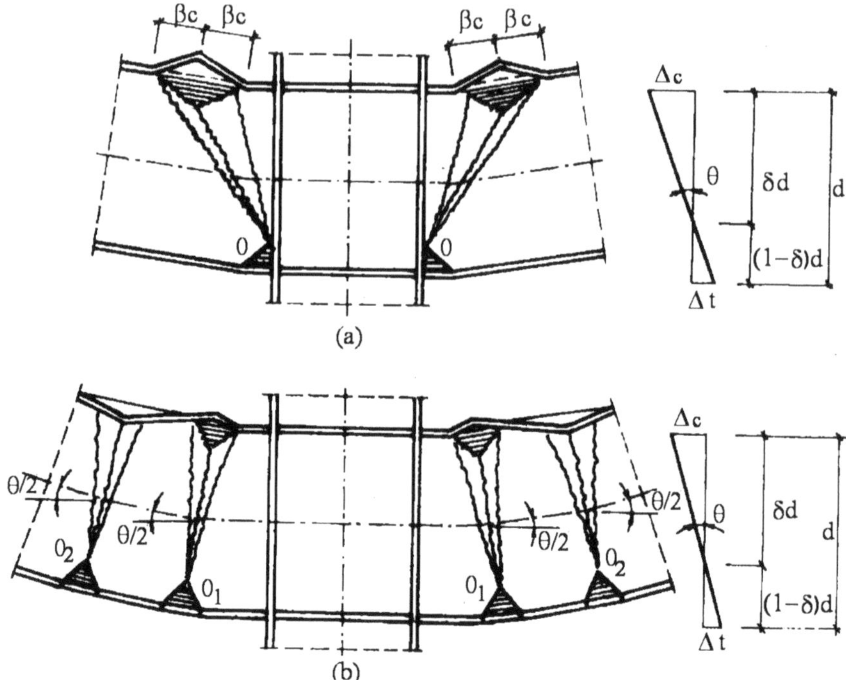

Figure 8.32: Local plastic mechanisms: (a) Gradient moment; (b) Quasi-constant moment

the displacements of compression and tension flanges:

$$\Delta_c = \delta d\theta \tag{8.40a}$$

$$\Delta_t = (1 - \delta)d\theta \tag{8.40b}$$

-for *quasi-constant moment*, the plastic mechanism is presented in Fig. 8.32b, being composed by four symmetrical local plastic mechanisms, having distinct tension zones. The plastic mechanism rotates around centers O_1 and O_2 with the angle $\theta/2$ and displacements of compression and tension flanges equal to the values obtained from (8.40).

The local plastic mechanism shape for gradient moments is shown in Fig. 8.33a. The plastic buckling is introduced by the instability of compression flange, which also involves the deformation of webs. The length of buckled flange is given by (see Section 7.2.2):

$$L_p = 2\beta c \tag{8.41a}$$

with

$$\beta = 0.713 \left(\frac{d}{b}\right)^{1/4} \left(\frac{t_f}{t_w}\right)^{3/4} \tag{8.41b}$$

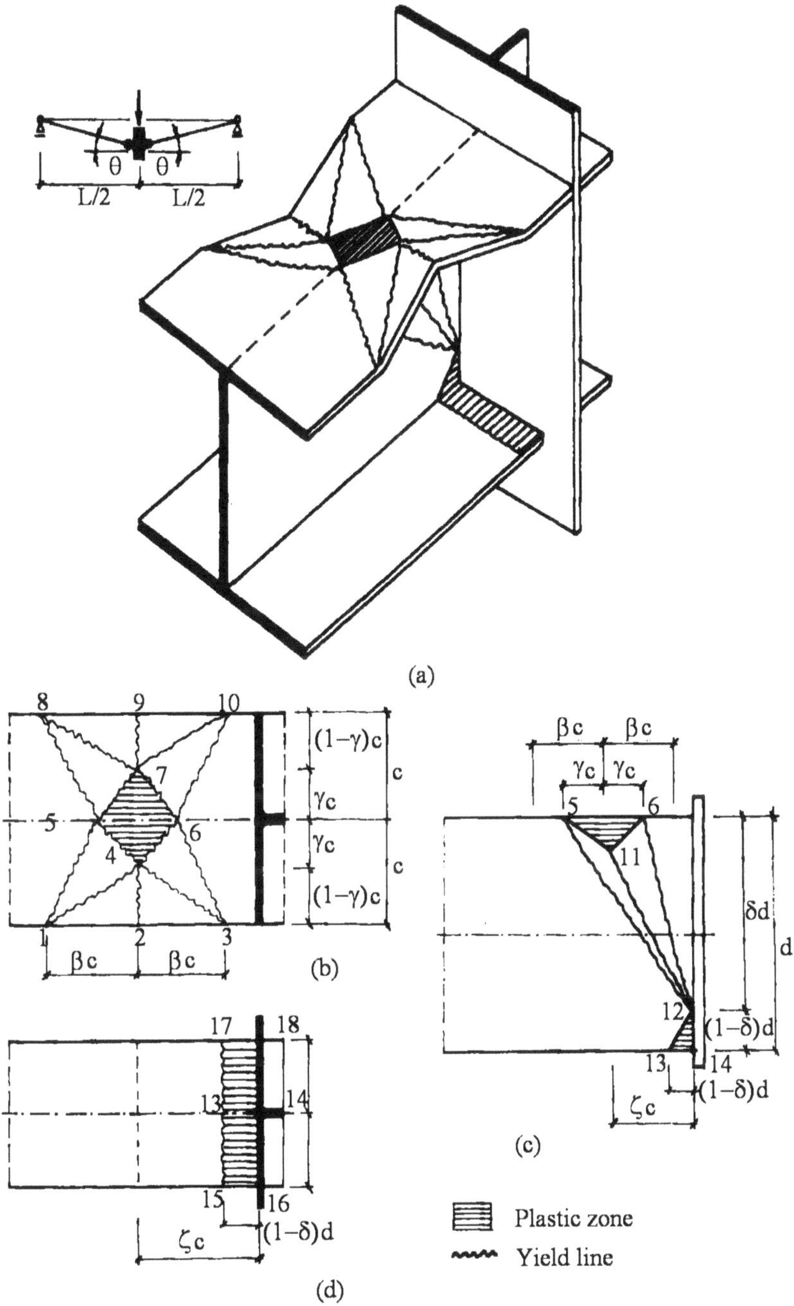

Figure 8.33: In-plane plastic mechanism: (a) General view; (b) Compression flange plastic mechanism; (c) Web plastic mechanism; (d) Tension flange plastic mechanism

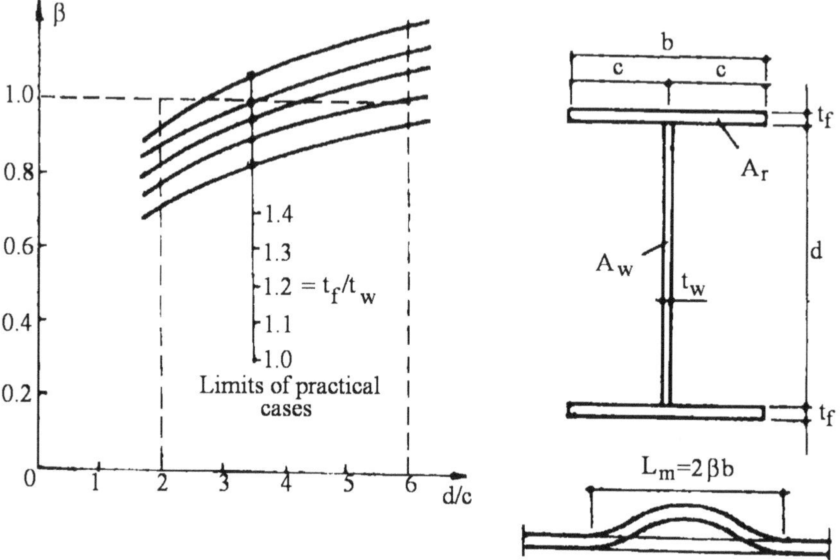

Figure 8.34: Length for plastic mechanism

For the usual I-section members Fig. 8.34 shows the values of the coefficient β. One can see that the practical range of these values is 0.8 to 1.2.

The actual shape of buckled flange is replaced by a plastic mechanism composed by plastic zones and yield lines (Fig. 8.33b). The plastic mechanism of web is composed by two yield lines (Fig. 8.33c). For tension flanges the mechanism is composed by one plastic zone and one yield line (Fig. 8.33d). The beam rotates with respect to the symmetry plane with the angle θ and the yield lines are travelling lines, only the mechanism length remaining constant (see Section 4.4.2). The dimensions, rotations and stresses of plastic zones and yield lines are presented in Table 8.4. For the yield lines the plastic moments are determined without considering the reduction due to simultaneous action of bending and compression, because this is compensated by the hardening effect which is neglected. For the stress in plastic zones an increased yield stress is considered in cases of gradient moment only, taking into account that the member works in a strain-hardening range (Fig. 8.29a):

$$\frac{f_{yf}^*}{f_{yf}} \simeq \frac{M_{sh}}{M_p} = \frac{1}{1 - \dfrac{2\beta c}{L}} \tag{8.42}$$

For constant moments this increasing of yield stresses is not necessary (Fig. 8.29b).

Using the rigid plastic analysis presented in Section 4.4.3 the variation of the internal potential energy U results:

Table 8.4: I-section: In-plane plastic mechanism

Plastic zones			
Zones	Volumes	Strains	Stresses
4-5-7-6	$2\gamma^2 c^2 t_f$	$\theta\delta d/(2\gamma c)$	f^*_{yf}
5-6-10	$\gamma^2 c^2 t_w$	$\theta\delta d/(2\gamma c)$	f_{yw}
11-12-13	$(1-\delta)^2 d^2 t_w$	$\theta/2$	f_{yw}
14-15- -16-17	$2(1-\delta)c_t t_{ft}$	$\theta/2$	f^*_{yf}
Yield lines			
Lines	Lengths	Rotations	Stresses
2-4-7-9	$(1-\gamma)c$	$2\left(\frac{\gamma}{\beta\chi}\right)^{1/2}\theta^{1/2}$	f_{yf}
1-4,3-4, 7-8,7-10	$[\beta^2+(1-\gamma)^2]^{1/2}c$	$\frac{1}{[\beta^2+(1-\gamma)^2]^{1/2}}\left[\frac{\beta}{1-\gamma}+\frac{\beta^2-3\gamma-\gamma\beta+1}{\gamma(1+\beta-\gamma)}\right]\left(\frac{\beta\gamma}{\chi}\right)^{1/2}\theta^{1/2}$	f_{yf}
1-5,3-6, 5-8,6-10	$[1+(\beta-\gamma)^2]^{1/2}c$	$\frac{[1+(\beta-\gamma)^2]^{1/2}}{\gamma(1-\beta-\gamma)}\left(\frac{\beta\gamma}{\chi}\right)^{1/2}\theta^{1/2}$	f_{yf}
4-5,4-6, 5-7,6-7	$2^{1/2}\gamma c$	$2^{1/2}\frac{1}{\gamma(1+\beta-\gamma)}\left(\frac{\beta\gamma}{\chi}\right)^{1/2}\theta^{1/2}$	f_{yf}
5-11	$2^{1/2}\gamma c$	$\left(\frac{2}{\chi}\right)^{1/2}\left[\frac{1-(1-\eta/\gamma)\chi}{1-(1+\eta/\gamma)\chi}\right]^{1/2}\theta^{1/2}$	f_{yw}
6-11	$2^{1/2}\gamma c$	$\left(\frac{2}{\chi}\right)^{1/2}\left[\frac{1-(1+\eta/\gamma)\chi}{1-(1-\eta/\gamma)\chi}\right]^{1/2}\theta^{1/2}$	f_{yw}
5-12	$[1+(1+\eta/\gamma)^2\chi^2]^{1/2}\delta d$	$\frac{1}{\chi^{1/2}}\left[\frac{1-(1-\eta/\gamma)\chi}{1-(1+\eta/\gamma)\chi}\right]^{1/2}\cdot\left[1+(1+\eta/\gamma)^2\chi^2\right]\theta^{1/2}$	f_{yw}
6-12	$[1+(1-\eta/\gamma)^2\chi^2]^{1/2}\delta d$	$\frac{1}{\chi^{1/2}}\left[\frac{1-(1+\eta/\gamma)\chi}{1-(1-\eta/\gamma)\chi}\right]^{1/2}\cdot\left[1+(1-\eta/\gamma)^2\chi^2\right]\theta^{1/2}$	f_{yw}
11-12	$[(1-\chi)^2+(\eta\chi/\gamma)^2]^{1/2}\delta d$	$\frac{(1-\chi)^2+(\eta\chi/\gamma)^2}{(1-\chi)^2-(\eta\chi/\gamma)^2}\theta^{1/2}$	f_{yw}
15-16, 17-18	b_t	θ	f_{yf}

* Increased yield stresses due to strain-hardening. For the other plastic zones and yield lines, only the randomness of yield stresses must be considered

-for gradient moment (two plastic mechanisms):

$$\frac{1}{cdt_f f_y}\frac{\partial U}{\partial \theta} = A + B\frac{1}{\theta^{1/2}} \tag{8.43}$$

where

$$A = 2\left[\gamma\delta\left(1+\frac{1}{2}\frac{t_w}{t_f}\frac{f_{yw}}{f_{yt}}\right)+\frac{1}{2}(1-\delta)^2\frac{dt_w}{ct_f}\frac{f_{yw}}{f_{yt}}+(2-\delta)+\frac{1}{2}\frac{t_f}{d}\right]\frac{f^*_{yf}}{f_{yf}} \tag{8.44a}$$

$$B = 2\left\{\frac{1}{2}\left[\frac{1-\gamma}{\beta}+\frac{\beta}{1-\gamma}+\frac{2+2\beta^2-3\beta\gamma+\gamma+\gamma^2}{\gamma(1+\beta-\gamma)}\right]\frac{t_f}{d}\frac{(\beta\gamma)^{1/2}}{\chi^{1/2}}+\right.$$
$$\left.+\frac{1}{2}\delta\frac{(2+\chi)(1+\eta^2/\gamma^2)\chi^2-\chi(2+\chi^2)}{\chi^{1/2}[(1-\chi)^2-\eta^2\chi^2/\gamma^2]^{1/2}}\frac{t_w^2}{bt_f}\frac{f_{yw}}{f_{yf}}\right\} \tag{8.44b}$$

with:

$$\chi = \frac{\gamma c}{\delta d} \tag{8.45}$$

is a geometric characteristic of the plastic mechanism and η a characteristic of mechanism asymmetry:

-for quasi-constant moment (four plastic mechanism):

$$\frac{1}{cdt_f f_y}\frac{dU}{d\theta} = A + 2B\left(\frac{2}{\theta}\right)^{1/2} \tag{8.46}$$

The variation of the external potential energy is (see equation (8.38)):

$$\frac{1}{cdt_f f_{yt}}\frac{\partial L_{ext}}{\partial \theta} = \frac{M}{M_p}C\frac{1+n_p\frac{M_p}{M}\frac{1}{\overline{e}_0}(\delta-0.5)}{1-0.822n_p\overline{\lambda}^{2}} \tag{8.47}$$

where, for gradient moment:

$$C = \left[4\left(1+\frac{t_f}{d}\right)+\frac{1}{2}\frac{dt_w}{ct_f}\frac{f_{yw}}{f_{yf}}\right]m_p \tag{8.48a}$$

and for quasi-constant moment:

$$C = \left[4\left(1+\frac{t_f}{d}\right)+\frac{1}{2}\frac{dt_w}{ct_f}\frac{f_{yw}}{f_{yf}}\right]m_p \tag{8.48b}$$

Equating the variations of internal and external potential energies, it results [(for $n_p = 0$):]

-for moment gradient:

$$\frac{M}{M_p} = \alpha_{1M} + \alpha_{2M}\frac{1}{\theta^{1/2}} \tag{8.49a}$$

-for quasi-constant moment:

$$\frac{M}{M_p} = \left[\alpha_{1M} + 2\alpha_{2M}\left(\frac{2}{\theta}\right)^{1/2}\right] \qquad (8.49b)$$

with the notations;

$$\alpha_{1M} = \frac{A}{C}; \qquad \alpha_2 = \frac{B}{C} \qquad (8.49c, d)$$

which represent the curves of plastic mechanism behaviour (Fig. 8.29c,d). The curves intersect the curve corresponding to the unbuckled section. An erosion of the cusp formed by this intersection occurs, giving rise to the maximum moment M_{max}. The second intersection with the horizontal line corresponding to M_p or $0.9M_p$ depends on the adopted ductility criterium.

The ultimate plastic rotation is given by:

-for moment gradient:

$$\theta_r = \left(\frac{\alpha_{2M}}{1-\alpha_{1M}}\right)^2; \quad \theta_{r0.9} = \left(\frac{\alpha_{2M}}{0.9-\alpha_{1M}}\right)^2 \qquad (8.50a, b)$$

-for quasi-constant moment:

$$\theta_r = \frac{1}{8}\left(\frac{\alpha_{2M}}{1-\alpha_{1M}}\right)^2; \quad \theta_{r0.9} = \frac{1}{8}\left(\frac{\alpha_{2M}}{0.9-\alpha_{1M}}\right)^2 \qquad (8.51a, b)$$

The ultimate rotation results from:

$$\theta_u = \theta_r + \theta_p \qquad (8.52)$$

where

-for moment gradient:

$$\theta_p = \frac{M_pL}{4EI} \qquad (8.53a)$$

-for quasi-constant moment:

$$\theta_p = \frac{M_pL}{3EI} \qquad (8.53a)$$

The geometrical parameters which govern the plastic mechanism are γ, δ and η. From the condition to obtain real solutions, the following relationship can be obtained:

$$\eta_{max} \leq (1-\chi)\frac{\gamma}{\chi} = \frac{\delta d}{c} - \gamma \qquad (8.54a)$$

This is a mathematical condition; for an actual pattern it is proposed that:

$$\eta_{max} \simeq 0.9\left(\frac{\delta d}{c} - \gamma\right) \qquad (8.54b)$$

Examining the equations (8.43) and (8.44), it results that the mechanism shape is dominated by the following variable geometrical parameters:

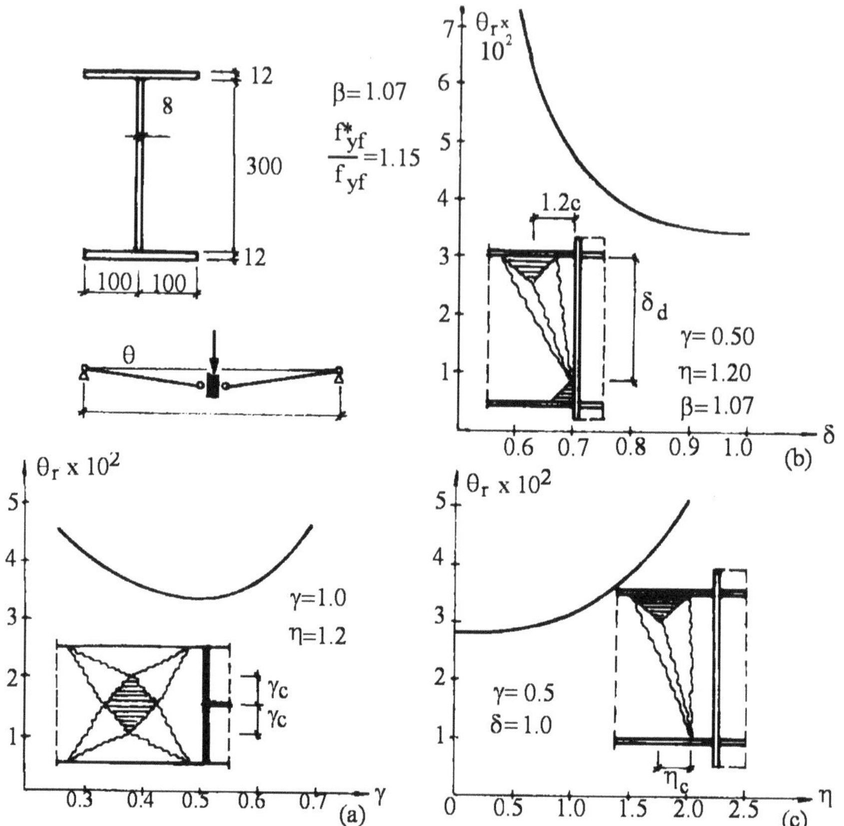

Figure 8.35: Geometrical parametrical study: (a) Influence of parameter γ; (b) Influence of parameter δ; (c) Influence of parameter η

-dimension of plastic zone of compression flange, γc;
-position of plastic mechanism rotation center, δd;
-asymmetry of plastic mechanism, ηc.

All these parameters must be treated as variables which need to be determined by minimizing the bending moment value or by imposing some constructional conditions. The only fixed geometrical characteristic is the buckled shape length.

The results of a parametrical study are presented in Fig. 8.35. One can see that a minimum value of plastic rotation is obtained by the variation of coefficient γ, the dimension of plastic zones in compression flanges being about half flange width (Fig. 8.35a). Relating the depth of web plastic mechanism the minimum value is obtained for $\delta = 1.0$, with the result that the rotation is produced around a point situated in tension flange (Fig. 8.35b). Concerning the mechanism asymmetry characterized by the coefficient η, the minimum of rotation capacity is obtained for a symmetrical shape (Fig. 8.35c). But this numerical finding is denied by the experimental evidence

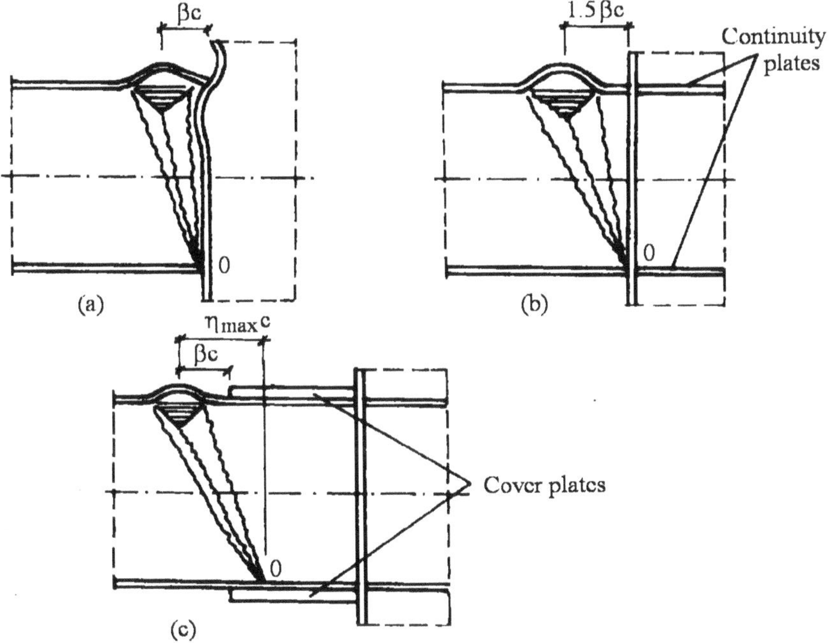

Figure 8.36: Asymmetry of plastic mechanism: (a) Column without continuity plates; (b) Column with continuity plates; (c) Beam with cover plates

where asymmetries in plastic mechanisms are observed. Concerning this aspect, there are two different situations:

-in case of *moment gradient* variation the parameter ηc does not result from a minimization. Indeed, the section rotation occurs around a point situated near the column flange. The asymmetry of a plastic mechanism is given by the position of a buckled flange. In case of a column without continuity plates, the position is situated near to the column flange and so $\eta = \beta$ (Fig. 8.36a), because the column is involved in flange rotation. In the presence of continuity plates, the position is remote from the column flange and so $\eta = 1.5\beta$ (Fig. 8.36b). In case of strengthened or weakened beam ends, when the position of buckled flange is fixed by constructional details, the asymmetry is given by distance $\eta_{max}c$ (see equation (8.54b)), (Fig. 8.36c);

-in the case of *quasi-constant moment*, the shape of plastic mechanism can be considered as a symmetrical one and, therefore, $\eta = 0$.

Fig. 8.37a shows the influence of the flange slenderness and of the ductility criterium. An important increase of ductility is obtained if the ductility is related to $0.9M_p$. In Fig. 8.37b the influence of moment variation is presented. The ductility of standard beam SB2 is higher than the ductility of standard beam SB1, showing the detrimental effect of moment gradient variation. In both cases an important increase of plastic rotation arises as far as the flange thickness increases.

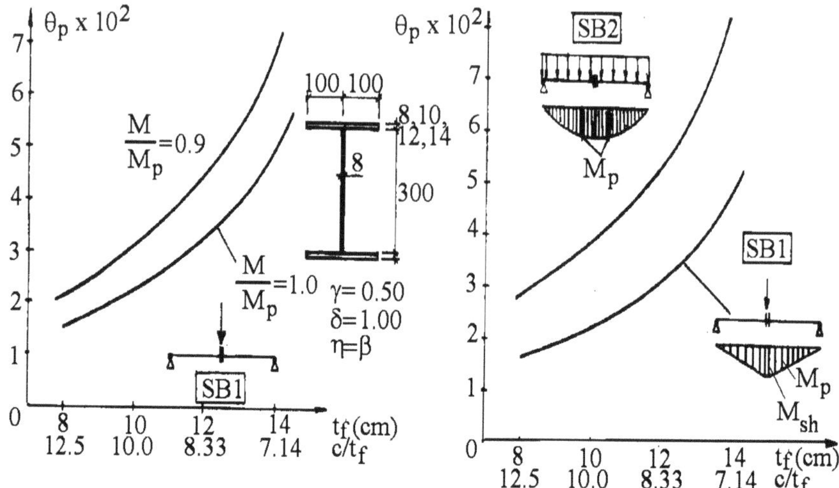

Figure 8.37: Influence of flange slenderness on plastic rotation: (a) Influence of ductility criterium; (b) Influence of moment variation

As it has been shown in Section 6.3.5, the member ductility is measured by the ultimate rotation capacity given by equations (6.47) and (6.48), as a function of the ductility criterium. So, in these cases, using the equations (8.50) to (8.53), it results:

$$\mu_\theta = \frac{\theta_r}{\theta_p}; \qquad \mu_{\theta 0.9} = \frac{\theta_{r0.9}}{\theta_p} \tag{8.55a, b}$$

Examining the ultimate plastic rotation and ultimate rotation capacity, it can be observed that the first concept is more tied to the section properties, while the second one is connected to the member properties, in which the influence of member span is very important (Fig. 8.38).

Fig. 8.39 shows the correspondence between theoretical and experimental results. One can see a satisfactorily concordance between these values, confirming that this methodology can be used for design practice.

(ii) *Out-of-plane plastic mechanism.* There are two out-of-plane mechanism types:

-*S-shaped plastic mechanism* is characterized by an asymmetrical shape around the plane of the mid-span stiffener (Fig. 8.40a). The compression flange behaves somewhat like a three-bar linkage (Earls and Galambos, 1998). This out-of-plane failure mode has been observed during the experimental tests performed by Lukey and Adams (1969). The corresponding plastic mechanism has been studied by Climenhaga and Johnson (1972) and Gioncu and Petcu (1997). This mechanism is typical for experimental SB1 beams, where there are no elements to prevent the rotation around beam mid-span. In a structure, this plastic mechanism can occur when the jointed column has a reduced torsional rigidity and in cases of lack of transverse beams;

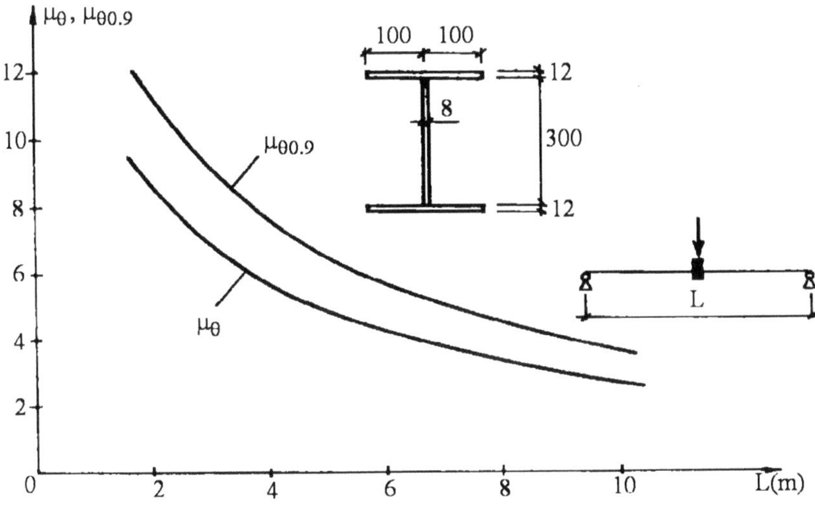

Figure 8.38: Influence of beam span on the rotation capacity

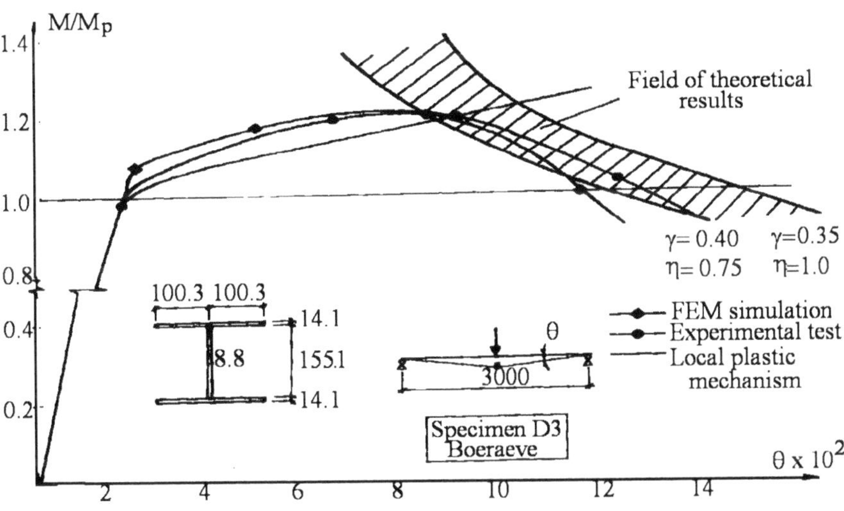

Figure 8.39: Correspondence between theoretical and experimental results

-*M shaped plastic mechanism* is characterized by a symmetrical shape around the plane of beam mid-span (Fig. 8.40b). The compression flange of beam behaves somewhat like a five-bar-linkage. This out-of-plane mechanism has been studied by Climenhaga and Johnson (1972), Ivanyi (1979) and Gioncu and Petcu (1997). The M shaped mechanism occurs when some elements (column or/and transverse beams) prevent the beam rotation around mid-span, being typical for the beam working in a complex structure.

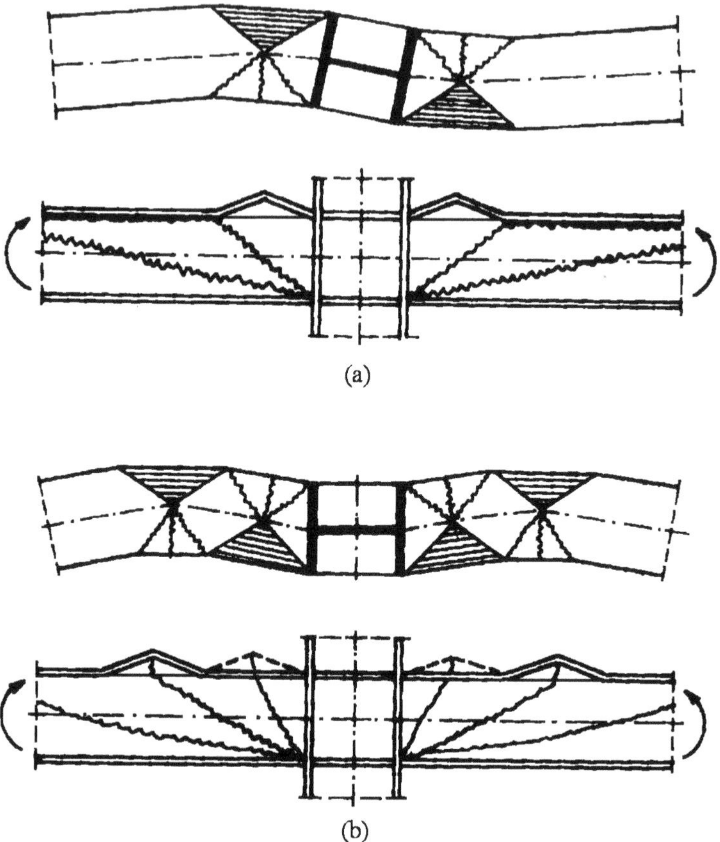

Figure 8.40: Out-of-plane plastic mechanisms: (a) S-shaped plastic mechanism; (b) M-shaped plastic mechanism

The characteristic dimensions of *S-shaped plastic mechanism* are presented in Fig. 8.41. The mechanism is composed by both plastic zones and yield lines for compression and tension flanges and yield lines only for web. The dimensions, rotations and stresses of these plastic zones and yield lines are presented in Table 8.5. The yield lines 11-12 and 11'-13 (being very long from the beam mid-span to beam ends) are incomplete yield lines (see Section 4.2.2), the plastic length being estimated to be equal to the beam depth. The lines 5A and 5B, having very small rotations, remain in the elastic range and do not dissipate energy.

Using the rigid plastic analysis presented in Section 4.4.3, the variation of internal potential energy results:

$$\frac{1}{cdt_f f_{yf}} \frac{\partial U}{\partial \theta} = A + B_1 \frac{1}{\theta^{1/2}} + B_2 \frac{1}{\theta^{3/4}} \tag{8.56}$$

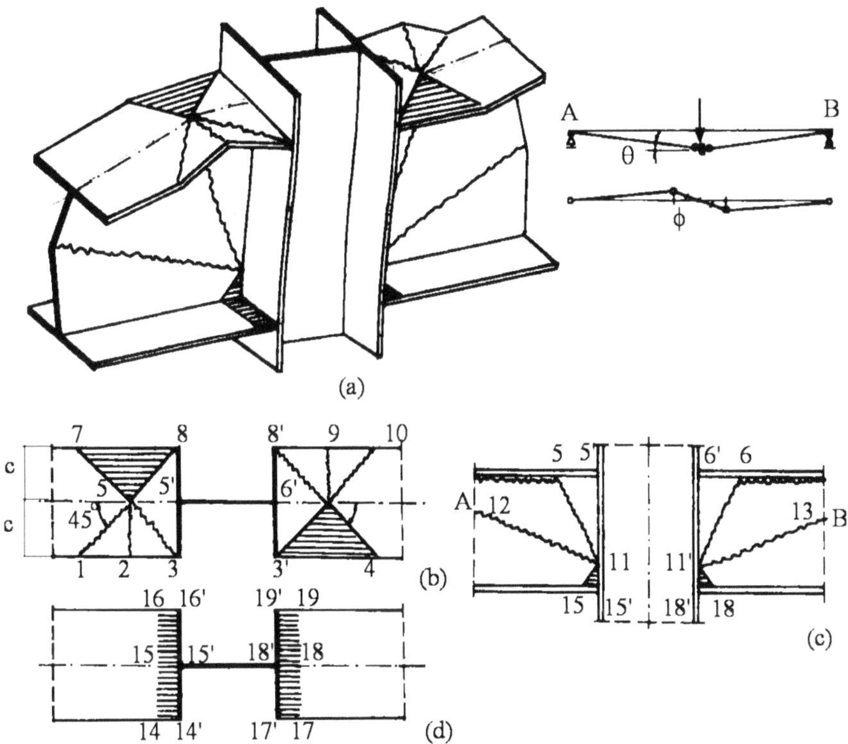

Figure 8.41: S-shaped plastic mechanism: (a) General view; (b) Compression flange plastic mechanism; (c) Web plastic mechanism; (d) Tension flange plastic mechanism

where

$$A = \left[\frac{t_{ft}}{d} + 2(1-\delta)\right] \frac{c_t}{c} \frac{t_{ft}}{t_f} \frac{f^*_{yf}}{f_y} + \frac{1}{2}(1-\delta)^2 \frac{d t_w}{c t_f} \frac{f_{yw}}{f_{yf}} \tag{8.57a}$$

$$B_1 = \frac{\sqrt{2}}{2}\left(\frac{c}{d}\right)^{1/2} \left\{ \delta^{1/2} \frac{f^*_{yf}}{f_{yf}} + \frac{1}{2}\left[\delta \frac{d}{c} + \right.\right.$$

$$\left.\left. + \frac{\left[1 + \frac{1}{\delta^2}\frac{c^2}{d^2} + \left(1 - \frac{2c}{L}\right)^2\right]^{1/2} + \frac{1}{\delta^{1/2}}\left(1 + \delta^2 \frac{4d^2}{L^2}\right)^{1/2}}{1 - \frac{2c}{L}}\right] \frac{t_w^2}{c t_f} \frac{f_{yw}}{f_{yt}} \right\} \tag{8.57b}$$

$$B_2 = \frac{3.2^{1/4}}{4} \delta^{1/4} \left(\frac{d}{c}\right)^{1/4} \frac{t_f}{d} \tag{8.57c}$$

Table 8.5: I-section: S-shaped plastic mechanism

Plastic zones			
Zones	Volumes	Strains	Stresses
3-4-6 5-7-8	$c^2 t_f$	$\frac{2^{1/2}}{2}\delta^{1/2}\left(\frac{d}{c}\right)^{1/2}\theta^{1/2}$	f^*_{yf}
11-15-15' 11'-18-18'	$(1-\delta)^2 d^2 t_w$	$\theta/2$	f_{yw}
14-14'-16-16' 17-17'-19'-19	$2(1-\delta) d c_t t_{ft}$	$\theta/2$	f^*_{yf}
Yield lines			
Lines	Lengths	Rotations	Stresses
1-5,3-5 6-8',6-10	$2^{1/2}c$	$2^{5/4}\delta^{1/4}\left(\frac{d}{c}\right)^{1/4}\theta^{1/4}$	f_{yf}
2-5 6-9	c	$2^{5/4}\delta^{1/4}\left(\frac{d}{c}\right)^{1/4}\theta^{1/4}$	f_{yf}
5-11 6-11'	$\left(1+\frac{c^2}{\delta^2 d^2}\right)^{1/2}\delta d$	$2^{1/2}\frac{1}{\left(1+\frac{c^2}{\delta^2 d^2}\right)^{1/2}}\left\{1+\frac{\left[1+\frac{c^2}{\delta^2 d^2}+\left(1-\frac{2c}{L}\right)^2\right]^{\frac{1}{2}}}{\delta\left(1-\frac{2c}{L}\right)}\cdot\frac{d}{c}\right\}\delta^{1/2}\left(\frac{d}{c}\right)^{1/2}\theta^{1/2}$	f_{yw}
11-12 11'-13	$\approx d$	$2^{1/2}\frac{1}{1-2c/L}\frac{1}{\delta^{1/2}}\left(\frac{c}{d}\right)^{1/2}\theta^{1/2}$	f_{yw}
5-A,6-B	$\frac{L}{2}\left(1-\frac{2c}{L}\right)$	≈ 0	Elastic range
14-16 17-19	$2c_t$	θ	f^*_{yf}

* Increased yield stresses due to strain-hardening. For the other plastic zones and yield lines, only the randomness of yield stresses must be considered

The variation of external potential energy is given by:

$$\frac{1}{cdt_f f_{yf}}\frac{\partial L_{ext}}{\partial \theta} = \frac{M}{M_p}C \tag{8.58}$$

where for only one plastic hinge:

$$C = 2\left(1+\frac{t_f}{d}\right) + \frac{1}{4}\frac{dt_w}{ct_f}\frac{f_{yw}}{f_{yf}} \tag{8.59}$$

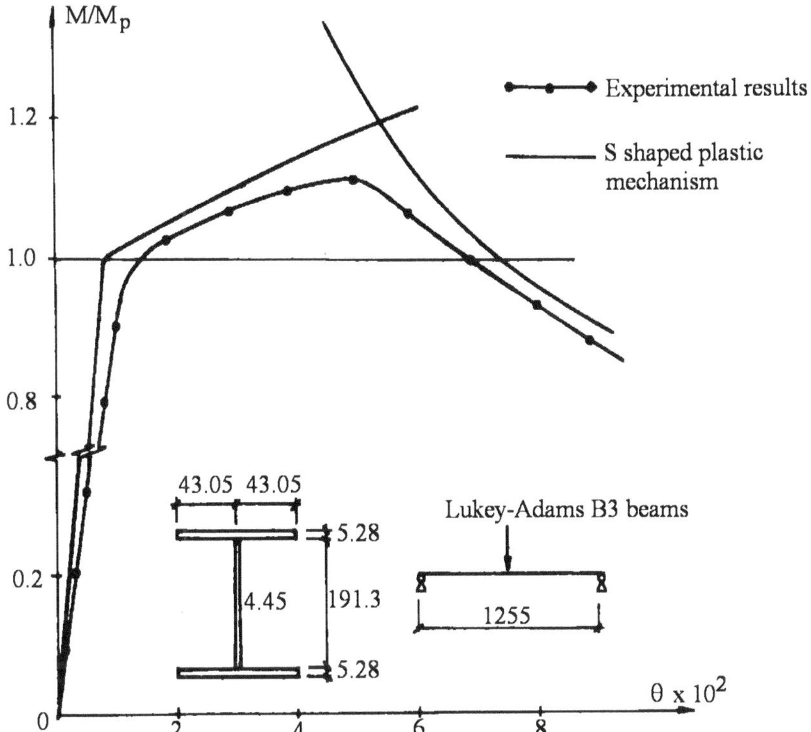

Figure 8.42: Correspondence between theoretical and experimental results

Equating the internal and external potential energies, it results (for $u_p = 0$):

$$\frac{M}{M_p} = \alpha_{1M} + \alpha_{2M}\frac{1}{\theta^{1/2}} + \alpha_{3M}\frac{1}{\theta^{3/4}} \tag{8.60a}$$

where

$$\alpha_M = \frac{A}{C}; \quad \alpha_{2M} = \frac{B_1}{C}; \quad \alpha_{3M} = \frac{B_2}{C} \tag{8.60b, c, d}$$

Equation (8.60a) represents the curve of the S-shaped plastic mechanism behaviour. This curve intersects the lines corresponding to M_p or $0.9M_p$, giving the ultimate rotation related to the adopted ductility criterion.

In order to verify the accuracy of this method, the experimental results obtained for specimen B3 of Lukey and Adams tests (1969) (see Fig. 7.3) has been used. The shape of buckled compression flanges shows that the collapse is due to an S-shaped out-of-plane mechanism. A very good correspondence exists between the experimental and theoretical results (Fig. 8.42).

The characteristic dimensions of *M-shaped plastic mechanisms* are presented in Fig. 8.43. This mechanism is similar to the S-shaped mechanism, but it is symmetric respect to the beam mid-span. The dimensions, rotations and stresses of the plastic zones and yield lines are presented in Table 8.6. The same approximations as for S-shaped mechanisms are used.

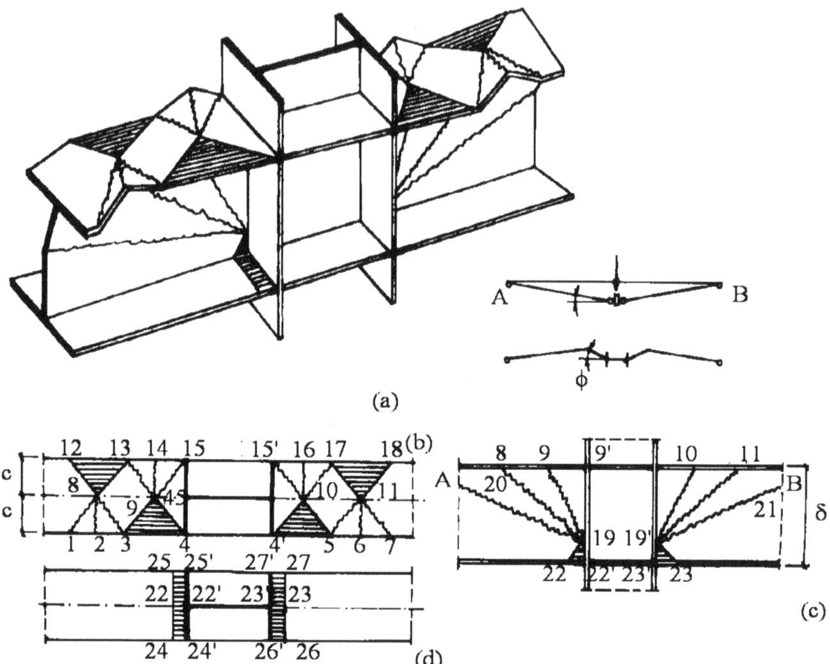

Figure 8.43: M-shaped plastic mechanism: (a) General view; (b) Compression flange plastic mechanism; (c) Web plastic mechanism; (d) Tension flange plastic mechanism

Similar equations as (8.56) and (8.58) are obtained, with the coefficients:

$$A = \frac{c_t}{c}\frac{t_{ft}^2}{dt_f}\frac{f^*_{yf}}{f_{yf}} + \frac{1}{2}(1-\delta)^2\frac{dt_w}{ct_f}\frac{f_{yw}}{f_{yf}} + 2(1-\delta)\frac{c_t}{c}\frac{t_{ft}}{t_f} \tag{8.61a}$$

$$B_1 = 2^{1/2}\delta^{1/2}\left(\frac{c}{d}\right)^{1/2}\frac{f^*_{yf}}{f_{yf}} + \left\{\frac{1}{4}\delta^{3/2}\left(\frac{1+\dfrac{1}{\delta^2}\dfrac{c^2}{d^2}}{1+\dfrac{9}{\delta^2}\dfrac{c^2}{d^2}} + 1 + \frac{3}{\delta^2}\frac{c^2}{d^2}\right)\left(\frac{d}{c}\right)^{1/2} + \right.$$

$$\left. +\delta^{1/2}\frac{\left[1+\left(\dfrac{3}{\delta}\dfrac{c}{d}+2\delta\dfrac{d}{L}\right)^2\right]^{1/2}}{1-\dfrac{6c}{L}}\left(\frac{c}{d}\right)^{1/2}\right\}\frac{t_w^2}{ct_f}\frac{f_{yw}}{f_{yf}} \tag{8.61b}$$

$$B_2 = \frac{3}{2}2^{1/4}\delta^{1/4}\left(\frac{d}{c}\right)^{1/4}\frac{t_f}{d} \tag{8.61c}$$

$$C = 4\left(1+\frac{t_f}{d}\right) + \frac{1}{2}\frac{dt_w}{ct_f}\frac{f_{yw}}{f_{yf}} \tag{8.61d}$$

A comparison between S-shaped and M-shaped mechanisms is presented in Fig. 8.44. One can observe that there are no important differences. The in-plane behaviour curve is also plotted, showing that, for the analyzed beam, this plastic mechanism decides the collapse mode. But the intersection between in-plane and out-of-plane curves in the post-critical range confirms the experimental observation that, in many experimental tests, buckling starts with an in-plane deformation and, in advanced post-critical behaviour, this buckling mode is transformed in an out-of-plane buckling, due to the weakening in lateral rigidity (Lay and Galambos, 1967).

(iii) *Rolled I-section members.* One of the main factors influencing member available ductility is the fabrication process. Hot-rolled profiles (as IPE, HEA, HEB), widely used in structural design, provide a larger rotation capacity than the welded sections (Anastasiadis, 1999, Anastasiadis and Gioncu, 1999). This increased ductility is due to the influence of the rigid zone created by the flange to web junction (Fig. 8.45a). The corresponding plastic mechanism is studied by Piluso (1995) and Anastasiadis (1999), by considering the flange to web junction as a plastic zone which does not allow any rotation (Fig. 8.45b). Fig. 8.46a,b shows the influence of this junction for different IPE and HEA profiles. One can see important increasing of plastic rotation capacity of hot-rolled sections as compared with the same sections in which the influence of the rigid zones are neglected (Anastasiadis and Gioncu, 1999). For IPE beams the increasing is about 64%, while for HEA beams it is about 38–48%. A simplified method to consider the effect of this junction is to correct the value of rotation capacity determined without junction with the coefficient c_j:

$$\theta_{rj} = c_j \theta_r \tag{8.62}$$

where (see Fig. 8.45a):

$$c_j = \left(\frac{c}{c_r}\right)^2 \tag{8.63a}$$

and

$$c_r = c - 0.5t_w - 0.8r \tag{8.63b}$$

r being the radius of flange to web junction. The correlation between the values obtained using the modified collapse mechanism and the ones obtained using equation (8.62) is presented in Fig. 8.46c, showing that this simple procedure allows us to determine the improved values of the rotation capacity of hot-rolled sections.

(iv) *Strengthened flanges.* In order to force the development of plastic hinges away from the column face, additional pieces as ribs, haunches, cover plates, sides plates, etc were proposed (see Section 6.4.3). The dimensions of the strengthened standard beam SSB1 are presented in Fig. 8.47a. For determining the required strengthened flange, it is recommended to consider different strain-hardening effects for beam and strengthened sections. The overstrength of beam can be of the order of 20–30% (Gioncu and Petcu, 1997), while for the strengthened section may be 5–10% only. The plastic

Table 8.6: I-section: M-shaped plastic mechanism

Plastic zones			
Zones	Volumes	Strains	Stresses
3-4-9 8-12-13 4'-5-10 11-17-18	$c^2 t_f$	$\frac{2^{1/2}}{2}\delta^{1/2}\left(\frac{d}{c}\right)^{1/2}\theta^{1/2}$	f^*_{yf}
19-22-22' 19'-23-23'	$(1-\delta)^2 d^2 t_w$	$\theta/2$	f_{yw}
24-24'-25-25' 26-26'-27-27'	$2(1-\delta)dc_t t_{ft}$	$\theta/2$	f^*_{yf}
Yield lines			
Lines	Lengths	Rotations	Stresses
1-8,3-8 9-13,9-15 5-11,7-11 10-15',10-17	$2^{1/2}c$	$2^{5/4}\delta^{1/4}\left(\frac{d}{c}\right)^{1/4}\theta^{1/4}$	f_{yf}
2-8,9-14 10-16,6-11	c	$2^{5/4}\delta^{1/4}\left(\frac{d}{c}\right)^{1/4}\theta^{1/4}$	f_{yf}
9-19 10-19'	$\left(1+\frac{c^2}{\delta^2 d^2}\right)\delta d$	$\frac{\left(1+\frac{c}{\delta^2 d}\right)^{1/2}}{1+\frac{9c^2}{\delta^2 d^2}}\delta^{1/2}\left(\frac{d}{c}\right)^{1/2}\theta^{1/2}$	f_{yw}
8-19 11-19'	$\left(1+\frac{9c^2}{\delta^2 d^2}\right)\delta d$	$\frac{1}{1+\frac{9c^2}{\delta^2 d^2}}\left\{1+\frac{3c^2}{\delta^2 d^2}+ +\frac{2}{\delta}\frac{c}{d}\frac{\left[1+\left(\frac{3c}{\delta d}+2\delta\frac{d}{L}\right)^2\right]^{1/2}}{1-\frac{6c}{L}}\right\}. \cdot\delta^{1/2}\left(\frac{d}{c}\right)^{1/2}\theta^{1/2}$	f_{yw}
19-A 19'-B	$\approx\left(1+\frac{9c^2}{\delta^2 d^2}\right)\delta d$	$\frac{2}{\delta^{1/2}}\left(\frac{c}{d}\right)^{1/2}\frac{\left[1+\left(\frac{3c}{\delta d}+2\delta\frac{d}{L}\right)^2\right]^{1/2}}{\left(1-\frac{6c}{L}\right)\left(1+\frac{9c^2}{\delta^2 d^2}\right)^{1/2}}$	f_{yw}
8-A,11-B	$\frac{L}{2}-3c$	≈ 0	Elastic range
24-25,26-27	$2c_t$	θ	f^*_{yf}

* Increased yield stresses due to strain-hardening. For the other plastic zones and yield lines, the randomness of yield stresses must be considered only

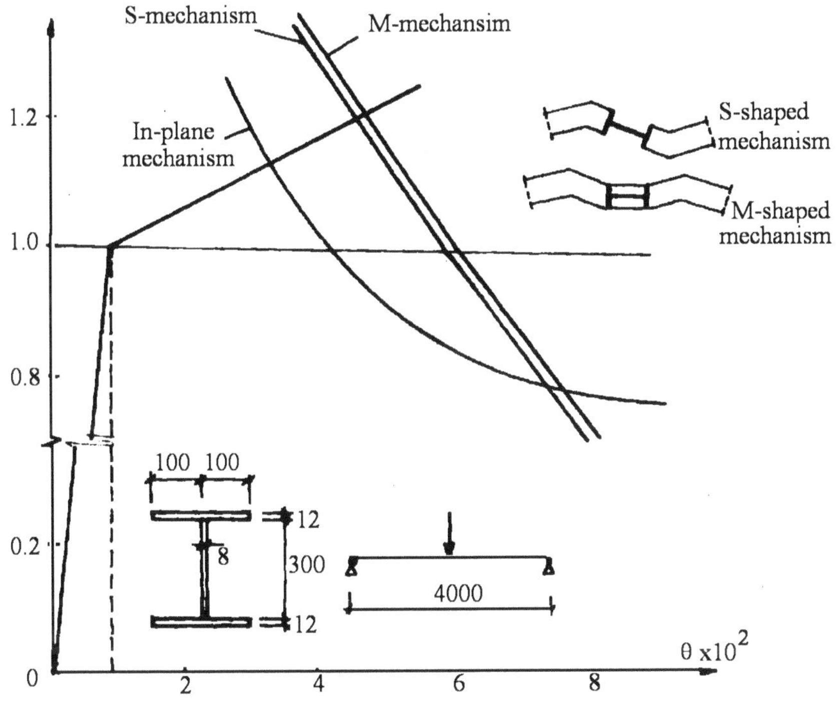

Figure 8.44: In-plane and out-of-plane S- and M-shaped plastic mechanisms comparison

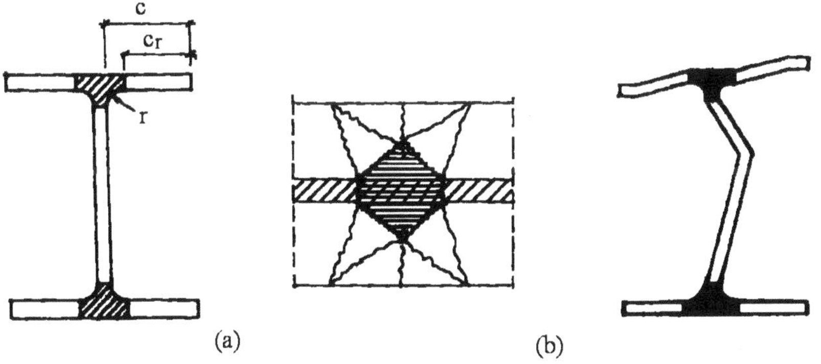

Figure 8.45: Hot-rolled I-section: (a) Flange-web junction; (b) Plastic mechanism with rigid junction

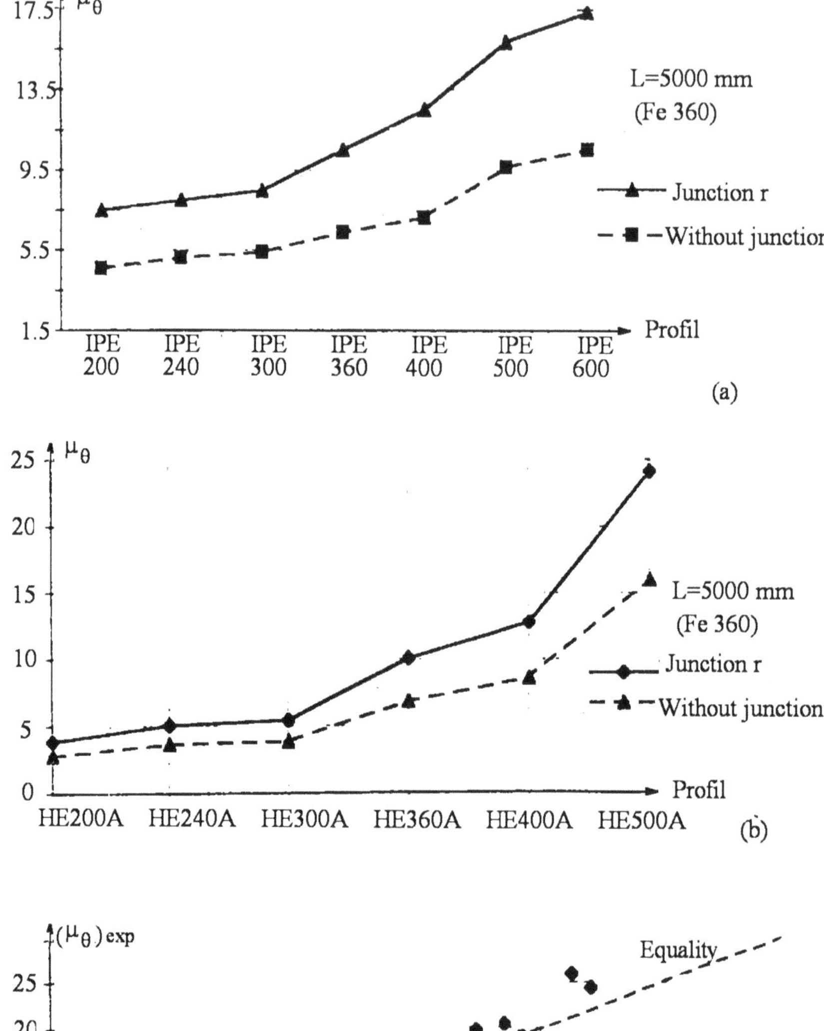

Figure 8.46: Influences of flange-web junction: (a) IPE-profiles; (b) HEA profiles; (c) Correlation between theoretical plastic mechanism values and simplified method

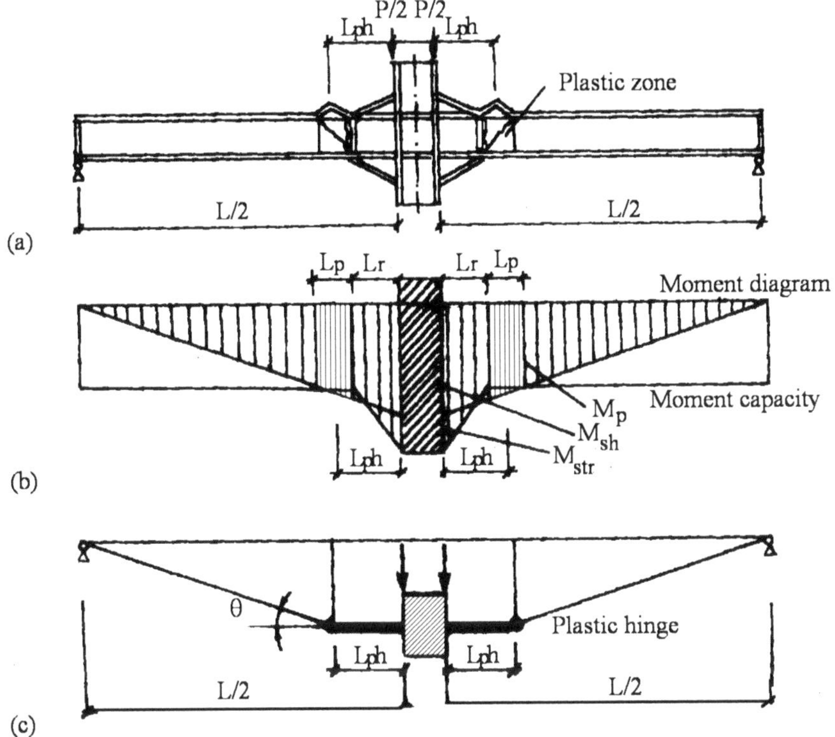

Figure 8.47: Strengthened flanges: (a) General view; (b) Bending moment diagrams; (c) Position of plastic hinges

moment capacity of strenthened section must be (Fig. 8.47b):

$$M_{str} > \frac{M_p}{s} \frac{1}{1 - 2\dfrac{L_r + L_p}{L}} \tag{8.64}$$

where the coefficient s considers the effect of strain-hardening: $s = 0.65$–0.75. The plastic hinge occurs at the distance $L_{ph} = L_p/2 + L_r$ from the column face (L_p is given in Fig. 6.48, as a function of the strengthening type). So, the rotation θ_p corresponding to the formation of plastic hinge is (Fig.8.47c):

$$\theta_p = \frac{M_p L \left(1 - 2\dfrac{L_{ph}}{L}\right)}{4EI} \tag{8.65}$$

The pattern of mechanism depends on the strengthening system. The asymmetry of a plastic mechanism in the presence of vertical ribs is not so important as in the case without these ribs (Fig. 8.48a). The influence of stiffener length on the ultimate rotation capacity is presented in Fig. 8.48b. One can see that the ductility of members without vertical ribs is

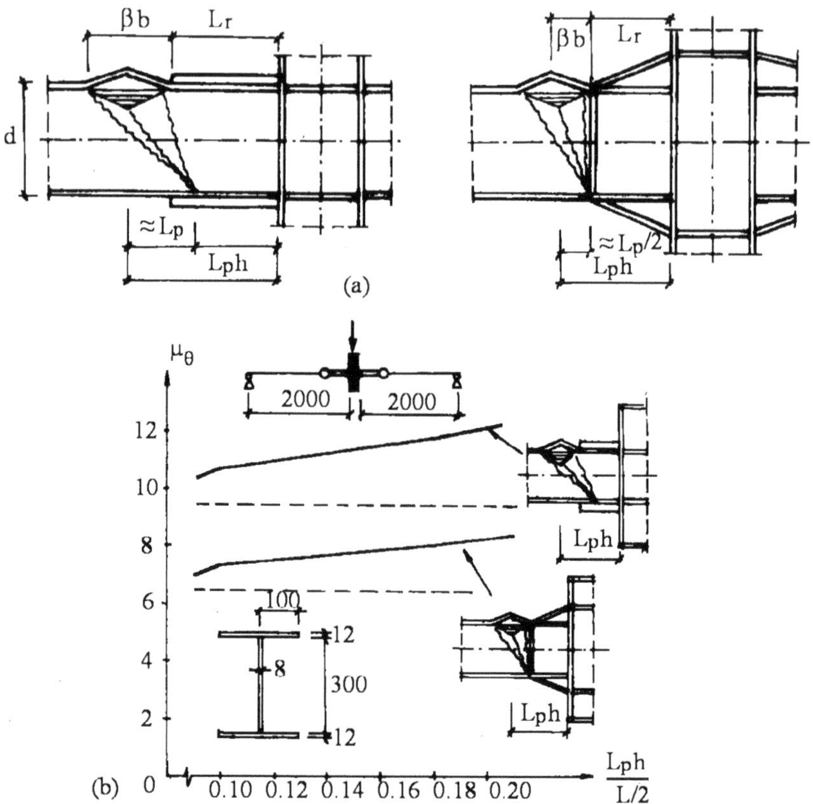

Figure 8.48: Influence of strengthening type: (a) Beam without and with vertical ribs; (b) Influence of rib lengths

higher than the solution with vertical ribs, because the presence of these ribs impedes the development of a more favorable plastic mechanism.

(v) *Weakened flanges.* Among the solutions for safeguard of brittle joints and for obtaining a collapse mechanism of global type, the idea of the weakening of flanges at the beam ends has been proved to be very effective. By cutting the beam flange near the joints in a specific zone (dog-bone), the formation of plastic hinges is assured due to the reduced moment capacity (Fig. 8.49). The constructional details are given in Section 6.4.3 (Fig. 6.44), where different solutions for member weakening are illustrated. Analyzing all these solutions, Plumier (1996) considers that the polygonal cuts with smoothed connecting curves between the straight lines would eliminate the stress concentration. The fully curved cut impedes the formation of an efficient plastic mechanism, while the tapered cut is not adequate when the bending moment diagram can have different shapes, as in the case of seismic actions. The weakened beam section with smoothed cuts is schematically illustrated in Fig. 8.50a. The distance L_1 must be positioned about $h/6$ to $h/3$ from the column face, in order to protect the connection as well as to

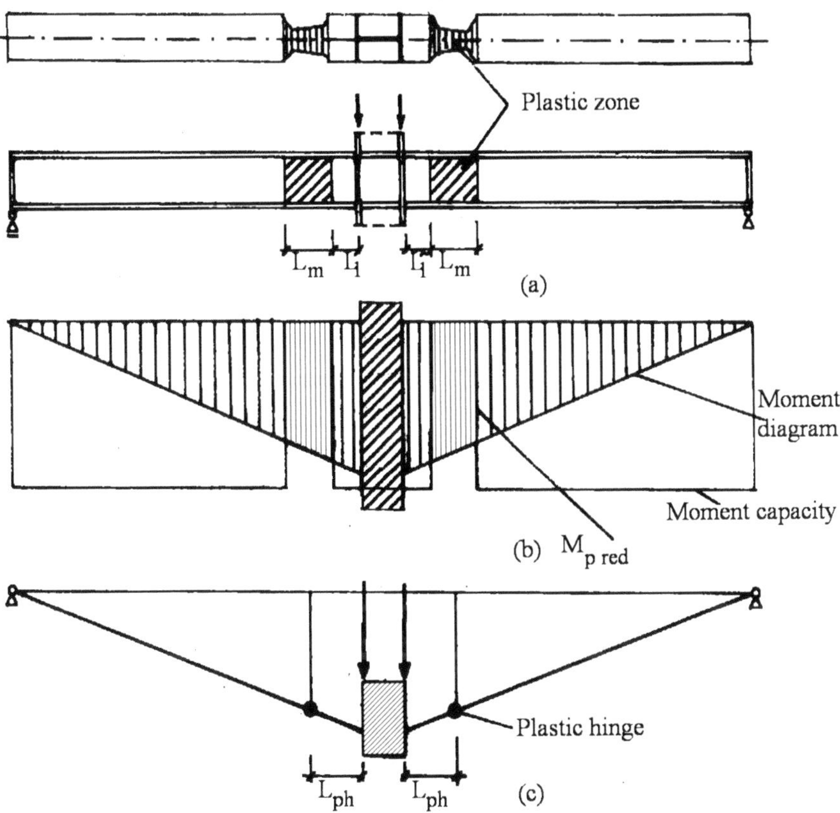

Figure 8.49: Weakened flanges: (a) General view; (b) Bending moment diagrams; (c) Position of plastic hinges

create a sufficient distance for the formation of plastic hinge. The transition zone L_2 must be sized avoiding stress concentration, having a length about 1.2 to $1.5c_r$, where c_r is the reduced flange width. Generally, a reduction of 35–45% is sufficient to provide an improved ductile behaviour. Finally, the critical weakened zone, L_w, is approximately positioned about $0.75h$ to h, but it must assure the formation of the plastic mechanism of buckled flange $L_w > \beta b$ (see relationship (8.41)) (Anastasiadis et al, 1999a,b). A short length impedes the formation of an adequate dissipative plastic hinge, while an overlong length may allow the occurring of two plastic hinges producing an unexpected local beam mechanism. A decreasing of moment capacity of 5–10% is proposed in order to ensure that yield occurs in the reduced section (Chen et al, 1996). So, the reduced plastic moment capacity, M_{pred} results (Anastasiadis et al, 2000):

$$\frac{M_{pred}}{M_p} = (0.90 \div 0.95)\left[\frac{2L_{ph}}{L} - 1 + \frac{1}{2\alpha}\frac{L_{ph}}{L}\left(1 - \frac{L_{ph}}{L}\right)\right] \tag{8.66}$$

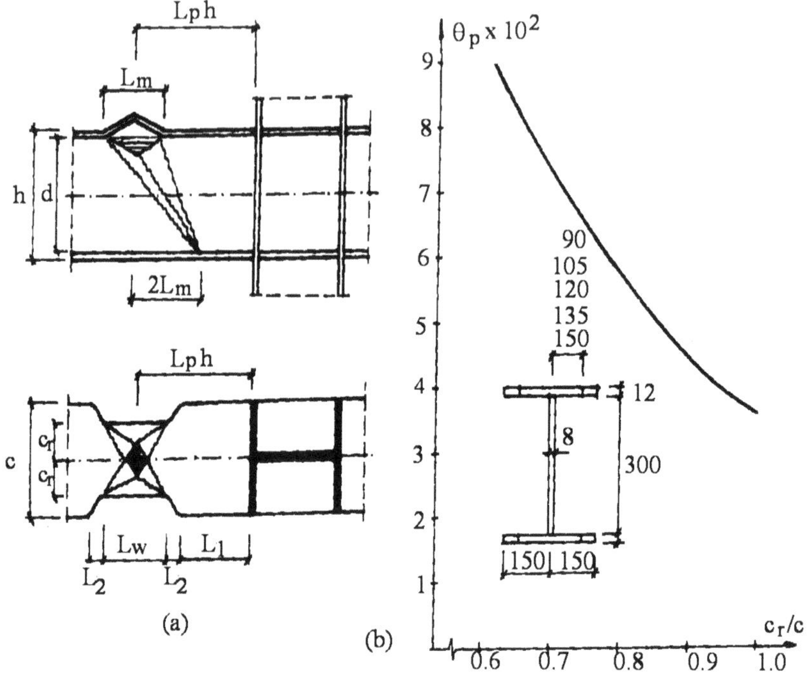

Figure 8.50: Influence of weakening: (a) Plastic mechanism shape; (b) Influence of reducing of flange width on plastic rotation

where L_{ph} is the distance from the column face to the assumed center of plastic hinge, L, the beam span, α, a coefficient introducing the influence of vertical loads $\alpha = M_p/qL^2$. Analyzing the above equation, a relation for direct sizing of the reduced width can be obtained:

$$\frac{c_r}{c} = \left[1 + \frac{d^2 t_w}{4ct_f(d+t_f)}\right] \frac{M_{pred}}{M_p} - \frac{d^2 t_w}{4ct_f(d+t_f)} \tag{8.67}$$

where the geometrical dimensions are illustrated in Fig. 8.50a. The increasing of ultimate rotation due to the flange reduction is presented in Fig. 8.50b. The effect is very important, especially due to the increasing asymmetry of plastic mechanisms. One must mention that, in addition to the protection of connections against brittle fracture and to the increasing of member ductility, the weakening procedure assures the attainment of global collapse mechanisms (Faggiano and Mazzolani, 1999, Anastasiadis et al, 1999, Montuori and Piluso, 2000).

(vi) *Stiffened webs.* In order to obtain an increasing of rotation capacity, the web can be provided with horizontal or vertical stiffeners (Fig. 8.51). The horizontal stiffeners change the point of rotation of the plastic mechanism. Using equation (8.50a), the effect of position of stiffeners is presented in Fig. 8.51a. One can see that the ultimate plastic rotation increases as far as the ratio d_s/d decreases. In the range $1.0 > d_s/d > 0.75$ the effect is quite

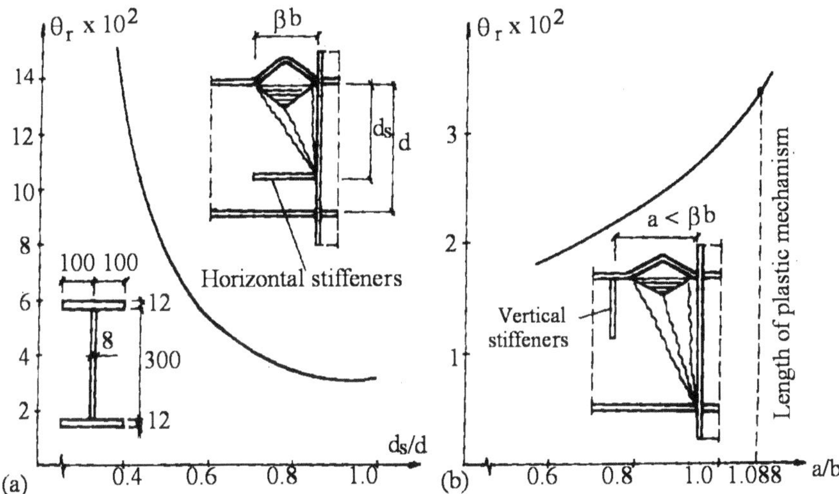

Figure 8.51: Influence of web stiffening: (a) Horizontal stiffeners; (b) Vertical stiffeners

insignificant, but for $d_s/d < 0.75$ the increasing of rotation capacity can be very important. The vertical stiffeners (Fig. 8.51b) reduce the length of the plastic mechanism. The effect of these stiffeners is detrimental for rotation capacity, the reduction of mechanism length producing a decreasing of the rotation capacity (Fig. 8.51b). Therefore, the use of vertical stiffeners is an unrecommended solution when the scope of design is for increasing ductility.

(vii) *Haunched or tapered members.* The use of haunched or tapered beams and beam-columns leads to more economic structures and, consequently, these solutions have wide use in moment-resisting frames. Generally, the haunching or tapering of members is achieved by using trapezoidal plates or by welding a cut half section. The corresponding standard beams are presented in Fig. 8.52a, with the haunches in the compression zone. The haunched or tapered portions of the beam are subjected to large in-plane moments, causing high compressive stresses and plastic buckling of inclined haunch flanges (Andrade and Morris, 1985, Szabo, 1990). By using the equation (8.49a), in which the geometrical characteristics of the two manufacturing solutions are used, the post-critical behaviour is obtained (Fig. 8.52b). The first solution has the web slenderness at the limit of ductility class 1 and, therefore, a reduced rotation capacity and very steep post-critical behaviour. Contrary to this, the second solution, having a reduced web slenderness, presents a larger rotation capacity and a reduced degradation in post-critical behaviour.

(viii) *Bended beams on the weak axis.* The experimental and numerical tests show that the plastic mechanism is composed by two distinct parts. The buckled shape of the compressed flange is composed by yield lines only, whilst in the tension part, plastic zones are formed. The web is bent

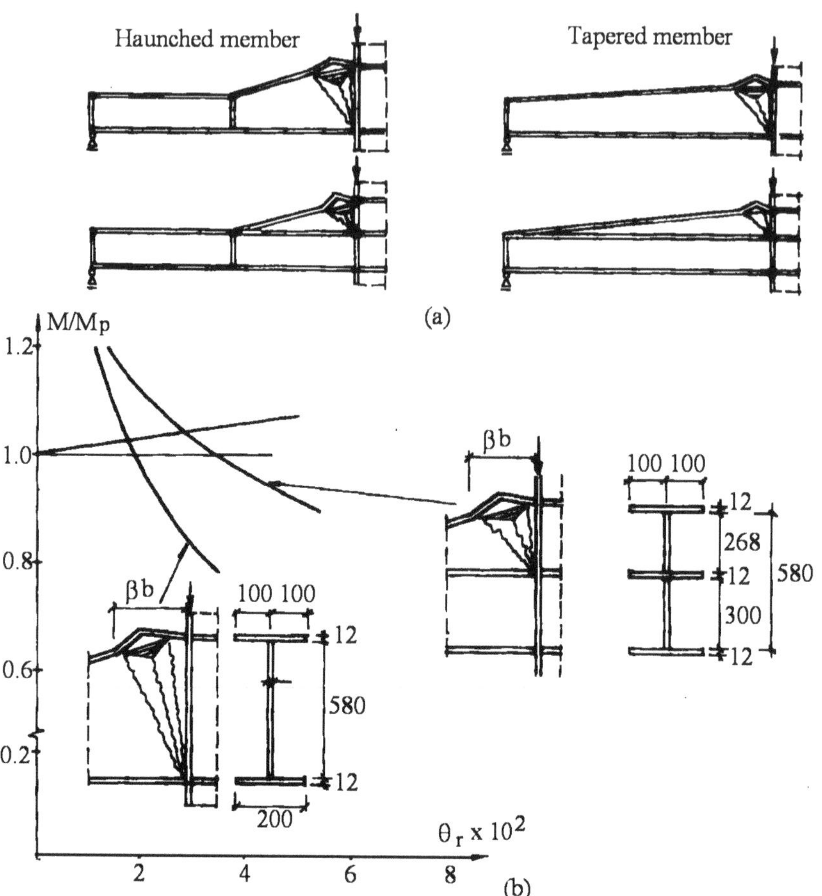

Figure 8.52: Haunched or tapered members: (a) General view with local plastic mechanism; (b) Rotation capacity

along a yield line (Fig. 8.53). The geometrical and numerical properties of this plastic mechanism are presented in Table 8.7. Using the rigid plastic analysis presented in Section 4.4.3 the variation of internal potential energy is given by:

$$\frac{1}{cdt_f f_{yf}} \frac{\partial U}{\partial \theta} = A + B \frac{1}{\theta^{1/2}} \tag{8.68}$$

where:

$$A = \frac{c}{d} + \frac{t_w^2}{ct_f} \frac{f_{yw}}{f_{yt}} \tag{8.69a}$$

$$B = \frac{3}{2} \frac{t_f}{d} \tag{8.69b}$$

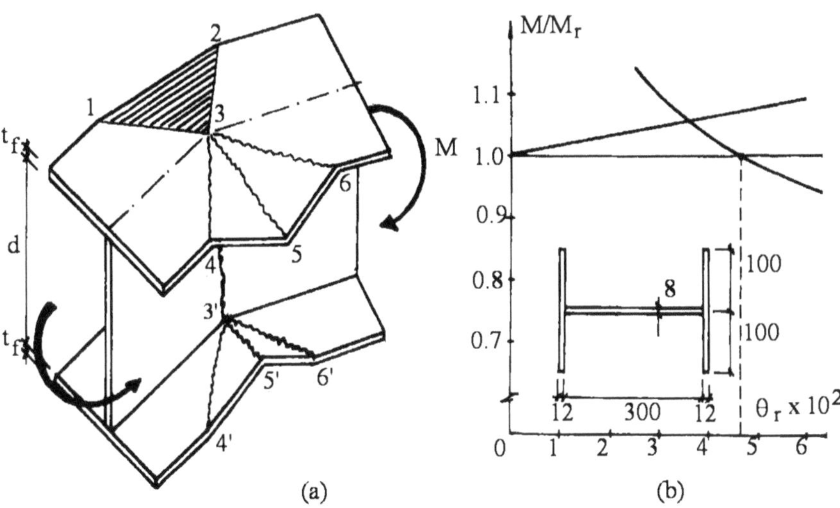

Figure 8.53: Bended beam on the weak axis: (a) Plastic mechanism; (b) Ultimate plastic rotation

Table 8.7: I-section: Members bended on weak axis

Plastic zones			
Zones	Volumes	Strains	Stresses
123 1'2'3'	$c^2 t_f$	$\theta/2$	f_{yf}
Yield lines			
Lines	Lengths	Rotations	Stresses
3-4,3-6 3'-4',3'-6'	$2^{1/2}c$	$2^{1/2}\theta^{1/2}$	f_{yf}
3-5,3'-5'	c	$2\theta^{1/2}$	f_{yf}
3-3'	d	2θ	f_{yw}

The external potential energy is:

$$\frac{1}{cdt_f f_{yf}} \frac{\partial L_{ext}}{\partial \theta} = \frac{M}{M_p} C \tag{8.70}$$

where

$$C = 2\frac{c}{d} \tag{8.71}$$

By equating the internal and external potential energy variations, it results:

$$\frac{M}{M_p} = \frac{1}{C}\left(A + B\frac{1}{\theta^{1/2}}\right) \tag{8.72}$$

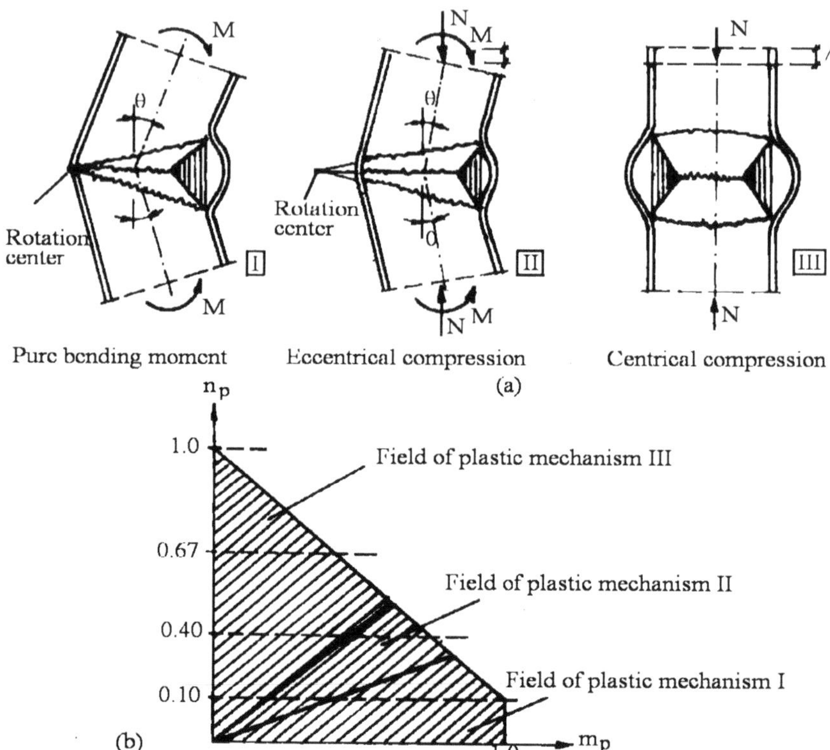

Figure 8.54: Beam-column members: (a) Plastic mechanism types; (b) Field of different plastic mechanism types

The ratio plotted in Fig. 8.53b allows us to determine the ultimate rotation for bending around the weak axis.

(ix) *Beam-column members.* The beam-column behaviour is characterized by the presence of axial forces, which can reduce the rotation capacity. The evolution of collapse mechanism from pure bending moment to simple axial compression is presented in Fig. 8.54a. Considering the plastic mechanism for a bending moment as the base for the evaluation of the ultimate rotation, the main aspect which must be lightened is the limit of eccentricity for which important changing in this plastic mechanism pattern occurs. If:

$$\frac{e}{d} = \frac{M}{Nd} = \frac{m_p}{n_p} e_0 \tag{8.73}$$

where $e_0 = M_p/N_p d$ and for I-sections:

$$e_0 = \frac{W_p}{Ad} \simeq \frac{1}{2} \frac{1 + \dfrac{1}{B}\dfrac{dt_w}{ct_f}}{1 + \dfrac{1}{4}\dfrac{dt_w}{ct_f}} \approx 0.35..0.45 \tag{8.74}$$

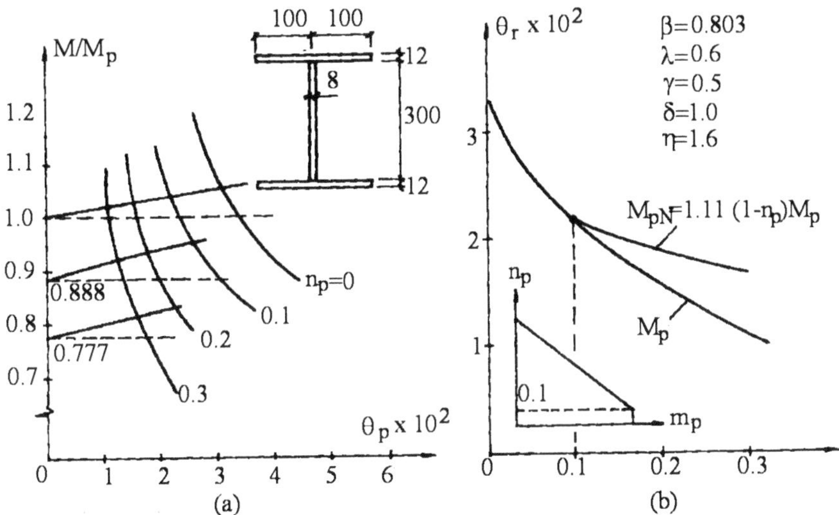

Figure 8.55: Influence of axial force level: (a) Post-critical behaviour; (b) Reduction of plastic rotation due to axial force

some change in plastic mechanism occurs for $e < d$ only, for which the range of axial level is:

$$n_p < (0.35..0.45)m_p \simeq 0.35..0.45 \tag{8.75a}$$

the value of rotation capacity being determined for $m_p = 1.0$. In this range, the collapse plastic mechanism I, determined for pure bending moment can also be used for beam-column analysis. The plastic mechanism III, corresponding to simple axial compression can be used for $e < 0.6d$, resulting

$$n_p < 0.67m_p \simeq 0.67 \tag{8.75b}$$

For

$$0.4 < n_p < 0.67 \tag{8.75c}$$

the plastic mechanism II must be used. Generally, the axial force level in moment-resisting frames is less than 0.4 and, therefore, the plastic mechanism determined for sections in pure bending can be used. The influence of the axial force level is plotted in Fig. 8.55a. One can see a very important degradation of the post-critical curves due to the effect of axial force. The ultimate plastic rotation decreases with the increasing of axial force level (Fig. 8.55b). The decreasing is less severe if the ultimate rotation is defined for reduced moment capacity, considering the effect of axial force.

(x) *Fracture behaviour*. As it has been shown in Section 7.2.6, the fracture of an I-section member can be due to cracks produced in the tension flange or by cracks in the buckled compressed flange (Fig. 8.56). The collapse by fracture of tension flanges occurs when the ultimate strain $\epsilon_{uf} = (\epsilon_u + \epsilon_t)/2$ for the couple of values (see Sections 4.4.4 and 5.2.2) (Fig. 8.56a):

$$\theta_{uf} = 2\epsilon_u; \quad \frac{M_u}{M_p} = \frac{1}{4}\left(1 + \frac{3}{\rho_y}\right) \tag{8.76a, b}$$

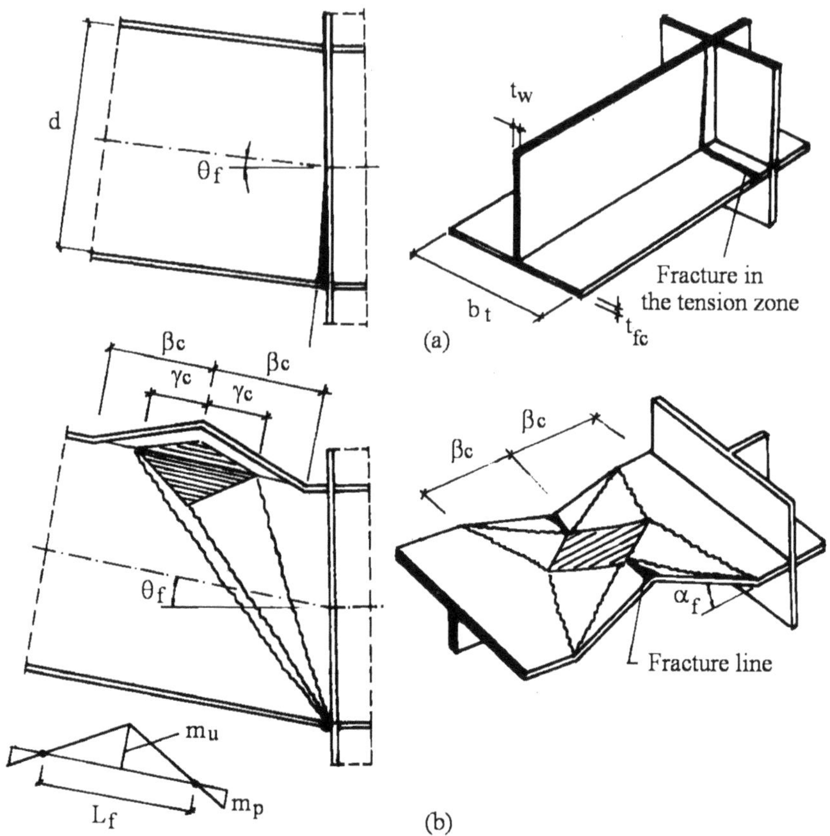

Figure 8.56: Fracture behaviour: (a) Fracture of tension flange; (b) Fracture of buckled compression flange

where ρ_y is the yield ratio (see Section 4.2.3). For ultimate strain an average value between uniform and total strains is proposed, considering that the value of ultimate strain ϵ_u obtained on a specimen is not the actual characteristic for the steel working in an element. The fracture of a compression flange occurs when some yield lines change in fracture lines (see Sections 4.5.2 and 7.2.6) (Fig. 8.56b). The fracture rotation is given by equation (4.59):

$$\alpha_f = \zeta \left(\frac{1}{\rho_y} - 1 \right) \frac{L_f}{t_f} \epsilon_{uf} \qquad (8.77a, b)$$

where L_f can be determined from (Fig. 8.56b):

$$L_f = \frac{1}{1 + \dfrac{m_p}{m_n}} b = \frac{2}{1 + \rho_y} c \qquad (8.78)$$

where m_p and m_u are the plastic and fracture moments of flange yield lines.

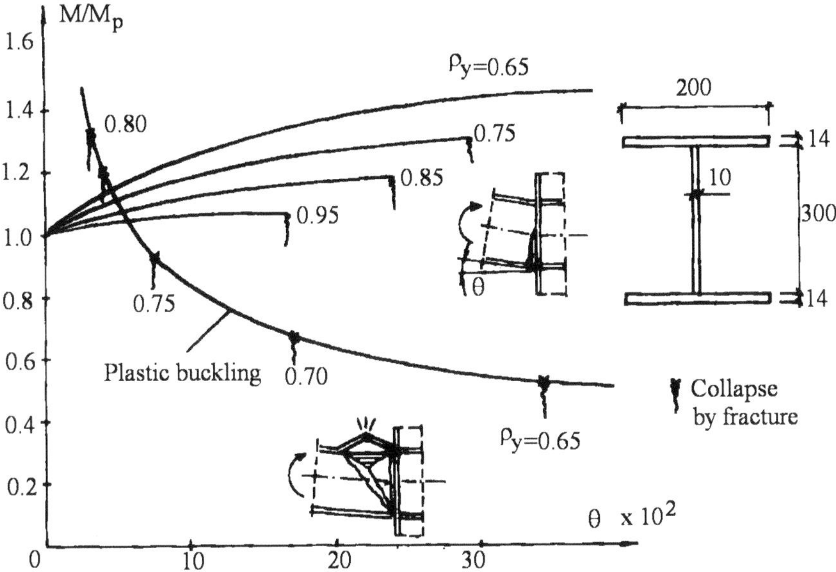

Figure 8.57: Influence of yield ratio on fracture rotation

The relation between fracture line rotation and member rotation results from Table 8.4 (lines 2–4 and 7–9):

$$\alpha_f = 2\left(\frac{d}{c}\right)^{1/2}\theta_f^{1/2} \tag{8.79}$$

where $\beta \simeq 1.0$ and $\delta \simeq 1.0$. Using equations (8.77) and (8.78), it results:

$$\theta_f = \zeta^2 \frac{\left(\dfrac{1}{\rho_y} - 1\right)^2}{(1+\rho_y)^3} \frac{c^3}{dt_f^2} \epsilon_{uf}^2 \tag{8.80}$$

One can see that both fracture types are influenced by the yield ratio ρ_y. For the profile presented in Fig. 8.57 both fracture types are determined. The tension flange fracture occurs in the primary unbuckled moment-rotation curve, while the compression flange fracture is produced during the post-critical plastic mechanism. For $\rho_y = 0.78$ the fracture rotation corresponds to the ductile rotation. For $\rho < 0.78$, the ductility is given by the ultimate plastic rotation, while for $\rho_y \simeq 0.78$, this ductile behaviour can be impeded by flange fracture. Therefore, the plastic buckling of compressed flange operates as a filter against high strain in tension flange. It is very clear that tension flange fracture is a more dangerous collapse than the buckling or fracture of compression flange. In order to eliminate this possible fracture type, a minimum value of flange slenderness

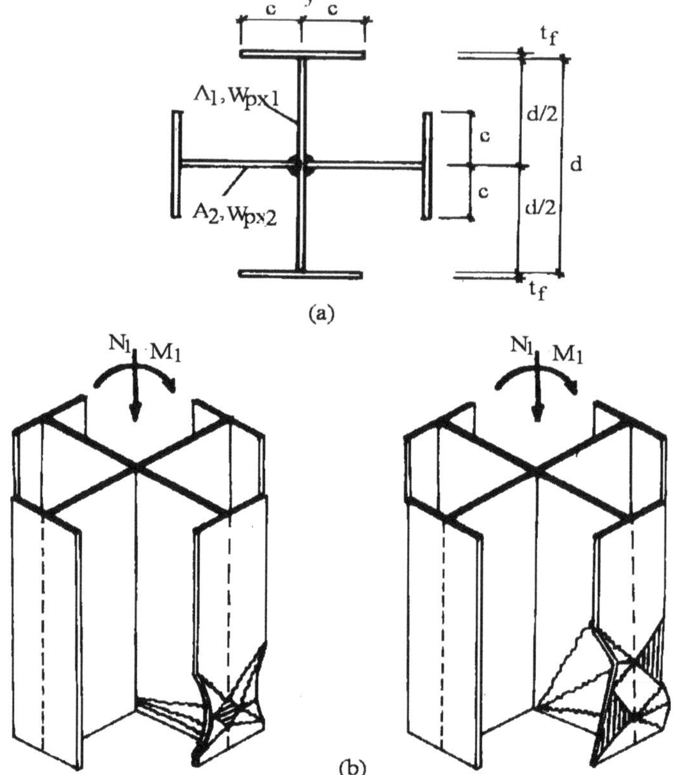

Figure 8.58: X-section member: (a) Cross-section dimensions; (b) Collapse modes

is proposed (Gioncu and Petcu, 1997):

$$\left(\frac{c}{t_f}\right)_{min} = \frac{1.126}{\left[\left(1 + 0.6\frac{f_{uf}}{f_{yf}}\right)\left(1 + 0.192\frac{E}{E_h}\right)\right]^{1/2}}\left(\frac{E}{f_{yf}}\right)^{1/2} \qquad (8.81)$$

resulting in the values $(c/t_f)_{min} = 8.38$; 7.27; 6.19 for Fe360, Fe430 and Fe510 steel qualities, respectively. For ratios above these, values for local plastic buckling will occur before the tension flange fracture, while for ratios less than these values, the fracture of a tension flange can occur before local buckling of the compression flange.

8.3.3. X-Section Beam-Columns

The X-sections (Fig. 8.58a) made from two I-profiles (especially IPE) are very useful for spatial frames, where the internal forces are almost equal in the two directions and the joints are spatial ones. The use of X-sections,

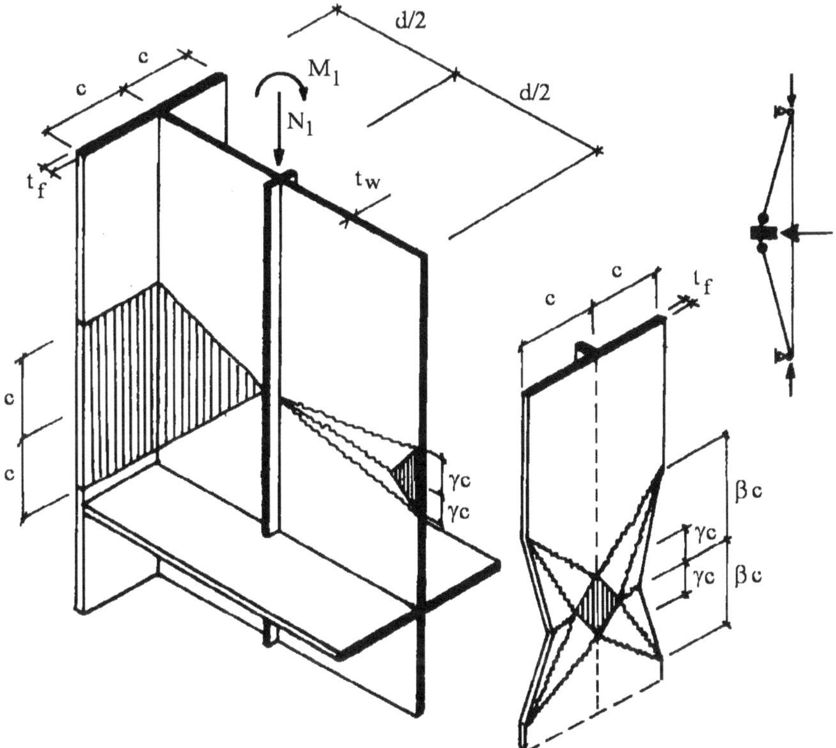

Figure 8.59: In-plane mechanism for X-section: (a) Tension flange and web plastic mechanism; (b) Compression flange plastic mechanism

(called also "austrian cross" shapes) avoids the connections of secondary beams directly to the column web. These sections are made from one I-profile, while the second is cut and welded to the first profile web. The axial force N and bending moment M are distributed on the two profiles in function of areas A_i and plastic section module W_{pxi} (i = 1,2):

$$N_{1,2} = \frac{A_{1,2}}{A} N; \qquad M_{1,2} = \frac{W_{p\times 1,2}}{W_{p\times}} M \tag{8.82a, b}$$

Fig. 8.58b shows the two plastic mechanisms considered in the determination of ultimate plastic rotation: the in-plane and out-of-plane mechanisms.

(i) *In-plane plastic mechanism.* The main characteristics of this plastic mechanism are shown in Fig. 8.59. The mechanism is the same as for I-sections, but the rotation of mechanism occurs around the middle of the section due to the presence of transverse profile. Therefore, the results for I-sections (see Section 8.3.2) can be used, considering $\delta = 0.5$ and $\eta = 0$ as fixed data. Fig 8.60 shows the plastic rotation capacity for a X-section beam-column as a function of axial force levels. Fig. 8.60b presents the

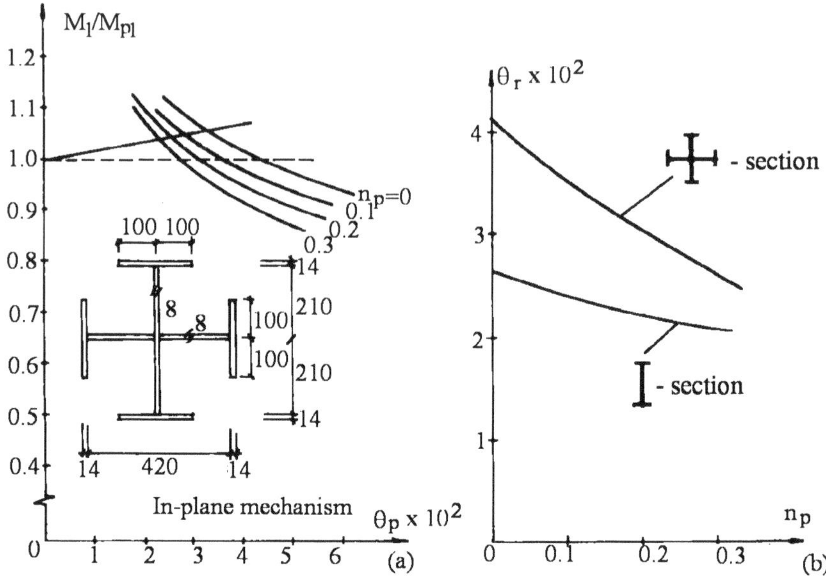

Figure 8.60: In-plane plastic rotation of X-section; (a) Influence of different axial level; (b) Comparison with I-section

reduction of ultimate rotation as a function of axial force levels for both X-section and I-section members. One can see an important increase in rotation capacity for X-section members due to the presence of the transverse I-profiles, which acts as a stiffener for web and influences the formation of the plastic mechanism pattern.

(ii) *Out-of-plane plastic mechanism.* Due to the fact that the plastic hinge occurs at the column base, the plastic mechanism corresponds to the M-shaped type. The characteristics of this plastic mechanism is shown in Fig. 8.61. One can see that it is possible to use the results obtained for I-sections considering $\delta = 0.5$. Fig. 8.62 shows the plastic rotation capacity as a function of axial force levels. In Fig. 8.62b the in-plane and out-of-plane mechanism behaviours are compared. It is clear that, for the analyzed section, the out-of-plane mechanism is the dominant one, but the influence of axial force level is smaller than for in-plane mechanisms.

8.3.4. Box-Section Members

In the following the local ductility of built-up box-sections made of welded plates (Fig. 8.63) is examined. Consequently, the section corners are sharp, contrary to hollow sections. Due to their great torsional rigidity, the lateral-torsional buckling is normally prevented, the in-plane plastic mechanism being the significant collapse mode.

(i) *Collapse plastic mechanism.* Contrary to the studies performed on I-section members, the results obtained for box-section members are less frequent:

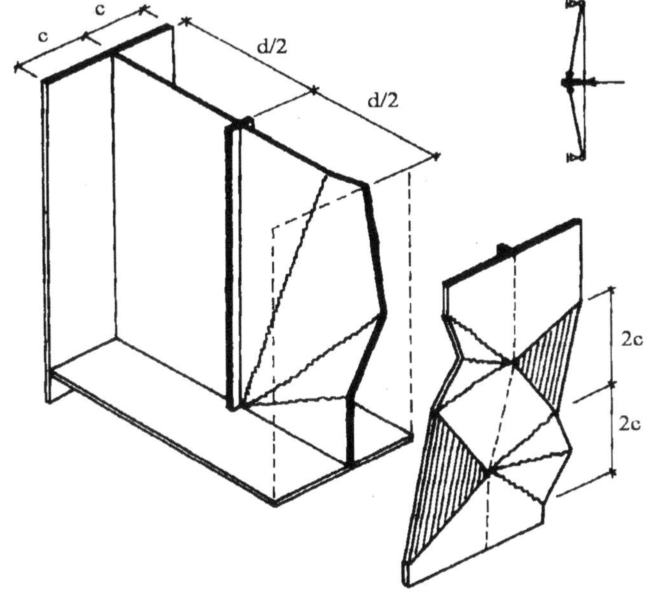

Figure 8.61: Out-of-plane mechanism for X-section; (a) Web plastic mechanism; (b) Compression flange plastic mechanism

-Gioncu and Petcu model (1995) characterized by plastic zones at the compression and tension flange corners and in webs (Fig. 8.63b). The yield lines form a plastic mechanism defined by the flange plastic zone length. The shortcoming of this model is related to the experimental results which have shown that the mechanism length also depends on web depth;

-Gioncu and Petcu improved model (2001) which considers both flange width and web depth as mechanism parameters (Fig. 8.63c).

The geometrical characteristics of the collapse mechanism are presented in Fig. 8.64. The length of buckled flange results from Lay's relationship (7.9a):

$$L_p = \beta b \tag{8.83a}$$

with

$$\beta = 0.713 \left(\frac{d}{c}\right)^{1/4} \left(\frac{t_f}{t_w}\right)^{3/4} \tag{8.83b}$$

The extension to box-section of the Lay's relation, determined for I-section, is possible considering that this section is formed by two U-sections (see Section 4.3.4 , Fig. 4.37). The geometrical and mechanical characteristics of the plastic mechanism are presented in Table 8.8. There are many similarities with the values obtained for I-sections (see Table 8.4). Using

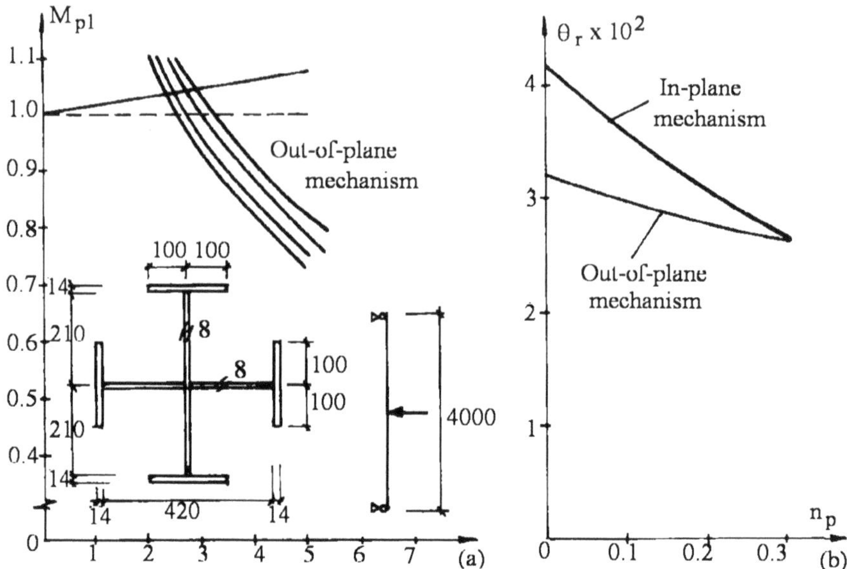

Figure 8.62: Out-of-plane plastic rotation for X-section: (a) Influence of different axial level; (b) Comparison of in-plane and out-of-plane plastic rotations

the rigid plastic analysis the post-critical curve results:

$$\frac{M}{M_p} = \frac{1}{C}\left(A + \frac{B}{\theta^{1/2}}\right) \tag{8.84}$$

where

$$A = 2\left[\gamma\delta\left(1 + \frac{t_w}{t_f}\frac{f_{yw}}{f_{yc}}\right) + (1-\delta)^2\frac{dt_w}{ct_f}\frac{f_{yw}}{f_{yf}} + 2(1-\delta) + \frac{1}{2}\frac{t_f}{d}\right]\frac{f^*_{yf}}{f_{yf}} \tag{8.85a}$$

$$B = \left(\frac{1-\gamma}{\beta} + \frac{\beta}{1-\gamma} + \frac{2+2\beta^2-3\gamma\beta+\gamma^2+\gamma}{\gamma(1+\beta-\gamma)}\right)\left(\frac{\beta\gamma}{\chi}\right)^{1/2}\frac{t_f}{d} +$$

$$+\frac{\delta}{\chi^{1/2}}\frac{(2+\chi)\left(1+\dfrac{\eta^2\chi^2}{\gamma^2}\right) - \chi(2+\chi^2)}{\left[(1-\chi)^2 - \dfrac{\eta^2\chi^2}{\gamma^2}\right]^{1/2}}\frac{t_w^2}{ct_f}\frac{f_{yw}}{f_{yf}} \tag{8.85b}$$

$$C = \left(1 + \frac{t_f}{d} + \frac{1}{4}\frac{dt_w}{ct_f}\frac{f_{yw}}{f_{yf}}\right)m_p \tag{8.85c}$$

Fig. 8.65 shows the evaluation of the ultimate plastic rotation. One can see that the minimum of this rotation is obtained for $\gamma = 0.35$ and $\delta = 1$, in very good concordance with the theoretical value obtained by Haaijer

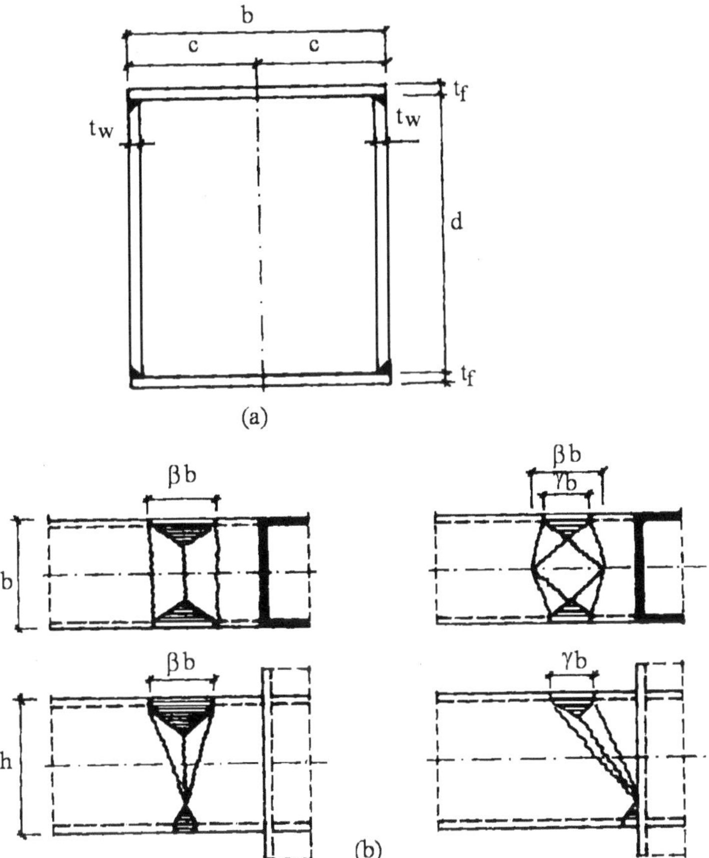

Figure 8.63: Welded box-section: (a) Cross-section dimensions; (b) Plastic mechanism types

(see Section 4.3.4). For the ductility criterion related to $0.9M_p$ results an increasing with 31% in comparison with the criterion related to M_p.

(ii) *Beam-columns.* This section type is very frequently used for frame columns and, therefore, the influence of axial forces is decisive. Fig. 8.66 shows the influence of axial force level. One can see that the minimum is obtained for $n_p = 0.1$. If the definition of ultimate rotation is related to M_p, a continuous decreasing in rotation capacity with the increasing of axial level is obtained. But if the rotation capacity is related to the reduced moment due to axial force, an increasing of rotation capacity is obtained, due to the fact that the reference level of bending moment decreases.

8.3.5. Hollow Section Members

Hollow rectangular sections are widely used as load-carrying members, especially for columns. There are two section types in function of manufacturing

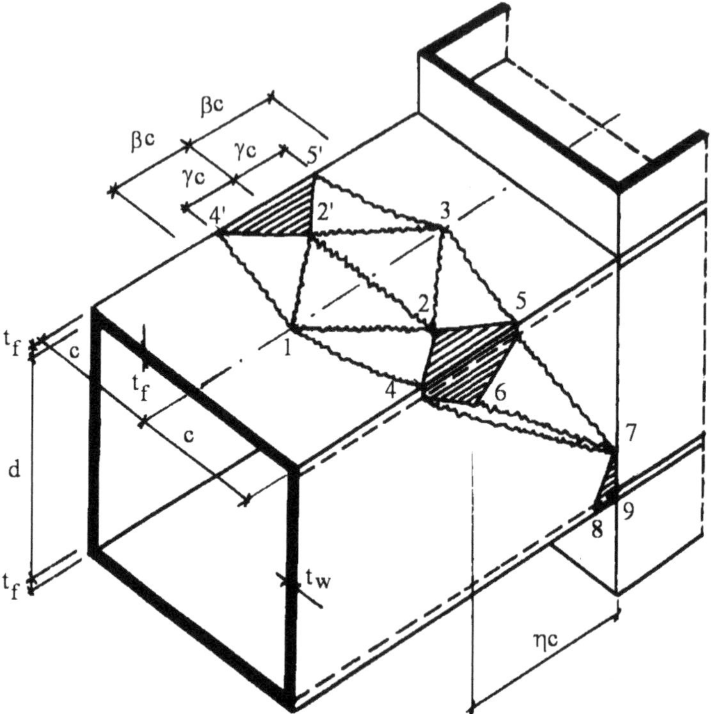

Figure 8.64: Plastic mechanism of box-section

process: hot-rolled and cold-formed profiles (Fig. 8.67). The main difference between these two profiles is the increase of yield stress in the corners for cold-formed profiles.

(i) *Collapse mechanism.* The difference between the plastic mechanisms for welded box-sections and hollow sections is related to the compressed corners deformation. For box-sections the corners remain straight, forming plastic zones, while for hollow-sections the corners are involved in the mechanism deformations. The main collapse mechanism types are presented in Fig. 8.68:

-Kecman (1983) and Kotelko (1996) model is a true mechanism (see Section 4.4.2), being composed by yield lines only. The main shortcoming of this plastic mechanism is related to the pattern of mechanism. In fact, the experimental tests show that the form of buckled flange is bounded by curved yield lines and not by straight lines. The pattern of deformed webs is characteristic only for constant moments;

-Stranhoner model (1995), in which the pattern of a plastic mechanism for compression flanges is the same as for the Kecman-Kotelko mechanism, but the rotation of mechanism occurs around of middle of the web depth. The shortcoming of this model is the same as for the first one;

-Gioncu and Petcu model (2001), in which the shortcomings of above models are corrected. The local plastic mechanism is bounded by broken

Table 8.8: Box-section: in-plane plastic mechanism

Plastic zones			
Zones	Volumes	Strains	Stresses
245,2'4'5'	$\gamma^2 c^2 t_f$	$\delta d/(2\gamma c)\ \theta$	f^*_{yf}
456,4'5'6'	$\gamma^2 c^2 t_w$	$\delta d/(2\gamma c)\ \theta$	f_{yw}
789,7'8'9'	$(1-\delta)^2 d^2 t_w$	$\theta/2$	f_{yw}
89'8'9'	$2(1-\delta)bdt_f$	$\theta/2$	f_{yf}
Yield lines			
Lines	Lengths	Rotations	Stresses
2-2'	$2(1-\gamma)c$	$2[\gamma/(\beta\chi)]^{1/2}\theta^{1/2}$	f_{yf}
1-2,2-3 1-2',2'-3	$[\beta^2+(1-\gamma)^2]^{1/2}c$	$\frac{1}{[\beta^2+(1-\gamma)^2]^{1/2}}\left[\frac{\beta}{1-\gamma}+ +\frac{\beta^2-3\gamma-\gamma\beta+1}{\gamma(1+\beta-\gamma)}\right]\left(\frac{\beta\gamma}{\chi}\right)^{1/2}\theta^{1/2}$	f_{yf}
1-4,5-3 1-4',5'-3	$[1+(\beta-\gamma)^2]^{1/2}c$	$\frac{[1+(\beta-\gamma)^2]^{1/2}}{\gamma(1+\beta-\gamma)}\left(\frac{\beta\gamma}{\chi}\right)^{1/2}\theta^{1/2}$	f_{yf}
2-4,2-5 2'-4',2'-5	$2^{1/2}\gamma c$	$2^{1/2}\frac{1}{\gamma(1+\beta-\gamma)}\left(\frac{\beta\gamma}{\chi}\right)^{1/2}\theta^{1/2}$	f_{yf}
4-6,4'-6	$2^{1/2}\gamma c$	$\left(\frac{2}{\chi}\right)^{1/2}\left[\frac{1-(1-\eta/\gamma)\chi}{1-(1+\eta/\gamma)\chi}\right]^{1/2}\theta^{1/2}$	f_{yw}
5-6,5'-6'	$2^{1/2}\gamma c$	$\left(\frac{2}{\chi}\right)^{1/2}\left[\frac{1-(1+\eta/\gamma)\chi}{1-(1-\eta/\gamma)\chi}\right]^{1/2}\theta^{1/2}$	f_{yw}
4-7,4'-7'	$\left[1+\left(1+\frac{\eta}{\gamma}\right)^2\right]^{1/2}\delta d$	$\frac{1}{\chi^{1/2}}\left[\frac{1-(1-\eta/\gamma)\chi}{1-(1+\eta/\gamma)\chi}\right]^{1/2}\cdot\left[1+\left(1+\frac{\eta}{\gamma}\right)^2\chi^2\right]\theta^{1/2}$	f_{yw}
5-7,5'-7'	$\left[1+\left(1-\frac{\eta}{\gamma}\right)^2\right]^{1/2}\delta d$	$\frac{1}{\chi^{1/2}}\left[\frac{1-(1+\eta/\gamma)\chi}{1-(1-\eta/\gamma)\chi}\right]^{1/2}\cdot\left[1+\left(1-\frac{\eta}{\gamma}\right)^2\chi^2\right]\theta^{1/2}$	f_{yw}
6-7,6'-7'	$\left[(1-\chi)^2+\frac{\eta^2\chi^2}{\gamma^2}\right]^{1/2}\delta d$	$\frac{(1-\chi)^2+\eta^2\chi^2/\gamma^2}{(1-\chi)^2-\eta^2\chi^2/\gamma^2}\theta^{1/2}$	f_{yw}
8-8,9-9'	b	θ	f_{yf}

* Increased yield stresses due to strain-hardening. For the other plastic zones and yield lines, the randomness of yield stresses must be considered only

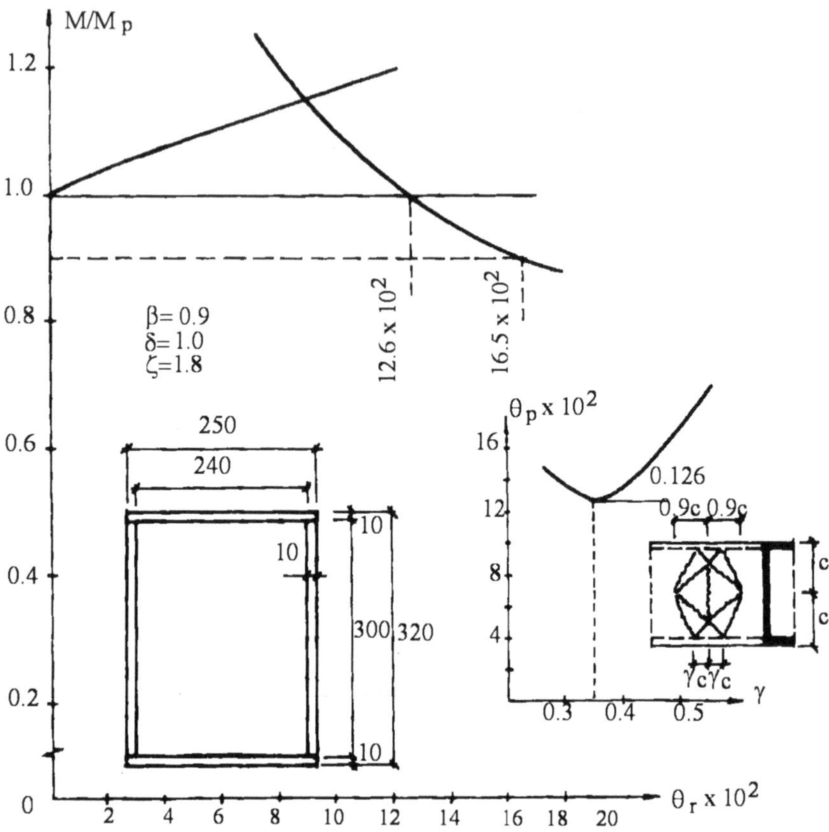

Figure 8.65: Plastic mechanism of box-section

lines and the deformations of the web are asymmetrical. The rotation of the plastic mechanism occurs around a line situated in the tension flange. The geometrical characteristics of this plastic mechanism are presented in Fig. 8.69 and Table 8.9. The following notations are used:

$$\Omega = \left\{(1+\beta^2)[1+(\beta-\gamma)^2]-\gamma^2\right\}^{1/2} \tag{8.86a}$$

$$\Phi_{1,2} = \left[1+(\eta\pm\gamma)^2\frac{c^2}{h^2}\right]^{1/2}\left\{1+\left[\eta^2-\frac{\gamma^2}{1\pm(\eta\pm\gamma)^2\dfrac{c^2}{h^2}}\right]\frac{c^2}{h^2}\right\}^{1/2} \tag{8.86b}$$

By means of the same procedure used for the previous cases, the following equation is obtained:

$$\frac{M}{M_P} = \frac{1}{C}\left(A+\frac{B}{\theta^{1/2}}\right) \tag{8.87}$$

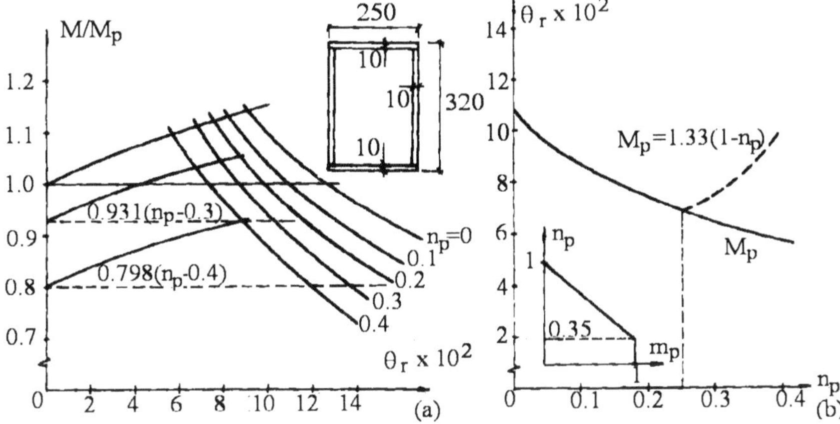

Figure 8.66: Influence of axial forces: (a) Post-critical behaviour; (b) Influence of axial force level

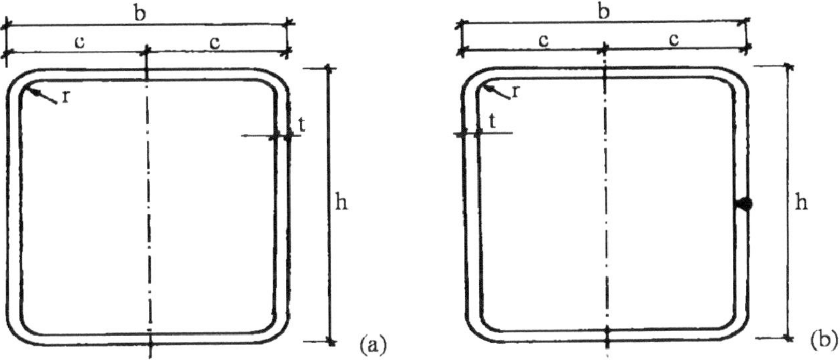

Figure 8.67: Hollow-section: (a) Hot-rolled; (b) Cold-formed

where

$$A = 2\left(\frac{\beta}{\gamma}\frac{t}{c}\frac{f_y^*}{f_y} + \frac{t}{h}\right) \tag{8.88a}$$

$$B = \sqrt{\beta}\left\{\frac{1}{\beta} + \beta + \frac{\Omega}{\gamma} + \left(1 + \frac{\Omega^2}{\gamma^2}\right)^{1/2} + \right.$$

$$\left. + \frac{1}{\gamma}\left[1 + (\eta^2 + \gamma^2)\frac{c^2}{h^2} + \frac{1}{2}(\Phi_1 + \Phi_2)\right]\frac{t}{h} + \frac{c}{h}\frac{t}{h}\right\}\left(\frac{h}{c}\right)^{1/2} \tag{8.88b}$$

$$C = 4\left(1 + \frac{t}{h} + \frac{1}{4}\frac{h}{c}\right) m_p \tag{8.88c}$$

The correlation of this relationship with an experimental test performed by Stronghoner (1995) is presented in Fig. 8.70. One can see that an acceptable agreement is obtained.

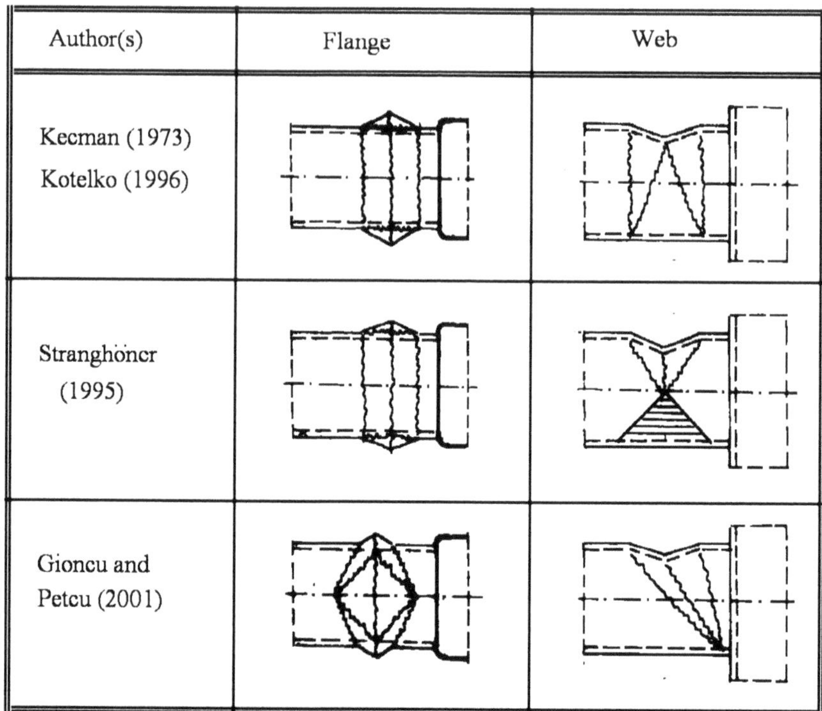

Author(s)	Flange	Web
Kecman (1973) Kotelko (1996)		
Stranghöner (1995)		
Gioncu and Petcu (2001)		

Figure 8.68: Plastic mechanism types for hollow-section

(ii) *Influence of axial force level.* For the section presented in Fig. 8.71, the reduction of ultimate plastic rotation as a function of the axial force level is shown. Two definitions of ultimate rotation are used, the first related to M_p, the second to M_{pN}, in which the interaction M-N is considered. If these results are compared with the ones obtained for welded box-section with the same geometrical dimensions (Fig. 8.66b), one can see an important reduction of ultimate rotation for the hollow sections, due to the detrimental effect of corner collapse.

8.3.6. Composite Section Beams

The main composite beam composite section types are presented in Fig. 8.72, the differences referring to the concrete slab: reinforced slab or concrete on corrugated sheets. To calculate the rotation capacity of composite section beams the member must be divided into two separate regions, namely regions of positive (sagging) bending moments and negative (hogging) bending moments. The ductility in positive bending is governed by the concrete crushing, while for negative bending by the local buckling of the steel beam. The ductility of composite section beams is studied considering that a secondary failure due to shear is prevented by a proper design of the connectors.

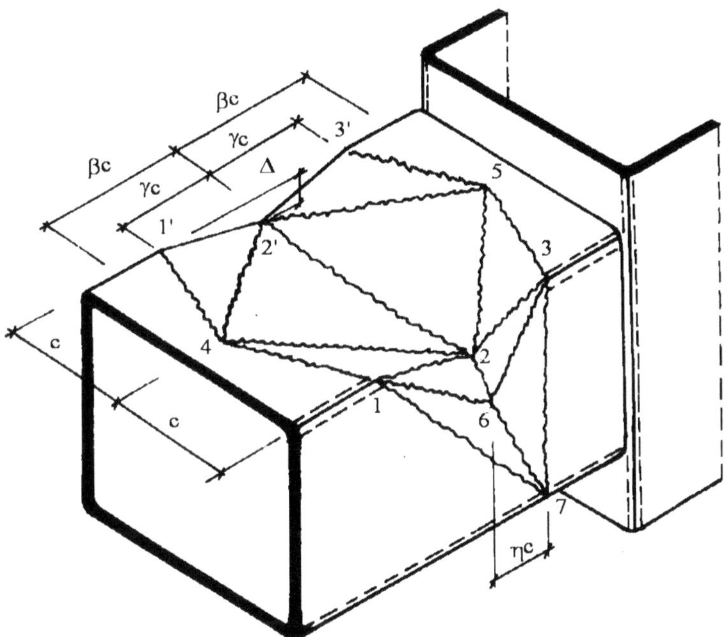

Figure 8.69: General view of plastic mechanism for hollow-section

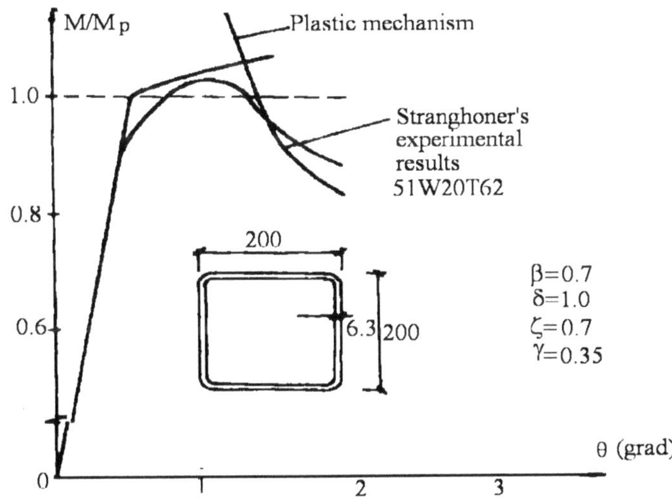

Figure 8.70: Correlation between theoretical and experimental results

(i) *Positive bending moment.* The collapse of a composite beam occurs by crushing of concrete, so the method of plastic mechanism composed by yield lines cannot be used. The results presented in Section 7.4.2 are helpful for evaluating the rotation capacity. The effective width results from (7.67a)

Table 8.9: Hollow-section: in-plane plastic mechanism

Lines	Lengths	Rotations	Stresses
1-4,3-5 1'-4,3'-5	$[1+(\beta-\gamma)^2]^{1/2}c$	$\frac{1}{4}\left[\frac{1+\Omega^2/\gamma^2}{1+(\beta-\gamma)^2}\right]^{1/2}\beta^{1/2}\cdot\left(\frac{h}{c}\right)^{1/2}\theta^{1/2}$	f_y
2-4,2-5 2'-4,2'-5	$(1+\beta^2)^{1/2}c$	$\frac{1}{4}\left(\beta+\frac{\Omega}{\gamma}\right)\beta^{1/2}\left(\frac{h}{c}\right)^{1/2}\theta^{1/2}$	f_y
2-2'	$2c$	$\frac{2}{\beta^{1/2}}\left(\frac{h}{c}\right)^{1/2}\theta^{1/2}$	f_y
1-2,2-3 1'-2',2'-3'	γc	$\pi/2$	f_y
2-6,2'-6'	Δ	$2\frac{\beta^{1/2}}{\gamma}\left(\frac{h}{c}\right)^{1/2}\theta^{1/2}$	f_y^*
1-7,1'-7'	$\left[1+(\eta+\gamma)^2\frac{c^2}{h^2}\right]^{1/2}h$	$\frac{\beta^{1/2}}{\gamma}\left[1+(\eta+\gamma)^2\frac{c^2}{h^2}\right]^{1/2}\cdot\left(\frac{h}{c}\right)^{1/2}\theta^{1/2}$	f_y
3-7,3'-7'	$\left[1+(\eta-\gamma)^2\frac{c^2}{h^2}\right]^{1/2}h$	$\frac{\beta^{1/2}}{\gamma}\left[1+(\eta-\gamma)^2\frac{c^2}{h^2}\right]^{1/2}\cdot\left(\frac{h}{c}\right)^{1/2}\theta^{1/2}$	f_y
6-7,6'-7'	$\left(1+\eta^2\frac{c^2}{h^2}\right)^{1/2}h$	$\frac{\beta^{1/2}}{\gamma}\frac{\Phi_1+\Phi_2}{\left(1+\eta^2\frac{c^2}{h^2}\right)^{1/2}}\left(\frac{h}{c}\right)^{1/2}\theta^{1/2}$	f_y
1-6,3-6 1'-6',3'-6'	γc	$\beta^{1/2}\left(\frac{c}{h}\right)^{1/2}\theta^{1/2}$	f_y
7-7'	$2c$	2θ	f_y

* Increased corner yield stress due to the forming process

(Plumier, 2000):

$$b_{eff} = 0.15L \tag{8.89}$$

The fully plastic moments are (Aribert, 1992):
-neutral axis in slab (Fig.8.73a):

$$M_p = A_s f_y \left(h_a - \frac{1}{2}\frac{A_s f_y}{b_{eff} f_c}\right) \tag{8.90a}$$

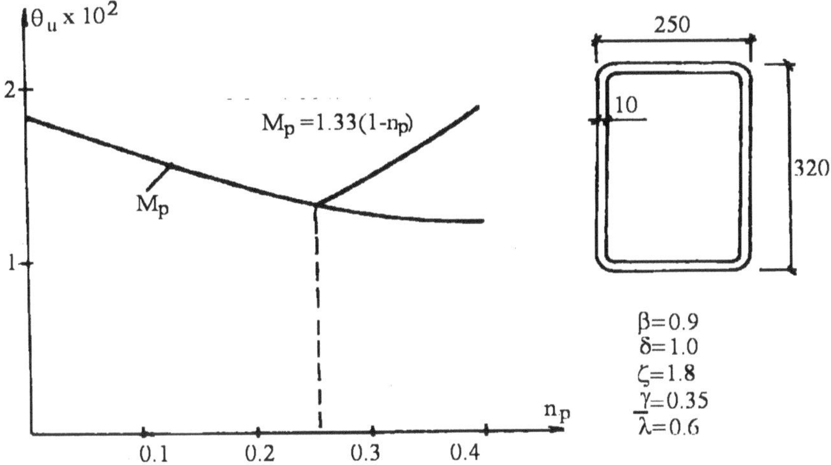

Figure 8.71: Influence of axial force level on plastic rotation

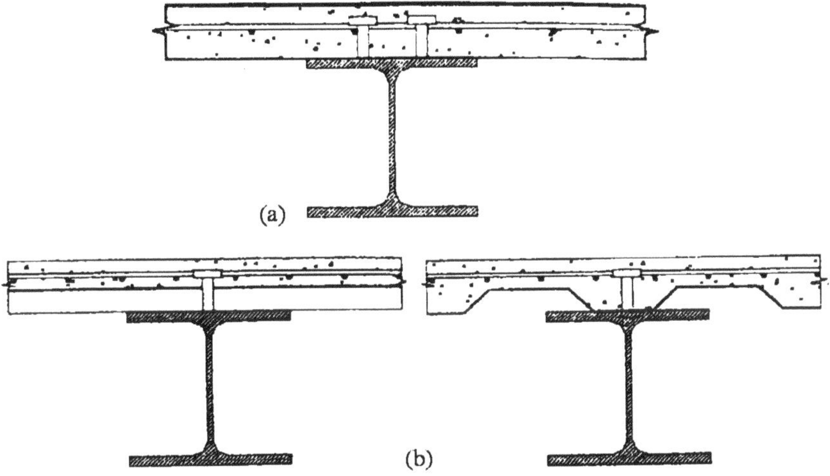

Figure 8.72: Composite section types: (a) Reinforced slabs; (b) Concrete on corrugated sheet

-neutral axis in steel profile (Fig. 8.73b):

$$M_p = A_s f_y \left(h_a - \frac{h_s}{2}\right) - 2A_s' f_y \left(h_a' - \frac{h_s}{2}\right) \tag{8.90b}$$

$$A_s' = \frac{1}{2}\left(A - s - b_{eff} h_s \frac{f_c}{f_y}\right) \tag{8.90c}$$

where A_s is the total area of profile, respectively. In order to obtain a ductile

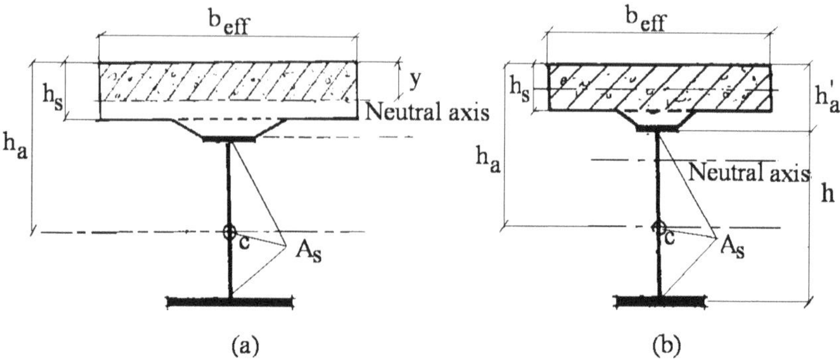

Figure 8.73: Composite section behaviour for positive moment: (a) Neutral axis in slab; (b) Neutral axis in steel profile

collapse, the condition:

$$\chi > 1.0 \tag{8.91}$$

is recommended by Ansourian (1982), where χ is the ductility parameter defined by equation (7.68). The minimum slab thickness h_s in order to obtain the concrete crushing after complete plasticization in tension of the steel profile can be obtained by the above condition:

$$\frac{h_s}{h} > 1.39 \frac{A_s}{b_{eff} h} \frac{f_y}{f_c} \left(1 + \frac{\epsilon_{sh}}{\epsilon_{cu}}\right) - 1 \tag{8.92}$$

where ϵ_{sh} is the strain at the start of strain-hardening, f_c, concrete cylindrical strength and ϵ_{cu}, the ultimate concrete strain:

$$\epsilon_{cu} = 0.0041 - 0.00000206 f_c \quad (\mathrm{N/mm^2}) \tag{8.93}$$

Fig. 8.74 shows the minimum slab thickness for different IPE profiles and concrete strength. The importance of concrete strength in ductility requirement clearly appears: for a thickness smaller than the value given by (8.92) a brittle beam collapse occurs. Concrete with cylindrical strength smaller than 20 N$/mm^2$ must be forbidden for composite beams and smaller than 25 N$/mm^2$ can be used only for small profiles.

The rotation capacity of composite beams for a positive moment is given by equation (7.69):

$$\frac{\theta_u}{\theta_p} = \left[1.7 \frac{b_{eff}(h + h_s)}{A_s \left(1 + \frac{\epsilon_{sh}}{\epsilon_{cu}}\right)} \frac{f_c}{f_y} - 1.85 \right] \tag{8.94}$$

Fig. 8.75 shows the moment-rotation curve of a composite beam, where the ultimate rotation capacity given by equation (8.94) is identified.

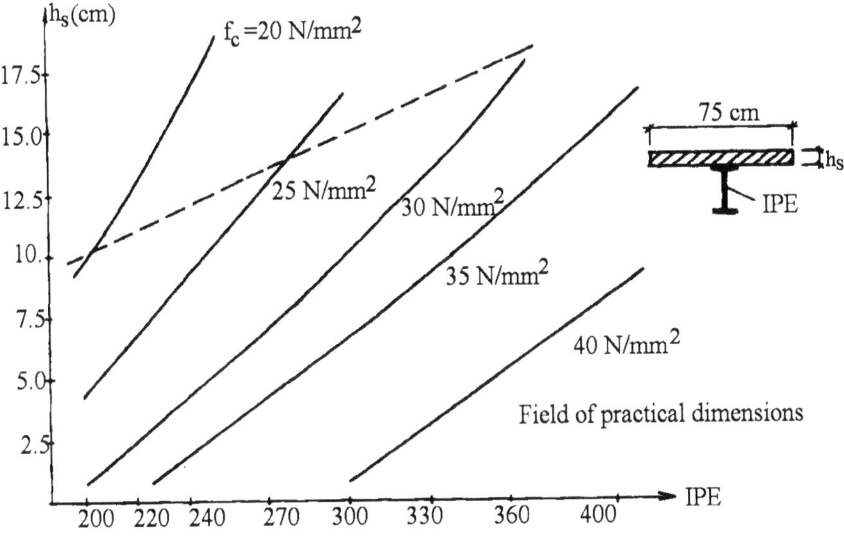

Figure 8.74: Minimum slab thickness for IPE profiles

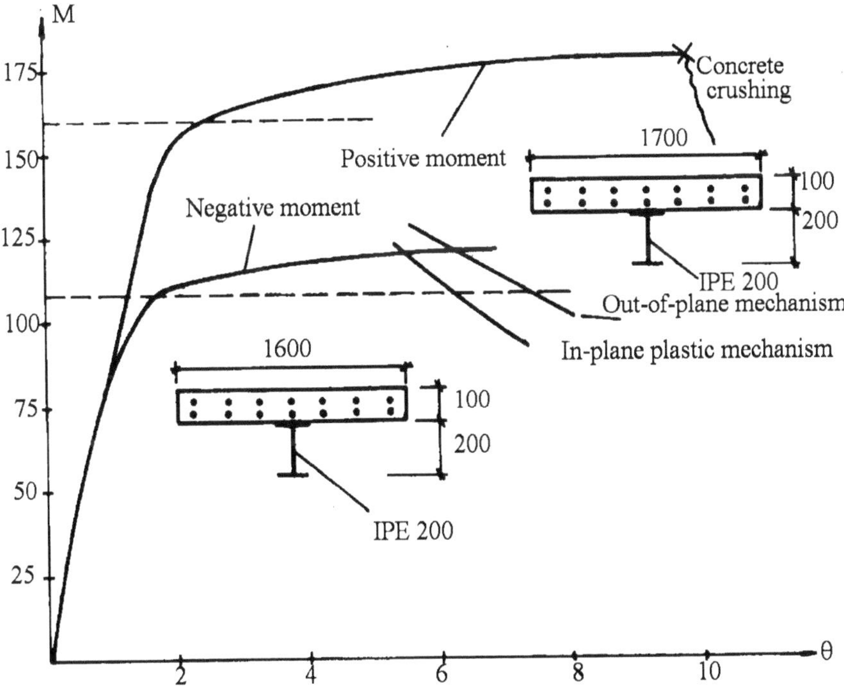

Figure 8.75: Composite section behaviour for positive and negative moments

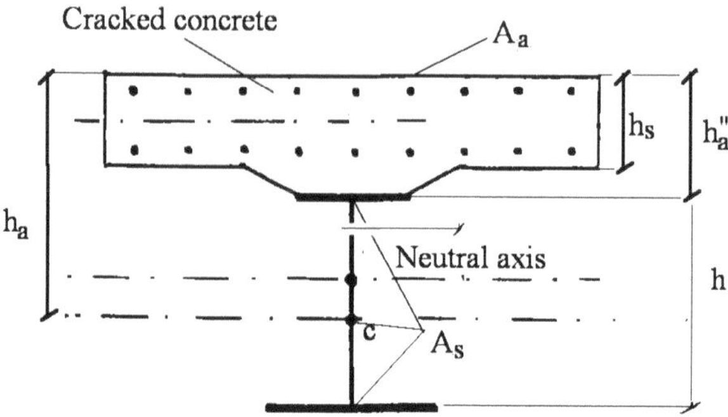

Figure 8.76: Composite section behaviour for negative moments

(ii) *Negative bending moment.* The effective width of slab is given by (see Section 6.62), (Plumier, 2000):

$$b_{eff} = 0.20L \tag{8.95}$$

and the plastic moment is (Aribert, 1992) (Fig. 8.76):

$$M_p = A_s f_y (h_a - h_s) - 2A"_s f_y (h"_a - h_s) \tag{8.96}$$

where

$$A"_s = \frac{A_s - A_a b_{eff}}{2} \tag{8.97}$$

The collapse of the composite beam occurs by plastic buckling of the compressed steel part during an in-plane or out-of-plane mechanism (see Section 7.4.3) (Fig. 8.77). In tension slab the concrete is cracked and only the steel rebars are active. The plastic mechanism of rotation includes the area of slab reinforcement. Therefore, in the equations (8.44a), (8.57a) and (8.61a) for I-section members, equation (8.81) for box-section members and equation (8.88a) for hollow section members the value (Fig. 8.78):

$$\Delta A = \frac{b_{eff} A_a}{c t_f} \left[(1 - \delta) + \frac{h_s}{2d} \right] \tag{8.98}$$

must be added, where A_a is the rebar area $cm^3/cm = cm^2$. In Fig. 8.75 the post-critical curves are plotted for the in-plane and out-of-plane plastic mechanism. One can see that the fully plastic moment for negative moment is lower than the one for positive moment. At the same time the minimum plastic rotation is obtained for negative moment and the in-plane plastic mechanism is the dominant mechanism.

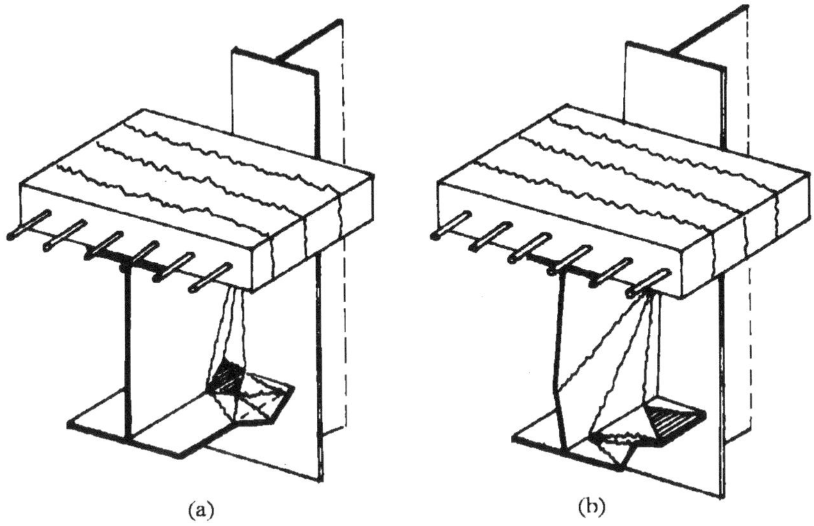

Figure 8.77: Plastic mechanism for negative moment: (a) In-plane mechanism; (b) Out-of-plane mechanism

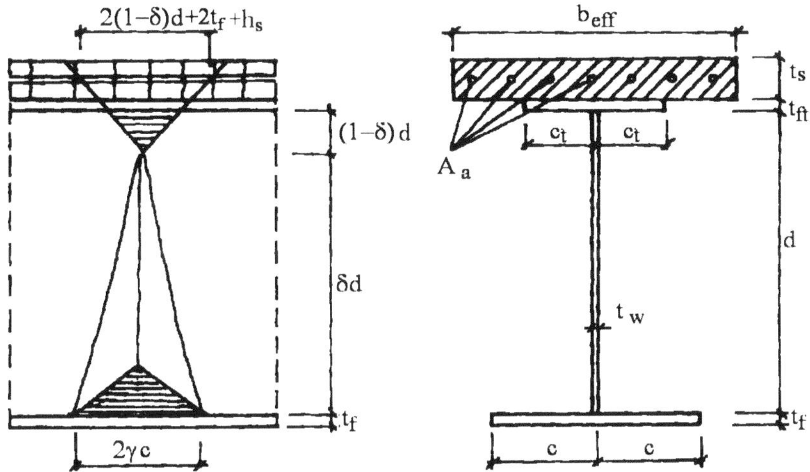

Figure 8.78: The geometrical characteristics of composite-section

8.4 Influence of Seismic Actions

8.4.1. Main Factors

The rotation capacity has been determined in the previous Sections exclusively under static and monotonic actions. In reality the response of a member in a building structure is greatly influenced by the dynamic feature of seismic actions, mainly characterized by velocity, amplitude and number

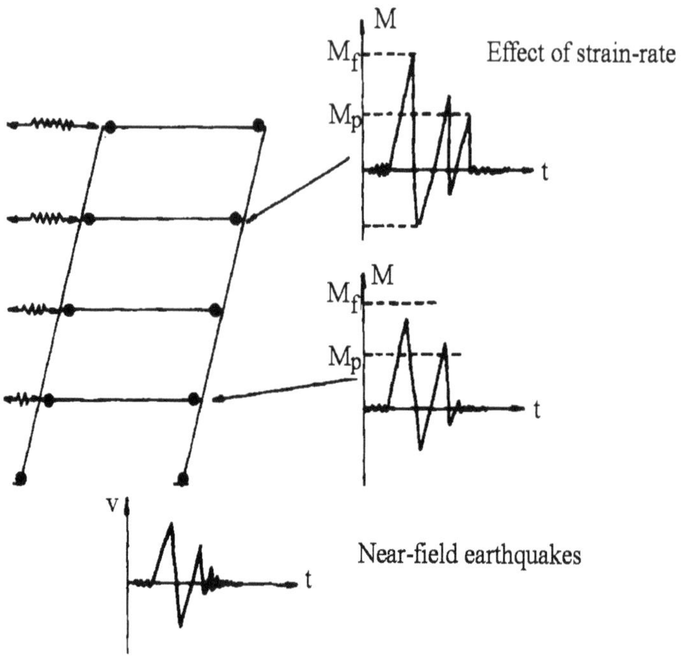

Figure 8.79: Behaviour of moment-resisting frame under near-field earthquakes

of cycles. At the level of structure these characteristics are turned in (see Section 6.5):

- strain-rate influence which can transform the plastic hinge behaviour from a ductile response to a brittle fracture;
- effect of repeated cyclic actions, which produces a reduction of rotation capacity due to accumulation of plastic rotations;
- structural damage produced by local fracture or accumulation of plastic deformations.

8.4.2. Influence of Strain-Rate

The greatest strain-rate influence occurs in the case of pulse seismic actions (especially for near-field earthquakes), when the first or second quake has a great velocity (Fig. 8.79), the following ground motions being much more reduced in intensity. The effect of strain-rate consists on both the increasing of plastic hinge moments over the design plastic moment M_p and the reaching, in some sections, of the fracture moment-rotation capacity. There are two main problems in evaluating the influence of strain-rate:

- determination of section strain-rate as a function of ground velocity, global behaviour of structure and local behaviour of plastic hinge;
- determination of fracture rotation as a function of section strain-rate.

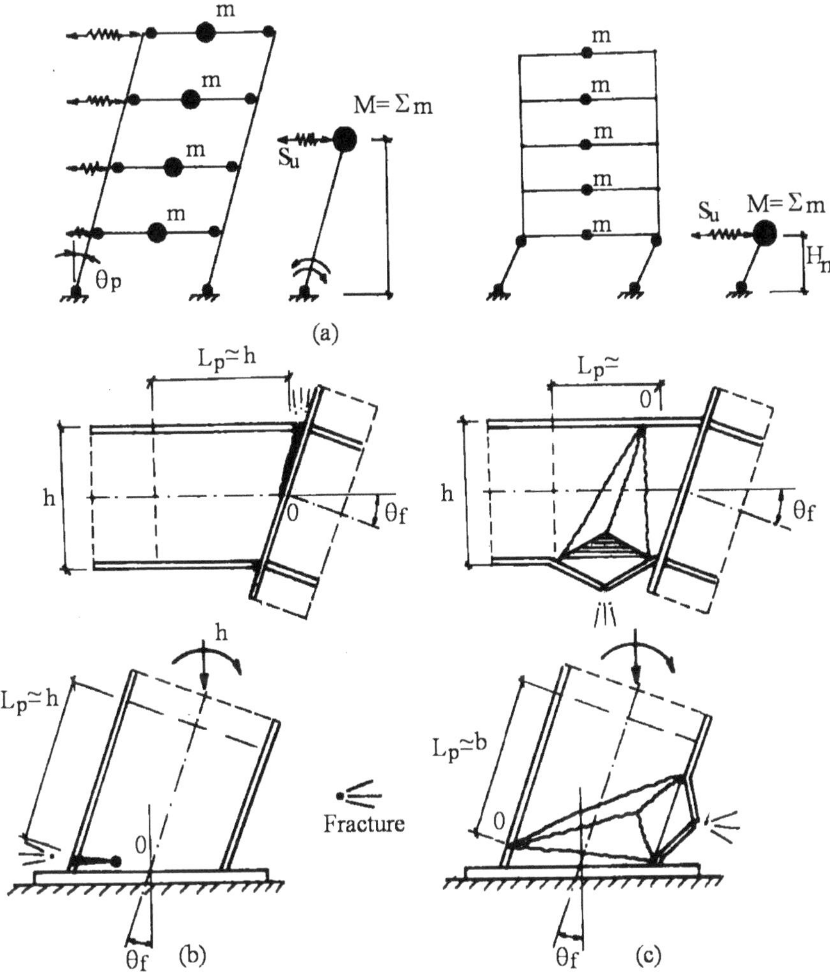

Figure 8.80: Influence of strain-rate: (a) Structure behavior; (b) Fracture of tension flange in the case of stocky flanges; (c) Fracture of buckled compression flange

(i) *Section strain-rate.* The elastic and plastic behaviour of a frame under seismic actions is presented in Fig. 8.80a. The rotation is given by equation (6.64):

$$\theta_p(1) = \frac{v_g}{H_m}\left(\bar{t} - \frac{\sin(2\pi t/T_g)}{2\pi T_g}\right) \tag{8.99}$$

where H_m is the height of center of mass involved in a plastic mechanism and v_g, the ground motion velocity. The velocity of the section rotation is:

$$\dot{\theta}_p(t) = \frac{d\theta_p(t)}{dt} = \frac{v_g}{H_m}\left(1 - \cos\frac{2\pi\bar{t}}{Tg}\right) \tag{8.100}$$

directly depending on ground motion velocity. It can be observed that the strain-rate developed in the case of a first storey mechanism is higher than in the case of a global mechanism. Therefore, even from this point of view, this mechanism type must be avoided in seismic areas with high velocity potential. As it is shown in Section 6.5.3, there are two different situations:

-in case of *stocky flanges*, when the plastic buckling of compression flange is avoided, the rotation of section occurs around a point situated at the middle of member depth (Fig. 8.80b). Considering that the length of plastic hinge is equal to the beam depth, the strain-rate results:

$$\dot{\epsilon}(t) \simeq \frac{1}{2}\frac{\dot{\theta}_p(t)h}{L_p} \quad \frac{1}{2}\frac{v_q}{H_m}\left(1 - \cos\frac{2\pi\bar{t}}{T_g}\right) \tag{8.101a}$$

with the maximum value for $t = T_g$:

$$\dot{\epsilon}_{max} = \frac{v_g}{H_m} \tag{8.101b}$$

The main effect is concentrated at the tension zone, by increasing the yield stress and, consequently, by increasing the fully plastic moment. If the velocity is very high, the fracture of a tension flange or beam-column connection occurs;

-in the case of *compression flange buckling*, the rotation of section occurs around a point situated in the tension flange and the length of plastic mechanism is approximately equal to the flange width (Fig.8.80c). So, the strain-rate results:

$$\dot{\epsilon}(t) \simeq \frac{\dot{\theta}_p(t)h}{b} = \frac{v_g h}{H_m b}\left(1 - \cos\frac{2\pi\bar{t}}{T_g}\right) \tag{8.102a}$$

with the maximum value for $\bar{t} = T_g$:

$$\dot{\epsilon}_{max} = \frac{2v_g h}{H_m b} \tag{8.102b}$$

The effect of strain-rate is concentrated in the compression zone, producing, in the case of very high strain-rate, a fracture of buckled flange, before the reaching of the ultimate plastic rotation.

(ii) *Increasing of yield ratio.* The effect of strain-rate produces the increasing of the yield ratio, due to the increase of yield stress, the ultimate strength remaining constant (see Section 4.2.5). In order to consider this effect, the Soroushian and Choi (1987) equation is used (see equation (4.12a)):

$$\frac{\rho_{ysr}}{\rho_y} = c_T c_w \frac{1.46 + 0.0925\log\dot{\epsilon}}{1.15 + 0.0496\log\dot{\epsilon}} \tag{8.103}$$

where the coefficient c_T considers the influence of low temperature if the structure is not protected ($c_T = 1.1$ for $T^0 =$ - 20^0C) and the coefficient c_w considers the influence of welding ($c_w = 1.15$). From the equation (4.8c) it

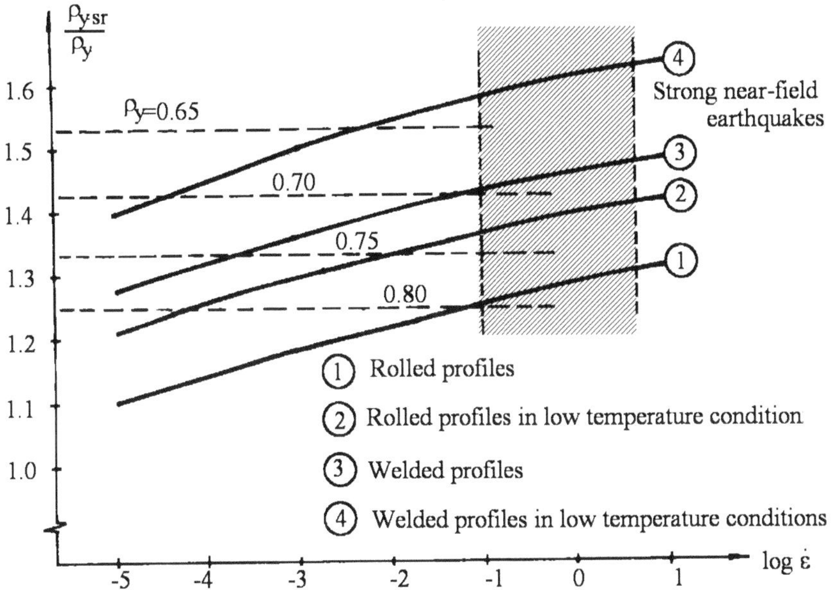

Figure 8.81: Influence of strain-rate on yield ratio

appears that the influence of steel grade is not so important that it has to be considered. In addition, this equation considers the random variation of yield stress and ultimate strength.

Fig. 8.81 shows the relationship (8.103) as a function of strain-rate for rolled and welded profiles in the room- and low-temperature conditions. The interval $\log\dot{\epsilon} = -1$ to 1 represents the field of strong near-field earthquakes. For this field the limits of the ρ_{ysr}/ρ_y ratios are plotted in Figure 8.81. One can see that for rolled profiles in room conditions the possibility of fracture exists only for $\rho_y > 0.80$, while for low temperature, the condition of fracture occurs for $\rho_y > 0.74$. The welded sections have more severe requirements to prevent the fracture: for normal conditions, $\rho_y < 0.70$, while for low temperature $\rho_y < 0.64$.

Considering that the ultimate strength remains unchanged, the increased plastic moment as the effect of the strain-rate results:

$$M_{psr} = \frac{\rho_{ysr}}{\rho_y} M_p \tag{8.104a}$$

A reduced rotation capacity is the consequence of the increasing of plastic moments. For instance, from (8.50a) it results:

$$\theta_{rsr} = \frac{m - \alpha_{1M}}{\rho_{ysr}/\rho_y - \alpha_{1M}} \tag{8.104b}$$

where m is the coefficient of ductility criterion (1.0 or 0.9).

(iii) *Fracture rotation of tension flange.* The effect of strain-rate on the M-θ curve is to increase the fully plastic moment and to reduce the plateau

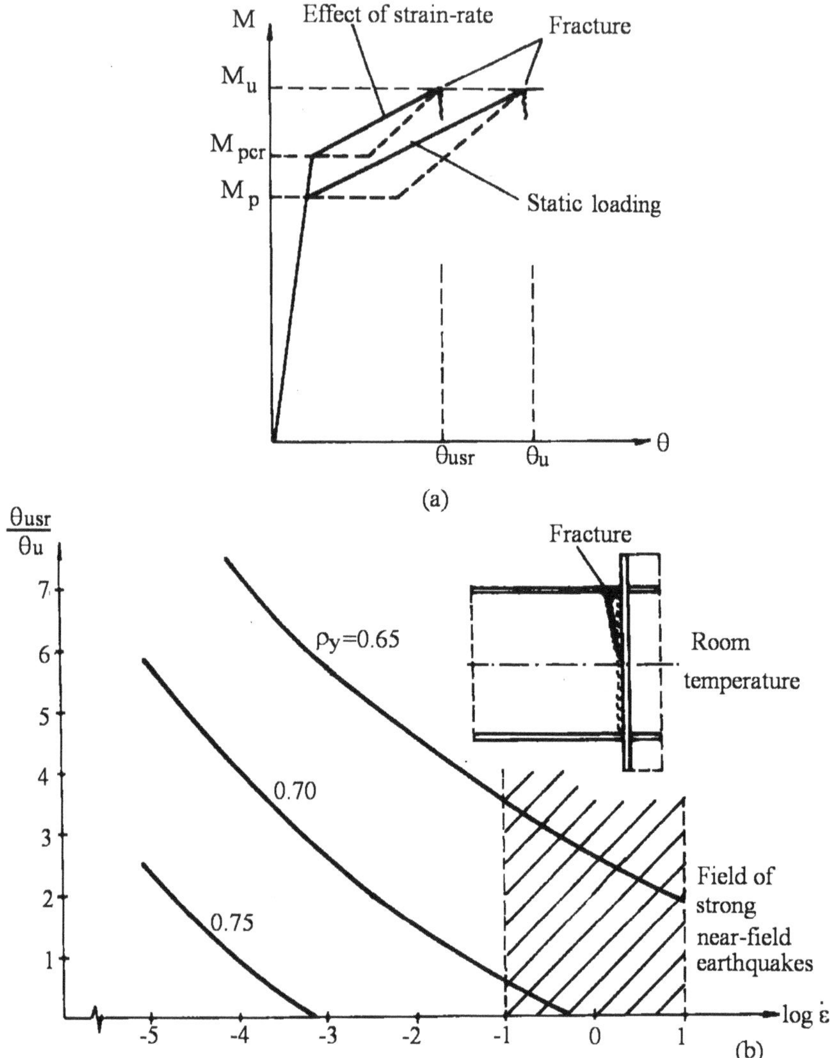

Figure 8.82: Influence of strain-rate on tension flange fracture: (a) Moment-rotation curve; (b) Influence of yield ratio

length (Fig. 8.82a). At the same time, the characteristics of strain-hardening are practically unchanged. In this condition, the plastic behaviour of members can be obtained by a translation of M-θ curve and the fracture rotation can be determined. After some algebra it results:

$$\frac{\theta_{usr}}{\theta_p} \simeq \frac{\rho_{ysr}}{\rho_y} + \frac{1-\rho_{ysr}}{1-\rho_y}\left(\frac{\epsilon_u}{\epsilon_y} - 1\right) \tag{8.105}$$

where the correspondence $\theta_u/\theta_p = \epsilon_u/\epsilon_y$ is considered, taking into account

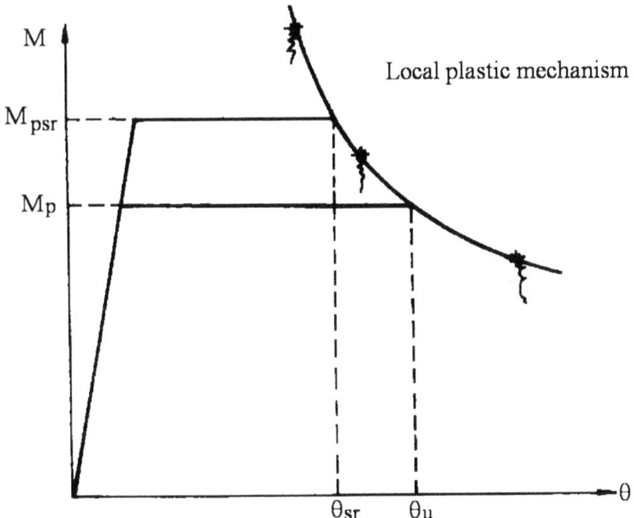

Figure 8.83: Fracture of compression-flange in the post-critical behaviour

the equation (4.40a). In equation (8.105) θ_{usr} is the plastic rotation corresponding to the bending producing the section fracture, when $\epsilon = \epsilon_u$ (see equation (4.15)). For a beam welded to a column and in conditions of room temperature, the fracture rotation is presented in Fig. 8.82b as a function of strain-rate and yield ratio. For $\rho_y = 0.65$ the section has a good behaviour even in the field of strong near-field earthquakes, but for $\rho_y > 0.70$, the danger of a premature fracture exists in these fields. For $\rho_y > 0.75$, it is possible, even for low strain-rates, for a fracture of tension flange to occur. Therefore, especially for near-field earthquakes, steels with $\rho_y > 0.75$ are not recommended to be used for welded steel frames.

(iv) *Fracture rotation of compression flange.* The fracture rotation of compression flange is given by the increasing of yield ratio ρ_y in equation (8.80):

$$\theta_f = \zeta^2 \frac{\left(\dfrac{1}{\rho_{ysr}} - 1\right)^2}{(1 + \rho_{ysr})^3} \frac{c^3}{dt_f^2} \epsilon_{uf}^2 \tag{8.106}$$

where ρ_{ysr} is determined from the equation (8.103). The following situations can be distinguished (Fig. 8.83):

-if $\theta_f < \theta_{sr}$, ζ, no fracture ($M > M_{psr}$) because the flange did not buckle;

-if $\theta_{sr} < \theta_f < \theta_u$, a reduction of rotation capacity arises due to the premature section fracture;

-if $\theta_f > \theta_u$, the fracture occurs after reaching the ultimate plastic rotation and so this fracture does not influence the structure ductility.

For some profiles the fracture rotation for different flange slendernesses is presented in Fig. 8.84. At the same time the ultimate plastic rotations

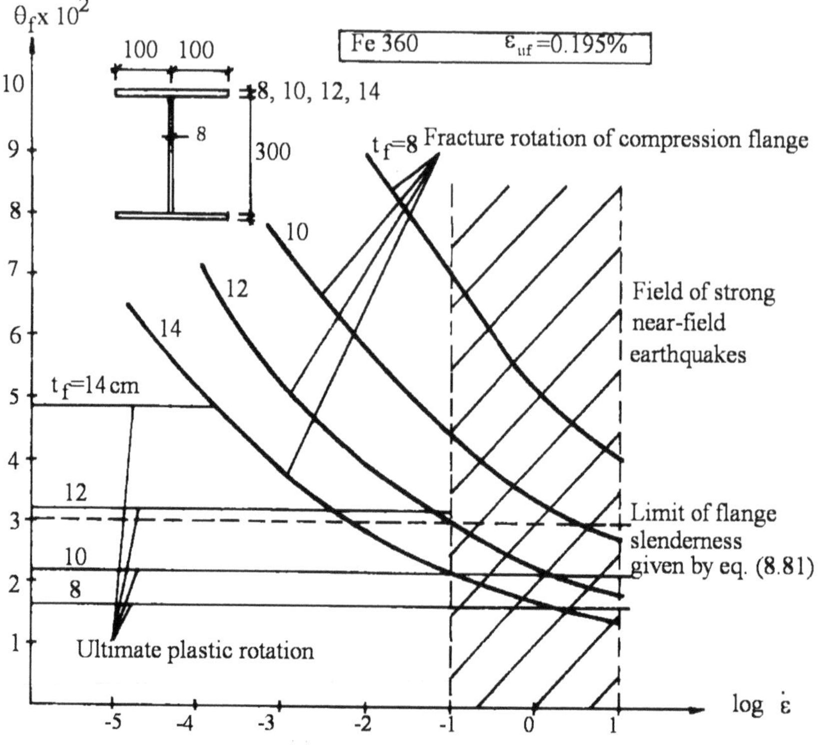

Figure 8.84: Influence of compression flange slenderness on fracture rotation

are plotted. The danger of a compression flange fracture exists for profiles with slenderness less than the limit given by equation (8.81). Contrary to this, for higher slendernesses, the collapse is produced by the ultimate plastic rotation, even for the strong near-field earthquakes. This observation underlines the importance of choosing of an adequate flange slenderness in order to prevent a brittle fracture.

8.4.3. Influence of Repeated Cyclic Actions

In the case of seismic actions the effect of repeated cyclic actions on the ductility of structural members becomes very important, in particular for the structures situated in intermediate or far-field areas and designed to dissipate an important amount of seismic energy. The main results are the reduction of rotation capacity due to the accumulation of plastic deformations, together with the occurrence of cracks and section fracture due to strain limits being reached.

(i) *Behaviour of plastic hinges under repeated cyclic actions.* The seismic actions for intermediate or far-field earthquakes are characterized by a first period with a slow increasing of accelerations, followed by a period with significant increasing of accelerations, the maximum being reached at

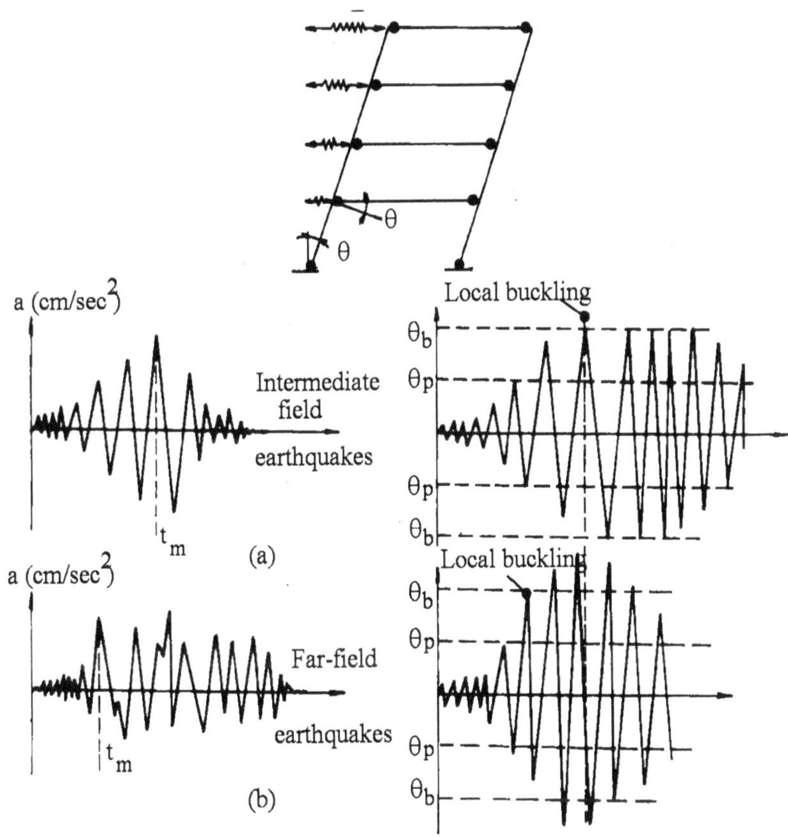

Figure 8.85: Frame behaviour under different earthquake types: (a) Intermediate-field earthquake; (b) Far-field earthquake

time t_m. After the culminating phase, decreasing ground motions occurs until movement completely stops. (Fig. 8.85). The effect of this action on the ultimate limit state is the formation of a global mechanism, composed of a sufficient number of plastic hinges. The behaviour of these plastic hinges can by very different. If θ_p is the rotation for which plastic rotation occurs and θ_b the rotation for which plastic buckling of compression flanges takes place, there are the following behaviour types:

-when sections work in the plastic range without flange buckling, an accumulation of plastic rotations occurs without any degradation in moment capacity of plastic hinges. In this case the collapse mode can be produced by tension flange fracture, due to accumulation of plastic deformations in this flange. So, this behaviour type must be avoided in design practice;

-when the flange buckling is produced at the maximum seismic action, the plastic hinge works with quasi-constant rotation amplitude (Fig. 8.85a);

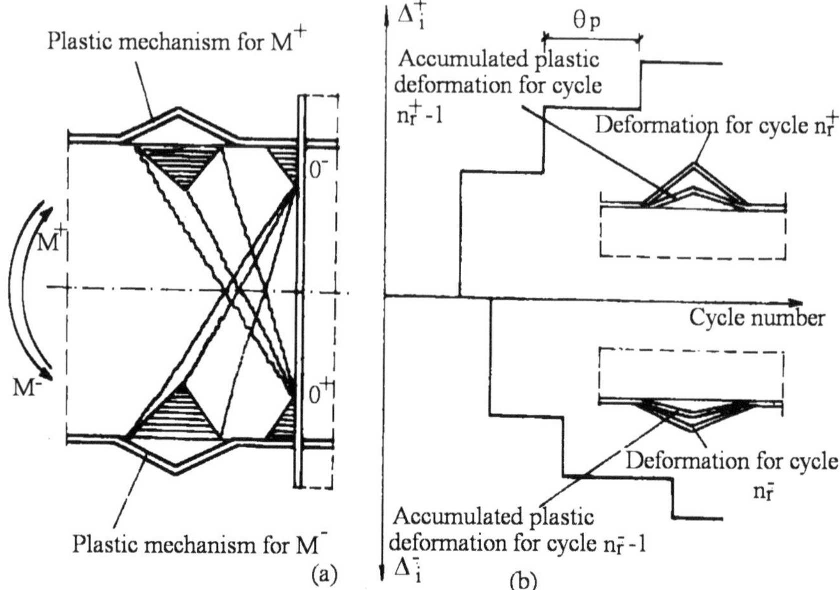

Figure 8.86: Influence of repeated cyclic actions: (a) Plastic mechanism; (b) Accumulation of buckled flange plastic deformations

-when the flange buckling is produced before reaching the maximum seismic action, the plastic hinge works with an increasing of rotation amplitute for each cycle (Fig. 8.85b).

Therefore, two types of movement under constant and increased amplitudes must be considered for study of the repeated cyclic actions.

(ii) *Local plastic mechanism under repeated cyclic actions.* The most important step for studying the member behaviour under cyclic actions is the choice of the characteristic control parameter (see Section 4.2.6). In the ECCS testing procedure (ECCS, 1986), the limit of elastic rotation θ_p is selected for this parameter. The first positive moment produces the buckling of a compression flange and the section rotates around the point O^+ located near the opposite flange. During the first negative moment the opposite flange also buckles and the rotation occurs around the point O^- (Fig. 8.86a). The process continues alternatively in the same way for the next cycles. Due to the fact that the buckled flange and the rotation points are in different positions, the tension strains are small and are not able to straighten the buckled flange. Therefore, the plastic collapse mechanism under cyclic actions is formed by superimposing, as in a mirror, two local plastic mechanisms. During the next cycle, the section works with an initial geometrical deformation which resulted from the previous cycle. Therefore, the cyclic behaviour produces an accumulation of plastic deformations in the buckled flange, which determines a deviation of the post-critical curve from the one corresponding to the static actions (Fig. 8.86b).

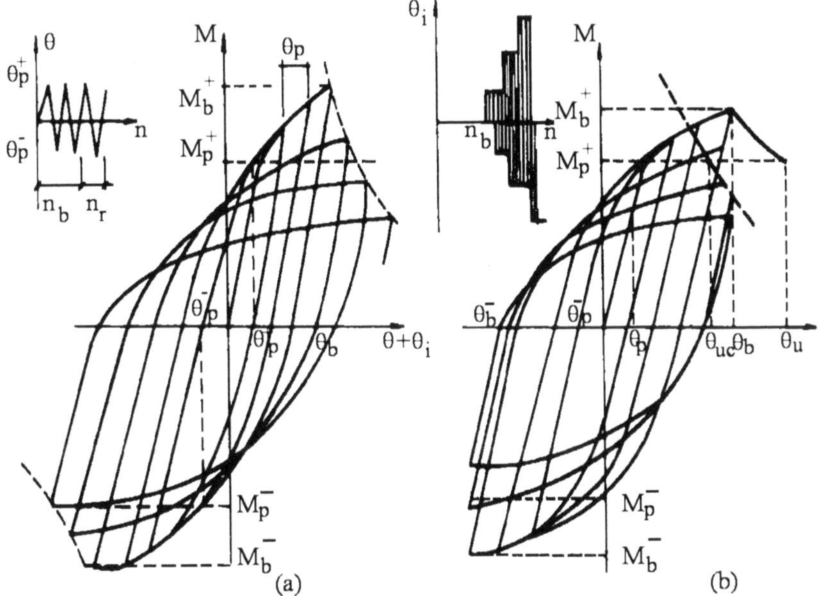

Figure 8.87: Constant rotation amplitude: (a) Moment-total rotation curve; (b) Moment-rotation curve

(iii) *Rotation capacity under cyclic actions.* The effect of cyclic actions is analyzed in the two above mentioned cases:

-*constant rotation amplitude*, presented in Fig. 8.87. There are no differences between monotonic and cyclic actions until the buckling of compressed flange and the skeleton curve can be used (see Section 4.2.6). The difference in behaviour begins to become significant only after plastic buckling occurs at the cycle $n_b = \theta_b/\theta_p$. There are two ways to plot the hysteretic curves. The first includes the accumulated plastic deformation (Fig. 8.87a). One can observe that the post-critical curve for the cyclic actions corresponds to the monotonic one, but with a sequential reducing of moment capacity. This observation allows for an extension of the stable part of moment-rotation curve to its unstable part. The second way refers to the plastic rotation only (Fig.8.87b), the sequential decreasing of moment capacity being noted after the local buckling of compression flanges;

-*increasing rotation amplitude*, presented in Fig. 8.88a, for $\theta + \theta_a$ rotation (θ_a being the accumulated plastic rotation) and in Fig. 8.88b for θ only. The main difference in comparison with the constant rotation amplitude consists in a more accentuated increasing of plastic rotation accumulations and a faster deterioration of moment capacity of members.

If n_r is the number of strong pulses produced after the local buckling:

$$n_r = n - n_b \tag{8.107}$$

where n is the total number of strong cycles, the post-critical curve for cyclic

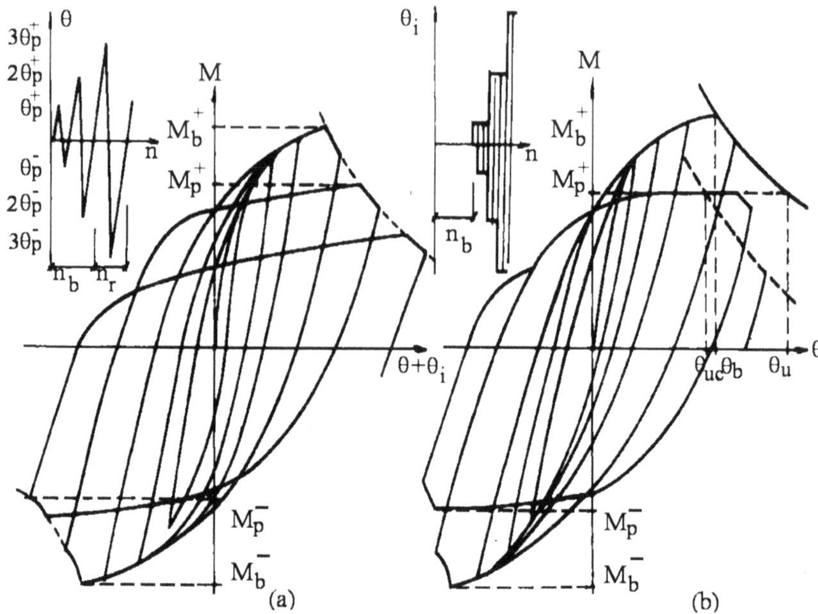

Figure 8.88: Increasing rotation amplitude: (a) Moment-total rotation curve; (b) Moment-rotation curve

actions is given by (see (8.49a)):

$$\frac{M}{M_p} = \alpha_{1M} + \alpha_{2M}\frac{1}{(\theta + \theta_{ar})^{1/2}} \qquad (8.108)$$

where θ_{ar} is the accumulated plastic rotation until cycle r, resulting from:
-for constant rotation amplitude:

$$\theta_{ar} = (n_r - 1)\theta_p; \quad \theta + \theta_{ar} = n_r\theta_p \qquad (8.109a, b)$$

-for increased rotation amplitute:

$$\theta_{ar} = \frac{n_r(n_r - 1)}{2}\theta_p; \quad \theta + \theta_{ar} = \frac{n_r(n_r + 1)}{2}\theta_p \qquad (8.109c, d)$$

Using the same criterion for determining the ultimate rotation capacity as for monotonic actions (the intersection of post-critical curve with the line corresponding to M_p or $0.9M_p$), from (8.49a) and (8.108) the ultimate rotation for cyclic actions θ_{uc} results:

$$\theta_{uc} = \theta_u - \theta_{ar} \qquad (8.110)$$

where θ_u is the ultimate rotation for monotonic actions. The rotation capacity $\mu_{\theta c}$ for cyclic action is given by:
-for constant rotation amplitude:

$$\mu_{\theta c} = \mu_\theta - (n_r - 1) \qquad (n_r \geq 2) \qquad (8.111a)$$

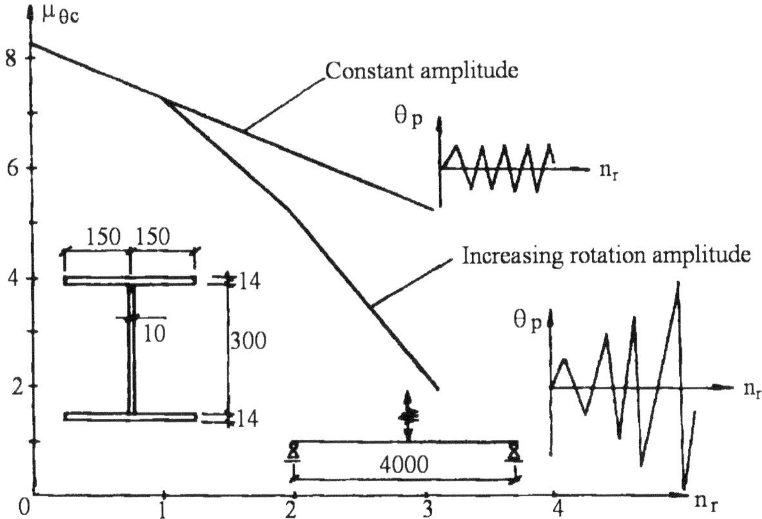

Figure 8.89: Influence of pulse number after local buckling on rotation capacity

-for increasing rotation amplitude:

$$\mu_{\theta c} = \mu_\theta - \frac{n_r(n_r - 1)}{2} \qquad (n_r \geq 2) \tag{8.111b}$$

where μ_θ is the rotation capacity for monotonic actions. The influence of the number of pulses for the two cyclic action types is presented in Fig. 8.89, where a very important decreasing of rotation capacity in the case of increasing rotation amplitude clearly appears.

Therefore, the problem of the evaluation of rotation capacity under repeated cyclic actions is based on the estimation of the number of pulses producing large plastic deformations. Taking into account that the number of pulses is higher for far-field earthquakes than for intermediate-field earthquakes, it is reasonable to consider:

-for intermediate-field earthquakes n_r = 1–2;

-for far-field earthquakes n_r = 2–4. That means that, in order to take into account the detrimental effects of cyclic actions, the local rotation ductility must be reduced by a value 1–2 for intermediate-field earthquakes and 3–4 for far-field earthquakes.

(iv) *Fracture under cyclic actions.* The number of pulses for the two cyclic types for different yield ratios is presented in Fig. 8.90. The fracture rotation is determined by using the equations (8.80) and (8.109b,d). One can observe that, for low yield ratios, the number of pulses is out of the range of practical values and, therefore, there is no danger of fracture. Contrary to this, for high yield ratios ($\rho_y > 0.80$), the number of pulses is in the range of possible realistic situations.

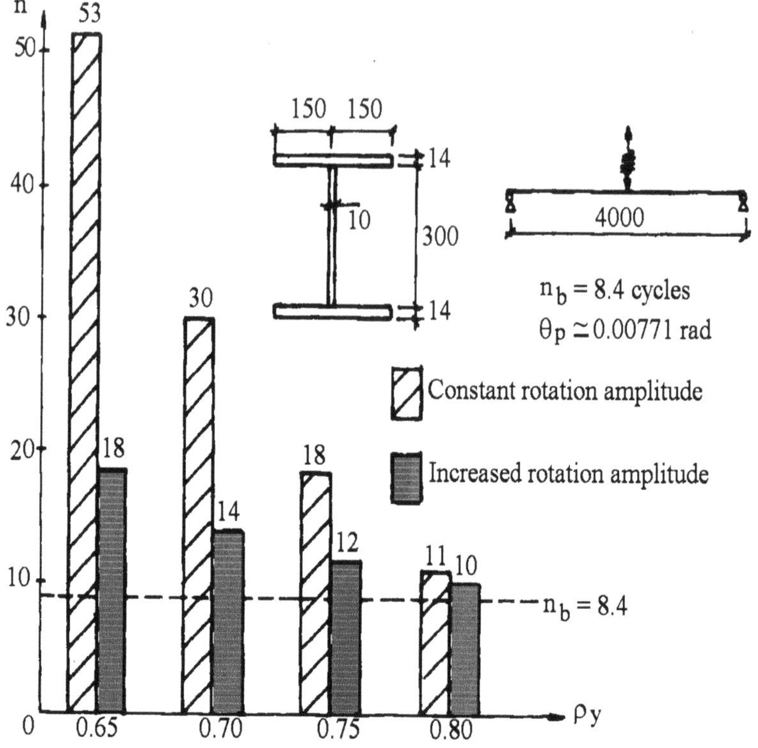

Figure 8.90: Number of fracture pulses

8.5 Structural Damage

8.5.1. Damage Index

The economic losses during the recent earthquakes were substantial. Therefore, the control of seismic damage has become a very important phase in design. Although the prediction of seismic damage is primarily a probabilistic problem, the ground motions being the most critical aspect, neverthless deterministic analyses can represent a valuable tool (Powell and Allahabadi, 1988). The prediction of the amount of seismic damage that a structure is likely to sustain during its lifetime is controlled by the damage index, defined at the level of rigidity, strength and ductility. At the level of ductility this parameter refers to the plastic rotation of plastic hinges, determined at the member level, storey level or global structure level:

-*member damage index* can be calculated as the ratio between the required plastic rotation and the available plastic rotation:

$$I_{dm} = \frac{\theta_r}{\theta_a} \qquad (8.112a)$$

-*storey damage index* refers to the damage of a storey:

$$I_{ds} = \frac{\sum_{1}^{n_b} I_{dm}^2}{\sum_{1}^{n_b} I_{dm}} \tag{8.112b}$$

where n_b is the number of frame bays;

-*global damage index* is a structure damage index which includes the global seismic behaviour of the structure:

$$I_{dg} = \frac{\sum_{1}^{n_b}\sum_{1}^{n_s} I_{dm}^2}{\sum_{1}^{n_b}\sum_{1}^{n_s} I_{dm}} \tag{8.112c}$$

where n_s is the number of frame storeys.

The equations (8.112b,c) are determined by using a weighting factor which reflects the replacement cost and/or the relative importance of the damaged member in maintaining the integrity of the structure (Powell and Allahabadi, 1988). For example, in a building, more importance might be assigned to the lower storeys than the upper ones. By assuming that the weight of each damaged section is proportional to its local damage index the above equations results.

Fig. 8.91 shows the damage index for members, storeys and global structure. Generally, the member damage index is the highest. The global damage index gives a more qualitative information on the structure damage and provides little or no information on the amount and location of member damage. Therefore, the damage index for members or storeys is suggested to be used in design practice.

8.5.2. Damage Levels

A correlation between damage levels and different limit states must be established. At the level of rotation capacity the identified damage levels are:

-*no damage* up to the elastic limit for:

$$\theta < \theta_p \tag{8.113a}$$

where the performance of structure is very similar for different earthquakes;

-*minor damage*, when the elastic range is exceeded somewhere:

$$\theta_p < \theta < 1.5\theta_p \tag{8.113b}$$

No interventions after earthquakes are necessary to assure that service, facilities and functions continously operate;

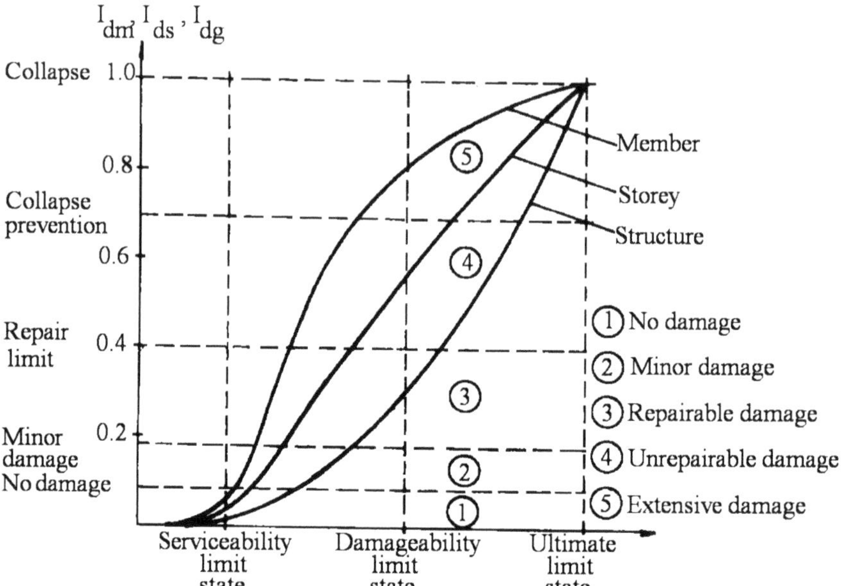

Figure 8.91: Damage index for member, storey and structure

-*easy repairable damage* for plastic rotation without local buckling;

$$1.5\theta_p < \theta < \theta_b \tag{8.113c}$$

when damage is light and repair is required to restore some structural sections. Structure is safe for occupancy immediately after earthquake and the structural interventions do not disturb the main activity in the building;

-*repairable damage*, for plastic rotation with local buckling:

$$\theta_b < \theta < \theta_u \tag{8.113d}$$

when the structure is locally damaged but remains stable. Repair is required to restore some zones of structure and the building activity is temporarily interrupted;

-*very difficult repairable* or *irreparable damage* for important plastic rotations or fractures

$$\theta > \theta_u \tag{8.113e}$$

when the structural damage is very severe and the building activity must be interrupted. The building must be demolished or strengthened, according to the decisions of owner and experts;

-*extensive damage* when a portion or the complete structure collapses during the earthquake.

At the level of global damage index some values may be considered:

-no damage if $I_{dg} < 0.05$;

-minor damage if $I_{dg} < 0.15$;

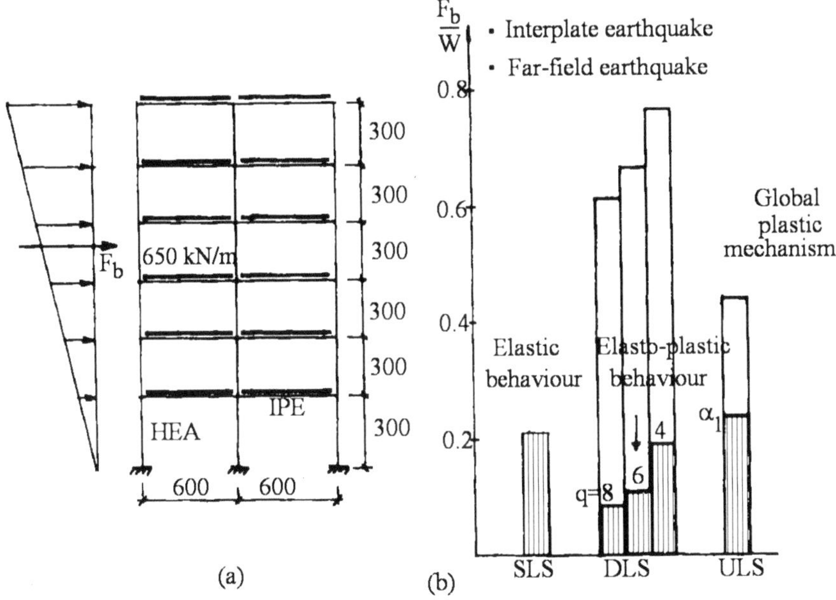

Figure 8.92: Worked example – structure subjected to far-field interplate earthquake: (a) Structure dimensions; (b) Seismic actions for the three limit states

-repairable damage if $I_{dg} < 0.50$;
-collapse prevention limit if $I_{dg} < 0.80$;
-near collapse if $I_{dg} > 0.80$;
-structure collapse if $I_{dg} > 1.00$

A similar classification (but with some differences in limit values) has been proposed by Ghobarah et al (1997).

Finally, it must be kept in mind that these values are informative only and each structural damage must be studied as a special case, in which the expert's experience plays the most important role.

8.6 Worked Examples

8.6.1. Main Characteristics of Structure

In order to show the applicability of the proposed methodology, the results for a moment-resisting framed building located at two different sites are herein presented. The structure is a six storeys-two bays frame (Fig. 8.92a). For each floor the value of dead load (including weight of external and internal walls) is 4.50 kN/m^2 and that of live load is 2.00 kN/m^2. These loads give rise to concentrate forces of 120 kN for external and 240 kN for internal columns. The total weight acting on the structure is W = 2880 kN. The beams are made of IPE profiles and the HEA profiles are used for column. The joints are made of welded connections, respecting the condition to have sufficient overstrength to not be included in dissipative zones. The fundamental vibration period is determined from equation (8.28b), giving $T = 0.66$sec.

8.6.2. Structure Subjected to Far-Field Interplate Earthquake

(i) *Main characteristics of site.* The building is placed in a seismic area with a source which produces interplate earthquakes with a ground acceleration $a_g = 0.36g$ corresponding to a return period of 475 years. The distance from the potential source being large, the ground motions have the characteristics of a far-field earthquake with a maximum of 8 big pulses. The subsoil corresponds to class C with $S = 1.35$ (see EC 8).

(ii) *Seismic actions.* The return period of the basic acceleration corresponds to the damageability limit state. Therefore, using equations (8.2), the accelerations corresponding to the three limit states are given by:

-serviceability limit state (SLS), $a_s = 0.15g$;
-damageability limit state (DLS), $a_d = 0.36g$;
-ultimate limit state (ULS), $a_u = 0.44g$

The earthquake type being an interplate one, the spectra given by equations (8.5a,b) are used, with $\eta = 1.2$ for SLS and $\eta = 1.0$ and $S = 1.35$ for DLS. The base shear forces corresponding to different limit states are determined using equations (8.8a,b). The results are presented in Fig. 8.92b, where it appears that SLS and ULS are the dominant limit states in design process. For DLS three values for q factor are considered; even for $q = 4$ the value of shear force is lower than that corresponding to SLS. This result gives a justification of the design strategy for moderate seismic actions presented in Table 8.1.

(iii) *Design for global plastic mechanism.* The formation of the global mechanism is performed for the ultimate limit state. From equation (8.8c) it results:

$$F_{bu} = 0.44 \times 2880 = 1270 \text{ kN}$$

The triangular distribution of lateral forces along the structure height is presented in Fig. 8.93a, giving rise to the distribution of the shear forces and the ratio m_h of eq. (8.11b) (Fig. 8.93b). The bending moments in the beams, as a sum of the effect of vertical loads $M_w \simeq qL^2/10$ and the one of horizontal forces $M_h = m_h M_w$ (see equations (8.11), are shown in Fig. 8.93c. For these values, IPE 450 (Fe 360) profiles are selected for the first three levels and IPE 400 (Fe 360) profiles for the last three. Using the equation (8.13b) the axial forces in internal and external columns are determined (Fig. 8.93d). In order to obtain an optimal global plastic mechanism, the plastic moment of first level columns is obtained from equation (8.14):

$$M_{pc,1} = \frac{\sum M_{pc,i1}}{3} = \frac{(0.7 + 0.15 \times 6) \times 4 \times 400 \times 1.1}{3} = 937 \text{ kNm}$$

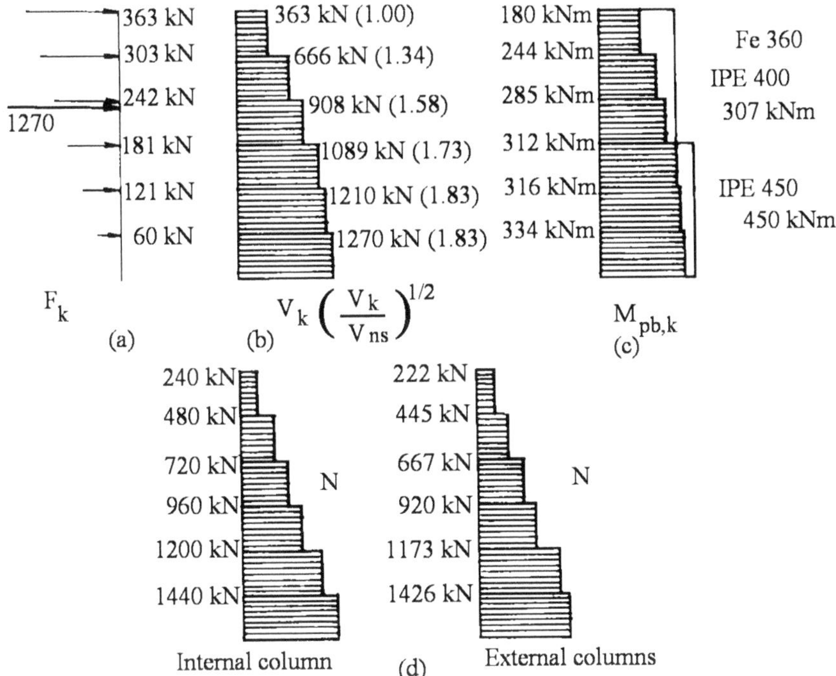

Figure 8.93: Actions and moments: (a) Loading distribution of lateral forces; (b) Shear force distribution; (c) Required and available bending moments; (d) Axial forces

in which an overstrength of beam plastic moments $m_y = 1.1$ is introduced. Considering the effect of axial forces $N = 1440$ kN (interaction M-N equation 5.4b), a HE 550A profile (Fe 430) results with $M_{pn} = 1066$ kNm. For these structure dimensions a global plastic mechanism is expected for the ultimate limit state.

(iv) *Required plastic rotation for static and monotonic actions* (Fig. 8.94). The simplified push-over method presented in Sections 8.2.2 to 8.2.3 is used for determining the required plastic rotation. From equations (8.23)–(8.25) it results:

$$\alpha_1 = \frac{3}{2}\frac{3 \times 1066 \times 1.1 + 6 \times 400 \times 1.1 + 6 \times 307 \times 1.1}{1270 \times 18.00} = 0.537$$

$$\alpha_2 = \frac{3}{4}\frac{1}{0.44} = 1.70$$

From equations (8.27) and (8.28), it results:

$$\theta_e = 0.537 \cdot 0.51 \times 10^{-2} \cdot 18^{0.6} \cdot 0.44 = 0.0068 \text{ rad}$$

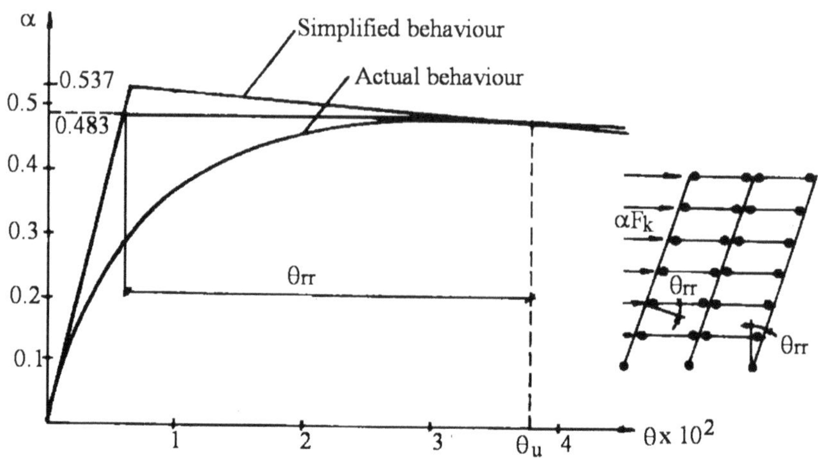

Figure 8.94: Required rotation under static and monotonic action

The required plastic rotation for $0.9\alpha_1$ results from equation (8.29c):

$$\theta_{rr} = 0.1\left(\frac{0.537}{1.7} + 0.0068\right) = 0.0323 \text{ rad}$$

(v) *Required plastic rotation for seismic actions.* The structure is situated in a far-field site and from equations (8.32) and Table 8.2 it results:
-for beams:

$$\gamma_a = 1.963$$

$$\theta_{rca} = 1.963 \times 0.0323 = 0.0634 \text{ rad}$$

-for column:

$$\gamma_a = 2.463$$

$$\theta_{rca} = 2.463 \times 0.0323 = 0.0796 \text{ rad}$$

(vi) *Standard beams.* The structure is subdivided in a sum of standard beams for which the ultimate plastic rotation and the rotation capacity are determined (Fig. 8.95), using the methodology presented in Section 6.3.2. Five standard beams must be considered for IPE 400 and IPE 450 beams (Fig. 8.95b,c) and for HE 550A column(Fig. 8.95d). In the case of beams, standard beam SB 1 type is considered for moment gradient variation and standard beam SB 2 type for quasi-constant moment. For columns, standard beam SB 1 type is used.

(vii) *Available rotation capacity for static and monotonic actions.* The moment-rotation curves for the three sections in the case of moment gradient are presented in Fig. 8.96, using the method of local plastic mechanism presented in Section 8.3.2. The $0.9M_p$ ductility criterion is used for determining the ultimate plastic rotation. It results:

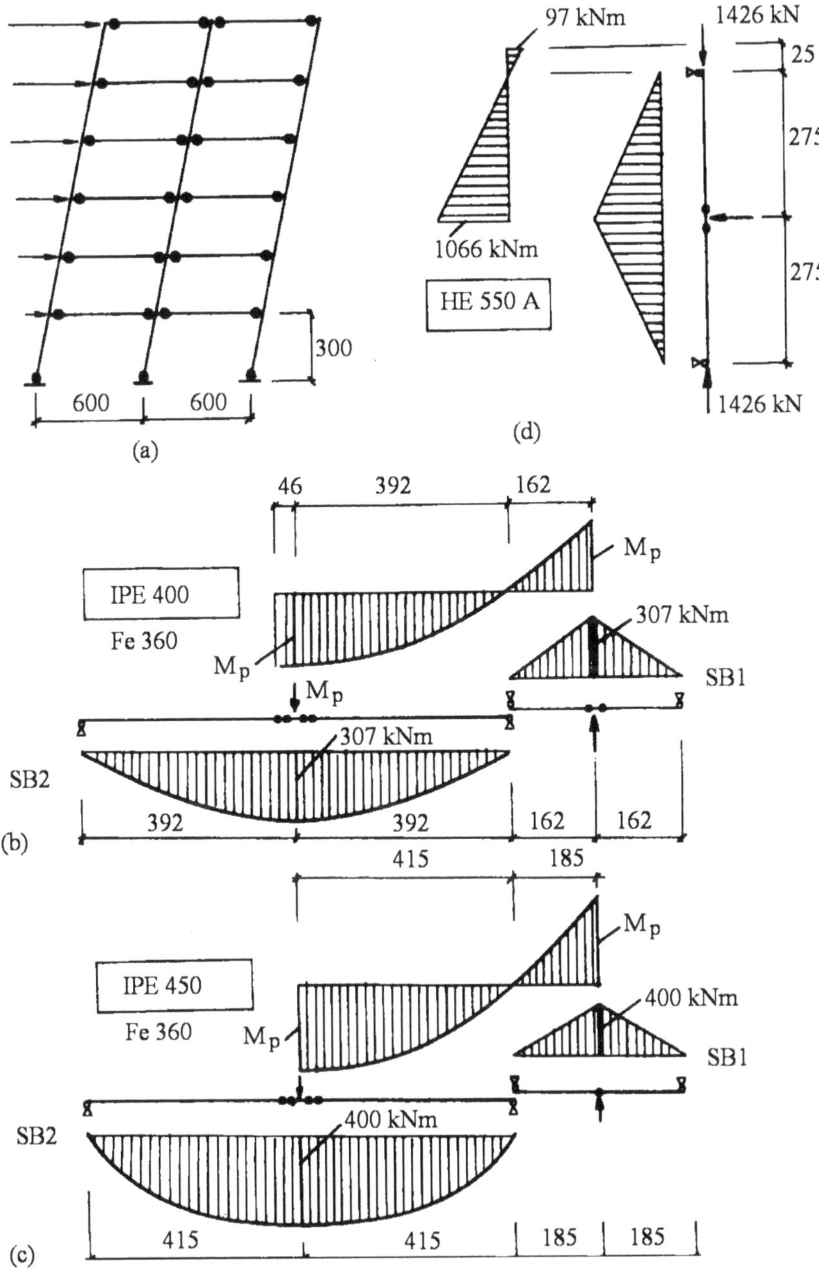

Figure 8.95: Determination of standard beam dimensions: (a) Structure configuration; (b) Standard beams for IPE 400 profiles; (c) Standard beams for IPE 450 profiles; (d) Standard beam for HE 550A profiles

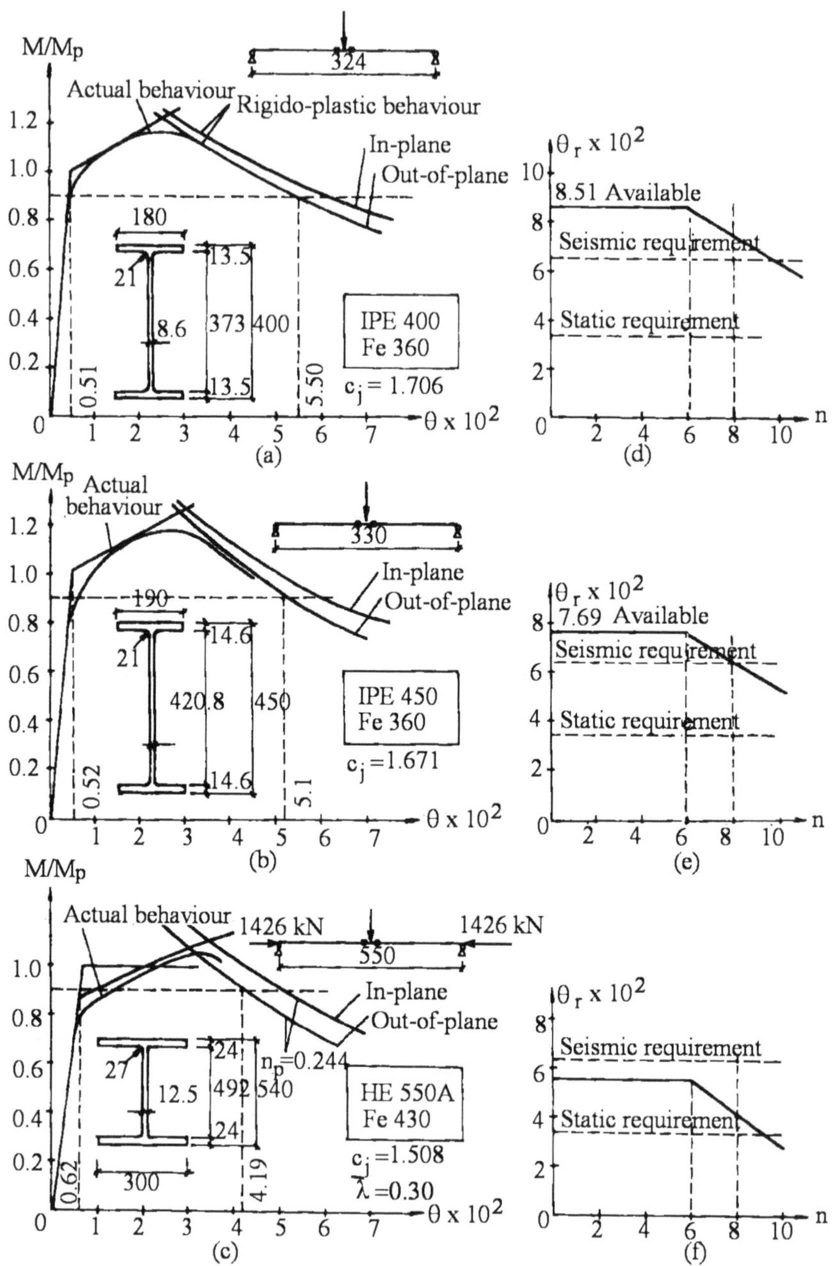

Figure 8.96: Available rotation capacity: (a) IPE 400 profiles; (b) IPE450 profiles; (c) HE 550A profiles

-IPE 400 profiles (Fe 360):
• in-plane mechanism, moment gradient: $\theta_r = 0.0582$ rad;
• in-plane mechanism, constant moment: $\theta_r = 0.1239$ rad;
• out-of-plane mechanism: $\theta_r = 0.0499$ rad.

These results show that the out-of-plane mechanism is responsible for the profile collapse. Considering the effect of flange-web junctions (relations (8.62), (8.63)), it results:

$$\theta_{rj} = 1.706 \times 0.0499 = 0.0851 \text{ rad } > 0.0323 \text{ rad}$$

(static required plastic rotation)

-IPE 450 profiles (Fe 360):
• in-plane mechanism, moment gradient: $\theta_r = 0.0570$ rad;
• in-plane mechanism, constant moment: $\theta_r = 0.1170$ rad;
• out-of-plane mechanism: $\theta_r = 0.0460$ rad.

Also in this case the out-of-plane mechanism determines the member collapse. The effect of a flange-web junction gives:

$$\theta_{rj} = 1.671 \times 0.0460 = 0.0769 \text{ rad } > 0.0323 \text{ rad}$$

(static required plastic rotation)

-HE 550A profiles (Fe 430, $n_p = 0.244$):
• in-plane mechanism, moment gradient: $\theta_r = 0.0462$ rad;
• out-of-plane mechanism: $\theta_r = 0.0357$ rad.

Even for this profile the out-of-plane mechanism is the dominant one. The effect of flange-web junctions gives:

$$\theta_{rj} = 1.508 \times 0.0357 = 0.0538 \text{ rad } > 0.0323 \text{ rad}$$

(static required plastic rotation)

By analyzing these results, one can conclude that for static and monotonic actions the structure has a sufficient rotation capacity to develop a good global plastic mechanism.

(viii) *Available rotation capacity for seismic actions.* The earthquake being of far-field type, the main seismic effect is due to accumulation of plastic rotation caused by the repeated cyclic actions. The reductions of rotation capacities due to cyclic actions are presented in Fig. 8.96d,e,f. One can see that for the first $n_b = 6$ pulses there is no reduction in rotation capacity. Considering that 8 has been assumed as the number of big pulses, only $n_r = 8 - 6 = 2$ pulses produce a reduction of rotation capacity due to accumulation of plastic rotations:

- IPE 400: $\theta_{rc} = 0.0851 - 2\times0.0051 = 0.0749 > 0.0634$ rad (seismic required plastic rotation);

- IPE 450: $\theta_{rc} = 0.0769 - 2\times0.0052 = 0.0665 > 0.0634$ rad (seismic required plastic rotation);

- HE 550A: $\theta_{rc} = 0.0538 - 2\times0.0062 = 0.0414 < 0.0796$ rad (seismic required plastic rotation).

It results that for beams there are no problems, the available rotation capacity being in both cases greater than the required one. Contrary to

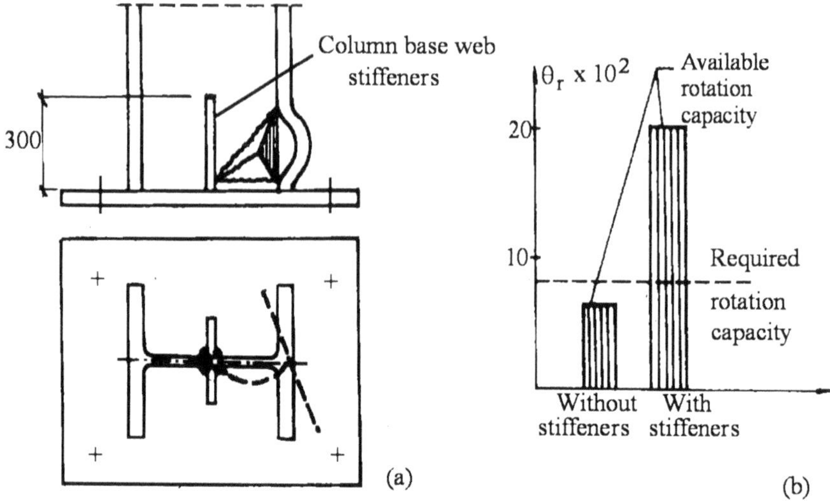

Figure 8.97: Strengthening of columns at the base: (a) Constructional details; (b) Effect of stiffeners on the rotation capacity

this, for columns, the requirement exceeds the available rotation capacity. In order to increase this rotation capacity, a pair of welded web stiffeners can be introduced (Fig. 8.97a). With this solution, the rotation of a plastic local mechanism occurs around the mid section ($\delta = 0.5$) and a very important increasing in rotation capacity is obtained (Fig. 8.97b):

- HE 550A: $\theta_{rc} = 0.2051 - 2\times0.0062 = 0.1927 > 0.0794$ rad.

So, the structure now has a sufficient rotation capacity to develop a good global plastic mechanism.

The member damage index results from equation (8.112a):

- IPE 400: $I_{dm} = 0.841$;
- IPE 450: $I_{dm} = 0.947$;
- HE 550A: $I_{dm} = 0.413$.

The global damage index results from equation (8.112c):

- $I_{dg} = 0.806$.

Therefore the structure is saved from collapse, but the earthquake produces very severe damage and the cost of repairing is very high. The expert viewing the damaged structure has to decide between two possible choices: strengthening or demolition.

8.6.3. Structure Subjected to Near-Field Intraplate Earthquake

(i) *Main characteristics of the site.* The same building (Fig. 8.98a) like the one presented in the previous Section is now placed in a seismic area characterized by intraplate earthquakes. The building is situated close to the epicenter in the near-field area, where the earthquakes have the ground

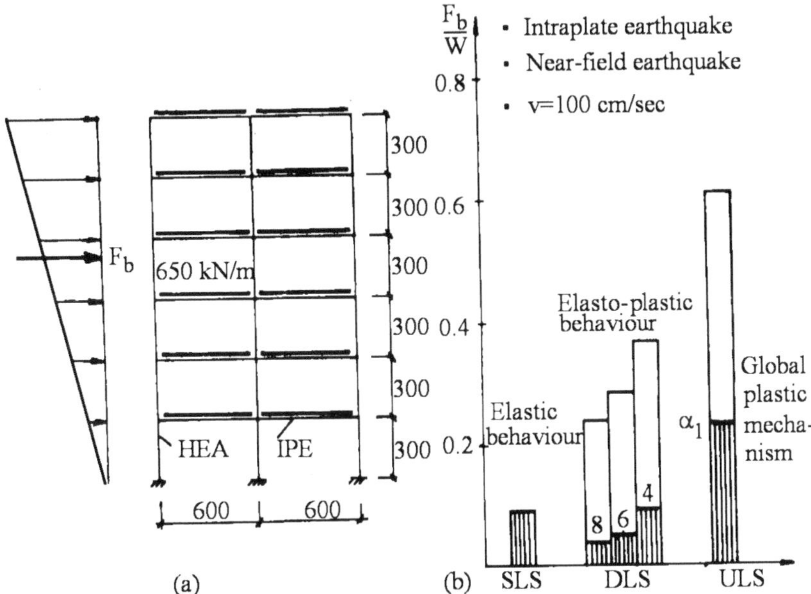

Figure 8.98: Worked example – structure subjected to near-field intraplate earthquake: (a) Structure dimensions; (b) Seismic actions for the three limit states

acceleration $a_g = 0.50g$, corresponding to a return period of 475 years, with a maximum ground velocity of $v = 100$ cm/sec and no more than two big pulses.

(ii) *Seismic actions.* The accelerations corresponding to the limit states are:

-serviceability limit state (SLS) $a_s = 0.21g$;
-damageability limit state (DLS) $a_d = 0.50g$;
-ultimate limit state (ULS) $a_u = 0.61g$.

To determine the seismic actions the equations (8.6) for intraplate earthquakes are used. The results are presented in Fig. 8.98b. One can observe that, in comparison with the previous case of interplate earthquake, the seismic loads for serviceability and damageability limit states are less important due to spectral values, but for the ultimate limit state the loads are increased. Also in this case, the dominant limit states are the serviceability and ultimate ones.

(iii) *Design for global plastic mechanism.* This corresponds to ultimate limit state. From equation (8.8c),

$$F_{bu} = 0.61 \times 2880 \simeq 1760 \text{ kN}$$

with a triangular distribution along the structure height. The dimensions of structure remain the same as in the previous case.

(iv) *Required rotation capacity for static and monotonic actions.* From the equations (8.23) to (8.25),

$$\alpha_1 = \frac{3}{2}\frac{3 \times 1066 \times 1.1 + 6 \times 400 \times 1.1 + 6 \times 307 \times 1.1}{1760 \times 18.00} = 0.387$$

$$\alpha_2 = \frac{3}{4}\frac{1}{0.61} = 1.23$$

From the equations (8.27) and (8.28), it results:

$$\theta_e = 0.387 \times 0.51 \times 10^{-2} \cdot 18^{0.6} \cdot 0.61 = 0.0068 \text{ rad}$$

The required kinematic ductility for $0.9\alpha_1$ is obtained from the equation (8.29c):

$$\theta_{rr} = 0.1\left(\frac{0.387}{1.23} + 0.0068\right) = 0.0321 \text{ rad}$$

which is practically the same value obtained in the previous example.

(v) *Required rotation capacity for seismic actions.* The structure being situated in a near-field site, it follows from equations (8.102b) and (8.103):

$$\text{IPE 400:} \quad \dot{\epsilon} = \frac{2 \times 100 \times 40}{900 \times 18} = 0.494 \text{ /sec;} \ \frac{\rho_{ysr}}{\rho_y} = 1.388$$

$$\text{IPE 450:} \quad \dot{\epsilon} = \frac{2 \times 100 \times 45}{900 \times 19} = 0.526 \text{ /sec;} \ \frac{\rho_{ysr}}{\rho_y} = 1.389$$

$$\text{HE 550A:} \quad \dot{\epsilon} = \frac{2 \times 100 \times 54}{900 \times 30} = 0.364 \text{ /sec;} \ \frac{\rho_{ysr}}{\rho_y} = 1.384$$

and from (8.32c) the required rotation capacities:

$$\text{IPE 400:} \quad \theta_{rrsr}\frac{1.388}{1.1} \times 0.0321 = 0.0405 rad$$

$$\text{IPE 450:} \quad \theta_{rrsr}\frac{1.389}{1.1} \times 0.0321 = 0.0405 rad$$

$$\text{IPE 550A:} \quad \theta_{rrsr}\frac{1.384}{1.1} \times 0.0321 = 0.0404 rad$$

where the overstrength of yield stress $m_y = 1.1$ is eliminated, being included in the evaluation of coefficient α_1.

(vi) *Available rotation capacity for static and monotonic actions.* The values of rotation capacity remain the same as the ones obtained in the previous example.

(vii) *Available rotation capacities for seismic actions.* Due to the fact that the ground motions have a reduced number of cycles, the effect of plastic rotation accumulation is greatly reduced. The rotation capacity is influenced only by the strain-rate effect to increase the fully plastic moment (see equation (8.104a,b)):

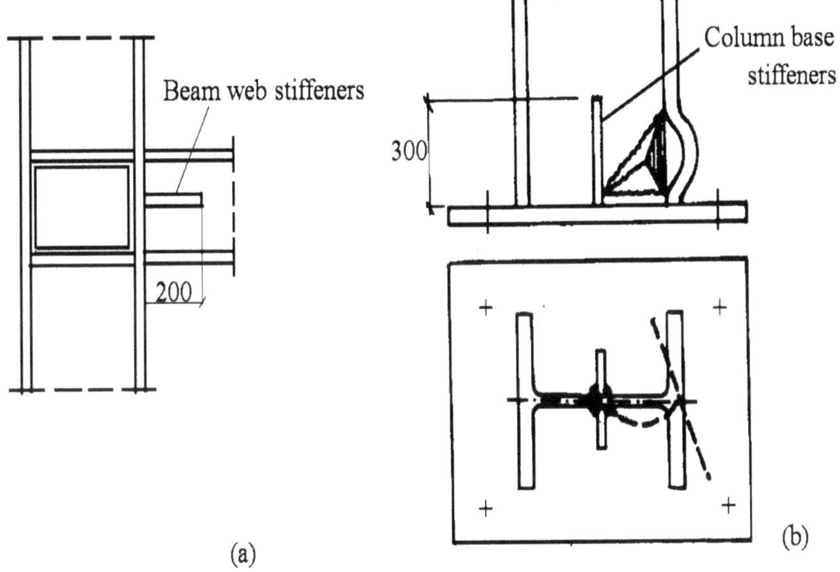

Figure 8.99: Measures for rotation capacity increasing: (a) Beam web stiffeners; (b) Column base web stiffeners

-IPE 400: $\theta_{rsr} = 0.414 \times 0.0851 = 0.0352 < 0.0405$ rad;
-IPE 450: $\theta_{rsr} = 0.415 \times 0.0769 = 0.0319 < 0.0405$ rad;
-HE 550A: $\theta_{rsr} = 0.408 \times 0.0538 = 0.0220 < 0.0404$ rad.

One can observe that the influence of strain-rate is very important, the available rotation capacity for all profiles being less than the required rotation capacity.

Concerning the fracture of tension flanges, the strain-rate results from equation (8.101b) with $H_m = 900$ cm:

$$\dot{\epsilon} = \frac{100}{900} = 0.111 \text{ /sec}; \quad \frac{\rho_{ysr}}{\rho_y} = 1.359$$

and from equation (8.105), it can be pointed out that there is no danger for fracture in tension flanges.

At the level of compression flanges, from the equation (8.106) it results (Fig. 8.83):

- IPE 400: $\theta_f = 0.0020 < \theta_{sr} = 0.0190$ rad;
- IPE 450: $\theta_f = 0.0018 < \theta_{sr} = 0.0187$ rad;
- HE 550A: $\theta_f = 0.0032 < \theta_{sr} = 0.0335$ rad.

The fracture cannot occur.

So, it results that the rotation capacity of all elements does not satisfy the condition of a good plastic ductility. In order to increase the rotation capacity, a pair of welded web stiffeners can be introduced (Fig. 8.99).

8.7 Conclusions

This Chapter presents a comprehensive methodology for the seismic design practice in which the ductility control is performed to the same degree of accuracy as the rigidity and strength checking:

-the methodology is based on multi-level criteria, which consider three limit states: serviceability, damageability and ultimate;

-four phases are developed during the design process: seimic input determination, global analysis, local analysis and required-available checking;

-design seismic inputs like return periods, ground motion accelerations, corner periods, spectra and base shear force are characteristics to be determined first of all;

-for the global analysis a simplified push-over method is set up in order to evaluate the required plastic rotation, both for static and monotonic actions and seismic actions;

-available plastic rotation capacity is determined for I, X, box, hollow and composite section members, using the local plastic mechanism methodology, for static and monotonic actions and seismic actions;

-the main factors influencing the available ductility are considered: collapse type, fabrication, strengthening or weakening member ends, stiffening of web, haunching or tapering, axial forces, member slenderness, fracture behaviour;

-the effects of seismic actions are considered, in order to correct the values determined from static and monotonic actions: the influence of strain-rate, cyclic actions and structural damage are taken into account;

-the worked examples show the applicability of proposed methodology;

-the local plastic mechanism method has been used to develop a computer program DUCTROT M (DUCTility for ROTation of Members) for practical use. This program is illustrated in the Appendix. A complimentary diskette for this computer program is enclosed in the book.

8.8 References

Alavi, B., Krawinkler, H. (2000): Consideration of near-fault ground motion effects in seismic design. In 12th World Conference on Earthquake Engineering, Auckland, 30 January-4 February 2000, CD-ROM 2665

Allen, H.G., Bulson, P.S. (1980): Background to Buckling. McGrow-Hill Book Company, London

Anastasiadis, A. (1999): Ductility Problems of Steel Moment Resisting Frames. Ph Thesis, Politechnica University Timisoara

Anastasiadis, A., Gioncu, V. (1998): Influence of joint details on the local ductility of steel MR frames. In 3rd National Conference on Steel Structures, Tessaloniki, 30-31 October 1998, 311-319

Anastasiadis, A., Gioncu, V. (1999): Ductility of IPE and HEA beams and beam-columns. In Stability and Ductility of Steel Structures, SDSS 99 (eds. D. Dubina and M. Ivanyi), Timisoara, 9-11 September 1999, Elsevier, Amsterdam, 249-257

Anastasiadis, A., Mateescu, G., Gioncu, V., Mazzolani, F.M. (1999a): Reliability of joints for improving the ductility of MR frames. In Stability and Ductility of Steel Structures, SDSS 99 (eds. D. Dubina and M. Ivanyi), Timisoara, 9-11 September 1999, 229-268

Anastasiadis, A., Gioncu, V., Mazzolani, F.M. (1999b): New upgrading procedures to improve the ductility of steel MR-frames. In XVII Congresso CTA, Napoli, 3-7 October 1999, Vol. 1, 193-204

Anastasiadis, A., Mateescu, G., Gioncu, V. (2000): Improved ductile design of steel MR-frames based on constructional details. In 9th International Conference on Metal Structures, (ed. M.Ivan), Timisoara, 19-22 October 2000, Editura Orizonturi Universitare, Timisoara, 367-376

Anderson, J.C., Bertero, V.V. (1987): Uncertainties in establishing design earthquakes. Journal of Structural Engineering, Vol. 113, No. 8, 1709-1724

Andrade, S.A.L., Morris, L. J. (1985): Residual stresses in portal frames haunched and their influence on member stability. Universidade Catolica de Rio de Janeiro, Techical Report AT 17/85

Ansourian, P. (1982): Plastic rotation of composite beam. Journal of the Structural Division, Vol. 108, ST 3, 643-659

Aribert, J.M. (1992): Notions de calcul des poutres mixtes acier beton aux states limites ultimes. Stage de perfectionnement, CTICM, 24-25 November 1992

ATC-Applied Technology Council (1987): Evaluating the seismic resistance of existing buildings. Report ATC No. 14

Chai, Y.H., Fajfar, P., Romstad, K.M. (1998): Formulation of duration-dependent inelastic seismic design spectrum. Journal of Structural Engineering, Vol. 124, No. 8, 913-921

Chen, S.J., Yeh, C.H., Chu, J.M. (1996): Ductile steel beam-connections for seismic resistance. Journal of Structural Engineering, Vol. 122, No. 11, 1292-1297

Climenhaga, J.J., Johnson, R.P. (1972): Moment-rotation curves for locally buckling beams. Journal of the Structural Division, Vol. 98, ST 6, 1239-1254

De Matteis, G., Landolfo, R., Dubina, D., Stratan A. (2000): Influence of the structural typology on the seismic performance of steel framed buildings. In Moment Resistant Connections of Steel Frames in Seismic Areas. Design and Reability, RECOS, (ed. F.M. Mazzolani), E&FN Spon, London, 513-538

Driver, R.G., Kennedy, D.J.L., Kulak, G.L. (2000): Establishing seismic force reduction factors for steel structures. In Behaviour of Steel Structures in Seismic Areas, STESSA 2000 (eds. F.M. Mazzolani and R. Tremblay), Montreal, 21-24 August 2000, Balkema, Rotterdam, 487-494

ECCS-TC 13 (1985): Recommended testing procedure for assessing the behaviour of structural elements under cyclic loads. Doc. 45/86

Earls, C., Galambos, T.V. (1998): Inelastic failure of high strength steel wide flange beams under moment gradient and constant moment loading. SSRC Annual Technical Session, Atlanta, 21-23 September 1998, 131-152

EUROCODE 8, EC 8 (1994): Design provisions for earthquake resistance of structures. ENV 1933, 1-1

Faggiano, B., Mazzolani, F.M. (1999): Proposals for improving the steel frame ductility by weakening. In XVII Congresso CTA, Napoli, 3-7 October 1999, Vol. 1, 269-280

Fajfar, P., Krawinkler, H. (eds)(1997): Seismic Design Methodologies for the Next Generation of Codes. Bled, 24-27 June 1997, Balkema, Rotterdam

Feldmann, M. (1994): Zur Rotationkapazitat von I-Profilen statisch and dynamisch belasteter Trager. Ph Thesis, RWTH Universitat Aachen

FEMA-Building Seismic Safety Council (1997): NEHRP guideline for the seismic rehabilitation of buildings. FEMA Report No. 273

Galambos, T., Ravindra, M.K. (1978): Properties of steel for use in LRFD. Journal of the Structural Division, Vol. 104, ST 9, 1459-1468

Gioncu, V. (2000): Design criteria for seismic resistant steel structures. In Seismic Resistant Steel Structures (eds F.M. Mazzolani and V. Gioncu), CISM courses, Udine 18-22 October 1999, Springer, Wien, 19-99

Gioncu, V., Petcu, D. (1995): Numerical investigations on the rotation capacity of beams and beam-columns. In Stability of Steel Structures (ed. M. Ivanyi), Budapest, 21-23 September 1995, Akademiai Kiado, Budapest, Vol. 1, 163-174

Gioncu, V., Petcu, D. (1997): Available rotation capacity of wide-flange beams and beam-columns. Part 1. Theoretical approaches. Part 2. Experimental and numerical tests. Journal of Constructional Steel Research, Vol.43, No. 1-3, 161-217, 219-244

Gioncu, V., Petcu, D. (2001): Ductility of steel members. Part I. Static and monotonic ductility. Part II. Seismic ductility (manuscript)

Gioncu, V., Tirca, L., Petcu, D. (1996): Rotation capacity of rectangular hollow section beams. In Tubular Structures VII (eds. J. Farkas and K. Jarmai), Miskolc, 28-30 August 1996, Balkema, Rotterdam, 387-395

Gioncu, V., Mateescu, G., Petcu, D., Anastasiadis, A. (2000): Prediction of available ductility by means of local plastic mechanism method: DUCTROT computer program. In Moment Resisting Connections of Steel Frames in Seismic Areas. Design and Reliability, (ed. F.M. Mazzolani), E&FN Spon, London, 95-146

Ghersi, A., Marino, E., Neri, F. (1999): A simple procedure to design steel frames to fail in global mode. In Stability and Ductility of Steel Structures (eds. D. Dubina and M. Ivanyi), Timisoara, 9-11 September 1999, Elsevier, Amsterdam, 377-384

Ghobarah, A., Aly, N.M., El-Attar, M. (1997): Performance level criteria and evaluation. In Seismic Design Methodologies for the Next Generation of Codes (eds. P. Fajfar and H. Krawinkler), Bled, 24-27 June 1997, Balkema, Rotterdam, 207-215

Goel, R.K., Chopra, A.K. (1997): Period formulae for moment-resisting frame buildings. Journal of Structural Engineering, Vol. 123, No. 11, 1454-1461

Ivanyi, M. (1979): Moment rotation characteristics of locally buckling beams. Periodica Polytechnica, Civil Engineering, Vol. 23, Nos 3/4, 217-230

Kecman, D. (1983): Bending collapse of rectangular and square section tubes. Journal of Mechanical Science, Vol. 25. No. 9-10, 623-636

Kennedy, D.J.L., Medhekar, M.S.(1999): A proposed strategy for seismic design of steel buildings. Canadian Journal of Civil Engineering, Vol. 26, 564-571

Kim, S., D'Amore, E. (1999): Push-over analysis procedure in earthquake engineering. Earthquake Spectra, Vol. 15, 417-434

Kotelko, M. (1996): Ultimate load and postfailure behaviour of box-section beams under pure bending. Engineering Transactions, Vol. 44, No. 2, 229-251

Kuhlmann, U. (1986): Rotationskapazitat biegebeanspruchter I-Profile under Berucksichtung des plastischen Beulens. Technical Report, Bochum University, Mitteilung Nr. 86-5

Kuwamura,H., Sasaki, M. (1990): Control of random yield-strength for mechanism-based seismic design. Journal of Structural Engineering, Vol.116, No. 1, 98-110

Lam, N., Wilson, J., Hutchinson, G. (1996): Building ductility demand: Intreplate versus intraplate earthquakes. Earthquake Engineering and Structural Dynamics, Vol. 25, 965-985

Lay, M.G., Galambos, T.V. (1967): Inelastic beams under moment gradient. Journal of the Structural Division, Vol. 93, ST 1, 381-400

Lee, L.H., Lee, H.H., Han, S.W. (2000): Method of selecting design earthquake ground motions for tall buildings. The Structural Design of Tall Buildings, Vol. 9, 201-213

Leelataviwat, S., Goel, S.C., Stojadinovic, B. (1999): Toward performance-based seismic design of structures. Earthquake Spectra, Vol. 15, No.3, 435-461

Lukey, A.F., Adams, P.F.(1969): Rotation capacity of beams under moment gradient. Journal of the Structural Division, Vol. 95, ST 6, 1173-1188

Mazzolani, F.M. (2000): Design of moment resisting frames. In Seismic Resistant Steel Structures (eds. F.M. Mazzolani and V. Gioncu), Udine CISM courses, 18-22 October 1999, Springer, Wien, 159-240

Mazzolani, F.M., Piluso, V. (1995a): Failure mode and ductility control of seismic resistant MR-frames. Costruzioni Metalliche, No. 2, 11-28

Mazzolani, F.M., Piluso, V. (1995b): A new method to design steel frames failure in global mode including P-Δ effects. In Behaviour of Steel Structures in Seismic Areas, STESSA 94 (eds. F.M. Mazzolani and V. Gioncu), Timisoara, 26 June-1 July, 1994, E&FN Spon, London, 300-309

Mazzolani, F.M., Piluso, V. (1996): Theory and Design of Seismic Resistant Steel Structures. E&FN Spon, London

Mazzolani, F.M., Piluso, V. (1997a): A simple approach for evaluating performance levels of moment-resisting steel frames. In Seismic Design Methodologies for the Next Generation of Codes (eds. P. Fajfar and H. Krawinkler), Bled, 24-27 June 1997, Balkema, Rotterdam, 241-252

Mazzolani, F.M., Piluso, V. (1997b): Plastic design of seismic resistant steel frames. Earthquake Engineering and Structural Dynamics, Vol. 26, 167-191

Mazzolani, F.M., Piluso, V., Rizzano, G. (1998): Design of full-strength extended end-plate joints accounting for random material variability. In Control of the Semi-Rigid Behaviour of Civil Engineering Structural Connections, COST C1, Liege, 17-19 September 1998, 405-414

Mazzolani, F.M., Montuori, R., Piluso, V. (2000): Performance based design of seismic-resistant MR-frames. In Behaviour of Steel Structures in Seimic Areas, STESSA 2000 (eds. F.M. Mazzolani and R. Tremblay), Montreal, 21-24 August 2000, Balkema, Rotterdam, 611-618

Montuori, R., Piluso, V. (2000): Plastic design of steel frames with dog-bone beam-to column joints. In Behaviour of Steel Structures in Seismic Areas, STESSA 2000, (eds. F.M. Mazzolani and R. Tremblay), Montreal, 21-24 August 2000, Balkema, Rotterdam, 627-634

MTC (1997): A study of provisions for seismic regulations. MTC Report No. 97

Piluso, V. (1995): Post-local buckling behaviour of rolled steel beams subjected to nonuniform bending. In XV Congresso CTA, Riva de Garda, 15-18 October 1995, 542-562

Plumier, A. (1996): Reduced beam section: A safety concept for structure in seismic zones. Buletinul Stiintific al Universitatii Politehnica Timisoara, Transaction on Civil Engineering, Architecture, Tom 41(55), No. 2, 46-60

Plumier, A. (2000): Seismic resistant composite structures. In Seismic Resistant Steel Structures (eds. F.M. Mazzolani and V. Gioncu), Udine CISM courses, 18-22 October 1999, Springer, Wien, 289-347

Plumier, A., Boushaba, B. (1988): Relation entre la ductilite locale et le facteur de comportement sismique de structures en acier. Construction Metallique, Nr. 2, 59-70

Powell, G.H., Allahabadi, R. (1988): Seismic damage prediction by deterministic methods: Concepts and procedures. Earthquake Engineering and Structural Dynamics, Vol. 16, 719-734

Prakash, V., Powell, G.H., Campbell, S.(1993): Drain-2DX. Base program description and user gide, University of California

Priestley, M.J.N. (1997): Displacement-based seismic assessment of reinforced concrete buildings. Journal of Earthquake Engineering, Vol.1, No. 1, 157-192

Sarma, S.K., Srbulov, M. (1998): An uniform estimation of some basic ground motion parameters. Journal of Earthquake Engineering, Vol. 2, No. 2, 267-287

Soroushian, P., Choi, K.K. (1987): Steel mechanical at different strain rate. Journal of Structural Engineering, Vol. 113, No. 4, 863-872

Spangemaher, R. (1992): Zum Rotationnachweis von Stalkonstruction, die nach Traglastverfahren berechnet werden. Ph Thesis, Aachen University

Szabo, B. (1990): Local buckling of frame corners with semi-rigid members. In Stability of Steel Structures (ed. M. Ivanyi), Budapest, 25-27 April 1990, Akademiai Kiado, Budapest, 731-737

Tehami, M. (1997): Local buckling in class 2 continuous composite beams. Journal of Constructional Steel Research, Vol. 43, No. 1-3, 141-159

Teran-Guilmore, A. (1998): A parametric approach to performance-based numerical seismic design. Earthquake Spectra, Vol. 14, No. 3, 501-520

Tso, W.K., Zhu, T.J., Heidebrecht, A.C. (1993): Seismic energy demands on reinforced concrete moment-resistant frames. Earthquake Engineering and Structural Dynamics, Vol. 22, 533-545

Zhu, T.J., Tso, W.K., Heidebrecht, A.C. (1988): Effect of peak ground a/v ratio on structural damage. Journal of Structural Engineering, Vol. 114, No. 5, 1019-1037

Appendix

DUCTROT M Computer program

Dana Petcu and Victor Gioncu

A.1 General description

A.1.1 Name of application: **DUCTROT M** (**DUCT**ility of **ROT**ation for **M**embers)
A.1.2 Version: 2001
A.1.3 Objective: the DUCTROT M computer program provides the designer with the rotation capacity of steel members in order to verify the structure ductility
A.1.4 Used method: the method of local plastic mechanism is used to determine the rotation capacity of steel members. This method has been proved to be the most effective one for practical design
A.1.5 Program characteristics: the DUCTROT M computer program is an interactive tool for computing rotation capacity of steel beams and beam-columns, elaborated as a result of many years of research works in this field by the research teams of Politechnica University Timişoara, Building Research Institute Timişoara and Western University Timişoara.

The program is designed to run on any PC with a version of Windows 95/98 or Millenium. A version for Windows 2000 and NT is also available. Its small dimension makes it easy to move it from one workstation to another.

The graphical interface is based on a number of computational panels, each one having a specific meaning (cross-section types, material characteristics, etc.). A panel can have four types of field: user input, results, buttons, and explanation figures. Fields for data input (in white) are provided for the user to describe his problem (yield stress, flange width, axial force, etc.). Messages are supplied when the user gives some wrong data, i.e. out of normal range. The fields with results (in grey) are filled with information provided by the program according to the input data. Note that each modification of input data automatically modifies the result values and also the effect of changing one datum can be easily seen on the results that are on the same computational panel and depend on that datum. The buttons (with text or pictures on top) are used to select one from many options or to navigate between panels (possible also using the scroll-bar).

The pictures can be classified into static and dynamic ones. Suggestive static pictures help the user to understand the notations used by the program. Note that each data field has associated with it a clear explanation on the bottom line (status line), which will appear when the user clicks in this field. If the user needs some supplementary information, he can activate a help window, where he can find proposed values for the respective characteristic or relationship used to determine the results. Dynamic pictures (diagrams) are those which depend on the input data. A simple modification of the data on a figure panel or some previous data can affect the shape of the curves represented in that picture.

The program execution is user driven, i.e. the computational panels depends on the data given by the user. So, according to the user inputs, different branches of the program can lead to different panels. In order to help the user to fill input data fields, each datum has an implicit value which will be displayed when the resistant panel is activated for the first time by the user. Any modification of a data field will influence the results on the current panel and those of the next panels. An order is established between panels: a panel will not appear until the previous one has been deactivated.

The program menu helps the user to save his current input (numerical values only) in a text file which can be used by another program or in another working session with DUCTROT. The implicit extension of the output file is “loc”. The information to be printed can be selective: the printed information can be a table with inputs or only those of specific panel results, dynamical diagrams (all or a specific one), or all information (inputs, numerical results in tables and diagrams). A toolbar allows rapid access to the menu facilities (new session, open session, save session, print and exit program).

The program was written in Microsoft Visual C++ version 5.0. Therefore the program interface is specific for Windows applications, the open, save and print interfaces correspond to that of the specific Windows version on which the program is running. The diagrams are constructed using the Visual C tools for two-dimensional graphics.

A.2 User Guide

The user must examine and complete 11 pages containing input data and results.

Page 1 *Section types.* The designer can select the cross-section of member between the followings:

- I-section;
- I-monosymmetrical section;
- X-section;
- box-section;
- hollow-section;
- composite-section.

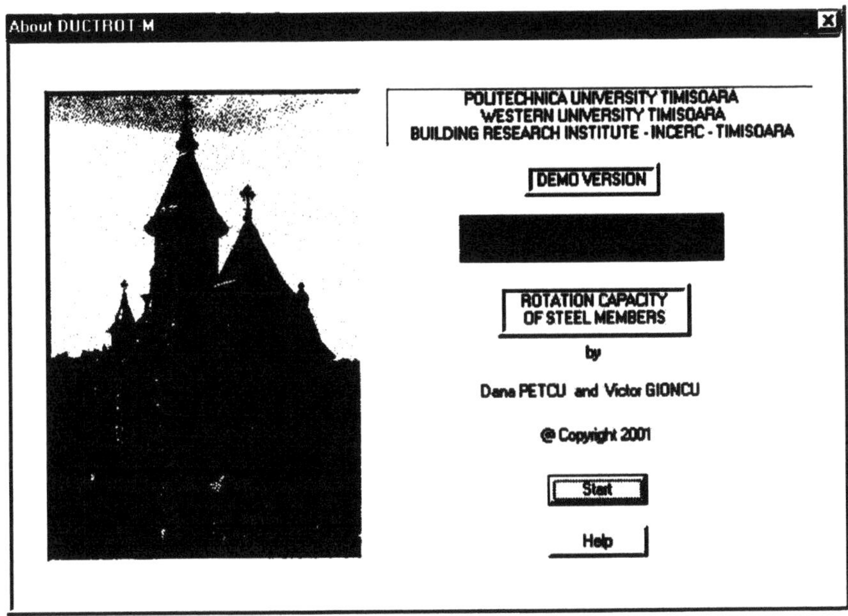

Presentation page

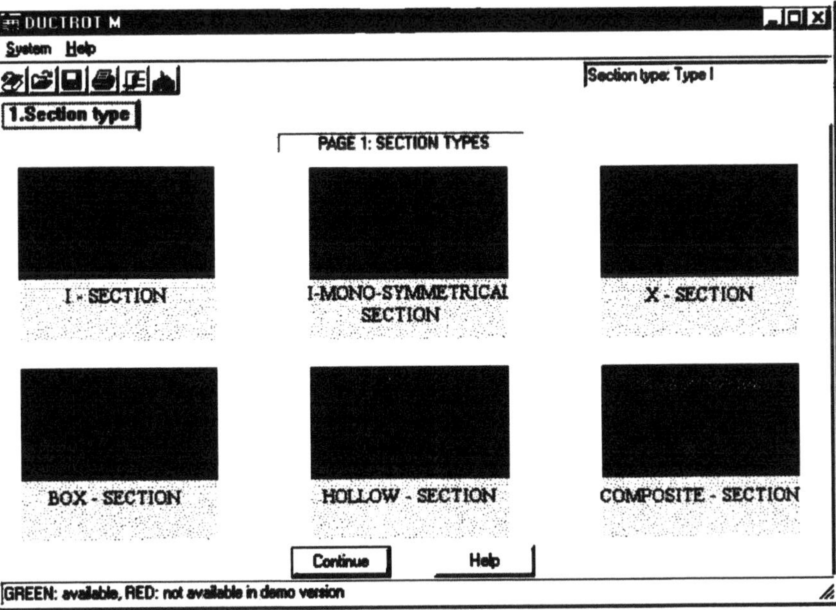

Page 1: Cross-section types

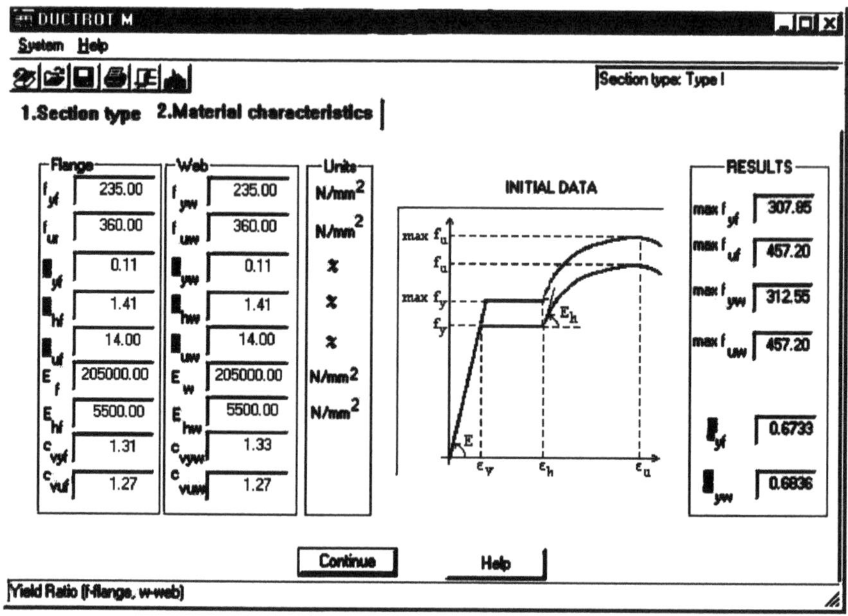

Page 2: Material characteristics

New section types will be provided in the next version of the computer program, as angle, channel, built-up, open web sections. The enclosed DEMO CD ROM refers only to the I-section. Therefore this user guide the figures refers only to these sections, in order to show the practical uses of the program.

Page 2 *Material characteristics.* Some input data required by the program are common for all section types. Therefore, the main mechanical properties for flanges and webs are required: yield stress, ultimate strength, yield strain, outset strain-hardening, total strain, modulus of elasticity, strain-hardening modulus, coefficient to determine the actual yield and strength stresses. To help the designer, supplementary information is provided in the form of values of characteristics for most steels used or values of possible random material variability. The program determines the yield ratios for common steels in order to verify their framing in recommended limits.

Page 3 *Geometrical section characteristics.* The designer must complete the required cross-section dimensions. The results refer to area, position of centroid, moment of inertia, elastic and plastic section modulus, characteristic values of selected cross-section. The program determines the cross-section slenderness, compares the values with the limits, and warns if some limits are exceeded. The designer can change the cross-section dimensions to satisfy this limit or ignore the notice and continue the calculation with unmodified geometrical characteristics.

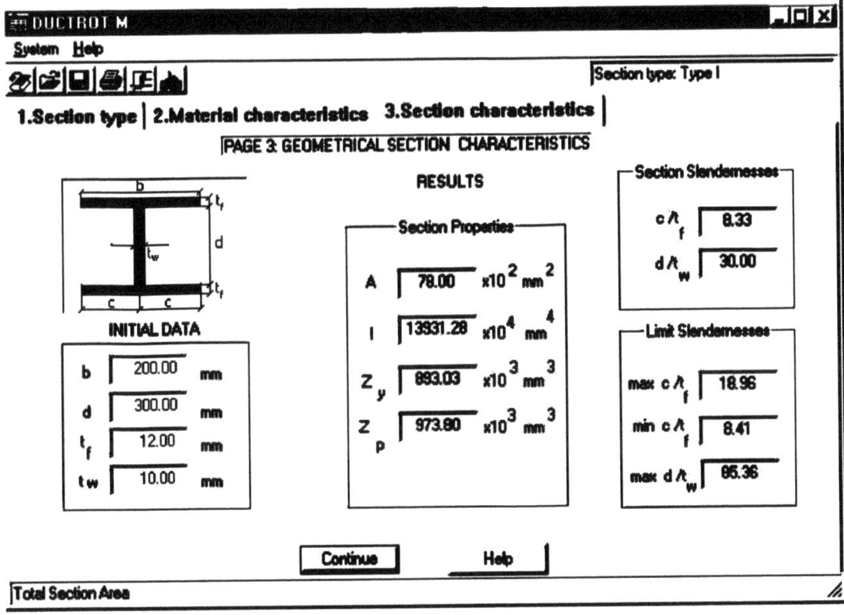

Page 3: Geometrical section characteristics

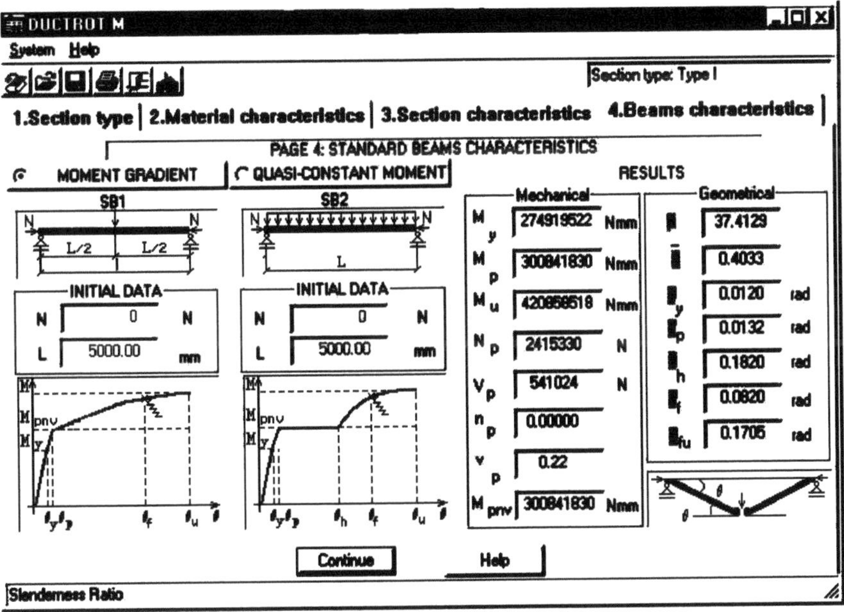

Page 4: Standard bean characteristics

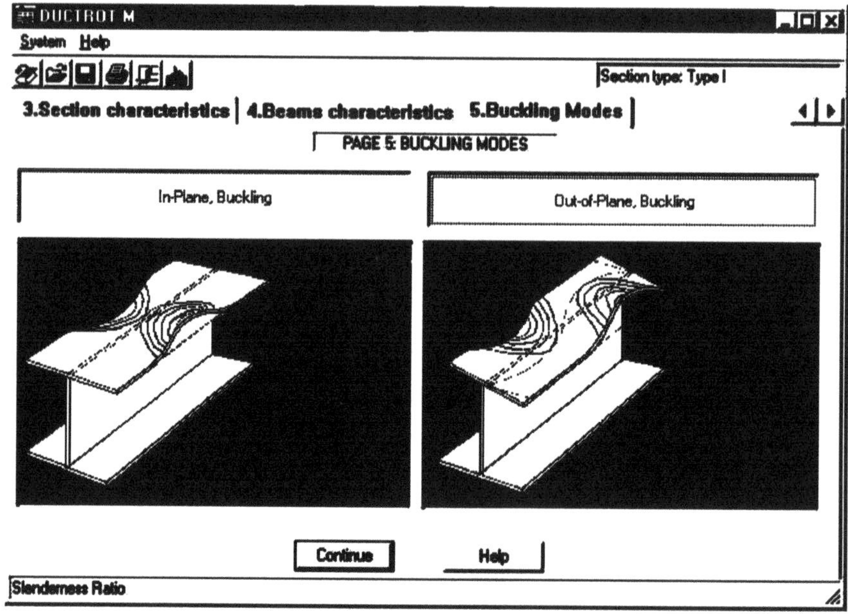

Page 5: Buckling modes

Page 4 *Standard bean characteristics.* Two standard beam types are available, SB 1 for gradient moment and SB 2 for quasi-constant moment. The required input data are the axial force (zero for beams) and beam span, determined from the structure by using the position of the inflection point. The program determines the mechanical properties of the section: first yield moment, full plastic moment, ultimate moment, plastic and ultimate axial forces, plastic shear force, reduced moments due to interaction with axial and shear forces. The geometrical characteristics refer to member slenderness and normalized slenderness and the beam rotations corresponding to first yield, full plastic moment, outset of strain-hardening, fracture of tension flange or compressed buckled flange.

Page 5 *Buckling mode.* The designer can select the buckling mode for which the ultimate plastic rotation is determined: in-plane or out-of-plane. The designer can ask, as help, for supplementary information about these local plastic buckling types.

Page 6a,6b *Rotation capacity under monotonic load.* The designer must provide data concerning the geometry of buckled shape. Information on these parameters can be obtained by using a help window. The ductility criterion must be chosen. The results refer to the rotations corresponding to maximum moment, the slope of post-critical behaviour, the rotation capacity corresponding to maximum moment and ultimate rotation for in-plane (page 6a) and out-of-plane (page 6b) buckling modes. The designer can determine these values for all plastic buckling modes. Supplementary information is provided on request by using help windows.

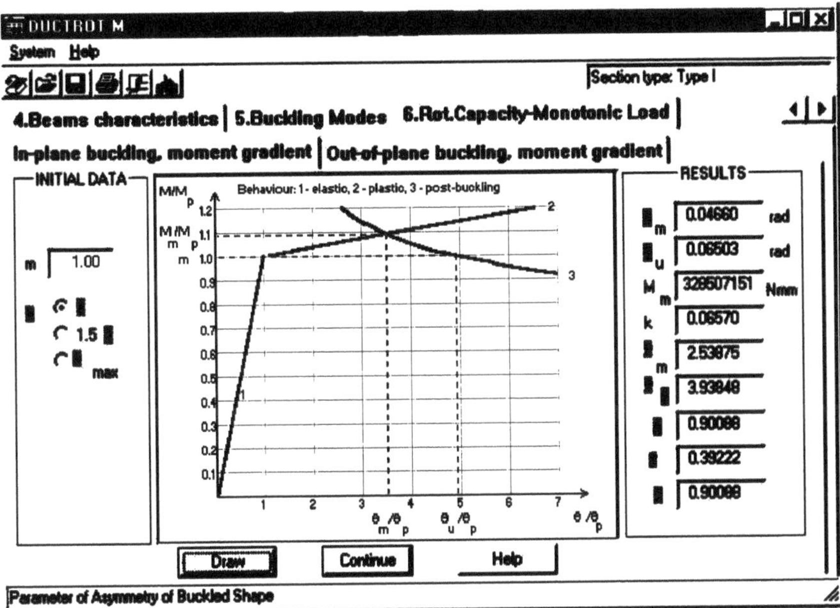

Page 6a: Rotation capacity for monotonic load: in-plane buckling

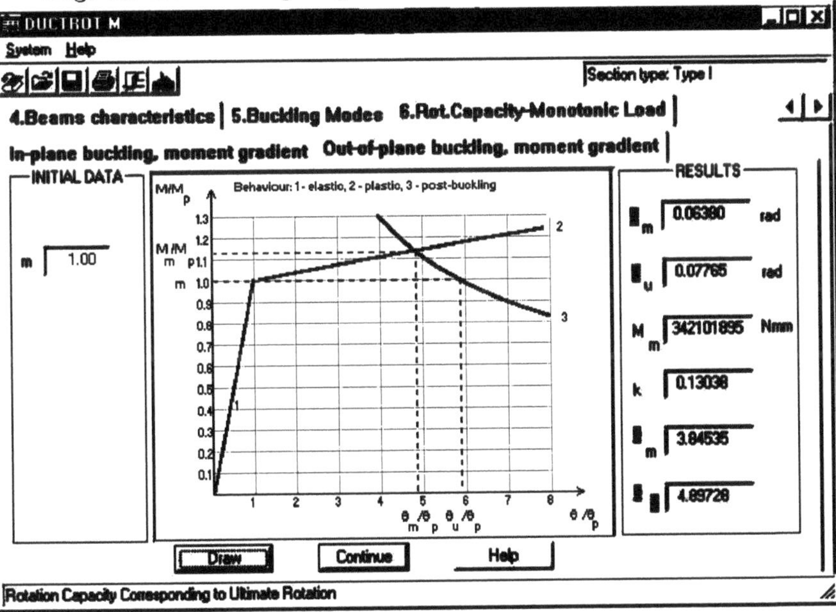

Page 6b: Rotation capacity for monotonic load: out-of-plane buckling

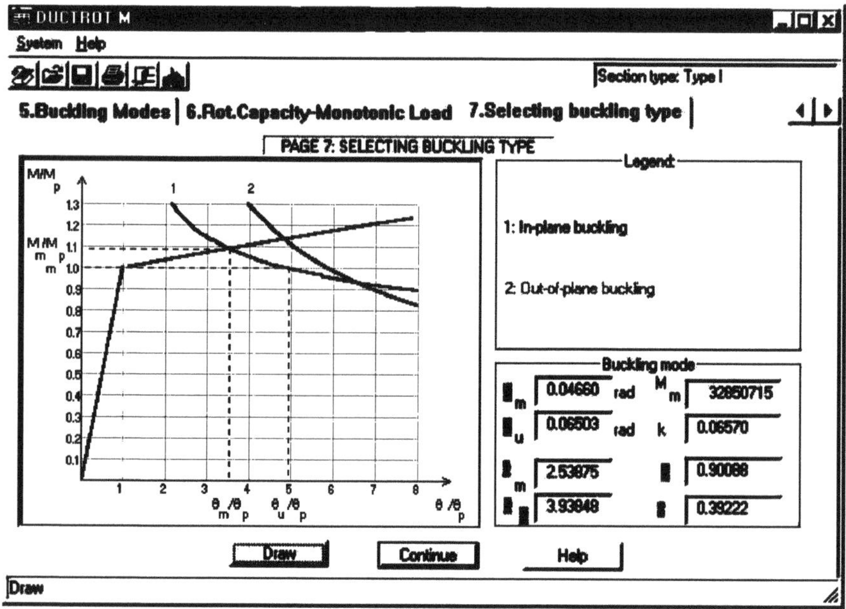

Page 7: Select buckling type

Page 7 *Selecting buckling type.* When the plastic buckling is evaluated for two modes, the minimum rotation capacity determines the selected mode. The corresponding characteristic values are kept for the following design.

Page 8 *Fabrication types.* The rotation capacity is determined using the cross-section dimensions from considering rectangular elements. For built-up sections with fully penetrated welds, the obtained values can be used in design. But for hot-rolled profiles or built-up sections with fillet welds the determined rotation capacity must be corrected taking into account the reduced slenderness of walls due to flange-to-web junctions. The designer must provide the dimensions of junction radius or welds.

Page 9 *Parametric studies for monotonic action.* In order to improve the local ductility a parametric study can be performed by selecting the examined characteristics and variables, giving the lower and upper limits and the step of variation. A graphic is obtained for the chosen characteristic as a function of two variables. Using these graphics the designer can decide on the most efficient modification in section geometrical or mechanical properties, in order to obtain increased ductility.

Page 10 *Seismic action.* When the influence of seismic action is evaluated, the designer must select the case of strain-rate (near-field earthquakes) (Page 10a) or cyclic loading (far-field earthquakes) (Page 10b). In case of near-field earthquakes the designer must estimate the value of strain-rate as a function of ground motion velocity and structure characteristics. A help window is presented on designer's request. The increased yield ratio and yield and strength stresses are determined. The reduced rotation capacity

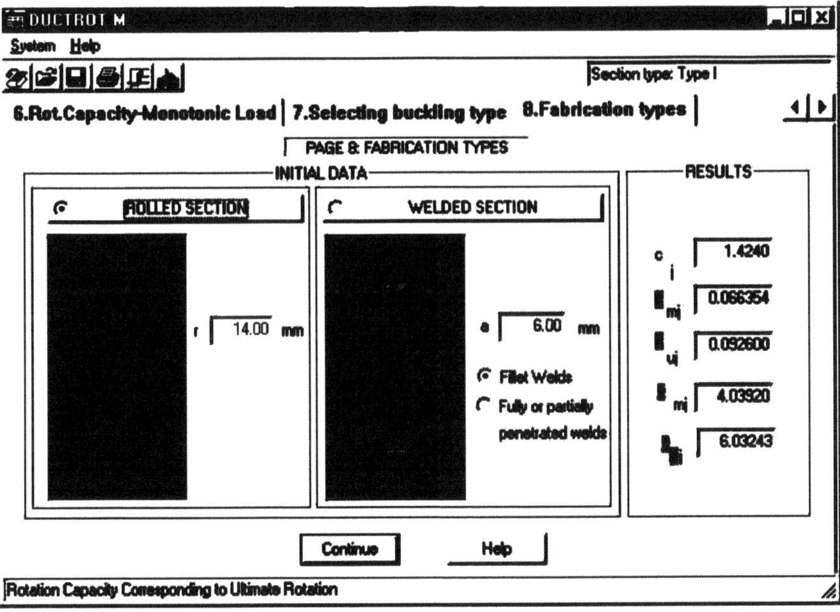

Page 8: Fabrication types

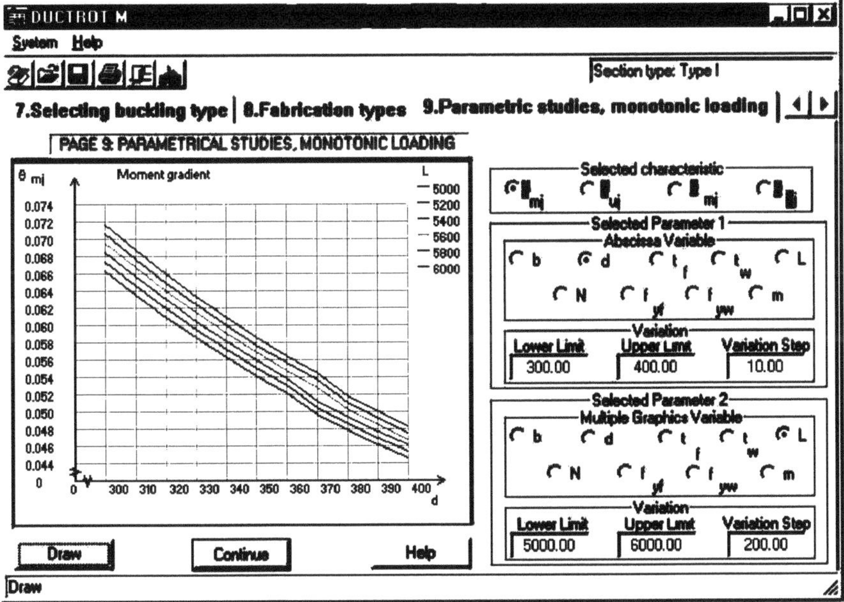

Page 9: Parametric study for monotonic actions

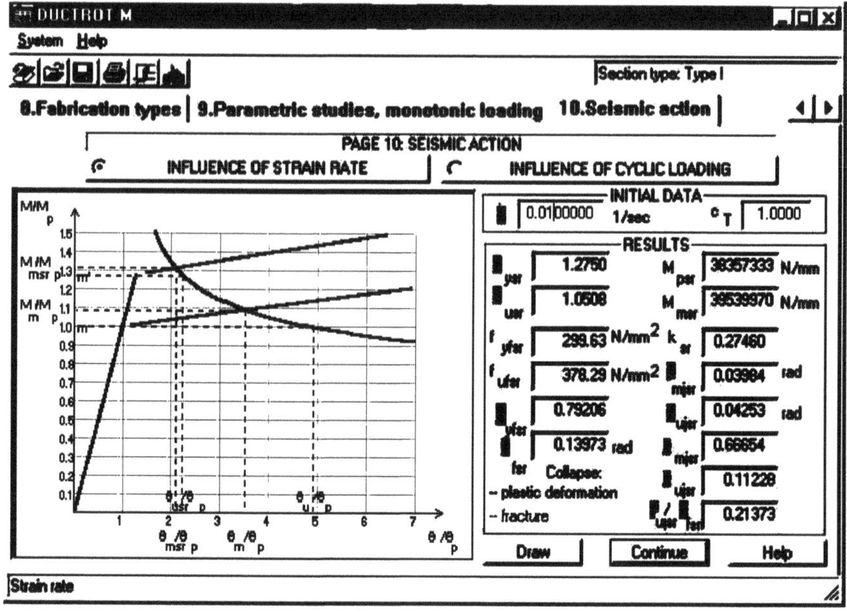

Page 10a: Seismic action influence of strain-rate

and the fracture rotation are also determined.

Page 10b *Influence of cyclic action.* In the case of intermediate and far-field earthquakes the cyclic action must be characterized by the designer specifying the main characteristics: number and amplitude of big pulses, increasing, constant or decreasing amplitudes. Using these values, the reduction of rotation capacity is determined.

Page 11 *Parametric study for seismic actions.* In order to improve the rotation capacity of members under seismic actions, a parametric study can be performed by the designer. The selected characteristics are plotted as a function of two selected variables. Using these graphics the designer can decide on the most efficient improvement in cross-section characteristics, in order to obtain an increased member ductility.

A.3. Acknowledgment

With a view to developing a new and improved version, the program authors would be grateful to all those who make suggestions or comment on this program at one of the following addresses:

Dana Petcu
Western University Timişoara
Department of Computer Science
B-dul Vasile Pârvan nr. 4
1900 Timişoara, Romania
E-mail: petcu@info.uvt.ro

Victor Gioncu
Politechnica University Timişoara
Department of Architecture
Str. Traian Lalescu 2
1900 Timişoara, Romania
E-mail: histruct@mail.dnttm.ro

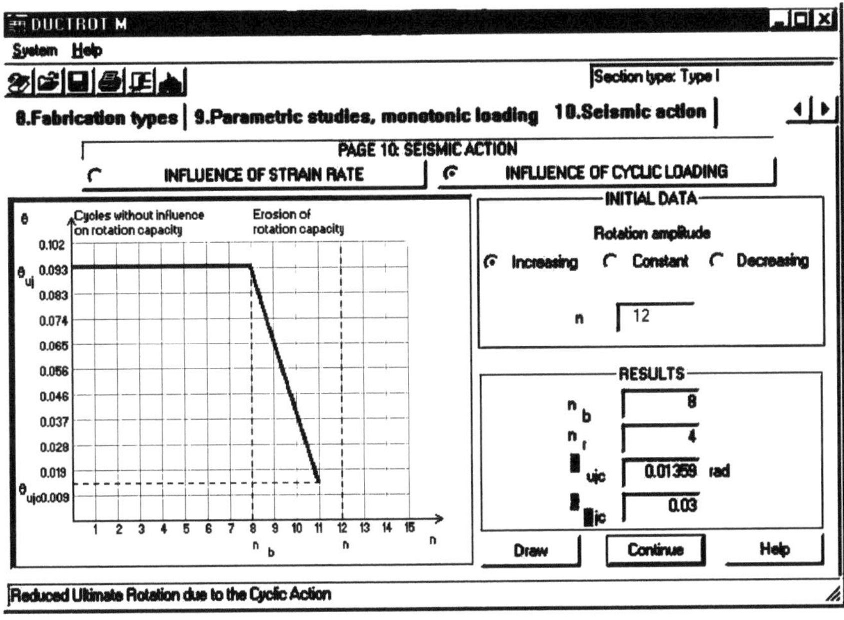

Page 10b: Influence of cyclic action

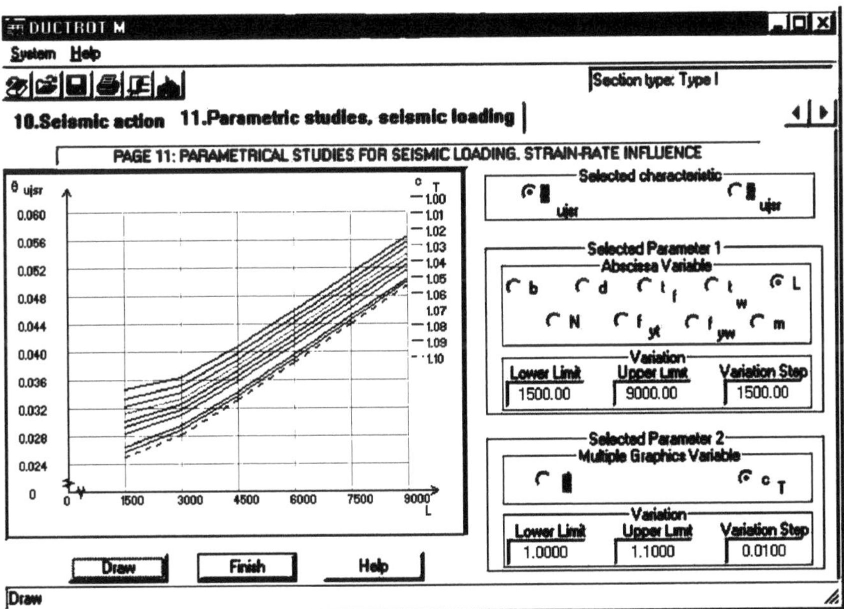

Page 11: Parametrical study for seismic actions

Index

CD-ROM Single-User Licence Agreement

We welcome you as a user of this Spon Press CD-ROM and hope that you find it a useful and valuable tool. Please read this document carefully. This is a legal agreement between you (hereinafter referred to as the 'Licensee') and Taylor and Francis Books Ltd., under the imprint of Spon Press (the 'Publisher'), which defines the terms under which you may use the Product. By breaking the seal and opening the package containing the CD-ROM you agree to these terms and conditions outlined herein. If you do not agree to these terms you must return the Product to your supplier intact, with the seal on the CD case unbroken.

1. Definition of the Product

   The product which is the subject of this Agreement, *DuctRot-M Demo Version 2002* (the 'Product') consists of:

   1.1 Underlying data comprised in the product (the 'Data')

   1.2 A compilation of the Data (the 'Database')

   1.3 Software (the 'Software') for accessing and using the Database

   1.4 A CD-ROM disk (the 'CD-ROM')

2. Commencement and licence

   2.1 This Agreement commences upon the breaking open of the package containing the CD-ROM by the Licensee (the 'Commencement Date').

   2.2 This is a licence agreement (the 'Agreement') for the use of the Product by the License, and not an agreement for sale.

   2.3 The Publisher licenses the Licensee on a non-exclusive and non-transferable basis to use the Product on condition that the Licensee complies with this Agreement. The Licensee acknowledges that it is only permitted to use the Product in accordance with this Agreement.

3. Installation and Use

   3.1 The Licensee may provide access to the Produce for individual study in the following manner. The Licensee may install the Product on a secure local area network on a single site for use by one user.

   3.2 The Licensee shall be responsible for installing the Product and for the effectiveness of such installation.

4. Permitted Activities

   4.1 The Licensee shall be entitled:

      4.1.1 to use the Product for its own internal purposes;

      4.1.2 to download onto electronic, magnetic, optical or similar storage medium reasonable portions of the Database provided that the purpose of the Licensee is to undertake internal research or study and provided that such storage is temporary;

      4.1.3 to make a copy of the Database and/or the Software for back-up/archival/disaster recovery purposes.

   4.2 The Licensee acknowledges that its rights to use the Product are strictly set out in the Agreement, and all other uses (whether expressly mentioned in Clause 5 below or not) are prohibited.

5. Prohibited Activities. The following are prohibited without the express permission of the Publisher:

   5.1 The commercial exploitation of any part of the Product.

   5.2 The rental, loan, (free or for money or money's worth) or hire purchase of this product, save with the express consent of the Publisher.

   5.3 Any activity which raises the reasonable prospect of impeding the Publisher's ability or opportunities to market the Product.

   5.4 Any networking, physical or electronic distribution or dissemination of the products save as expressly permitted by this Agreement.

   5.5 Any reverse engineering, decompilation, disassembly or other alteration of the Product save in accordance with applicable national laws.

   5.6 The right to create any derivative product or service from the Product save as expressly provided for in this Agreement.

   5.7 Any alteration, amendment, modification or deletion from the Product, where for the purposes of error correction or otherwise.

6. General Responsibilities of the License

   6.1 The Licensee will take all reasonable steps to ensure that the Product is used in accordance with the terms and conditions of this Agreement.

   6.2 The Licensee acknowledges that damages may not be a sufficient remedy for the Publisher in the event of breach of this Agreement by the Licensee, and that an injunction may be appropriate.

   6.3 The Licensee undertakes to keep the Product safe and to use its best endeavours to ensure that the product does not fall into the hands of third parties, whether as a result of theft or otherwise.

   6.4 Where information of a confidential nature relating to the product of the business affairs of the Publisher comes into the possession of the Licensee pursuant to this Agreement (or otherwise), the Licensee agrees to use such information solely for the purposes of this Agreement, and under no circumstances to disclose nay element of the information to any third party save strictly as permitted under this Agreement. For the avoidance of doubt, the Licensee's obligations under this sub-clause 6.4 shall survive the termination of this Agreement.

7. Warrant and Liability

   7.1 The Publisher warrants that it has the authority to enter into this agreement and the Authors warrant that they have secured all rights and permissions necessary to enable the Licensee to use the Product in accordance with this Agreement.

   7.2 The Publisher and the Licensee acknowledge that the Publisher supplies the Product on an 'as is' basis. The Publisher gives no warranties:

      7.2.1 that the Product satisfies the individual requirements of the Licensee; or

      7.2.2 that the Product is otherwise fit for the Licensee's purpose; or

      7.2.3 that the Data are accurate or completely free of errors or omissions; or

      7.2.4 that the Product is compatible with the Licensee's hardware equipment and software operation environment.

   7.3 The Publisher hereby disclaims all warranties and conditions, express or implied, which are not stated above.

   7.4 Nothing in this Clause 7 limits the Publisher's liability to the Licensee in the event of death or personal injury resulting from the Publisher's negligence.

   7.5 The Publisher hereby excludes liability for loss of revenue, reputation, business, profits, or for indirect or consequential losses, irrespective of whether the Publisher was advised by the Licensee of the potential of such losses.

7.6 The Licensee acknowledges the merit of independently verifying Data prior to taking any decisions of material significance (commercial or otherwise) based on such data. It is agreed that the Publisher shall not be liable for any losses which result from the Licensee placing reliance on the Data or on the Database, under any circumstances.

7.7 Subject to sub-clause 7.4 above, the Publisher's liability under this Agreement shall be limited to the purchase price.

8. Intellectual Property Rights

8.1 Nothing in this Agreement affects the ownership of copyright or other intellectual property rights in the Data, the Database of the software.

8.2 The Licensee agrees to display the Authors' copyright notice in the manner described in the Product.

8.3 The Licensee hereby agrees to abide by copyright and similar notice requirements required by the Authors, details of which are as follows: '©2002 D. Petcu and V. Gioncu. All rights reserved. All materials in *DuctRot-M Demo version 2002* are copyright protected. No such materials may be used, displayed, modified, adapted, distributed, transmitted, transferred, published or otherwise reproduced in any form or by any means now or hereafter developed other than strictly in accordance with the terms of the licence agreement enclosed with the CD-ROM. However, text and images may be printed and copied for research and private study within the preset program limitations. Please note the copyright notice above, and that any text or images printed or copied must credit the source.'

8.4 This Product contains material proprietary to and copyedited by the Publisher, Authors and others. Except for the Licence granted herein, all rights, title and interest in the Product, in all languages, formats and media throughout the world, including copyrights therein, are and remain the property of the Publisher or other copyright holders identified in the Product.

9. Non-assignment

This Agreement and the licence contained within it may not be assigned to any other person or entity without the written consent of the Publisher.

10. Termination and Consequences of Termination

10.1 The Publisher shall have the right to terminate this Agreement if:

- 10.1.1 the Licensee is in material breach of this Agreement and fails to remedy such breach (where capable of remedy) within 14 days of a written notice from the Publisher requiring it to do so; or
- 10.1.2 the Licensee becomes insolvent, becomes subject to receivership, liquidation or similar external administration; or
- 10.1.3 the Licensee ceases to operate in business.

10.2 The Licensee shall have the right to terminate this Agreement for any reason upon two month's written notice. The Licensee shall not be entitled to any refund for payments made under this Agreement prior to termination under this sub-clause 10.2.

10.3 Termination by either of the parties is without prejudice to any other rights or remedies under the general law to which they may be entitled, or which survive such termination (including rights of the Publisher under sub-clause 6.4 above).

10.4 Upon termination of this Agreement, or expiry of its terms, the Licensee must:

- 10.4.1 destroy all back up copies of the product; and
- 10.4.2 return the Product to the Publisher.

11. General

11.1 Compliance with export provisions
The Publisher hereby agrees to comply fully with all relevant export laws and regulations of the United Kingdom to ensure that the Product is not exported, directly or indirectly, in violation of English law.

11.2 Force majeure
The parties accept no responsibility for breaches of this Agreement occurring as a result of circumstances beyond their control.

11.3 No waiver
Any failure or delay by either party to exercise or enforce any right conferred by this Agreement shall not be deemed to be a waiver of such right.

11.4 Entire agreement
This Agreement represents the entire agreement between the Publisher and the Licensee concerning the Product. The terms of this Agreement supersede all prior purchase orders, written terms and conditions, written or verbal representations, advertising or statements relating in any way to the Product.

11.5 Severability
If any provision of this Agreement is found to be invalid or unenforceable by a court of law of competent jurisdiction, such a finding shall not affect the other provisions of this Agreement and all provisions of this Agreement unaffected by such a finding shall remain in full force and effect.

11.6 Variations

This agreement may only be varied in writing by means of variation signed in writing by both parties.

11.7 Notices

All notices to be delivered to: pdana@mail.dnttm.ro or histruct@mail.dntt

11.8 Governing law

This Agreement is governed by English law and the parties hereby agree that any dispute arising under this Agreement shall be subject to the jurisdiction of the English courts.

CD-ROM information

Minimum System Requirements for PCs
Pentium Processor
Windows 95 or later operating system (Designed for Windows 95/98/Me and Windows NT/2000)
16MB of RM
10MB of available hard disk space
400 MHz processor speed
CD-ROM drive
Mouse

Double-click on InstallD.exe to start and follow on-screen instructions.